Some of the foremost authorities in contemporary Earth Sciences have contributed to *Understanding the Earth* to provide a highly accessible and stimulating account of the recent rapid evolution of knowledge about the processes that drive and shape the Earth. Well-illustrated and extremely readable accounts cover many aspects of the Earth Sciences that are at the forefront of research and often confined to the more esoteric literature, from magma generation in the Earth to the formation of sedimentary basins. *Understanding the Earth* reflects the increasingly integrated and multi-disciplinary research that has fed our understanding of the Earth's origin and its internal and surface processes, and provides an indispensable guide to students of modern Earth Sciences.

Starting with the exploration of the Solar System and the study of planet formation, the reader will then find novel contributions on the Earth's interior and its dynamic evolution. Synthesis of the related topics of magma generation, plate tectonics, volcanic, hydrothermal and mineralisation processes, and crustal evolution leads the reader to a new understanding of lithospheric tectonics. Data from seismic reflection studies, earthquake focal plane solutions and from metamorphic belts are shown to underpin these new discoveries. The relation between the surface environment and the evolving stratigraphic record is looked at through the consideration of extensional tectonics. Finally, the evolution of the biosphere and its interactions with the atmosphere and hydrosphere are discussed, together with the natural and human-driven perturbations to the environment.

Understanding the Earth

Understanding the Earth

Edited by

GEOFF BROWN,
CHRIS HAWKESWORTH
and CHRIS WILSON

Department of Earth Sciences
The Open University, Milton Keynes, UK

Published by the Press Syndicate of the University of Cambridge
The Pitt Building, Trumpington Street, Cambridge CB2 1RP
40 West 20th Street, New York, NY 10011–4211, USA
10 Stamford Road, Oakleigh, Victoria 3166, Australia

First published 1992

Printed in Great Britain at the University Press, Cambridge

A catalogue record for this book is available from the British Library

Library of Congress cataloguing in publication data

Understanding the Earth/edited by Geoff Brown, Chris Hawkesworth, and Chris Wilson.
p. cm.
Includes bibliographical references.
ISBN 0 521 37020 5 (hb) ISBN 0 521 42740 1 (pb)
1. Earth sciences. I. Brown, G. C. (Geoff C.) II. Hawkesworth, C. J. III. Wilson, R. C. L.
QE26.2.U53 1992
550—dc20 91–22281 CIP

ISBN 0 521 37020 5 hardback
ISBN 0 521 42740 1 paperback

UP

CONTENTS

LIST OF EDITORS AND CONTRIBUTORS

The editors:

Geoff C. Brown, Chris J. Hawkesworth and Chris Wilson, Department of Earth Sciences, The Open University, Walton Hall, Milton Keynes MK7 6AA, UK.

The contributors:

Don L. Anderson, Seismological Laboratory, California Institute of Technology, Pasadena, California 91125, USA

Eric J. Barron, Earth System Science Center, Pennsylvania State University, University Park, Pennsylvania 16802, USA

Sierd Cloetingh, Institute of Earth Sciences, Vrije Universiteit, De Boelelaan 1085, 1081 HV Amsterdam, The Netherlands

Simon Conway Morris, Department of Earth Sciences, University of Cambridge, Downing Street, Cambridge CB2 3EQ, UK

Keith G. Cox, Department of Earth Sciences, University of Oxford, Parks Road, Oxford OX1 3PR, UK

Philip England, Department of Earth Sciences, University of Oxford, Parks Road, Oxford OX1 3PR, UK

Peter W. Francis, Department of Earth Sciences, Open University, Walton Hall, Milton Keynes MK7 6AA, UK

Ian G. Gass, Department of Earth Sciences, Open University, Walton Hall, Milton Keynes MK7 6AA, UK

Chris J. Hawkesworth, Department of Earth Sciences, Open University, Walton Hall, Milton Keynes MK7 6AA, UK

Simon L. Klemperer, British Institutions Reflection Profiling Syndicate, Bullard Laboratories, Madingley Rise, Cambridge CB3 0EZ, UK. Now at Department of Geophysics, Mitchell 366, Stanford University, Stanford, California 94305–2215, USA

Ralf Littke, Institut für Chemie, Kernforschungsanlage Jülich GmbH, Postfach 1913, D-5170 Jülich, Germany

Keith O'Nions, Department of Earth Sciences, University of Cambridge, Downing Street, Cambridge CB2 3EQ, UK

Carolyn Peddy, British Institutions Reflection Profiling Syndicate, Bullard Laboratories, Madingley Rise, Cambridge CB3 0EZ, UK

Michael R. Rampino, Earth Systems Group, Department of Applied Science, New York University, New York, NY 10003, USA

Michael J. Russell, Department of Geology and Applied Geology, University of Glasgow, University Avenue, Glasgow G12 8QQ, UK

Adolf Seilacher, Institut und Museum für Geologie und Paläontologie, Universität Tübingen, Sigwarstrasse 10, D-7400 Tübingen, Germany

Peter W. Skelton, Department of Earth Sciences, Open University, Walton Hall, Milton Keynes MK7 6AA, UK

Alan G. Smith, Department of Earth Sciences, University of Cambridge, Downing Street, Cambridge CB2 3EQ, UK

Stephen J. Sparks, University of Bristol, Queens Building, University Walk, Bristol BS8 1TH, UK

Stuart Ross Taylor, Research School of Physical Sciences, Australian National University, Canberra, Australia 2601

Alan B. Thompson, Institut für Mineralogie und Petrographie, Swiss Federal Institute of Technology (ETH), CH-8092 Zürich, Switzerland

Maurice E. Tucker, Department of Geological Sciences, University of Durham, Durham DH1 3LE, UK

Tjeerd H. van Andel, Department of Earth Sciences, University of Cambridge, Downing Street, Cambridge CB2 3EQ, UK

Peter van Calsteren, Department of Earth Sciences, Open University, Walton Hall, Milton Keynes MK7 6AA, UK

Roger G. Walker, Department of Geology, McMaster University, Hamilton, Ontario L8S 4M1, Canada

Tony Watts, Department of Earth Sciences, University of Oxford, Parks Road, Oxford OX1 3PR, UK

Dietrich H. Welte, Institut für Chemie, Kernforschungsanlagelich GmbH, Postfach 1913, D-5170 Jülich, Germany

Chris Wilson, Department of Earth Sciences, Open University, Walton Hall, Milton Keynes MK7 6AA, UK

Peter J. Wyllie, Division of Geological and Planetary Sciences, California Institute of Technology, Pasadena, California 91125, USA

FOREWORD

Just over twenty years ago I helped to write and edit the first version of *Understanding the Earth*. The book-was specifically designed to support the Earth Sciences component of the Open University Science Foundation Course, which was presented for the first time in 1971. When the first preface was written, The Open University was in its infancy, having only received its Charter in 1969. The book would have to be at the right level, contain no higher mathematics or advanced physics but have plenty of ideas that were mentally stimulating and conceptually demanding. Perhaps more than anything else we needed a book which would emphasise that the Earth Sciences were alive, vigorous and intellectually fascinating. Our contributors were all scientists of international repute and standing whose articles were authoritatively written and completely up to date. The timing was extremely fortunate, for the Earth Sciences had just passed through the plate tectonic revolution. This paradigm presented, for the first time, a global model wherein all geological, geochemical and geophysical data concerning the Earth could be fitted. As the discovery of the periodic table was to chemistry, so plate tectonics was to the Earth Sciences – what previously had been random collections of data now became part of a coherent whole. *Understanding the Earth* was not a textbook; you could not, even by reading it from cover to cover, get a complete survey, even at the shallowest level, of all aspects of the Earth Sciences. It was a 'reader' providing up-to-date and well-informed synopses of various aspects of our science for a wide audience among first-year undergraduates and geological enthusiasts. The new volume of *Understanding the Earth* is the same in that it looks at a wide range of issues and some of the major areas of Earth Sciences research in the modern context.

Despite the vast wealth of ideas, theories and models derived therefrom, plate tectonics still retains its pride of place as the all-embracing, global model. However, despite its unchallenged status, it has become a background against which other, often cross-disciplinary aspects of the Earth Sciences have become more prominent. Recently, seven themes and two geographic areas of research priority were identified in the UK by the Royal Society and the Natural Environment Research Council. The themes are palaeoclimatology, biomolecular palaeontology, neotectonics, fluid processes, sedimentary basin evolution, the Earth's deep interior, and space geochemistry and planetary evolution. The geographic areas are the ocean floor and Antarctica. Palaeoclimatology has become critically important for there is now grave concern that we are so polluting the Earth that this will have a markedly detrimental effect on global climate. Of particular concern is the greenhouse effect and the polar gaps in the ozone layer. Only 10 000 years ago, without any human intervention, the Earth's climate suddenly (on a geological time scale) got some 3–5 °C warmer. Although we now know that climatic changes are linked to variations in the Earth's orbit around the

Sun, a detailed study of the change to the last warm period could provide a clue as to the Earth processes that amplify this weak astronomical signal into one of major global significance. In the last decade or so it has been shown that virtually intact molecular structures exist in sedimentary rocks of the Earth's crust and that these were formed as far back as Precambrian (> 570 Ma) times. The study of these fossil molecules is biomolecular palaeontology and it has become possible through recent technological developments in molecular biology and organic geochemistry. It is a huge and complex field of study but offers much to our understanding of evolution, sediment diagenesis and oil generation. Neotectonics is the study of current deformation of the Earth. By relating seismic events to geomorphological changes in the Earth's surface, it is possible to quantify the processes causing these changes and thereby to understand, more precisely, older deformation events. fluids within the Earth, from groundwater to oil and gas are essential to human survival and civilisation. Many of these and other physical resources are located within sedimentary basins, so the evolution of these basins, as well as the nature and behaviour of fluids within the Earth, is an important area of frontier research. The last two themes, involving the Earth's deep interior, space geochemistry and planetary evolution could be thought of as types of remote sensing bearing on planetary processes. For, although we known plate tectonics has worked for the last 2500 Ma of geological time, we still do not fully understand *how* it works. Seismological and, to a lesser extent, magnetic studies of the deep Earth should provide clues as to the operative processes. Lastly, the study of meteorites and other cosmic material is providing invaluable clues to the origin of the Earth, the Solar System and indeed, the universe – it is an invaluable area for academic study. The choice of the two geographic areas was obvious. Oceans cover over 70 per cent of the Earth's surface and, as we know little about the rocks beneath them, they need to be studied further. Similarly, Antarctica is the one great continental arena that has not been studied in detail. As it was at the heart of the once great continent of Gondwanaland, its investigation is critical to the understanding of global processes. The list is long and the subjects profound: a complex situation faced the editors of this book when choosing its contributors.

In 1970, the three members of the Open University's Earth Sciences academic staff then in post acted as the editors for the first version of *Understanding the Earth*. In the 1990s the choice is wider, and it is appropriate that the three editors are all senior members of departmental academic staff, one of whom (Chris Wilson) was one of the editors of the first version. The number of chapters is the same and two of the original authors, Keith Cox and Alan Smith, act again. Some titles remain closely similar to those of 1970 although the authorship has changed, and yet again there are now subjects relating to the new areas of topical research identified above. Indeed, the more interdisciplinary nature of modern Earth Sciences, reflected increasingly in the physical and mathematical rigour of undergraduate teaching, has meant that the level of presentation in this book has advanced beyond OU Foundation Level. Obviously, much care has gone into the choice of topics and authors. The value of this version of *Understanding the Earth* lies in providing easy access to concepts not always readily available in the undergraduate library. So, I wish this book every success and hope that it proves even more popular than its predecessor.

Ian G. Gass
The Open University

April, 1992

INTRODUCTION TO CHAPTERS 1–4

Enquiry into the origin of the Earth and Solar System is guaranteed to capture the imagination of scientists and philosophers alike. After centuries of astronomical observation, speculation and physical modelling it is intriguing that there is still no fully satisfactory unifying theory. Of course, the main problem is one of scale, and in our attempts to penetrate the Solar System with space missions, humans have travelled just 0.000 001 per cent and space probes only 0.01 per cent of the distance to the nearest star. Yet these missions have brought new observations, dramatically enhancing our knowledge of almost all the planets and their dynamics. No excuse is needed, therefore, for side-stepping the title of this book in the first chapter. Peter Francis, whose primary interests are in volcanological and planetary processes, discusses planetary exploration from historical and modern perspectives, so providing a backdrop against which we can develop our understanding of the Earth.

The growth of modern Earth Sciences has drawn heavily on interdisciplinary research with the application of broader scientific techniques and principles to geological problems. Nowhere is this better illustrated than in Chapter 2, which develops a geochemical basis for understanding the Earth's origin. The generation of chemical elements inside stars, their scattering into interstellar space during giant explosions, and their subsequent incorporation into new stars and planetary bodies provides a basis for debating the accretion of the Earth and the chemistry of its interior. We are fortunate that supporting evidence from meteorites, fragments of small planetary bodies from within the Solar System, indicates that the Earth's core essentially is made of metallic iron and is surrounded by a silicate mantle. Stuart Ross Taylor, who has also written several texts on lunar science, concludes his contribution in characteristic style with some new, exciting and increasingly popular ideas about the intimate relationship between the Earth and Moon which dates from the time of their formation.

Focussing on the Earth, Chapters 3 and 4 take complementary approaches to fundamental questions about the physical nature and chemical composition of the inaccessible deep interior. Don Anderson, a seismologist concerned with charting the passage of earthquake waves through the Earth, describes the way that the speed and direction of those waves are affected by the density and elastic properties of the material in their paths. Data on the travel times of different waves are used to construct geophysical models illustrating the variation of physical properties with depth; in turn, these models constrain the likely minerals present in the Earth, together with their properties and their chemistry. The problem can be approached in a different way by taking minerals and rocks to the high pressures and temperatures predicted for the Earth's interior and then comparing measured physical properties with those deduced from geophysics. The techniques of experimental petrology are some of the most sophisticated in contemporary Earth Sciences, and even then, some of the highest pressure minerals – representing the deepest mantle and core – are stable in the laboratory for only the fractions of a second during which their properties are measured. To understand the relevance of the results to the Earth's interior, some simple phase diagrams are introduced by Peter Wyllie, one of the foremost authorities on pressure–temperature–composition relationships in the Earth.

Chapter 3 also includes some of the newest seismic imaging techniques that reveal lateral variations in the Earth's mantle related to its dynamic evolution, while Chapter 4 begins to examine the products of internal dynamics in zones of crustal melting and rock metamorphism.

Peter W. Francis
The Open University

EXPLORATION OF THE SOLAR SYSTEM

We live in remarkable times. The last two decades have seen our knowledge of the Solar System grow far more than in the previous two millennia. In our lifetime, answers have been found to questions that have concerned people ever since the first truly sapient *homo sapiens* looked up at the night sky. In decades to come, we will find answers to more questions as spacecraft extend our knowledge to the furthest edges of the Solar System. This is a period unprecedented in the history of science; a glorious and continuing intellectual adventure whose equal will not come again.

Sir Isaac Newton (1643–1727) said: 'If I have seen further, it is because I have stood on the shoulders of giants'. We should not forget this in our exhilaration as we explore the Solar System. The spacecraft voyages of discovery of our era would have been impossible without the centuries of earlier work by astronomers and mathematicians, who solved the problems of orbital mechanics and determined the fundamental physical properties of the planets, such as their distances, sizes, masses and densities. One of the giants that helped Newton to see further was Nicolaus Copernicus (1473–1543), who invoked the wrath of the Church by overturning the paradigm of millenia that required the Earth to be at the centre of the Solar System. Another was Johannes Kepler (1571–1630), who used the pattern of Mars' movements to show that its orbit, and those of all the planets, is elliptical, and thereby derived his Laws of Planetary Motion. Since Newton, our knowledge of the Solar System has enormously expanded. It has not done so continuously, however, but in a series of quantum leaps separated by long periods of stasis.

Construction of the world's great telescopes during the late 19th and early 20th centuries dramatically changed our understanding of the size and evolution of the Universe, and added a further increment to our knowledge of the Solar System. Unfortunately, though, the Earth's atmosphere imposes severe limits on what one can observe telescopically on a planet millions of kilometres distant, so that the view obtained by a modest telescope is as good as that provided by the largest. Thus, construction of large telescopes provided an initial burst of new data, but subsequently did not advance Solar System studies much, and our knowledge of the Moon and planets remained static. For centuries, the only available data were sketch maps drawn by painstaking observers after many a long and cold night at the telescope. Galileo (1564–1642) saw the craters on the Moon the instant he turned his first telescope towards it in 1609. In the following centuries, the craters were minutely mapped, measured, photographed and named, but their origin could not be resolved by any amount of careful observation. Opposing hypotheses for the craters were still being advanced even as the first Apollo spacecraft headed towards the Moon in the late 1960s. Until the first spacecraft missions, study of the Moon and planets was the concern of a small, fringe group of astronomers. (*Serious* astronomers, as they thought of

Table 1.1. *Basic Solar System statistics*

	Mercury	Venus	Earth	Moon	Mars	Jupiter	Saturn	Uranus	Neptune	Pluto
Distance from Sun (10^6 km)	57.9	108.2	149.6	( )	227.9	778.3	1427	2870	4497	5900
Period of revolution	88 d	224.7 d	365 d	27.3 d	687 d	11.86 yr	29.46 yr	84 yr	165 yr	248 yr
Axial rotation period	58.6 d	243 d retro	23 hr 56 m	27.3 d	24 hr 37 m	9 hr 55 m	10 hr 40 m	17.3 hr retro	18 h 30 m	6 d 9 hr retro
Axial inclination	0°	3°	23°27'		25°12'	3°5'	26°44'	97°55'	28°48'	?
Inclination of orbit to ecliptic plane	7°	3.4°	0°	5°	1.9°	1.3°	2.5°	0.8°	1.8°	17.2°
Orbital eccentricity	0.206	0.007	0.017	0.06	0.093	0.048	0.056	0.047	0.009	0.254
Equatorial diameter (km)	4880	12 110	12 756	3476	6794	143 200	120 000	51 800	49 500	3000?
Mass (kg)	0.33×10^{24}	4.9×10^{24}	6.0×10^{24}	7.4×10^{22}	6.5×10^{23}	1.9×10^{27}	5.7×10^{26}	8.7×10^{25}	1.0×10^{26}	1.6×10^{22}
Density (Mg m^{-3})	5.4	5.2	5.5	3.3	3.9	1.3	0.7	1.2	1.7	1.5
Atmosphere (main components)	Virtually none	Carbon dioxide	Nitrogen, oxygen	None	Carbon dioxide	Hydrogen, helium	Hydrogen, helium	Hydrogen, helium, methane	Hydrogen, helium	None detected
Satellites	0	0	1	NA	2	16+	17+	5	2	1
Rings	0	0	0		0	1	1000?	10	?	?

themselves, concerned themselves solely with stars and galaxies.) Once spacecraft data became available, the situation changed overnight. The planets ceased to be the concern of astronomers, and abruptly became the focus of interest of *geologists*. Spacecraft made it possible for the first time to study the Moon, planets and all the swarms of planetary satellites as individuals, each with their own unique record of geological evolution.

Bulk compositions

Before examining some of these individuals, it is worth while exploring some factors that the members of the **Solar System** have in common. It is quite a varied assemblage (see also Chapter 2). Three different groups of bodies emerge from consideration of sizes and densities: first, there are the large, low-density 'gas giant' planets (Jupiter, Saturn, Uranus and Neptune); second, a group of medium-size, high-density 'rocky' planets, asteroids and satellites (Mercury, Venus, Earth, Mars, many asteroids and a few of the larger planetary satellites including our Moon); and third, a group of small 'icy' bodies, including poor Pluto, a tiny, icy planet on the frigid fringes of the Solar System, and most of the smaller satellites, and comets. (Table 1.1 summarises some of the key parameters of the planets; Tables 1.2–1.4 those of the main planetary satellites; unfortunately there is not space to include the innumerable asteroids and comets.) Recognition of these groups as gaseous, rocky or icy stems simply from a consideration of their density, and from inferences of what is plausible in cosmochemical terms. This is perhaps the most important unifying concept in Solar System studies: *all* the bodies in the Solar System condensed from the *same* **solar nebula** *c.* 4600 million years (Ma) ago, and thus their compositions can be simply understood in terms of the varying conditions of temperature and pressure under which each condensed. Sadly, there is no scope for moons made of green cheese in modern planetology.

Meteorites and asteroids

Long before the Apollo program provided the first samples of lunar rocks, there was only one *direct* source of compositional data on the Solar System: rocks that literally fell from the sky. **Meteorites** have been known since pre-history, but they are studied now more intensively than ever before, because they provide the only tangible samples of materials that date back to the earliest days of the Solar System. Meteorites provided the first clues to the composition of planets beyond the Earth, and it was from their chemistry and mineralogy that geologists were able to predict the likely structures and compositions of the rocky inner planets. A wide range of meteorite compositions is known, ranging from **carbonaceous chondrites**, which are primitive, and close to the composition of the Sun's atmosphere, right

Table 1.2. *Satellites of Jupiter*

Name	Letter (in discovery sequence)	Distance from Jupiter (R_J)*	Orbital period (hr)	Diameter (km)	Mass (10^{20} kg)	Density (Mg m^{-3})	Albedo
Adrastea	J14	1.76	8	40			< 0.1
Amalthea	J5	2.55	11.7	90**			0.05
Thebe	J15	3.11	16.1	80			< 0.1
Io	J1	5.85	42.5	3640	892	3.5	0.6
Europa	J2	9.47	85.2	3130	487	3.0	0.6
Ganymede	J3	15.1	171.6	5270	1490	1.9	0.4
Callisto	J4	26.6	400.8	4830	1075	1.8	0.2
Leda	J13	156	5760	~ 10			
Himalia	J6	161	6024	180			0.03
Lysithea	J10	164	6240	~ 20			
Elara	J7	165	6240	80			0.03
Ananke	J12	291	14808	~ 20			
Carme	J11	314	16608	~ 15			
Pasiphae	J8	327	17640	~ 20			
Sinope	J9	333	18192	~ 15			

* R_J refers to the radius of Jupiter (71600 km – see Table 1.1)
** Amalthea is highly irregular, 135 × 85 × 75 km.

Table 1.3. *Satellites of Saturn*

Name	Distance from Saturn (R_S)	Orbital period (hr)	Diameter (km)	Mass (10^{20} kg)	Density (Mg m^{-3})	Albedo
1980S28	2.28	14.4	30			0.4
1980S27	2.31	14.7	100			0.6
1980S26	2.35	15.1	90			0.6
1980S3	2.51	16.7	120			0.5
1980S1	2.51	16.7	200			0.5
Mimas	3.08	22.6	392	0.45	1.4	0.7
Enceladus	3.95	32.9	500	0.84	1.2	1.0
Tethys	4.88	45.3	1060	7.6	1.2	0.8
1980S13	4.88	45.3	30			0.6
1980S25	4.88	45.3	25			0.8
Dione	6.26	65.7	1120	10.5	1.4	0.6
1980S6	6.26	65.7	35			0.5
Rhea	8.74	108	1530	25	1.3	0.6
Titan	20.3	383	5150	1346	1.9	0.2
Hyperion	24.6	511	300			0.3
Iapetus	59.0	1904	1460	19	1.2	0.5
Phoebe	215	13211	220			0.06

Notes: 1980S28, 1980S27 and 1980S26 are 'Ring Shepherds'; 1980S13, 1980S25 and 1980S6 occupy Langrangian points on the orbits of larger satellites.

Table 1.4. *Satellites of Uranus*

Name	Distance from Uranus (R_U)	Orbital period (hr)	Diameter (km)	Mass (10^{20} kg)	Density (Mg m^{-3})	Albedo
1986U7	1.90	7.9	15			
1986U8	2.06	8.9	20			
1986U9	2.28	10.4	50			
1986U3	2.38	11.1	70			
1986U6	2.42	11.4	50			
1986U2	2.48	11.8	70			
1986U1	2.55	12.3	90			
1986U4	2.70	13.4	50			
1986U5	2.90	14.9	50			
1985U1	3.32	18.3	160 × 170			
Miranda	5.00	33.9	480	0.71	1.26	0.22
Ariel	7.38	60.5	1170	14.4	1.65	0.25–0.45
Umbriel	10.3	99.5	1190	11.8	1.44	0.15
Titania	16.8	209	1590	34.3	1.59	0.23
Oberon	22.5	323	1550	28.7	1.50	0.20

through to **basaltic achondrites**, which resemble terrestrial basalt lavas and are the results of complex geological fractionation processes. Meteorites also provided another vital fixed datum point: all but a trivial number formed about 4600 Ma ago, tying down the age of the entire Solar System. **Asteroids** are rather larger, kilometre-sized chunks of material that represent intermediate building blocks in the assembly of planets and occupy a distinct region in the Solar System, the **asteroid belt**, between the orbits of Mars and Jupiter. In Chapter 2, S. R. Taylor discusses further the compositions of meteorites and asteroids, and shows how they are critically important in understanding the origin of the planets.

The Moon

The key geological features of the **Moon** are visible to the naked eye: large, whitish areas surrounding darker grey, irregular circular blotches that make up the 'Man in the Moon'. Even the smallest telescope will show that the whitish areas are rather rugged and heavily cratered – they are **highland regions** – while the darker areas – known as 'seas' or **maria** to the earliest observers – are vast lowland plains, largely devoid of craters. By the time of the Apollo program, the front side of the Moon had been minutely mapped, and it was suspected that the dark plains were huge expanses of basaltic lava, but major geological problems remained. What was the nature of the light-toned highlands rocks? How old were they? How old were the lava plains? And what was the origin of all the thousands of craters that scarred the surface and splashed bright 'rays' hundreds of kilometres over its surface?

The 381.69 kg of rocks returned by the six Apollo landing missions solved all these problems – and a great many more – demonstrating how science advances in quantum leaps. Despite the drama inherent in returning rocks from the Moon to the Earth, many of the samples looked remarkably dull, just like dusty, terrestrial basalt lavas. They were indeed basalt lavas, though geochemically distinct from terrestrial ones. Much the most startling result from the Apollo missions emerged from radiometric dating of the first Apollo 11 lava samples. They were *3800 Ma* old. And it was clear from photo-interpretation studies that these lavas from the Sea of Tranquillity *are among the youngest rocks on the Moon!* This result was unexpected because, although geologists are used to dealing with ancient materials, rocks older than three thousand million years are exceedingly scarce on Earth, where the most familiar part of the geological record – the fossiliferous part – covers only the past 570 *million* years (see Chapters 22 and 23). An immediate implication of this result was that the lighter toned highlands must be much older, because it was clear from photogeological studies that material from the maria overlay highlands material. So it transpired. When samples from the highlands were obtained, it was found that these were a curious group of igneous rocks, dominated by plagioclase feldspar, whose nearest terrestrial counterparts are Archaean and Proterozoic rocks known as anorthosites. The lunar anorthosites yielded ages greater than 4000 Ma; some of them close to 4600 Ma.

Given the meteorite evidence that the Solar System formed 4600 Ma ago, and that the Moon's youngest lavas are 3200 Ma old, it is plain that the Moon remained geologically active for only a relatively short time after its formation. It is easy *now* to understand why the Moon should have had such a short geological life, while the Earth continues to be active. It is much smaller than the Earth – only about one-hundredth of the mass – and thus had initially a smaller total content of **radioactive heat-producing elements**. Furthermore, because it was so small, whatever heat was generated was lost much more quickly by radiation, because of the Moon's much larger surface area relative to its volume. This simple realisation was a very important lesson for planetary scientists, and it has underpinned much subsequent thinking about the geological evolution of the terrestrial planets.

Twenty years after Apollo 11, detailed petrological studies of the returned samples are still continuing, with no prospect of an end. Meticulous studies are now being made of individual fragments in breccias, and of the multitude of fine particles obtained from core samples through the lunar soil (Figure 1.1). A consistent overall picture of lunar evolution has emerged, which suggests that the Moon was formed *c.* 4600 Ma ago when massive impact took place between the primordial Earth and an impactor the size of the planet Mars. The material of the mantle of the impactor was largely vaporised, and re-condensed to form the Moon. (The geochemical evidence for this is discussed by S. R. Taylor in Chapter 2.) Shortly after formation, the Moon's core separated towards its centre, and the remainder was melted, forming a vast magma 'ocean', hundreds of kilometres deep. Olivine and then orthopyroxene began to crystallise out, and settle towards the core. Calcium and aluminium began to migrate outwards, crystallising as plagioclase feldspar. By about 4500 Ma ago, the plagioclase had floated upwards to form a continuous anorthositic crust, and the interior had largely crystallised. Partial melting of the interior by radiogenic and other sources of heat, however, produced basalts which found their way to

Figure 1.1 One of the returned lunar samples. This one was collected on the Apollo 16 mission to the lunar highlands. Its brecciated character is clear, as is the pale colour of the matrix. Clasts of many different types are present, but the most common are anorthosites, impact glasses, basalts, and fragments of older shocked breccias. The sample is about 5 cm across. □

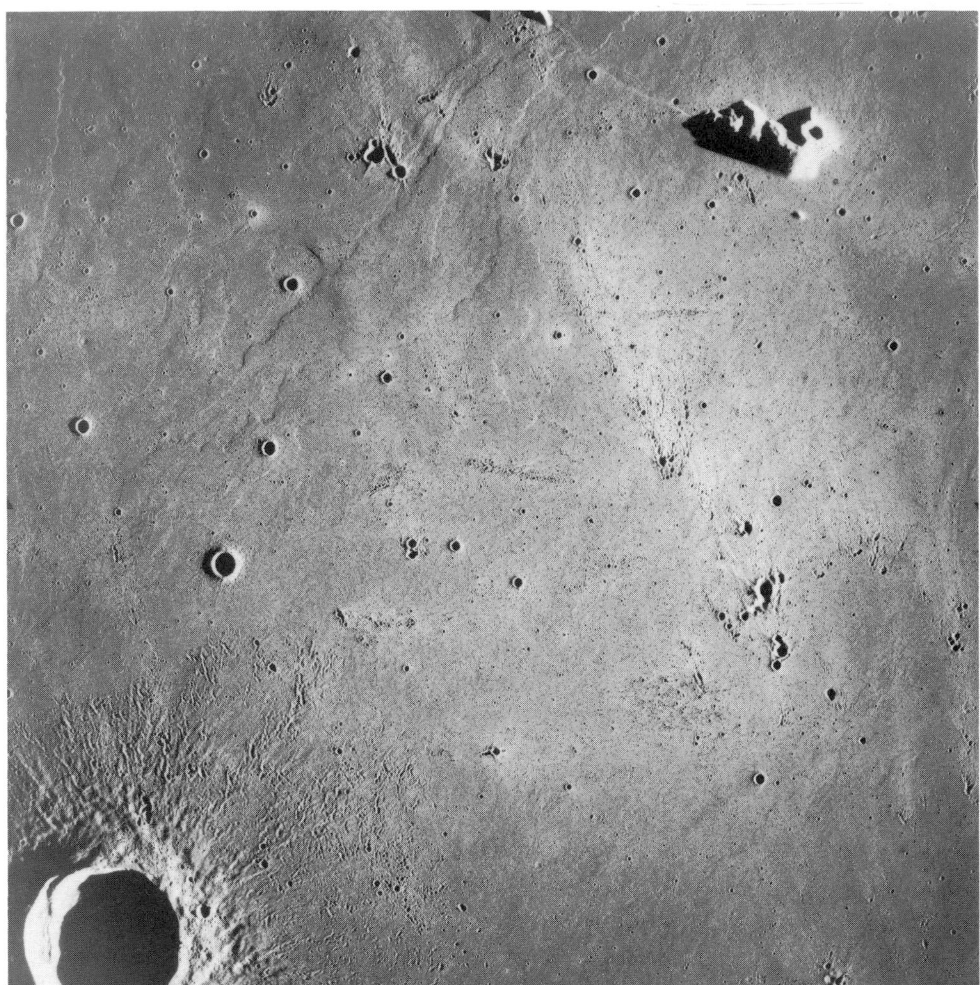

Figure 1.2 An Apollo orbital photograph of part of the Mare Imbrium, showing the crater Euler, and an isolated mountain called La Hire. To the left of La Hire, tongues of basalt lava can be seen extending over the surface of the mare. These flows are among the youngest on the Moon, about 3200 Ma old. Euler is approximately 27 km in diameter, and the mountain about 170 km distant. Note the 'aureole' of secondary craters surrounding the main crater. Between Euler and La Hire a number of small craters with 'V'-shaped appearance pockmark the smooth mare surface. These are secondary craters caused by ejecta thrown out by the impact that formed the 100 km diameter Copernicus, located well off the photograph towards the south (bottom). □

the surface. As the heat sources decayed away with the passage of time, the depth of partial melting steadily increased, with consequent changes in magma composition, until about 3200 Ma ago, surface volcanic activity died away, and the Moon became geologically extinct (Figure 1.2).

Although a consensus was emerging, the great controversy over the origin of the lunar craters had not been finally resolved before the Apollo program got under way. One school of thought had argued for decades that they were of volcanic origin. A conflicting school insisted that they must be of impact origin. The issue was soon decisively resolved. Studies of returned rocks showed that the Moon had been subjected to a massive asteroidal bombardment from its earliest days, which disrupted the anorthositic crust, probably even while it was still forming, and produced circular **impact structures** on scales ranging from thousands of kilometres (such as the Orientale basin on the far side of the Moon), through the familiar craters with diameters of the order of 100 km, visible from Earth with the most modest binoculars, down to submillimetric 'zap pits' gouged in the surface of rock samples (Figure 1.3). All but a handful of the tens of thousands of craters of all sizes on the Moon's surface are of impact origin. The very few that may be of volcanic origin are small, obscure features characterised by 'dark haloes' of ejected material. These may represent sites where Hawaiian-style fire-fountaining volcanic eruptions took place during the effusion of the basaltic lavas which accumulated within impact basins to form the basalt maria.

Studies of crater size/frequency statistics and radiometric age determinations show that the period of heavy bombardment of the Moon lasted from its formation until about 3800 Ma ago, and then faded away rapidly. Since 3800 Ma ago, impacts have occurred – and still continue – but only a few major craters have been formed during the period of the entire Phanerozoic record on Earth (i.e. the last 570 Ma). One of the most striking features of the full Moon is the 85 km diameter crater Tycho, from which brilliant white 'rays' of ejected material radiate for thousands of kilometres across the surface. Tycho was formed about 100 Ma ago, approximately at the beginning of the Cretaceous Period on Earth, while dinosaurs still roamed. Realisation of the antiquity of the era of heavy bombardment and of the decaying (but still continuing) flux of impacting bodies has provided a major insight into the geological record of the other planets in the Solar System, but profound problems remain over whether the impact flux on the other planets has varied through time in the same way as the lunar flux has. Tycho's existence on the Moon also raises a question of direct relevance to us here on Earth: did a similar impact wipe out all the dinosaurs and many other species at the end of the Cretaceous 65 Ma ago on Earth? Could another impact wipe *us* out?

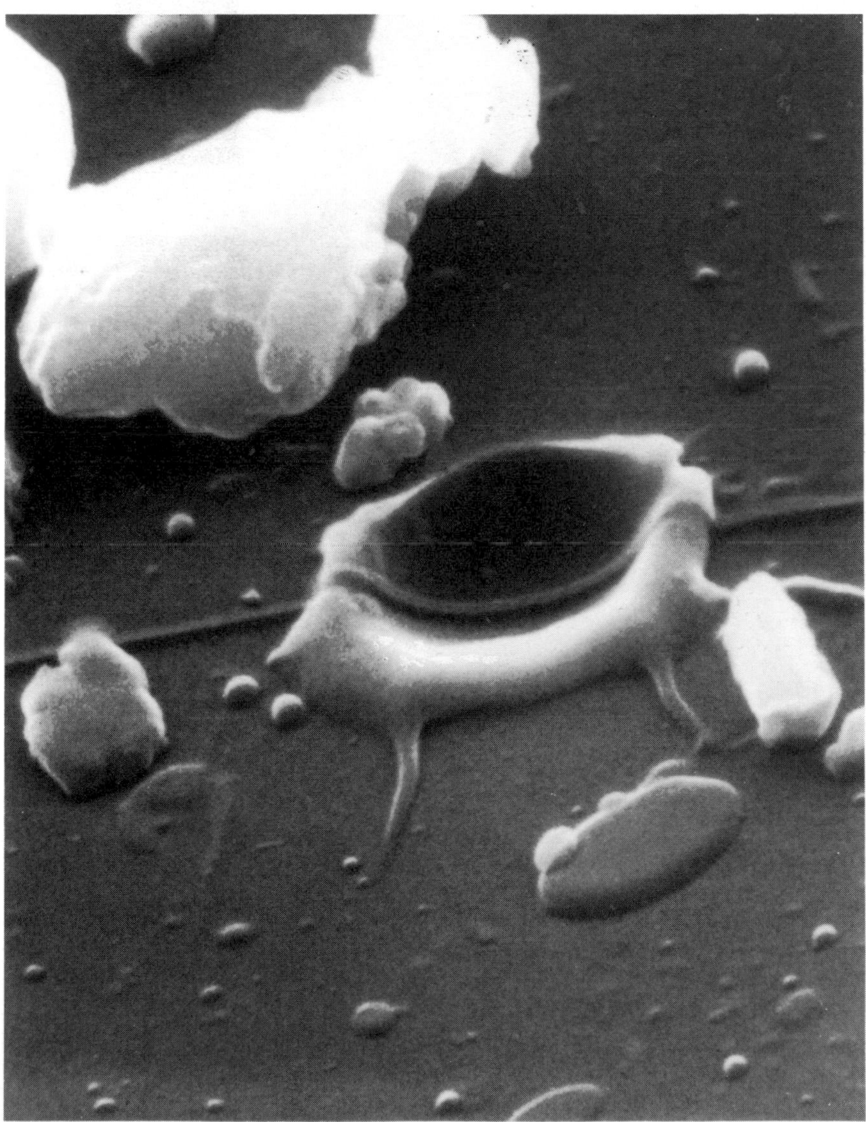

Figure 1.3 **A tiny 'zap pit' about one-tenth of a millimetre in diameter, caused by a micrometeorite impact on an Apollo 15 rock specimen.** □

Figure 1.4 A photomosaic of one hemisphere of Mercury, composed of pictures taken by Mariner 10 as it moved away from the planet after its first encounter. The similarities with the Moon are obvious. Notice in particular the concentric structure of a vast impact basin, just visible on the boundary between the sunlit and dark hemispheres of the planet. Mercury is 4880 km in diameter; the impact basin is about 1000 km across. □

Mercury

Mercury is a small planet, not much bigger than the Moon. Because it orbits so close to the Sun (only 58 million kilometres) it is usually lost against the solar glare, and is therefore exceedingly difficult to study telescopically. Thus, not much was learned about it in the era of visual observations. Remarkably, even its rotation period remained unknown until the 1960s, when radar studies showed that this period is 58.6 days, thereby revealing an elegant example of celestial mechanics. Mercury's rotation period is exactly two-thirds of its orbital period (88 days) so that the planet rotates three times around its axis while completing two orbits around the Sun. This spin–orbit coupling is a more complex example of the gravitational phenomenon that keeps the same face of the Moon always turned towards the Earth. Like the Moon, Mercury is not a completely symmetrical sphere, but has a slight bulge on one side. Because its orbit is also elliptical, tidal forces acting between Mercury and the huge mass of the nearby Sun keep the planet facing more or less towards the Sun while it is closest to it but, when further away, it makes a complete rotation.

One other important property of Mercury was learned early on: although small, it is a surprisingly dense planet. This means that it must have a massive metallic core, which is proportionally much larger relative to its diameter than that of any other planet. The reasons for this are not fully understood, but one suggestion is that Mercury may have lost much of its original 'mantle' material during a catastrophic impact event in the earliest days of the Solar System.

Almost all that we have learned about Mercury resulted from a single space mission, Mariner 10. The Mariner 10 spacecraft did not head directly to Mercury. Its trajectory was finely calculated so that it first swept close by Venus to gain acceleration before being flung off towards Mercury in what one writer described as 'an exquisite celestial slingshot'. The images obtained by Mariner 10 revealed that Mercury looks rather like the Moon – a barren sphere pockmarked with craters. There are even examples of rayed craters, splashing white streaks over the surface. At a casual glance, it is often difficult to distinguish between pictures of Mercury and the Moon (Figure 1.4). Closer inspection, however, reveals some important differences.

First, while the Moon's topography shows a clear division between rugged highlands and smooth, dark lava plains (maria), there are no obvious equivalents of the lunar maria on Mercury – the surface is dominated by heavily cratered 'highlands' material.

Second, while there is unambiguous evidence of relatively young volcanic activity on the Moon – most prominent in the lavas of the maria – the situation on Mercury is more complex, and there are *no* obvious lava-covered areas. There are, however, some 'intercrater plains' which *may* be of volcanic origin.

Third, there are subtle differences in the morphology of impact craters. Because of its greater size and density, surface gravity is two and a half times greater on Mercury than the Moon, and hence material ejected on impact behaves very differently – for example, **secondary craters**, formed by material falling back around primary impact sites, are much less widely distributed around the primary impact site.

Fourth, the surface of Mercury is broken by a number of long, low breaks or steps in the topography, known as **scarps**, which can be traced for hundreds of kilometres. Nothing comparable is known on the other planets. It is thought that the scarps were produced early in the history of Mercury, perhaps when the planet was entirely molten. As it cooled, the planet shrank and decreased in volume, causing the outer crust to wrinkle up and contract, forming the scarps. Tidal interactions with the Sun may also have been involved.

Finally, Mariner 10 also showed that Mercury has a significant magnetic field, with north and south poles aligned with the planet's rotation axis like the Earth's, but only about one-hundredth of the intensity. Mars and Venus both lack significant magnetic fields. The Moon has none at present, but returned samples of mare basalts reveal a remanent, or fossil magnetism. Why should this be? What is there about the Earth that is responsible for its powerful, changing field? Although the **Earth's magnetic field** has been closely studied for three centuries, its origin remains to be fully understood. Motions in the Earth's fluid metallic core and our planet's rapid rotation period are clearly involved in some way, but the details are controversial. Future studies of Mercury's somewhat similar field could provide insights into the origin of the Earth's field.

Although Mariner 10 was a remarkable mission, much remains to be learned about Mercury. Only about half of the planet was imaged during the mission; the rest remains as unknown now as it has always been. However, it seems unlikely that any bizarre anomalies are concealed, and so Mercury, like the Moon, is a barren, heavily cratered body that has been geologically extinct through most of the history of the Solar System.

Venus

Mercury is hard to see. Venus, by contrast, blazes so brightly in the evening (or morning) sky that it is hard to avoid, and the planet is reported with tedious frequency as an Unidentified Flying Object. Observed through a telescope, Venus appears as a disappointingly blank billiard ball, because all that can be seen is the uppermost surface of its dense, cloud-laden atmosphere which completely conceals the topography beneath. Venus owes its brilliance to the high albedo of its cloud cover, and also to its proximity – it often comes within 40 million kilometres of the Earth. In terms of size, mass and density, Venus is similar to the Earth, so similar that it is often thought of as the Earth's twin, and early astronomers postulated that Venus and Earth should therefore share similar properties.

For a long time the dense cloud cover precluded effective studies of Venus. In the three centuries since the invention of the telescope, no progress at all was made in determining so simple a parameter as the axial rotation period. All sorts of guesses were made, from 24 hours to 225 days (the orbital period). The problem was not finally resolved until radar astronomers turned their attention to Venus in 1962, and found that the planet turns on its axis once every 243 days – *backwards*! Venus lumbers slowly round on its axis in the opposite sense to every other major body in the Solar System. The 243 day period revealed another elegant but enigmatic Solar System statistic: the Earth's orbital period and Venus' axial rotation period are in the ratio exactly 3:2.

In the 1970s and 1980s, a positive shower of Soviet and American spacecraft descended on Venus. In December 1978, no less than seven spacecraft reached the surface within a few days of each other. These spacecraft had many different objectives: profiling through the atmosphere, mapping the surface from orbit using radar altimeters and imaging radars, and direct examination of the surface by landing craft. As a result, our knowledge of Venus has increased enormously, and continues to do so.

The orbital radar data show that Venus is distinctly different from the Earth, having far less topographic relief. Most of its surface is covered by endless rolling plains with rather uniform elevations. There are some higher regions, however, of which two (Ishtar Terra and Aphrodite Terra) have been compared with terrestrial continents, although they are much smaller. Several smaller elevated areas such as Beta Regio are thought to be large volcanic constructs. Other circular features have been interpreted as giant calderas. Complex sets of linear ridges in Ishtar have been interpreted in terms of collisional tectonics, similar to those in Tibet produced by the indentation of the African plate into the Asian plate (see Chapter 14). Features reminiscent of the symmetrical topography of terrestrial ocean ridges have also tentatively been identified. Perhaps most important of all, a number of large impact craters has been identified, suggesting that large parts of the Venusian surface are geologically ancient, although by no means as old as the lunar highlands.

Images from the first Soviet spacecraft (Venera 9) to survive landing on the planet revealed a rather grim prospect: a bleak, flat, rock strewn terrain stretching from horizon to horizon. The view from Venera 10, which landed 10 000 km away, was somewhat similar, although the rocks at that site were smaller and slabbier. Gamma-ray and X-ray fluorescence analyses of surface rocks carried out on board these and later spacecraft confirmed what had been suspected previously: the rocks were basalts, generally similar to terrestrial ocean-ridge basalts. Perhaps the key results from the landers, though, concerned surface conditions. Temperatures are searingly hot (around 500 °C), and the atmospheric pressure crushing; ninety times greater than the Earth's.

The atmosphere itself is no less unappealing, consisting of 95 per cent carbon dioxide and a few per cent of nitrogen with small traces of sulphur dioxide and water. Three well-developed decks are present, separated by clear layers. Of these, the lowest and thickest cloud layers are about as dense as clouds on Earth, but are much higher, with bases at 50 km altitude. The most likely constituent of the clouds is liquid droplets of sulphuric acid. Thus, as one distinguished American scientist wrote 'Venus is astonishingly hot, with an oppressively dense atmosphere containing corrosive gases, with a surface glowing dimly by its own red heat ... Venus ... seems very much like the classical view of Hell'.

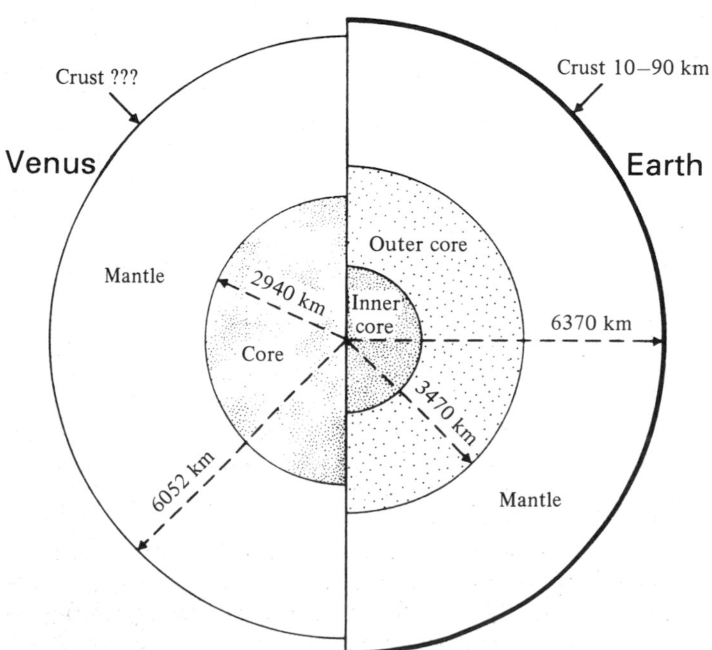

Figure 1.5 Internal structures of Venus and Earth, drawn to the same scale. Venus is closely similar to the Earth in many of its properties, but strikingly different in its surface geology. □

Hellish or not, having a closely similar size, mass, density and bulk chemical composition, Venus clearly started off in a similar state to the Earth, and probably experienced similar internal processes (Figure 1.5). The extraordinary divergence in the pattern of surface geological evolution of the two planets must be bound up with the quite different atmospheric evolution that they have experienced. Exactly how this evolution took place, and how their atmospheric and lithospheric evolutions were interlinked remains to be explored, and present some of the most challenging questions in planetary science today. One problem relates to how Venus dissipates its internal heat. On Earth, this takes place through the operation of plate tectonics: new oceanic crust is continually being created at spreading ridges and cooling via conduction and convection to the ocean. Because of its similar size and composition, Venus should have a similar thermal budget, but the evidence for terrestrial-type plate tectonics is slight. Furthermore, the high surface temperatures suggest that the Venusian lithosphere should be thinner and more buoyant than the Earth's, thus prohibiting subduction of terrestrial type. Some scholars think that it would be difficult to detect fast-spreading ridge features on Venus on radar images currently available, so the possibility remains. An alternative model suggests that Venus loses its internal heat through a 'hot spot' mechanism, that is, through a small number of large, central volcanic complexes, akin to the Hawaiian hot spot. Beta Regio may be such a hot spot. At the time of writing (1990), a NASA spacecraft (Magellan) had reached Venus, producing many high-resolution orbital radar images which are being used to unravel the complex geological history of this planet.

Mars

Mars moves rapidly against the background of fixed stars and in some years blazes a fiery red against the night sky, whereas in other years it is a feeble point of light. These variations, which are due to the changing relative positions of Earth and Mars in their orbits, were for millenia a rich source of myth for astrologers. In the days of telescopic observation, Mars was a natural target for study, and it still fires the imagination of scientists and public alike. Largely, this may be because *all* the other planets are unbearably hostile to life. *Only* on Mars – and perhaps Pluto – could lightly protected astronauts explore the surface and establish permanent bases. (Several planetary satellites, like the Earth's Moon, offer similar potential.)

Because Mars has easily visible surface features, basic parameters such as size and axial rotational period were easily measured telescopically; the latter as early as 1666, when Cassini found the period to be 24 hours 40 minutes, strikingly close to the Earth's. The earliest observations showed that Mars has polar ice caps, like the Earth, and diffuse dusky markings which, although essentially stable in overall outline, appeared to show seasonal variations, shrinking and expanding in rhythm with the polar caps. This naturally prompted speculation that the variable markings were huge tracts of vegetation and that Mars would be an agreeable abode for life. At the beginning of this century many observers, notably the American, Percival Lowell, believed that they could see spidery networks of **'canals'** on the surface, and thus the legend grew up that Mars was inhabited by an intelligent race, struggling to preserve life on their planet by building vast canal systems from the poles across the ochreous desert wastes, providing irrigation for scattered oases.

All this, of course, was merely the product of fevered imaginations. Telescopic observers such as Lowell were unwilling to accept the well-understood physical limitations of terrestrial telescopes. Because of Mars' small size and vast distance from the Earth (up to 100 million kilometres) and atmospheric turbulence, it is impossible to resolve features on Mars which are less than hundreds of kilometres across, even with the best telescopes. Thus, Mars' topography and geology remained unknown until the era of spacecraft exploration, which revealed that the intricate networks of 'canals' drawn by Lowell and others simply did not exist.

Lacking canals, the first spacecraft views of Mars were not impressive. In 1965, Mariner 4 transmitted twenty-two television pictures, which revealed a rather dull, cratered surface, depressingly like that of the Moon. Subsequently, Mariner 9 and the Viking 1 and 2 missions obtained far more data, and showed that Mars is a richly diverse planet, geologically far too multifaceted to encapsulate adequately in a brief review (Figure 1.6).

In summary, Mars exhibits a dichotomy between its northern and southern hemispheres, the southern being elevated and rugged, the northern lower and smoother. The southern hemisphere is heavily cratered, contains large impact basins, such as Hellas (hundreds of kilometres in diameter), and resembles the lunar highlands in many ways. The Martian highland rocks may be comparably old. This is a first key point of Mars' geology: much of the crust is ancient, perhaps 4000 Ma or more. The northern hemisphere is less heavily cratered and therefore less old, but craters are

Figure 1.6 This oblique view from the Viking I orbiter over the surface of Mars reveals many of the planet's complexities. Cloud or haze layers in the atmosphere are visible on the horizon and are about 25–40 km above the planet's surface. In the middle distance is part of a large circular impact basin known as Argyre Planitia, ringed by mountains and hills. Argyre is over 700 km in diameter. In the foreground at bottom left, a multitude of small channels is visible. These are thought to have been formed by run-off of surface waters. □

none the less numerous enough to emphasise that these surfaces too are ancient by terrestrial geological standards.

The second key point is that several giant volcanoes rise high above the surface, and that their lava flows – clearly visible on high-resolution Viking images – are almost uncratered, and therefore must be geologically young. Much the largest volcano is Olympus Mons, which rises no less than 26 km above the surrounding plains, and has a summit caldera complex 70 km across, which alone could enclose a dozen terrestrial volcanoes of the size of Vesuvius. Until samples are returned from Mars for age dating, it will be difficult to pin down exactly how old – or young – Mars' volcanoes are. Most estimates suggest that 'active' volcanism had ceased by about 1000 Ma ago. This accords reasonably well with estimates of the likely thermal evolution of Mars deduced from knowledge of its mass, probable bulk composition and content of heat-producing radioactive isotopes. The vast size of Martian volcanoes allows us to infer something about Mars' lithosphere: in order for volcanoes to grow to such large sizes in fixed positions, the present Martian lithosphere must be thick and rigid, at least 200 km thick. This, of course, precludes the possibility of plate-tectonic processes of terrestrial style.

The third key point is perhaps the most intriguing of all: Mars shows unambiguous evidence of dramatic changes in climate through its geological history. At present, the atmospheric pressure is tenuous, corresponding to about 6 terrestrial millibars (standard atmospheric pressure = 1113.25 mb), and surface conditions are extremely cold and dry. Specifically, liquid water would not be stable at this extremely low pressure, and would rapidly boil off. Mariner and Viking images, however, revealed remarkable meandering channels, valleys and canyons, all pointing to an earlier period – or periods – when Mars' climate was less harsh, and liquid water could exist on the surface. These features showed that the surface of Mars is much more Earth-like than the other planets, and that water has played a major role in its geological history (Figure 1.7). Impact crater morphology provides a second major pointer to this. The ejecta blankets surrounding many of Mars' craters are quite different from those on the Moon, and *superficially* give the appearance that the impact took place into a wet, plastic medium like mud (Figure 1.8). These distinctive ejecta blankets may have been formed by impacts striking areas of Mars' surface where there was a deep permafrost layer, containing large volumes of volatiles which were liberated by the impact event.

The vast areas of Mars underlain by permafrost ice represent an enormous reservoir of water, and much present research is focussed on how this might have been liberated in the past. Some studies have suggested that variations in Mars' orbital parameters (similar to those used in the Croll–Milankovich hypothesis to account for the Earth's Ice Age – see Chapter 24) could have caused large-scale transfer of water from permafrost ice to the atmosphere, increasing atmospheric pressure drastically, and in turn raising surface temperatures and enabling liquid water to flow in channels. Much of this is still controversial. Some scientists believe that the erosional features observed on Mars were not eroded by *flowing* water, but by sapping of ground water liberated by melting of permafrost (Figure 1.7). There are good analogues of this kind of erosion in the Colorado Plateau of the USA. Regardless of the details, the questions raised are fundamental, and show that if we are to understand Mars' geological history, we have also to understand its atmospheric history, and the complex interactions between atmosphere and crust.

Public interest in Mars has, of course, centred around the possibility of life existing there, and the discovery of water-sculpted topographic features naturally gave these speculations a fresh impetus. The two Viking surface landers were designed specifically to answer the question 'Is there life on Mars', and they did: *they found no evidence whatsoever of any form of life*. This has not ended the debate, however. Perhaps prompted more by wishful thinking than the true scientific spirit, various groups have postulated a range of reasons of varying degrees of plausibility as to why the Vikings did not detect life. Future surface missions

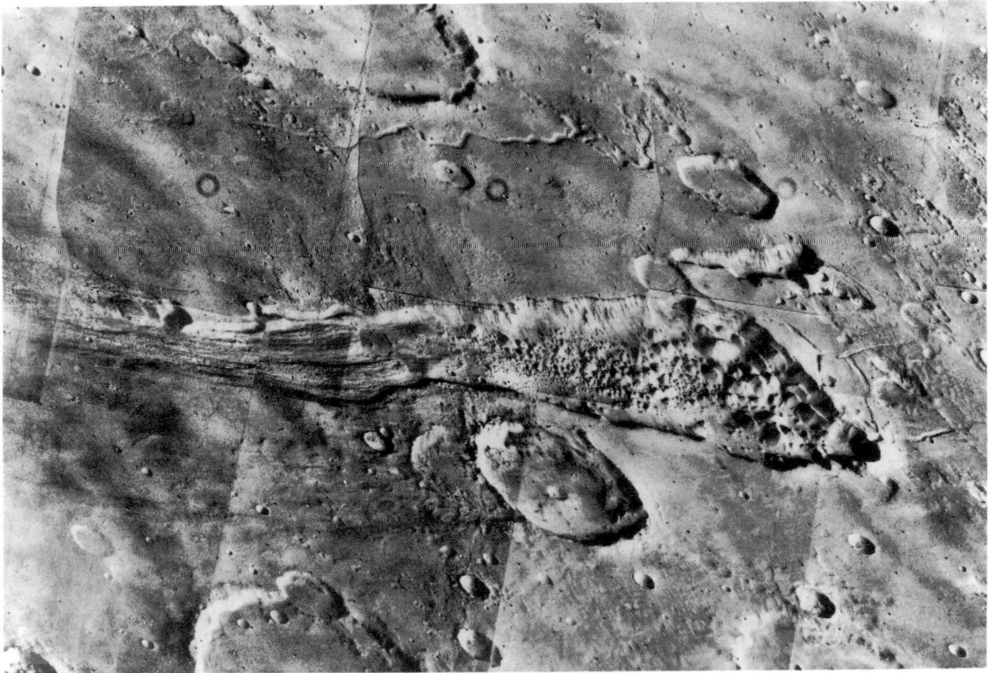

Figure 1.7 A great collapse feature on Mars, probably triggered by melting of permafrost ice at the head of the gorge; slumping of the walls has yielded chaotic mounds on the floor (right), while lower down, well-defined stream features suggest erosion by torrents of water released by melting ice (left). Ground water sapping may have enlarged the valley. The picture covers an area about 300 km from side to side. □

Figure 1.8 Like many lunar craters, Arandas crater on Mars (25 km in diameter) has a pronounced central peak, but it also displays spectacular 'ramparts' of material which appear to have splayed out radially in a sludge-like fashion. Ramparts such as these have been used to argue that the Martian crust contains a high proportion of volatiles in the form of permafrost ice. □

will doubtless continue the search. For geologists, perhaps the most interesting question is, given the absence of life today, could it have evolved earlier in Mars' history, and are there traces of it (fossils) preserved in the rocks?

To conclude this brief review of Mars, it is essential to recall the case of the Egyptian dog. In this case, the killer is much more interesting than the dog who, so far as we know, died nameless and unmourned. There are two remarkable things about the dog that died at Nakhla, Egypt, in 1911. The first is that the dog was apparently killed by a falling meteorite. The second, much more remarkable, is that *the meteorite probably came from Mars*. This surprising deduction arises from the facts that the Nakhla meteorite has an obvious igneous texture, far different from the majority of meteorites, and that it has an apparent crystallisation age of about 1300 Ma. Where in the Solar System could there have been a crystallisation event 1300 Ma ago? As Sherlock Holmes would have argued, once the impossible has been eliminated, whatever remains, however improbable, is the solution. Thus, Mars seems to be the only possible source of the Nakhla meteorite (and a handful of others known as the **SNC**s, for **Shergotty–Nakhla–Chassigny**). Glancing blows from impacting asteroids would jettison small quantities of material from the surface of Mars into orbits that would eventually intersect the Earth's. These samples are being minutely studied to determine what they can reveal about Mars' geological history, without the expense of actually going there.

Phobos and Deimos, and the asteroids

Some of the fortunate by-products of the remarkably successful Viking missions were the first detailed images of Phobos and Deimos, Mars' pair of tiny moons. Viking 2 swooped past Deimos at a range of only 26 km, a stunning piece of celestial navigation. Phobos is an ellipsoid, with a greatest dimension of 27 km (Figure 1.9), while Deimos is more spherical and only 15 km in diameter. Both have heavily battered, cratered surfaces, are very dark, and have low densities, suggesting that they are made of material akin to that of carbonaceous chondrite meteorites. Because their orbits are not stable, it seems clear that these tiny satellites have not orbited Mars since the origin of the Solar System. They are most likely

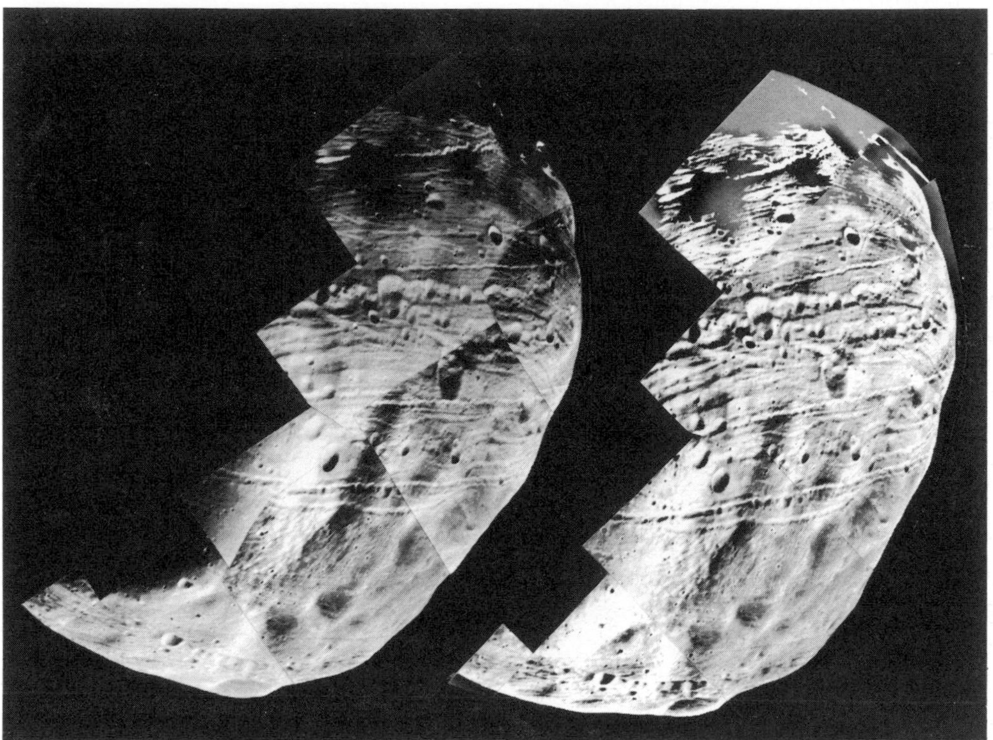

Figure 1.9 A view of Phobos obtained by the Viking I Orbiter when it flew past at a distance of less than 300 km. Numerous craters and linear grooves are conspicuous. Phobos is probably typical of hundreds of asteroids in the asteroid belt, and may consist of a jumble of fragments produced by impacts which have re-assembled together. (The right-hand picture is an enhanced view of the left hand one.) □

asteroids, captured somehow from the asteroid belt between Mars and Jupiter and, as such, they provide us with the only available close-up views of asteroids. Telescopic studies of asteroids show that there are several thousands in the asteroid belt, ranging in size from respectable bodies with diameters of hundreds of kilometres, down to small tumbling boulders, smaller even than Phobos and Deimos. Spectroscopic studies show that several different classes are present, which have been interpreted as corresponding to carbonaceous, metallic and rocky types, similar to the known range of meteorite types. Indeed, it is thought that almost all meteorites are derived from the asteroid belt, although the issue is not fully resolved. (S. R. Taylor discusses meteorites and their role in the assembly of the Earth in Chapter 2.)

Gas giants: Jupiter to Neptune

Once beyond the asteroid belt, we enter a different realm. Jupiter and the other more distant planets are not even vaguely Earth-like, but are large, low-density balls of compressed gases, mostly hydrogen and helium, so they are often grouped together as the 'gas giants'. In detail, there are two pairs, Jupiter and Saturn, and Uranus and Neptune. Jupiter and Saturn are true gas giants and are composed, respectively, of 97 and 70 per cent hydrogen and helium, whereas Uranus and Neptune are composed of only 10–20 per cent hydrogen and helium, most of their mass being icy and rocky material. On all four planets, however, we can observe directly only the extreme outermost fringes of their atmospheres, and must hypothesise on conditions in their interiors, where pressures are so great that we have limited knowledge of the physics that prevails. Exploration of the Solar System has led in many strange directions, and in the case of Jupiter has led to studies of the properties of hydrogen when confined in a diamond cell at pressures equivalent to 360 000 atmospheres, at which pressure the element has many of the properties of a metal. At the centre of Jupiter, temperatures and pressures may both be so high that hydrogen behaves as an electrically conducting, liquid metal.

Both Jupiter and Saturn are easy to observe, and their principal telescopic features were learned early on. Their masses, densities and rotation periods were easily measured, but are still among their most interesting parameters: although Jupiter is by far the most massive planet in the Solar System, it also has the shortest axial rotation period, whipping around in a mere 9 hr 55 m. The rotation period is so fast that the planet is visibly flattened by the strong centrifugal force. The degree of flattening – or oblateness – however, is not as much as would be expected from moment of inertia considerations for a homogeneous body, suggesting that the planet must have a concentration of mass towards its centre, probably a small amount of rocky material.

Telescopic studies of Jupiter necessarily focussed on its outer atmosphere, and detailed maps were made of various equatorial belts or zones of clouds. A few persistent spots were also observed, including the famous 'Great Red Spot' which has come and gone intermittently for 300 years. Although not fully understood, it was readily perceived to be an atmospheric phenomenon, because its position varied relative to other features. Two Pioneer and two Voyager spacecraft missions during the 1970s and 1980s provided stunning new details of the atmospheric circulation, and a wealth of other data (Figure 1.10). Voyager 2 has been a particularly historic mission, providing brilliant views not only of Jupiter, but also of Saturn, Uranus and Neptune during its unique 'grand tour' of the Solar System. It is a remarkable tribute to the design of the spacecraft that all of the software for the on-board computers was re-written and 'uplinked' to the spacecraft while it was *en route* from Saturn to Uranus, many years and millions of kilometres since it left Earth. Voyager 2's last encounter was with Neptune in August 1989. It is now heading out into interstellar space.

At present, it is thought that the uppermost clouds of Jupiter consist of tiny crystals of **ammonia ice** (akin to terrestrial cirrus) and that layers of ammonium hydrosulphide (NH_4SH) and water exist at deeper levels. The top of the ammonia cloud deck probably has a temperature of −113 °C, and is at a pressure of about one atmosphere. One intriguing problem is that while ammonia cirrus should be white, Jupiter's clouds are visibly tinged with yellows, oranges and reds. Many suggestions have been put forward to account for these colours. Organic polymers, perhaps formed when

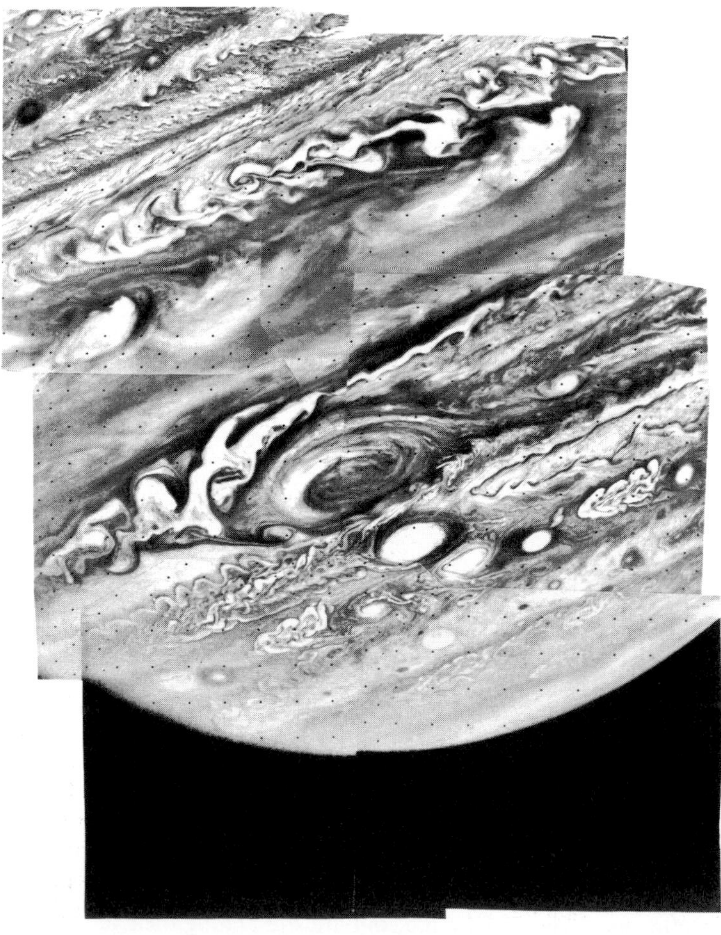

Figure 1.10 Complex swirls and eddies abound in this composite photograph of Jupiter obtained by Voyager 1. The Great Red Spot is prominent towards the bottom, with a complex atmospheric circulation pattern surrounding it. The white circular patches are believed to be enormous convection cells, where warm gases (pale tones) rise in a spiral fashion at the centre and turn downwards to form the outer, darker rings. Most of the spiral cells also have an anticlockwise rotation sense. □

lightning flashes cause chemical reactions between ammonia and methane, may be one possibility. According to another theory, the Great Red Spot may owe its colour to red phosphorus, formed by breakdown by ultraviolet radiation of phosphine molecules derived from deep within the atmosphere. This begs the question of course, of where the phosphine came from. Unfortunately, detailed studies of Jupiter's atmosphere will have to await the Galileo mission, part of which will include a probe that will descend deep into the atmosphere, sampling as it goes. This mission was intended for launch several years ago; was delayed by the disaster to the Shuttle Challenger; was finally launched in 1989, and will reach Jupiter in 1996, after looping once round Venus and twice round the Earth.

Apart from details of its upper atmosphere, the Pioneer and Voyager missions provided new insights into some other aspects of Jupiter. It was discovered that the planet has an intense magnetic field, ten times more powerful than the Earth's. The field is broadly dipolar, but is much more complicated near the surface, where quadrupolar and octopolar fields were detected. This observation has two implications: to generate a magnetic field, there should be an electrically conducting medium within the planet. In the Earth, this is iron; in Jupiter it must be a form of metallic hydrogen. Secondly, the conducting material should be in motion, and this implies in turn a source of energy to provide movement.

Jupiter's vast size and exotic internal processes can best be understood by considering it from a different perspective: it is more of a failed star, like the Sun, than a 'real' planet. Early in its life, Jupiter shone like a star, about 1 per cent as luminous as the Sun is today, heated by the accretion of nebular material. Had it been seventy times more massive than it is, gravitational contraction would have caused a further temperature rise, until self-sustaining nuclear reactions could ignite in its interior. If this had happened, the Sun would have been a double star, and the Earth and other planets might not have formed. Jupiter was too small, however, and within 10 Ma of its formation had shrunk to its present size without igniting, and today its luminosity is only one ten thousand millionth of the Sun's. Its internal energy is still very large, however: its internal temperature is probably 30 000 K, sufficient to keep it entirely molten, with no solid core whatever. And about 10^{17} watts of power reach the surface from the interior, roughly comparable to that received by Jupiter from the Sun.

Much of that heat is probably pumped to the surface by convection currents, carrying warm, relatively low-density hydrogen upwards, while cooler, denser hydrogen sinks. Such convection currents, taking place within the liquid metallic hydrogen part of Jupiter, would provide an elegant source of energy to generate the magnetic field: they are analogues of the circulating currents in the liquid iron of the Earth's core (see Chapter 3).

For present purposes we can consider Saturn as sharing most of Jupiter's properties. Uranus and Neptune present more challenging problems. Although it is termed a 'gas giant', Uranus is rather an odd-ball. For one thing, its axis of rotation lies almost in its orbital plane, rather than being roughly perpendicular to it, as all the other planets are. One consequence of this is that when the Voyager spacecraft encountered Uranus, it dived headlong *through* the plane passing through the equator of the planet and all its satellites, like an arrow through an archery target, rather than curving *around* it. Uranus also has a much less massive hydrogen–helium atmosphere relative to its dense core than Jupiter or Saturn. This is true also of Neptune. An important problem therefore emerges: why should the dense cores of the four giant planets vary in mass by a factor of only three or four, while their gaseous envelopes vary by a factor of ten to twenty? No simple answer to this question is likely to emerge in the short term. It is clear, though, that the internal structures of Uranus and Neptune are different from Jupiter and Saturn. One model for Uranus invokes three layers: a dense, rocky core, a 'mantle' of liquid water around the core, forming an ocean thousands of kilometres deep, and a dense atmosphere of hydrogen and helium. The 'ocean' probably contains methane and ammonia as well as water, is electrically conducting, and may even be metallic near its centre. Motions in this conducting fluid are responsible for generating Uranus' magnetic field, which is comparable in intensity to the Earth's. Voyager observations of Uranus' magnetic field, however, revealed a further astonishing Uranian anomaly: the axis of the field is inclined at an angle of 60° to the rotation axis! (Recall that the Earth's field is tilted only 11° to the rotation axis.)

Uranus and Neptune are rather similar in size, mass and rotation period, so they are sometimes termed twins. They also have some startling differences, however, so they are best thought of as cousins rather than twins. Uranus is rather a featureless object, as the Voyager images revealed. Neptune is much more diverse, and rapidly evolving cloud features scud across its surface. Some of the clouds are so large that they can be detected with Earth-based telescopes which show that the apparent brightness of the planet varies rapidly, whereas Uranus' is constant. Unlike Uranus, Neptune radiates far more heat outwards than falls on it from the Sun. Its atmosphere also contains much less methane than that of Uranus. One explanation of all these important differences is that Uranus has a much more stable internal stratification than Neptune. Thus, convection currents arising from heat sources deep within the core are confined to great depths. In Neptune, which is less stably stratified, convection not only carries much heat upwards and outwards to the atmosphere, but also transports huge volumes of methane upwards, where it condenses to form clouds in the frigid hydrogen–helium atmosphere.

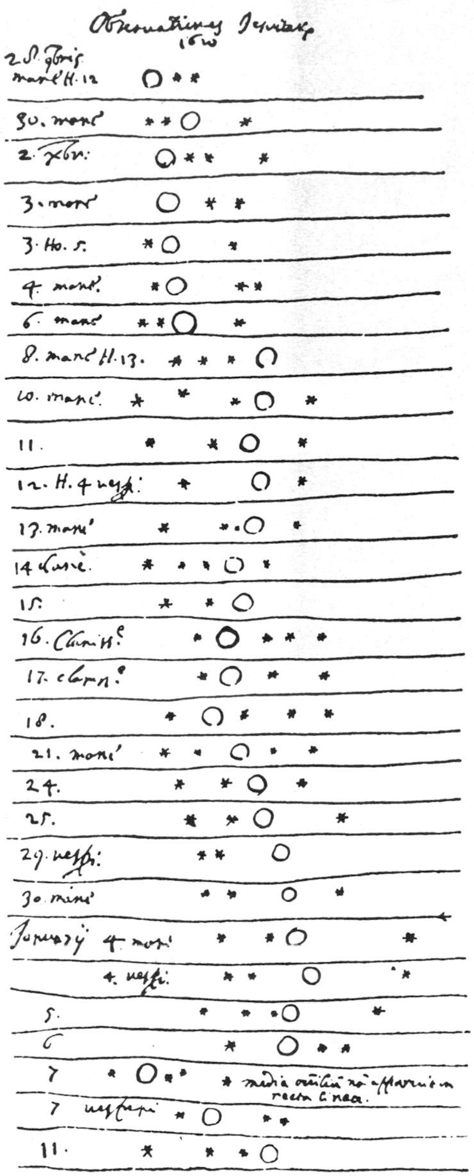

Satellites of the outer planets

An expected but still significant result of the spacecraft exploration of the outer planets was the discovery of swarms of satellites; satellites so numerous that names are no longer sought for all of them, and numbers are now formally assigned (Tables 1.2–1.4). The major satellites, of course, had long been known. The four Galilean satellites of Jupiter (Io, Europa, Ganymede and Callisto) can easily be seen with a pair of binoculars, and because they are so easily observed, they have played a major role in the history of science: Galileo used them to show for the first time that objects in the Solar System do not all revolve around the Sun (Figure 1.11), and Romer used timings of their orbital motions to make the first accurate measurement of the velocity of light. Another success of the Voyager missions were superb images of the surfaces of the satellites, providing many clues to their compositions and evolution.

The Galilean satellites are large – Ganymede and Callisto are as big as the planet Mercury, while Io and Europa are about the same size as the Moon. Each has had a distinctly different geological evolution. In the case of Io, this has been dominated by its proximity to the enormous mass of Jupiter, its complex orbital resonances with Europa and Ganymede, and the huge internal tidal stresses that result. In a brilliant piece of scientific theorising, two American scientists, Peale and Casson, predicted that Io was so close to Jupiter that dissipation of tidal energy would generate sufficient heat to melt the interior, and thus that volcanism should be vigorously active. The Voyager spacecraft proved the accuracy of their predictions, with stunning pictures of eruptions actually in progress (Figure 1.12). Now the problem is to determine the nature of the materials erupted. There are some reasons for believing that the 'lavas' of Io's volcanoes are liquid sulphur, but this is controversial, and conventional silicates may be involved as well.

Figure 1.11 Galileo's meticulous records of his first observations of the relation positions of Jupiter's four large satellites, made in December 1609 and January 1610. Observation dates appear in his handwriting on the left; on each date Galileo made a sketch of the relative positions of Jupiter and the four satellites. Jupiter is the large circle, the satellites small stars. He even recorded details of 'seeing' conditions – e.g 'clarissime' (exceptionally clear) on 16 December. □

Io and Europa are unusual among the satellites in that they have densities comparable to the Moon, and are therefore composed dominantly of silicate materials. The remaining scores of satellites have much lower densities, and must be made largely of ice, mostly

Figure 1.12 **One of the most extraordinary scientific photographs ever taken, this Voyager view of Io shows two volcanic eruptions in process. One is seen on the limb of the satellite, spraying particulate material over 200 km into space, which then falls back along parabolic ballistic paths. The second is on the boundary between light and dark areas, and is seen as a bright patch because it is high enough to catch the rays of the Sun. Io is 3640 km in diameter.** □

ordinary water ice, but with other phases such as methane and ammonia also present. They may also have small rocky cores. The surface features of all of these satellites are dominated by the interplay of two processes: impact cratering, and 'volcanic' resurfacing. Most of the satellites have heavily cratered surfaces, which are clearly ancient, but many of those which are close enough to their parent planet show evidence that tidal energy dissipation has caused melting of the interior sufficient for the cratered surface to be wholly or partially smoothed over or replaced by new ice. Ganymede even appears to exhibit a form of icy plate tectonics (Figure 1.13). The details of such extraordinary icy processes are, of course, far outside our experience, but are the subject of much ongoing research.

Two of these outer planetary satellites are exceptional in that they are large enough to possess their own atmospheres. Saturn's Titan has a methane atmosphere so dense that the surface is invisible to spacecraft. Early in the next century, the joint NASA/European Space Agency Cassini mission is planned to drop a probe through the atmosphere and investigate the surface. Neptune's satellite Triton has a much more tenuous atmosphere consisting mostly of nitrogen with smaller amounts of methane. It is particularly interesting, because it appears to show seasonal variations, with shifting 'frosts' of frozen nitrogen ice forming at the poles and nitrogen geyser-like volcanoes that form where liquid penetrates through the ice cover. Triton may resemble the outermost (tiny, low-density planet) Pluto in many ways. In studying the remarkable Voyager 2 images of Triton, we have obtained the best idea of what Pluto is like that will be possible for many decades to come.

Rings

Saturn's elegant system of rings has provided something approaching a caricature of a planetary object since their discovery in the 17th century. For a long time, they were thought to be unique, but in 1977 careful telescopic studies of Uranus as it passed in front of a star showed a curious series of blips in light from the star *before* it passed behind the planet, and again afterwards. The pattern of blips could only be explained by a series of rings around the planet. Two years later, when Voyager 1 reached Jupiter, its pictures revealed a previously unsuspected narrow ring system around it too. Voyager 2 found a complex series of rings around Uranus, and when it arrived at Neptune, three rather tenuous rings were even found there. Far from being unique, ring systems therefore are common features, and will probably be found

Figure 1.13 Ganymede, a satellite of Jupiter which is larger than the planet Mercury. The bright patches are impact craters, like those on the Moon, but the surface is probably dust-covered ice rather than rocks. Large dark regions are separated by lighter toned ice. It has been suggested that some sort of icy 'plate-tectonic' process has operated, causing chunks of the satellite's crust to move apart. Each of the dozens of icy satellites in the outer Solar System appears to have experienced a unique geological history, dominated by impacts and icy volcanism. □

around planets in other solar systems.

Saturn's rings, of course, are much the best known. The Voyager images showed them to be remarkably complicated in detail, with many individual rings separated by gaps (Figure 1.14). The spacing of the gaps is in some cases controlled by orbital resonances with tiny, so called 'shepherd satellites'. Perhaps the most striking feature of Saturn's ring system is how thin it is. Although some 27 000 km across, the rings are only about a kilometre thick. They probably

Figure 1.14 A Voyager image of Saturn, processed to enhance subtle details of the rings. A puzzling aspect of this image is the dusky 'spokes' visible in the middle region of the rings. Like many other aspects of the rings, the nature of these spokes remains to be resolved. □

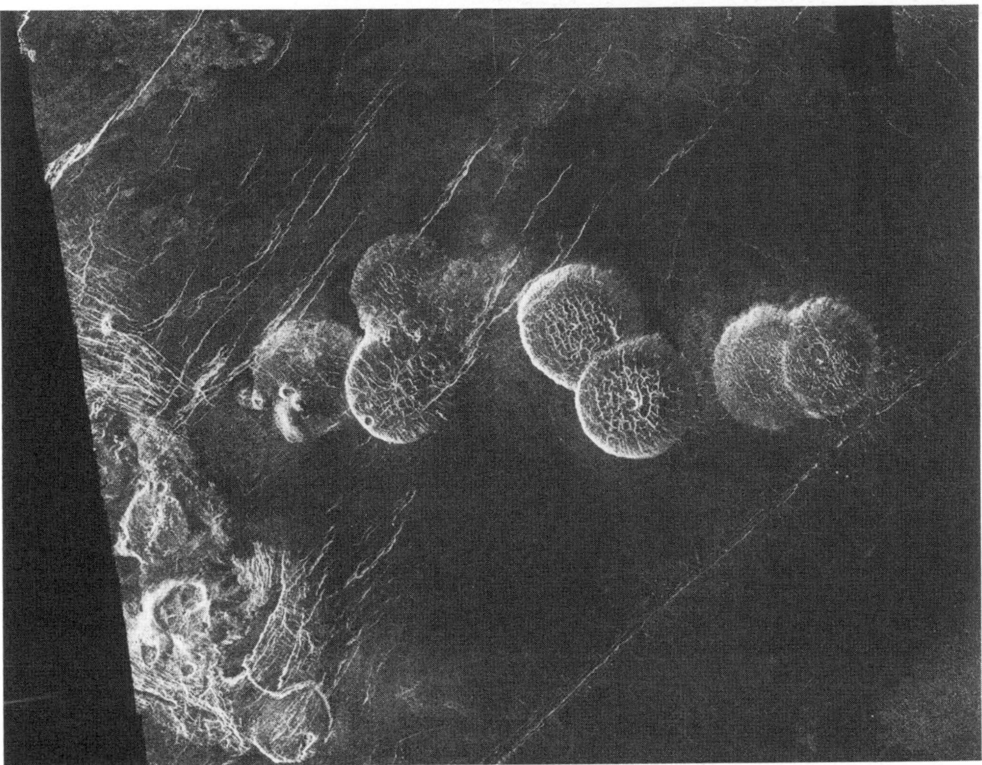

Figure 1.15 A single example of the thousands of extraordinary images of the surface of Venus obtained by the Magellan spacecraft. This radar image shows seven circular, dome-like hills averaging 25 km in diameter and reaching 750 metres in height, thought to be extrusive lava flows which oozed out onto relatively level ground. Whereas most volcanic features on Venus are of basaltic composition, the viscous nature of the lava implied by the shapes of these domes may suggest that more silicic magmas are involved. □

consist of myriads of tiny metre-sized chunks of ice, perhaps impregnated with 'dirt' of silicate or carbonaceous material. The rings may represent debris from a satellite or satellites that disintegrated with Saturn's massive gravitational field. It has been known for a long time that if a satellite comes closer than a certain distance to its parent planet – known as the **Roche limit** – it will be disintegrated because of the huge gravitational forces imposed upon it. It will be almost literally pulled apart. Further out, beyond the Roche limit, some larger satellites also appear to have disintegrated early on in their history as a result of massive impact events, and the resulting debris subsequently assembled together again. If such a disintegration had taken place nearer to Saturn, the debris would have distributed itself in the form of a ring around the planet.

Conclusions: Looking forward

At the time of writing, several major spacecraft missions were in progress or in prospect. Since 10 August 1990 Magellan has been in orbit around Venus, and returning superb radar images of the surface, revealing features less than 200 metres in size (Figure 1.15). Collectively, the Magellan images reveal that Venus, so similar in size to the Earth, has evolved into an entirely different geological entity. Rather than losing its internal heat through plate recycling and plate tectonics, most of Venus' internal heat leaks slowly away by lithospheric conduction. At the same time, Galileo was heading towards Jupiter, following a convoluted trajectory in order to build up sufficient velocity. On its looping path, Galileo has already provided unique perspectives of Venus, the Earth and Moon, and will fly by an asteroid before going into orbit around Jupiter in December 1995. If all goes well, it will provide three years' worth of data on the giant planet and its satellites. An American Mars Observer mission is being planned for a launch in the mid 1990s, and designs for a Mars Rover-Sample Return mission are also on the drawing board. Mars will also be a target for Soviet planetary scientists in 1994, notwithstanding the failure of both their *Phobos* missions there in 1989. There are ambitious plans for this mission to deploy a balloon which will drift gently across the face of Mars, periodically descending to sample the surface. In the mid 1990s, Cassini will head towards Saturn and dispatch a probe to Titan. Despite its embarrassing and well-publicised technical defects, the Hubble Space Telescope, launched in 1990, will provide a further unique source of data on the Solar System, enabling us to obtain superb views of planets where no spacecraft is operating. We live in remarkable times.

Further reading

There are a great many books dealing with the exploration of the Solar System. In increasing order of detail, the following are recommended:

Beatty, J. K. & Chaikin, A. (eds.) (1990) *The New Solar System*, Cambridge University Press, 326 pp.
A general, highly readable introduction.

Hartmann, W. K. (1983) *Moons and Planets*, Wadsworth Publishing Co, Tarrytown on Hudson, New York, 404 pp.
An excellent undergraduate textbook.

Taylor, S. R. (1982) *Planetary Science, a Lunar Perspective*, Lunar and Planetary Institute, Houston, 481 pp.
A detailed geochemical appraisal of the Solar System.

For the most detailed studies of individual planets in the Solar System, the following, published by the University of Arizona Press, are highly recommended:

Burns, J. A. (ed.) (1977) *Planetary Satellites*, 598 pp.

Gehrels, T. (ed.) (1976) *Jupiter*, 1254 pp.

Gehrels, T. (ed.) (1979) *Asteroids*, 1181 pp.

Greenberg, R. & Brahic, A. (eds.) (1984) *Planetary Rings*, 784 pp.

Hunten, D. M. *et al.* (ed.) (1984) *Venus*, 1143 pp.

Morrison, D. (ed.) (1982) *Satellites of Jupiter*, 972 pp.

Wilkening, L. L. (ed.) (1982) *Comets*, 766 pp.

CHAPTER 2

THE ORIGIN OF THE EARTH

Stuart Ross Taylor
Australian National University

Introduction

In this chapter, I have endeavoured to set forth the current thinking on the origin of the Earth, a topic which encompasses most scientific disciplines. Although the Earth is a unique planet, it is not possible to discuss the origin of the Earth separately from that of the other 'terrestrial' planets, or indeed that of the entire Solar System, and one must also account for the nearby presence of that unique object, the Moon. Evidence bearing on the origin of the Earth is to be sought not so much on the planet itself, but from considering the information from other planets, satellites, and the meteorites. After several centuries in which the discussion has been dominated by dynamical considerations, recent exploration of the Solar System has revealed not only an astonishing diversity among the planets and some sixty satellites, but also a new set of chemical constraints to add to the formidable problems addressed by Copernicus, Kepler, Newton, Kant, Laplace, Chamberlin, Moulton, Jeans and Jeffreys.

The first question to be addressed concerns the origin and primordial composition of the rotating disc of dust and gas, the **solar nebula**, from which both the Sun and the planets were derived. The **meteorites** provide many insights into the nature of the early Solar System, particularly with respect to nebula conditions and time scales. However, the currently available population, apart from the rare lunar and martian meteorites, is derived mainly from the inner asteroid belt, and differs in minor, but significant aspects from the composition of the terrestrial planets. The chief importance of meteorites as a source of information about the origin of the Earth is twofold: firstly they gave unique age information on early Solar System processes and secondly, compositional information about the asteroid belt and the early solar nebula, prior to the accretion of the Earth and the other terrestrial planets.

The inner Solar System with its small rocky terrestrial-type planets is distinct from the realm of the giant planets which dominate the outer reaches of the system. A basic cause of this division was the loss of gaseous elements and depletion of volatile elements in the inner nebula, probably associated with early violent solar activity. Jupiter and Saturn appear to have formed earlier than the inner planets while the gas was still present; the terrestrial planets accreted later in a gas-free environment.

Several lines of evidence, including the ubiquitous ancient cratered surfaces, point to the former presence of a hierarchy of **planetesimals**. (These were bodies which formed initially in the nebula and ranged in size from a few metres up to Mars-sized objects.) The asteroids are our best currently available examples. The formation of the Earth, and the other inner planets, appears to have occurred principally through the accretion of those planetesimals which grew in the inner nebula after the gas was gone. The large planetesimals were probably melted early in Solar System history and differentiated into metallic cores and silicate mantles. The terrestrial planets differ in composition among themselves, so that they collected somewhat differing populations of planetesimals. Following the accretion of the Earth, the separation of the metallic core from the silicate mantle appears to have occurred promptly. Possibly it was effectively coincident with accretion. No early sialic (high silica, high aluminium) crust formed.

The Earth–Moon system is unique in the Solar System; the formation of the Moon is inextricably linked to that of the Earth. The evidence appears decisive that the Moon formed towards the end of the period of planetary accretion through the collision of a Mars-sized impactor with the Earth, and that the material forming the Moon came not from the terrestrial mantle, but from the metal-poor silicate mantle of the impactor. Such an event would have removed any pre-existing atmosphere, and melted the terrestrial mantle, with consequences for mantle crystallisation and evolution which have yet to be worked out in detail.

The Earth in relation to the Solar System

There are many unusual features about the Solar System. It is probably unique, and other planetary systems must be expected to differ substantially from our own. The planets are usually divided into three major groups: the small, **inner** or **terrestrial** rocky **planets** (Mercury, Venus, Earth, Mars) the **gas giants** (Jupiter and Saturn), and the smaller **ice giants** (Uranus and Neptune), but all are different in size and composition. The ninth member, Pluto, is similar to Neptune's satellite, Triton, and is most probably a left-over planetesimal. It is only called a planet by courtesy, since it is much smaller than the Moon, and has an eccentric and inclined orbit around the Sun, far removed from the plane of the **ecliptic**. (This is the plane of the rotation of the Earth about the Sun, and all the planets except Pluto (with a 17° inclination) and the innermost planet, Mercury (with a 7° inclination), lie within a few degrees of it.) The plane is a consequence of the formation of the Sun and planets from a rotating disc of dust and gas, called the solar nebula, a concept which goes back 200 years to the French natural philosopher, Pierre-Simon, Marquis de Laplace (1749–1827).

The current scenario began about 4600 million years (Ma) ago. The nebula formed when a mass of gas and dust became detached from a larger molecular cloud in a spiral arm of the Milky Way galaxy and collapsed under gravitational attraction into a disc. Mass flowed inwards, the Sun formed in the centre, and angular momentum was transferred outwards, so that most now resides in the planets. Small metre to kilometre-sized bodies began to grow in the nebula. At a later stage in the early history of the Sun, the flow of material was reversed and violent solar flares and winds (T Tauri and FU Orionis stages) drove hydrogen, helium, the noble gases and many of the volatile elements out beyond 4–5 AU (1 **AU** (astronomical unit) = Earth–Sun distance) where they were accreted to the giant planets. All these events occurred probably within a million years. Water was able to condense in the nebula at a temperature of 160 kelvin (K) as ice at a 'snow line' at 4–5 AU and be retained in the icy satellites of the giant planets.

This **nebula stage** was shortlived, taking perhaps 10^5–10^6 years. Jupiter grew rapidly, perhaps as a consequence of the massive outflow of gas. In the inner nebula, only bodies large enough (metre size) to survive the early intense heating episodes from the early Sun were left. These aggregations grew by collisions into a hierarchy of planetesimals, a few of which reached the size of Mars before finally accreting into the Earth, Venus, Mars and Mercury, a process taking about 100 Ma.

The planets acquired somewhat differing suites of planetesimals, accounting for their variations in composition, although these do not necessarily represent any strict original zonation. Late in this **accretional stage**, perhaps about 100 Ma after the system separated from the molecular cloud, a massive Mars-sized planetesimal crashed into the Earth. The mantle of the impactor was ejected into orbit and formed the Moon. The core of the impactor accreted to the Earth. Other higher velocity collisions stripped much of the silicate mantle from Mercury, and pushed Venus into rotating backwards. Sweep-up of residual planetesimals continued down to 3800 Ma ago, producing the observed battered surfaces of the Moon and Mercury, the ancient cratered terrain on Mars, and destroying any early crust on the Earth.

One basic question about this model for the accretion of the Earth and the other inner planets concerns the size of the accreting material. Did the planets accrete directly from dust and the dispersed material of the nebula, or are they the end product of the accretion of a hierarchical succession of bodies? Several observations seem to point to the latter process. Firstly, there is ample observational evidence from the battered surfaces of planets and satellites throughout the Solar System that they were impacted by innumerable large (> 100 km) bodies. Secondly, the obliquities (tilts) of the planets are consistent with the collision of very large objects (> 1000 km). Thirdly, the most reasonable hypothesis for the origin of the Moon demands that the Earth was hit by an object of 0.1–0.2 Earth masses at a rather late stage in accretional history, when both the impactor and the Earth had already formed a core. The high density of Mercury is likewise best explained by removal of much of its silicate mantle by collision with a massive body. Fourthly, accretion from a dusty nebula might be expected to lead to uniform planetary compositions rather than the observed chemical and isotopic diversity, both of meteorites and planets.

These observations point to **planetary accretion** from a hierarchy of massive objects rather than from the infall of dust. The meteoritic evidence is consistent with this: individual meteorites do not consist of single phases but are always mixtures of several components (e.g. mineral grains, Ca,Al-rich inclusions (CAI), chondrules or metallic particles), or are fragments of larger bodies.

One corollary of the accretion of the terrestrial planets from sizeable objects is that internal fractionation may occur in such bodies before their final sweep-up into the planets, so that separation of metal, silicate and sulphide phases may occur in small objects before planetary accretion. Thus the Earth may have formed from bodies which had already differentiated into metallic cores and silicate mantles.

The solar nebula

The Sun and the planets formed about 4560 Ma ago from a rotating disc of dust and gas which had become separated from a larger molecular cloud in a spiral arm of the galaxy. (The rotation is an inherent feature of the Galaxy.) The Universe was already old. Some 12–15 thousand million years had elapsed since the **Big Bang** (the currently accepted model for the origin of the Universe, though our concepts may change following the deployment of the Hubble Space Telescope). Countless stars had formed and died since that event and had enriched the interstellar medium in the heavier elements. Only hydrogen, helium and a little lithium and beryllium were produced in the Big Bang; the heavier elements, from which the Earth was constructed and which comprise 2 per cent of the Solar System were formed by nuclear reactions in stellar interiors and supernovae. These stellar explosions both produced many of the heavier elements and dispersed them and the pre-existing stellar material into the gas and dust of interstellar space to provide material from which new stars were born. The age of the Solar System, given by the meteoritic data as 4560 Ma, is thus only about a quarter of the age of the visible Universe. This fact, a relatively recent discovery in scientific history, separates the origin of the Earth and the Solar System from that of the Universe. Just as Copernicus dethroned the Earth from a central position, more recent discoveries in this century have in turn relegated the Sun to a position in a spiral arm of our Galaxy, about 30 000 light years from the centre. The familiar Milky Way Galaxy is, of course, only one among a mighty host of some 10^{11} galaxies, each containing about 10^{11} stars. This placing of the Earth and the Solar System, itself possibly unique, in a rather distant corner of the Universe, has implications for the position of *Homo sapiens* in the Universe which have not yet been accommodated in most philosophical, mythological or religious systems.

Following its separation as a fragment of a molecular cloud, the primitive solar nebula consisted mainly of hydrogen and helium, which are the principal constituents of the Sun, Jupiter and Saturn. It is generally assumed that the composition, for the non-gaseous elements, of the primordial solar nebula is similar to that of the **Type 1 carbonaceous chondrites** for which the abbreviation **CI** (Carbonaceous Ivuna) is commonly used. The rationale for this is the similarity between the abundances of the non-gaseous chemical elements in the outer layer of the Sun (the photosphere, in which element concentrations can be obtained from solar spectra) and in the CI chondrites (Figure 2.1).

In the present discussion, the term 'gaseous' is used to refer to H, C, N, O and the noble gases. The other **elements** are **classified** as 'very volatile' (e.g. Bi, Tl), 'volatile' (e.g. Rb, Cs), 'moderately volatile' (e.g. K, Mn), 'moderately refractory' (e.g. V, Eu), 'refractory' (e.g. Ca, Al, U, La) and 'super-refractory' (e.g. Zr, Sc), as well as being termed as lithophile (e.g. Si, Al, Mg, K, U), chalcophile (e.g. S, Cu) and siderophile (e.g. Fe, Ni, Co, Ir) from their entry into silicate, sulphide or metal phases in meteorites.

This similarity between the composition of the Sun and the meteorites out at 3 AU implies a broad uniformity in the composition of the nebula. However, there are some differences in detail and it is becoming clearer that the nebula was neither well mixed nor particularly homogeneous.

There are many second-order heterogeneities, for example, in **oxygen isotopes** and K/U ratios among the planets and meteorites, while the asteroid belt shows compositional zones. Thus there appears to have been incomplete mixing or some fractionation in the nebula.

It has been clear for a long time that the compositions of the inner planets are depleted in a wide range of elements which are volatile below about 1000 K and that they differ in detail from those of the CI meteorites. The terrestrial planets are thus different in composition to the primitive nebula: it appears that substantial chemical fractionation occurred in the inner nebula before the final accretion of the terrestrial planets.

What was the initial state of material in the nebula? Two extreme models exist, one hot and the other cold. In the first, condensation from a hot nebula produces with decreasing temperature a sequence of oxides, silicates, Fe metal and sulphide (Figure 2.2). In the second, a lower temperature scenario, the dust in the rotating disc of gas and dust was oxidised.

A major paradox is that the astrophysical and isotopic evidence suggests a cool nebula (a few hundred kelvin at most at the asteroid belt), whereas the mineralogical evidence from the meteorites indicates high temperatures (>1500 K), at least locally at the site of the inner asteroid belt at distances from the Sun of 2–3 AU.

Some resolution of this paradox may be possible, if the heating events recorded by the meteorites were local, perhaps above the mid-plane of the nebula, rather than nebula-wide. Heating of chondritic material could occur during discrete local events, on short time scales (years rather than 10^5 years for hot nebula models), mostly connected with the infall of material into the mid-plane of the nebula. This topic is discussed in the section on chondrules, which provide significant evidence for local, rather than nebula-wide high (>1500 K) temperatures.

One of the more cogent arguments for a cold heterogeneous nebula comes from the oxygen isotope evidence, which clearly indicates the existence of various oxygen isotope reservoirs for the Earth and

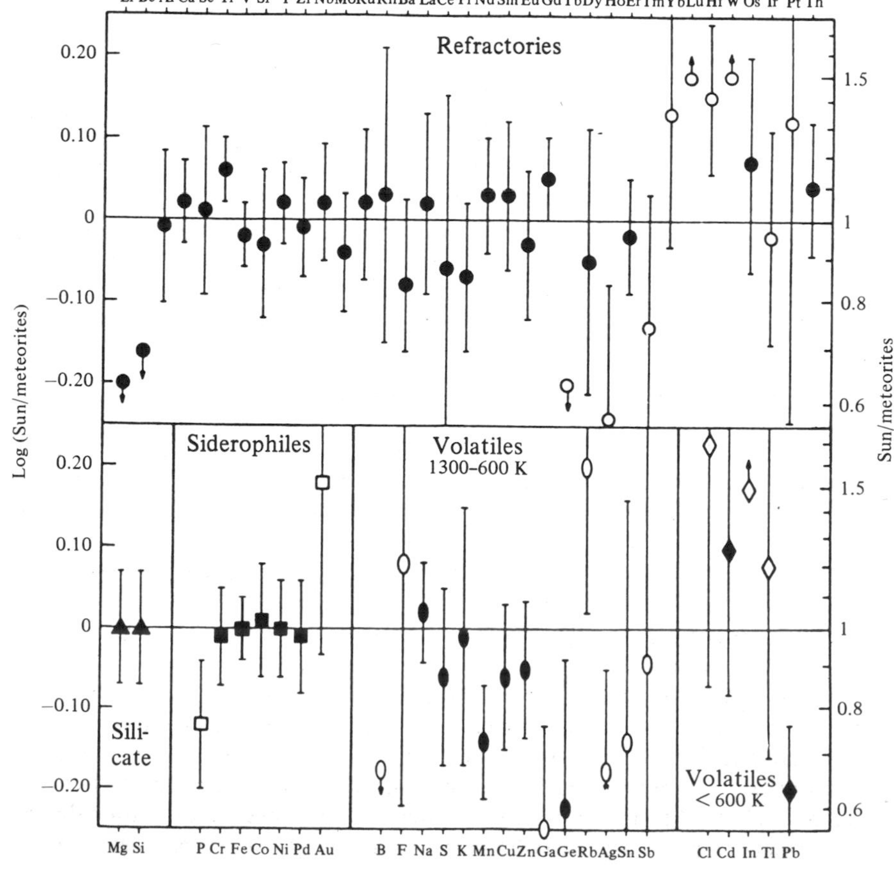

Figure 2.1 The close correspondence between the element abundances in the solar photosphere and in the Type I carbonaceous chondrites (CI) is taken as evidence that the CI compositions are representative of the primordial solar nebula. New determinations of the iron abundance in the sun now match those in the CI meteorites, removing an earlier discrepancy. The CI meteorites can of course be analysed accurately in the laboratory, compared to the inherently less precise spectroscopic determinations of solar abundance. □

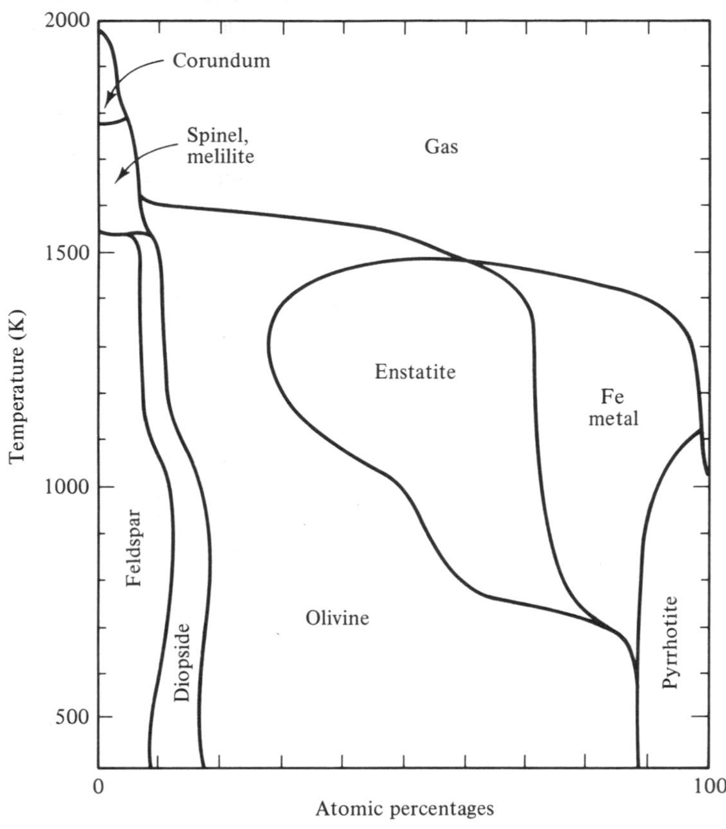

Figure 2.2 Simplified mineral condensation sequence in the solar nebula. This assumes that most of the gaseous components (H and He) have been swept away so that the dust/gas ratio is enriched a hundred times over the primordial solar nebula value and that the nebula pressure is 10^{-5} atm (approx. 10^{-8} kbar). These conditions are judged appropriate to the region of meteorite formation. The base line gives the atomic percentages of the abundant elements Si, Mg, Fe, Al, Ca, Na and Ti in the various phases. □

different classes of meteorites (Figure 2.3). These would be homogenised in a hot (> 1500 K) nebula. In the present context, two observations may be made. The oxygen isotopic signatures of the meteorite groups are mostly distinct, and they do not coincide with the values for the Earth or Mars. One group of meteorites which probably come from Mars (the shergottites, nakhlites and chassignites or SNC: see Chapter 1) form a separate group, which also do not overlap with any of the other meteorite groups or with that of the Earth–Moon system. This is additional evidence for lack of homogeneity in the nebula, and hence a cool nebula at distances of 1–3 AU.

It should be realised that the nebula is not a static entity, but evolves with time. During the early collapse phase, material flowed inward to the forming Sun. Temperatures in the nebula were high, reaching 1500 K in the inner regions (where the terrestrial planets later formed) during this brief hot inflow stage, which lasted perhaps 10^5 years. However, no solid condensed material was present at this time. When the growing Sun reached critical mass and thermonuclear burning began, violent T Tauri and FU Orionis activity swept out the remaining gas from the inner nebula. At this stage, condensed material in the inner nebula in metre-sized planetesimals survived and subsequently accreted into the terrestrial planets. Nebula temperatures were much cooler, a few hundred kelvin at 2–3 AU at the distance of the Earth. Accordingly, much of the debate over hot versus cold solar nebula models refers to differing stages of nebula evolution. It is becoming clearer that the early solar nebula was a much more turbulent environment than commonly realised.

The meteoritic evidence

Meteorites present us with tantalising information about the early Solar System. In 1492 a meteorite fall was witnessed in Ensisheim, Alsace, then part of the Holy Roman Empire, but now, after a chequered political history, a province of France. The stone was recovered, and placed in the church with the following inscription: 'Many know much about this stone, everyone knows something, but no one knows quite enough'. This comment from the Middle Ages summarises much about our current understanding.

Meteorites have long provided us with analogues for planetary compositions. The internal structures of the terrestrial planets are explained most rationally by metallic cores overlain by silicate mantles and the

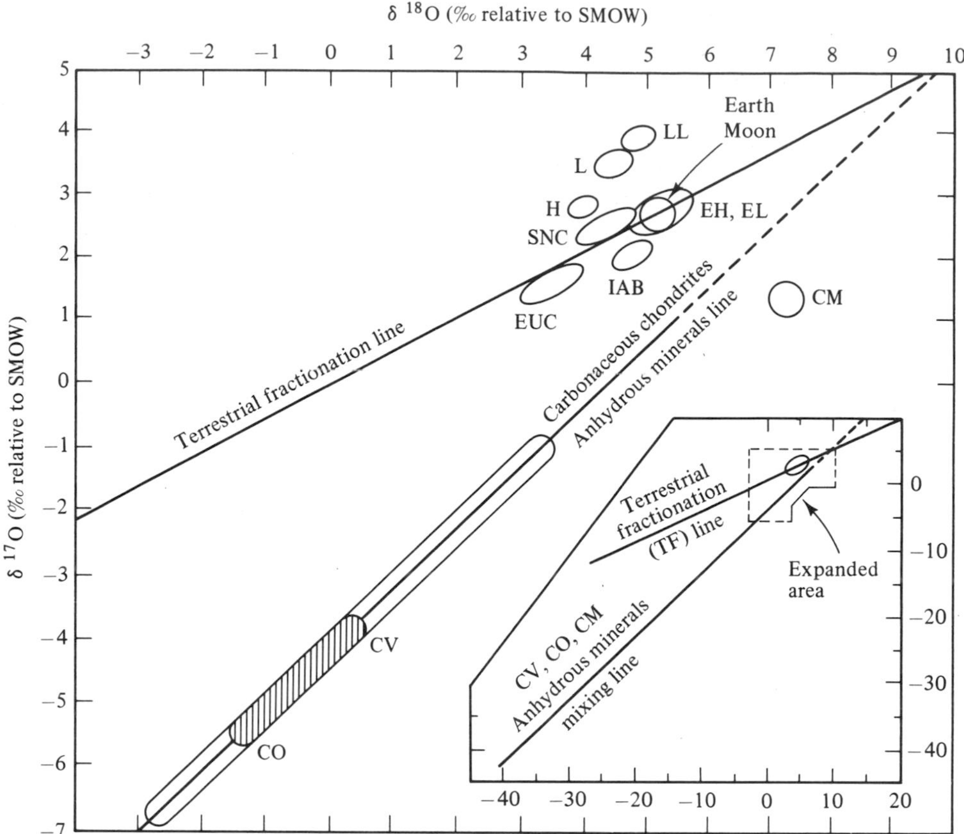

Figure 2.3 The abundances of the three oxygen isotopes ^{16}O, ^{17}O and ^{18}O show differences due not only to isotopic fractionation, which cause the data to lie along lines with a slope of about 0.5, but also indicate the existence of separate reservoirs with oxygen isotopic compositions distinct from terrestrial. Most of the Solar System units sampled come from different oxygen isotope reservoirs, indicating that the nebula was not well mixed with respect to oxygen isotopes. The delta notation refers to the difference in the ratio of ^{17}O or ^{18}O to ^{16}O in parts per thousand (or per mil) between the sample and SMOW = Standard Mean Ocean Water; LL, L, H, EH, EL, IAB, EUC, CM, CO and CV refer to the various meteorite classes. □

presence of metallic and silicate phases in meteorites has lent credence to these models.

Meteorites contain unique and crucial information about the dates of early events in the Solar System. The oldest reliably dated objects in meteorites are the millimetre to centimetre-sized refractory **calcium–aluminium inclusions** (CAl). These give an age of 4559 ± 4 Ma. The ordinary **chondritic meteorites** (H, L, LL and E classes) all have similar whole rock ages, the best estimate being 4555 ± 4 Ma. This is a little younger than the ages recorded by the CAl inclusions.

Formation of igneous rocks in asteroidal bodies began within a few million years of the canonical age of the Solar System or T_0 (4560 Ma). Angra dos Reis (ADOR) is an igneous meteorite with an age of 4551 ± 2 Ma. The **basaltic achondrites** are true basalts similar to lunar basalts. They have an average crystallisation age of 4539 ± 4 Ma, notionally only 20 Ma younger than the oldest material dated by the Allende CAl. Clearly the achondrites reveal that planetary type differentiation and the eruption of basalts was occurring on a small but extensive scale in asteroids very close to T_0. **Iron meteorites**, perhaps the ultimate in differentiated objects, come from at least sixty separate bodies, providing evidence for many episodes of differentiation in precursor bodies.

Chondrules, which make up the bulk of ordinary chondritic meteorites, are millimetre-sized silicate spherules, depleted in siderophile and chalcophile elements. They formed, in the words of Sorby over 100 years ago, as 'molten drops in a fiery rain'. Our understanding of the place and conditions of their origin has advanced only a little since that time.

Chondrule production occurred very early and the 'chondrule factory' must have been efficient, since they are so abundant (Figure 2.4). Chondrules formed by melting of pre-existing dust in the nebula: volcanic and impact origins can be ruled out. Also they did not condense from a hot vaporised nebula, since that is ruled out also by the preservation of local isotopic anomalies in the refractory inclusions (CAl), and by the ubiquitous oxygen isotopic variations: all would have been homogenised in a hot nebula. Chondrules are mainly composed of silicates and are depleted in siderophile elements, metals, sulphides and some volatile lithophile elements. The most likely scenario is that they were melted from oxidised pre-existing dust. Although some reduction of iron, and depletion of siderophiles, chalcophiles and some volatiles occurred during chondrule formation, the major formation of these phases and elements apparently occurred prior to the chondrule melting event. There seems to have been preferential aggregation of silicate lumps while metals and sulphides remained dispersed. Perhaps silicate dust was separated from metals and sulphide, either by differential gravitational settling or magnetically in the case of the metal, or perhaps the silicates just stuck together more efficiently. In any event, metal–sulphide–silicate fractionation in meteorites occurred effectively at T_0. Cooling times of chondrules (< 1 hour) are very fast. It is not possible to cool molten drops so rapidly in a hot nebula; the process must have been highly localised in an overall cool environment.

The chondrules are individually variable in elemental and oxygen isotopic composition, but their mean compositions for all meteorite groups are relatively uniform: the matrix in the differing groups of meteorites accounts for the intragroup variations in composition. The chondrule-forming process probably occurred on a brief time scale of a million years or less. The chondrules preserve rare evidence of the refractory millimetre to centimetre-sized inclusions (CAl) among their precursors, so that the formation of these

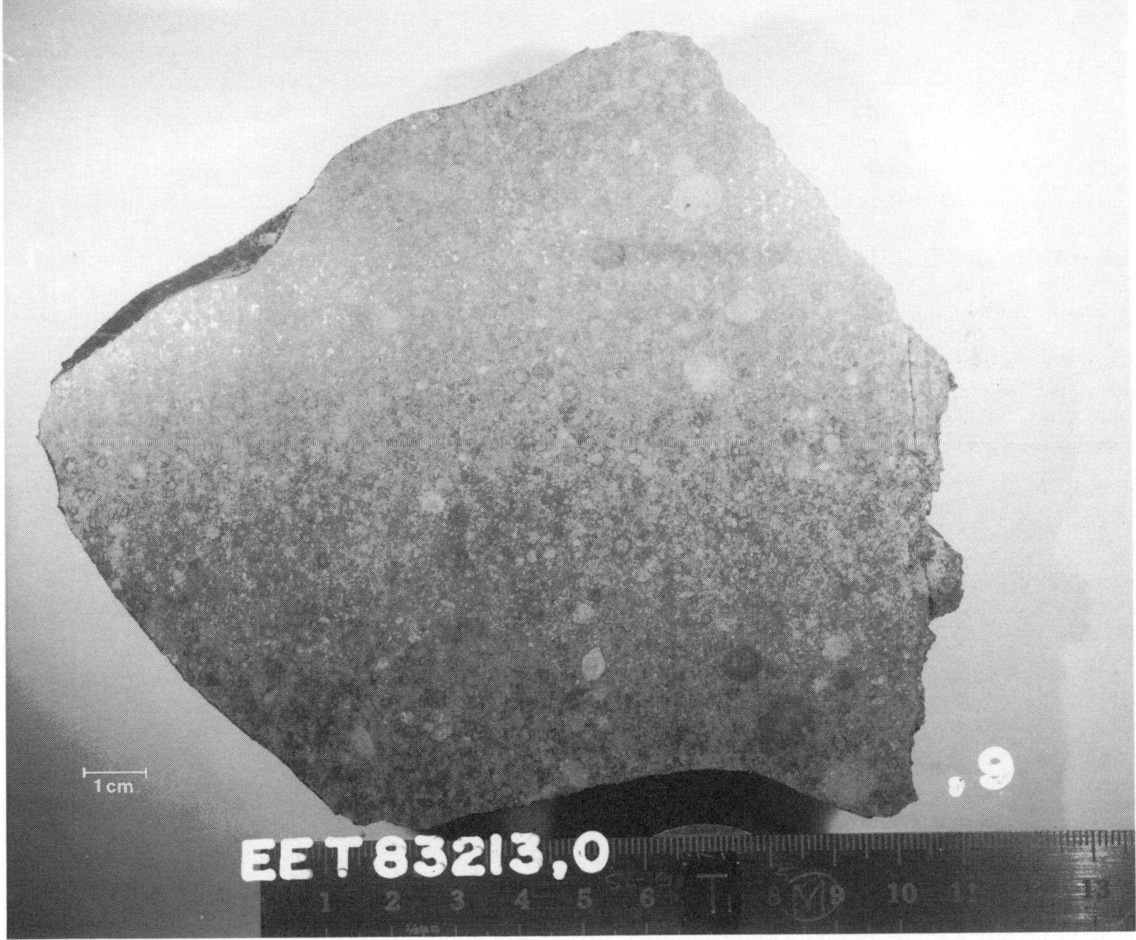

Figure 2.4 Chondritic texture revealed in a saw-cut slab of the L3 chondrite named Elephant Moraine, Antarctica, 83213. Chondrules comprise the light, oval-shaped objects set in a dark, fine-grained matrix. □

refractory inclusions pre-dates chondrule formation, consistent with the dating record.

Many origins have been suggested, but most fail to explain the evidence. One current plausible scenario forms chondrules by flash melting of clumps of silicate dust by nebula flares (analogous to solar flares) during an early turbulent period of nebula history.

Refractory inclusions also preserve many examples of variations in isotope abundances (referred to as isotopic anomalies) different from the familiar terrestrial patterns. These record a pre-solar memory of incomplete mixing of isotopes from different element-forming events, principally in supernovae. These refractory inclusions also record trace element evidence of multiple episodes of evaporation and condensation due to complex heating episodes in the early nebula. None of this evidence would have survived in a hot nebula, which would have homogenised such anomalies.

Meteorites are mostly derived from the **asteroid belt**. This shows a zonation with distance from the Sun. Telescopic studies of asteroid surfaces measuring reflectance spectra reveal that differentiated objects containing metal and silicates occur in the inner belt, while primitive carbonaceous compositions dominate in the outer belt. The presently observed zones have been broadened by collisions from initial widths of perhaps 0.1 AU. The asteroid belt represents a planetary accretion zone arrested at an early stage of development, most likely by early growth of Jupiter. It also records the existence of an early heating episode which decreased in intensity with distance from the Sun.

Many scenarios for the origin of the terrestrial planets have been based on the two components, fine-grained matrix and coarser metal particles, chondrules etc., which are intimately mixed in many meteorites. Sometimes these were characterised, respectively, as volatile-rich and volatile-poor, or refractory. The philosophy behind this approach is that the variations in the compositions of the inner planets are similar to those observed in chondrites.

Recent models employing this general scenario have tended to simplify the components to two; one a volatile-rich, and the other a high-temperature volatile-poor, refractory and metal-rich (and so reduced) component. Such models attempt to account for the obvious presence of reduced metal, oxidised phases and some volatiles in the Earth, but do not shed much light on the source or origins of the postulated two components in the nebula.

Several properties (e.g. oxygen isotopes, noble gas contents etc.) in the Earth indicate that neither the volatile nor the refractory component can be matched with a specific meteorite class. Although the volatile component is commonly identified with CI, this class of meteorites is ruled out on many grounds (e.g. oxygen isotopes). Many of the volatile constituents in the Earth may have been derived from late-accreting comets. The E chondrites are the most obvious candidates for the refractory component but, although they are reduced (so that all iron is present as metal), they are not volatile depleted. One might expect that, if the planets were built out of these two components, both planetary and meteoritic compositions should show considerably more homogeneity than is the case. Such models merely rationalise what is observed, rather than shed light on the processes involved.

The present population of meteorites is therefore not particularly representative of the building blocks of the terrestrial planets. There are many differences in detail, e.g. volatile/refractory element ratios such as K/U, oxygen isotopes, noble gases, density and other characteristics. These rule out any of the known meteorite classes as having appropriate chemical or isotopic compositions to make them potential candidates for the source material for the inner planets. For example, the common classes of chondrites (E, H, L, LL and CI) all have higher K/U ratios than the terrestrial planets and so are unsuitable building blocks for them (Figure 2.5), as discussed later.

Gas loss and volatile depletion in the inner nebula

One of the most significant geochemical observations about the terrestrial planets, is their **depletion** in **noble gases** and volatile elements relative to the outer planets and even most meteorites. Probably this is due to early intense solar activity (e.g. solar flares, strong solar winds) as the Sun settled onto the main sequence, and cleared the inner nebula of volatiles and water out to a 'snow line' at about 4 AU in the outer reaches of the asteroid belt. This process is likely to have taken about a million years.

The meteorites provide us with the time of volatile depletion in the inner nebula, since the Pb–Pb and Rb–Sr ages give the time of separation and depletion of volatile lead and rubidium relative to refractory uranium and strontium. All the meteorite data give effectively the same age for this event, indicating that this depletion in volatile elements occurred at about 4560 Ma. Accordingly, volatile depletion in meteorites, and presumably also in the inner nebula closer to the Sun, in the region of the inner planets, occurred effectively at T_0.

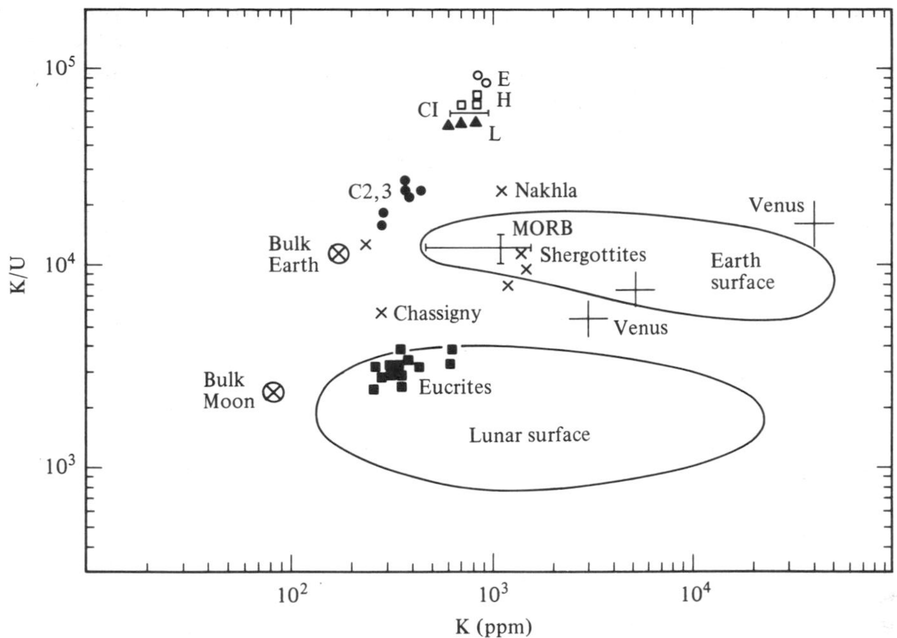

Figure 2.5 Fractionation between volatile elements, here represented by potassium (K), and refractory elements, of which uranium (U) is a typical example, is widespread in the terrestrial planets. Information on the abundances of these two elements is obtainable on planetary surfaces by gamma-ray spectrometry. The primordial solar nebula value is given by the abundances in the carbonaceous chondrites [CI]. Earth, Mars (represented by the Shergotty, Nakhla, Chassigny meteorite data), and Venus are depleted in K relative to U, compared to this value. Most classes of meteorites have higher K/U ratios than the planets. The Moon is more highly depleted in K, and other volatile elements, than the terrestrial planets. MORB refers to Mid-Ocean Ridge Basalts, the commonest variety of terrestrial volcanic rock. Both K and U are incompatible elements, excluded from common mantle minerals, and concentrated together in residual melts during fractional crystallisation. Thus these two elements, although dissimilar chemically, preserve their original bulk planetary abundance ratios. □

The depletion of potassium (a moderately volatile element) compared with uranium (a refractory element) is well shown in Figure 2.5. Since these elements are both gamma-ray emitters, this is one of the few geochemical measurements available for Earth, Venus and Mars, as well as for the meteorites, and it provides some crucial information. Potassium and uranium, although distinctly different in chemical properties, ionic radius and valency, nevertheless share a common characteristic which makes them useful as geochemical indices of planetary compositions; they are both excluded from the common rock-forming minerals in basalts (i.e. are 'incompatible' elements; see Chapter 6). They are therefore concentrated together in residual melts during crystallisation of basaltic silicate melts, and so they tend to preserve their bulk planetary ratios during planetary differentiation. Thus K/U ratios can provide information about the depletion of volatile relative to refractory elements in their parent body.

The primordial ratios, as given by the CI meteorites, are approximately 60 000, while terrestrial ratios are 10 000. An initial question is whether the depletion of potassium observed in terrestrial samples is a bulk planetary effect or is local and surficial. This question can be addressed only indirectly from the information available on the Earth. However, potassium appears to be depleted, relative to uranium, in the surface rocks of Mars and Venus, as well as on the Earth. Measurements by Soviet landers of K/U ratios for Venus indicate an overlap with those for the terrestrial surface. The K/U ratios for the SNC meteorites, almost certainly derived from Mars, appear to be somewhat higher (15 000), consistent with a somewhat higher volatile content for that planet, but still a factor of four lower than the primordial values. The higher volatile content of Mars is also supported by the Nd and Sr isotope data, assuming that the SNC meteorites come from Mars. Since the trapped gases in the SNC meteorites match the unique composition of the Martian atmosphere as measured by the Viking Landers, this assumption is nearly a scientific consensus.

Accordingly, it is a firm conclusion that potassium, and presumably the other volatile elements are de-

Figure 2.6 Mare Orientale, 900 km in diameter, is a type example of a multi-ring basin formed by impact of a planetesimal or asteroid perhaps 100–200 km in diameter. The concentric rings of mountains were formed within a few minutes, 3800 Ma ago. During the period 4400–3800 Ma, at least 200 similar structures were formed on the Earth by bombardment by the tail of the planetesimal swarm, destroying any primitive crust. □

pleted in the surficial rocks of Mars, Earth and Venus relative to the primordial nebula values. Apart from an initial depletion in the incoming planetesimals, are there other possible explanations to account for this depletion in volatile elements?

For example, it has been suggested that potassium could behave as a metal at high pressures and be buried in the planetary cores. However, the central pressure in Mars (r = 3390 km) is only 400 kbars. This pressure is insufficient to allow potassium to enter a Martian core. Thus the depletion of potassium in Mars must be due to some other process. Many other volatile elements in addition to potassium are also depleted in the Earth relative to primordial (CI) abundances. Most of them have chemical properties which make it unlikely that they would enter into metallic phases.

Potassium is a moderately volatile element. Could it have been boiled off during a high-temperature stage of planetary accretion? It turns out that elements of the atomic weight of potassium cannot be lost from the terrestrial planets, even at elevated temperatures, once these bodies have reached their present size. If they were boiled off in some manner during accretion, the K/U ratio should vary with planetary size. This is not apparent.

In addition to the chemical evidence, the Rb/Sr isotope data also indicate that the Earth is depleted in volatile Rb relative to refractory Sr. Samples derived from the mantle indicate that the mantle Rb/Sr ratio has always been much lower than the primordial nebula values given by the CI meteorites. Since rubidium has closely similar properties to potassium, it is unlikely that either potassium or rubidium are present in the mantle in their primordial solar nebula concentrations, and probable that rubidium and the other volatile elements were already depleted in the precursor material from which the Earth accreted. The most reasonable explanation is that these volatile elements were depleted in the precursor planetesimals from which the inner planets accumulated.

The inner nebula was thus cleared of the gaseous, and depleted in volatile elements very early, probably within about one million years of the arrival of the Sun on the main sequence. The cause of this clearing of the inner regions of the nebula is most probably connected with early intense solar activity, with strong stellar winds, and flare outbursts of the type observed with the very young T Tauri and FU Orionis stars.

Planetesimals

According to the current version of the planetesimal hypothesis, at a very early stage the dust in the rotating disc of the solar nebula began to clump together, so that a hierarchy of bodies formed, beginning with grains and proceeding through metre-sized lumps to objects of kilometre size, finally reaching dimensions of hundreds to thousands of kilometres for the final pre-planetary stage. These objects are termed planetesimals, and current thinking regards them as the building blocks of the planets.

What sort of evidence do we have for these now vanished objects? There are several different converging lines.

The asteroids are left-over planetesimals. Phobos (see Figure 1.9, p. 17), one of the Martian satellites, appears to be a primitive object and may be a captured asteroid. Thus it may provide us with an analogue for a planetesimal. The absence of a planet in the asteroid belt, in which over 4000 small bodies have been numbered (apart from countless smaller ones), is probably due to the influence of massive Jupiter, which swept up, or ejected many of the bodies. The total mass of the myriad of small objects in the belt is less than 5 per cent of the mass of the Moon. (The small size of Mars is probably due to a similar cause; starvation caused by massive Jupiter, which formed earlier and depleted neighbouring areas.)

A direct piece of evidence for the former existence of bodies up to a hundred kilometres diameter comes from the observation that all of the older preserved surfaces on planets and satellites are saturated with craters. The lunar surface is the classic example, but from Mercury, close to the Sun, out to the satellites of Uranus, photographs reveal that a massive bombardment struck planets and satellites. Craters of all sizes are present, from micrometre-sized pits due to impact of tiny grains on lunar samples, up to giant ringed basins over a thousand kilometres in diameter. Like the smile of the Cheshire Cat in *Alice in Wonderland*, the craters record the previous existence of now vanished objects, in this case, the planetesimals.

The extent of this bombardment on the Moon after it had reached its present size *and after the lunar crust had formed*, is revealed by the evidence that at least eighty basins with diameters greater than 300 km and 10 000 craters in the size range from 30 to 300 km formed before the main bombardment ceased about 3800 Ma ago. Figure 2.6 shows the last major ringed basin, Mare Orientale, which formed on the Moon at about 3800 Ma ago. Since a similar but probably more intense barrage struck the Earth, this accounts for the absence of identifiable rocks older than that age. Current estimates indicate that 200 ringed basins with diameters greater than a thousand kilometres formed due to the impact of bodies a few hundred kilometres in diameter on the Earth between 4400 and 3800 Ma; i.e. following the accretion of the Earth. On the Earth, in contrast to the Moon, the evidence has been removed, since the smashed-up breccias would have been easy meat for the terrestrial agents of erosion. Although this bombardment was catastrophic for any early crust, it would have added only a minor amount of material to the surface of the Earth.

A few survivors of the planetesimals are still around. There are over 120 identified asteroids in Earth-crossing or Earth-approaching orbits. Between 100 and 1000 tonnes of meteoritic material, mainly as dust, falls on the Earth every day. About every 20–30 Ma an asteroid large enough to form a 20 km diameter crater hits the Earth. The extinction of the dinosaurs 65 Ma ago at the end of the Cretaceous Period was probably connected with the impact of an asteroid some 10 km in diameter. The evidence for such a collision includes, at the exact Cretaceous–Tertiary boundary stratum, a world-wide 'spike' in iridium (rare in the Earth's crust but more abundant in meteorites), quartz grains shocked by pressures of hundreds of kbars, and soot from massive wildfires. No single geological process internal to the Earth is capable of accounting for these facts. The youngest large lunar crater, Tycho, 85 km in diameter, formed by a similar impact about 100 Ma ago indicating that relatively large impacts have still occurred rather recently in geological time in the terrestrial neighbourhood. (A near miss occurred in March 1989, when a 100 metre asteroid, capable of forming an impact crater of kilometre dimensions, passed within twice the Earth–Moon distance.)

Accordingly, there is plenty of evidence for the existence of planetesimals up to 100 km or so in diameter in the early Solar System, together with a few left-over survivors. Were the planets assembled from these, or were there larger intermediate-sized bodies (Moon–Mercury–Mars size) in the hierarchy of objects which accreted to form the terrestrial planets?

The major piece of evidence for the previous existence in the early solar nebula of very large objects (of lunar, Mars and Earth-sized masses) comes from the tilt or inclination of the planets to their axis of rotation (see Table 1.1, p. 4). The largest impact is required to account for Uranus. Calculations show that a body the size of the Earth crashing into the planet would be needed to tip it through 90°. Smaller collisions are needed to account for the tilts of the other planets, but a few very large objects at least as large as Mars (1/10 Earth mass) must have been responsible, since the impacts of a host of smaller (Phobos-size) bodies will average out.

How many of these very large objects were there and what was their size? Calculations have been carried out only for the region of the inner planets (Mercury–Venus–Earth–Mars). Computer simulations of the accretion process show that about a hundred Moon-size bodies, ten Mercury-size and three to five Mars-size bodies would have formed the final population of planetesimals existing just before the final sweep-up. Venus and the Earth acquired most of them (Mars is only a tenth of Earth's mass while Mercury is only about a twentieth Earth's mass).

A large impact is probably responsible for the strange facts that Mercury has such a small rocky mantle, such a large iron core, and an inclined orbit so close to the Sun. Two explanations are current. The first proposes that the silicate was boiled away in some early high-temperature event, connected with early solar activity (Mercury is close to the Sun and the surface temperature on the sunlit side at present is 425 °C, hot enough to melt lead). However, extremely high temperatures of several thousand degrees are required to boil off the rocky mantle.

The alternative, and currently favoured explanation is that Mercury was struck by a body about a sixth of its mass, at a late stage in its accretion. The collision fragmented the planet: most of the silicate was lost to space but the iron core reaccreted together with a smaller silicate mantle.

Finally, the current explanation for the origin of the Moon (see later in this chapter) involves a collision of an object larger than Mars with the Earth.

In summary, there is ample evidence for the existence of large precursor bodies, or planetesimals in the early Solar System.

The accretion of the Earth

The currently favoured hypothesis is that the Earth, and the other terrestrial planets accreted from a hierarchy of planetesimals of varying sizes, a process taking perhaps 100 Ma. Very early formation of the inner planets (within a million years or so of T_0), which is the only time during which accretion in the primordial hydrogen and helium gas-rich solar nebula could occur, seems to be ruled out. Thus the noble gases (He, Ne, Ar, Kr and Xe) and hydrogen are strongly depleted in the Earth relative to solar abundances. By the time the Earth, Venus, Mars and Mercury accreted, the gaseous components of the nebula were gone. As discussed earlier, the cause of this early volatile loss in the inner portions of the solar nebula must be connected with intense solar activity (T Tauri and FU Orionis stages) as the early Sun joined the main sequence stage of stellar evolution.

A significant question is the relative sizes of the planetesimals, which in the currently preferred scenario accreted to form the terrestrial planets. Large precursor objects were relatively common; probably 50–75 per cent of present Earth mass was originally present in large planetesimals (lunar mass or larger). These bodies may already have been melted, and formed metallic cores and silicate mantles. The addition of such differentiated bodies to the Earth adds a complicating factor to our present limited understanding of terrestrial core–mantle relationships, since, in this view, much of the core and mantle arrived prepackaged.

If there is little identifiable input into the inner planets from the meteorites now coming from the asteroid belt, then this has important implications for planetary accretion models. One is that there was little lateral mixing in the nebula and that the terrestrial planets either accumulated from rather narrow (perhaps < 0.5 AU) concentric zones in the solar nebula or from different populations of planetesimals. The present asteroid belt has a zoned structure, ranging from apparently differentiated objects in the inner belt, to apparently primitive ones which dominate in the outer reaches. Although the zones in the asteroid belt have been broadened through time by collisions, they may be an analogue for the original structure of the nebula. Temperature, and hence fractionation, is expected to increase sunwards of the inner edge of the belt. This view receives some support from the differences in composition between the inner planets, particularly Earth and Mars, which differ in K/U ratios, Mg/(Mg + Fe) ratios, V–Cr–Mn abundances, and oxygen isotopes, all of which indicate accretion from somewhat differing populations of planetesimals. Whether these represented some initial heliocentric zoning or were merely a random assortment, the message is clear that the planetesimals in each zone from which the inner planets accreted were not uniform in composition.

When did the accretion of the largest planetesimals occur? Most probably during the closing stages of the formation of the planets about 4500 Ma ago, with a tail of 100-km-sized and smaller objects being swept up in the next 500 Ma. Mostly the sweep-up of these bodies was completed by about 3800 Ma ago. (We know this date from the age of the last great collisions on the Moon, dated from the returned samples from the Apollo missions.) This bombardment would have disrupted any early crust.

The Solar System is now reasonably clean. For example, only one tiny object, the asteroid (most probably a comet) Chiron, 350 km in diameter, has been detected wandering alone in the immense space between Saturn and Uranus (ten times the Earth–Sun distance separates these two planets). The objects which now hit the Earth are either comets coming from the outer reaches of the Solar System, or asteroids and meteorites perturbed from their orbits in the asteroid belt, by collisions or gravitational effects of Jupiter. Occasional meteorites, knocked off from Mars or the Moon by large impacts, are also swept up by the Earth, and have been discovered on that excellent collection surface for meteorites, the Antarctic ice plateau.

Although, as noted earlier, the inner planets are chondritic in a broad sense, it does not appear possible to construct the Earth and the other terrestrial planets out of the building blocks as supplied by the currently sampled population of meteorites. If, as appears reasonable, these provide us with an adequate sample of the inner asteroid belt, then there were substantial differences between that area and the zone sunwards of about 2 AU. The most significant difference appears to have been a greater depletion of the volatile elements in the region in which the terrestrial planets accumulated.

A consequence of the planetesimal hypotheses is that planetary melting occurs due to the stored kinetic energy from the large impacts; in scenarios involving accretion from fine dust the heat from gravitational infall is mostly re-radiated to space. This question will be addressed in the later section on the Earth–Moon relationships.

Metallic core and silicate mantle formation

The scenario in which the terrestrial planets accrete from planetesimals which were already mostly differentiated into metallic and silicate phases implies little further reaction between metal and silicate once these bodies accreted to the Earth. **Core formation** is expected to occur catastrophically once melting temperatures are reached, and metallic iron in the incoming planetesimals can sink rapidly through the silicate mantle. Possibly the iron cores in the planetesimals were still molten when they accreted to the Earth. Thus core formation is effectively instantaneous and coeval with accretion, rather than occurring over a lengthy period (hundreds of millions of years) of geological time.

However, the present upper mantle did not achieve equilibrium with the core. There are high abundances in the upper mantle of the Earth of Re, Au, Ni, Co and the **platinum group elements**. (Pd, Ir, Pt, Rh, Os, Ru) (Figure 2.7). The distribution of the platinum group elements appears to be rather uniform, and they are present in approximately primordial (CI) proportions, although depleted relative to Fe. Their abundances are determined principally from analysis of mantle fragments (xenoliths) carried up in basaltic lavas. It is worth remembering that these are samples only from the top 100–200 km of the mantle, and that the composition of the lower mantle is essentially unknown. Thus, current estimates are very model-dependent. However, it seems clear that the highly siderophile elements listed above would have been extracted into the metal phase under equilibrium

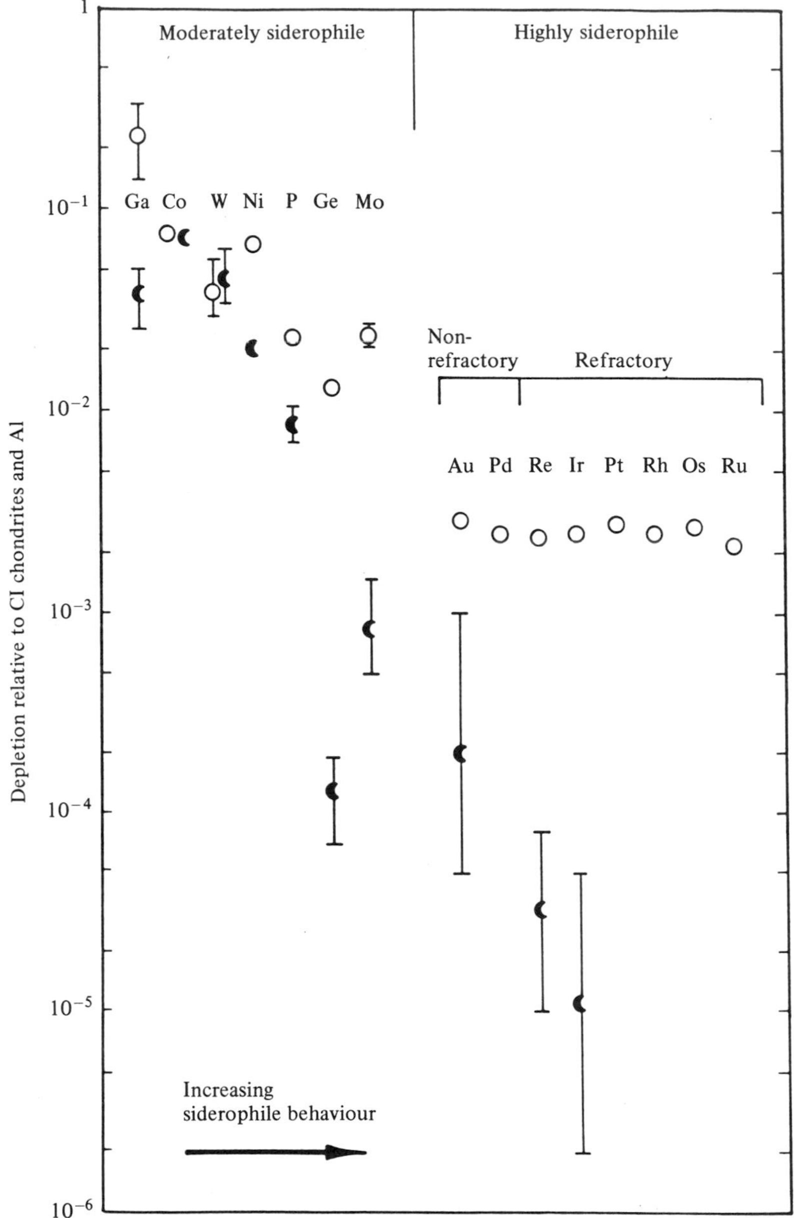

Figure 2.7 The abundance of siderophile (metal-loving) elements in the upper mantle of the Earth (open circles), and in the Moon (filled crescents), normalised relative to primitive CI concentrations. The relatively high levels of Ni and Co, and the uniform 'chondritic-type' abundance pattern of the highly siderophile elements, indicates a lack of equilibration between the silicate mantle and the metallic iron core. □

conditions. Thus the present upper mantle was apparently never in equilibrium with the core. The variety of explanations offered for the relative abundances of siderophile elements in the terrestrial upper mantle, and their resemblance to CI patterns negates claims that there is a unique terrestrial siderophile element signature.

Late accretion of CI planetesimals rich in platinum group elements is a common explanation for their over-abundance in the upper mantle. A similar scenario of terminal accretion of volatile-rich planetesimals is often invoked to account for the volatile (e.g. H_2O) inventory of the Earth. A cometary influx could be an equally viable source. However, late infall of planetesimals to planets which are essentially complete is like adding icing to a cake: the decoration may give little insight about the composition of the interior.

A further consequence may be noted. The metallic outer core of the Earth contains about 10 per cent of some light element. The two current contenders are oxygen and sulphur. If high-pressure core–mantle equilibrium was not attained in the early Earth, then it seems unlikely that oxygen entered the core, since this requires megabar pressures, as is the case for potassium. Sulphur then becomes the most viable candidate for the light element in the Earth's core. The often-repeated argument that the Earth must be depleted in sulphur since that element is more volatile than potassium, fails to recognise that most of the sulphur accreting to the Earth in planetesimals is stabilised by combination with iron as troilite (FeS).

Atmosphere, hydrosphere and early crusts

There is no sign of any primitive atmosphere on the Earth which might have accreted after the hydrogen and helium, the principal components of the primordial solar nebula, had been dissipated. Most of the evidence for this comes from the abundance and isotopic composition of the noble gases. There appears to be no trace of any primitive solar nebula gases.

If the Earth had been immersed in the gas-rich solar nebula during accretion, it is sufficiently massive to have captured a primary atmosphere which would have had a surface pressure of about 1 kbar, with a surface temperature of 4000 K. Much absorption of gas into the molten crust from such an atmosphere would result, so that, for example, neon abundances would be ten to one hundred times greater than the present atmospheric content. Neither the low abundance nor the isotopic composition of the rare gases in the atmosphere fit such a model.

An additional factor is that large collisions in the final stages of accretion are likely to have removed any primitive atmosphere which might have formed. Thus the present atmosphere and hydrosphere of the Earth appears to be entirely secondary in origin, formed by degassing from the interior or by late accretion from comets and asteroids from beyond Mars.

When did this occur? Was the atmosphere slowly evolved from the mantle, or was there a sudden early degassing or outgassing event. The isotopic composition of the noble gases helium, argon and xenon have provided crucial evidence. The current outgassing rate, to judge from the rate at which helium is leaking from the mantle, is very low and could not be responsible for the present atmosphere. Early extensive degassing is indicated by the argon data. The mean atmospheric age from the noble gas data is greater than 4000 Ma. Most of the primitive volatiles were degassed from the mantle in the first 500 Ma after accretion, before there was significant addition to the mantle of ^{40}Ar from the decay of radiogenic ^{40}K. Xenon data also indicate that up to 80 per cent of the degassing occurred within about 50 Ma following accretion. This early rapid degassing would be consistent with a molten mantle, resulting both from the accretion of large planetesimals and from the formation of the Moon by a massive collision.

The source of water in the Earth remains as an interesting problem. Little water was available in the zone from which the Earth and the other inner planets were formed since water was never condensed or was lost along with the other volatiles in the early heating event. Some water, perhaps present in hydrated minerals in already formed planetesimals, might survive the early intense heating which drove the volatiles out of the inner Solar System, but probably much of the water came as a late-accreting veneer from beyond Mars. Water-ice is expected to be a stable phase in the nebula only at temperatures below 160 K at nebula pressures. This means that water-ice will occur only beyond 4–5 AU from the Sun, in the outer reaches of the asteroid belt. This is consistent with the observation that icy satellites are restricted to the region of the giant planets. Carbonaceous chondrites, probably typical of asteroid compositions beyond about 3 AU, contain up to 20 wt% water. Thus most of the terrestrial water was possibly derived from planetesimals or comets from beyond Mars, perhaps late in the accretional history of the Earth. If comets comprised 10 per cent of the bodies responsible for the bombardment between 4400 and 3800 Ma they could supply the appropriate amount of water for the terrestrial oceans. However, such a model is not without problems, since comets impact at high velocity, and so may remove earlier atmospheres and hydrospheres.

Crustal development really lies outside the main topic of this chapter (see Chapter 8), but since it is sometimes imagined that an early sialic crust was formed 'in the beginning', some comment is necessary. Studies of the development of the silicic continental crust show that it grew gradually throughout geological time. There is considerable evidence for an episodic crustal growth, with major crust-forming events in the late Precambrian (3000–2500 Ma ago) and other minor episodes (e.g. at 1800 Ma).

No rocks older than 3960 Ma have been identified, and they are unlikely to be found. The massive pre-3800 Ma bombardment will have broken up any early crust. The resulting breccias would have been easily removed by erosion (unlike the Moon, where the smashed up highland crust is preserved).

Although no massive world-encircling sialic crust formed in the beginning, did other crusts develop? Although the Moon formed a massive early anorthositic crust, this was a consequence of its more aluminous composition, lower pressures and bone-dry nature, which allowed a plagioclase-rich crust to float. This combination of circumstances was absent on the early Earth. Plagioclase is stable only in the upper 40 km or so, and it would sink in a terrestrial hydrous magma ocean, converting to denser phases (e.g. garnet).

There is evidence from the Sm–Nd isotope data that some extraction of elements from the mantle occurred before 4000 Ma ago. The most likely candidate is formation of an early crust of basalt, the primary partial melt from planetary mantles. This crust, which probably resembled the present basaltic crust of Venus, was removed by a combination of subduction and destruction by the meteoritic bombardment.

The Earth–Moon system

The Earth's Moon is a unique satellite; the satellites of the other outer planets are mainly rock–ice mixtures, formed by accretion around their parent planets, or by subsequent capture. The origin of the Moon has been an outstanding problem. It is in plain sight, accessible even to naked-eye observation, yet it has remained until recently one of the most enigmatic objects in the Solar System.

None of the other terrestrial planets possess comparable moons: Phobos and Deimos, the tiny Martian moons, are probably captured asteroids. The angular momentum of the Earth–Moon pair is anomalously high compared with that of the other inner planets. Some event or process spun up the system, although it is not rotating rapidly enough for classical fission to occur.

The lunar orbit is also strange. It is neither in the equatorial plane of the Earth nor in the plane of the ecliptic, but is inclined at 5.1° to the latter. Compared with the satellites of the giant planets, the Moon has a high mass relative to that of its primary planet. The bulk density of the Moon (3.34 g/cm^3), attributable to a low metallic iron content, is much less than that of the Earth (5.54 g/cm^3) or of the other inner planets.

The lunar samples returned by the Apollo and Luna missions added further complexities to these classical problems, revealing an unusual composition by either cosmic or terrestrial standards. The Moon is bone-dry, no indigenous water having been detected at parts per billion (10^9) levels. It is strongly depleted in volatile elements (e.g. K, Pb, Rb) by a factor of about fifty compared to the Earth, or 200 relative to primordial solar nebula abundances, and is probably enriched in refractory elements (e.g. Ca, Al, Ti, U) by about a factor of 1.5 compared with the Earth. The bulk lunar composition contains about 50 per cent more FeO than current estimates of the terrestrial mantle. It is perhaps not surprising that previous theories for the origin of the Moon failed to account for this diverse set of properties and that only recently has something approaching a consensus been reached.

Hypotheses for lunar origin can be placed into five categories, although they are not all mutually exclusive, elements of some hypotheses occurring in others:

1 Capture of an already formed Moon from an independent orbit has been shown to be improbable on dynamic grounds, and it provides no explanation for the compositional peculiarities, since the Moon might be expected to be an example of a common early Solar System object.

2 Formation as a double planet system immediately encounters the density and compositional problems. Attempts to overcome the density problem led to co-accretion scenarios in which the Moon formed from a ring of low-density silicate debris produced by disruption of incoming differentiated planetesimals. Their silicate mantles disintegrated as they came within about three Earth radii, but the tougher, denser iron cores survived to accrete to the Earth. This attractive scenario has been shown to be unlikely since the proposed breakup of planetesimals close to the Earth is unlikely to occur. In addition, the required angular momentum is difficult to achieve.

3 Fission hypotheses, deriving the Moon from the terrestrial mantle as proposed by George Darwin (son of Charles) over a hundred years ago were popular since they provided a low-density metal-poor Moon. They were testable following the lunar sample return, since they predict that the bulk composition of the Moon should match that of the terrestrial mantle. However, they failed to account for significant chemical differences between the composition of the Moon and that of the terrestrial mantle, or to provide a unique identifiable terrestrial signature in the lunar samples. These differences between the chemical composition of the Earth's mantle and the Moon remain fatal to theories which wish to derive the Moon from the Earth.

4 A modification of the fission hypothesis places terrestrial mantle material into orbit via multiple small impacts. Apart from the chemical difficulties inherent in the hypothesis, it is exceedingly difficult to obtain the required high angular momentum by such processes; multiple impacts should average out. A further objection turns on the uniqueness of the Earth's Moon; this scenario might be expected to produce moons as a

common accompaniment to the formation of the inner planets.

None of these theories accounted for the lunar orbit and for the high angular momentum (relative to the other terrestrial planets) of the Earth–Moon system, a rock on which all early hypotheses foundered. By specifying processes which might have been common in the early Solar System, so that similar satellites might be expected around the terrestrial planets, these hypotheses generally failed to account for the unique nature of the Moon.

5 A giant collision is now perceived to be the most likely explanation for the origin of the Moon. This single impact theory resolves many of the problems associated with the origin of the Moon and its orbit. Initially it was proposed to resolve the angular momentum problem but, in the manner of successful hypotheses, has accounted for other properties as well, and has become virtually a consensus since a 1984 conference on the origin of the Moon. The theory proposes that during the final stages of accretion of the terrestrial planets, a body somewhat larger than Mars collided with the Earth, and spun out a disc of material from which the Moon formed.

In more detail, the following scenario is the current view of lunar origin. When the Earth was close to its present size, it suffered a grazing impact, at about 5 km/s, with an object about 0.14 Earth masses (i.e. over 30 per cent larger than Mars). This body, as well as the Earth, is assumed to have differentiated by that time into a metallic core and silicate mantle. The

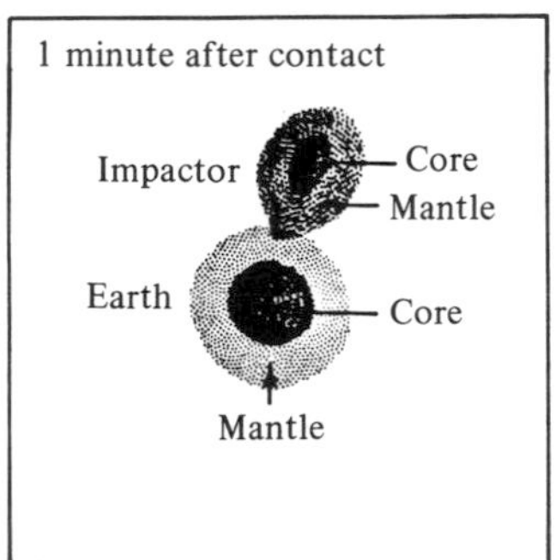

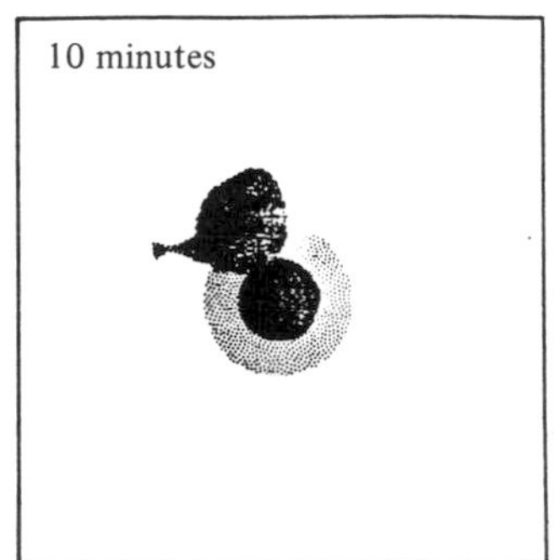

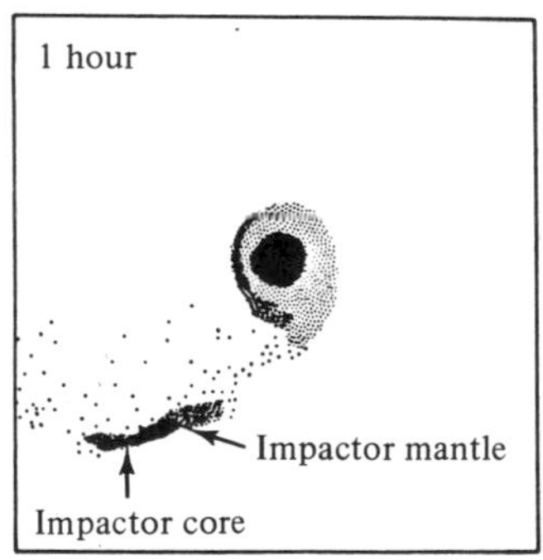

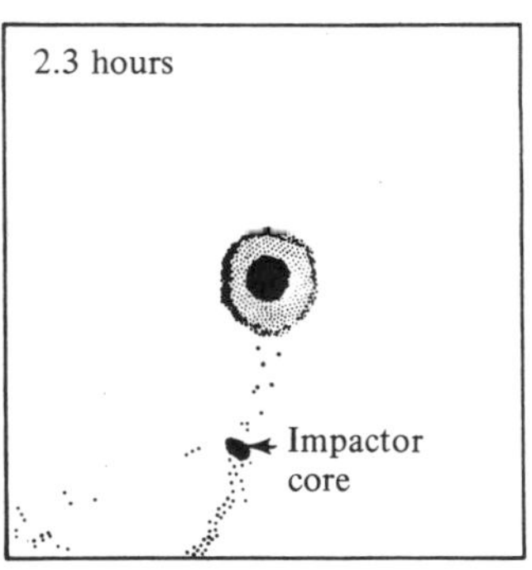

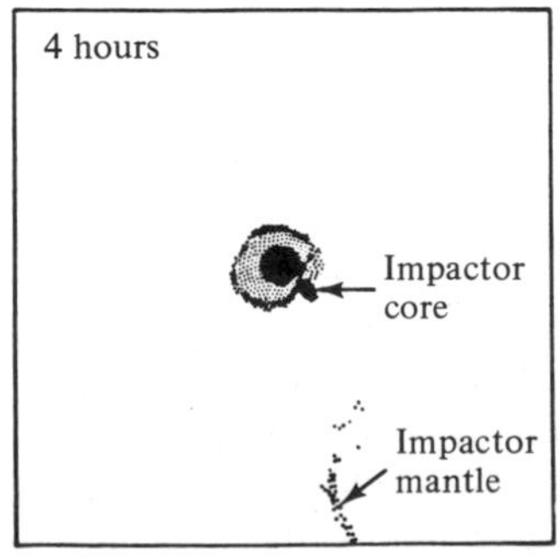

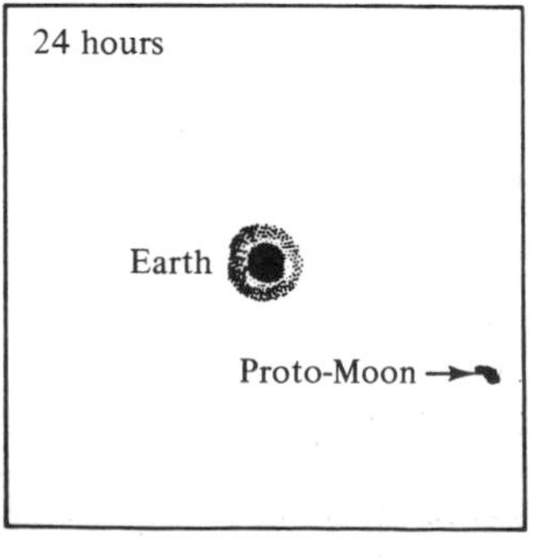

Figure 2.8 A computer simulation of the origin of the Moon by the glancing impact of a Mars-sized body with the Earth, in the final stages of planetary accretion. Both bodies have differentiated into a metallic core and a silicate mantle. Following the collision, the mantle of the impactor is ejected into orbit and accretes to form the Moon. The core of the impactor is swallowed up by the Earth. All the diagrams are not to the same scale. □

collision disrupted the impactor, much of which went into orbit about the Earth. Gravitational torques, due to the asymmetrical shape of the Earth following the impact, were sufficient to accelerate material into orbit. Another effect which promoted material into orbit was the acceleration away from the Earth due to expanding gases from the vaporised part of the impactor. An additional process occurred following the impact. The core of the impactor separated, the mantle material was accelerated and the core decelerated relative to the Earth. The core accreted to the Earth within about 4 hours.

Whether the material immediately coalesced to form a totally molten Moon, or broke up into several moonlets that subsequently accreted to form a partly molten Moon is uncertain. Geochemical studies indicate that at least half the Moon was molten shortly after accretion, with the feldspathic highland crust crystallising from this 'magma ocean'. The second scenario, of a partly rather than fully molten Moon, is probably most consistent with geochemical and geophysical constraints.

The impact event was sufficiently energetic to vaporise much of the material which subsequently recondensed to make up the Moon. This naturally explains such unique geochemical features as the bone-dry nature of the Moon, the extreme depletion of very volatile elements and the enrichment of refractory elements in the Moon. Figure 2.8 illustrates several stages of a computer simulation of the formation of the Moon according to one version of the single giant impact hypothesis. When we see the Moon in the sky, we are looking at the results of a chance collision about 4400 Ma ago.

The implications for the Earth of the single impact origin of the Moon are considerable. The event probably triggered or enhanced complete mantle melting. The accretion of the Earth from large planetesimals, rather than fine dust, virtually guarantees planetary melting in any event, since the heat dumped in by such events cannot readily be lost. In addition to the accretion of the impactor's core, about 10 per cent of the mass of the Earth's mantle would be added from the impactor's mantle.

Most of the metal core ends up in the Earth, with the metal penetrating the mantle and ending up wrapped about the Earth's core. Such an event would not disturb siderophile abundance patterns already present in the Earth's mantle. However, a significant amount of material from the impactor's core, enriched in siderophile elements, would probably be vaporised and redistributed into the mantle. The detailed implications for siderophile elements depends on the fraction of the accreting metal core which has a small enough grain size to equilibrate with the mantle. Another variation on this theme is that the present abundances of highly siderophile elements (e.g. the late veneer) in the Earth's mantle are due to the addition of a small portion of the impactor's core, while the rest of the impactor's core accreted to the Earth's core without significant interaction with the Earth's mantle. The Moon shows little evidence of a late veneer of siderophile elements, supporting a unique event involving the Earth as the explanation for the abundances of the highly siderophile elements in the Earth's mantle.

Clearly the Moon has a composition that cannot be made by any single-stage process from the material of the primordial solar nebula. The compositional differences from that of the primitive solar nebula, the Earth, from Phobos and Deimos (almost certainly of CI carbonaceous chondritic composition), and from the satellites of the outer planets (rock–ice mixtures with the exception of Io), thus call for a distinctive mode of origin.

Unique events are difficult to accommodate in most scientific disciplines. The Solar System, however, is not uniform. All nine planets (even such apparent twins as the Earth and Venus) and over sixty satellites are different in detail from one another. Bodies as massive as Mars existed during the accretional stages but have now been removed. The planets all possess varying obliquities. In contrast to the Earth, Venus rotates slowly backwards, has a low magnetic field, no oceans, a thick (100 bar) atmosphere mainly of carbon dioxide, and no Moon. All this diversity makes the occurrence of single events more probable in the early stages of Solar System history.

A significant feature of this model for the formation of the Moon is that the event is sufficiently energetic to melt the terrestrial mantle. This has important implications for the subsequent crystallisation and possible differentiation of the mantle, which cannot be pursued further here (see Chapter 3). Such melting is predictable during planetesimal accretion, since the energy is buried, rather than being re-radiated as in models which accrete the Earth from fine dust.

Conclusion: a broader perspective

The origin of the Earth and of the Solar System are of course among the great philosophical questions addressed by mankind. This chapter has concentrated on what appears to be the current scientific consensus for the origin of the Earth, and its satellite.

Rather recently, as the complexity of the Solar System has been revealed, there has been a shift in emphasis away from attempts to find some general and universal solution to the problem of its origin. It is not at all clear that the system as we know it could be modelled in a computer from the immutable operation of the laws of physics and chemistry acting on a fragment of a molecular cloud. Too many chance factors intervene. These include the initial size of the nebula, its detailed evolution, the early formation of Jupiter (a crucial part of the scenario), and the many random collisional events, one of which tipped Uranus on its side, another of which stripped off a major part of the silicate mantle of Mercury, while a third formed our unique satellite, the Moon.

It is just as unlikely that such a sequence of events will occur elsewhere as it is that the path of evolution, which led to *Homo sapiens*, would be duplicated in another planetary system. An unpredictable and random event, such as the Cretaceous–Tertiary boundary impact, was probably crucial in clearing the scene for mammalian evolution to proceed unhindered by the giant reptiles of the Mesozoic.

It is worth reflecting that of the four giant planets, Jupiter, Saturn, Uranus and Neptune, only the first three possess substantial regular satellite systems. Although these miniature solar systems might be expected to be similar, they are all distinct. Thus, even formation of satellite systems within our own Solar System does not lead to a uniform product. Other solar systems doubtless exist; they will almost certainly be different in the numbers and types of planets. That a large element of chance has entered in the evolution of our present system has led to the realisation of the inherent difficulties in constructing general theories for the origin of solar systems. The concept of 'clockwork' solar systems, popular since the time of Newton, has now generally been abandoned.

Further reading

Binzel, R. P., Gehrels, T. & Matthews, M. S. (eds.) (1989) *Asteroids*, University of Arizona Press, 1258 pp.

Black, D. C & Matthews, M. S. (eds.) (1985) *Protostars and Planets*, University of Arizona Press, 1293 pp.

Burns, J. A. & Matthews, M. S. (eds.) (1986) *Satellites*, University of Arizona Press, 1021 pp.

Kerridge, J. F. & Mathews, M. S. (1988) *Meteorites and the Early Solar System*, University of Arizona Press, 1269 pp.

Taylor, S. R. (1982) *Planetary Science: A Lunar Perspective*, Lunar and Planetary Institute, Houston, 481 pp.

Taylor, S. R. (1987) The origin of the Moon, *American Scientist*, **75**, 469–77.

Wasson, J. T. (1985) *Meteorites: Their Record of Solar-System History*, Freeman, NY, 267 pp.

CHAPTER

3

Don L. Anderson

California Institute of Technology

THE EARTH'S INTERIOR

Introduction

Seismologists have developed many methods for studying the structure and composition of the Earth's interior. Tens of thousands of **earthquakes** occur every year and each one sends out seismic waves in all directions. The bigger events send waves into the deep interior and these are used to infer the three-dimensional structure of our planet. Large earthquakes also generate surface waves which travel repeatedly around the surface and constructively interfere to generate the free oscillations, or normal modes, of the Earth. The velocities of seismic waves depend on composition, mineralogy, temperature and pressure and are useful as tools for studying these parameters. The new science of **seismic tomography** (meaning to 'slice the Earth') provides images which are useful in mapping convection.

While we can sample the crust of the Earth by a variety of techniques, the rest of the interior is largely inaccessible. Geophysical data, such as the mass and moment of inertia of the Earth and the velocities of seismic waves, are not adequate unambiguously to infer the chemistry of the mantle and core. We know from the distribution of elements in the Solar System which elements are likely to be most important in the Earth. We also obtain valuable clues by studying meteorites, the properties of the other planets in the Solar System, and the Sun (see also Chapters 1 and 2).

Nebular condensation and planet formation

It is commonly assumed that the Sun and planets formed more-or-less contemporaneously from a common mass of interstellar dust and gas, known as the **solar nebula**. There is a close similarity in the relative abundances of the condensable elements in the atmosphere of the Sun, in **chondritic meteorites** and, probably, in the Earth. To a first approximation we can assume that the planets incorporated the condensable elements in the proportions observed in the Sun and the chondrites. On the other hand, the differences in the mean densities of the planets, corrected for differences in pressure, show that they cannot all be composed of materials having exactly the same composition. Variations in iron content and in the oxidation state of iron can cause large density variations among the terrestrial planets, whereas, the giant, or Jovian planets, must contain much larger proportions of low atomic weight elements than Mercury, Venus, Earth, the Moon and Mars.

The equilibrium assemblages of solid compounds that exist in a system of solar composition depend on temperature and pressure, and, therefore, on location and time relative to nebular condensation (Figure 3.1). At a nominal nebular pressure of 10^{-4} bar the material would be a vapour at temperatures greater than ~ 1800 K. The first solids to condense at lower tem-

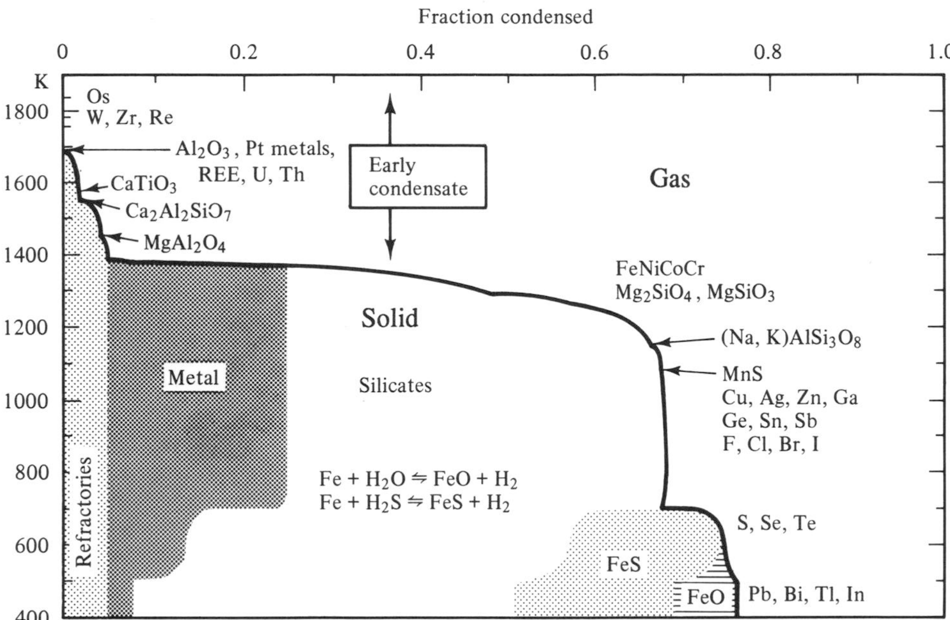

Figure 3.1 Condensation of solar gas at 10^{-4} bar represented in terms of the fraction condensed against absolute temperature (see also Chapter 2). REE = rare earth elements.

perature or higher pressure are the refractory metals such as W, Re, Ir and Os. Below about 1700 K refractory oxides of more abundant elements such as Al, Ca, Mg and Ti condense, and metallic Fe condenses near 1400 K. Below ~ 1150 K, Na and K form feldspars and a portion of the Fe is stable in olivine and pyroxene with the proportion increasing with a further decrease in temperature. FeS forms below ~ 750 K. Hydrated silicates condense below ~ 300 K. Differences in planetary composition may thus depend on the location of the planet, the location and width of its feeding zone and the effects of other planets in sweeping up material or perturbing the orbits of **planetesimals**, the small bits of pre-planetary material. In general, one would expect planets closer to the Sun to be more refractory rich than the outer planets. On the other hand, if the final stages of accretion involved coalescence of large objects with solar orbits of different eccentricities then there may be little correspondence between bulk chemistry and the position of the terrestrial planets.

There are several ways in which interactions between gaseous nebula and solid condensate particles might have controlled the composition of the planets. At one extreme, all or most of the material joining a planet may have equilibrated in a relatively narrow range of temperatures peculiar to that planet. Alternatively, one mineral after another condenses as required to maintain thermodynamic equilibrium in the nebula, and immediately accretes into a planet, or into a planetesimal which eventually joins a planet. At some point the process is interrupted by dissipation of the uncondensed nebular remnants. Differences in the mean compositions of the planets would result when the gaseous nebula, with any remaining uncondensed elements, was removed. If temperatures declined outward in the nebula at the time when condensation ended, this could account qualitatively for the density differences in the planets. The planets formed may thus have been layered from the outset, having the highest-temperature condensates at their centres, and successively lower-temperature condensates closer to their surfaces. Such planets would contain substantial metallic cores mainly as a result of accretion rather than subsequent interior melting and differentiation. This is termed **heterogeneous accretion** and is in contrast to the situation where all the solids have completely condensed before serious planet building starts: **homogeneous accretion**.

Modern theories of planetary accretion are of hierarchical nature. There is assumed to be a range of sizes in the pre-planetary body population and some large objects were involved in the formation of each planet. For example, an Earth-sized body may have coalesced from one Mars-sized and ten Moon-sized objects and numerous smaller bodies during its accretionary stage. In fact, current theories for the origin of the Moon have the Moon resulting from the splash between the proto-Earth and a Mars-sized impactor (details in Chapter 2). Large impacts tend to melt, or even vaporise, the colliding objects and to throw a large amount of material out into space. The Earth may have been extensively melted several times before the major part of the impacting population was used up and incorporated into the surviving planets. The present internal structure of the Earth and its chemical stratification may be a result of these early processes occurring during accretion. Where gas pressures are high, for example near a growing planet, or immediately after a giant impact, solids, melts and gases can coexist. At low nebular pressures most gases condense directly to a solid as indicated in Figure 3.1.

Cosmic element abundances and the composition of the Earth

The planets and the Sun formed from a diffuse nebula of gas and dust that collapsed sometime just after 4600 million years (Ma) ago, the age of the Earth (see Chapter 2). The **terrestrial or 'rocky' planets** and the **meteorites** can be thought of as rocky debris from this event, most of the mass having entered the giant planets and the Sun. Since the deep interior of the Earth is inaccessible we use data from the Sun and meteorites, as well as geophysical constraints, to infer what the Earth may be made of.

The term **volatile elements** is often used to refer to elements such as H, He, C and S and compounds containing these elements which are gaseous until very low temperatures and which are not easily accreted into planets. In contrast, compounds containing the **refractory elements***, Mg, Si, Ca, Al, Fe etc., condense as solids at relatively high temperatures (Figure 3.1) and are easily incorporated into growing planets. Elements such as K, Na, Rb and Pb are of intermediate volatility and apparently may be partially lost during the high-temperature accretional stages of a planet.

Of course, the Sun is mainly H, He and C, which are very volatile elements, but it also contains small amounts of Mg, Si, Fe, Ca, Al and O, the main constituents of the meteorites and rocky planets. After the Sun formed, most of the more volatile elements were swept out of the vicinity of the inner Solar System, probably at an early period when the solar wind was much more intense. The terrestrial planets are therefore composed primarily of the more refractory non-volatile or condensable elements, whereas the volatile elements are the primary building blocks of the Sun, the giant planets, and some of the satellites in the outer part of the Solar System. Comets are also composed primarily of volatile elements and compounds. Through solar spectra the Sun provides us with estimates of the ratios of the refractory elements such as Fe/Si, Mg/Si, Al/Si and so on. These ratios are likely to be close to those in the Earth as a whole.

Table 3.1. *Solar and meteoritic (CI carbonaceous chondrites) atomic abundances of some elements (relative to Si)*

	Sun*	CI†
Na	0.067	0.0574
Mg	1.089	1.074
Al	0.0837	0.0849
Si	1	1
P	0.0049	0.0010
S	0.242	0.0515
K	0.0039	0.00377
Ca	0.082	0.0611
Ti	0.0049	0.0024
Fe	1.270	0.900
Ni	0.0465	0.0493

* Breneman, H. H. & Stone, E. C. (1985) *Astrophysics, J.*, **299**, L57.

† Anders, E. & Grevesse, N. (1989) *Geochim. et Cosmochim. Acta*, **53**, 197.

Table 3.1 gives the composition of the Sun, as derived from spectroscopic studies and analysis of particles emitted by the Sun. Also given is the composition of a class of meteorites called the **carbonaceous chondrites**, Type 1 (CI). These meteorites have compositions close to the refractory portion of the Sun (i.e. the refractory elements occur in solar ratios) and also contain a large fraction of volatile elements (such as C and S).

Various attempts have been made to estimate the composition of the solar nebula by combining solar and chondritic chemical composition data. These are the so-called **'cosmic abundances'** which are the reference composition for the planets and other meteorites. Materials which have nearly cosmic abundances of a group of elements are called 'primitive' or 'undifferentiated' or 'chondritic'. The CI carbonaceous chondrites are the most solar-like in their composi-

* Refractory elements are those that condense at fairly high temperatures, say 1400 K or more. We define volatile elements as those that condense at lower temperatures, say 1200 K or less.

tions and have higher abundances of volatile elements than other types of meteorites. They are therefore called 'primitive' even though there is reason to believe they have experienced low-temperature alteration, possibly on the surface of a small Solar System object, or inside of a comet. Other meteorites appear to have been processed at higher temperatures, either in space or in small bodies such as asteroids, satellites or near the surface of small planets.

With some idea of the relative importance of the various chemical elements in our Solar System we can construct crude models of the Earth and other terrestrial planets. The Earth has a heavy core which is 32.5 per cent by mass of the planet. A glance at the cosmic abundance table (Table 3.1) shows that iron (Fe) is the only abundant heavy element so it is reasonable to assume that most of the core is Fe. It probably also contains about 6 per cent Ni. On the other hand, the core is a little less dense than pure Fe or Fe–Ni. There is a variety of elements, the **siderophile** (iron-loving) **elements** (see Figure 3.2), which may be mixed with Fe in the core, but these are unlikely to reduce its density. However, oxygen is an abundant element which dissolves in molten Fe at high pressure. FeO may also be an important component of the lower mantle. We therefore assume that the core is:

$$Fe + FeO = \text{'}Fe_2O\text{'}.$$

This has about the right density and there is enough 'Fe_2O' in cosmic abundances to make a core about the right size. We write Fe_2O in quotes since there is no guarantee that there is such a compound. There is probably some sulphur in the core as well, but sulphur is a volatile element and may not have been accreted by the Earth in cosmic abundances (see also discussion in Chapters 2 and 4). But there is a possibility that **chalcophile** (sulphur-loving) **elements** are concentrated in the core.

The major rock-forming refractory elements are Mg, Si, Al, Ca, Fe and O and their abundances in the silicate Earth are estimated in Table 3.2. In minerals these occur as MgO, SiO_2, Al_2O_3, CaO and FeO, Fe_2O_3 or Fe_3O_4. For example, the minerals enstatite (pyroxene) and forsterite (olivine) can be written:

$$MgO + SiO_2 \rightleftharpoons MgSiO_3 \text{ (enstatite)}$$

and

$$2\,MgO + SiO_2 \rightleftharpoons Mg_2\,SiO_4 \text{ (forsterite).}$$

We assume that these elements are fully oxidised in the crust and mantle (with Fe as FeO). Fe can substitute in these minerals for Mg, and the FeO-rich end-members are ferrosilite and fayalite respectively. On

Table 3.2. *Estimates of the composition of the silicate Earth (i.e. mantle plus crust)*

	(1)	(2)	(3)
SiO_2	45.0	44.2	44.9
Al_2O_3	3.2	2.0	3.1
FeO	15.7	8.3	13.5
MgO	32.7	42.3	33.0
CaO	3.4	2.1	3.8

(1) Computed from solar abundances, with 32.2 wt % core (1 mole Fe + 1 mole FeO = Fe_2O).
(2) Average peridotite from kimberlite pipe.
(3) Komatiite melt.

this basis, Table 3.2 (column 1) shows the oxide composition of 'primitive' mantle (mantle plus crust). Note that the remaining cosmic Fe that is not needed in the core is enough to make 15 per cent FeO for the mantle. This is much greater than the proportion found in most rocks and magmas which occur in the upper mantle, so this suggests that upper mantle rocks may not be typical of the whole mantle (see upper mantle peridotite composition, column 2 in Table 3.2). Some ancient lavas, called **komatiites**, do have a very high FeO content (~14 per cent – see Table 3.2, column 3).

Another way to estimate the composition of primitive mantle uses rocks and magmas which are exposed at the surface and which were brought up from great depth by volcanic eruptions. Most basalts come from the mantle, particularly those at ocean ridges, oceanic islands and volcanic arcs. There are also large basaltic lava flows on land, called continental flood basalts, which originate in the mantle. **Basalt** is the most abundant material from the mantle but it probably is only a partial melt so that a large amount of more refractory material is left behind. Rocks called **peridotites**, harzburgites and lherzolites are composed primarily of the refractory crystals olivine and orthopyroxene and have only a small amount of calcium, aluminium, sodium and titanium, elements which are important in basalts. Some peridotites may represent the material left behind after basalt extraction. The 'bulk silicate Earth' also includes the crust so this must be mixed back in when estimating the composition of primitive mantle. The crust is only a small fraction of the Earth but it contains a large fraction of some of the so-called 'incompatible' or **'lithophile'** (rock-loving) **elements**.

The relative portions of basalt and peridotite in the Earth are unknown but we can use chondritic or solar ratios based on the refractory elements to decide how much of each is required. Table 3.3 gives an estimate

Table 3.3. *Terrestrial abundance table*

	Mantle and crust	Relative to CI
Li	2.1 ppm	0.87
Na	2040 ppm	0.26 v
Mg	20.52 %	1.46
Al	2.02 %	1.57
Si	22.40 %	1.44
P	57 ppm	0.05 s,v
S	48 ppm	0.0025 s,v
K	151 ppm	0.17 v
Ca	2.20 %	1.58
Ti	1225 ppm	1.86
Fe	6.1 %	0.22 s
Ni	1961 ppm	0.13 s
Rb	0.39 ppm	0.11 v
Sr	16.2 ppm	1.42
Th	0.0765 ppm	1.50
U	0.0196 ppm	1.40

s = these elements may occur primarily in the core.
v = volatile, these elements are depleted in the Earth compared with CI meteorites or the Sun (relative to the refractory elements).

of the composition of the Earth based on these considerations and an extended version with more elements appears as Figure 3.2.

The refractory elements (Mg, Al, Si, Ca, Ti, Th, U ...) all occur at a level of about 1.5 times chondritic (CI meteorites) in the mantle–crust system. This is the level expected if these elements have been excluded from the core and occur only in the silicate part of the Earth. The volatile elements (Na, P, S, K, Rb ...) are depleted in the mantle and crust, relative to chondrites and the Sun, and the elements which are depleted the most are the most volatile or the most siderophile. This depletion probably occurred during the accretion of the Earth, and during core formation. The Fe in the Earth is primarily in the core along with other siderophile elements such as Ni, Co, Re, Os ..., and also, possibly, P and S. The mantle–crust system is extremely deficient in siderophiles. Having developed broad constraints on the chemistry of the Earth's interior the next step is to introduce geophysical data that allow us to construct an amplified picture.

Seismology and the Earth's interior

The Earth's interior is illuminated by **seismic rays** radiating outward in all directions from the numerous earthquakes that occur in tectonically active regions of the world. The wavefronts developed at right angles to these ray paths are refracted (bent) and reflected by discontinuities and gradients in material properties in the Earth's interior and are recorded at seismic stations on the Earth's surface. From analysis of such waves seismologists have divided the Earth into a crust, mantle and core and into numerous smaller subdivisions such as the upper mantle, transition region, lower mantle, outer core and inner core. These will be discussed later. Box 3.1 discusses the math-

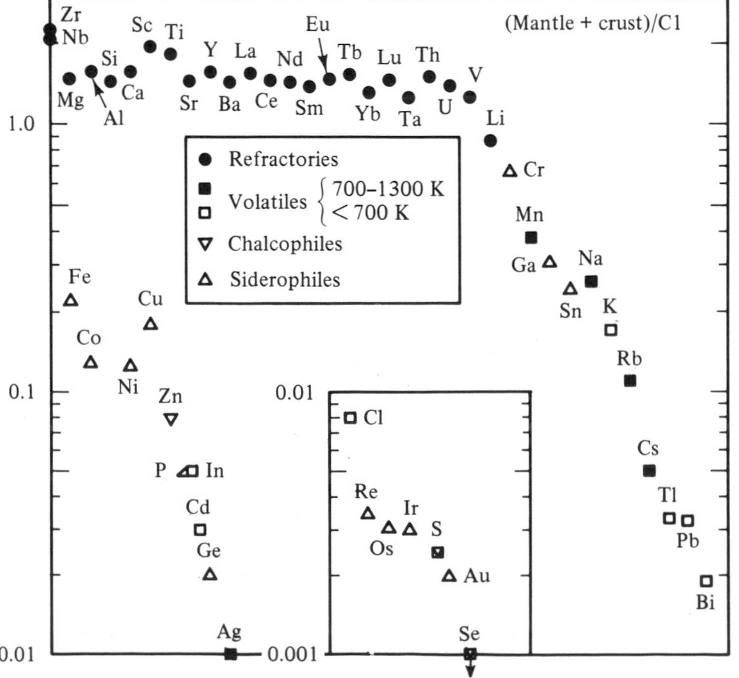

Figure 3.2 Chemical composition of the silicate Earth (mantle plus crust), sometimes called the bulk silicate earth (BSE) or Primitive Mantle relative to the compositions of CI carbonaceous chondrite meteorites.

BOX 3.1

Seismic wave velocities, elastic moduli and density

An isotropic solid exhibits both a **compressibility** and a **rigidity**. The former describes the resistance of the solid to a change in volume due to a change in confining pressure. The rigidity expresses the ability to withstand a shape change, with no change in volume, due to an imposed shear stress. Liquids and gases are relatively compressible and have no rigidity.

Sound waves in fluids travel with the compressional velocity, V_P:

$$V_P = \sqrt{K_S / \rho}\,, \qquad (1)$$

where K_S is the adiabatic **bulk modulus**, the reciprocal of the compressibility, and ρ is the density. In a solid the compressional velocity is:

$$V_P = \sqrt{\frac{K_S + (4/3)G}{\rho}}\,, \qquad (2)$$

where G is the **rigidity modulus**.

The shear velocity is:

$$V_S = \sqrt{\frac{G}{\rho}}\,. \qquad (3)$$

The first wave to arrive from an earthquake travels at velocity V_P and is compressional; it is often called the P- or primary-wave. The particles in such a wave exhibit a push–pull motion, just as do sound waves in the air (Figure 3.3). The shear, secondary or S-wave, travels much more slowly, although it usually has a larger amplitude because of the shearing nature of motion along earthquake faults. The motion associated with the shear wave is a side-to-side shaking motion. The particle motion, in fact, is perpendicular to the direction of wave propagation (Figure 3.3).

Seismic compressional waves and most sound waves depend on the adiabatic bulk modulus, K_S. It is called 'adiabatic' because there is little time during the passage of the wave for heat to flow. The compressed parts of the wave warm up the solid but the dilatational part of the wave soon arrives and cools it off. In a static experiment a solid can be compressed very slowly and one can wait until the temperature of the solid equilibrates with the surroundings. In this case the compression of the solid is controlled by the isothermal bulk modulus, K_T. The two are related by a well-known thermodynamic identity:

$$K_S = K_T(1 + \alpha\gamma T) \qquad (4)$$

where α is the coefficient of thermal expansion and γ is the Grüneisen parameter, a small number usually between 1 and 2. The coefficient of thermal expansion is typically between 20 and 30×10^{-6} K^{-1}. For a temperature, T, of 1000 K and with $\gamma = 1$ and $\alpha - 20 \times 10^{-6}$ K^{-1}:

$$K_S/K_T = 1.02. \qquad (5)$$

ematical details of seismic wave velocities, elastic moduli and density, while some of the seismic wave paths which are used are shown in Figure 3.4.

The nomenclature of Figure 3.4 is quite simple. The primary (**compressional**) and (**shear**) secondary **waves** in the mantle are P-waves and S-waves. Compressional waves in the outer core are K and in the inner core are I. Waves reflected at the boundary of the outer core include the letter c in their designation (e.g. PcP, ScS). Waves reflecting off the inner core boundary are PKiKP, SKiKS and so on. A shear wave through the inner core is called J but evidence for its existence is indirect because no shear waves occur in the outer core. Waves can reflect off the surface of the Earth many times (e.g. Figure 3.4: SS, SSS, PP ...). They also bounce around in the outer core (PKKP, PKKKP, ...).

The travel times of these various waves tell us the depths to **seismic discontinuities** (important physical and chemical boundaries in the Earth) and the variation of seismic velocity with depth and from place to place. The amplitudes of these waves tell us about the size of the seismic event, the reflection coefficients of seismic discontinuities and the attenuation of seismic waves. In general, the velocities of seismic waves increase with depth but in tectonic and volcanic regions there is often a region in the shallow mantle where particularly the shear velocity decreases with depth, due to high temperature gradients near the Earth's surface and the onset of partial melting (Figure 3.5). Pressure compresses the rocks and minerals of the

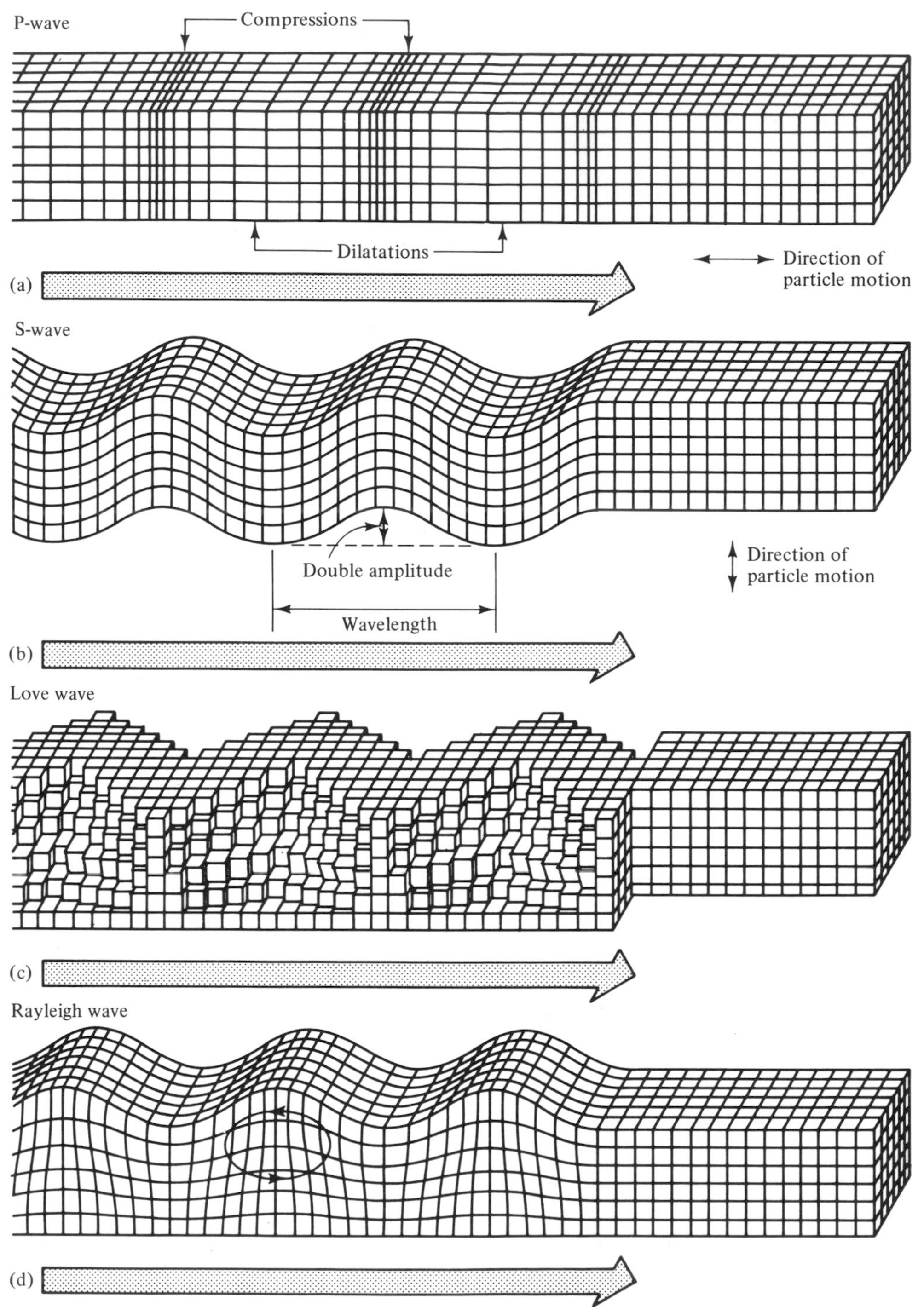

Figure 3.3 The forms of the motion and ground deformation due to (a) P-waves, (b) S-waves, (c) Love waves and (d) Rayleigh waves. The particle motion of P and S-waves is transmitted through the Earth, so these are known as body waves, whereas the motion due to Love and Rayleigh waves diminishes with depth, so these are known as surface waves. □

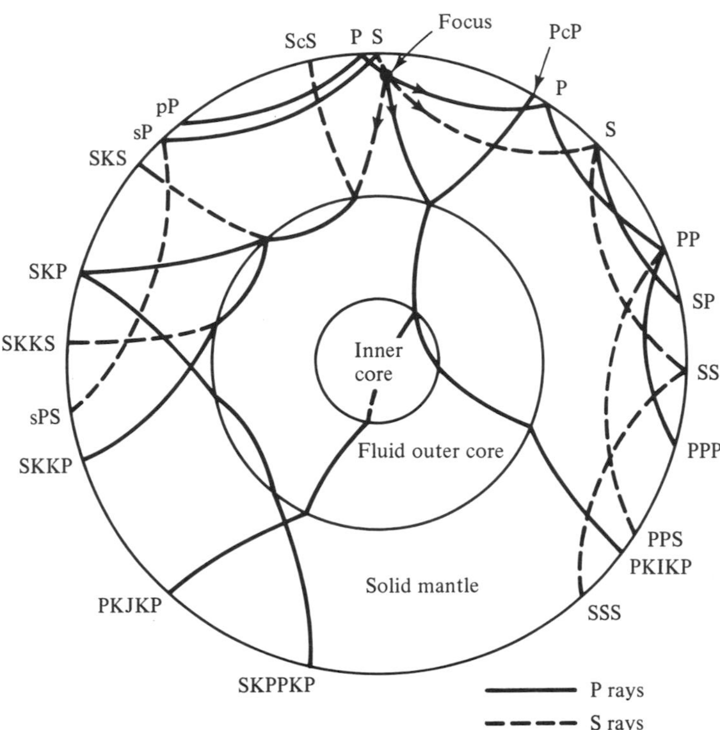

Figure 3.4 Representative seismic wave paths in the Earth all arising from the same focus. P, K and I are compressional waves; S and J are shear waves (see text for further details). □

mantle and causes the density and seismic velocities to increase. Temperature has the opposite effect, and partial melting also decreases density and seismic velocity. In the upper part of the mantle the temperature gradient is high and the tendency for velocity to decrease with temperature is comparable to or greater than the tendency of pressure to increase velocity. At greater depth, pressure is the dominant influence.

The temperature gradient is also high across major chemical interfaces such as the **core–mantle boundary** (CMB). The reason is that heat cannot be convected across such boundaries and a steep temperature gradient – across a 200–300 km thick zone (discussed in more detail later) – is established in order to conduct the heat away from the interior. The shear velocity may also decrease at the base of the mantle, a result of the high thermal gradient. P-waves are somewhat less sensitive to temperature and more sensitive to pressure than S-waves.

Seismic velocities vary with depth, and from place to place at the same depth. For example, stable shield areas not only have thick crusts they also have high velocities in the underlying mantle, a combination of cold temperatures and high velocity refractory minerals. Volcanic regions and regions of young lithosphere, such as near oceanic ridges, have very low seismic velocities, primarily the result of the velocity reduction due to the presence of magma in the shallow mantle.

Figure 3.5 shows the variation of shear velocity with depth in a stable shield area, in an old oceanic area and in a tectonically active area. Note the large variation of velocity above 200 km depth. The seismic lithosphere, or lid, is thick (~ 150 km) under shields and much thinner under the other regions. The near-surface high-velocity regions may correspond to the minimum thickness of the plates that are involved in plate tectonics (see also Chapter 8). The velocities under the lid in tectonic and oceanic regions are so low that no combination of the abundant upper mantle minerals can account for them unless the temperature is so high that partial melting occurs. The presence of a melt phase is implied down to a depth of at least 300 km in these regions. The relatively high velocities under shield areas between about 150 and 400 km have been used by some to argue that continental roots extend this deep and that the continental plate extends deep into the mantle. The alternate point of view is that shield mantle is normal and it is the presence of an enhanced melt phase under oceans and tectonic areas that makes them differ from normal colder, or average temperature, mantle.

The lateral variation of seismic velocity decreases markedly below 200 km (Figure 3.5) and again below 400 km. In fact, the high-resolution **body wave** (P and

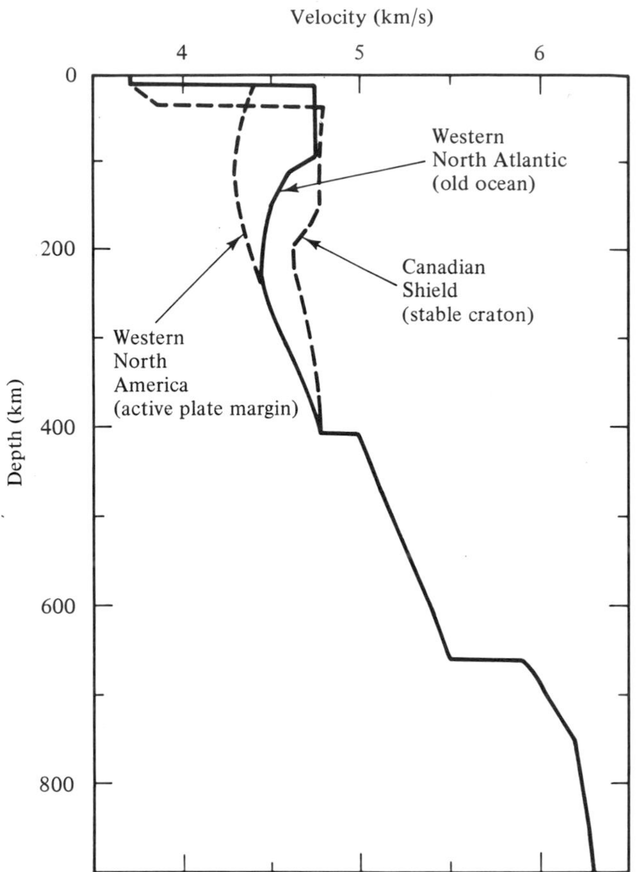

Figure 3.5 Seismic shear velocity vs depth in three different tectonic provinces as determined by Helmberger and co-workers at Caltech by comparing theoretical seismograms with observed seismic records. The theoretical mantle model is adjusted until there is good agreement with the data. The travel times of seismic waves at different distances from a seismic source are also used to construct velocity–depth functions. □

S-waves travelling through the Earth) studies of Professor Helmberger and his colleagues (Caltech) show little if any lateral variation below 400 km in the regions studied. In other studies there are slight variations below this depth. However, the new science of seismic tomography has resolved slight variations down to the core–mantle boundary; these studies will be discussed later.

There are large and abrupt increases in seismic velocity at 5–40 km depth, also near 400 and 650 km depth (Figures 3.5 and 3.6). These are associated with changes in mineralogy as, for example, at the shallowest discontinuity, the crust–mantle boundary or **Mohorovicic discontinuity** (Moho). A change in mineralogy can be due to a change in chemistry, or a change in temperature or pressure with no change in chemistry. For example, the common minerals in the continental crust are quartz and feldspar, a result of the high SiO_2, Al_2O_3, Na_2O and K_2O content of the crust. The main stable minerals at the top of the mantle are olivine and orthopyroxene, a consequence of the higher MgO and lower SiO_2 etc. content. Because of this change in mineralogy there is a large change in density and seismic velocity at the Moho.

A change in chemistry can be more subtle. The large jump in seismic velocity at 400 km is primarily due to the effect of increasing pressure which causes the collapse of olivine and orthopyroxene to denser minerals with the same chemistry (i.e. **phase changes**). At temperatures of 1400–1500 °C these phase changes occur with increasing pressure equivalent to 400 km depth. The discontinuity may therefore be an equilibrium phase boundary between low and high density phases (for details, see Chapter 4). However, it has been suggested by some that the region between 400 and 650 km depth may be richer in a basaltic fraction, and therefore poorer in olivine, than the mantle above 400 km. If there is such a change in chemistry at this boundary, an increase in garnet and clinopyroxene for example, it would be hard to detect because of the large phase change effect. Another large seismic discontinuity occurs at about 650 km depth. It is also primarily due to a change in mineralogy and may also be accompanied by a chemical change. The high velocity gradient region just below the discontinuity (Figure 3.6) may represent a spread-out phase change.

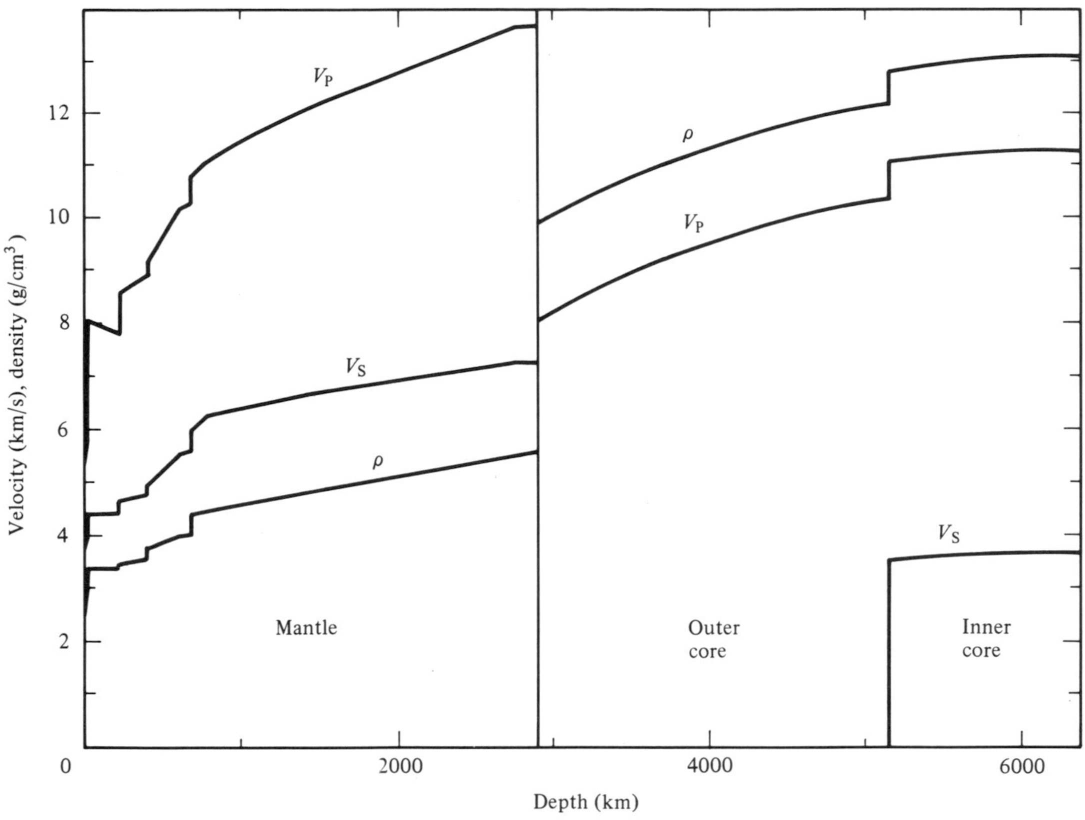

Figure 3.6 Seismic velocities V_P and V_S throughout the Earth and density (ρ) deduced as explained in Box 3.2. □

The density and seismic velocity throughout the Earth are summarised in Figure 3.6 and a summary of the way in which we derive density data from observed quantities is given in Box 3.2. The boundary between the mantle and the core (CMB) occurs at a depth of about 2900 km. There is a marked decrease in compressional velocity, V_P, and an increase in density as we go from the mantle to the core; moreover, shear waves are absent. These observations are consistent with a molten iron-rich core. The subsequent sections of this chapter amplify details of the structure, composition and mineralogy of the Earth, building on the outline introduced above using Figures 3.5 and 3.6.

Properties of the Earth's layers: some further details

At the centre of the Earth is the solid **inner core**. It is about the size of the Moon but has approximately the density of pure iron or an iron–nickel alloy. It is generally thought to have frozen out of the liquid outer core. The outer core is molten but is probably close to the freezing point. The inner core is therefore suspended in a relatively low-viscosity fluid and is fairly isolated from the rest of the Earth.

The **outer core** extends from the boundary of the inner core (5155 km depth) to the core–mantle boundary which is at a depth near 2900 km, about half of the radius of the Earth (Figure 3.6). The outer core does not transmit shear waves and is therefore a fluid.

The adiabatic bulk modulus, K_S of the top of the core is about the same as that at the base of the mantle but since there is no rigidity in the outer core ($G = 0$), the value of V_P actually drops when crossing from the mantle to the core (Box 3.1, Equation 2). The density of the core is about twice the density of the mantle and this also contributes to the velocity drop. Seismic waves entering the core from the mantle therefore refract downwards (e.g. at B and E, Figure 3.7). They are eventually refracted upwards by the increase of seismic velocity with depth but there is a gap at the Earth's surface in which no direct waves through the top of the core arrive. This is called a **'shadow zone'**

BOX 3.2

Determination of density in the Earth

Zones within the Earth are considered to be homogeneous if there are no chemical variations or phase changes. The change of density, ρ, with radius, r, in any such region is given by:

$$\frac{d\rho}{dr} = \left(\frac{\partial\rho}{\partial P}\right)_T \frac{dP}{dr} + \left(\frac{\partial\rho}{\partial T}\right)_P \frac{dT}{dr}, \qquad (6)$$

where the subscripts refer to the parameter (pressure, P, or temperature, T) being held constant. For a homogeneous self-compressed region in which temperature increases because of this compression (i.e. under an **adiabatic temperature gradient**):

$$\frac{dP}{dr} = -g\rho, \text{ where :} \qquad (7)$$

$$g = \frac{GM(r)}{r^2}. \qquad (8)$$

g is the acceleration due to gravity, G is the gravitational constant and $M(r)$ is the mass inside radius r.

In a convecting mantle the mean temperature gradient, away from thermal boundary layers, is close to adiabatic, denoted by the subscript S.

$$\frac{dT}{dP} = \left(\frac{\partial T}{\partial P}\right)_S = \frac{T\alpha}{\rho Cp}, \qquad (9)$$

where Cp is the specific heat at constant pressure and α is the volume coefficient of thermal expansion (cf. Box 3.1) defined as:

$$\alpha = -\frac{1}{\rho}\left(\frac{\partial\rho}{\partial T}\right)_P. \qquad (10)$$

It is therefore convenient to write the temperature gradient in the Earth as:

$$\frac{dT}{dr} = \frac{T\alpha}{\rho Cp}\frac{dP}{dr} - \tau, \qquad (11)$$

where τ is the departure from the adiabatic gradient (required by the isobaric dT/dr in equation 6). Adiabatic compression of a material is given by the adiabatic bulk modulus, K_S:

$$K_S = \left(\frac{\rho\partial P}{\partial\rho}\right)_S, \qquad (12)$$

where S means constant entropy.

Seismic waves are also adiabatic, and hence we can use (from equations 2 and 3, Box 3.1):

$$V_P^2 - (4/3)\, V_S^2 = K_S/\rho = (\partial P/\partial\rho)_S = \phi \qquad (13)$$

to calculate the variation of density with radius in a homogeneous region of the Earth for which we have seismic data. Combining equation 6 with equations 7 and 13, and equations 10 and 11, for a homogeneous adiabatic self-compressed region, we have:

$$\frac{d\rho}{dr} = -g\rho/\phi + \alpha\rho\tau. \qquad (14)$$

This is known as the **Williamson–Adams equation**.

Much of the upper mantle and transition region are not homogeneous or close to adiabatic because of large temperature gradients, partial melting and phase changes. In such regions, the term $\alpha\rho\tau$ can be used to correct for non-adiabatic gradients provided there is no chemical change or melting, and, of course, provided α and τ are well known (which is often not the case). Most of the lower mantle, however, appears to be close to homogeneous and adiabatic, so that the equation holds well. The Williamson–Adams equation also cannot be used in parts of the Earth where the chemistry is variable, and here we resort to producing Earth density models by trial-and-error that are constrained by seismic travel times, free oscillation data, and other properties such as the Earth's mass and moment of inertia. The simplified results shown in Figure 3.6 were produced using a combination of these techniques.

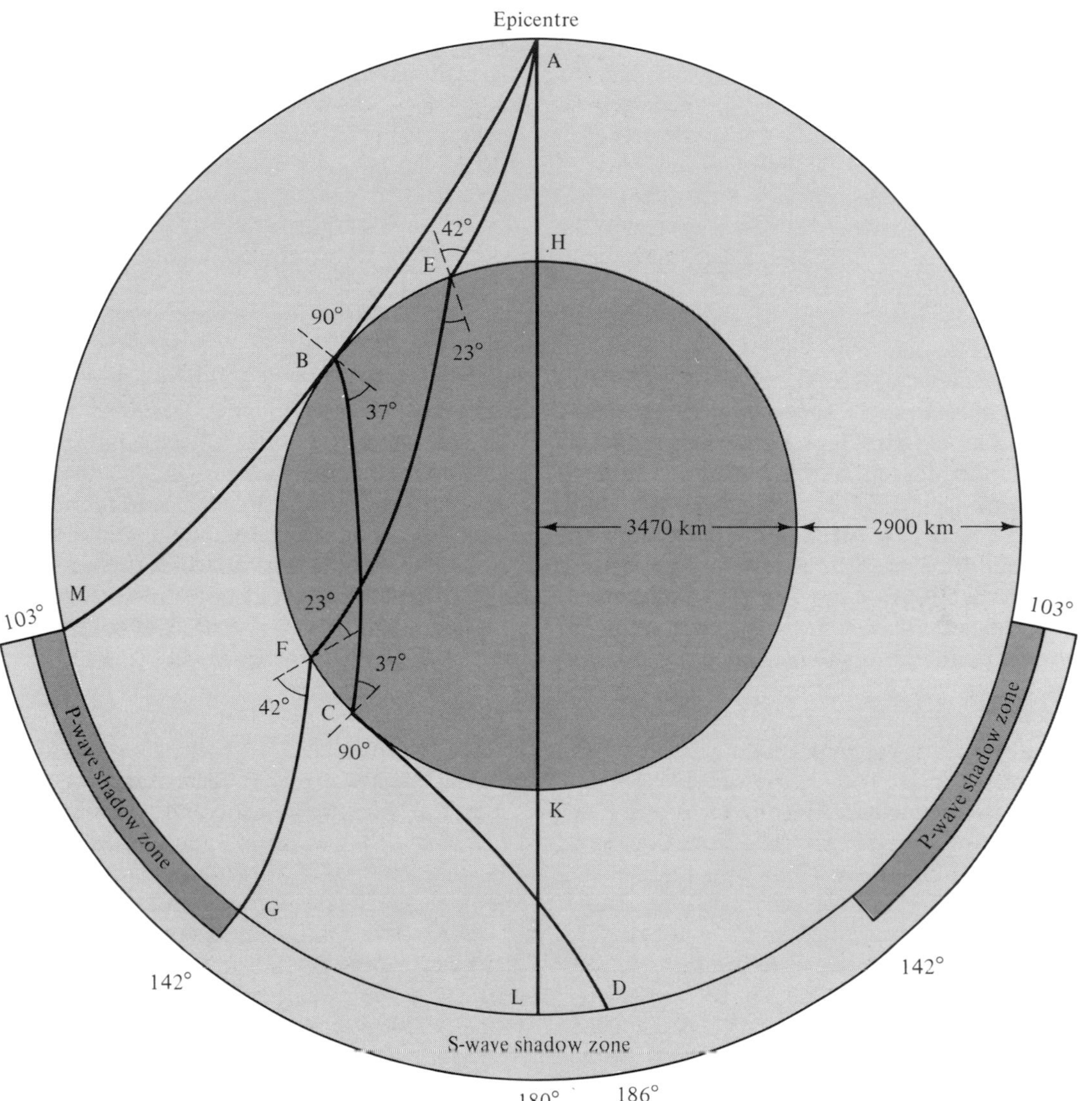

Figure 3.7 The origin of the P-wave shadow zone due to the outer core. A mantle wave path AM just grazes the core and a path only minimally steeper than AB causes the ray to be strongly refracted into the core and out again at C, arriving at D. Waves reaching the core between B and E arrive at the surface between D and G, but as the initial path (e.g. AH) becomes steeper than AE further arrivals at the surface move from G to L. Thus no waves arrive between M and G – the shadow zone between 103° and 142° around the Earth from the epicentre (known as the epicentral angle). This area does, however, receive some seismic energy deflected strongly from the inner core (see Figure 3.4, path PKIKP) but the strength of these arrivals is nowhere near so strong as between 142° and 180°. □

(Figure 3.7) and its size is the primary evidence of the radius of the core.

Various theoretical attempts have been made to estimate the viscosity of the outer core and these generally conclude that it is not much more viscous than water, and may even be more than twenty orders of magnitude less viscous than the mantle. It therefore convects readily and cannot support large internal temperature or chemical gradients. It is probably the most homogeneous part of the Earth but this does not rule out light layers at the top and dense layers at the bottom where material may have settled out. In fact, the inner core may have formed from iron particles settling through the outer core. The core has probably cooled over time and, as it cools, crystals of pure iron may form from the iron–oxygen (and/or sulphur) mixture which has a lower freezing temperature. Production of these crystals will induce a form of chemical convection and this may help stir the core and provide some of the energy to drive the terrestrial 'dynamo' and generate the **Earth's magnetic field**. The other observable feature about the outer core is that it transmits seismic waves with very little attenuation (i.e. loss of energy). Once a seismic wave gets into the core it can bounce around for a long time. In contrast, the inner core attenuates seismic waves very rapidly.

Seismic waves reflect well off the core–mantle boundary (CMB), good evidence that the boundary is sharp and exhibits a large change of velocity and density over a fraction of a seismic wavelength. There may be long wavelength bumps on the CMB, of the order of 0.5 to 10 km, but this is still uncertain. However, topographic coupling is one way in which stresses may be transmitted across the CMB. For example, motions in the core could affect the rotation of the Earth by applying pressure to the bumps and valleys. The bumps themselves would be maintained by convection in the high viscosity mantle where relatively highly viscous and rigid, yet lower density material floats on the high density fluid below.

There is an irregular solid layer just above the CMB which has a thickness of 200 to 300 km. This is called the D" layer, a leftover from the nomenclature of Keith Bullen of Australia who proposed a letter designated for each region of the Earth. This layer is variable in both velocity and thickness and in some places is separated from the overlying mantle by a sharp but small discontinuity. In some places the shear velocity gradient in this region is negative, i.e. the velocity *decreases* with depth. These features are best explained by the existence of a chemically distinct layer at the base of the mantle and this layer is also a **thermal boundary layer**. The core is convecting and is losing heat to the overlying cooler mantle (see Chapter 4 for temperatures). However, since little or no material is crossing the boundary this heat must be lost by conduction. There is therefore a region of high thermal gradient just above the CMB which may extend for several hundred kilometres. A high thermal gradient usually means low seismic velocity and density gradients. In a chemically uniform mantle the material in such a thermal boundary layer would become buoyant and the layer would become unstable and rise. Thus it is likely that D" is a chemically distinct layer that is intrinsically much denser, so that the material may be trapped. Nevertheless, it is believed that some sustained mantle plumes may originate in this layer. However, there is no clear evidence that such plumes rise to the surface of the Earth, or even into the upper mantle.

The CMB represents the largest density contrast in the Earth. It is therefore a natural collection point for any light material leaving the core or dense material settling out of the mantle. Thus D" may be a remnant of the original differentiation of Earth, containing material intermediate in density between the core and the bulk of the mantle. Candidate materials of high density are stishovite, a high-pressure form of SiO_2, $(Mg,Fe)SiO_3$ (perovskite) and (Mg,Fe)O (magnesiowüstite). If D" represents early condensing refractory material, or even deeply subducted oceanic lithosphere, it would also be rich in CaO, Al_2O_3 and TiO_2. D" is only about 2 to 3 per cent of the mass of the Earth.

The **lower mantle** is the largest subdivision of the Earth. It extends from the top of D" (about 2740 km depth) to the major seismic discontinuity at a depth near 650 km. The lower mantle is 49.2 per cent by mass, of the Earth and 72.9 per cent of the mantle plus crust. The predominant mineral in the lower mantle is $(Mg,Fe)SiO_3$ in the dense **perovskite structure**, a high pressure form of the chemically identical lower density pyroxene (see Box 3.3 for details of mantle minerals). The second most abundant mineral is probably (Mg,Fe)O in the NaCl or rock-salt structure. Other possible minerals are Al_2O_3 (corundum), different perovskite-structure minerals containing CaO and Al_2O_3, and stishovite. Seismic velocities increase gradually with depth throughout most of the lower mantle and there is no evidence for discontinuities or changes in chemistry. It is thought therefore that the lower mantle is relatively uniform in composition and that the density and seismic velocities increase with depth primarily due to compression – thus the Williamson-Adams equation applies, cf. Box 3.2. Some lateral variations occur, however, primarily because of temperature differences.

The next important feature of the mantle is the **650 km discontinuity** (sometimes called the 670 km discontinuity, but it may vary in depth from place to place by 50 to 100 km). This represents a change in mineralogy between intermediate density minerals, such as γ-spinel and majorite, and high density phases such as perovskites and magnesiowüstite (see Box 3.3 and the next section of this chapter). It may be simply an equilibrium phase boundary, or the two sides of the boundary may also differ chemically in, for example, the contents of FeO, CaO, Al_2O_3 or SiO_2. Phase changes in mantle minerals are generally spread out over at least tens of kilometres depth (see Figure 3.8). But the 650 km discontinuity at 22–24 GPa pressure is a good reflector of seismic energy and this requires that its surface is sharp to about 4 km. It is also close

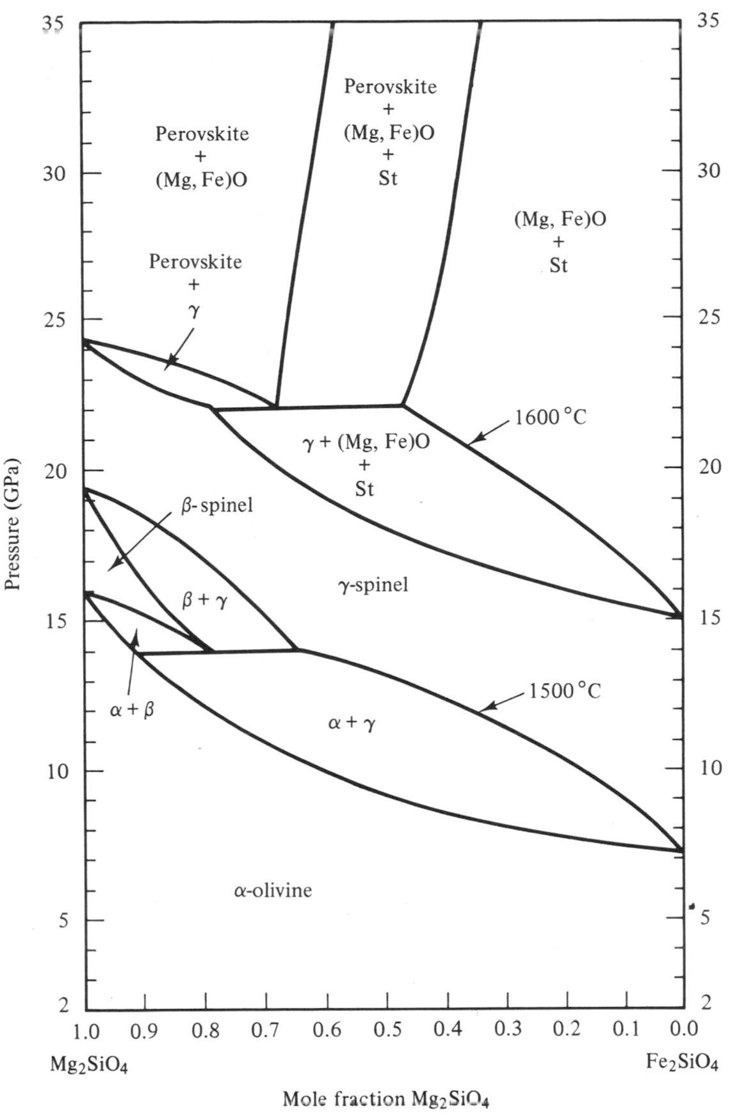

Figure 3.8 **Phase diagram of the olivine system $(Mg,Fe)_2SiO_4$, showing the stability fields of the various stable phases, St–stishovite (SiO_2). The α phase is olivine, and both the β and γ phases are forms of spinel. Notice how both the olivine to spinel and spinel to perovskite etc. phase changes occur at greater depths in Mg-rich than in Fe-rock systems. These two sets of phase changes are believed to be important, respectively, at the 400 and 650 km discontinuities.** □

to the maximum depth at which earthquakes occur. These features are sometimes taken as evidence that this is a chemical boundary but these arguments are not conclusive. In some Earth models there is a high seismic velocity gradient for 100 km below the discontinuity, and this suggests that the solid–solid phase changes do occur over a broad depth interval even though the upper surface of this zone is a sharp discontinuity.

The region of the mantle between the 650 km discontinuity and another mantle discontinuity at 400 km is known as the **transition region**, being the transition between the upper and lower mantle. The **400 km discontinuity** at 13–15 GPa pressure is due, in part, to the transformation of magnesium-rich olivine to β-spinel (Figure 3.8) and orthopyroxene to a garnet-structured mineral called **majorite**. However, the jump in seismic velocity at 400 km is much less than the jump predicted for these phase changes (see Figure 3.10 and related discussion in the next section) so there must be other minerals near this depth that do not transform. Garnet and clinopyroxene, the main minerals in eclogite, are candidates. **Eclogite** is the high pressure form of basalt and may be more abundant in the mantle than generally thought. The mineralogy of the transition region is therefore thought to include β- and γ-spinel, garnet and majorite. Clinopyroxene is stable at the top part of the transition region but it eventually collapses to a garnet-like phase, probably at about 500 km depth. Indeed, some seismic models have a discontinuity, or at least a high-velocity gradient region, near 500 km.

In the **upper mantle**, down to about 400 km depth, the seismic velocities are consistent with a mixture of

BOX 3.3

Mantle minerals

Atoms and elements (Table 3.3) are the basic building blocks of minerals. Minerals are the basic building blocks of rocks, both in the crust and mantle. The most abundant minerals, in turn, may be considered as composed of the oxides of the elements (Table 3.2). The important upper mantle minerals are olivine, orthopyroxene, clinopyroxene and garnet. These form **solid solutions** between relatively simple end-members:

forsterite Mg_2SiO_4	+	fayalite Fe_2SiO_4	=	olivine, $(Mg,Fe)_2SiO_4$
enstatite $MgSiO_3$	+	ferrosilite $FeSiO_3$	=	orthopyroxene, $(Mg,Fe)SiO_3$
diopside $CaMgSi_2O_6$	+	jadeite $NaAlSi_2O_6$	=	clinopyroxene, $(Ca,Mg,Na,Al)Si_2O_6$
pyrope $Mg_3Al_2Si_3O_{12}$	+	almandine $Fe_3Al_2Si_3O_{12}$	=	garnet $(Mg,Fe)_3Al_2Si_3O_{12}$

Note that diopside and pyrope can be expressed as $CaO + MgO + 2SiO_2$ and $3MgO + Al_2O_3 + 3SiO_2$ respectively. The above reactions are oversimplified but, nevertheless, by comparing the elements in the above minerals with the abundances in Table 3.3 we see that the most abundant elements are all accounted for. Estimated abundances of major minerals in the mantle lie in the following ranges (weight per cent): olivine 37–51, orthopyroxene 26–34, clinopyroxene 12–17, garnet 10–14.

The above minerals are typical of upper mantle assemblages. At high pressure these minerals convert to denser forms with different crystal structures, a process known as a solid–solid phase change (see Chapter 4 for definitions). For example, diamond and graphite are two solid phases of carbon; diamond is the high pressure and low temperature phase. Natural diamonds were made at depth in the mantle and then were brought rapidly to the Earth's surface in kimberlite eruptions. The important mantle minerals also undergo these pressure-induced phase changes. The mineral forsterite (olivine), Mg_2SiO_4, for example, converts to a dense cubic mineral having a structure similar to normal **spinel**, Al_2MgO_4, at a depth of about 400 km (13–15 GPa, Figure 3.8). This is referred to as the olivine–spinel phase change. There are two spinel-like forms of $(Mg,Fe)_2SiO_4$, the β-phase structure is about 7.5 per cent denser than olivine (the α-phase) and the γ-phase is about 10 per cent denser. The first phase change (α to β) is probably responsible for the seismic discontinuity in the mantle at a depth of 400 km; at greater depths (pressure) β-spinel converts to γ-spinel and then a further sharp density change to the perovskite structure (see below) occurs at 650 km.

The effects of temperature and pressure on the crystal structure of $MgSiO_3$ are shown in Figure 3.9. Note that in this illustration the chemical composition is constant; the phase boundaries move if impurities such as Al_2O_3 and FeO are introduced. Orthopyroxene, $(Mg,Fe)SiO_3$, is also unstable below about 400 km depth. It recrystallises to a garnet-like phase and can be thought of as dissolving in garnet at high pressure. Note the similarity in the formula of orthopyroxene and garnet, when written in the following form:

orthopyroxene	garnet
$(Mg,Fe)_4Si_4O_{12}$	$(Mg,Fe)_3Al_2Si_3O_{12}$

The garnet form of orthopyroxene is called majorite. Clinopyroxene undergoes a similar transformation at somewhat greater pressure. This mineral is not as well studied at high pressure and it may require the presence of garnet in order to transform to a garnet-like structure. At a depth of 500 km in the mantle all of the familiar low-pressure minerals have disappeared, to be replaced by β- and γ-phase spinels and a complex garnet (majorite) solid solution (see Table 3.4).

At still higher pressure the spinel form of olivine decomposes to an ultradense assemblage containing $(Mg,Fe)SiO_3$ in the perovskite (named for $CaTiO_3$) structure and (Mg,Fe)O, known as magnesiowüstite, in the rock salt (NaCl) structure. The garnet and majorite structure also collapse to perovskite-like minerals (Figure 3.9). These phases are probably stable throughout the whole lower mantle where there are fewer changes and lateral variations in velocity are very small; this is because variations in temperature and pressure occur without changes in mineralogy.

3.3 continued

Table 3.4. *Mineralogy of the mantle*

Depth (km)	Olivine	Orthopyroxene
0	α-phase $^{VI}(Mg,Fe)_2{}^{IV}SiO_4$	orthopyroxene $^{VI}(Mg,Fe)^{IV}SiO_3$
400	β-phase $^{VI}(Mg,Fe)_2{}^{IV}SiO_4$	majorite $^{VIII}(Mg,Fe)_2{}^{VI}[(Mg,Fe)Si]^{IV}Si_3O_{12}$
500	γ-phase $^{VI}(Mg,Fe)_2{}^{IV}SiO_4$	
650	perovskite + magnesiowüstite $^{VIII-XIII}(Mg,Fe)^{VI}SiO_3$ $+{}^{VI}(Mg,Fe)O$	perovskite $^{VIII-XII}(Mg,Fe)^{VI}SiO_3$

In summary, Table 3.4 shows the formulae and coordination numbers (roman numerals) of the cations of the various phases of the two most important mantle minerals, olivine and orthopyroxene. The **coordination number** is the number of cations of a given type which occur as nearest neighbours to each oxygen atom. For example, in normal silicate minerals at the Earth's surface, there are four oxygens around each silicon ion and six oxygens around each magnesium or iron ion. Coordination number is conserved across some phase boundaries but increases with pressure or decreases with temperature across other boundaries. An increase of coordination is generally associated with an increase of density and seismic velocity and this is a particularly important effect at the 650 km discontinuity.

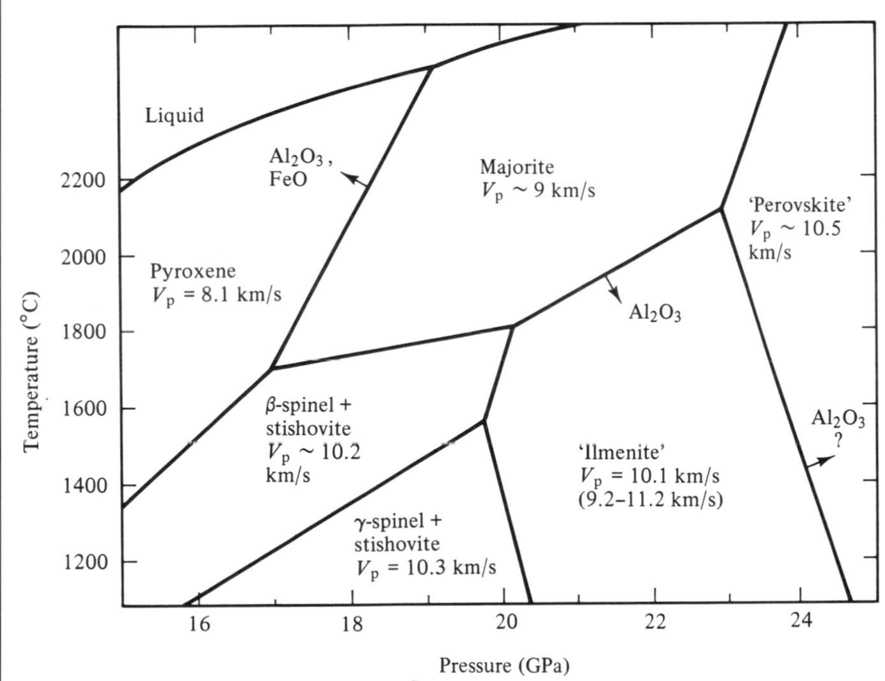

Figure 3.9 **Phase relations in pure $MgSiO_3$, showing the effect of temperature on the phase boundaries. Also given are estimates of the compressional velocity at zero pressure.** □

olivine and orthopyroxene (peridotite), or of garnet and clinopyroxene (eclogite), or a mixture of peridotite and eclogite (so-called fertile peridotite). The deepest samples, from kimberlite pipes, come from as deep as 200 km and these are primarily peridotite although eclogites are not uncommon. In the upper mantle it takes a relatively small change of temperature or pressure to change the stable mineral assemblage. Lateral variations in seismic velocity are primarily due to the solid–liquid phase changes caused by such fluctuations in temperature and/or composition. When eclogite melts it turns into a basaltic magma (for details, see Chapters 4 and 5). Basalt is the predominant material that reaches the surface from the interior of the Earth. A variety of evidence suggests that basalts separate from their immediate parent at depths shallower than about 100 km, but the ultimate source region may be much deeper, perhaps in the transition region. Buoyant upwelling in the mantle is probably responsible for much of the lateral heterogeneity of the upper mantle which is being mapped by seismologists using the methods of seismic tomography (see later).

Seismic velocities of mantle minerals

Seismology is used in several entirely different ways in studying the Earth's interior. It is used as a structural tool and, in this mode, as we have seen, has led to the discovery of the Moho, the outer core and the inner core, and several discontinuities in the mantle. The velocities of seismic waves are sensitive to chemistry and mineralogy and, as will by now be increasingly clear, seismology is also useful as a petrological probe.

Crustal rocks are composed of relatively low density minerals and yet seismic velocities are nevertheless low compared with mantle rocks. The velocities are low (Box 3.1, Equation 2) because the large interatomic spacing has a greater effect on K_S, the adiabatic bulk modulus, than it does on density. The rigidity, G, also generally increases with density, unless the density increase is due to an increased FeO content, or a high concentration of other high atomic weight ions.

Table 3.5 gives some of the properties of common crustal and upper mantle minerals. Note that typical crustal minerals (the upper half of the table) have densities less than about 3 g/cm^3 and compressional velocities less than about 6.5 km/s. The rock types including these minerals have similar properties; for example, quartz and feldspar-bearing granites have low velocities and densities. Basalt has intermediate seismic velocities because it is made up of minerals which, together, have intermediate velocities (clinopyroxene, feldspar). At high pressure basalt converts to eclogite and this is composed of garnet and

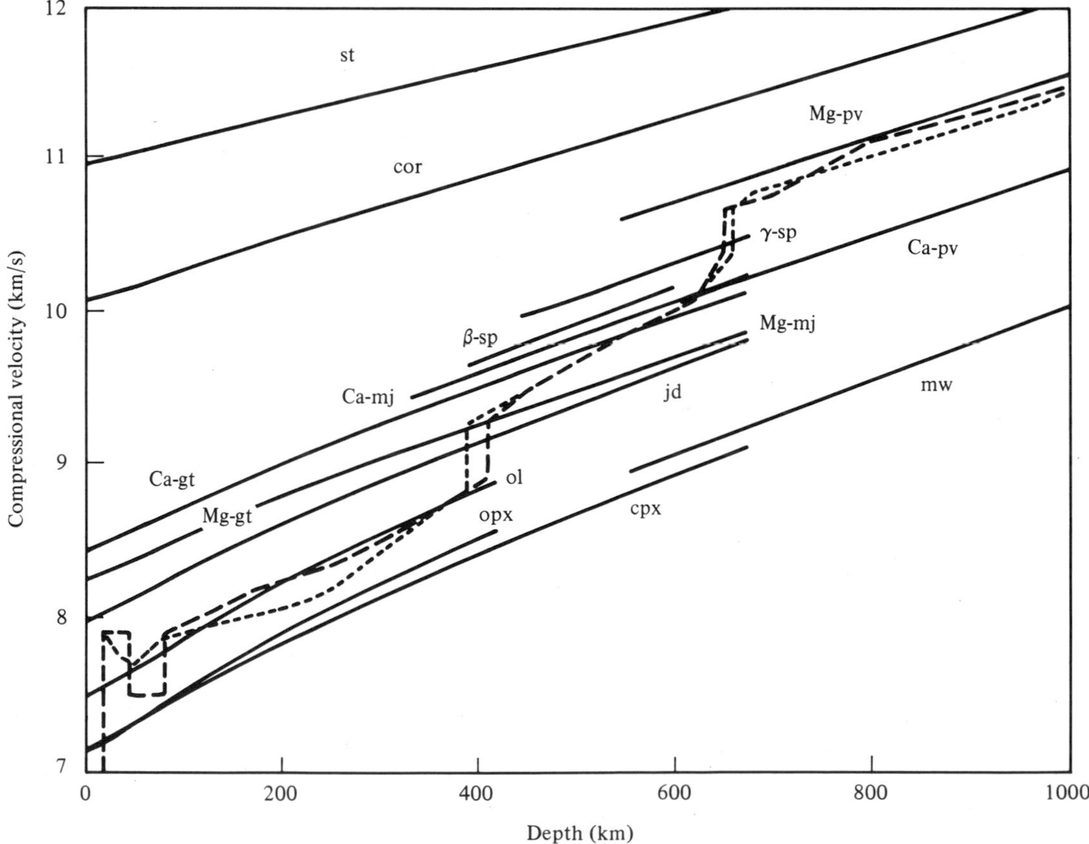

Figure 3.10 Calculated compressional velocities plotted against depth for a variety of mantle minerals. The dashed lines are typical seismic profiles. The calculations are done along a 1400 °C adiabat. (st = stishovite, cor = corundum, pv = perovskite structure, mw = magnesiowüstite, sp = spinel structure, mj = majorite, gt = garnet, jd = jadeite, ol = olivine, opx = orthopyroxene, cpx = clinopyroxene.) Ca- and Mg-rich end-member composites are plotted separately where these strongly influence the compressional wave velocity. □

Table 3.5. *Properties of crustal and upper mantle minerals at typical depths for these minerals (see also Figure 3.10)*

Mineral	Density (g/cm³)	V_P (km/s)	V_S (km/s)
Quartz	2.65	6.05	4.09
K-feldspar	2.57	5.88	3.05
Plagioclase	2.64	6.30	3.44
Mica	2.8	5.6	2.9
Amphibole	3.2	7.0	3.8
Clinopyroxene	3.3	7.80	4.51
Orthopyroxene	3.35	7.84	4.73
Olivine	3.3	8.4	4.9
Garnet	3.56	8.96	5.05

clinopyroxene which, together, yield relatively high densities and seismic velocities. This is one reason why the extent of the eclogite component in the upper mantle is difficult to resolve, geophysically, from that of the olivine-rich peridotite. Equations 2 and 3, Box 3.1, can be used to convert the parameters in Table 3.5 to elastic moduli, e.g. K_S and G.

Figure 3.10 shows the compressional wave velocities in various minerals as a function of pressure or depth, calculated along an adiabat assuming a 1400 °C surface temperature. The dashed lines show a range of typical mantle compressional velocity profiles for several regions. Note that the upper mantle, shallower than 400 km, has velocities close to olivine but ortho- and clinopyroxene can also be present as long as there is enough garnet to counteract the low velocities.

The minerals which are stable below 400 km (Table 3.4) are β-spinel and majorite, the high-pressure forms of olivine and orthopyroxene respectively, and clinopyroxene and garnet. Below about 500 km the stable minerals are a garnet-form of clinopyroxene, majorite, garnet and γ-spinel. Below about 750 km the stable minerals are various forms of perovskite ($MgSiO_3$, $CaSiO_3$ ...) and magnesiowüstite (Mg,Fe)O. There is also some FeO in the perovskites. Stishovite and corundum may also be stable under lower mantle conditions, depending on the average composition.

In order to refine estimates of mantle mineralogy we need to improve our knowledge of mineral seismic velocities and their variation with temperature and pressure. Mineral physics is that branch of geophysics which concerns itself with the measurements and extrapolation of physical properties under extremes of temperature and pressure. One of the most important goals is the determination of accurate 'Equations of State' which will allow the computation of physical properties under the range of conditions found in planetary interiors.

Seismic tomography

Lateral heterogeneity of the Earth's interior is studied by seismic tomography. Earthquakes are used as seismic sources and seismic stations on the Earth's surface serve as receivers. The idea of tomography can be illustrated very simply. The inside of a cave can be studied by a group of people all using flashlights as a source of illumination and their eyes as receivers. By combining their information they can draw three-dimensional maps of the cave. Seismic tomography works in much the same way. Computers replace the mental processing and convert the arrival times of seismic waves to velocity anomalies. Both **surface waves** (see Figure 3.3) and body waves can be used in tomography. After a large earthquake the surface waves travelling around the world constructively interfere and set up standing wave patterns called normal modes or **free oscillations**. The properties of these waves, their periods and spatial patterns, can be used to infer radial and lateral velocity structure; the longer the period of the standing wave the more effectively it 'sees' to a greater depth in the Earth because of the way in which the amplitude of the disturbance decreases with depth (Figures 3.3c and d). However, some of the normal modes are also due to constructive interference of body waves which travel through the interior and reflect off interfaces.

Figure 3.11 shows maps at various depths of the shear wave velocity determined from surface waves and long-period shear waves by Toshiro Tanimoto. Figure 3.12 shows cross-sections of the upper mantle derived from the surface wave tomographic results of Henri-Claude Nataf, Ichiro Nakanishi and myself. It turns out that there is a remarkable correlation between upper mantle seismic velocities and tectonics, at least down to a depth of 300 km. Active tectonic and oceanic regions have very low shear velocities, in many cases so low that partial melting is implied. In particular, extensional regions (ocean ridges, rifts, back-arc basins) tend to have very low velocities whereas stable continental regions such as shields and platforms have high velocities. Below 300 km, convergence regions tend to have high velocities, evidence for cold subducted slabs.

There are several reasons why seismic velocities vary from place to place at the same depth. The most important are isobaric phase changes, including partial melting, and changes in mineralogy due to chemical changes. Upper mantle minerals are also anisotropic and tend to be aligned by flow processes in the mantle. Changes in mineral orientation therefore can also cause lateral changes in velocity. The total range of shear velocity variation is about 10 per cent above 100 km, 7 per cent between 100 and 220 km, 4 per cent between 220 and 400 km, 3 per cent between 400 and 670 km and 1 per cent in the middle of the lower mantle. The variation picks up again in D", being about 2 per cent in that region.

Tectonic and young oceanic regions have high mantle temperatures and exhibit volcanism, evidence for melts in the mantle. In the course of continental drift, continents tend to move away from hot areas and come to rest over cold areas of the mantle (see also Chapters 8 and 10). In the process they override cold oceanic lithosphere. This may explain why continents generally have high seismic velocities between 200 and 400 km depth compared to tectonic regions (Figure 3.11a). However, below 400 km convergent bounda-

220–400 km

(a)

(b)

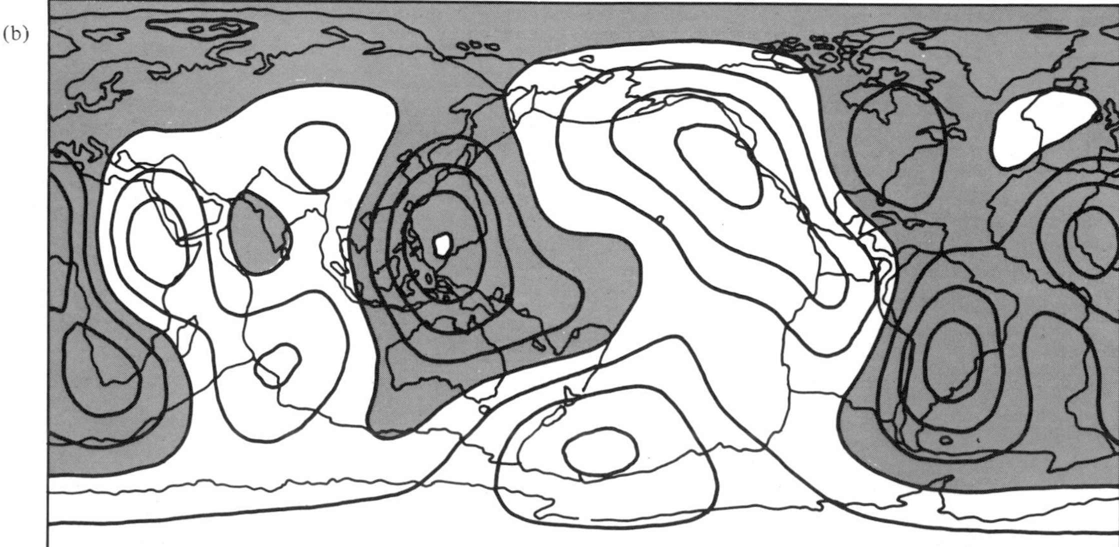

Figure 3.11 Tomographic maps. Global distribution of shear velocity as a function of depth (a–d) as determined by Toshiro Tanimoto of Caltech. The contour interval is 0.5 per cent. Dark regions are faster than average. □

ries and older oceanic regions also have fast seismic velocities (Figure 3.11b). Fast velocities between 400 and 650 km generally correlate with the places where oceanic lithosphere has been overridden by continents since the Mesozoic breakup of the supercontinent **Pangaea**.

Slow velocities under tectonic regions and young oceans can generally be traced to deeper than 300 km (Figure 3.12a), sometimes to 400 km (e.g. the eastern Pacific and northeast Africa, Figure 3.12b), although they are not long linear features (e.g. contrast in the three sub-Atlantic ridge zones crossed in Figure 3.12a–c). This suggests that the approximate linearity of ocean ridges is due to the fracture properties of the lithosphere rather than to the constancy of the magma source region. Thus, it is not surprising that the deep slow, presumably hot, regions are sometimes offset from the surface expressions of ridges or rifts.

Tomography has fairly poor resolution currently because of the locations and spacings of seismic stations and earthquakes. Features having dimensions of several thousand kilometres can be resolved with

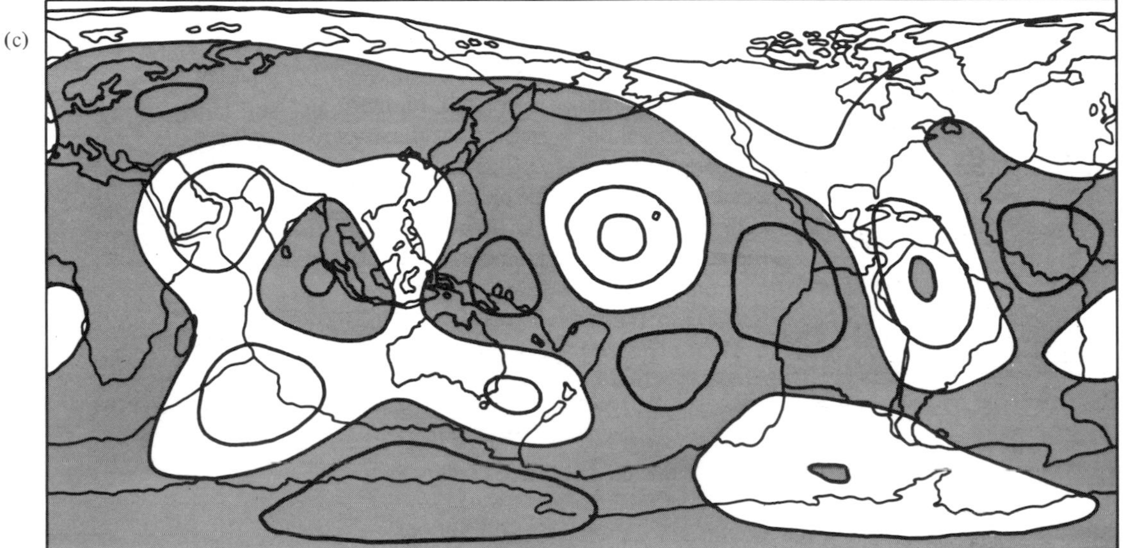

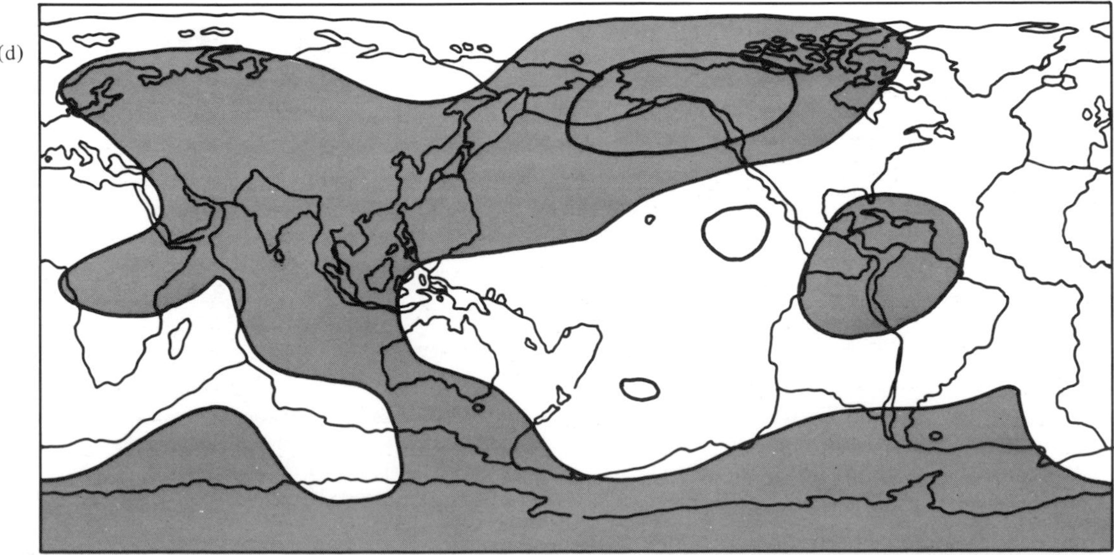

global surface wave tomography and, in some regions, body wave tomography can resolve features of a few hundred kilometres in size. With this qualification in mind one gets the impression that mantle upwellings are of large dimension. Upper mantle velocities are very low for large distances away from spreading centres, not just the hundred or so kilometres usually envisaged in cartoons of ocean ridge magmatism, and this is particularly noticeable where the Indian and Pacific ridges are crossed by the Figure 3.12 sections. Oceanic ridges and continental extension regions therefore appear to be embedded in very large regions underlain by seismically slow and, probably, upwelling material. Downwelling slabs are thought to be no thicker than about 100 km and these are not resolved tomographically in the upper 300 km of the mantle. However, below 300 km the shear velocity near subduction zones is higher (e.g. Figure 3.11b near southeast Asia and Figure 3.12b under New Guinea and South America). This suggests that large amounts of cold material are present in the upper mantle in these regions.

Seismic tomography also shows that there are large-scale, low-amplitude lateral variations of seismic velocity in the lower mantle. The lower velocity regions are centred under Africa and its antipode, the central eastern Pacific (Figure 3.11c and d). The low velocities are probably due to high temperatures and these are therefore thought to be upwelling regions of the lower mantle. High velocities occur in a band surrounding the Pacific, including the Pacific rim continents. Most of the continents (except Africa) and most of the subduction zones (except Tonga-Fiji) are above the long-wavelength high-velocity regions of the lower mantle. This may be because continental drift has brought the continents and the associated subduction zones towards downwelling parts of the mantle. If so, prior to the breakup of Pangaea, the supercontinent may have been sitting on top of a hot zone of mantle upwelling.

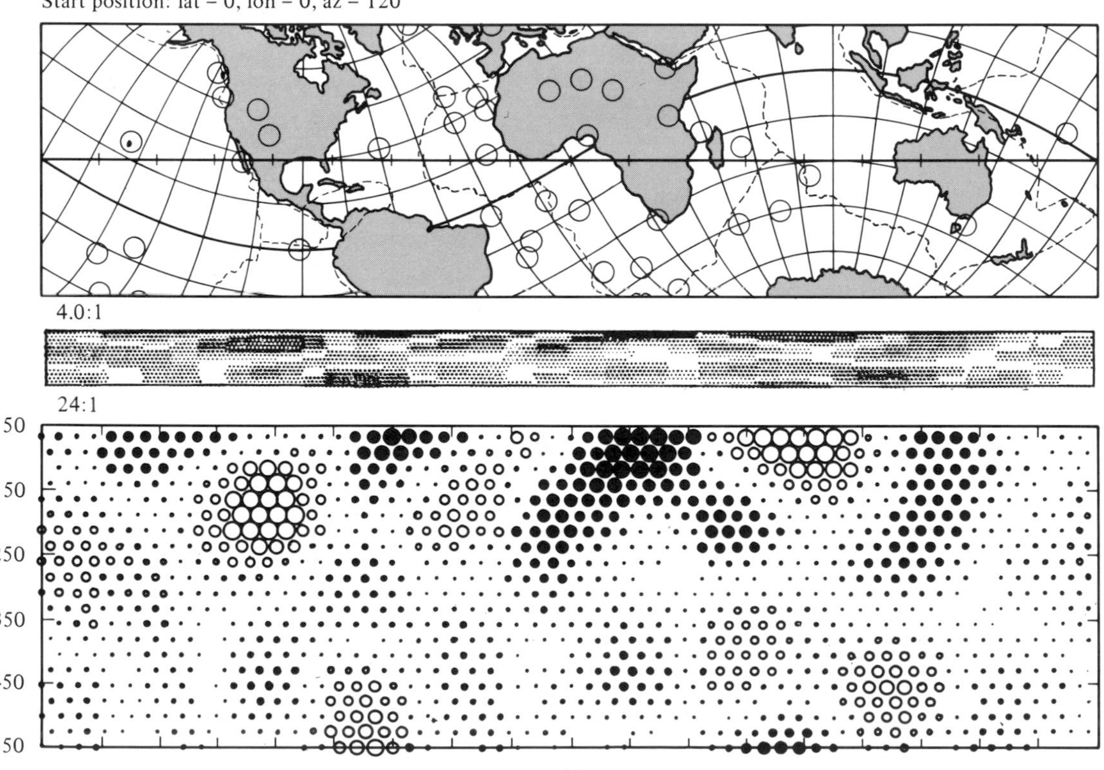

Figure 3.12 a–c Tomographic cross-sections. The shear velocity from 50 to 550 km depth determined from surface waves by H. C. Nataf, I. Nakanishi and Don L. Anderson. The cross-sections are along the great circle path shown in the centre of the upper panel. Plate boundaries (dashes) and hot spots (open circles) are also shown. Cross-sections are given with two different vertical exaggerations (4:1 and 24:1). Open symbols are slower than average velocities; closed symbols are faster than average velocities. □

Start position: lat = 0, lon = 125, az = 120

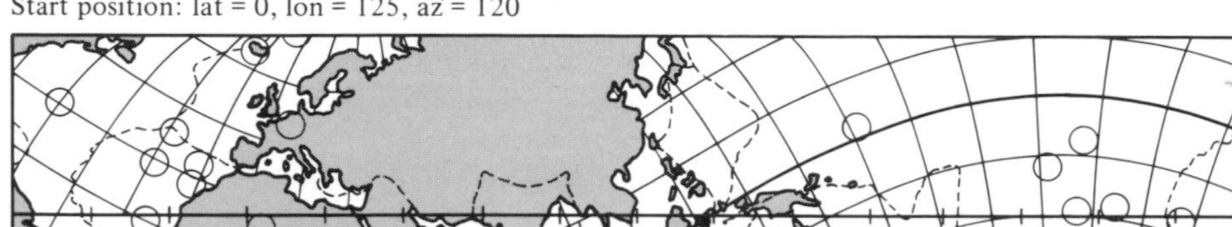

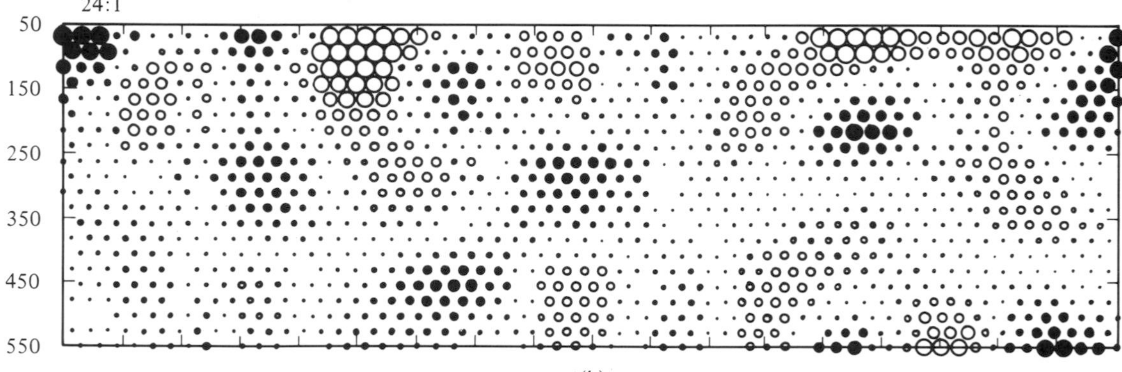

(b)

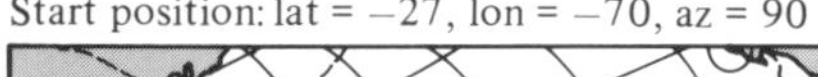

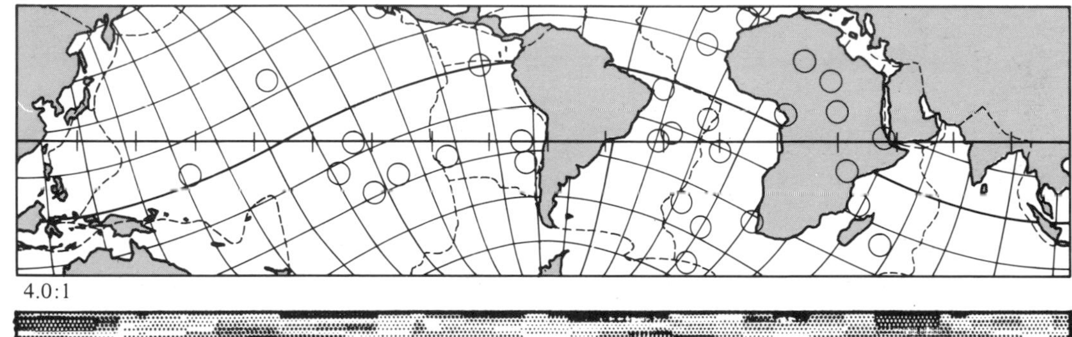

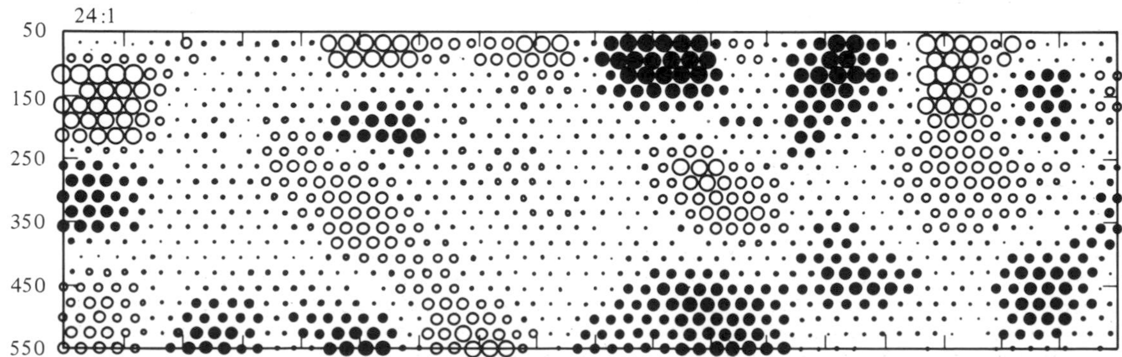

(c)

The slower than average, upwelling parts of the lower mantle, centred under the central Pacific and Africa, also correlate with highs in the geoid. The **geoid** is literally 'the shape of the Earth'. Technically, it is the gravitational equipotential surface of the Earth, like the surface of the ocean. The surface of the water is high in the vicinity of excess mass in the crust and mantle. Bumps on the core–mantle boundary also cause highs in the geoid, or equipotential surface. Regions of the mantle which have slow seismic velocity are probably hotter than average and therefore low density and buoyant. This ordinarily would reflect a mass deficiency and would cause a geoid low. However, in dynamic Earth, the buoyant upwelling causes the surface of the Earth and the surface of the core to rise. This elevation of dense material causes the equipotential surface to rise and a geoid high develops. This correlation between tomography and gravity adds confidence to the interpretation of the deep slow zones as regions of upwelling. Hot-spots are generally above these slower regions of the deep mantle. This does not necessarily mean that hot-spots represent material coming from the lower mantle since many hot-spots are underlain by faster than average, presumably colder than average, shallow mantle. Even if the mantle convects as a layered system there will be coupling between the upper and lower mantles, either thermal or mechanical. Temperatures in the lower mantle do seem to affect temperatures in the upper mantle, and may therefore have some influence on plate tectonics even if the lower mantle is chemically isolated from the upper mantle.

Conclusions

While we have come a long way from the initial resolution of seismic discontinuities to define the Earth's layered internal structure, much more research is needed if the new results of seismic tomography are to be fully interpreted in terms of the Earth's deep internal dynamics. However, we do have a good picture of the broad compositional state of the Earth's layers and it is now the fine detail of this 'static' view of its interior that needs to be refined with further study; meanwhile, the results from the Earth will provide important lessons if and when future exploration of the planets permits the acquisition there of similarly high quality seismic data.

Further reading

Anderson, D. L. (1989) *Theory of the Earth*, Blackwell Scientific Publications, London, 366 pp.

Bolt, B. A. (1982) *Inside the Earth; Evidence from Earthquakes*, W. H. Freeman, San Francisco, 191 pp.

Brown, G. C. & Mussett, A. E. (1981) *The Inaccessible Earth*, Unwin Hyman, 235 pp. (2nd edition, Chapman & Hall, in press.)

CHAPTER

4

Peter J. Wyllie
California Institute of Technology

EXPERIMENTAL PETROLOGY: EARTH MATERIALS SCIENCE

Introduction

Petrology is the science of rocks. Geologists map rocks in the field, and bring selected samples back to the laboratory for detailed petrographic analysis, mineralogical study and chemical analysis. On the basis of these studies, existing hypotheses for the origin of the rocks are tested, or new hypotheses are erected. From examination of the rocks in field and laboratory, geologists then attempt to deduce their histories. Experimental petrology involves further laboratory experiments which reproduce the conditions within the Earth during the generation and evolution of a rock or rock suite. This involves subjecting minerals and rocks, or their synthetic equivalents, to high pressures and temperatures under varied but precisely controlled conditions. Determination of the reactions which occur under the known conditions in the laboratory provides calibrations for the processes involved in formation of the rocks in nature, defining the actual conditions, and facilitating selection among competing hypotheses of origin. Furthermore, exploration of reactions under various conditions within the laboratory may reveal processes operating within the Earth which were previously unsuspected.

Experimental petrology had its beginnings in adventurous experiments on minerals and rocks using furnaces or cannon barrels. It became a force in Earth sciences starting in the early 1900s with the systematic determination of high-temperature phase equilibria involving the crystallisation of synthetic silicate liquids, which included representatives of the common rock-forming minerals. These investigations brought the rigour of thermodynamics to the processes of partial melting of rocks and the crystallisation of magmas, and elucidated many problems in igneous petrology. Only in the 1950s did phase equilibrium experiments with simultaneously maintained high temperatures and pressures become routine. At first the experiments reproduced conditions only within the continental crust, then during the 1960s the experimental range was extended to high-pressure conditions within the mantle, equivalent to about 100 km depth. During the 1980s one type of large-volume apparatus reproduced conditions down to 650 km in the Earth, and a miniature device has reproduced conditions corresponding to 2000 km depth, not far short of the mantle–core boundary of the Earth. The equipment mentioned above is static, with samples being held at constant pressure and temperature as reactions occur in the samples. A dynamic experimental approach reproduces conditions to the centre of the Earth; a sample is shattered by the passage of a shock wave, and its properties under extremely high pressures are measured during the last nanoseconds before its destruction.

Experimental petrology is the materials science of the Earth and other planets. The experiments define how rock materials behave under various conditions. If masses of rock are moved by processes such as mantle convection or plate tectonics, they experience changes of pressure or temperature or both. If they are transported to greater depths, the minerals may experience **phase transitions**, forming new minerals with denser structures. If they are transported to higher temperatures, the minerals may be dehydrated or decarbonated, they may react with each other to form new minerals, and the rock may begin to melt. Determination of the array of **phase equilibrium boundaries** for the reactions that can occur in planetary materials provides the framework of experimental petrology. It is necessary to determine also the variations in compositions of minerals, melts and vapours within the framework. Of growing importance are other experiments, such as the determination of the physical properties of the materials as a function of pressure and temperature, and kinetic studies of reactions. The results of experimental petrology are most informative when used in conjunction with results from thermodynamics, tectonics, geochemistry and geophysics.

The variables: materials, pressure and temperature

The main variables in experimental petrology are the compositions of the materials, and pressure and temperature. The structure of the Earth is revealed by the velocities of seismic waves, with results shown in Figure 4.1a (see also Chapter 3). Discontinuities in the seismic wave velocity mark the depths of changes in physical properties, which may be caused by phase transitions or changes in composition, or both. The discontinuities divide the Earth into core, mantle and crust, with the core subdivided into solid inner and molten outer core. The core is an alloy of FeNi, with density reduced by a light element, sulphur or oxygen (see discussions in Chapters 2 and 3). The mantle corresponds in composition to a peridotite, with debated volumes of eclogite. The crust, a very small fraction of the mantle, from which it was derived by chemical differentiation, is extremely heterogeneous, with mineralogy dominated by feldspars and quartz, and with hydrous minerals (such as clays) and carbonates (limestones) concentrated in the outer, sedimentary

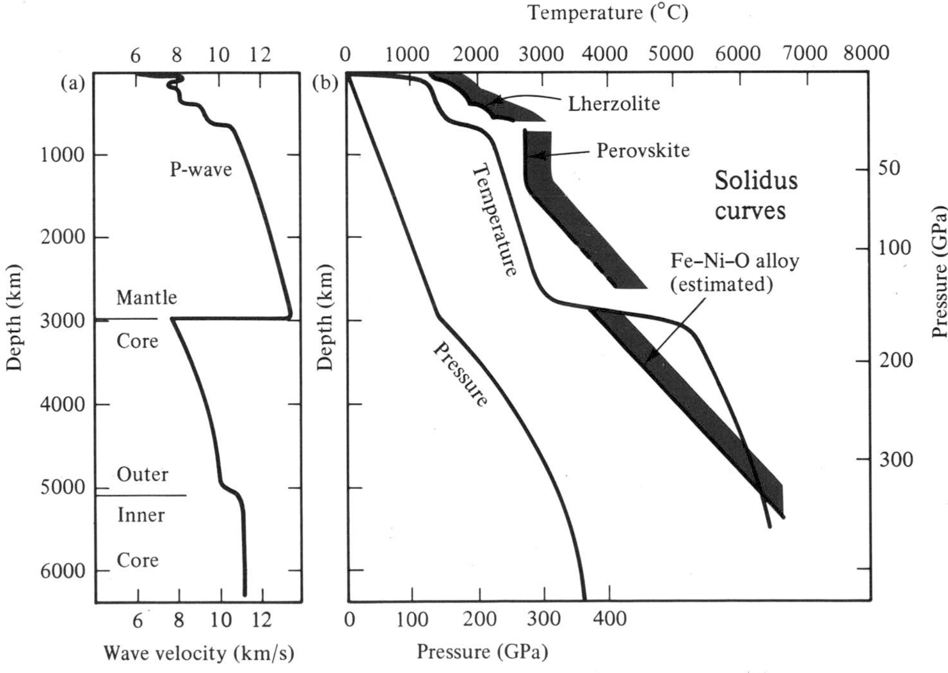

Figure 4.1 (a) P-wave seismic velocity profile through the Earth, defining the major sections: core and mantle. (b) The variation of temperature (upper axis) and pressure (lower axis) with depth in the Earth are shown together with the solidus (i.e. beginning of melting – see later in chapter) curves for the Fe–Ni–O alloy of the core, about 1000 °C below the fusion curve for Fe (Figure 4.7), for lherzolite to 650 km (Figure 4.8a), and for synthetic perovskite ($Mg_{0.9}Fe_{0.1}SiO_3$) between 25 and 100 GPa. Equivalent scales of depth and pressure appear to the left and right of this figure.

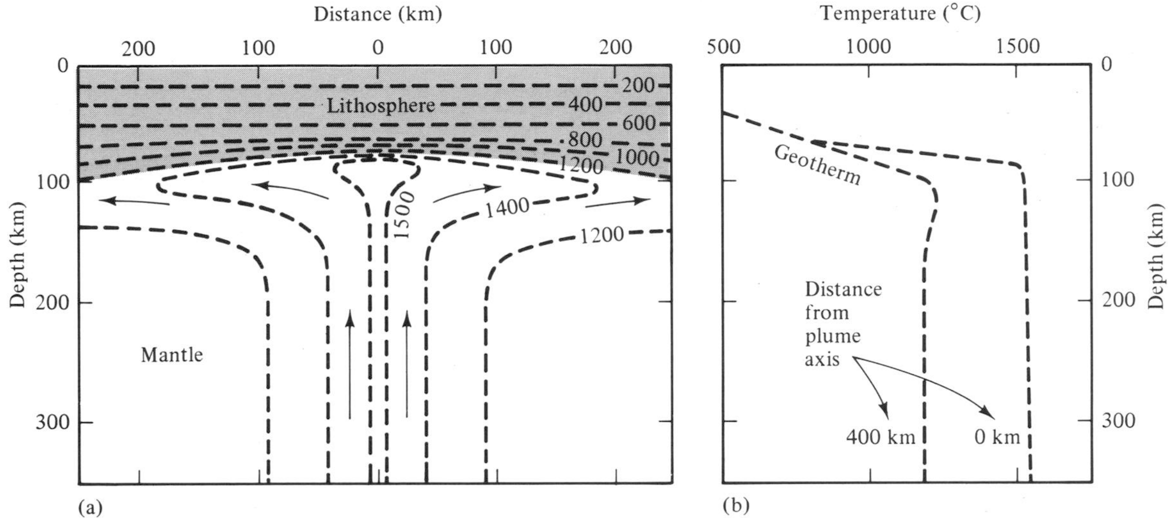

Figure 4.2 Effect of rising plume on mantle temperatures. (a) Isotherms (in °C), (b) geotherms from centre and edge of plume.

layers. In addition, Earth materials include melts, vapours, and the two fluid envelopes of hydrosphere and atmosphere. The crust is in contact with the oxidised volatile components H_2O and CO_2, and the lower mantle is in contact with the reduced iron of the outer core.

Pressure within the Earth has been calculated, with results shown in Figure 4.1b. The relationship with depth is almost linear, with a change in slope at the mantle–core boundary resulting from the very large density difference between the silicates and metallic alloy. Pressures in current literature are expressed as bars (atmospheres, kilobars), or as **pascals.** One **gigapascal** (GPa) equals 10 kilobars (kbar).

Temperatures within the Earth are commonly expressed in two ways. A **geotherm** for a particular location gives the increase in temperature as a function of depth (see Chapter 12). An average geotherm for the whole Earth based in part on recent experimental results at mantle and core pressures (discussed below) is given in Figure 4.1b. The corresponding **isotherms** would be a series of lines parallel with the surface. Uprise of hot material within the Earth generates a thermal structure mapped in cross-section by the isotherms with the mushroom shape depicted in Figure 4.2a. Specific geotherms through different parts of the thermal structure are illustrated in Figure 4.2b. Sinking of cold lithosphere generates a similar arrangement with the plume or mushroom of isotherms upside down compared with Figure 4.2a. Each point within such a thermal structure is characterised by a specific pressure and temperature, which provides the basis for calibration by phase boundaries.

Experimental methods and apparatus

The idea of experimental petrology is to reproduce conditions within the Earth's interior. High temperatures at one atmosphere are easily achieved, and high pressure experiments at low temperatures are not much more difficult. High pressures may be achieved by pumping a gas into a strong vessel containing the sample, or by squeezing the sample between a pair of solid anvils which are driven toward each other by a hydraulic press. It is technically more difficult to apply pressure with the sample being simultaneously maintained at high temperatures. Techniques involve either heating the whole pressure vessel, or heating the sample inside the apparatus. This provides essentially four types of apparatus, externally or internally heated pressure vessels, with the pressure being imparted either by a fluid medium, or by a compressed, solid medium.

A sample can be reacted under desired conditions of pressure and temperature, and the phase assemblage produced can commonly be preserved by quenching from the experimental 'run' conditions by rapid cooling and release of pressure. Many silicate liquids quench to glasses which are easily identified by petrographic microscope. The microscope was the main tool for identification of phases until X-ray diffraction became available, but modern analytical techniques have greatly enhanced the study of fine-grained phase assemblages from experimental runs. The scanning electron microscope has revealed textures previously unseen, and the electron microprobe makes identification of phases, through their chemical signature, unambiguous. The ion microprobe promises an exciting new era for analysis of trace elements in the experimental charges.

Techniques have been developed for some high-pressure apparatus to measure the properties of the phases *in situ*, under the conditions of the run, and this is particularly important when transient high-pressure phases are involved. Spectroscopic and X-ray techniques have been used successfully. Indeed, intense X-radiation sources have been interfaced with some high-pressure apparatus, permitting the study of kinetics of phase transitions as they occur, and observations on the geometry and distribution of melts among minerals.

For the first half of this century, most of what is now called experimental petrology consisted of the determination of liquidus phase diagrams at one atmosphere (i.e. phase diagrams with boundaries between partially and totally liquid conditions). Few high-pressure experiments of petrological interest were conducted before the design of the simple 'cold-seal test-tube vessel', or **Tuttle bomb**, in 1949. The vessel is a thick-walled metal tube with the pressure seal made by a screw cap. The sample, in a sealed gold capsule, is placed at the closed end of the metal test-tube which is positioned at the hot-spot of an external furnace; the sealing nut remains outside the furnace where its cool strength is maintained. A fluid is pumped in at the desired pressure, and the temperature of the sample can be precisely controlled. At the end of a run the sample is quenched by plunging the vessel into cold water (which can be very exciting for high-temperature runs when the vessel is cherry red), or by cooling it with a blast of compressed air. This simple apparatus opened up a whole range of experimental studies relevant to metamorphism in the crust, and the generation of migmatites and solidification of granites and pegmatites. Furthermore, the presence of water vapour under pressure enhances reaction rates tremendously. A rather cumbersome variant of this apparatus with even thicker walls and an internal heat source attains conditions corresponding to the uppermost mantle, with temperatures high enough to investigate the crystallisation of basaltic magma. This is an important range of geological conditions, but there are not many laboratories that have consistently produced petrological results.

Compression of a sample between opposing anvil faces of two pistons can generate very high pressures. The pistons are driven by a hydraulic ram with large radius, and the pressure is multiplied as it is transmitted to the small-diameter faces of the opposing pistons. This **piston-cylinder apparatus**, an internally heated uniaxial press, routinely reproduces conditions corresponding to depths of more than 100 km in the Earth, the average depth of subducted oceanic crust beneath a volcanic island arc, and deeper than the thickness of the oceanic lithosphere. This important geological range, together with the reliability of the apparatus and its relative simplicity of operation, have made it one of the most productive items in experimental petrology since the 1960s. Moreover, another class of apparatus using an array of four to six pistons to compress a deformable solid assembly which encloses sample, furnace and thermocouple has generated conditions equivalent to 650 km deep in the Earth. The disadvantages are the larger size, high cost, complexity of alignment and operation, and the difficulty of calibrating for precise temperatures and pressures; these are offset by the desirability of the data.

Even higher pressures and temperatures have been reproduced in the **diamond-anvil apparatus**, a miniature version of a simple uniaxial press. Small anvils made from diamonds are squeezed together in a holder with great care taken to maintain proper alignment, compressing a small sample in wafer form secured within an annular gasket. Internal heating is accomplished by directing a laser beam through the transparent diamond. Because the diamond is transparent, the sample can be studied *in situ* under high-pressure, high-temperature conditions, using an optical microscope, spectroscopic techniques and X-ray diffraction. This apparatus can reproduce conditions at depths of 2000 km, approaching the mantle–core boundary.

Data about materials at very high pressure are obtained by **shock-wave experiments** in which the sample is blasted to destruction. The blast is delivered by a large cylindrical plastic projectile, shaped like an enormous bullet, which is launched through a tube resembling a gun barrel toward the target which is suspended in an evacuated impact tank. The impact generates a shock wave which passes through the sample. Pressures in excess of 400 GPa have been achieved in this way. The projectile velocity and the transit time of the shock wave are measured, and the transient pressure and density of the sample can be calculated. Shock-wave techniques are routinely used for accurate determinations of the density of solid

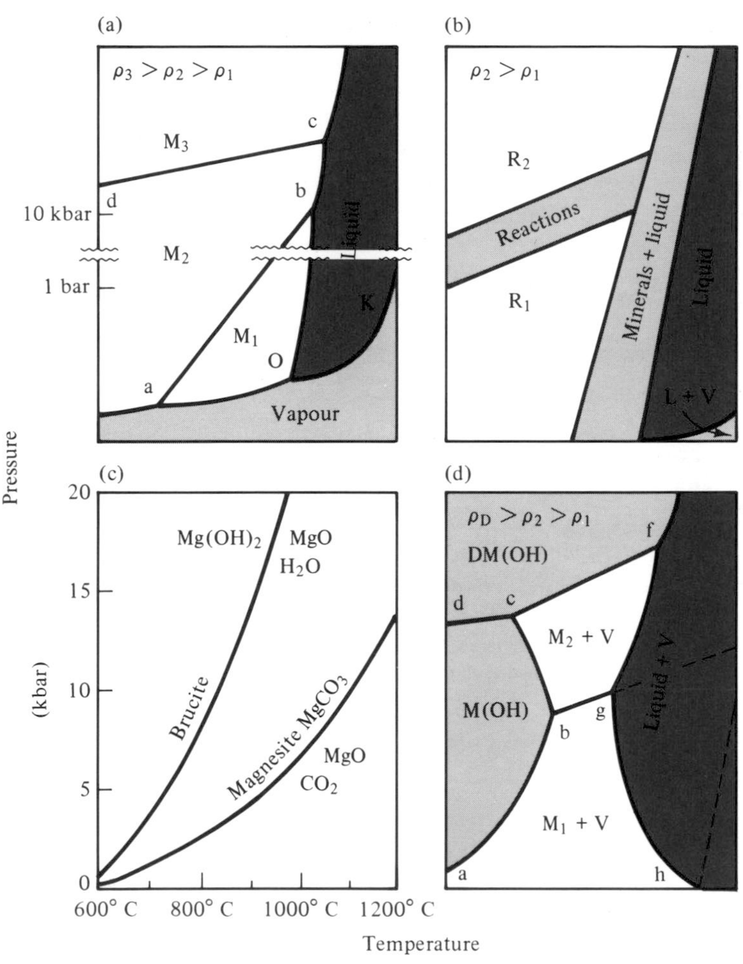

Figure 4.3 Reactions in minerals and rocks (see text for full description). (a) Solid, liquid and vapour in a mineral (M) system. (b) Solid and liquid in a rock (R) system. (c) Solid–vapour dissociation reactions. (d) Hypothetical mineral–H_2O system (DM = dense mineral). Note that diagrams (a), (b) and (d) are schematic, with pressure increasing vertically and temperature horizontally; L = liquid, V = vapour, ρ = density. □

materials as a function of pressure, and have been successfully modified for study of molten silicates.

Phase relationships and physical properties

Phase equilibrium experiments on rock materials and volatile components (e.g. H_2O, CO_2, F) provide the depth–temperature framework for phase transitions and for the distribution of various rock–fluid assemblages in the Earth, as well as the percentages of liquid produced as a function of temperature during progressive melting of a rock. This information alone is not sufficient to describe the physical behaviour of rock–fluid masses. We need to know in addition the fluid distribution and transport mechanism within a rock host.

Rock materials experience many types of reactions, and some of them are illustrated in Figure 4.3 in terms of the variables, pressure and temperature. Reactions of minerals with few components may be expressed as simple lines (**univariant curves**), but the corresponding reactions for rocks with many components occur through finite bands of pressure and temperature (**multivariant intervals**).

Phase transitions in minerals and rocks

Figure 4.3a illustrates the reactions that could occur in a simple system. The mineral M_1 **sublimes** to yield a vapour phase (V) at very low pressures (much less than 1 bar for silicate minerals) along the curve Oa, with the reaction: $M_1 \rightleftharpoons V$. With increasing pressure, the mineral is transformed along the univariant curve ab into a denser **polymorph** (i.e. same chemical composition, but a different structure) mineral, M_2, by the reaction: $M_1 \rightleftharpoons M_2$. Additional transitions may occur at higher pressures as shown by the reaction to an even denser mineral, M_3, along the curve cd.

The minerals melt as shown along the univariant

curve rising steeply from O, with inflections occurring at the triple points b and c: $M_1 \rightleftharpoons L$, $M_2 \rightleftharpoons L$, $M_3 \rightleftharpoons L$. The normal effect of increasing pressure is to raise the melting temperature.

The liquid phase **vaporises** (boils) along the evaporation curve OK, terminating at the **critical point**, K, where the physical properties of liquid and vapour become coincident and the meniscus between them disappears. With only a single homogeneous fluid phase remaining at higher pressures and temperatures, the univariant curve for reaction: $L \rightleftharpoons V$ can no longer exist.

The corresponding reactions for an assemblage of silicate minerals constituting a rock are illustrated in Figure 4.3b. If rock R_1 is subjected to high pressures by burial, it experiences a group of related solid–solid phase transitions through a **transition interval** forming another mineral assemblage, rock R_2, of higher density than rock R_1. The minerals may change composition progressively during the reactions. The distribution of elements among the coexisting minerals, either during the transition or within the separate fields for R_1 and R_2, may vary sufficiently to provide a means of calibrating the pressure and temperature at which a mineral assemblage of particular composition was formed (**geobarometry** and **geothermometry**; see below and Chapters 6 and 12).

Each rock also melts through an interval with coexisting minerals + liquid, and within that interval the liquid phase changes composition progressively from the **solidus**, where the first trace of liquid appears, to the **liquidus**, where the rock has completely melted to a liquid of its own composition. It follows that once we have measured the changes in the liquid composition, the paths of crystallisation, at any depth, or of progressive fusion of the rock, can be defined from the results.

With a pressure scale of geological relevance, the sublimation reactions are not distinguishable from the abscissa, and a vapour phase is shown in Figure 4.3b only at very high temperatures. Elsewhere in the universe and during the formation of Earth, however, minerals were formed at extremely low pressures either by direct precipitation from a dispersed gaseous phase, or via high-temperature precipitation of melt from a gas phase, under conditions corresponding to parts of the curves Oa and OK in Figure 4.3a (see Chapter 2).

Systems with volatile components

If carbonates and hydrated minerals are heated, they dissociate to yield carbon dioxide (CO_2) or water (H_2O) vapour. These reactions are of fundamental importance during progressive **regional metamorphism,** and heating of subducted ocean crust. The reverse reactions also have importance, because a flux of volatile components into a rock under appropriate conditions will generate hydrous minerals or carbonates. Figure 4.3c shows the univariant **dissociation reactions** for brucite and magnesite. Metamorphic dissociation reactions generally have fairly steep positive slopes on pressure–temperature diagrams at crustal pressures. In general, the phase assemblage on the high-pressure side of a reaction curve has higher density than that on the low-pressure side. In the hypothetical system $M–H_2O$ in Figure 4.3d, the mineral M(OH) dissociates along the univariant curve ab (which has a positive slope) to form the lower density (lower pressure) assemblage $M_1 + V$. The dehydration product, mineral M_1, experiences a transition at higher pressures to a denser polymorph, M_2, along bg (compare Figures 4.3a and b). Along the dissociation curve bc, where M(OH) dehydrates now to $M_2 + V$, the density of the dissociation products, including vapour, has become higher than that of the hydrated assemblage being dissociated. Therefore, the slope of the dissociation reaction has changed to negative.

There are dense hydrous minerals that are generated only at very high pressures, as illustrated in the hypothetical system of Figure 4.3d. The mineral M_2 in the presence of H_2O vapour becomes hydrated along the reaction curve cf, and the low-pressure hydrated mineral M(OH) reacts along the line dc to form the denser, hypothetical hydrated mineral DM(OH). Looking ahead to Figure 4.9c, there are two similar reactions occurring at high pressures; water reacts with peridotite to form DHMS (dense hydrous magnesian silicate), and H_2O and CO_2 react with olivine and orthopyroxene to form brucite and magnesite. These minerals may be important storage sites for H_2O and CO_2 in planetary interiors.

Volatile components such as H_2O and CO_2 dissolve in silicate melts under pressure, and their solubilities increase with increasing pressure. Since the crystallisation temperature of silicate melts is reduced by the solution of additional components, the solidus curves for minerals in the presence of vapour are depressed to lower temperatures with increasing pressure. The extent of the depression is a function of the solubility, and H_2O is most influential because it has the highest solubility. The effect is illustrated in Figure 4.3d for the reaction along hg: $M_1 + V \rightleftharpoons L$. Along curve bg, M_1 transforms to the denser polymorph, M_2, and at the triple point g the slope of the solidus curve changes, becoming more nearly parallel with the fusion curves

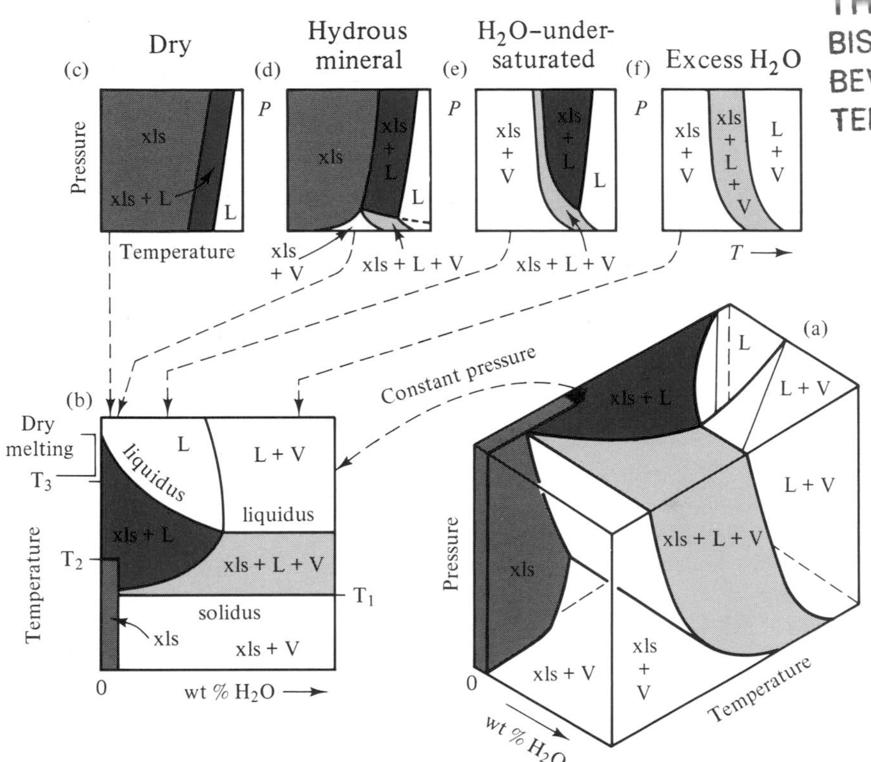

Figure 4.4 Phase relationships for a rock with H_2O, represented within a *P–T–X*(H_2O) model, with phase fields illustrated by isobaric (constant pressure, b) and isoplethal (constant H_2O content, d–f) sections. (a) Three-dimensional model of the entire system. (b) Isobaric section at high pressure (the top surface of (a)). (c), (d), (e), (f) *P–T* sections for the specific H_2O contents indicated. Note that (c) represents the rear of model (a) while (f) represents the front. xls = crystals; L = liquid; V = vapour.

for the minerals and rocks in Figures 4.3a and b. The change in slope is again caused by a change in density: the assemblage M_1 + V is less dense than the liquid, but the assemblage M_2 + V is more dense than the liquid.

The dehydration curves remain subsolidus for the conditions given in Figure 4.3d, but if the curve abc overlapped hgf then the hydrous mineral, M(OH), would be involved in the partial melting; similarly, crystallisation of the melt would precipitate the hydrous mineral directly. There are many natural rock compositions with one or more hydrous minerals, such as hornblende and biotite, which are stable above the solidus in the presence of H_2O.

Figure 4.4 is a convenient geometrical representation showing the various effects of H_2O on the phase relationships of rocks in pressure–temperature (*P–T*) space. Phase fields are illustrated in Figure 4.4 using sections of constant pressure (**isobaric** sections) and others at constant H_2O content (**isoplethal** sections). Note the field at low percentage H_2O (xls) in which all H_2O is stored in hydrous minerals. Figure 4.4b is an isobaric T–X(H_2O) section at high pressure. Note the three different solidus temperatures, depending on the H_2O content. The lowest melting temperature, T_1, is for the reaction with excess vapour; for a rock with all H_2O stored in the hydrous mineral the solidus is T_2; and for an anhydrous rock with no hydrous minerals, it is T_3. The boundary between the fields for L and L + V defines the maximum solubility of H_2O in the melt at this pressure. Note the large field for **water-undersaturated liquid** with crystals (xls + L). With decreasing temperature, liquids in this field crystallise, causing the residual liquid to change composition towards an increasing concentration of H_2O; only as the temperature approaches the solidus does the liquid become saturated with H_2O, and where the system crosses the phase boundary into the field crystals + L + V, a vapour phase separates from the now-saturated liquid. This kind of process is important for interpretation of explosive volcanism, the generation of ore deposits, and the formation of **pegmatites**, rocks with giant crystals and sometimes gemstones, fluxed by the volatile components.

The other sections in Figure 4.4 are *P–T* sections for different H_2O contents. Figure 4.4c is for the anhydrous rock, corresponding to Figure 4.3b. This kind of diagram is relevant for partial melting of the Earth's mantle, and the crystallisation of basalts. Figure 4.4f is for the extreme situation with enough H_2O present to saturate the liquid. Figure 4.4d represents a rock without pore fluid, but with some H_2O stored in a hydrous mineral. At lower pressures, where the hydrous mineral dissociates at subsolidus temperatures, the solidus is the same as in Figure 4.4f, but at higher pressures the solidus corresponds to the reaction where the hydrous mineral breaks down providing H_2O directly for solution in H_2O-undersaturated liquid, without formation of a separate vapour phase. The solidus in Figure 4.4e is the same as that in Figure 4.4f, but the system contains insufficient H_2O to saturate the completely molten rock, so the melting interval contains a large field for crystals with H_2O-undersaturated liquid. Figures 4.4d and e are relevant to partial melting and questions of magma production in the crust (see Chapters 12 and 14 where these data are applied).

The sections in Figure 4.4 cover only the pressure interval below the conditions for the transformation of the crustal silicate rock into a denser mineral assemblage. That is, they correspond to the region below the reaction curve bg in Figure 4.3d. Figure 4.3d shows that the slopes of dehydration and solidus reactions may change sign at pressures above such transitions. These reactions are relevant for processes in the Earth's mantle.

Similar diagrams can be constructed for silicate–CO_2 systems if their compositions are appropriate for the formation of carbonates by subsolidus reactions. There are **decarbonation reactions** involving (1) the silicate minerals of the mantle with (2) the carbonates dolomite and magnesite, which intersect the solidus curve for peridotite–CO_2 (xls + L) at pressures greater than about 21 kbar. At higher pressures, the melting of peridotite–CO_2 is controlled by silicate–carbonate reactions, yielding liquids with very high concentrations of CO_2. This provides the key to the origin of kimberlites, and the CO_2-rich magmas associated with continental rift valleys.

It follows from Figures 4.4c–f that the composition of the first liquid derived at the solidus by partial melting of a silicate assemblage changes as a function of pressure, and as a function of dissolved volatile components such as H_2O and CO_2. The magnitude of the change has been determined experimentally for **peridotite**, the major magma source rock in the upper mantle. Figure 4.5 shows the measured solidus curve for the type of peridotite that is likely to occur in the upper mantle, known as **lherzolite**; this is the solidus curve without volatile components (compare Figures 4.1b and 4.8a). Also shown, for different pressures (depths), are the compositions of the melts derived through a moderate temperature range above the solidus. The compositions, given as names of the corresponding lavas, show a trend with increasing pressure of decreasing SiO_2 and increasing MgO.

In the presence of volatile components, the solidus temperature for peridotite is lowered compared with that given in Figure 4.5. Liquids derived from peridotite are then enriched in SiO_2 when H_2O is present, trending toward **andesite** instead of basalt. This is particularly important in a subduction-related environment where basaltic andesites and andesites derived by partial melting of volatile-fluxed and possibly hydrated peridotite above a down-going slab of oceanic lithosphere, form the most significant component of island arcs and active continental margin magmatism (see Chapters 8, 9 and 14). In the presence of CO_2, the liquid from peridotite migrates away from SiO_2; at pressures above 21 kbar it reaches the composition of a **carbonatite**, a melt dominated by calcic dolomite, with alkalis to the extent that they are available in the peridotite, and with only about 10 per cent silicate components. Melts of this kind are probably influential in **mantle metasomatism**, a process which causes redistribution of trace elements within the mantle.

The properties of melts and vapours

The chemical differentiation of the Earth and other planets is controlled largely by the migration of silicate melts, and thus by the properties of the melts. In order to understand the processes, it is essential to know the composition, density and viscosity of melts, and the chemical diffusivity of elements in melts at high pressures and temperatures. It is believed that these properties are controlled by atomic or molecular structural changes within the melts.

It is well established that the **four-fold (tetrahedral) coordination** of silicon and aluminium in silicate minerals at low pressures is changed to **six-fold (octahedral) coordination** through phase transitions to denser minerals at higher pressures. It is expected that with compression, similar changes in packing will occur in silicate melts. Spectroscopic measurements on silicate glasses (supercooled, quenched melts) of several compositions in diamond-anvil apparatus have provided direct observations on the coordination of Si and Al at high pressures. Changes occurred continuously through the pressure range up to 40 GPa, in contrast to the changes that occur discontinuously in solids, at phase boundaries. The initial changes at pressures up to 10 GPa involved decreases in the average Si–O–Si angle between tetrahedra. The tetrahedra became distorted between 10 and 20 GPa, and at higher pressures the distorted tetrahedra were converted to distorted and then to regular octahedra.

Silicate melts exhibit a wide range of viscosities, varying as a function of composition, pressure and

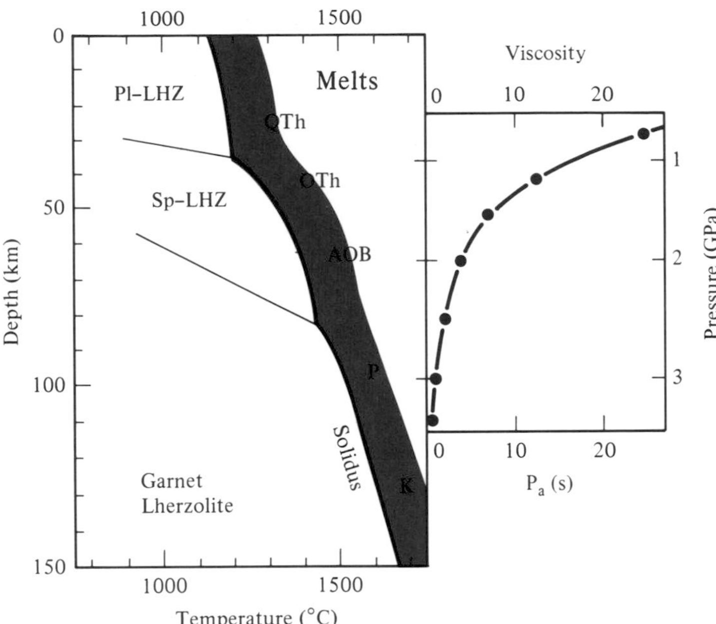

Figure 4.5 Phase relationships and properties of melts derived from a natural lherzolite. Pl-LHZ = plagioclase lherzolite; Sp-LHZ = spinel lherzolite. The solidus curve and the compositions of melts derived just above the solidus as a function of pressure; QTh = quartz tholeiite, OTh = olivine tholeiite, AOB = alkali olivine basalt, P = picrite, K = komatiite. The right-hand inset gives experimentally measured viscosities of these melts at indicated pressures. □

temperature. Viscosity decreases as melt compositions change from granitic through basaltic to picritic, with decreasing SiO_2 and increasing MgO. Moreover, the viscosity decreases (1) with increasing pressure at constant temperature, (2) with increasing temperature, and (3) with increasing solution of H_2O (see also Chapter 5). Figure 4.5 shows the decrease in viscosity of the liquid produced by partial melting of peridotite, which is more a function of the compositional change with depth than a function of pressure or temperature on melt properties. These results demonstrate that magmas in the mantle must be extremely fluid, even when anhydrous. The viscosity of the carbonatitic melts that may be generated under suitable conditions deeper than 75 km is less than 10^{-2} Pa s, approaching the value for liquid water. In contrast, the viscosity of granitic magmas in the crust is orders of magnitude higher. A value typical for a hydrous granitic melt is 10^6 Pa s, far more thixotropic than cold treacle.

The systematics of the densities of silicate melts at 1 bar are well known, and measurements at high pressures and temperatures have been made up to 5 GPa using a technique in which spheres of known density are placed in silicate samples within a pressure vessel. The relative density of the melt is noted according to whether the spheres sink or float. The densities of molten silicates up to 32 GPa have been measured for the first time recently using shock-wave techniques. It has been found that the melt density increases smoothly with pressure up to 25 GPa because of the change in coordination of Si and Al from four-fold to six-fold at higher pressures.

The results for a synthetic model basalt composition were compared with the density–depth relations for the Earth derived from seismological models (see Chapter 3). The implications are that melts become denser than host mantle at pressures of 6–10 GPa (200–300 km in the Earth), and that this is a maximum pressure from which basaltic or picritic melts can rise within planetary interiors.

The local distribution of a small percentage of fluid within a rock with which it is in chemical and mechanical equilibrium is controlled by thc condition of minimum interfacial energy between mineral–mineral and mineral–fluid contacts, which controls the **'wetting angle'** or **dihedral angle** Θ formed where three interfaces meet along a line. The principle is illustrated in Figure 4.6, which shows three representations of cross-sections through a fluid-filled channel between three minerals (see also Chapter 5). The surface energy, or surface tension, controls the curvature of the boundaries between mineral and fluid. The value of the dihedral angle determines whether a fluid is continuous in three dimensions, as indicated in the sketch of a rock section for $\Theta = 30°$, or whether the fluid is contained within closed pores or blebs, as indicated in the rock sketch for $\Theta = 90°$. If $\Theta < 60°$, even for a small amount of fluid, grain surfaces are 'wetted' and the fluid penetrates along grain edges making the rock permeable. For values of $\Theta > 60°$, the rock remains impermeable with dry grain surfaces until the amount of fluid reaches 6–8 per cent. Transport of the fluid with lower concentrations and high dihedral angle can proceed only by hydrofracture

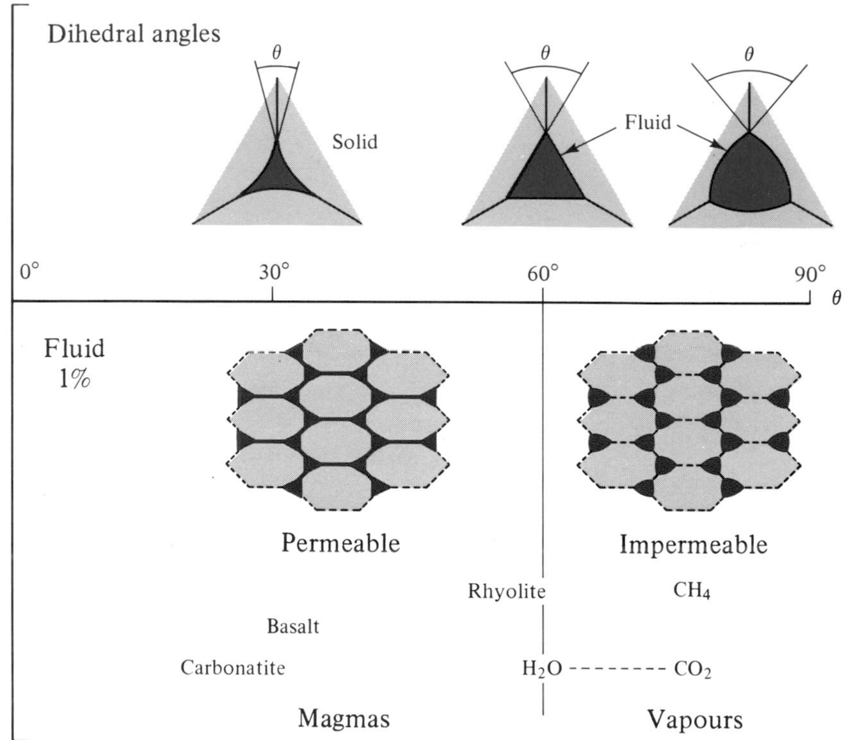

Figure 4.6 Illustrations of the dihedral angles between fluids and crystalline host. For angles < 60°, 1 per cent fluid forms an interconnected network of channels making the system permeable. For angles > 60°, the fluid forms closed blebs and the system is impermeable for fluid percentages up to 6–8 per cent. Most magmas form channels; most vapours form blebs. □

in which fluid pressure creates a microfracture network that allows subsequent migration.

Experiments at high pressures have provided information about melts and vapours summarised at the bottom of Figure 4.6. For up to 1 per cent fluid, basalt and carbonatite melts have low dihedral angles in assemblages of peridotite minerals, and therefore form continuous networks of interconnected channels. The dihedral angle for rhyolite melts is higher, 45–60°, indicating that at equilibrium they would only just form interconnected channels in crustal rocks. Vapours, CO_2, CO_2–H_2O mixtures, and probably CH_4, occupy closed cavities and cannot permeate rocks along grain boundaries. These properties, along with the viscosities of the fluids, are clearly of paramount importance in controlling the conditions of transport of magma or vapour within or through a rock.

Experimental petrology of the Earth's core and mantle

The range of our experimental knowledge about phase transitions and melting curves for mantle materials now covers the range of conditions throughout the Earth, although many details remain to be determined.

Melting of Fe and geotherm for the Earth's core

Figure 4.7 shows a compilation of results for the fusion curve of pure iron with data up to 100 GPa from large presses and diamond-anvil apparatus. In dynamic shock-wave experiments, melting occurred at 243 GPa along the **Hugoniot** of iron. (The Hugoniot curve is the locus of shock-induced final thermodynamic [pressure–temperature] states. Its trajectory shifts where the iron melts.)

Iron is the primary constituent of the Earth's core, but the density of the outer core indicates the presence of a lighter component in the Fe–Ni alloy (see Chapter 3). It is estimated that the core alloy melts at about 1000 °C below pure iron. The solidus for the core given in Figure 4.1b is therefore drawn 1000 °C lower than that determined for pure Fe, as in Figure 4.7. The intersection of this solidus curve with the boundary between the inner and outer core, which coincides with the change from solid to molten alloy, gives a value of 6400 °C for the temperature at this depth. A calculated **adiabat** (i.e. change of temperature due only to material compression) for the outer molten core passing through the intersection point gives the geotherm for the core, a maximum temperature of about 6700 °C for

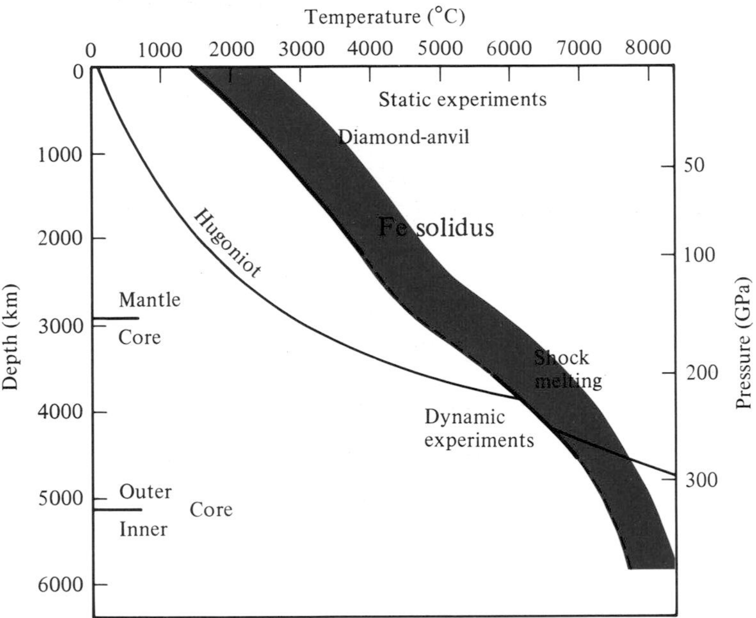

Figure 4.7 The fusion curve for iron. Results from diamond-anvil uniaxial device to 100 GPa. Results from shock-wave apparatus at 243 GPa. □

the centre of the Earth, and an estimated temperature of about 5000 °C for the alloy at the core–mantle boundary (considerably above the core's solidus curve). As shown in Figure 4.1b, this temperature is much higher than current temperature estimates of about 3000 °C in the convecting mantle just above the core–mantle boundary. This has significance for the energy source and dynamics of the Earth's convecting mantle, for interpretation of the seismic data in this region, and for the exchange of heat and perhaps matter between the core and mantle (see discussion in Chapter 3).

Phase transitions and melting in the mantle

Most of the different types of reactions investigated in experimental petrology occur within the Earth's mantle, and therefore a selection of mantle materials and mantle processes provides an informative guide to the applications of experimental petrology. The mantle is somewhat heterogeneous (see tomographic data summarised in Chapter 3), but for simplicity we shall consider initially a mantle of uniform composition, which exists as garnet–lherzolite at upper mantle pressures, and shall neglect the effect of volatile components. The solidus curve is shown in Figure 4.1b, reaching the 650 km seismic discontinuity, and some details are given in Figure 4.8a. Experiments in diamond-anvil apparatus, using the octahedrally coordinated silicate mineral **perovskite**, $(Mg,Fe)SiO_3$, as a model for the lower mantle composition, have provided the solidus given in Figure 4.1b between 650 km and 2000 km.

Figure 4.8a shows the phase relationships for lherzolite determined to pressures approaching 25 GPa, corresponding to depths of about 650 km. Consider the subsolidus phase fields. The mineral assemblages change at each phase boundary shown. The minerals present in each phase field are given in Figure 4.8b, which was determined experimentally at temperatures corresponding to the geotherm in Figure 4.8a, reproduced from that plotted in Figure 4.1b. The results in Figure 4.8b were based on experiments with natural lherzolite, and with simpler synthetic systems selected to investigate the mineralogical phase transitions encountered. The phase assemblages in Figure 4.8b represent a petrological cross-section through a mantle with this composition, and with the geotherm given.

The peridotite composition consists of lherzolite up to about 15 GPa (*c.* 400 km depth). The rock consists essentially of olivine, two aluminous pyroxenes, and a fourth aluminous mineral. The fourth mineral is plagioclase at low pressures, and this is transformed successively into spinel and garnet. At pressures greater than about 3 GPa, the peridotite consists of garnet–lherzolite. The changing proportions of pyroxenes and garnets involve redistribution of Al_2O_3 among the minerals, and progressive solution of the pyroxenes into **majorite**, a garnet solid solution. The garnet lherzolite is transformed successively into a garnet-spinel rock and then into a **perovskitite**, through two major phase transitions. The olivine is converted to a denser, spinel structure starting at 15 GPa (400 km depth), and then near 25 GPa (650 km depth) most of the minerals are transformed to perovskite, with Si in six-fold coordination. The gamma-spinel transforms sharply at 23 GPa, whereas the majorite transformation occurs through a wider pressure interval between 23 GPa and 26 GPa (Figure 4.8b).

These changes in physical properties of mantle materials are indicated clearly by the two discontinuities in seismic velocity profiles illustrated in Figure 4.8c and discussed in Chapter 3. There has been a long history of experimental studies attempting to correlate these seismic discontinuities with phase transitions in mantle minerals. In turn, these correlations can constrain mantle chemistry. It has been calculated that the density increases associated with these two phase transitions (about 10 per cent) are appropriate to account for the observed seismic discontinuities, and many scientists accept this evidence as support for an upper mantle with mineralogy corresponding to a lherzolite, dominated by olivine, converted to perovskitite at 650 km. Others conclude that the 400 km discontinuity and the physical properties between 400 km and 650 km may be better matched by a mantle with a smaller percentage of olivine (perhaps as low as 35–45 per cent) and a correspondingly higher percentage of the eclogite minerals (garnet + clinopyroxene), constituting a rock termed **piclogite**.

The densities and elastic properties of the mineral assemblages at high pressures are not yet known well enough for an unambiguous discrimination between the peridotite and piclogite models for the upper mantle. There are few measurements of the pressure and temperature derivatives of elastic moduli with direct relevance to the Earth, but estimates of these derivatives are now being made by new calculations based on the structural and chemical trends evident in the large elasticity data sets. The exercise of correlating high-pressure phase equilibrium experiments, mineral physics, and seismology continues to be a profitable approach.

Melting occurs in the upper mantle when the geotherm rises high enough to cross the solidus (Figures 4.1a and 4.8a), and relatively low-density, low-viscosity melt rises buoyantly through interconnected channels (Figures 4.5 and 4.6), associated with compaction of the rock matrix. These melts may reach the surface in basalt eruptions. This simple picture is modified in the presence of volatile components, as shown in Figure 4.4 and developed in an earlier section of this chapter. Evidence outlined earlier suggests that the density of mantle melts exceeds that of the crystalline host at depths of 200–300 km. Such melts

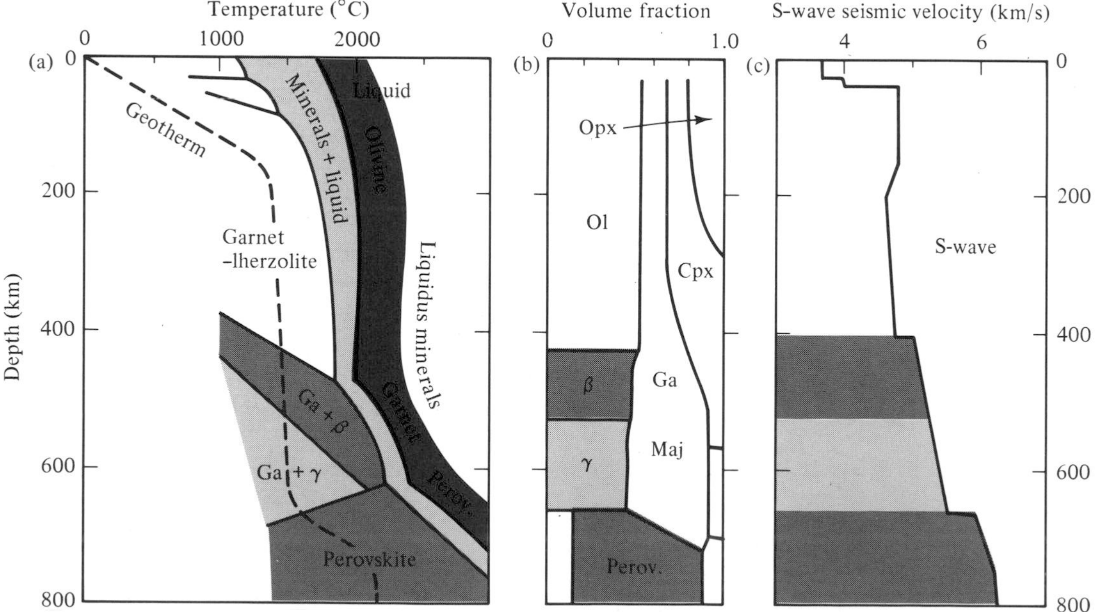

Figure 4.8 Phase diagram for lherzolite compared with seismic velocity profile. (a) Phase boundaries and geotherm. Ga = garnet; β and γ are forms of spinel. (b) Modal percentages of minerals in lherzolite composition measured at temperatures corresponding to the geotherm in (a). Ol = olivine, Opx = orthopyroxene, Cpx = clinopyroxene, Maj = majorite, Perov = perovskite. Further details of these minerals are given in Chapter 3. (c) S-wave seismic velocity profile beneath a continent (compare Figure 4.1a). □

would be unable to rise, which supports the concept of a **komatiite** (high-Mg) magma ocean at an early stage in the evolution of the Earth. This has important implications for the Earth's thermochemical evolution and the development of stratification in the mantle.

The solidus curve for peridotite increases steadily in temperature with increasing pressure (Figure 4.8a), with well-defined curves associated with the changes in slope that occur with each change in subsolidus assemblage. The melting interval, which is quite wide at low pressures, becomes narrower with increasing pressure as the solidus and liquidus approach each other. The liquidus mineral (i.e. the last mineral to melt) is olivine up to about 15 GPa, garnet through a moderate pressure interval, and perovskite above about 22 GPa, where the solidus and liquidus become less steep (dP/dT). At higher pressures, Figure 4.1b shows additional changes in slope for the solidus of synthetic perovskite.

Perovskite is the most abundant mineral in the lower mantle, and the ratio of Mg/Fe at depth is probably about 9. Figure 4.1b shows a recent experimental determination of the solidus curve for perovskite of composition $Mg_{0.9}Fe_{0.1}SiO_3$, which remains at essentially constant temperature from 25–60 GPa. This indicates that the solid and liquid must have similar densities, probably due to the close approach in coordination of silicon and aluminium in both the solid and liquid phases over a broad interval of pressure. Compression of oxygen atoms in the melt from four-fold to six-fold coordination around silicon would occur at these pressures, the same kind of change that occurred in the minerals at shallower depth, where perovskite was formed.

An extrapolation of this solidus curve toward the mantle–core boundary would intersect the geotherm (see Figure 4.1b), indicating either that melting should be widespread in the lower mantle, or that lower mantle temperatures are lower than currently estimated. According to results during the late 1980s on synthetic perovskite, which extend the solidus from 60 to 100 GPa, there is a change in the solidus slope with a distinct decrease in dP/dT for this high-pressure portion of the solidus (Figure 4.1b). Thus temperatures in the lower mantle could be considerably higher than plotted in Figure 4.1b without causing partial melting. The explanation proposed is that by 60 GPa, all silicon in the liquid has reached a coordination of six so that the liquid becomes less compressible with increasing pressure, with corresponding change in slope of the solidus curve.

Partial melting in a thermal plume

Major chemical differentiation of the Earth is associated with uprise of solid mantle material in convective motions beneath oceanic ridges, or in thermal plumes associated with hot-spots, such as the Hawaiian Islands, or the igneous complex at Yellowstone Park in the USA. The geotherm in Figures 4.1b and 4.8a does not reach the solidus curves for peridotite or perovskite, and no melting occurs under these conditions. Melting will occur if physical conditions raise the geotherm across the solidus as shown in Figure 4.9a, or if addition of volatile components lowers the solidus curve to temperatures below the geotherm, as shown in Figure 4.9c.

Consider an oceanic environment, uncomplicated by the presence of continental crust, with uprise of a thermal plume in the mantle generating a hot-spot at the base of the lithosphere. The hot plume causes some thinning of the lithosphere, but because the plate moves continuously across the plume, the asthenosphere does not rise to shallow levels as it does with the extreme upwelling that may be associated with seafloor spreading at ocean ridges (see Chapter 9).

Figure 4.2a shows a specific arrangement of isotherms. Geotherms corresponding to positions at the centre and toward the margin of a plume are given in Figures 4.9a and 4.9c. Each point in the plume is characterised by a specific pressure and temperature. Any phase boundary from a pressure–temperature phase diagram may therefore be plotted on the mantle isotherms, providing a cross-section showing the distribution of different petrological phase assemblages (e.g. subsolidus assemblages in Figure 4.8b, and those with melt in Figures 4.9a and 4.9c).

Superposition of the phase boundaries in Figure 4.8a onto the plume isotherms outlines the melting region M mapped in Figure 4.9b. Melting begins in plume material as it crosses the solidus boundary and enters the region M. As the pressure decreases and the rock temperature rises further above the solidus, the percentage of melt increases, reaching perhaps 25 per cent, the amount depending on the maximum temperature within M (Figures 4.9a and 4.2a). If equilibrium were maintained, the amount of liquid would decrease as the plume is carried out of the melting region M to either side. In fact, some melt rises through the rock matrix and either forms magma chambers at the base of the lithosphere, or rises directly into shallower magma chambers beneath the volcanoes. The physical details are not well established (see Chapters 5 and 8 for some discussion). As shown in

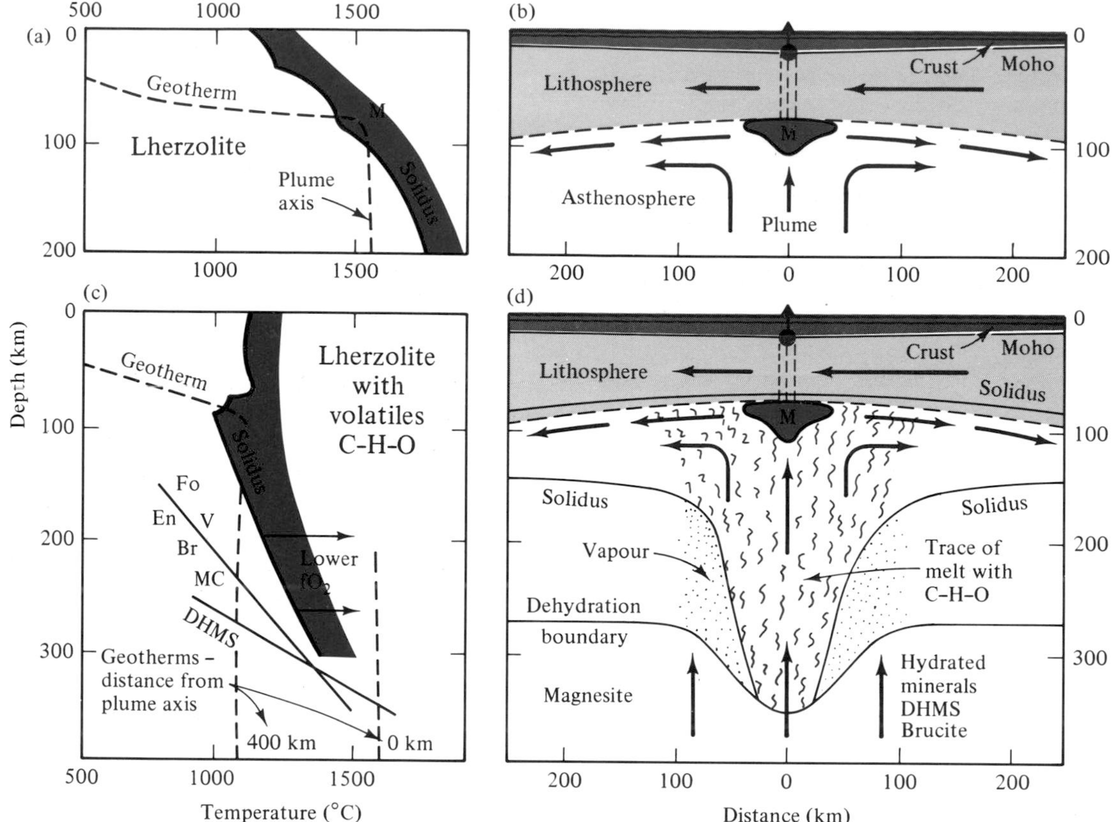

Figure 4.9 (a) Lherzolite solidus from Figure 4.8a, with the geotherm elevated to the temperature in the centre of a thermal plume (Figure 4.2). (b) The melting region (M) in volatile-free mantle with a rising thermal plume, based on specific isotherms in Figure 4.2a, and the solidus in Figure 4.9a. Melting occurs where the isotherms rise above the solidus. The line near 100 km depth represents the lithosphere–asthenosphere boundary layer. (c) Solidus for lherzolite–CO_2–H_2O. The solidus moves to higher temperature with lower oxygen fugacity (fO_2). Note two curves where, with increasing pressure, H_2O and CO_2 react with olivine (Fo) and pyroxenes (En) of lherzolite to form brucite (Br), magnesite (MC), and DHMS (dense hydrous magnesian silicates). V = vapour. (d) Map of phase boundaries, vapour and melt distribution in mantle with a trace of volatile components, based on isotherms in Figure 4.2a and the solidus and dissociation curves in Figure 4.9c. The shapes of the phase fields vary as isotherms and phase boundaries shift (with changing physical conditions, or changing compositions of rock and volatile components), but the general arrangement persists. The line for the typical position of the lithosphere–asthenosphere boundary layer (Figure 4.9b) is given, but note that the shallowest solidus curve associated with the plume axis is situated a few kilometres above this boundary. □

Figure 4.5, the initial melt composition in the melting region M is **picrite**; its viscosity is very low, and the dihedral angle is low enough (Figure 4.6) that percolation should occur at a relatively rapid rate. Uprise through the lithosphere may include both percolation and transport through cracks. Heating of the lithosphere by magmas from M may cause additional partial melting. Most magmas reaching the surface have had their compositions modified to less magnesian basalts.

Of course, even the most primitive of lavas contain traces of dissolved H_2O and CO_2 derived from the mantle sources. What proportion of these volatile components are primordial and what have been recycled via subduction remains to be determined. Yet, the presence of volatile components lowers the solidus curve, changes the composition of the melts, introduces new hydrated or carbonated minerals, and thus modifies the structure depicted in Figure 4.9b.

Let us assume that traces of volatile components composed of C–H–O are uniformly distributed through the mantle. There is debate about the oxidation state of the mantle and the compositions of vapours at different depths, which could vary from $CO_2 + H_2O$ to $CH_4 + H_2$ depending on the oxygen fugacity, and this contributes to uncertainty about the position of the solidus for lherzolite–C–H–O. The effect of adding volatile components is to lower the solidus temperature (cf. Figures 4.4, 4.9a and 4.9c); under less oxidising conditions, the solidus curve is lowered by a smaller amount. The distinctive feature is the ledge on the solidus at about 80 km depth which is involved with the formation of dolomite and amphibole in the presence of CO_2 and H_2O. Two curves are plotted in Figure 4.9c for deep hydration and carbonation reactions. Lherzolite minerals will absorb all H_2O at these reactions to generate brucite and DHMS (dense hydrous magnesian silicates), and all CO_2 to generate magnesite. These reactions thus place a maximum depth limit on the volatile-enriched solidus, under oxidised conditions. If conditions are reducing, however, then CH_4 and perhaps H_2 can exist as a vapour phase to deeper levels.

The composition of the liquid generated by partial melting of peridotite–CO_2–H_2O at depths greater than 75 km is carbonatitic, dominated by calcic dolomite. This liquid composition may become progressively enriched in silicate components with increasing depth, and with increasing temperature.

From the plume isotherms in Figure 4.2a and the phase boundaries in Figure 4.9c, the distribution of vapour and liquid is easily constructed (Figure 4.9d). Consider the processes occurring when the mantle plume rises through the phase fields in Figure 4.9d. The main effect is the generation of relatively large amounts of picritic melt in the region M where temperatures exceed the volatile-free solidus for lherzolite (Figure 4.9a), just as depicted in Figure 4.9b. However, associated with the rising plume is a sheath of volatile components, very small in quantity, which are transported with the plume as vapour. The vapour either rises from great depths as CH_4 + H_2 with some H_2O, or it is H_2O + CO_2 released from the solids as the plume crosses the dehydration/decarbonation boundary near 300 km depth. The vapour will be replaced by traces of interstitial melt when the plume crosses the solidus for peridotite–C–H–O (i.e. that in Figure 4.9c). This then grades into the more extensive melting that occurs in region M (Figure 4.9d).

The physical behaviour of the vapour and melt in a rising plume is controlled by the factors discussed in the previous section. The viscosity of all liquids associated with the plume is very low, and the interfacial angles for the volatile-rich liquids are well below 60° (Figure 4.6) so, again, they would form interconnected channels and flow easily. The vapours, in contrast, have interfacial angles greater than 60° and would remain sealed in blebs unless they were released through fracture, or by melt formation where temperatures exceed the solidus. It is likely, therefore, that vapours would rise at the same rate as the plume, but as soon as they crossed the solidus boundary, the trace of melt formed would rise faster than the rock matrix of the plume. Differential movement of material, such as escape of lava from the top of the plume, will change the bulk composition of the melt zone and thus the positions of phase boundaries. The framework of phase boundaries shown in Figures 4.9a and 4.9c appears to be robust enough to accommodate these and other adjustments, and it is precisely through such adjustments arising from the interplay of geochemistry, fluid dynamics and experimental petrology that we may advance our understanding of processes.

The volatile-rich partial melts are strongly enriched in incompatible trace elements including radioactive isotopes. They constitute the most likely cause of mantle metasomatism, the process that may re-enrich mantle rocks in incompatible elements, long after extraction of lava has depleted the rocks in these elements along with the more fusible basaltic components. The volatile-rich melt in the central part of the plume is incorporated into the major melting episode, whereas that in the outer sheath of the plume percolates directly into the lithosphere. What happens to it there depends on the physics of rock–fluid systems and the thickness of the lithosphere.

The ultimate products in ancient continental cratonic regions with thick lithospheres include **kimberlites** and carbonatites. The history of the traces of vapour and volatile-rich melt in a rising plume thus play an important role in determining the distributions of isotopes and trace elements in rocks of the upper mantle and crust. Interpretation of these isotope and trace element systematics in turn is critical for understanding rock reservoirs in mantle convection, source materials for igneous processes, and the evolution of continents (see Chapters 6 and 8).

Finally, the cross-section of a thermal plume developed in Figure 4.9 has all the phase fields relevant to processes of sea-floor spreading at ocean ridges. The main difference is that in this environment the lithosphere is much thinner and the geotherm may well cross the solidus curves near 10 kbar or 35 km depth (Figure 4.8a). A narrow ribbon of partially molten mantle extends down to at least 60 km, but the probability is that magmas escape rapidly once they rise to 30 km because of fracturing in this strongly extensional regime.

Experimental petrology of the continental crust

The continental crust, which is the product of chemical differentiation of the Earth's mantle through time, has itself differentiated further by processes of partial melting and melt migration, and by the weathering of rocks and the redistribution of the reaction products by atmosphere and hydrosphere to form new sedimentary rocks. When sedimentary rocks become buried and involved in mountain-building processes, they are subjected to increased pressure and temperature which transports them across reaction boundaries (see also Chapters 12 and 14). New mineral assemblages are formed at each reaction and the rocks are thus metamorphosed. If temperatures become high enough, the metamorphic rocks are partially melted.

The most dramatic changes in the conditions of continental rocks occur when continents collide and the crust is thickened, either by partial subduction of one continental mass beneath the other, or through compression at the convergent plate boundary. A zone of thickened continental crust is heated and thinned from below, yielding granitic melts which may rise through the crust. The continent rises higher, being less dense than the mantle, and thinning from above occurs through rapid erosion and extension on brittle faults (see Chapter 14). These processes cause each package of rock within the crust to follow a path of

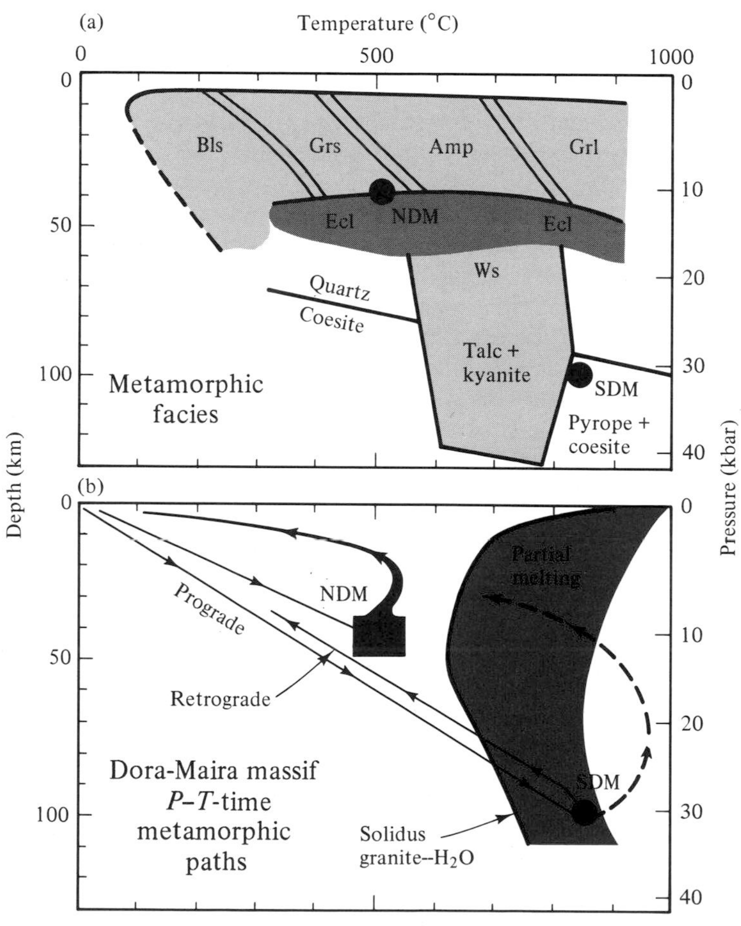

Figure 4.10 (a) Metamorphic facies for crustal rocks, based on experimental calibration of reactions. Bls = blueschist, Grs = greenschist, Amp = amphibolite, Grl = granulite, Ecl = eclogite, Ws = whiteschist. The eclogite facies has been regarded as the deepest, but note the additional high-pressure whiteschist facies which is not commonly shown on metamorphic facies diagrams. Estimated conditions of metamorphism of rocks from the Dora-Maira massif in the Italian Alps are given by NDM, northern region, and SDM, southern region. See text for detailed discussion. (b) Estimated depth–temperature paths for the NDM and SDM rocks during prograde and retrograde metamorphism. The dashed line might be considered a normal retrograde path for SDM, but retrograde minerals indicate that the rocks remained at high pressures during cooling. The solidus curve for granite–H_2O marks the position for the formation of migmatites (see H in Figure 4.11). □

changing depth (pressure) and temperature with time, known as a ***P–T–t* path**. At ocean–continent plate boundaries the basaltic ocean crust is subducted to depths of at least 650 km, causing extreme metamorphism compared with that of continental crust, and initiating magma generation in subducted oceanic crust and the overlying mantle wedge. All of these processes have been calibrated by experimental petrology and the evidence from this approach appears in several other chapters of this book: a few processes are outlined below.

Petrogenetic grid for metamorphic rocks: metamorphic facies and geothermobarometry

At crustal pressures, the slopes (dP/dT) of the reactions for the dehydration and decarbonation of sedimentary rocks, as illustrated in Figure 4.3c, are normally steeper than those of solid–solid reactions, as illustrated in Figure 4.3b. The array of intersecting reaction curves through the range of depth–temperature conditions corresponding to the continental crust forms a **petrogenetic grid**, dividing *P–T* space into a series of pigeon-holes, each characterised by a set of mineral assemblages. **Metamorphic facies** are defined in terms of the characteristic mineral assemblages. One of the triumphs of experimental petrology has been calibration of the reactions separating the metamorphic facies.

Figure 4.10a shows the standard metamorphic facies (cf. Chapter 12) up to about 12 kbar (1.2 GPa, about 50 km depth), somewhat deeper than the base of the average continental crust, with higher pressures being represented as **eclogite facies**. It has been determined experimentally that rocks expected in the deep crust, **amphibolites**, mafic **granulites**, and metamorphosed gabbros and diorites, are indeed transformed to varieties of eclogite (i.e. rocks with dense pyroxenes and garnet) at the pressures indicated. Most known metamorphic rocks fit neatly into this facies classification.

The system $MgO–Al_2O_3–SiO_2–H_2O$ has now been studied under mantle conditions because it includes minerals important in mantle petrology. These experiments revealed a large number of minerals that are only stable at high pressures. For example, the assemblage talc + kyanite was found to be stable at pressures between 6 kbar and 40 kbar (0.6–4 GPa), in the temperature interval shown in Figure 4.10a. This assemblage was subsequently identified in metamorphosed shales, termed **whiteschists**, which are now recognised as an extremely high-pressure metamorphic facies for crustal rocks. Whereas most crustal rocks when metamorphosed transform to greenschists, amphibolites, and granulites or rocks with eclogitic mineralogy, some shales at high pressures pass from greenschist to whiteschist.

Another product of experimental petrology is that the compositions of minerals between reactions provide information about the pressures and temperatures at which the minerals last equilibrated. In combination with the thermodynamics of mineral solid solutions and mineral assemblages, this has proved to be extremely effective in providing **geothermometers** and **geobarometers** for metamorphic mineral assemblages. For example, the use of pyroxenes in peridotite xenoliths (metamorphic rocks of the mantle) brought to the surface by kimberlite or alkali basalt provides fossil geotherms (temperature as a function of depth) down to depths of 250 km in ancient lithosphere, and this is one way in which the shallow geotherms depicted in Figures 4.1, 4.8 and 4.9 were determined. In the eclogite facies (Figure 4.10a), the compositions of clinopyroxenes and garnets vary with temperature and therefore provide a geothermometer. Similarly, geobarometers have been calibrated for many mineral assemblages in the other facies (see Chapter 12 for further discussion).

A rock being transported to greater depths by tectonic processes experiences a series of reactions, but relicts of the earlier minerals are commonly preserved as inclusions in the new minerals, providing information about the depths and temperatures that these **prograde rocks** (i.e. with advancing metamorphism) have experienced. During their transportation back to the surface, as the surface layers are being removed by erosion, the rocks experience **retrograde metamorphism** as they pass back through metamorphic facies and reaction boundaries, which may destroy the evidence for their earlier passage through higher temperatures at greater depths. However, the fact that metamorphic rocks of different grade do persist at the surface demonstrates that these retrograde reactions do not go to completion. The kinetics for reactions proceeding in the direction of decreasing temperature are less favourable than for the prograde reactions, and the abundant pore fluid expected during prograde reactions may not be available for retrograde reactions. It is the study of the mineral inclusions and relicts of former reactions in metamorphic rocks, along with the pressure–temperature calibrations of experimental petrology and various measurements of time that reveals the *P–T–t* paths of rock masses during various tectonic processes. These paths provide limits for the calculations of thermal regimes and are adding a new dimension to our understanding of mountain-building processes (see Chapters 12 and 14). There is an opportunity here for experimental kinetic studies of metamorphic reactions, and of the effect of pore fluids on the mechanisms and reaction pathways in heterogeneous rocks, an approach still in its infancy.

The Dora-Maira massif: a case study in pressure – temperature – time paths

The crystalline rocks of the Dora-Maira massif in the western Italian Alps include a series of schists and gneisses enclosing eclogites. They have been interpreted in terms of normal regional metamorphism passing through the facies greenschist to amphibolite, with some peak metamorphism reaching the eclogite facies. The physical conditions of metamorphism in the eclogites of the northern part of the massif (NDM) have been determined from detailed study of the mineralogy and petrography, using all available mineral geobarometers and geothermometers. The peak of metamorphism is estimated to have occurred at 500 ± 50 °C between 9 and 13 kbar pressure, as shown by the plotted box NDM in Figure 4.10b. These results indicate that the rocks were metamorphosed at depths of 40–50 km, somewhat deeper than the normal thickness of continental crust. The eclogites have been extensively recrystallised in the greenschist facies, and study of the retrograde mineral reactions indicates that the eclogites rose up to shallower levels under isothermal conditions, or even under conditions of slightly increasing temperatures. The retrograde path then continued through the greenschist facies with decreasing temperature. The prograde and retrograde paths for these rocks are represented by the path through NDM in Figure 4.10b. This kind of retrograde path has been determined in other parts of the Dora-Maira massif, and in many other metamorphic belts. It is explained in terms of burial of the rocks along a geotherm, with slow heating due to their low thermal conductivity. The rocks then experience relatively rapid decompression during uplift in response to erosion at the surface, while cooling is a relatively slow process.

There is evidence that quite a different history was experienced by rocks in the southern part of the Dora-Maira massif. Some of the rocks in this region yield spectacular large pyrope (Mg-garnet) crystals which are characteristic of the eclogite facies. Only in 1984 was it discovered that the pyrope is associated with a talc–kyanite whiteschist, indicating unusually high metamorphic pressures. It was also discovered that the garnets contain tiny silica crystals which are now of quartz, but to judge from their crystal shapes were originally high-pressure **coesite** crystals. There are also some crystals of unaltered coesite. Coesite is only stable at pressures above 20 to 30 kbar (depending on temperature), as shown in Figure 4.10a. Furthermore, the experimental studies have demonstrated that the assemblage pyrope + coesite is produced by the dehydration of talc + kyanite, and that this reaction occurs near 800 °C, and only at depths of at least 100 km. So, some Alpine crust was buried as deeply as 100 km.

According to these mineralogical calibrations, the coesite–pyrope and whiteschist must represent a large sample of the continental crust which was buried to a remarkable depth of about 100 km at a temperature near 800 °C in the region marked SDM in Figure 4.10. Even more remarkable is the fact that the mineralogical signal was preserved, despite some retrograde reactions.

The evidence is strong that during Alpine mountain-building (further details in Chapter 14), some of the continental rocks of the Dora-Maira massif were subducted beneath the African continent to depths previously unsuspected for buoyant continents. In order to preserve its mineralogy the rock mass must have returned to the surface without further heating into the crustal facies of amphibolite and granulite; thus it avoided the more normal retrograde path indicated by the dashed line in Figure 4.10b. Mineral reactions indicate that on the way back to the surface the mass returned along the relatively low-temperature retrograde path identified by the solid line, which corresponds closely to the cool, prograde 'subduction' path. Working out the tectonic details of this process remains as a challenge for structural geologists.

Melting in the crust, and at convergent plate boundaries

Much experimental attention has been devoted to the conditions of partial melting of possible magma source rocks and to crystallisation of the melts, but only a brief outline can be presented here. Figure 4.11 shows the rock masses present to a depth of about 150 km in a location about 100 km from a convergent plate boundary, where the subducted oceanic lithosphere has reached a depth of about 100 km. The continental crust overlies mantle and, about 60 km deeper, a 5 km-thick layer of oceanic crust is intersected. The temperatures for melting in the continental crust, the mantle and the subducted oceanic crust are given by the solidus curves (for the beginning of melting) and the shaded intervals at temperatures above them. These solidus curves, determined experimentally either dry or with known contents of volatile components, provide a calibrated depth–temperature framework useful for considering the processes of magma generation and the interactions of magmas in various tectonic environments.

The characteristic magma from the mantle is basalt, generated when hot mantle material rises above its solidus (e.g. AB in Figure 4.11 for dry conditions; see also Figures 4.1b, 4.5, 4.8a and 4.9a). The characteristic magma generated within the crust is intrusive **granite**, and this may have erupted as rhyolite, obsidian or pumice, depending on the dissolved volatile components. The phase diagram controlling basalt magmatism verges on the 'dry' type (Figure 4.4c), whereas that for granite is of the hydrous type (K or H in Figure 4.11, see also Figures 4.4d or 4.4e). Granitic magmas can be generated by partial melting in the continental crust, but melting is often initiated at H or K (Figure 4.11) by heat and material transported into the crust by higher temperature basalts from the mantle (cf. below and Chapter 5)

The situation is more complicated at ocean–continent convergent plate boundaries. The magmatic process starts with dehydration or partial melting of the subducted oceanic crust at a depth near 100 km. The vapours or melts rising from FG (Figure 4.11) into the overlying mantle wedge initiate melting in peridotite in the shaded mantle above the wet solidus curve CD (Figure 4.11). Again, the effect of water is to make the melt somewhat richer in SiO_2 than it would be under dry conditions. The overlying crust is then invaded by vapours and basic melts. The characteristic magma of this tectonic environment is andesite, intermediate between basalt and rhyolite. Granite production represents a further stage in the evolution of magmatic and thermal processes involving melting at crustal levels as heat and volatiles arrive from the mantle. A granitic pluton thus represents a frozen stage near the end of a complicated, dynamic, and long-duration magmatic process initiated in the mantle. Basaltic magmas provide thermal energy for these processes and either cause partial fusion of the crust, or become converted to more siliceous melts by reaction with crustal rocks and magmas, with concomitant fractional crystallisation. Interaction between magma, mantle and crust during upward passage to eventual

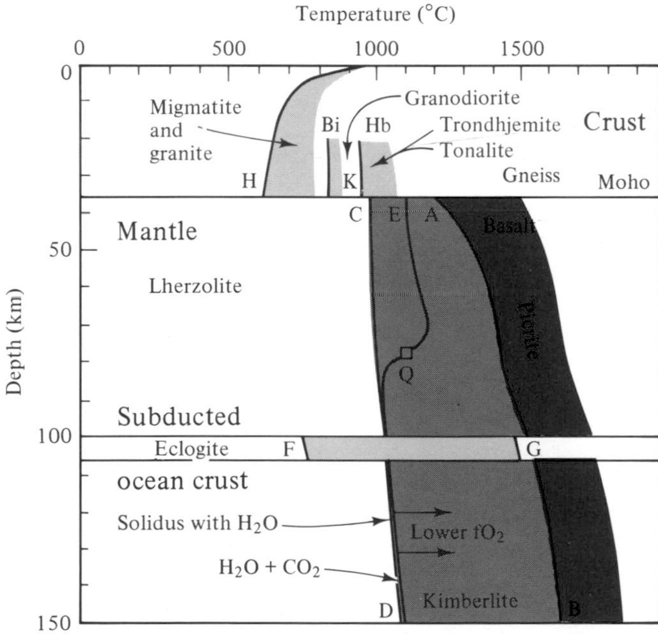

Figure 4.11 Conditions for partial melting (shaded areas) of rock masses present in cross-section about 100 km from a convergent plate boundary. Continental crust overlies mantle peridotite, and the oceanic crust layer of the subducted oceanic lithosphere is intersected at a depth of about 100 km. The curve AB is the dry solidus for peridotite (Figure 4.8a); CD is for peridotite–H_2O; EQD is for peridotite–C–H–O under oxidised conditions (Figure 4.9c); H is for granite–H_2O; K is for dehydration melting of biotite (Bi) or amphibolite (Hb); FG is the melting interval for eclogite with 5 per cent H_2O. The compositions of melts produced vary as a function of source rock composition, depth and volatile components, as explained in the text. □

eruption, and fractionation *en route* or in magma chambers, yields a melt situated on the liquid crystallisation path for alkali feldspar, plagioclase feldspar, and quartz; precipitation of these minerals along with some mafic and opaque accessory minerals forms a granite.

Let us take a look at melting in the crust. The minimum temperatures for partial melting of common crustal rocks are given in Figure 4.10b and at point H in Figure 4.11 by the experimentally determined solidus curve for granite in the presence of water, which has been measured to pressures of 35 kbar. Rocks other than those with granitic mineralogy will begin to melt at somewhat higher temperatures, but the difference is small for the common metamorphic rocks in the continental crust between 20 and 40 km depth. The granite solidus curve passes through the middle of the amphibolite facies (cf. Figure 4.10a), so we expect **migmatites** (i.e. mixed rocks with granitic veins) to form in amphibolites if pore fluid is available. According to Figure 4.10, the subducted metasediments of the Dora-Maira massif (SDM) would have melted partially in the presence of the pore fluid released as the whiteschists dehydrated.

If no pore fluid is available, melting may begin by **dehydration melting**, when a hydrous mineral such as muscovite, biotite or amphibole reacts directly to yield a H_2O-undersaturated melt with a phase diagram resembling Figure 4.4d, and represented by curves near K in Figure 4.11. For dehydration melting of amphibolite (Hb), the solidus curve is displaced towards higher temperatures than the boundary between the amphibolite and granulite metamorphic facies, almost 300 °C higher than the H_2O-saturated solidus at 10 kbar (Figures 4.10 and 4.11). The dry solidus temperature for basalt at 10 kbar (near to that of peridotite, A in Figure 4.11) is about another 300 °C higher. Evidence from field relations, geochemistry, and experimental petrology supports the conclusion that magmas generated at a range of temperatures in both mantle and crust are involved in the evolution of many granitic complexes.

The framework of petrogenesis based on experimental petrology has been built from equilibrium phase diagrams of simple systems which provide the principles and details of reactions and processes of melting and crystallisation, and of complex rock systems which provide a closer approach to reality. Given the complexity of dynamic processes in crustal melting, especially in subduction zone environments, one may question the applicability of phase equilibrium experiments. The response is that the equilibrium phase diagrams provide a firm basis for developing an understanding of these processes. Indeed, equilibrium phase diagrams are now being supplemented by experiments simulating reactions between appropriate rocks and melts, or between two dissimilar melts.

It has been shown experimentally that melts with compositions from granite to granodiorite may be produced by partial fusion of the crust during regional metamorphism, and the dihedral angles measured in experiments indicate that continuous networks of melt channels should form (Figure 4.6). However, experiments have also shown that the melts tend to collect in interstitial pools rather than to percolate through the crystalline matrix, because the viscosity of this silica-rich melt is so high. It has been calculated that percolation of melt by compaction of the rock matrix is too slow to account for the segregation of granitic magmas in the crust, although migration of melt through fractures does occur, as shown by the structures of migmatites.

The most likely mechanism for generating large, kilometre-size granitic plutons is for sufficient melt to be generated in a segment of the crust that the whole mass begins to convect. Estimates of the **critical melt fraction** required for mobilisation of the entire mass of a partly molten crustal rock range from 30 to 50 per cent. The effective viscosity of a partly melted zone is sensitively dependent on the H_2O content, because this influences both the percentage of melt produced at a particular depth and temperature (Figure 4.4a) and its viscosity. Given large fluxes of externally derived aqueous vapour, as is possible in subduction zones, the condition for mobilisation could be achieved at about 750 °C. However, given vapour-absent conditions and dehydration melting of basement gneisses, temperatures of at least 950 °C are required for segregation of a granodioritic melt. This requires influx of heat, and basalt is the only likely provider. The problem of producing and mobilising granite melts in the crust is addressed, in a similar vein, in Chapters 5, 8 and 12.

Conclusions

There are many additional topics in experimental petrology for which there is no room in this chapter. We have reviewed the framework of phase boundaries through which pass mobile masses of rock, melt or vapour, and some aspects of the physical properties of planetary materials, especially of rock–fluid systems. The phase boundaries calibrate the processes which cause the formation of magmas and rocks. They provide constraints to complement geochemical, geophysical and geological approaches to the materials science of the Earth.

Further reading

Akimoto, S. & Manghnani, M. H. (eds.) (1982) *High Pressure Research in Geophysics*, Reidel, Dordrecht, The Netherlands.

Holloway, J. R. & Wood, B. J. (1988) *Simulating the Earth: Experimental Geochemistry*, Unwin Hyman, London.

Morse, S. A. (1980) *Basalts and Phase Diagrams*, Springer, New York.

Ulmer, G. C. (ed.) (1971) *Research Techniques for High Pressure and High Temperature*, Springer, New York.

Ulmer, G. C. & Barnes, H. L. (eds.) (1987) *Hydrothermal Experimental Techniques*, Wiley, New York.

Wyllie, P. J. (1988) Magma genesis, plate tectonics, and chemical differentiation of the Earth, *Reviews Geophysics*, **26**, 370–404.

INTRODUCTION TO CHAPTERS 5–8

The study of geology has traditionally been that of the rocks, minerals and fossils which are readily accessible on the Earth's surface. However, one of the exciting developments in recent years has been how the advent of the unifying theory of plate tectonics has made Earth scientists much more aware of the wider implications of their detailed studies. Volcanic eruptions are among the most spectacular and awesome expressions of natural forces, and since the Earth's crust consists largely of igneous rocks, volcanoes are also the natural laboratory in which we can study how the crust itself was formed.

Igneous rocks are the products of a complex sequence of events and processes which include the partial melting of a variety of possible source rocks, and the crystallisation of the resultant magmas under a range of different conditions of pressure and temperature. One important approach has been to study the conditions under which naturally occurring rocks and minerals can be reproduced in the laboratory (Experimental Petrology, Chapter 4), but two others which are at the forefront of current investigations are first, the physical dynamics of melting and magma movement, and second, the chemistry of igneous rocks and how that can be used to determine the sorts of processes that have been in operation. Stephen Sparks is one of the leading exponents of the physical dynamics approach, and in Chapter 5 he provides a powerful statement of just why the observations and knowledge of petrology and geochemistry cannot be fully used unless they are accompanied by a proper sense of magmas as dynamic systems. In Chapter 6 Keith Cox then offers a pleasingly straightforward insight into the complex subject of how the many currently available geochemical and experimental techniques can be used to unravel the history of igneous rocks from their chemical compositions.

While plate tectonics has been the major conceptual revolution in Earth sciences this century, one of the most influential technological advances has been in the development of robust high-precision mass-spectrometers for the routine measurement of isotope ratios. Some isotopes are produced as the products of radioactive decay with half lives which are similar to the age of the Earth, and these are now widely used to tackle problems across a wide spectrum of subjects within the Earth sciences. The most dramatic result has been the measurement of the age of the Earth, and the realisation of the sheer immensity of Geological Time (Chapter 7). However, isotopes produced as a result of radioactive decay are also the cornerstone of any attempt to understand how the Earth was formed and how it has subsequently evolved. Keith O'Nions has a worldwide reputation for his work on how the outer portions of the Earth have differentiated, with the generation of the continental crust leaving behind a complementary depleted reservoir, probably in the upper mantle. In Chapter 8 he reviews current ideas of how and when the continents were generated, and how much of the Earth's mantle may have been processed in the formation of continental crust.

CHAPTER

5

Stephen J. Sparks
University of Bristol

MAGMA GENERATION IN THE EARTH

Introduction

Volcanism is the most spectacular manifestation of the internal dynamics of the Earth, and its causes have aroused scientific curiosity ever since Empedocles took the unwise step of jumping down the crater of Mount Etna. In the first half of this century enormous strides in understanding the origin of magmas were achieved by applying the principles of physical chemistry and phase equilibria to synthetic and natural silicate melts. These studies enabled geologists to constrain the conditions of temperature and pressure which were necessary to melt rock and form magma and to understand how a wide diversity of different magmas could be generated by differentiation during their solidification. However, it was not until the theory of plate tectonics was developed in the last twenty years that magmatism could be considered in the context of a general dynamic model of the Earth. Many volcanoes occur at major plate boundaries and consequently the formation of magmas is intimately connected with the processes that generate and destroy tectonic plates. There are in addition other volcanoes (hot-spots) which occur away from plate boundaries which reflect convection within the Earth's mantle. Ultimately, igneous processes are the principal mechanisms by which the Earth has differentiated.

While plate tectonics provide a conceptual framework for understanding volcanism, developments in geophysics, geochemistry and mineralogy provide the observational data on which hypotheses for the origin of magmas can be tested. Progress in these fields has been prodigious in recent decades and the quality and quantity of information on igneous rocks and volcanic systems has increased dramatically. Studies of seismicity, ground deformation, electrical conductivity and gravity in active volcanic areas have enabled geophysicists to identify source regions where rocks are melting, the sizes and shapes of magma chambers and to monitor the subterranean movement of magma. Analyses of radiogenic and stable isotopes allow geochemists to identify where melting takes place and what the source rocks are like. Studies of variations in major, minor and trace elements in igneous rocks and minerals allow geologists to recognise and distinguish different processes in the generation and evolution of magmas. Much can now be learnt about the conditions of melting and crystallisation from studies of the structure, compositions and thermodynamic stability of minerals. Estimates of temperature, pressure, volatile partial pressures and kinetic parameters, such as cooling rate, can be determined from studies of minerals in igneous rocks.

Plate tectonics and the wealth of sophisticated observational data on igneous rocks has provided a fertile environment in which to make progress in understanding magmatism. However, the central feature of magmatism is that rocks melt and the melt is then transported from one place to another. Magmatism is a process of heat and mass transfer by fluid flow in the Earth. The most important problems in igneous petrology require understanding of the physics of magmatic processes and in particular their fluid dynamics. This aspect of magmatism has developed into a major field in geology. Research in the last decade has resulted in a much more complete understanding of magmatism and in some instances has caused geologists to reconsider fundamental axioms.

This chapter discusses magmatism in terms of the physical dynamics of melting, magma movement and crystallisation in magma chambers. It is no longer adequate for students of igneous petrology to learn only about phase equilibria, mineralogy, geochemistry and classical petrography. These are all essential ingredients in understanding magmas, but without a sense that magmas are dynamic systems the observations and knowledge of these areas cannot be fully used. It is of equal importance to any of these other topics that simple principles of fluid flow and heat transfer are understood.

Melting the mantle

The Earth's mantle is essentially a crystalline solid. There appears to be no large permanent region of abundant melt in the Earth and so early ideas of a molten interior to the Earth are untenable. The region in the mantle between 150 and 400 km depth has anomalously low seismic velocities and could contain a very small fraction of permanent melt, but interpretation of this **'low velocity zone'** is still not certain and the amount of melt present could not account for most volcanism. However, although the interior of the Earth is solid, it is also quite near to its **solidus** temperature (i.e. the temperature at which it will start to melt), so that substantial melting can occur in suitable circumstances. The association of many volcanoes with plate boundaries demonstrates that substantial melting occurs where plates are created and destroyed. Intra-plate volcanoes also show that local melting is possible due either to convection in the interior or to other internal effects. The principles of melting and their effects are now discussed in terms of major tectonic environments.

Ocean ridges

At constructive plate margins the cold lithosphere is pulled apart at rates of centimetres per year and hot mantle from the asthenosphere flows up continuously towards the **ocean ridge** axis. The pattern of flow in the mantle beneath a spreading ridge is shown in Figure 5.1. The fundamental feature of this flow is that the upwelling mantle experiences a large pressure decrease which is a maximum immediately beneath the ridge axis. Experimental studies and principles of thermo-

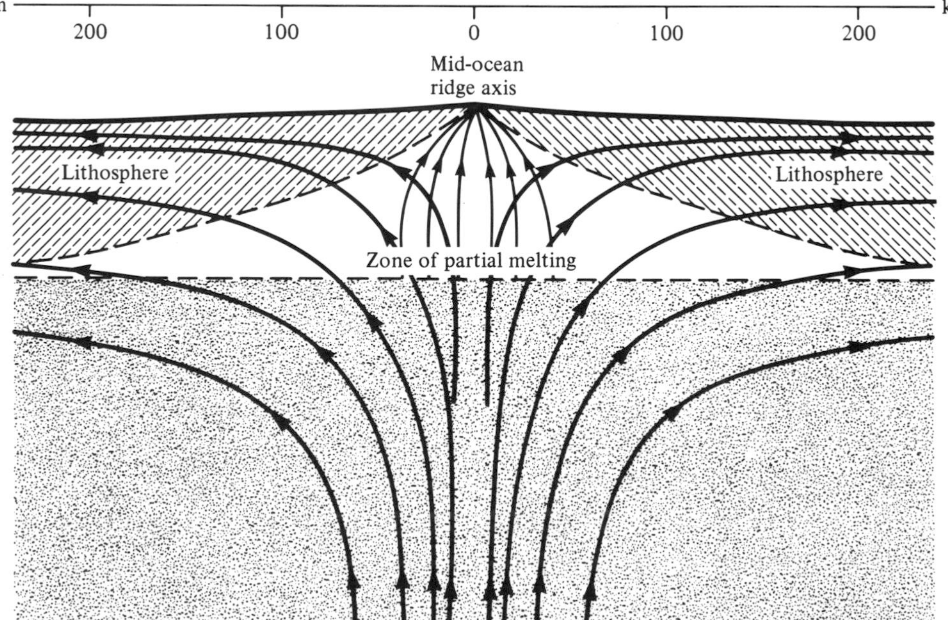

Figure 5.1 Schematic cross-section of a mid-ocean ridge. The thick solid lines show the symmetrical diverging flow-paths of the mantle beneath the ridge axis. The thickening of the lithosphere (diagonal ornament) due to cooling as the plates move away from the ridge is indicated. A central region of partial melting beneath the ridge is depicted. The thin solid lines show the flow-paths of the partial melts in this region converging on the ridge crest. □

dynamics require that the solidus and liquidus temperatures of the mantle decrease as the pressure drops. Thus, if the mantle in the asthenosphere is close to its solidus temperature, decompression must cause melting.

Figure 5.2 illustrates the principles of **decompression melting**. Imagine a small parcel of the mantle that rises along a particular flow line. While a solid the parcel will decrease in temperature due to adiabatic decompression, corresponding to about 0.6 °C per kilometre of ascent. Eventually the parcel of mantle intersects the solidus curve and melting begins. On further ascent, melting continues so that the proportion of partial melt increases with decreasing pressure. When melting is in progress energy is required to provide the latent heat of fusion of the crystals, causing a greatly enhanced temperature decrease of about 4 °C per kilometre during further ascent. From a thermodynamic point of view the parcel of mantle behaves as a closed system with regard to heat loss. For ridges being pulled apart at a few centimetres per year, the ascent velocity of the upwelling mantle is typically a few millimetres per year and this is sufficiently fast that heat loss to the Earth's surface is negligible.

In oceanic regions a 6 km thick oceanic crust is typically observed overlying the mantle. This oceanic crust is thought to represent the basalt extracted from the region of partially molten mantle beneath the ridge. Figure 5.3 shows the amount of melt produced by decompression melting of the mantle expressed as an equivalent thickness of a melt layer that would overlie the mantle if all the melt is completely extracted. Results are shown for different upper mantle temperatures.* The calculations require basic information on the thermodynamic properties of the crystalline and melt phases in the mantle and on the melting behaviour of peridotite. For a mantle temperature of 1350 °C a melt layer of 7 km is implied. Independent estimates of mantle temperature from considerations of heat flow and convection in the mantle indicate that this temperature is reasonable. The mean temperature of this melt layer can also be estimated at 1210 °C for a mantle temperature of 1350 °C (Figure 5.2). Thus decompression melting can generate the observed amounts of magma required to form the oceanic crust. Ocean ridge volcanism accounts for about 90 per cent of magma production in the Earth.

The flowlines in the mantle beneath a ridge show a symmetrical distribution about the central axis (Figure 5.1). With distance from the central axis the flowlines diverge sideways at increasing depth. In Figures 5.1 and 5.2 the melting behaviour along different flowlines away from the axis can be compared and the diagrams show that the amount of melting decreases with distance from the axis. In addition each parcel of mantle eventually cools back to beneath the solidus as it is transported laterally away from the ridge into the region where surface cooling becomes important and the cold plate forms (Figure 5.1). The history of melting and cooling of the mantle in this environment thus defines a **region of partial melting** in which the degree of partial melting and the depth of melting vary from place to place. The magmas generated during

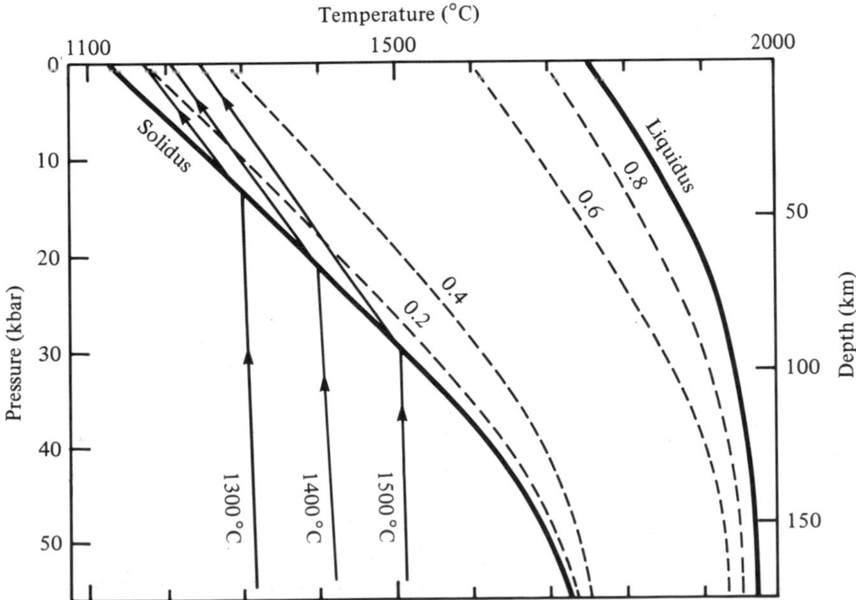

Figure 5.2 **Decompression melting of the mantle. A parcel of mantle with a particular temperature ascends and intersects the solidus of peridotite and melting begins. Further ascent results in extensive cooling of the partially molten mantle as energy is utilised to provide the heat of fusion. The melting behaviour is shown for three different mantle temperatures. The dashed curves show appropriate fractions of partial melt.** □

* The mantle temperature is defined as its equivalent temperature at a pressure of one atmosphere. In the convecting region of the mantle temperature varies with depth because of adiabatic effects.

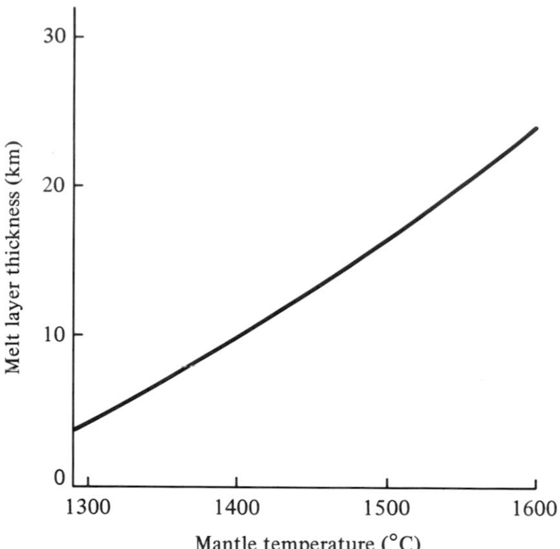

Figure 5.3 **The amount of melt generated at an ocean ridge is shown as a function of mantle temperature. The amount of melt is represented by the equivalent thickness of a melt layer calculated by assuming that all the melt formed in the region of partial melting (Figure 5.1) is extracted to form the overlying oceanic crust.** □

decompression melting should vary greatly in composition and temperature, because of the wide ranges of melting histories that are possible as the mantle is processed through the partially molten zone. Mantle rising close to the central axis will partially melt by relatively large amounts at low pressure. Mantle which follows flow-paths close to the edge of the partially molten region will experience small degrees of melting at high pressure. Experimental data and observations indicate, for example, that the melt generated at the ridge axis will be silica-saturated **tholeiitic basalt**, whereas the melt generated at the edge of the melting zone should be silica-undersaturated **alkali basalt**, rich in incompatible trace elements such as potassium.

The geometry of the partially molten zone in a normal ridge is an important consideration. If the underlying mantle is well mixed by convection and has a uniform temperature then melting along each flowline must initiate at the same depth (Figure 5.1). Consequently in normal regions of the mantle, away from ascending or descending convective plumes or jets, the partially molten zone will have a flat base (Figure 5.1). As the mantle moves laterally away from the ridge, surface cooling by conduction forms the lithosphere plate which grows as the square root of time, until the plate reaches a thickness of about 80 km, after which other factors associated with convection of heat from the mantle become important. The partially molten region is bounded by convex curving surfaces which represent the position of the mantle solidus (Figure 5.1). In normal oceanic ridges the depth of melting is thought to be limited to about 80 to 100 km, with most melting taking place at depths less than 50 km.

The calculations of the temperature of the magma erupted at oceanic ridges (Figure 5.2) assume that melt and matrix do not separate, and remain in thermodynamic equilibrium along each flowline. Another possibility, however, is that melt can be extracted at depth and can rise independently (for example in dykes) without reacting with the crystalline mantle. In this case the pronounced cooling of the melt caused by further fusion of the mantle crystalline matrix is not involved and the ascending melt only cools adiabatically at about 1.0 °C per kilometre. Thus the melts that reach the surface if they are able to separate at depth can be considerably hotter. For example a melt segregating from a mantle at 1400 °C and at 50 km depth would erupt at about 1350 °C. Current views on ridges favour decompression melting with melt being liberated from the mantle at shallow levels. However, the difference between the two cases is considerable and illustrates that the mechanisms of segregation of melt and ascent are important in determining the compositions and temperatures of magmas.

Mantle plumes (hot-spots)

Convection in the mantle requires that there are regions of the mantle where hot jets or **plumes** rise **(hot-spots)** and other areas where cold material sinks. Areas of the most active volcanism on Earth, such as Hawaii or Iceland, are thought to represent the ascent of unusually hot mantle in plumes. As in the case of Hawaii, some plumes occur beneath the interior of plates. Current modelling of convection in the mantle indicates that the mantle is as much as 200 °C hotter in the interior of a hot rising plume.

The decompression melting model can be applied to hot plumes. However, the temperature across the plume varies from normal background mantle temperatures to some much higher temperature in the interior (Figure 5.4). Thus mantle rising along the plume axis will intersect the solidus at greater depth than cooler mantle rising at the side of the plume. The shape of the region of partial melting is consequently more like a mushroom with a stalk tapering down symmetrically about the plume axis (Figure 5.4).

The amount of melt generated and the temperature of the melt both increase as mantle temperature increases. For example if the mantle temperature increases by 200 °C above presumed normal values of

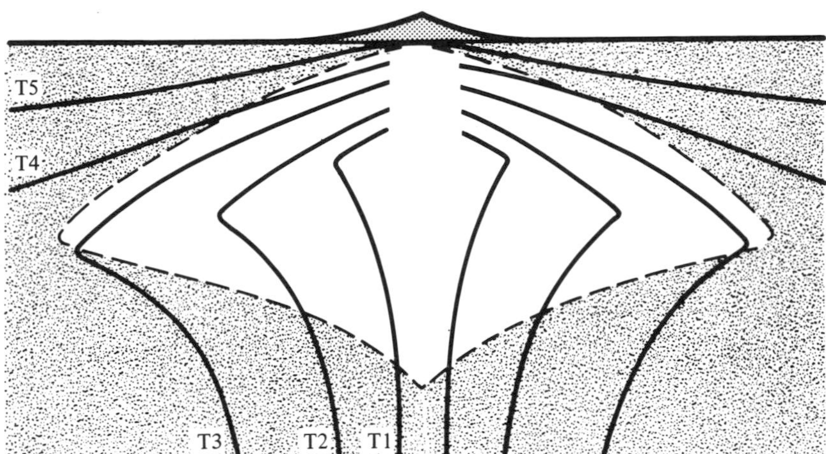

Figure 5.4 Schematic picture of melting zone structure in a mantle plume beneath a hot-spot such as Hawaii. In contrast to a normal mid-ocean ridge (Figure 5.1), the mantle becomes hotter at the centre of the plume so melts at a greater depth. Isotherms with arbitrary values are shown to illustrate the thermal structure. □

about 1350 °C then the rate of magma generation and hence the potential thickness of oceanic crust above the plume would increase threefold (Figure 5.3). The temperature of the extracted magma would also increase by 80 °C (Figure 5.2). In Iceland the oceanic crust is 15 to 20 km thick and the increased magma production rate can be simply explained by invoking a hot mantle plume. Observations and models of plume dynamics indicate that their influence can extend over regions with diameters of 1000 to 2000 km.

Diagrams such as Figure 5.3 invite speculation on volcanism during the early history of the Earth when the mantle should have been hotter. For example if the normal mantle were 200 °C hotter in the Archaean (3000 Ma ago) the oceanic crust at that time might have been 15 to 30 km thick. In the Archaean mantle plumes could have generated oceanic crust 30 to 40 km thick if they had temperatures 100 °C to 200 °C hotter than background.

Subduction zones

In **subduction zones** cold lithosphere sinks back into the mantle and so it is therefore somewhat surprising that volcanism is associated with subduction. The critical factor appears to be water. The subducted plate consists of sediment and metamorphosed basaltic ocean crust overlying the mantle. Water has been added to the oceanic crust by interaction with sea-water in circulating hydrothermal systems close to the ridge. Observations on the sea-floor indicate that sea-water penetrates several kilometres into newly formed oceanic crust and is heated to at least 350 °C. Thus the oceanic crust and uppermost mantle of the subducted plate contains hydrous minerals, such as chlorites, amphiboles and serpentines, which will progressively dehydrate as they are heated up on the upper surface of the subducting slab.

Water dramatically depresses the melting point of silicate materials (Figure 5.5). Water dissolves in silicate melts as the water molecule (H_2O) and as hydroxyl (OH) groups. The effect of dissolution is to decrease the **chemical activity** of other components (notably anhydrous silicate species) in the melt phase. Water becomes increasingly soluble with increasing pressure and so there is a corresponding dramatic decrease in solidus temperature with increasing pressure if abundant water is available (Figure 5.5). The depth to the subduction zone from the volcanoes is typically 100 to 200 km and averages 112 km. At these pressures the solidus temperatures of peridotite and basalt can be lowered by several hundred degrees centigrade. Thus the release of water by **dehydration reactions** in subducted rocks could trigger melting in both the slab and overlying mantle (Figure 5.6).

The exact role of water in the production of arc magmas is still uncertain. Several possibilities have been suggested. Metamorphic hydrous fluids at high pressure can dissolve large amounts of silicate material. For example at 30 kbars (about 100 km depth) experiments suggest that water can dissolve up to 20 per cent silica, whereas silica is virtually insoluble in water at the Earth's surface. Such hydrous fluids could be released into the overlying mantle, and the transported silica and water could react with the mantle to form orthopyroxene and hornblende. These reactions make the mantle considerably lower in density and so blobs or diapirs of mantle enriched in orthopyroxene and hydrous minerals ascend and eventually melt due to decompression and the depressed solidus temperature (Figure 5.5). Currently many workers favour the idea of hornblende-bearing mantle above the slab being transported downwards by the drag of the slab. At pressures equivalent to about 100 km depth hornblende breaks down and triggers melting. Another possibility is that the

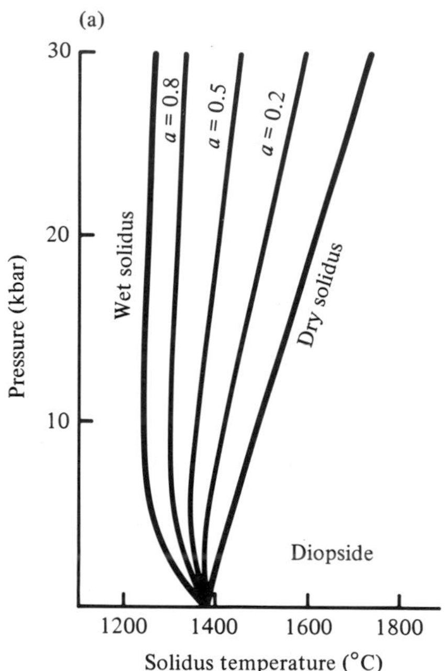

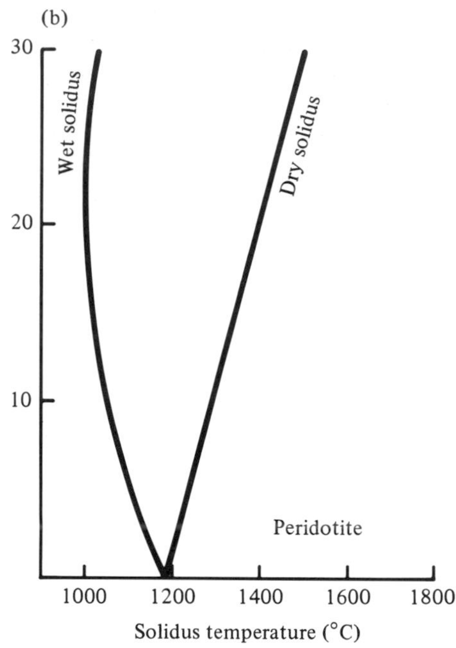

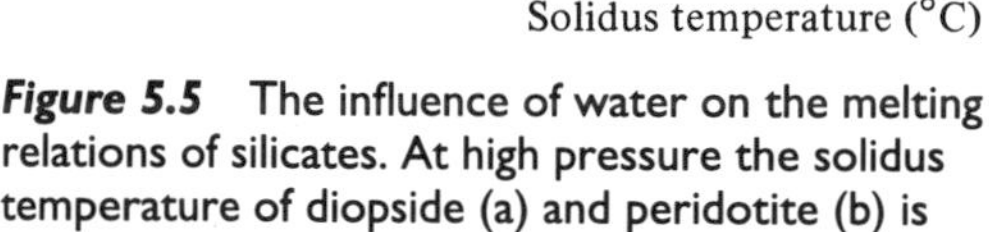

Figure 5.5 The influence of water on the melting relations of silicates. At high pressure the solidus temperature of diopside (a) and peridotite (b) is depressed by a few hundred degrees in the presence of water. For diopside approximate positions of the solidus for different activities (*a*) of water are shown. □

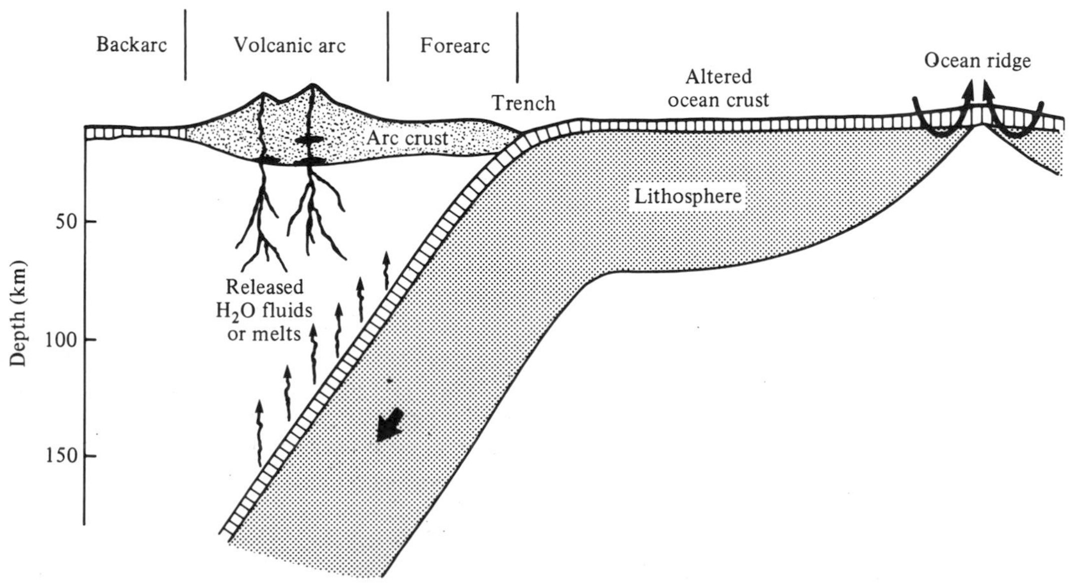

Figure 5.6 Model of a subduction zone and magma genesis. □

subducted oceanic crust itself melts as water is released, but recent thermal models suggest that temperatures are not high enough at the mantle-slab interface unless the lithosphere is young and hot. Experiments under the appropriate conditions indicate that rather silica-rich melts (**andesites** and **dacites**) form. It is most unlikely that such melts get to the surface. A more probable circumstance is that these hydrous silicic melts react with the overlying mantle, which then melts during ascent. In oceanic regions, where oceanic lithosphere is subducted beneath oceanic lithosphere, basalt is the predominant volcanic rock. In these areas the best available evidence is that basaltic magmas are generated from the mantle **peridotite** wedge, which has been variably contaminated with hydrous melts and/or hydrous fluids from the slab.

Other melting mechanisms

Decompression melting and freezing-point depression by volatiles appear to be the major mechanisms of melting in the mantle. Other volatiles such as carbon dioxide and halogens can also depress the solidus temperature of mantle rocks and a number of explanations of intra-plate volcanism have invoked the influx of H_2O or CO_2-rich fluids into the upper mantle from some much deeper source.

Heating is of course a simple way to melt rocks and there may be circumstances in the mantle where this is important. However, as convection is thought to be the major heat transfer mechanism in the mantle it is hard to see how heating could be readily distinguished from decompression melting. Locally heating may be important if magmas separate from their source and are then emplaced in a cold region where they lose their heat and melt surrounding rock. This can be seen as a second-order effect, but is important in the generation of magmas in continental crust and lithosphere and is discussed in a later section.

Frictional heating along subduction zones has been proposed, but detailed calculations do not indicate that large amounts of melt can be produced. Another possibility is the burial of rocks deep in the mantle which are rich in radioactive elements. Sediments are one possible example of suitable rocks, but no evidence has yet emerged to suggest that this is of more than local significance.

Melt segregation and ascent

A major challenge is to understand how partial melts separate from their mantle source. Separation is clearly accomplished very efficiently. The production of large volumes of crystal-poor basaltic magma is the most compelling observation. Normal upper mantle is thought to consist of peridotite, a mixture of olivine (60–70 per cent), orthopyroxene (10–15 per cent), clinopyroxene (10–15 per cent) and an aluminous phase (garnet, spinel or plagioclase depending on pressure). When basalt is formed by melting of peridotite, clinopyroxene and the aluminous phase are selectively consumed into the melt leaving a residue of olivine, orthopyroxene and some clinopyroxene. Many samples of the rock-type **harzburgite**, which consists of only olivine and orthopyroxene, have been found in places where there is good reason to suppose that they represent the uppermost parts of the mantle. For example, harzburgite is often the main rock type in the mantle part of ophiolites and is also a common rock-type dredged from fracture zones in the oceans. Harzburgites are widely interpreted as mantle from which substantial amounts of basaltic partial melt has been extracted. The low to negligible contents of clinopyroxene and an aluminous phase in many of these rocks strongly suggest that the melt extraction process can be extremely efficient.

The cause of **melt separation** is also well understood. The basaltic partial melts of the upper mantle have a lower density than the residual crystalline matrix. There is thus a buoyancy force which acts on the fluid and enables segregation to occur. Unfortunately, the actual physics of partially molten systems and melt segregation has turned out to be complex and progress in understanding has only been made recently.

The geometry of partially molten rock

The geometrical structure of partially molten materials is crucial to understanding their behaviour. When melting initiates in a mantle containing four principal minerals it must do so at grain corners where all four minerals touch. However, this configuration does not minimise the **energy associated with surfaces** between the melt and crystals. The equilibrium configuration which has the minimum possible energy is determined by the grain boundary energies both between the grains and between the melt and grains. Figure 5.7 shows the definition of the **dihedral angle**, θ, between two solid grains and melt. The value of θ is critical in controlling the distribution of melt in a partially molten rock. If θ is less than 60° then melt at the corners of grains will become interconnected with

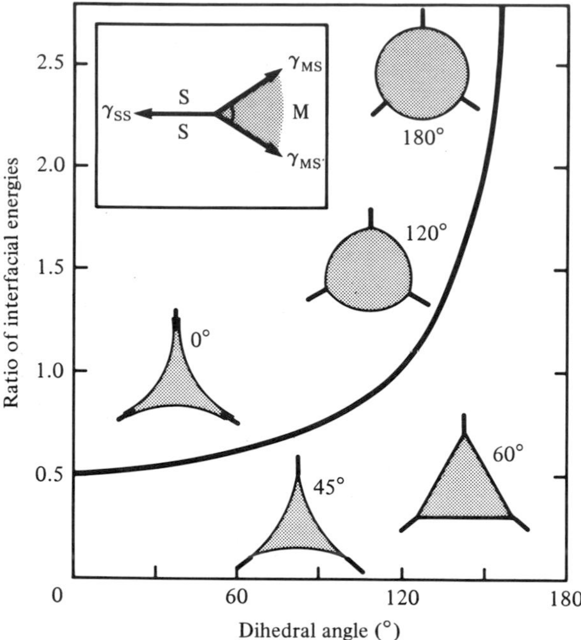

Figure 5.7 **Relationship of the dihedral angle, θ, to the ratio of interfacial energies between melt and crystalline solid. The shapes of melt at the corners of solid grains are shown for different dihedral angles. The dihedral angle can be obtained by resolution of the tensional forces relating to the difference in energies between melt–solid and solid–solid boundaries. The angle is defined: $\cos {}^1/_2\theta = \gamma_{SS}/2\gamma_{MS}$, where γ_{SS} is the surface energy between two solid grains and γ_{MS} is the surface energy between grains and the melt.** □

each other along the edges of grains. However low the fraction of melt there will be a continuous interconnection (Figure 5.8). Experiments have found that in partially molten peridotite θ is less than 50°. This is an extremely important result, because it means that melt can flow through these interconnected pores and escape, even if the proportion of partial melt is very small.

If θ is greater than 60° there is a minimum porosity (fraction of partial melt) required to form interconnected networks. There is some experimental evidence indicating that in granites θ can be considerably greater than 60° and so melt in these circumstances cannot easily escape unless there is a substantial amount of partial melting.

Porous flow

The flow of a fluid through an interconnected porous material is governed by **D'Arcy's Law**:

$$V = \frac{k\phi}{\mu\phi}\frac{dP}{dz}, \quad (1)$$

where $k\phi$ is the permeability, μ is the viscosity of the fluid, ϕ is the porosity (fraction of melt) and dP/dz is the pressure gradient driving flow. The permeability is determined by the porosity and grain size. The ability of a partial melt to separate from its source rock is thus strongly dependent on the viscosity of the melt, the amount of melt formed as measured by porosity and by the grain size.

A brief digression is now ncccssary on the **viscosity** of the basaltic melts. This property is of fundamental importance to understanding the flow of magma. The viscosity of basaltic liquid is strongly dependent on temperature (Figure 5.9). Values of about 50 Pa s characterise basalt lavas at the Earth's surface, but in the hotter interior, where partial melting takes place and segregation occurs, values between 1 and 10 Pa s are more typical. For comparison water has a viscosity of 0.001 Pa s and engine oil has a viscosity of about 1 Pa s. Basalt is therefore a rather viscous sticky liquid in comparison to our every day experience, but is actually very fluid for a geological material. Basalt lava can flow at speeds of a few metres per second down the slopes of a volcano, and even in the tiny channelways of partially molten mantle basalt liquid can flow at velocities which are geologically fast. For example calculations indicate that melt flows in partially mol-

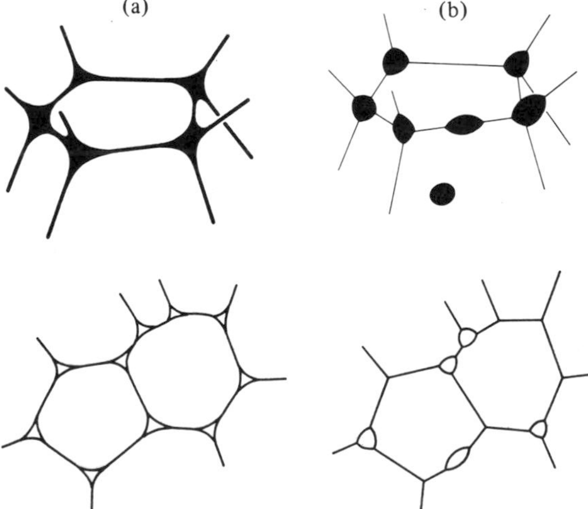

Figure 5.8 **The three-dimensional distribution of melt around grain boundaries is illustrated for (a) a dihedral angle less than 60° and (b) a dihedral angle of greater than 60°. The lower diagrams show the distribution of melt in a two-dimensional section. Note that for small dihedral angles there is a continuous interconnected network of melt.** □

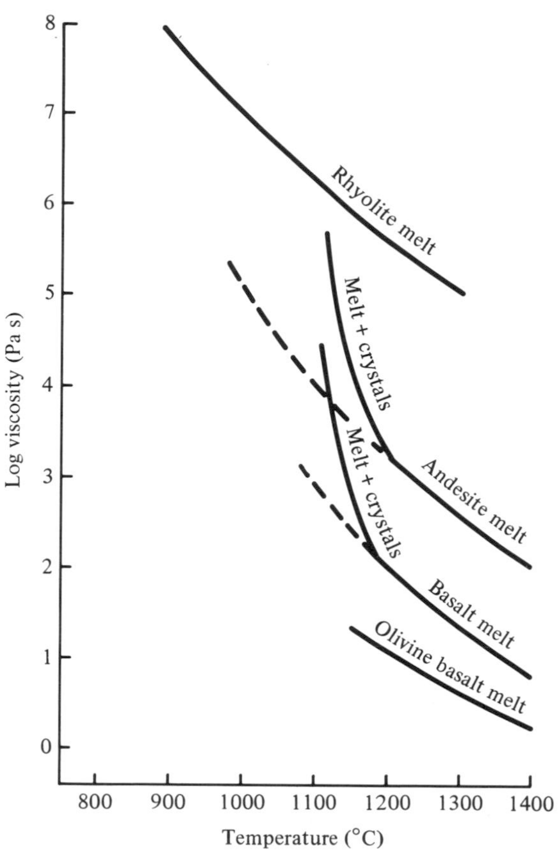

Figure 5.9 The variation of viscosity with temperature for different kinds of magma. The effect of crystallisation is to greatly increase the rate of viscosity change with decreasing temperature.

ten mantle at velocities of several metres per year, which is much greater than the velocities of plates.

The escape of basalt melt requires a driving force to cause the pressure gradient in D'Arcy's Law (equation 1). The main force is thought to come from the difference in density between the melt and the crystalline matrix in partially molten mantle. The melt has a significantly lower density by about 5 to 10 per cent and so gravity will act to force the melt upwards and the crystalline matrix downwards. If the crystalline matrix of the mantle were a rigid solid then no flow could occur. However, when any crystalline material is near to its melting temperature it can behave as a very viscous fluid. A good example of this behaviour is seen in ice which, although it is a crystalline solid, also flows readily in glaciers. The hot mantle also flows and the interior of the mantle could not convect and plate movements could not occur if this were not so.

Progress in understanding melt separation has recently been made by considering a simple situation in which a layer of partially molten rock with a constant initial fraction of melt (the porosity) is considered (Figure 5.10). The crystalline matrix deforms and flows downwards and the melt moves upwards until the melt forms a separate layer overlying an entirely crystalline layer. This process is called **compaction** and is analogous to the separation of oil and vinegar into two layers when they are mixed together in salad dressing. However, the crystalline component in the mantle has a very high viscosity and so in the geological system the separation takes place much more slowly. Estimates of the viscosity of the crystalline matrix of the mantle at temperatures appropriate for melting indicate a value of about 10^{18} Pa s.

In analysing any dynamic flow it is useful to define scales of length and time over which the process is likely to be important. Compaction provides a good example of this general approach. In Figure 5.10 the variation of matrix velocity and melt velocity with depth in the layer of partially molten material is shown. The melt velocity increases with height and is zero at the base of the layer. At some height the velocity of the melt reaches a constant value, W_o, where the pressure of the upward flow balances the weight of overlying matrix. In this system the compaction rate is a maximum at the base of the layer. Theoretical studies suggest that a useful length-scale to take is the distance over which the rate of compaction decreases by a factor of e (2.718) and is defined as:

$$\delta_c = \left(\frac{\eta}{\mu} k\phi \right)^{1/2} \qquad (2)$$

where η is the viscosity of the crystalline matrix. This is an extremely useful parameter. If the actual thickness of the layer is small compared to this **compaction length** then the effective viscosity of the matrix is large and compaction proceeds very slowly. If the thickness of the layer is much greater than the compaction length then the compacting region will be small and the escape of melt at higher levels in the layer will depend on the melt viscosity and permeability.

For the mantle, if $\mu = 10$ Pa s, $\eta = 10^{18}$ Pa s and $k\phi$ (permeability) varies from 10^{-14} to 10^{-12} for melt fractions of 1 to 10 per cent, the compaction length varies from 100 to 1000 m. The important point is that partially molten regions of the mantle are likely to be tens of kilometres in height (Figures 5.1 and 5.4), and therefore very much larger than the compaction length. The ability of the melt to escape will therefore be strongly dependent on the viscosity and the fraction of melt.

Progressive compaction of a partially molten rock layer

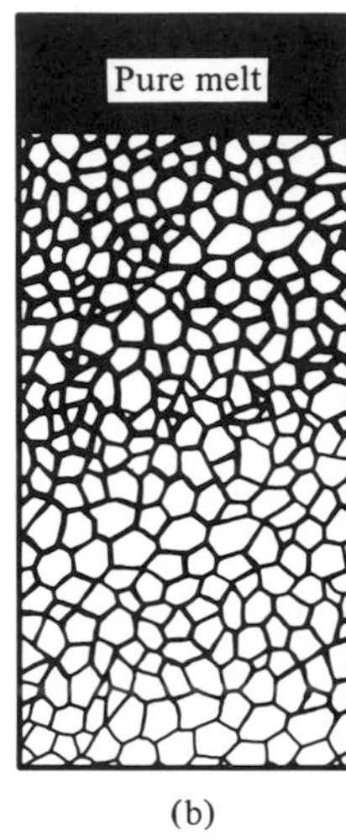

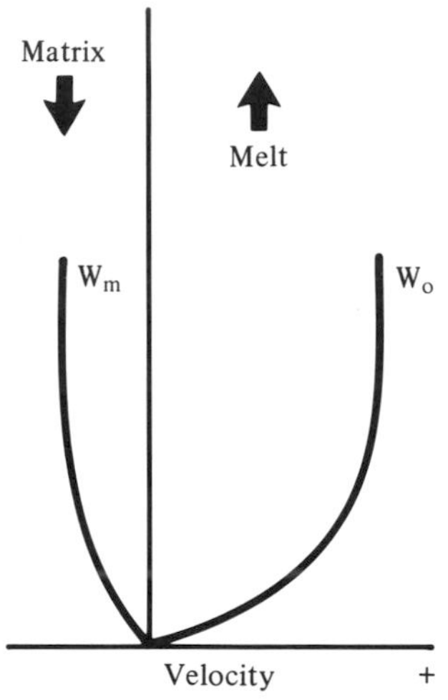

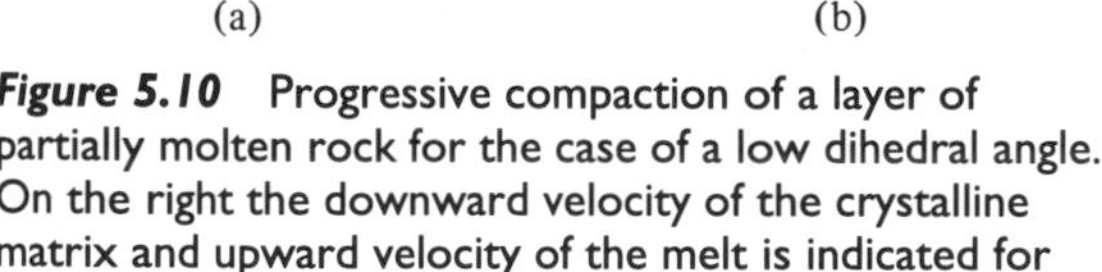

Figure 5.10 Progressive compaction of a layer of partially molten rock for the case of a low dihedral angle. On the right the downward velocity of the crystalline matrix and upward velocity of the melt is indicated for compaction at stage (c). W_o is the asymptotic value of the melt velocity at a height where the flow velocity is sufficient to balance the weight of overlying solid matrix. The matrix also reaches a steady asymptotic value, W_m. □

Whether this physical process is important also depends on the time available. The compaction timescale required to expel 1/e of the melt from a compacting layer of thickness h, τ, is defined as:

$$\tau = \left(\frac{h}{\delta_c} + \frac{\delta_c}{h} \right) \frac{\delta_c}{W_o(1 - \phi)} . \tag{3}$$

For layers of partially molten rock with thicknesses of the same order as the compaction length the timescales are only hundreds or thousands of years, a time which is very short compared with the millions of years characteristic of large-scale geological processes.

The analysis of the compaction of a partially molten layer is of course a much simpler situation than the environments of melt segregation within the mantle. The mantle itself is flowing so there will be complicated interactions between compaction-driven flow of the melt and the overall mantle flow. Complete physical descriptions of the melt flow are not yet available, but two further theoretical developments give insights into how melt will segregate.

Investigations of the equations which describe the flow in compacting regions have found that such systems should progressively break up into waves of high and low porosity. The regions of high porosity can develop into layers of melt which ascend as solitary waves. These waves of magma have been called **magmons** and could be the principal way in which magma is separated from the mantle and then feeds into magma chambers or develops into vertical flow in dykes. Calculations have also been made on how melt would flow in a region of constant porosity with the geometry of the melting zone beneath a ridge. The pressure gradients induced by flow in the mantle drive the melt towards the ridge axis (Figure 5.1), although the crystalline matrix itself moves symmetrically away from the axis.

Ascent of basalt magma

Magma must traverse through the lithosphere and crust to reach the surface or form crustal magma chambers and intrusions. In this region the rocks are normally well below their solidus temperature. The main mechanism of magma ascent is thought to be propagation of magma-filled cracks, which of course are better known to geologists as dykes.

The heights of volcanoes

The cause of magma ascent can be understood in terms of principles of hydrostatics. In Figure 5.11 a reservoir of magma is envisaged at the interface between a region of partial melting in the mantle and the lithosphere. A **lithostatic pressure** acts on the reservoir. A column of magma connects the reservoir with the surface. Flow will occur until the pressure at the base of the column (the **hydrostatic pressure**) equals the lithostatic pressure. If the density of the magma in the column is less than the mean density of overlying rock the height of the column has to be above the Earth's surface. If the column is less than this height then there is a pressure imbalance and magma should flow to the surface. The flow will cease when a volcanic edifice of sufficient height has been constructed. The pressure gradient causing flow can then be defined as:

$$\frac{dP}{dz} = (\bar{\rho}_L - \bar{\rho}_M)g, \quad (4)$$

where $\bar{\rho}_L$ is the mean density of the overlying lithosphere and $\bar{\rho}_M$ is the mean density of the magma column. The height of the volcanic edifice, h, at hydrostatic equilibrium is given as:

$$h = H\left(\frac{\bar{\rho}_L - \bar{\rho}_M}{\bar{\rho}_M}\right), \quad (5)$$

where H is the depth of the magma reservoir. If $\bar{\rho}_L$ = 3200 kg m^{-3}, $\bar{\rho}_M$ = 2900 kg m^{-3}, and H = 60 km, then the volcano height will be 6 km above the surface. The Hawaiian volcanoes are over 6 km above the sea-floor and their height can be considered consistent with these simple hydrostatic considerations.

Ascent in dykes

Figure 5.12 shows possible ways in which magma-filled cracks may ascend through the crust of lithosphere from a magma reservoir. When the magma pressure exceeds the strength of surrounding rocks in tensile failure, a fracture filled with magma can propagate towards the surface. The velocity of propagation of a crack in rock is typically a few kilometres per second. However, basalt magma is much too viscous to flow at such speeds and so the velocity is governed by the rate at which magma can flow into the crack tip. The velocity of ascent can be described adequately by the same relationships that describe the flow of fluid between two parallel plates:

$$V = \frac{d^2}{64\mu}(\rho_L - \rho_M)g, \quad (6)$$

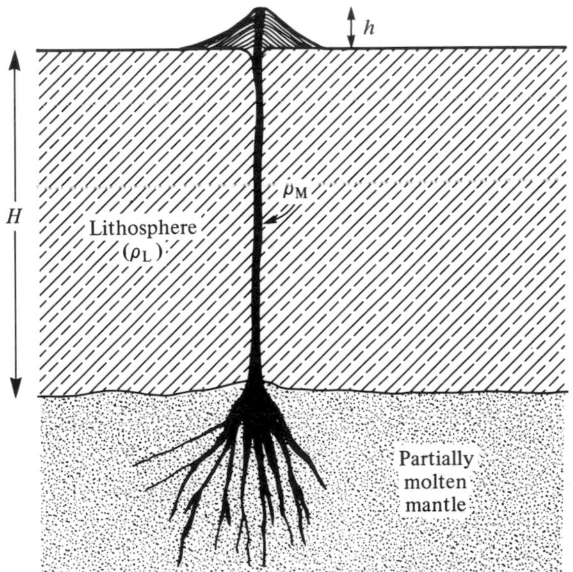

Figure 5.11 Hydrostatic model for the height of volcanoes and transport of magma through the lithosphere. Melt accumulates at the base of the lithosphere and ascends to an excess height, h, so that there is a pressure balance between the column of magma and the lithosphere. □

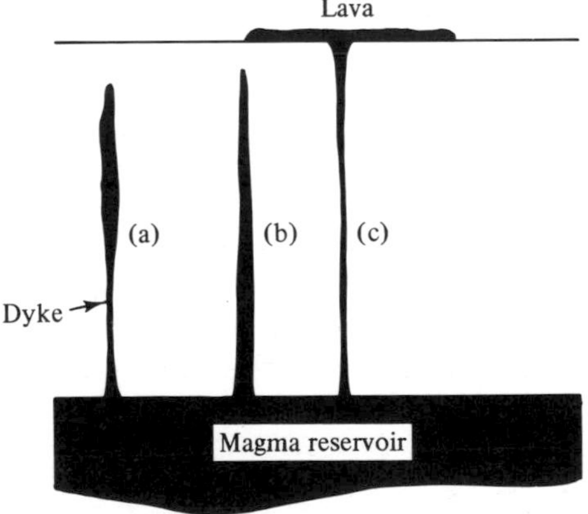

Figure 5.12 Dyke propagation through the lithosphere or crust from a magma reservoir. In (a) a volume of magma is intruded into the crust as an elliptical crack which leaves behind a trail of congealed magma as a dyke. In (b) the pressure in the chamber is sufficient to keep the crack open at the bottom end. In (c) the magma reaches the surface to form a lava, leaving a dyke behind. Note that the width of the moving magma-filled crack can be considerably greater than the dyke left behind. □

where d is the width of the fracture, μ is the viscosity, and the pressure gradient, dP/dz, is considered to be due to buoyancy. The width of the crack is governed by the elastic properties of surrounding rock and the length of the crack. If values of $\mu = 30$ Pa s, $d = 1$ to 10 m and $(\bar{\rho}_L - \bar{\rho}_M) = 300$ kg m^{-3} are substituted into equation (6), then ascent velocities of 0.1 to a few metres per second are estimated. In the rift zones of Kilauea volcano, Hawaii, and of Krafla volcano, Iceland, earthquake studies have recorded the propagation of dykes at velocities of 0.5 to 1.0 m s^{-1}. In the 1976 eruption of Tolbachik in the Kamchakta peninsula, USSR, an earthquake swarm was observed to initiate at 30 km and ascend to the surface at 0.1 m s^{-1}. A basaltic eruption began when the swarm reached the surface, and this is perhaps the best example of magma ascending through the crust in a dyke. As magma flows up through the fracture it will freeze against the walls. During flow the fracture is probably wider than the eventual dyke (Figure 12a). As the pressure in the source reservoir declines and the congealed walls of frozen magma thicken the fracture will close, eventually causing the flow to cease. Pressure in the source reservoir and tectonic stresses must now build up again before another eruption can occur.

Magma chambers

The concept of the **magma chamber** is fundamental in most interpretations of magmatism and volcanic activity. Studies of igneous rocks have demonstrated conclusively that most magmas have been chemically differentiated either on their way to the surface or at their final level of intrusion. Studies of phase equilibria elegantly explain how magmas with wide ranges of composition and temperature can be produced from a single starting composition by **fractional crystallisation**. When this process is combined with other processes such as magma mixing and assimilation the potential diversity of magmas becomes even greater. These processes are thought to occur principally in large reservoirs of magma situated beneath volcanoes. The fundamental cause of differentiation is heat loss from such chambers. Information on their size, shape and physical condition comes from geophysical observations on active volcanoes and from geological observations of igneous intrusions.

Geophysical observations

Kilauea volcano on Hawaii is the world's most active volcano and geophysical studies have identified the magma reservoir. When magma moves from the mantle towards the surface beneath Kilauea the rocks surrounding the conduit system are repeatedly fractured causing earthquakes. The depth and position of these events can be precisely located outlining the shape of the plumbing system. Data from the 1969–74 period (Figure 5.13) define a central cylindrical conduit about 1 km in diameter feeding a central reservoir with a capacity of 10 km^3 at between 2 and 6 km beneath the summit crater. A subsidiary conduit feeds magma directly to the flanks of the volcano.

Before an eruption, the summit region of Kilauea swells (tumescence), producing a circular uplift pattern (Figure 5.14). This is caused by magma being fed from the conduit system into the reservoir beneath the summit. The magma pressure increases and deforms and uplifts the overlying roof by as much as a metre. Eventually the strength of the surrounding rock is exceeded and dykes are propagated to the summit or are injected laterally into the rift zone. Most magma in fact is injected laterally and can flow to over 40 km in subsurface dykes to erupt as flank lavas. As magma is withdrawn from the reservoir the pressure drops and the summit area subsides. In some cases the roof rocks collapse enlarging the summit caldera.

A similar pattern of behaviour was observed in the basaltic volcano of Krafla in northern Iceland. The volcano consists of a central 6-km-diameter caldera region and a rift zone extending both north and south of the caldera. Between 1976 and 1984 there were over twenty cycles of inflation and rapid deflation in which magma was either erupted or injected as a dyke into the rift zone. Inflation periods typically lasted a few months in which time the centre of the caldera uplifted at 6 mm per day. Calculations indicate that basalt magma was being fed from depth at a rate of 5 m^3 s^{-1} in a flat thin magma chamber (a sill) at a depth between 3 and 7 km. Eruption of lava was accompanied by large deflations of the summit of tens of centimetres to a couple of metres in a few hours. Evidence from Kilauea and Krafla show that the magma chambers beneath active volcanoes are open-systems in which influx of new magma from depth triggers eruption.

Many other magma reservoirs have been identified by seismic methods. Magma will not transmit shear (S) waves, so that ray paths that cross a chamber will lose the S-wave component. In some cases there is evidence of both deep and shallow reservoirs. For example in the Kamchatka volcanic arc Soviet seismologists have identified seismically anomalous regions at mid-crustal (10 to 15 km) levels and deeper in the mantle. Often these anomalies are consistent with large regions of partially molten rock rather than discrete large magma

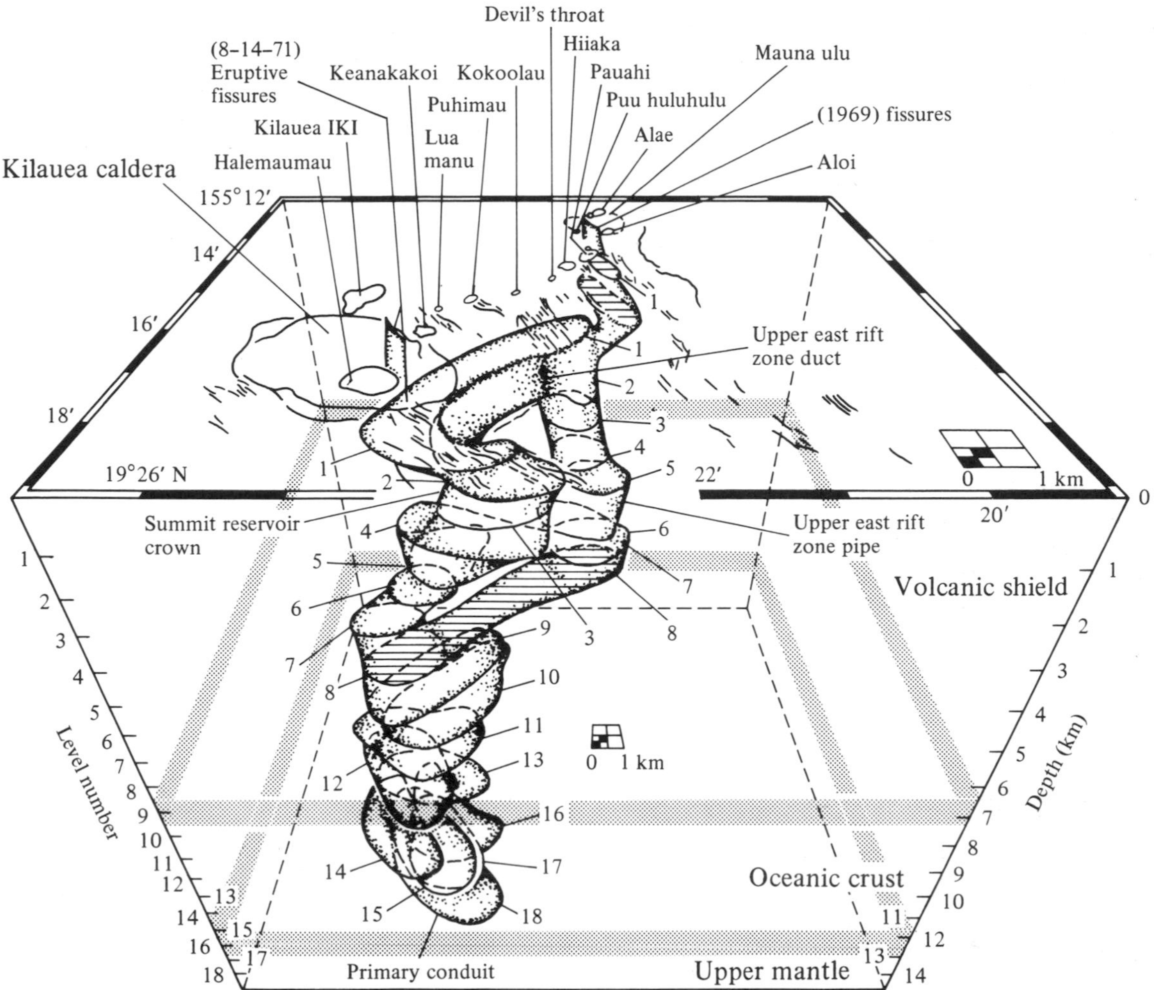

Figure 5.13 Three-dimensional model of the conduit and magma reservoir system of Kilauea volcano, Hawaii. The model is based on the location of small earthquakes around the conduit system and surface deformation of the volcano. This system has fed magma from the mantle to the surface at an almost constant rate of 5 $m^3 s^{-1}$ during this century. □

chambers. The size and shape of smaller discrete magma chambers within such anomalous regions is not certain. The data indicate that magma chambers can form at a wide range of depths in the upper mantle as well as in the crust.

Geological evidence

Intrusions provide compelling evidence for the existence of magma chambers and for very efficient chemical differentiation processes within them. A common form of basic intrusion has a sill-like or laccolithic shape. Other geometries include cylindrical intrusions and dyke-like bodies. Sizes range from a few hundred metres in width, diameter or length to huge intrusions such as the Bushveld in South Africa which has an area of 300 000 km^2 and a thickness of several kilometres. The final dimensions of such intrusions do not necessarily mean that a body of magma of the same size ever existed. Most magma chambers are open-systems connected to volcanoes and likely go through a complicated history of crystallisation, magma influx and magma loss as at Kilauea. The dimensions of the magma body present at any time could be much less than the final dimensions of the intrusion.

Geological and geophysical observations have led to the concept of the level of neutral buoyancy in controlling the depth of magma chambers. Magmas

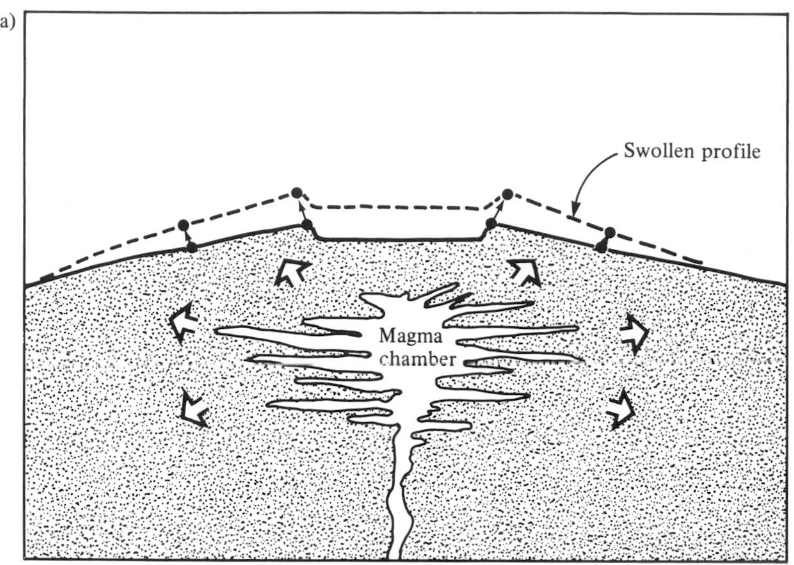

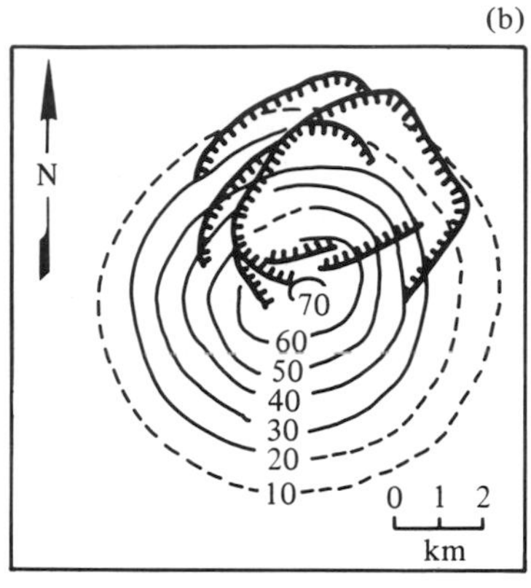

Figure 5.14 Surface deformation of Kilauea volcano during flow of magma into the reservoir near the summit. The ground surface is lifted up by as much as a metre (a). A concentric pattern of uplift of the ground around the caldera occurs before a major eruption (b) with the contours in centimetres. Rapid subsidence of the ground occurs during an eruption as pressure decreases in the chamber.

ascend to the levels in the crust where they have the same density as surrounding rock and spread out to form magma chambers. There are consequently favoured sites of magma chamber formation such as the Moho and at shallow depths where rocks decrease their density due to porosity increased fractures.

Differentiation mechanisms

Igneous rocks show overwhelming evidence that magmas have been chemically differentiated. In most basic intrusions the rocks are composed of concentrates of crystals precipitated from magma. For example olivine is commonly the first major crystal to form in basalt. In many intrusions rocks composed entirely of olivine (dunite) are found. Lavas in basaltic volcanoes such as Kilauea in Hawaii show ranges of composition which can only be explained by efficient separation of olivine crystals before eruption. The layers of dunite found in intrusions cannot represent a magma, because impossibly high temperatures are required to melt olivine (1800 to 1900 °C). The dunite layers are interpreted, therefore, as layers formed by the concentration of olivine from basaltic magma. Most intrusive basic rocks, although they contain several minerals, cannot be magma compositions, but represent the crystals extracted from magma. Likewise many volcanic rocks require very efficient processes of separation of crystals from liquid in magma chambers. Three major mechanisms have been identified which can cause efficient separation of liquid and crystals in magma chambers: mechanical sedimentation, compositional convection and compaction. There is considerable uncertainty in the relative importance of these processes.

(i) Crystal sedimentation

There can be large density differences between crystals suspended in the interior of magma and gravitational **crystal settling** is a widely invoked mechanism of differentiation. In a static liquid crystals settle or float according to **Stokes's Law**:

$$V = \frac{\Delta\rho g d^2}{18\mu}, \qquad (7)$$

where $\Delta\rho$ is the density difference between crystal and liquid, d is the grain diameter, g is acceleration due to gravity, and μ is the liquid viscosity. For olivine settling in basaltic magma values of $d = 0.5$ cm, $\Delta\rho = 500$ kg m^{-3} and $\mu = 30$ Pa s can be chosen as typical. These values give a velocity of 0.5 metres per hour. Settling velocities of crystals in basaltic magmas are thus fast for a geological process and would produce rapid separation. Concentration of olivine at the bottom of lava flows has been observed and is readily explained by settling. Some crystals can be lighter than magma. For example plagioclase feldspar is marginally lighter than many basaltic magmas and should float. In basalt, for example, experimental studies indicate that olivine, plagioclase and pyroxenes co-precipitate at low pressure in the approximate proportions 0.1:0.6:0.3. The proportions of phenocrysts in many porphyritic basalt lavas, however, are strongly enriched in plagioclase which can be explained by the settling out of the dense ferromagnesian minerals and retention of buoyant feldspar.

Basic intrusions often show strong layering with concentrations of particular minerals in individual layers. These features have been interpreted as due to differential crystal settling. However, layered intrusions also pose problems for simple crystal settling. In many cases the rocks formed at the bottom of basaltic magma chambers have approximately the correct proportions of plagioclase and ferromagnesian minerals for co-precipitation. If segregation according to density were the major factor then one might expect the rocks formed on the floor to be composed of olivine and pyroxene and those at the roof to be composed of plagioclase. One possible solution to this difficulty is that crystals commonly aggregate together so that most feldspars are in fact dense due to the attachment of denser minerals.

Magma chambers, however, do not consist of bodies of static liquid. Theoretical and experimental studies indicate that magmas should convect vigorously. Simple thermal convection should oppose settling by keeping crystals in suspension. Experimental studies so far indicate that convection delays but does not stop crystal settling. In dilute turbulent suspensions crystal concentration decreases exponentially with time due to settling at the floor. It is also possible that in systems consisting of solid crystals dispersed in a liquid that new and different forms of convection may arise where sedimentation is enhanced. Experiments have been carried out where light and heavy particles are mixed together into a liquid of intermediate density. The particles are observed to organise

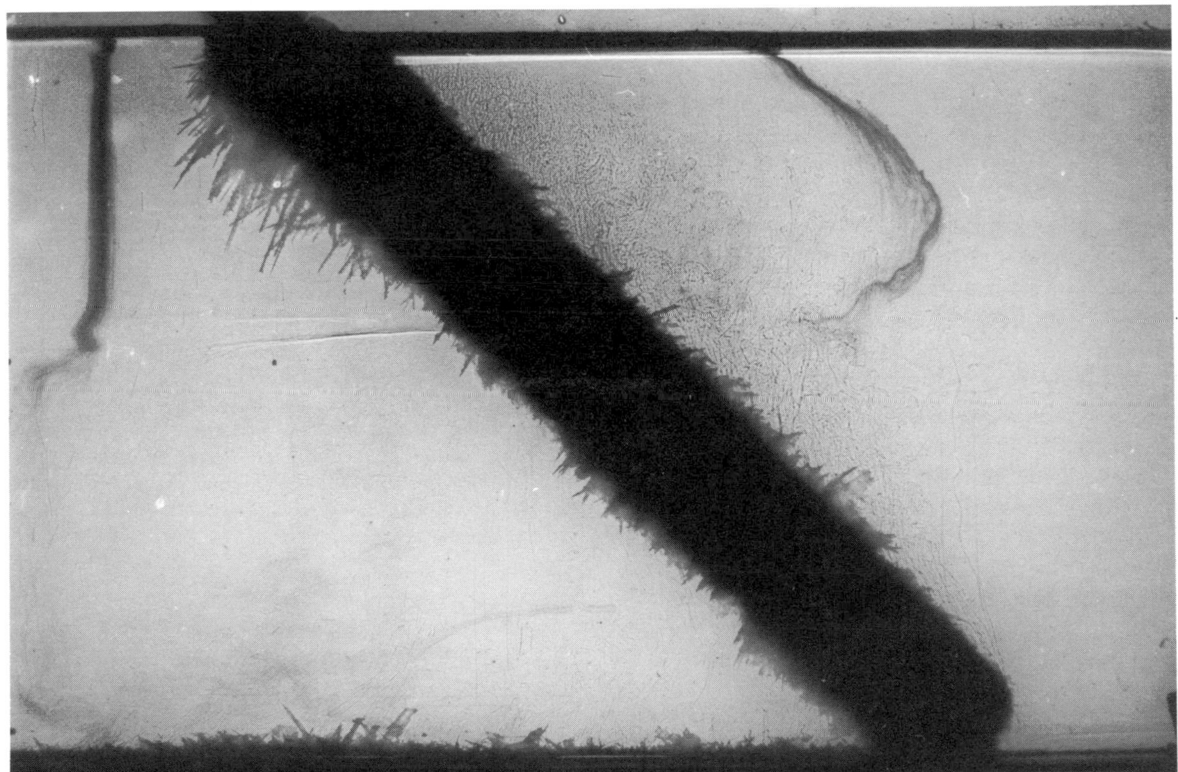

Figure 5.15 A solution of Na_2CO_3 has been cooled from an inclined plane through which alcohol at −21 °C is circulated. The photographs show compositional convection above the plate as thin vertical plumes. Below the plate convection of light fluid as a thin flow at the plate has created a compositional zoned upper region separated by a sharp interface from homogeneous solution below. With time the compositional convection causes both sides of the plane to become compositionally and thermally stratified. This mechanism provides an explanation of the compositional zoning that occurs in many magma chambers. □

rapidly into alternating columns of rising fluid containing the light particles and dense fluid containing the heavy particles. The two particle types are separated at rates which are much faster than predicted by Stokes's Law.

(ii) Compositional convection

Experimental studies on the crystallisation of **saturated solutions** have recently shown another mechanism of separating crystals from liquid. Figure 5.15 shows an experiment where a saturated solution of Na_2CO_3 is cooled along an inclined plane, within which alcohol at –21 °C is circulated. The inclined plane divides the experimental tank into two parts so that two experiments can be carried out. In one-component fluids cooling increases density so that cold fluid adjacent to the slope would be expected to form a descending flow. However, nucleation of crystals against the plane produces thin plumes of buoyant fluid and a circulation pattern with the opposite sense of motion is actually observed (Figure 5.15).

The motion shown in the experiment is known as **compositional convection**. The fundamental cause is crystal growth which locally depletes the solution in Na_2CO_3. The decrease in density due to removal of Na_2CO_3 component from the solution is much greater than the increase in density due to cooling, so the solution is buoyant and convects away from the crystals. Figure 5.16 compares the change in density with decreasing temperature in residual liquids for Na_2CO_3 solutions and basalt crystallising olivine. In both cases cooling and crystallisation produces residual liquids of *decreasing* density.

Compositional convection occurs on a very small scale. When a crystal grows from a melt those components not included in the crystal structure are concentrated in a thin (typically tens of micrometres) region adjacent to the crystal. This thin film usually has a different density to adjacent fluid and **buoyancy forces** cause the film to continue rising away from the growing crystal as a thin plume (Figure 5.15). The ascent of buoyant residual fluid eventually leads to the development of compositional and thermal layering throughout the tank with the coolest and most depleted fluid occurring at the top. The same phenomenon should occur in magma chambers in which crystals grow along the walls and floor.

Figure 5.16b shows schematically the variation of density in residual melts during the progressive fractional crystallisation of an ocean ridge basaltic magma. A primitive basalt, P, directly derived from the mantle crystallises olivine first in a shallow magma chamber

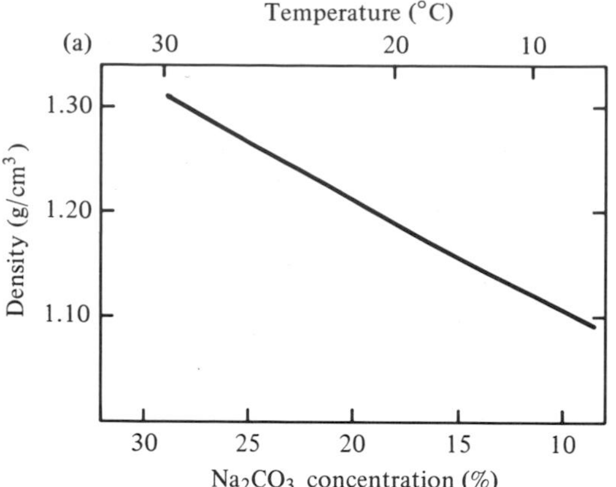

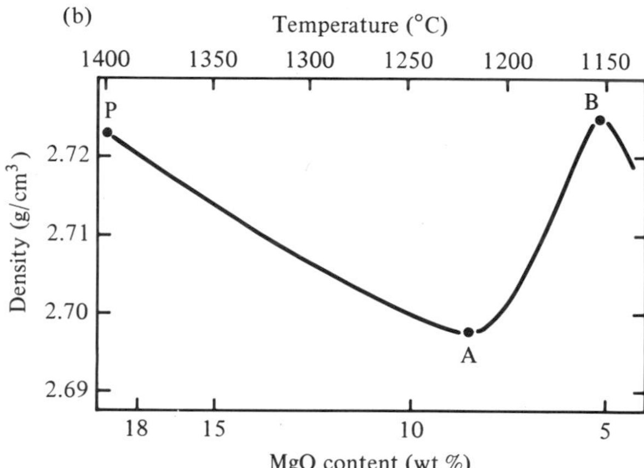

Figure 5.16 The variation of density with temperature in saturated solutions of Na_2CO_3 and in basaltic magma during differentiation. In the aqueous solution density decreases with temperature due to the crystallisation of Na_2CO_3. The same feature is seen in basalt magmas where cooling and crystallisation of olivine decreases the density and MgO content of residual liquids (P to A). When plagioclase begins to crystallise (A to B) the residual melt density can increase again. At B magnetite begins to crystallise and once again residual melt density decreases. □

and density decreases because olivine removes dense components from the melt. The density reaches a minimum, A, when plagioclase begins to crystallise with olivine and sometimes pyroxene. Plagioclase crystallisation removes light components from the magma. At the maximum, B, magnetite begins to crystallise and the density of residual liquids increases rapidly as the residual melts evolve to andesite and ultimately rhyolite. These relationships are analogous to the Na_2CO_3 solution in which the changes in composition that occur during crystallisation have a

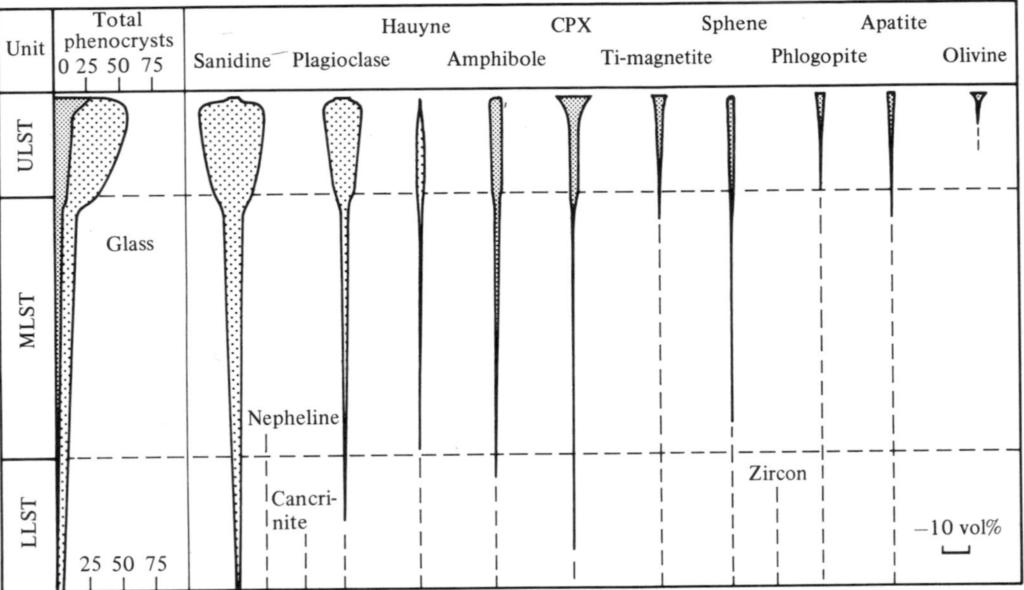

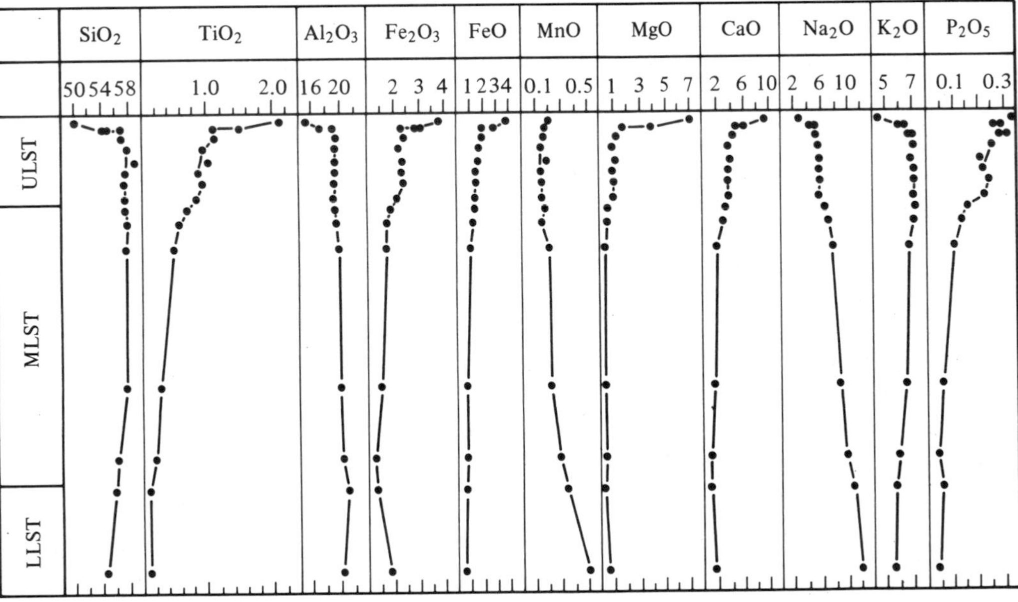

Figure 5.17 Variation of the mineralogy and chemical composition in the Laacher See pyroclastic deposits of the Eifel volcanic district. The eruption at 10 700 years BP discharged sequentially volcanic layers named LLST, MLST and ULST (Lower, Middle and Upper Laacher See Tephra), probably in only a few days. With time the magma that was erupted changed in composition, mineralogy and crystal content. The data show that a chemically and thermally zoned magma chamber had formed with cooler, crystal-poor, more differentiated magma at the top grading downwards to magma which was hotter, more crystal-rich and less differentiated. Note that the top of the magma chamber erupted first to form the lowest (LLST) layer in the deposit. □

much larger effect on density than thermal contraction. In many situations the residual melt density can decrease substantially despite the accompanying fall in temperature. Compositional effects usually dominate density variations and thus compositional convection may be an important process in magma chambers. The density variations during fractional crystallisation will vary considerably from one magma to another. Figure 5.16b shows the effects for crystallisation of mid-ocean ridge basalt at low pressure with little water present.

A common feature of differentiated magmas is the occurrence of compositional zoning in magma chambers. Figure 5.17 shows profiles of the mineralogy and chemistry of pumice fragments through the deposits of the Laacher See volcano in the Eifel District of Germany which were erupted about 10000 years ago. During a major explosive eruption the different levels of this chamber were systematically tapped and allowed this reconstruction to be made. The chamber contained **phonolitic** magmas formed by the extensive fractional crystallisation of basanite (a silica-undersaturated basic magma). The top of the chamber was occupied by water-rich, highly differentiated phonolite, with high concentrations of incompatible trace elements. With depth the magma became hotter, richer in ferromagnesian components, increasingly rich in crystals, less differentiated and consequently denser. Primitive **basanite** was present beneath this zoned chamber. This kind of zoning pattern has been found in many volcanic systems. Commonly the zoned magmas are crystal-poor and it has proved difficult to explain the zoning by crystal settling. The zoning also can develop remarkably quickly. For example in the volcano Hekla in Iceland zoning from basaltic andesite to dacite can be established in a few tens of years. A very efficient fractionation mechanism is indicated. The zoning patterns of magma chambers are reminiscent of the patterns observed in the experiment with aqueous solutions (see Figure 5.15). The idea of marginal crystallisation and upwelling of buoyant differentiated melt is an attractive explanation.

(iii) Compaction

The ideas on compaction discussed for separation of partial melt from the mantle are equally applicable to thick piles of hot igneous rock in intrusions. When crystals settle to the floor of a magma chamber or crystallise *in situ* some liquid is trapped between the crystals. In small rapidly cooled intrusions some of this liquid is frozen in pore spaces, but in large slowly cooled intrusions the residual melt is extracted with remarkable efficiency. The common occurrence of monomineralic rock layers such as dunite or anorthosite require a process that removes almost all the liquid. Compositional convection can provide some closure of pore space by circulating melt between the chamber and the pores. However, if the porosity is too small convection is not possible. The time periods over which significant compaction can occur appear to be in the range of hundreds to perhaps a few thousands of years. Large magma chamber can take thousands to tens of thousands of years to solidify. Thus this will be a very efficient process of completing the efficient separation of crystals and liquids (Figure 5.10).

(iv) Other differentiation mechanisms

The discussion so far has assumed that the fundamental cause of magma differentiation is the separation of melts from crystals, either during partial melting or crystallisation in magma chambers. Other processes have been invoked to cause magma differentiation.

Liquid immiscibility involves the spontaneous disassociation of a magma into two liquid phases, usually during cooling. Immiscibility has been observed in a number of silicate systems, notably those which are unusually rich in high field strength cations such as Fe^{2+} + and Ti^{4+} alkalis or dissolved CO_2. Immiscibility has been invoked in a number of cases, usually in magmas with exotic or unusual compositions. The process, for example, provides an explanation of the formation of carbonate melts (**carbonatites**) which can form by immiscibility from alkali-rich, silica-deficient silicate melts which contain abundant dissolved CO_2.

Vapour transfer involves the transport of chemical components in gas bubbles from one place to another. Elements such as the alkalis appear to be particularly susceptible to movement by vapour.

Diffusion is another process which can cause differentiation. Some components, such as H_2O and the alkalis, diffuse much more rapidly than the main components in magmas, such as silica and alumina. Diffusion on its own in a static magma is a very slow and inefficient process and is more likely to have an influence in combination with other processes, such as convection.

Open-system behaviour

The processes of differentiation that have been discussed thus far involve closed-system behaviour in which a batch of magma cools and crystallises in a chamber. However, there is overwhelming evidence that the chambers beneath volcanoes are open-systems in which new magma is repeatedly introduced from

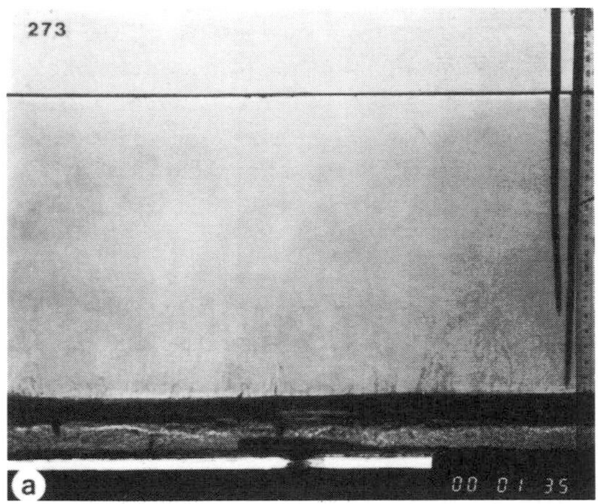

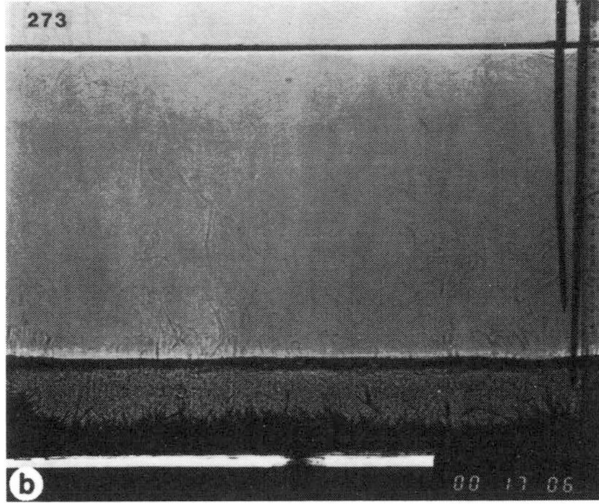

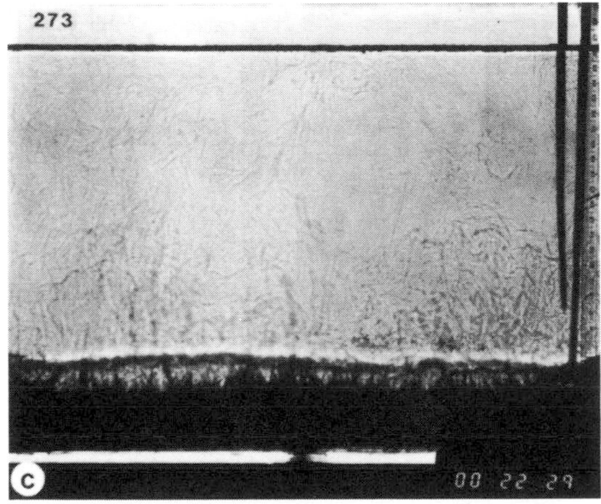

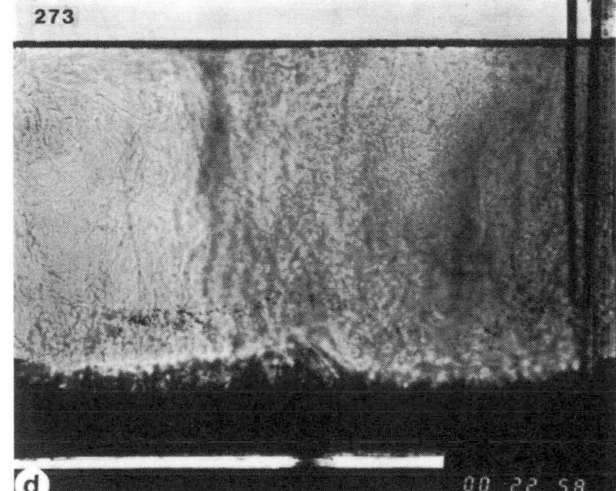

Figure 5.18 The photographs show various stages in an experiment in which a layer of KNO_3 solution at 65 °C is emplaced beneath a layer of $NaNO_3$ solution at 11 °C (a). The time is shown in hours, minutes and seconds in the lower right-hand corner. The dark rods on the right are thermistors. The lower layer cools and crystallises (b and c) and then overturns and mixes with overlying solution when the residual liquid has become less dense (d). □

depth and differentiated resident magma is discharged onto the surface. **Magma recharge** into a chamber is probably the major cause of volcanic eruptions. Influx of new magma increases the fluid pressure in the chamber and eventually the roof breaks and magma can erupt.

A common situation in magma chambers involves the emplacement of hot primitive magma from the mantle into a chamber containing more differentiated magma. As can be deduced by inspection of Figure 5.16b, the hot primitive magma can be either denser or lighter than the resident magma. Figure 5.18 shows the results of an experiment which replicates some of the essential physics of the input of a hot dense fluid. A saturated solution of hot KNO_3 at 65 °C has been introduced into the bottom of a tank of cold $NaNO_3$ solution at 11 °C. The KNO_3 is much denser despite its high temperature due to its composition. A sharp interface is formed between the layers and heat is transferred rapidly from the lower to upper layer causing vigorous convection in both layers. A very important feature of the experiment is that there is negligible transfer of composition (K and Na), because of the low diffusivity of the chemicals compared to heat. The lower layer cools and crystallises KNO_3 so that the residual solution in the layer decreases in density. Eventually the lower layer decreases to the same density as the upper layer and rapid mixing occurs (Figure 5.18).

The experiment illustrates some important effects

that can occur when primitive basalt replenishes a chamber of differentiated magma. The hot magma should pond at the base of the chamber and would cool and crystallise producing a layer of olivine crystals. If the residual liquid becomes lighter than the more differentiated magma above (see Figure 5.18) then mixing can occur and a new layer of crystals, such as gabbro, will form above the olivine layer. Further replenishment events would produce a cyclic layering of different rock-types, which is in fact commonly observed in intrusions. If a magma chamber exists then the primitive magma will not be able to reach the surface and only differentiated magma will be erupted.

Figure 5.19 shows a different situation in which the influx is of fluid with a lower density. A plume of buoyant fluid rises into the tank and mixes turbulently with its environment. Mixing between the different composition fluids causes crystals to form. This process can also be anticipated in a basaltic magma chamber, for inspection of Figure 5.16 indicates that it is sometimes possible for hot primitive magmas to be less dense than differentiated basaltic magmas.

These experiments illustrate the two cases of replenishment. However, the details of the behaviour depend on a great many factors such as the chamber geometry, the temperatures, densities and viscosities of the magmas and the replenishment rate. In the first case replenishment can be seen as a mechanism of producing a zoned or layered magma chamber. In both cases replenishment provides a mechanism for mixing magmas together. Magma mixing is also possible by associated processes. For example, eruption of magma along conduits or dykes can cause magmas from different levels in a chamber to be forcefully mixed together.

Another important open-system effect is the assimilation of surrounding wall-rocks into magmas. This

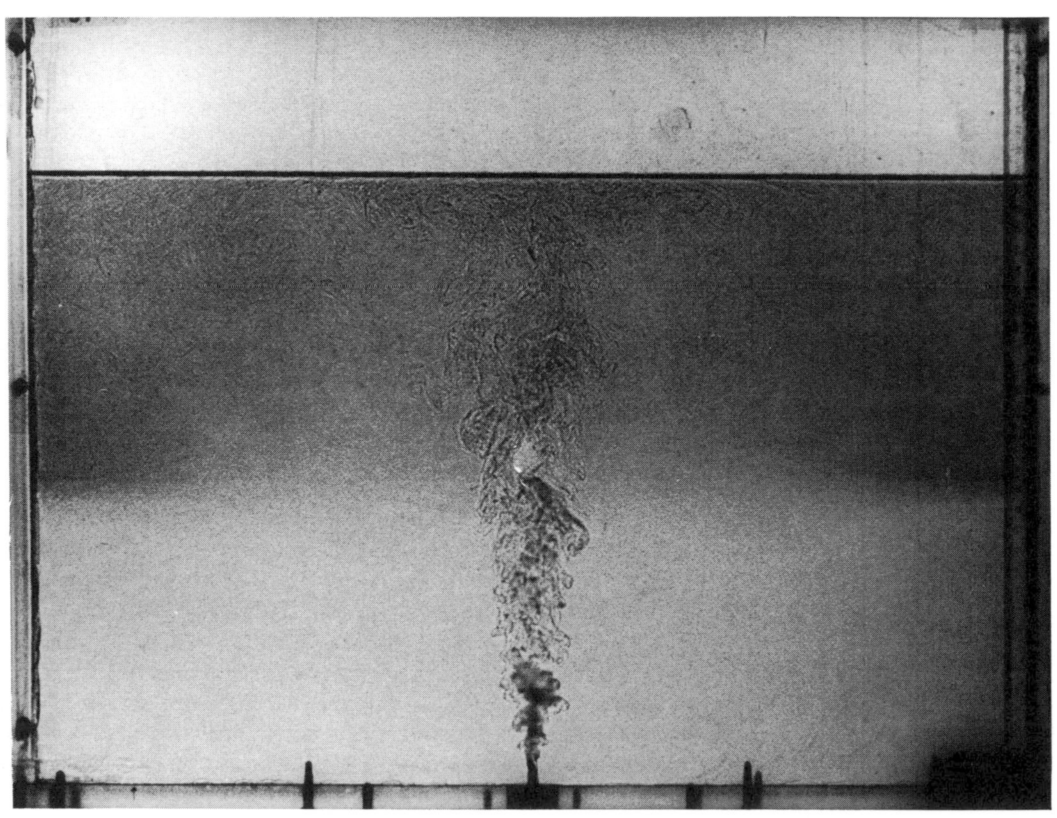

Figure 5.19 A flow of fresh water into a tank of salty water develops into a typical turbulent plume. The plume rises and mixes with its environment to form the dark layer of mixed fluid. The clear layer below is the original salty water. □

can occur by a variety of mechanisms both during magma ascent in dykes or in magma chambers. If the surrounding rocks have a low melting temperature then the wall-rocks can melt and be mixed into the magma. The roof of a chamber can also collapse causing blocks to fall into the magma, where they can be assimilated or where chemical reactions can take place between magma and wall-rock. Heat is generally required to cause melting and this is provided by the cooling and crystallising magma. There is thus a widespread view that assimilation and fractional crystallisation are usually closely associated.

Magma generation in continents

In the oceans magmatism is dominated by the generation of basalt. The relatively minor amounts of intermediate to felsic or silicic magmas that are found at ocean ridges, on oceanic islands and in oceanic island arcs are ultimately linked to the basalts by differentiation processes. In the continents magmatism is much more diverse and complex. A coherent classification of **continental magmatism** has not emerged and there are still quite fundamental disagreements about the significance of the different magma types. For the purposes of discussion three major categories of continental volcanism can be identified, although the reader should be aware of a certain degree of arbitrariness and oversimplification in this scheme.

(i) Flood basalt volcanism

Certain volcanic provinces are characterised by the formation of large numbers of basalt lava flows. Individual lavas can have very large volumes (hundreds of cubic kilometres) and can extend distances of a few hundred kilometres. Examples of **continental flood basalt** provinces include the Deccan Plateau of India and the Columbia River basalts of the north west United States. The basalts are usually tholeiitic, suggesting extensive partial melting of the mantle, but have geochemical features which imply contamination with material from the continental lithosphere or from the crust. Silicic volcanic rocks are usually subsidiary in volume. Flood basalt provinces have been interpreted as the consequence of rapid stretching of the continental lithosphere over mantle plumes which may be over 200 °C hotter than normal mantle. The large amounts of melt formed can be viewed as a consequence of decompression melting of hot mantle.

(ii) Alkaline magmatism

A wide and bewildering diversity of alkaline and usually silica-deficient igneous rocks occur in continental regions. These include **alkali basalts**, **basanites** and **nephelinites**, their differentiated derivatives and plutonic equivalents. In this category can also be grouped **carbonatites** and **kimberlites**, although this should not be construed as implying a similar origin for such diverse igneous rock types. Numerous hypotheses have been put forward to explain these different kinds of alkaline igneous rock. A common theme to many models, however, is the idea that these magmas owe many of their characteristics to the mantle part of the continental lithosphere.

There is now evidence that parts of the mantle of the continental lithosphere are very ancient and that in places it has been invaded by volatile components, notably water and carbon dioxide. Garnet inclusions in diamonds from kimberlite pipes yield ages of over 3000 Ma and must originate from depths of 150 to 200 km. The addition of fluids into the mantle beneath continents is also indicated by veins from nodules brought to the surface during volcanic eruptions and by the high contents of either H_2O or CO_2 in various kinds of alkaline magmas. The cause of melting in the sub-continental mantle is unclear, particularly if it is old and cold. Influx of volatiles from some deeper and mysterious source could conceivably trigger small amounts of melting in the most fusible components. However, many examples of alkaline igneous activity are associated with rifting (e.g. the East African Rift and Rhine Graben) and so magmatism could simply represent small amounts of decompression melting in the asthenosphere coupled with contamination of material from the ancient sub-continental lithosphere during ascent.

(iii) Silicic magmatism

The most prominent category of continental magmatism, however, involves the development of large volumes of intermediate and silicic magmas, which usually dominate over basaltic products. In plutonic environments granites are formed, using the term granite in a broad sense to include a spectrum of intermediate to silicic rocks such as **diorites**, **tonalites**, **granodiorites** and true **granite**. In the volcanic environment rocks in the **andesite–dacite–rhyolite** range are formed. The origin of these magmas is of fundamental importance to understanding how the continental crust has formed, since the bulk of the upper and middle crust is composed of either granite or granite-gneisses.

In the present-day Earth silicic magmas are generated in several different tectonic environments. Where oceanic lithosphere is subducted beneath continental margins large batholithic plutonic complexes are developed in long linear belts, such as in the Sierra Nevada Cordilleran belt in the western United States. These plutons were inevitably associated with medium to large volume volcanic centres, typically characterised by calderas and ignimbrites. Large-scale plutonism and volcanism is also associated with regions of substantial stretching of the continental lithosphere. An example of such a province is in the Tertiary Basin and Range Province of the western United States. Some silicic systems appear to be associated with hot-spots, such as the Yellowstone Caldera in Idaho. Finally, large amounts of granites are associated with continental collision zones, such as the Himalayas and the Alps. The rest of this section is concerned with the origin of silicic magmas in continents.

Origins of silicic magma

Two major hypotheses for the origin of silicic magma have emerged. First there is the idea that since the continental crust is essentially composed of granitic materials then the magmas are simply the result of remelting the crust. Second there is the view that the silicic magmas are formed by differentiation of basaltic magma from the mantle. Most discussions of the petrogenesis of silicic magmas centre around the issue of the relative importance of these two sources. Since the continental crust is old and the upper parts have been progressively enriched in radiogenic components such as ^{87}Sr with time, study of the isotopic compositions of these rocks has contributed greatly to the resolution of this problem. Many silicic magmas are formed by melting upper crustal rocks, whereas others are related either to melting the mafic lower crust or are formed directly by differentiation of basaltic magma. A significant number are best interpreted as mixtures of material from the two sources.

From the standpoint of the physical mechanisms of magma generation the most significant question to ask is how the crust can be raised to a temperature which is sufficiently high to melt and generate magma. The normal geothermal gradient in the crust is about 20 °C per kilometre. Although crustal rocks begin to melt at temperatures of 700–750 °C, extensive melting can only occur when the temperature is high enough for the major hydrous minerals (biotite and hornblende) in the crust to break down and release their water. The temperatures required for large-scale melting are thus typically 800 °C to over 1000 °C. Thus a great deal of heat usually has to be supplied to raise crustal rocks from normal temperatures to temperatures at which substantial melting can take place.

There are a number of arguments which lead to the conclusion that intrusion of basalt into the continental crust (sometimes called **underplating**) is often the major heat source. The major tectonic environments where silicic magmas are formed are also situations where substantial melting of the mantle would also be expected. In subduction zones beneath continents basalt magma is generated. Even in the central Andes where the crust reaches 70 km thickness small amounts of basalt and basaltic andesite reach the surface and gabbros and diorites are found in association with large cordilleran silicic plutons. Intermediate and some silicic igneous rocks are typically complex mixtures of crustal magmas and basalt differentiates. In regions of crustal extension or where hot-spots impinge beneath the continents decompression melting would be expected and the silicic magma systems in these settings are almost always intimately associated with mafic magma. In continental collision zones decompression melting of the mantle is also feasible if the thickened lithosphere root zone becomes unstable and sinks into and is replaced by asthenosphere. Basalt intrusions provide an effective way of heating up the crust, since basalt has a temperature of 1200 to 1300 °C and crustal rocks melt extensively at temperatures of about 1000 °C or less.

Figure 5.20 shows a model of how the penetration of basaltic magma into and through the continental crust can lead to the large-scale melting of the crust, the formation of large calderas and the emplacement of large granitic plutons. The model is an amalgam of observations on volcanic and plutonic complexes and theory. In the earliest stage basalt and derivative mafic magmas are intruded into the crust and erupt at the surface to form early lava shields and high-level basic intrusions. At this stage the crust is cold and dykes can easily penetrate to the surface. Little melting occurs, although the basaltic magmas can differentiate or can become contaminated by crust on the way to the surface. With time the deeper levels of the crust heat up sufficiently that basalt intrusions can start to generate an increasing amount of crustal melt. Eventually the crust at some level becomes so hot and ductile and basaltic dykes can no longer penetrate to the surface and are intruded to generate large volumes of silicic magmas, which ascend to form large high-level plutons and erupt as voluminous ignimbrites. Repeated injection of basalt into the source region provides the opportunity for extensive differentiation of the basalt, melting of crustal rocks and hybridism

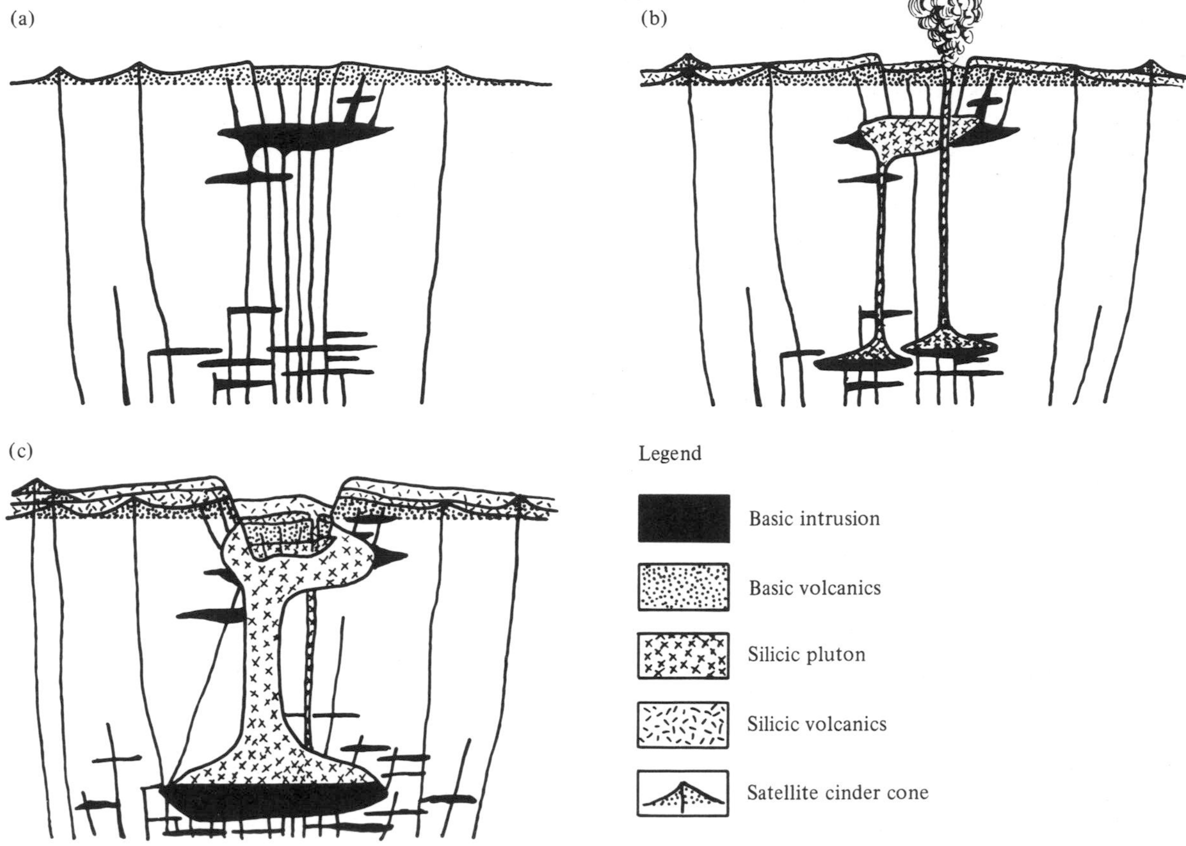

Figure 5.20 **A model for the development of granite magmas by emplacement of basalt intrusions into the continental crust.** □

between the magmas. In this model silicic magmatism is really seen as a second-order effect of mantle magmatism. The continental crust acts as a physical barrier to basalt reaching the surface, because of the low density and low fusion temperatures of crustal rocks.

Other mechanisms of melting continental crust can be envisaged. Fluxing of water into the crust could trigger melting by the freezing-point depression mechanism. The water could be supplied from the surface by deep penetration in hydrothermal systems or from depth, for example, by progressive dehydration of deeper rocks or by release of water from subduction zones. However, this mechanism is unlikely to heat the crustal rocks up in the first place, so the effect is more plausibly lined to basalt intrusion. In regions of continental collision, crustal rocks, particularly those with high contents of radioactive elements, can be buried by tectonic thickening. Eventually these rocks could be heated up and melt. Some of the granites in the Himalayas, for example, have been formed by melting pelitic metamorphic rocks, and magmas appear to have been formed next to a major thrust fault where two different slices of continental crust have been superposed.

Segregation and ascent of silicic magmas

Silicic magmas have much higher viscosities than basalt magmas (Figure 5.9). Typically their viscosities range from 10^6 to over 10^{10} Pa s for temperatures of 800 to 1000 °C. These much higher viscosities have major implications for the mechanisms of magma segregation and ascent. In the case of partial melting it is very difficult to separate efficiently melt from residual solid. This is seen in the geological record by the widespread preservation of regions of partially molten crustal rocks (or migmatites). Theoretically a mechanism such as compaction does not seem viable and the compaction time-scale (equation 3) is hundreds of millions or even thousands of millions of years for dry granitic melts. There is also some evidence that the dihedral angle of silicic melts with their crystalline matrix can be greater than 60°, which inhibits the formation of an interconnected partial melt until high degrees of melting are achieved (Figures 5.7 and 5.8). The formation of large bodies of granitic magma is, however, common and highlights the question of whether these magmas represent partial melts or bulk melts of crustal rocks.

One view of granite magmas is that they represent mixtures of partial melt and solid residue (or restite) rather than liquids. Thus when the melting source region reaches a critical amount of melting the region converts to a crystal-rich magma. The heterogeneities in the original rock are destroyed by convection. Another view is that granite magmas represent large degrees of partial fusion of crustal rocks. Basalt intrusion provides an effective way of forming high-temperature silicic magmas by providing the heat to drive convection and melt the overlying crustal rocks (Figure 5.20).

The high viscosity of silicic magmas causes them to ascend as large masses rather than in narrow cracks. Granites are typically found in large elliptical intrusions which are thought to represent the buoyant ascent of voluminous balloons or diapirs of magma. One idea is that their ascent is assisted by heating the rocks around them so that they become ductile and flow around the rising pluton. Another possibility is that the magmas ascend by melting their roof and crystallising at their floor. Once in the upper crust silicic magmas can rise further by stoping off fragments of the roof in piecemeal fashion or by the collapse of large coherent cauldron blocks.

Differentiation and zoning of silicic magmas

Large volumes of silicic magma can be emplaced in the upper crust where cooling and crystallisation in magma chambers over long periods of time can cause differentiation. Many silicic volcanic rocks, in particular ignimbites, show evidence that these magma chambers are zoned, becoming typically hotter and more mafic with depth. Crystallisation along the margins is a plausible way of differentiating and zoning silicic magmas (Figure 5.16). Addition of basic magma from depth probably accentuates and complicates zoning. It is also possible to envisage melting of different rock types in a source region forming magmas with different densities, which become organised and zoned as they ascend and coalesce.

Conclusions

Magmas are generated in the Earth due to convective motions in the mantle and plate movements. These processes cause hot mantle close to its solidus temperature to ascend and partially melt. This mechanism of pressure release melting probably accounts for most global magmatism. Plate movements also lead to other melting mechanisms. Subduction of hydrated oceanic crust and sediment can release water and trigger melting. Thickening of continental crust in collision zones can also bury rocks rich in radioactive elements which can then heat up enough to melt. Injection of basalt into the continental crust is a major cause of remelting crustal rocks.

Basaltic magmas can segregate from the mantle with remarkable efficiency. This is because the melt phase is always interconnected in partially molten mantle and the buoyancy forces are large enough to allow the basalt melt to flow and segregate at rates which are geologically rapid. Basalt magma typically ascends through the crust and lithosphere along fractures. Due to their high viscosity granitic melts are much more difficult to segregate from partially molten rock and the mechanisms are not yet fully understood.

Magmas often accumulate in the crust to form chambers where they cool and differentiate. Processes such as crystal settling, crystallisation along the chamber walls with accompanying compositional convection and compaction in crystal–liquid mushes leads to strong chemical differentiation. Magma generation is much more complex in the continental crust where assimilation, mixing and remelting of crustal rocks can occur.

Further reading

McBirney, A. R. (1984) *Igneous Petrology*, W. H. Freeman and Co., San Francisco.

Nicholls, J. & Russell, J. K. (eds.) (1990) Modern methods of igneous petrology, in *Understanding Magmatic Processes*, Reviews in Mineralogy volume 24, Mineralogical Society of America.

Press, F. & Siever, R. (1984) *The Earth*, 3rd edition W. H. Freeman and Co., San Francisco.

Williams, H. & McBirney, A. R. (1979) *Volcanology*, W. H. Freeman and Co., San Francisco.

CHAPTER 6

Keith G. Cox
University of Oxford

THE INTERPRETATION OF MAGMATIC EVOLUTION

Introduction

In the previous chapter the physical processes involved in igneous activity were outlined. Now we turn to the chemical aspects of the same topic, particularly how chemical methods can be used to determine what sorts of process have operated, and how the large diversity of observed igneous rock compositions may be explained. There are essentially three methods of approaching these tasks, via experimental petrology, via petrography and mineralogy, and via geochemistry. Thus the following discussion will touch on aspects of experimentally determined phase diagrams, the mineralogy and textures of igneous rocks, and the interpretation of analytical data. It is assumed that the reader has some knowledge of the compositions of the common rock-forming minerals, and of rock classification and nomenclature.

Most igneous rocks contain substantial quantities of the elements O, Si, Ti, Al, Fe, Mn, Mg, Ca, Na, K and P, usually termed the **major elements**. Apart from O, these elements are sometimes also referred to as the cations, because their valency can be expressed in terms of positive charges. Chemical analyses of igneous rocks are expressed as lists of major element oxides, because in most cases, the positive charges of the cations are balanced by equal amounts of oxygen, although other anions such as CO_3, OH, F, Cl and S may also be present, usually, however, in very small quantities. The use of the terms cations and anions is of course potentially misleading, because much of the bonding in silicate minerals is not ionic. However, it is a useful way of keeping an eye on charge-balance. Note that analysis for oxygen itself is hardly ever carried out, it being assumed that enough oxygen is present to satisfy the charge requirements.

The less-abundant constituents of igneous rocks are termed **trace elements** and their concentrations are usually expressed as parts per million (ppm) of the element itself, rather than as oxides. Among the trace elements for which analytical data are often obtained are Rb, Ba, Sr, Nb, Cr and Ni, though the full list is very much longer.

Table 6.1 gives some typical major element analyses and illustrates the considerable variation within the igneous rock spectrum. With the exception of the **peridotite** in column 1 all the analyses are of volcanic rocks and can be taken to represent approximately the compositional range of magmas.

The concept of fractionation

In the face of compositional variation as illustrated in the table, petrologists always start thinking in terms of how it might be possible for one magma, or perhaps one rock which is undergoing melting, to give rise to compositionally different magmas. The term **fractionation** in the broadest sense means evolution of a single starting composition into a number of different products, fractions of the original. The term is also

Table 6.1. *Analyses representing the range of variation (as wt %) in igneous rocks*

	Peridotite	Basalt	Andesite	Rhyolite	Trachyte	Phonolite	Nephelinite
SiO_2	42.26	49.20	57.94	72.82	61.21	56.19	40.60
TiO_2	0.63	1.84	0.87	0.28	0.70	0.62	2.66
Al_2O_3	4.23	15.74	17.02	13.27	16.96	19.04	14.33
Fe_2O_3	3.61	3.79	3.27	1.48	2.99	2.79	5.48
FeO	6.58	7.13	4.04	1.11	2.29	2.03	6.17
MnO	0.41	0.20	0.14	0.06	0.15	0.17	0.26
MgO	31.24	6.73	3.33	0.39	0.93	1.07	6.39
CaO	5.05	9.47	6.79	1.14	2.34	2.72	11.89
Na_2O	0.49	2.91	3.48	3.55	5.47	7.79	4.79
K_2O	0.34	1.10	1.62	4.30	4.98	5.24	3.46
P_2O_5	0.10	0.35	0.21	0.07	0.21	0.18	1.07

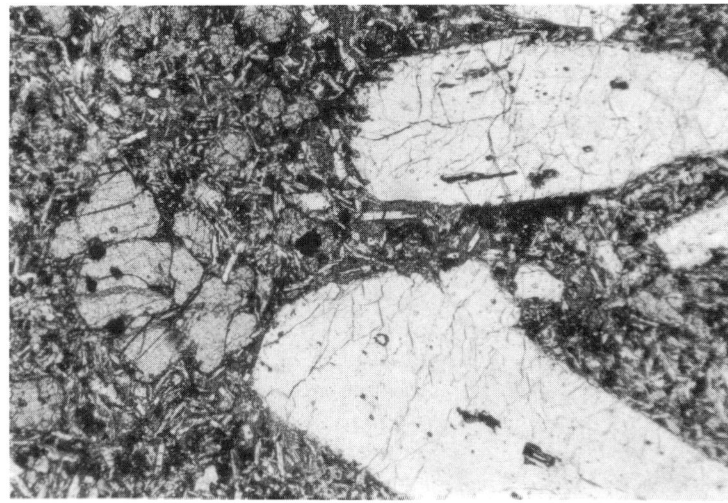

Figure 6.1 Photomicrograph of thin-section containing phenocrysts of feldspar (large almost colourless crystals) and olivine (moderate sized, medium grey) in a very fine-grained groundmass. The field of view is 6 mm wide. □

applied to individual elements to express the range of variation they show. For example Table 6.1 shows that of all the elements included, it is Si which shows the least fractionation, in that all the concentrations, from about 40 per cent to a little over 70 per cent, lie within a factor of two. Al also shows relatively little fractionation in most of the spectrum. By contrast, the strongest fractionation is exhibited by Mg, as comparison of the concentrations in the peridotite and the rhyolite shows (a factor of *c.* 70 difference). The remainder of this chapter will be devoted to trying to explain how this variation has come about, and a discussion of the lines of evidence which are employed in petrogenetic studies.

Fractional crystallisation

As argued in the previous chapter, the process of **fractional crystallisation** is one of the most important specific mechanisms by which compositional variation may be generated, and much of what follows will be devoted to the investigation of this topic.

Fractional crystallisation takes place when crystals which appear in a magma as a response to cooling become separated from the parent liquid. The mechanism by which this takes place is not however the present concern (see Chapter 5). For present purposes the most important point is that the composition of the crystalline material is only under exceptional circumstances the same as that of the liquid from which it crystallises. It follows that crystal-removal will leave a liquid which is compositionally different from the starting liquid. It is convenient to refer to these as the **residual liquid** and the **parent liquid** respectively.

We can surmise for a good reason that fractional crystallisation might be very important. Many, probably most, volcanic magmas are erupted to the surface as liquids carrying crystals. In rapidly cooled lavas the liquid fraction crystallises rapidly into a fine-grained mass (termed the **groundmass**) while the original crystals are visible as much larger inclusions (termed **phenocrysts**) (see Figure 6.1). Clearly such magmas have undergone cooling in the relatively cold near-surface environment of the Earth, and if suitable conditions for separating crystals from liquid are encountered, fractional crystallisation can take place.

Binary phase diagrams

The way silicate melts crystallise is best studied experimentally (see Chapter 4), and even if very simplified compositions are investigated a large number of general principles can be clarified. The experimental technique mainly employed is termed the **quenching method**. A powdered rock composition or simplified analogue is heated in a furnace, usually in a platinum or similarly non-reactive capsule, held at a desired temperature long enough for equilibrium to be established, and then cooled rapidly (quenched). Experiments can be carried out at atmospheric pressure or at higher pressures which simulate those of the deeper parts of the Earth.

Ideally, any liquid present in the capsule will be retrieved as a glass, and any crystals which were present will be suspended in it (Figure 6.2). The glass and the crystals are directly analogous to the groundmass and the phenocrysts in a volcanic rock. If, for example, the charge is held at a temperature above its **liquidus temperature** it will be completely molten, and nothing but glass will be found in the quenched charge. Conversely, below the **solidus temperature** no liquid is present and the charge is completely crystalline. At temperatures between the liquidus and the solidus both glass and crystals are found in the charge. This is analogous to the state in which the majority of volcanic liquids are erupted, and it is the first necessary requirement for fractional crystallisation.

Figure 6.3 shows a simple, experimentally-determined, **binary phase diagram**, so termed because all the compositions used were binary mixtures of two components, $CaMgSi_2O_6$, representing the mineral diopside, and $CaAl_2Si_2O_8$, representing the mineral anorthite. The diagram was made by studying a large number of different compositions and temperatures of equilibration. The labelled fields describe what the experimenter found in the quenched charges, given that glass is interpreted as representing former liquid. Thus, for example, a composition such as A, which consists of 80 per cent diopside + 20 per cent anorthite (usually expressed by the notation $Di_{80}An_{20}$), when equilibrated at the temperature of 1450 °C proved to be all liquid (glass). The same composition held at 1300 °C contained diopside crystals as well as glass, while at 1250 °C no glass was found, the charge consisting only of diopside and anorthite crystals. The field labels do not reveal the relative proportions of the phases when two are present, but this is readily determined from the diagram – for example at 1250 °C, composition A is very rich in diopside crystals while composition B is very rich in anorthite crystals

This diagram can be used to illustrate some very fundamental features of magmatic crystallisation. First, consider what happens to a liquid represented by point A, as it is cooled. Note that this liquid represents a magma which is above its liquidus, and in nature this is referred to as **superheated**. We will worry at a later stage about where such magmas come from, but for the moment we will take it as an example of a parental liquid, formed somewhere deeper in the Earth and now ready to undergo crystallisation at a relatively low pressure.

As A is cooled, diopside begins to crystallise at 1350 °C (point P). Clearly this follows from the experimental observation that between 1350 °C and 1274 °C diopside is always found coexisting with liquid. Cooling through this temperature interval obviously must involve the formation of more and more diopside crystals. Hence the *liquid* fraction is not only cooling but also getting richer in the anorthite component as diopside is removed from it into the crystals. The track

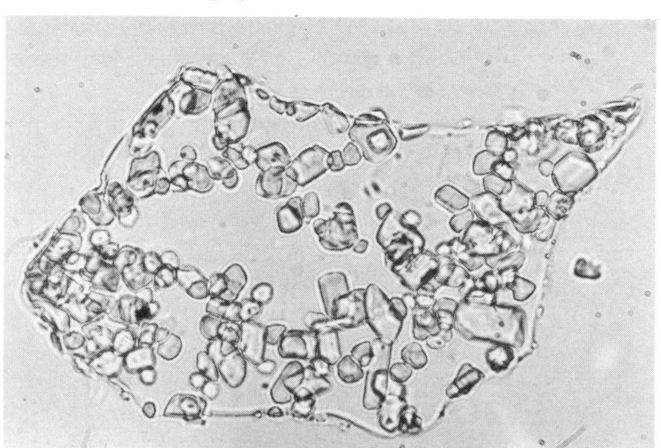

Figure 6.2 Photomicrograph of an experimental charge (composition 90 per cent diopside) from the system diopside–silica. Tiny crystals of diopside are set in a glass fragment. The charge was quenched from a temperature of 1378 °C and helps to map out the diopside + liquid field of the phase diagram. The field of view is 0.4 mm wide. □

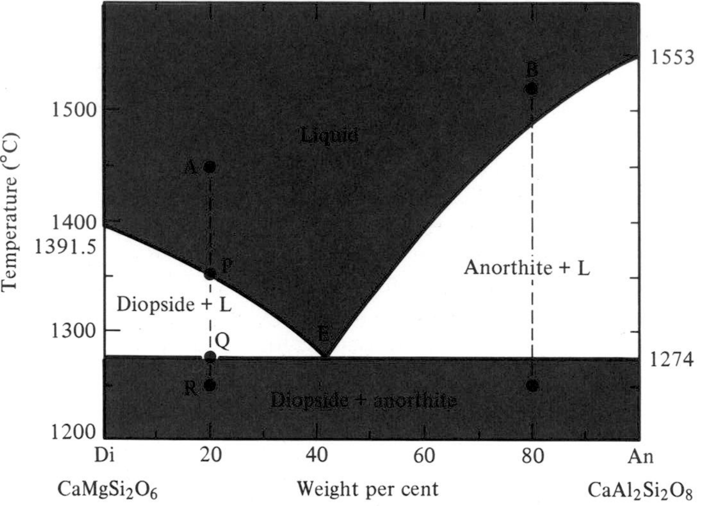

Figure 6.3 Experimentally determined phase diagram of the system diopside–anorthite at 1 atmosphere (after N. L. Bowen). Composition is measured on the horizontal axis (numbers refer to the anorthite content). □

followed by the liquid is along the curve separating the diopside + liquid field from the all-liquid field. As the charge reaches 1274 °C (point Q) the liquid reaches point E, a so-called **eutectic**, and anorthite begins to crystallise along with diopside. The liquid composition stays constant while the two phases crystallise, because the amounts of diopside and anorthite appearing collectively have the same composition as the liquid. When all the liquid has disappeared the charge simply cools in the subsolidus as a solid mixture of diopside and anorthite crystals, e.g. point R, which has the two phases present in the same proportion as the two components were present in the original liquid.

General conclusions that can be extracted so far include:

1 The first phase to crystallise on cooling of a superheated magma will depend on the composition of the magma. For example, all compositions more diopside-rich than the eutectic (to the left of E) will crystallise diopside first. All compositions to the right of E will crystallise anorthite first. This is a very important concept, because one sometimes sees statements like '... during the crystallisation of basaltic magma the first phase to appear on cooling is olivine...'. This is true for *many* basaltic magmas, but is not a valid generalisation. What crystallises first depends on the composition.

2 In general magmas will show a crystallisation interval rather than having a fixed melting or solidification temperature. The only composition in this system which will crystallise at a fixed temperature is the very special one which already (for whatever reason) has the eutectic composition.

3 Even this simple system suggests that the general style of crystallisation of magmas consists of the crystallisation of one phase which at a lower temperature will be joined by another. In more complex systems the number of phases which can crystallise together may become quite large, e.g. volcanic rocks with phenocryst assemblages consisting of olivine, clinopyroxene, plagioclase, magnetite and apatite, are not uncommon. However, more-complex systems also show that down-temperature reactions can take place so that a phase ceases to crystallise. For example, some basaltic magmas crystallise olivine at a high temperature, but crystallisation ceases at a lower temperature and any olivine present is redissolved.

4 The lowest temperature liquids in this system always have the composition of the eutectic liquid *irrespective of the starting composition.* If we consider heating solid charges, the converse follows – mixtures of diopside and anorthite, irrespective of bulk composition (i.e. the diopside and anorthite proportions) *all* melt initially to the same liquid, that is the eutectic liquid. Translating these conclusions into natural rock terms, they suggest that magmas of different starting compositions may produce low-temperature liquids which converge in composition. From the point of view of melting, the important conclusion is that the composition of the first liquid to form will depend on the mineral assemblages being melted, but not on the proportions in which the minerals are present. This follows from the fact that there is only one liquid composition in the system, the eutectic liquid, which is simultaneously in equilibrium with both diopside and anorthite. Obviously melting a mixture of these two phases must at least initially give rise to that liquid.

Fractional crystallisation and phase relations

Fractional crystallisation simply involves the removal of crystals as they are formed. Hence in our simple system we can imagine that we might have a magma chamber at some higher level in the Earth's crust, which was initially filled by our parent magma A. Subsequent fractional crystallisation could then give rise to liquid fractions representing all the compositions lying along the liquidus curve P–E. Depending on the volcanological circumstances, any of these might be erupted as individual lava flows.

Although this is a very simple system it is easy to check whether natural magmas can fulfil the necessary conditions imposed by these phase relations. If we were to collect the individual lava flows referred to above and carry out fresh experiments on them, we should find that they all produced diopside as the first phase to crystallise (the terminology used usually expresses this by saying 'they all have diopside on the liquidus'), but in general all at different temperatures. However, the temperature of diopside entry should correlate with the composition of the individual lavas, as, in our simple analogue, the temperature of diopside entry correlates with the increasing anorthite component in the liquids. Finally, although diopside entry is over a range of temperatures, all the experiments should show the entry of the second phase at the same, fixed, temperature. This would be true in any system, however complex, because the initial parent under given conditions can only give rise to residual liquids following a specific compositional evolution.

Let us look at a real case. Table 6.2 gives chemical analyses of five picrite basalts (very magnesium-rich basalts) from the Deccan Trap province of India, on which quenching experiments have been carried out to determine the temperature of entry of different phases during cooling (an average analysis of olivine phenocrysts from the same rocks is also included, for later discussion). Investigation of the compositional characteristics, a topic which we shall study next, had suggested that the different lavas were related to each other by the fractional crystallisation of olivine alone. Hence we should expect all of them to have olivine on the liquidus, we should expect the olivine entry temperature to correlate with chemical composition, and we should expect other phases to enter at fixed temperatures irrespective of composition. In this case the phases that appear are olivine, augite and plagioclase, and since crystal fractionation appears to have involved olivine we can use MgO as an index of compositional change, because olivine is very rich in this component and successive residual liquids should show progressive MgO-depletion. Augite and plagioclase are, respectively, naturally occurring Fe-bearing and Na-bearing analogues of diopside and anorthite (Figure 6.3).

Figure 6.4 shows the different compositions plotted with their mineral entry-temperatures plotted against MgO. Given that there are error brackets on the temperatures (typically ± 7–10 °C), and this is a compositionally complex suite of natural rocks, the correspondence with theory is very striking. Notice the gradual decline of the olivine liquidus temperature with falling MgO, and the relatively constant temperatures for augite entry (the second phase to crystallise) and plagioclase entry. Clearly all the magmas show a temperature interval, a large one in the case of the more magnesian compositions, over which olivine crystallises alone. The phase equilibrium requirements for fractional crystallisation of this phase on its own are clearly fulfilled perfectly.

Table 6.2. *Analyses (as wt %) of picrite basalts and their olivine phenocrysts*

	Rock analyses					
Sample	W-1	D-11	D-12	D-6	D-10	Olivine
SiO_2	44.15	46.44	46.60	46.90	46.44	40.03
Al_2O_3	6.35	9.28	9.90	9.59	11.82	0.25
Fe_2O_3	2.96	4.80	3.98	6.28	5.63	–
FeO	7.38	6.37	6.94	4.57	5.81	11.53
MgO	22.37	15.23	12.98	12.78	10.42	47.16
CaO	8.33	11.11	12.97	12.22	12.11	0.53
Na_2O	0.64	1.75	1.62	1.42	2.08	–
K_2O	0.40	0.67	0.90	0.80	1.11	–

Note: the olivine analysis is the average of six phenocryst analyses from different picrite basalts within the sequence.

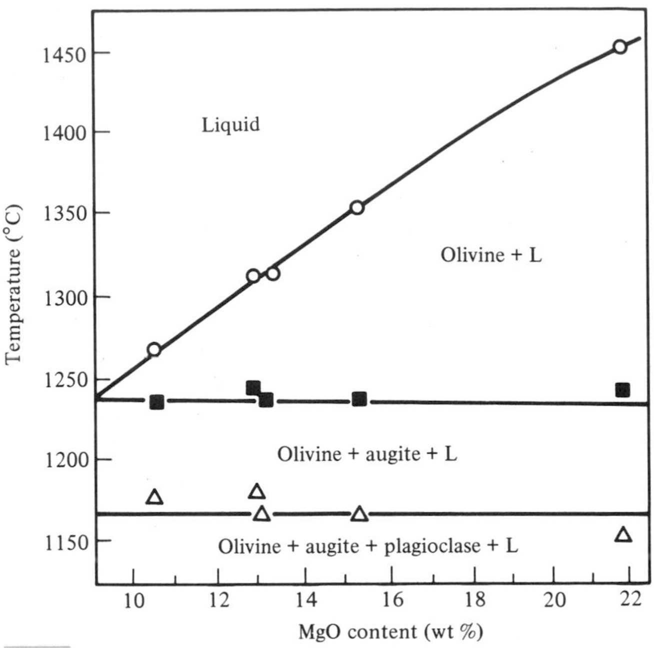

Figure 6.4 Experimentally determined crystallisation sequences at 1 atmosphere for the picrite basalts of Table 6.2, plotted against MgO content. Circles represent olivine entry temperatures, squares – augite, triangles – plagioclase. The approximate form of the phase diagram of these natural rocks is sketched in. □

Fractional crystallisation and compositional change

We now look at the same suite of lavas from the point of view of compositional change. How was the conclusion that the different lavas were related by olivine fractionation reached in the first place? To study this we need the concept of the **variation diagram**. Variations diagrams take many different forms but one of the simplest types, and most useful, simply involves plotting two compositional parameters against each other. These are often known as two-element or element–element plots, though in fact the data plotted are frequently oxides.

Fractional crystallisation involves the removal of crystals from liquids, and mass balance has to be maintained for the system as a whole. Thus if we consider any particular component, e.g. MgO, the amount of it in the crystals removed, plus the amount in the residual liquid, must add up to the amount in the parental liquid. In a two-element variation diagram these three compositions must always plot on a straight line, with the parental composition lying between the crystals and the liquid (see Figure 6.5). Moreover, simple geometrical consideration of the **'lever rule'** (see figure caption) allows the determination of the relative amounts of crystals and liquids, that is to answer, for example, questions like 'how much olivine must crystallise to give rise to this particular residual liquid?'

Returning to the natural rocks under study. Figure 6.5 also shows the analytical data for MgO plotted against CaO and Al_2O_3. Suppose we take the most magnesium-rich composition as a possible parent (P), and the other rocks as representing residual liquids of fractional crystallisation, leading to the most evolved liquid E. Clearly if olivine fractionation was responsible, then olivine compositions should plot on the extension of E–P into the more MgO-rich part of the figure. There is some scatter in the natural data, but again the hypothesis passes with flying colours, and everybody would agree that the average olivine fits the back-projection of E–P rather well. Satisfying as that is, it is important to realise that the hypothesis has been tested so far *only* for the relationships of MgO, CaO and Al_2O_3. These are important components of the rocks, but nevertheless the full test has to consist of checking all the components against each other. If analyses have been carried out for ten major elements, for example, no less than 45 separate diagrams are required. It is no problem, of course, with a computer.

In the case discussed, since compositional change has evidently been generated by the fractionation of olivine by itself, it follows that the fractionation process must have taken place at temperatures above the entry of augite and plagioclase into the crystallisation sequence. But if the magmas had cooled more before eruption the possibilities would certainly have existed for the fractionation first of olivine and augite together, and later for all three phases, olivine + augite + plagioclase. Obviously we need a method for unravelling these more complex cases, when the extracted crystalline material consists of a number of different minerals.

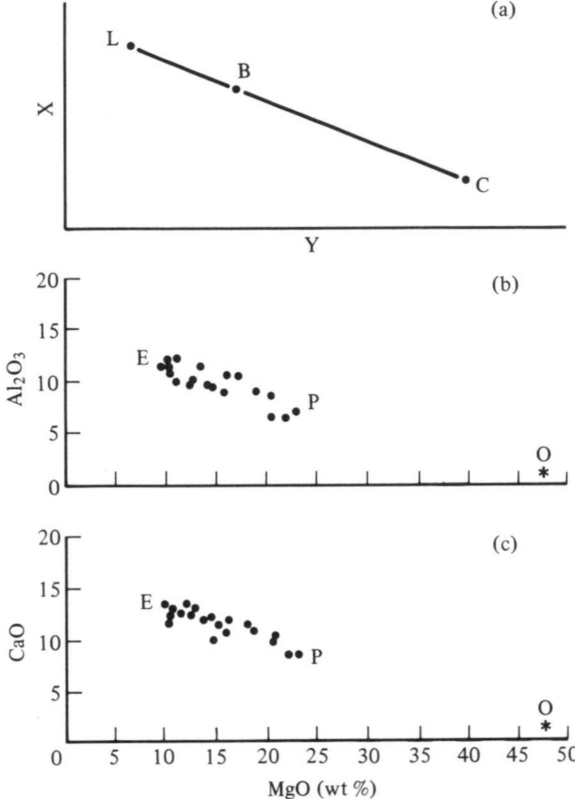

Figure 6.5 (a) Illustration of the lever rule on a diagram where concentrations of two elements (X and Y) are plotted against each other. Bulk composition B consists of crystals (C) and liquid (L). Relative proportions are determined by measurement of the distances BL, BC and LC. The fraction of the sample which is composed of liquid is given by BC/LC, the fraction of crystals by BL/LC, respectively 0.67 and 0.33 (or 67 per cent and 33 per cent). (b) and (c) These graphs show analytical data for the picrite basalts of Table 6.2 (filled circles) and other analysed rocks from the same suite, and the average olivine phenocryst composition (asterisk–O). E is the least Mg-rich sample, P the most Mg-rich. E–P–O approximates to a straight line, showing, for example, that P *minus* olivine can give rise to E. □

This can be carried out graphically by an extension of the lever-rule argument illustrated in Figure 6.5, but it has to be treated numerically if the number of minerals concerned is more than three. The calculation, usually termed a **mixing calculation**, requires as input the assumed composition of the parent magma, the composition of the supposed residual liquid, often termed the daughter (in the case already discussed these would be represented by analyses P and E), plus the compositions of all the minerals which might be suspected of involvement. Often it is the petrography of the rocks, that is the phenocryst types present, which suggests which minerals should be included. The calculation then determines what mixture of the given minerals, or some of them, can be removed from the parent to give the closest fit to the composition of the daughter. The investigator has to judge for him or herself whether the fit is satisfactory enough to confirm a given hypothesis.

Phenocryst assemblages are often a useful guide to the minerals which might be responsible for fractionation, but it is also extremely useful to gain independent insights by inspection of the analyses of parent and daughter. After all, some rocks do not contain phenocrysts, for example volcanic suites in which phenocryst removal has been completely effective before eruption. Additionally, and these are very interesting cases which give information about more complex modes of magmatic evolution, some suites quite clearly have undergone fractionation processes which can *not* be interpreted in terms of the fractionation of the phenocrysts actually present.

The principles are very simple. If we look at the analyses in Table 6.3, which are arranged in pairs of supposed parents and daughters, we inspect the changes of concentration in each element. In all cases we shall assume that the more magnesian analysis of the pair is the parent, and the daughter has been formed by the removal of some likely mineral or mixture of minerals. What is being removed will be termed the **'extract'**. The extract, even though a mixture of minerals, has to have a definite bulk composition.

We can think of three cases:

1 If an element shows no change, or very little change, from parent to daughter, it follows that the extract contains the same amount of that element as the parent and daughter do.
2 If an element is enriched in the daughter, the extract must contain less of that element than the parent.
3 If an element is depleted in the daughter, the extract contains more of that element than the parent.

All the rock analyses are essentially basaltic in nature, so in Table 6.3 some analyses of typical minerals which may be found as phenocrysts in basaltic compositions are also given. With these data a surprisingly useful number of provisional conclusions can be reached about the likely fractionating assemblages.

Looking at the first pair, A1 and A2, it is clear that the extract has to be (relative to the parent A1) poor in Si, Ti, Al, Ca, Na, K and P, and rich in Mg. The Fe content of the extract is about the same as that of the parent, and we can also suspect that although the

Table 6.3. *Analyses of parent and daughter compositions, mineral analysis (wt %)*

	A1	A2	EF	B1	B2	C1	C2	Olivine	Augite (clino-pyroxene)	Plagioclase	Ti-magnetite	Apatite
SiO_2	47.37	49.13	–	49.09	48.91	48.54	51.09	39.87	52.92	49.06	0.51	–
TiO_2	1.49	1.68	1.13	2.18	3.49	2.28	1.73	0.03	0.30	–	50.02	–
Al_2O_3	7.53	8.76	1.16	14.02	12.56	13.66	13.00	–	2.80	32.14	–	–
FeO	8.80	8.33	–	12.04	14.94	14.30	12.34	14.00	6.50	0.20	45.18	0.21
MgO	16.11	10.63	–	6.77	5.40	6.67	5.88	45.38	16.40	0.20	0.46	0.54
CaO	11.21	12.67	1.13	11.72	9.62	10.60	10.27	0.25	19.97	15.38	0.71	52.40
Na_2O	0.94	1.05	1.12	2.23	2.58	2.31	2.41	0.04	0.15	2.57	–	–
K_2O	0.22	0.25	1.14	0.24	0.32	0.14	0.25	0.01	0.01	0.17	–	–
P_2O_5	0.11	0.12	1.09	0.18	0.30	0.19	0.10	–	–	–	–	44.8

Note: Column EF gives enrichment factors for A2 relative to A1.

extract is poor in Si, it is only a little depleted relative to the parent, because the relative change in Si is much smaller than the other elements showing enrichment in the daughter. Looking at the minerals in Table 6.3 we can immediately begin to suspect the role of olivine. Certainly none of the other minerals acting alone could conceivably have produced this result. Clinopyroxene removal, for example, would have generated Ca-depletion in the daughter, plagioclase removal would have generated Al-depletion, titanomagnetite Ti-depletion, and apatite P-depletion.

Looking at the second pair, B1 and B2, this time it is clear that plagioclase has to be involved, because it is the only mineral containing enough Al to account for the Al-depletion of the daughter. But, there has to be a magnesium-rich mineral in the extract as well, to account for the simultaneous Mg-depletion. Whether the Mg-bearing mineral is olivine, or clinopyroxene, or both, would remain to be seen by using graphical or mixing calculation methods. However we now have some good ideas to investigate.

The third pair of analyses, C1 and C2, are very similar to the second pair, but this time the daughter shows depletion in Fe, Ti and P. Obviously that is consistent with the probability that the fractionating assemblage now contains titanomagnetite and apatite, in addition to plagioclase and one or more magnesium-rich minerals.

The discussion illustrates the extreme usefulness of simply inspecting analyses with some idea of likely mineral compositions in mind. A little bit more can, however, easily be done if we look at the idea of incompatible elements.

Incompatible elements

An **incompatible element** is one which does not enter a particular crystal lattice (mineral) in significant quantities. The term only means something when the mineral or mineral assemblage is specified. Hence from Table 6.3 we can say that Al and Ca are effectively incompatible in olivine, Al is reasonably incompatible in clinopyroxene (the concentration is much lower than in the liquids from which pyroxenes normally crystallise), and P is incompatible in olivine, clinopyroxene, plagioclase and titanomagnetite. Conversely, it is highly compatible in apatite.

The distribution of an element between crystals and coexisting liquids is conveniently expressed by the **distribution coefficient (K_D)** defined under the specific circumstances (e.g. liquid composition, temperature and pressure) as:

$$K_D = \frac{\text{concentration in the solid}}{\text{concentration in the liquid}}.$$

The distribution coefficient for a particular element and a particular mineral, can be quite variable, especially for major elements, depending on the conditions. But incompatible elements have distribution coefficients approaching zero, and these remain very small irrespective of the circumstances. Take rubidium in olivine for example. Rb is an extremely large univalent ion. Neither in terms of charge nor valency can it find a site in olivine, which has cation sites suitable for the much smaller, divalent ions, Mg and Fe.

This is an extremely useful concept because over a wide range of compositions it enables us to guess quite

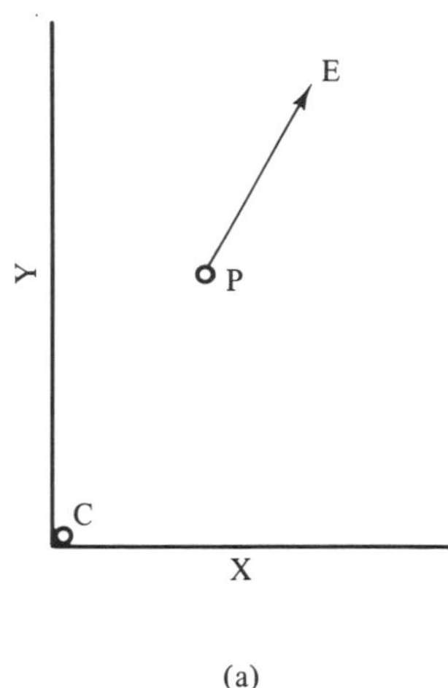

(a)

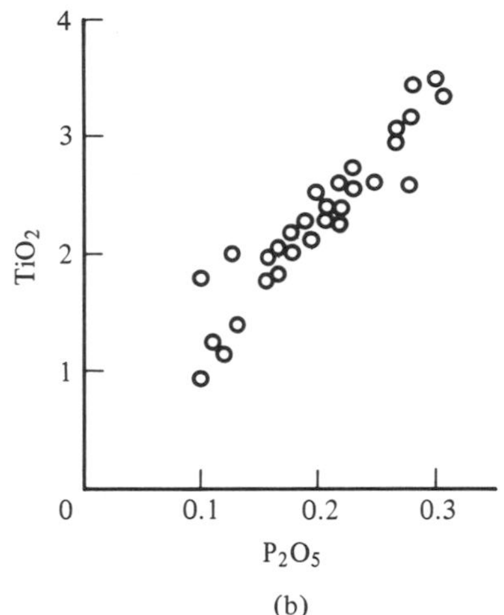

(b)

Figure 6.6 (a) Behaviour of two incompatible elements (X and Y) in a two-element plot. The extracted crystals plot at or near the origin (C), and their removal from the parent magma (P) drives residual liquids towards E. Analytical data for rocks on the P–E trend should back-project towards the origin. (b) P_2O_5 plotted against TiO_2 for a series of basalts fractionating olivine, plagioclase and clinopyroxene. Both elements are partitioned strongly into the liquid. □

accurately which mineral phases are *not* involved in the fractionating assemblage. The key relationship is that if two elements are incompatible in the assemblage, then they must retain the same ratio to each other in the daughter as they do in the parent. Figure 6.6 illustrates the principle in a variation diagram, and also gives a set of data illustrating a natural case. Diagramatically, any two incompatibles plotted against each other must define a straight line projecting back more-or-less towards the origin. This is the geometrical expression of constant ratio.

With the analyses in Table 6.3 it is easy to do a quick check. All elements which show more-or-less the same **enrichment factor** (daughter/parent) can be suspected of being the set of incompatible elements for the fractionation process concerned. For A1 and A2 the enrichment factors are given. The great similarity between the numbers for Ti, Al, Ca, Na, K and P suggests that these are the incompatibles, and thus the hypothesis that the two compositions are related by olivine fractionation is confirmed. The enrichment factors clearly excluded the possibility that other phases such as augite, plagioclase, titanomagnetite, and apatite, were involved in any significant capacity.

Trace element evolution

The concept of distribution coefficient, K_D, is very valuable in predicting the behaviour of trace elements during fractional crystallisation. The general equation for perfect fractional crystallisation, so-called Rayleigh fractionation, is:

$$\frac{C_L}{C_O} = F^{(K_D - 1)},$$

where C_L is the concentration in the residual liquid after fractional crystallisation, C_O is the concentration in the original magma (the parent), and F is the fraction of liquid remaining (e.g. if a liquid loses 10 per cent of its original mass as olivine then $F = 0.9$).

For many trace elements K_D with respect to a particular mineral remains relatively constant over fairly small degrees of fractional crystallisation, and such K_D's also remain constant during small changes in the degree of partial melting (see later in this chapter). This fact enables predictions to be made of changes in both concentrations and the relative concentrations of pairs of trace elements, which has important implications for interpretations of observed radiogenic isotope variations (see Chapters 5 and 8). For example, during plagioclase crystallisation in basaltic magma Sr has a K_D of *c.* 2 while the K_D of Rb is *c.* 0. Sr thus becomes depleted in the residual liquid while Rb is enriched. Clearly strong changes in the ratio Rb/Sr can be generated by this mechanism. Partial melting can be handled similarly, though using a different equation.

Cumulates and cumulus enrichment

The examples chosen for discussion so far have been deliberately simple cases in which compositional changes have been interpreted in terms of straightforward fractional crystallisation models. But before going on there are some further important questions to ask. First of all, in the case of the picrite basalts (Figure 6.5) what was the justification for choosing the composition P as the parent? It does have the highest liquidus temperature, so in that sense is perfectly allowable. But, and a fairly large but, we must also consider the process known as **cumulus enrichment**. The crystals removed obviously have to go somewhere, and they form rocks which are called **cumulates**. These may form as a heavy sludge in the bottom of the magma chamber, or they may crystallise on magma chamber walls or roofs. In any case, cumulates are geochemically rather different from residual liquids. They are formed by the accumulation of crystals and thus sometimes have compositions which could never be achieved by liquids. For example there is no known liquid equivalent in composition to a pure olivine cumulate (the plutonic rock-type known as dunite). The temperatures required to produce molten olivine are simply never achieved within the Earth at any depth where olivine is stable.

Cumulus-enriched liquids can form in a number of ways, but all involve the acquisition of crystals which have originated elsewhere and then been incorporated in the magma. One of the most obvious mechanisms affects liquids which lie in the lower parts of magma chambers cooling from above and undergoing crystal settling. Crystals of olivine, for example, originating at the top of the chamber can be added to the liquid lower down, and can be erupted with it. Compositionally this process enriches the bulk composition of the magma in the components of the cumulate. Hence in the case of the picrite basalts we discussed, we could conceivably have chosen liquid E as the initial magma, then all the other compositions would be generated by *adding* various amounts of cumulus olivine to it. The phase relations would remain the same, and the compositional variation would be much the same as (though not absolutely identical to) that generated by fractional crystallisation of P. Alternatively, and in fact detailed studies have shown this probably to be the case for the picrite basalts under discussion, we could choose a parent liquid lying somewhere between E and P. Then magmas could evolve in the Mg-poor direction, i.e. towards E, by fractional crystallisation, and in the opposite direction, towards P, by the accumulation of olivine. It is not always easy to decide on questions like this, but in the present case a detailed study of the olivine compositions has resolved the problem.

Partial melting trends

The above discussion illustrates one of the problems – that of deciding between fractional crystallisation and cumulus enrichment trends, even in a simple case. However there is another potentially important ambiguity in the evidence we have looked at so far. In the discussion of the liquid evolution path in the system diopside–anorthite (Figure 6.3) we worked throughout on the assumption that the parent magma was hot and that magma compositions were then able to evolve by fractional crystallisation in response to cooling. We touched briefly on melting, pointing out that all diopside–anorthite mixtures gave rise to the same, eutectic, liquid at the beginning of melting. Further heating of a composition such as A, which when completely molten must consist of a liquid of composition A, takes place by evolution of the liquid up the liquidus curve to P after all the anorthite has been melted out of the source material. If we were able to extract batches of liquid at various times during the partial melting process we might expect to produce magmas showing variations very similar to those of the magmas we previously ascribed to fractional crystallisation. In practice partial melting trends and fractional crystallisation trends are usually not exactly the same, but in many cases they may be so similar as to be difficult to distinguish. All the arguments we have made about phase relation requirements, changing compositions, and the correlation between these two phenomena could be essentially similar in the two cases.

In the case of the picrite basalts, an interpretation in terms of partial melting takes the form of saying that we are looking at a series of melts extracted at different temperatures from a source rock in which every mineral except olivine had already been melted and entered the liquid. Hence proceeding to a higher temperature involves the melting of more olivine (the only mineral there) with the addition of more of the olivine component to the liquid. This results in the composition trend we see, and the phase relations observed experimentally. In this case they are all almost exactly the same as what we should expect from fractional crystallisation of olivine or fractional crystallisation combined with cumulus enrichment.

How then may fractional crystallisation and partial melting trends be distinguished? The answer is sometimes they can't be, though probably, in *basaltic* rocks at least, partial melting trends are rarely preserved. Basaltic magmas are generated by partial melting, usually quite deep within the Earth (in the mantle), and then as they proceed upwards, entering cooler regimes, the possibility of fractional crystallisation rapidly becomes very high. Now we come to the question of pressure and its effects on phase diagrams. A magma can be thought of as having to start life in a phase diagram for a specific series of components. It dies (solidifies) in the same phase diagram. But although the components are the same, the effect of pressure on the geometry of the diagram, and even the mineral species involved, may be dramatic. Hence the evolution of basaltic liquids during melting is not likely to be the same as their evolution during crystallisation – the pressure change from the high-pressure environment of melting to the low-pressure environment of cooling is sufficient to ensure that. The compositional effects of low-pressure fractional crystallisation are likely to obliterate any partial melting trends represented in the sequence of erupted magmatic compositions (the principles are illustrated in Figure 6.7). Thus in the absence of other evidence it is reasonably safe to try to interpret most observed trends in basaltic rocks in terms of fractional crystallisation processes.

Acid magmas, which give rise to granites and rhyolites, present a somewhat more difficult problem. While a few are demonstrably generated by the advanced fractional crystallisation of basaltic parental magma, many are clearly produced by the melting of existing crustal rocks. They are cooler than basaltic magmas, and thus lose less heat as they approach the surface, a factor which may help to suppress fractional crystallisation, and allow the preservation of partial melting trends. On the other hand, compared with basalts they undergo relatively little pressure change during ascent, which implies that fractional crystallisation and partial melting trends are rather similar to each other.

Soluble and insoluble minerals

The remainder of this chapter will work towards trying to synthesise the evidence from petrological, petrographic and experimental studies, into a general explanation of how magmas evolve to give the variety of igneous rocks we see. However, there are one or two important points to clear up, which in the past have been a frequent source of misconception.

Clearly one of the most general points of interest is the sequence in which minerals crystallise with falling temperature. We know that high-temperature magmas such as basalts tend to crystallise minerals such as olivines and pyroxenes, and calcic plagioclase. Conversely, quartz, potassium feldspars and sodic plagioclases are characteristic crystallisation products of low-temperature magmas like rhyolites. It is often thought that this contrast is something to do with the melting points of the different minerals. In fact the relationship between the melting point of a pure mineral and the temperature at which it characteristically crystallises from a multi-component silicate melt is very far from direct. It is better to think in terms of relative solubilities – an insoluble mineral crystallises early (at high temperatures) and a soluble mineral late. Sodic-plagioclases, potassium feldspar, and quartz are all highly soluble in typical magmas – that is why they crystallise from low-temperature melts. To reinforce the point it is worth noting that the melting point of pure SiO_2 is 1713 °C, which is a lot higher than that of pure diopside (1391 °C) or pure anorthite (1550 °C).

The concept of **relative solubilities** has very important consequences for magmatic evolution in general. Consider the diagrammatic binary phase diagram in Figure 6.8. The eutectic point in this diagram is very asymmetrically placed. Notice that the eutectic liquid

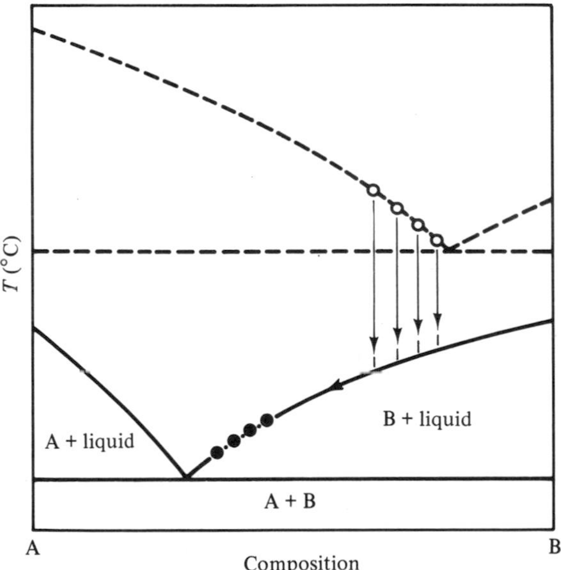

Figure 6.7 A hypothetical binary system in which two minerals A and B crystallise. High-pressure phase relations are shown by dashed lines, low-pressure by solid lines. Note that pressure release causes a large expansion in the compositional stability field of B + liquid, together with a general fall of liquidus temperatures. Partial melting is imagined to produce a series of high-pressure liquids (open circles) in equilibrium with residual solid A. Sudden pressure-drop and cooling allows these to evolve (arrowed paths) and fractionate B to produce liquids shown by solid circles. The evidence of high-pressure evolution is completely obscured by low-pressure effects. □

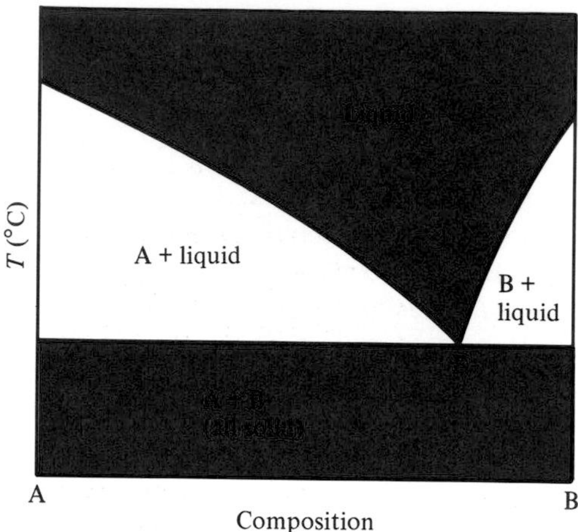

Figure 6.8 Hypothetical binary phase diagram for the system A–B. The eutectic liquid E is very rich in the B component. The phase B is highly soluble relative to A in melts consisting of A–B mixtures. □

is very rich in the component B, and that all the liquids which can crystallise the mineral B (lying to the right of the eutectic) are already very rich in this component. B is an example of a highly soluble mineral. It only crystallises from this system when the concentration of the necessary components is very high. Putting it another way, a highly soluble mineral has, in terms of compositional range, a very small stability field in coexistence with liquid. Conversely, a highly insoluble mineral such as A has a very large stability field.

How do residual liquids undergoing fractional crystallisation evolve in such a system? They obviously all eventually converge on the eutectic, irrespective of starting composition. Because the eutectic is so asymmetrically placed, it is possible to produce liquids which are very rich in the component B from a wide variety of starting compositions, many of which could have been initially very B-poor. It is *not* possible to produce very A-rich liquids by fractional crystallisation.

If we examine a few more complex (ternary) phase diagrams representing the phase relations of some common minerals we can extract the general pattern from this. **Ternary phase diagrams** are made from experiments on three-component mixtures and are normally presented as triangular plots in which composition and the compositional range over which each mineral is the first to crystallise are represented. Because, compared with a binary diagram, an extra dimension is needed for the representation of composition, temperatures cannot be shown directly in a two-dimensional plot. However, one can think of the temperature axis as being at right angles to the page, so temperature can be represented by contours or spot heights.

In Figure 6.9 the phase relations for the system diopside–silica–leucite are shown. The details of the

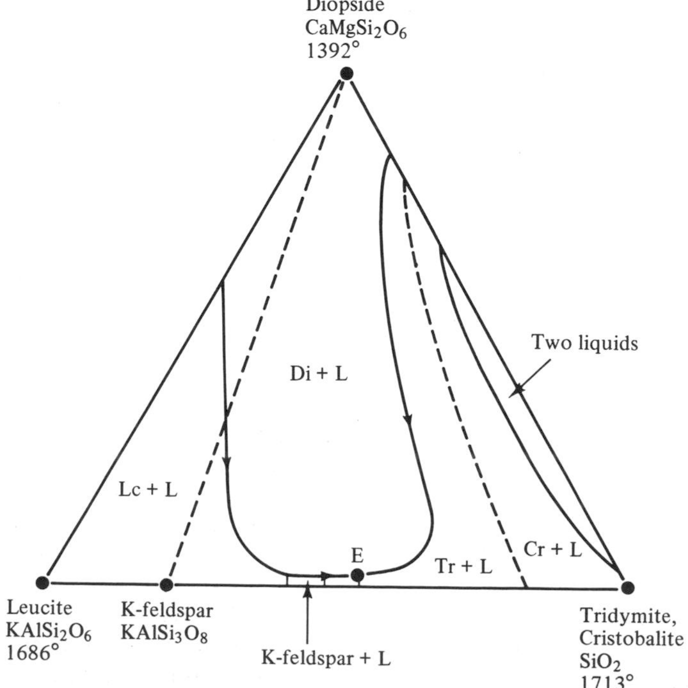

Figure 6.9 Phase diagram for the system diopside–silica–leucite at 1 atmosphere. Temperatures at the corners represent melting points of the pure phases. Labelled fields show compositional ranges over which each mineral crystallises first during cooling. Boundaries between fields are loci of liquids in equilibrium with two crystalline phases. Arrows show directions of falling temperature. All fractionating liquids in this system eventually converge on the eutectic E and crystallise to mixtures very rich in potassium-feldspar and silica minerals. Note that the very small potassium-feldspar field indicates high solubility of potassium-feldspar in melts of these compositions. It is this feature which allows residual liquids to become so depleted in diopside. □

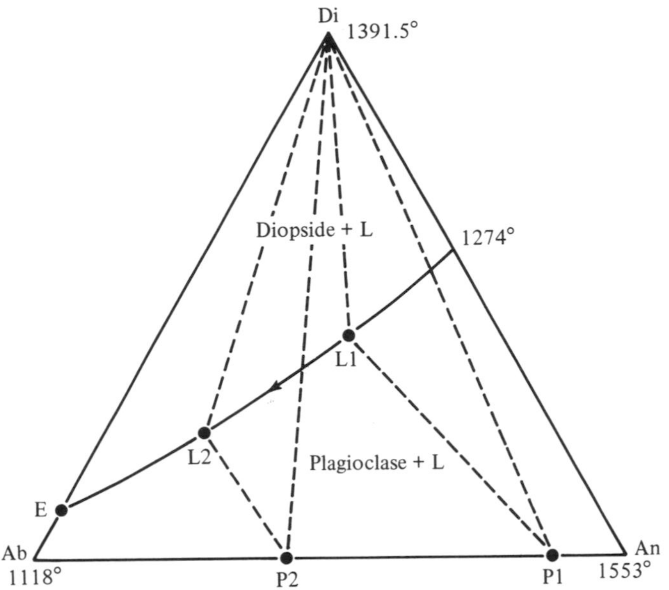

Figure 6.10 The system diopside–albite–anorthite at 1 atmosphere. Temperatures at the corners represent melting point of the pure phases. The curve separating the two fields shows the track taken by liquids fractionating diopside and plagioclase. A liquid such as L1 is in equilibrium with diopside and a calcic plagioclase P1. Removal of these solid phases drives the residual liquid down the curve to the left (arrow shows direction of falling temperature). Plagioclase compositions change continuously so that a lower-temperature liquid (L2) is in equilibrium with a more sodic plagioclase (P2). The lowest temperature liquids in the system are at E, a composition very rich in albitic plagioclase with only a little diopside. □

diagram do not matter except to note that a potassium-feldspar field is present, it is extremely small, and it lies in the lowest temperature part of the diagram. Diopside by contrast has an extremely large stability field. This is just one example to illustrate the general point that an initial liquid rich in the calcium and magnesium component diopside can, by fractional crystallisation, give rise to residual liquids very rich in potassium feldspar. Many other experimentally determined phase diagrams add to the general impression that initial magmas rich in the components of olivine and pyroxene (e.g. basalts) can fractionate to give very feldspathic liquids akin to phonolites, trachytes or rhyolites.

The second example (Figure 6.10) looks at the case of the plagioclase feldspars. This diagram shows phase relations for the system diopside–albite–anorthite. The plagioclase feldspars form complete solid solutions, so there are only two different mineral + liquid stability fields. Liquids on the boundary between the two fields crystallise diopside and a highly calcic plagioclase at high temperatures, and fractionate down the boundary curve towards the albite corner. The important point to note is that liquids on the boundary curve near the albite–diopside join (the low-temperature end of the curve) not only crystallise albite-rich plagioclase but have bulk compositions very rich in albite. High-temperature liquids at the other end of the boundary curve are simple analogues of basalt (diopside plus Ca-rich plagioclase in approximately equal proportions), while their low-temperature fractionation products are like trachytes (a small amount of diopside with a large amount of albite, which for taxonomic purposes is an alkali feldspar). This example again illustrates the capacity of basalt-like liquids to fractionate towards feldspathic end-products, and also gives an explanation of why basic rocks contain calcic-plagioclase while more-evolved magmas crystallise albitic plagioclase.

Primary magmas from the mantle

The range of compositions of igneous rock types is large but not infinite. Perhaps the most fundamental thing to understand is that the spectrum of compositions we see is controlled in the first place by the composition of the Earth itself. Just as in the simplest phase diagram the course of evolution of residual liquids of fractionation, or the course of liquid evolution during partial melting, is controlled by the starting composition, so in the Earth the range of igneous rocks is constrained by the original materials. As petrologists interested in the rocks we see today at the surface of the Earth, we may skip over some of the earlier parts of Earth history, the original accretion, and the segregation of the core. We come in on the story after the Earth has acquired its mantle, because it is in the mantle, especially in the upper part (see Chapter 3), that most magmas originate by decompression in response to convection.

The composition of the upper mantle is well known from studies of inclusions brought up in diamond pipes and basalts, and from the study of ophiolite complexes. The dominating rock type is peridotite, consisting largely of olivine and orthopyroxene, with subordinate amounts of clinopyroxene, and either garnet or spinel, depending on the depth from which the samples were derived (the latter being from higher levels in the upper mantle, the former from deeper). When the mantle melts it gives rise to basaltic, albeit according to the experimental evidence, rather Mg-rich melts, which are considerably enriched in Ca, Al, Na and K compared with their source rocks (compare the analyses of basalt and peridotite in Table 6.1).

In a mantle peridotite the minor phases act as repositories for some of the important constituents required for the formation of basaltic magmas. Clinopyroxene is the main repository of Ca and Na, while garnet or spinel is the principal repository of Al. Potassium is highly incompatible in all the major minerals of the mantle, and hence is contained in very minor amounts of accessory potassium minerals, e.g. micas or amphiboles.

During progressive partial melting, unless the minor phases are exceptionally insoluble, it is difficult to escape the conclusion that there soon comes a point when phases present in small amounts have completely melted (i.e. their constituents have been entirely contributed to the liquid) while phases originally present in large amounts remain, as the unmelted residue, though in reduced amounts. This is the case with peridotite melting. Relatively small amounts of partial melting result in the complete disappearance of garnet or spinel, clinopyroxene, and minor potassium-rich phases, into the liquid. It follows that the liquids are, relative to the source rock, enriched in Ca, Al, Na and K, and concomitantly depleted in Mg, which is partially retained in the residue of unmelted olivine and orthopyroxene. This is why partial melts of the mantle are more basaltic in composition than peridiotitic.

Magmas which originate directly from partial melting, and which are subsequently unmodified by fractional crystallisation *en route* to the surface are known as primary. **Primary magmas** from the mantle rarely reach the surface because they rapidly fractionate olivine during cooling and decompression. However, they are quite readily identifiable when they do, because they not only have high contents of MgO but they also have characteristically high Mg/Fe ratios, like picrite basalt D-11 in Table 6.2, which is a good candidate for a primary magma. The Mg/Fe argument depends on the fact that primary magmas from the mantle must originate in equilibrium with mantle olivines. The latter, from studies of mantle samples, typically have Mg/Fe ratios of about 9:1 or 10:1 (expressed in atomic ratios). Experiments show that liquids coexisting with olivine have Mg/Fe ratios close to one-third of that of the olivine. Hence primary magmas from the mantle should have Mg/Fe ratios of about three. Most basalts erupted at the surface have ratios substantially lower than this, showing that they have lost Mg by fractionation, probably mainly of olivine, on their way to the surface.

The evolution of basaltic magmas

Basaltic magma may be taken as the starting point for the evolution of most of the world's igneous rocks. Olivine-removal from picritic precursors eventually generates liquids which are sufficiently rich in Ca, Al, Ti and Fe to start crystallising and fractionating augitic clinopyroxenes and calcic plagioclases, and often a little later, opaque phases such as titanomagnetite.

A typical fractionation trend of residual liquid composition might start with quite strong Ca-depletion (the combined effect of clinopyroxene and plagioclase removal), very modest Al-depletion (because although plagioclase is very aluminous, clinopyroxene is Al-poor, so the two effects almost balance). At this stage Fe enrichment is likely because the fractionation of ferromagnesian minerals has very little effect on Fe (crystallising olivines or pyroxenes tend to contain about the same amount of Fe as their liquid) but plagioclase removal obviously enriches Fe in the liquid. K is an incompatible element, and so becomes enriched. Na is enriched less rapidly because some Na is removed in the plagioclase. During this stage Si is likely to remain almost constant. All the phases being removed are silicates, and their collective Si content is much the same as that of the liquid.

As fractionation continues the opaque phase is likely to appear, and being a non-silicate, forces the liquid evolution towards silica enrichment. The next possibility, though it is not inevitable, is that there is enough build-up of water in the melt (H_2O has acted as an incompatible so far) to stabilise hydrous ferromagnesian minerals such as hornblende, or if K has built up enough, biotite. At any rate, the increase of K frequently results in the stabilisation of K-feldspar, and if silica enrichment has been sufficient, quartz may also start to crystallise. By this stage magmatic temperatures have become so low that little further evolution is possible before ultimate solidification. But what sort of a magma have we got?

Typically the answer is a rhyolitic magma. Because of the phase relations discussed earlier it is relatively rich in the components of the alkali feldspars, probably both albitic and potassic, it is quite silica-rich, and it is thoroughly depleted in Mg, Fe, Ca and Ti. The effect of a long history of plagioclase fractionation will ensure that it is probably a little Al-poor compared with a basalt, and has a substantially higher K/Na ratio.

In transit, the liquid has passed from basaltic composition to rhyolitic via andesite and dacite. If we prefer to look at this in terms of plutonic rocks we are seeing the sequence from gabbro to granite via diorite and quartz diorite. In volumetric terms the enormous majority of the world's surface igneous rocks belong to this series (see Figure 6.11).

However, all the above discussion has been in terms of a fractional crystallisation model. Does this mean that all granites are the fractional crystallisation products of basaltic magmas? Emphatically not.

In an earlier section we discussed the difficulties of distinguishing partial melting trends from fractional crystallisation trends. The great series of magma compositions from basalt to granite *can* be, and sometimes is, produced by fractionation of basaltic magma. But the augmentation of the general fractionation trend by episodes of re-melting at crustal levels is probably far more important in granite production.

Imagine an island arc built up by the effusive products of andesitic volcanoes. It contains pristine igneous rocks and sediments derived from them. The magmas are andesitic because they are mantle melts showing reasonably advanced fractionational crystallisation before eruption. The whole complex is eventually incorporated into new continental crust and at some subsequent time is remelted, perhaps by the injection of fresh basaltic magma (see Chapter 5). By this time the rocks concerned are probably a collection of metamorphic assemblages which will in places contain alkali feldspar and quartz. These are the soluble minerals in silicate melts, so as melting commences they contribute strongly to the liquid and produce granitic magma. This demonstrates that basaltic magma can evolve into granitic magma by a series of stages involving fractionation part-way, aided by remelting episodes. The constraints of the phase relations cannot be avoided, and it is these that impose a relatively restricted series of evolutionary paths on the liquids. However, in the study of granites and similar rocks it is obviously important to study the experimental data with partial melting trends firmly in mind.

The evolution of more alkalic magmas

The distribution of rock types is illustrated in Figure 6.11 on a diagram in which silica is plotted against the total Na_2O+K_2O. This is known as the **TAS diagram (total alkalis–silica)** and is the basis of some recent classifications of volcanic rocks. The main fractionation

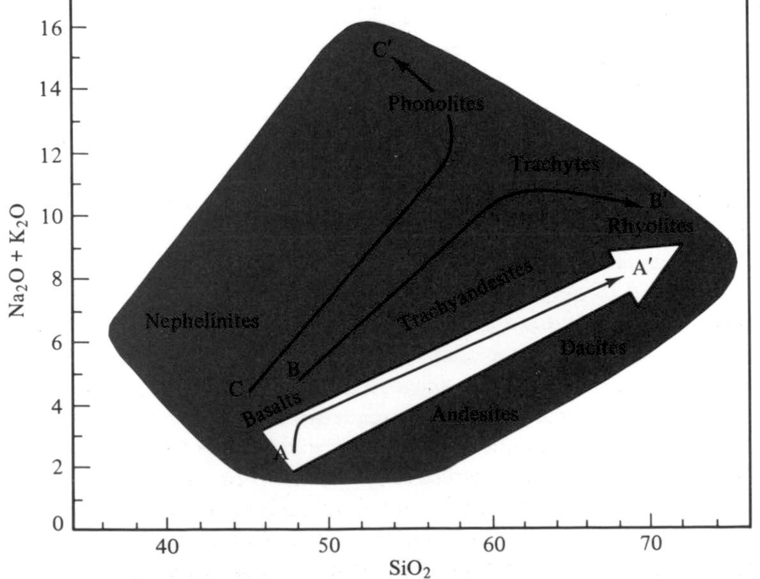

Figure 6.11 Distribution of volcanic rock types on the TAS diagram. The most abundant igneous rocks fall on the basalt–rhyolite trend shown by the broad arrow. A–A', B–B' and C–C' are specific fractionation trends discussed in the text. □

trend of the common rock types just discussed is indicated.

Relative to this main trend, other rocks are richer in alkalis at any given silica content. These types may also be seen as originating by fractional crystallisation from essentially basaltic magmas, but from parents which are already unusually enriched in either Na or K or both. Phonolites, for example, are in many important respects similar to rhyolites. They are both rock types rich in feldspathic constituents and highly depleted in Mg, Fe and Ca. However, because their parental magmas are basalts already enriched in Na, plagioclase fractionation involves on average more albitic plagioclases than in more normal magmatic suites. Albite contains less Al than anorthite, and hence phonolites are Al-rich (*c.* 17–20 per cent) relative to rhyolites (occasionally as low as 10 per cent). The high content of alkalis and the relatively low silica contents also ensure that quartz is not present, but is replaced by the feldspathoid nepheline.

Relatively Na-rich basaltic magmas, such as give rise to phonolites on fractionation, can be generated by lower-than-usual degrees of partial melting of mantle peridotites. If this effect is very slight (i.e. the parental magmas are only slightly more sodic than usual), the fractionation products are likely to be trachytes. On the other hand, magmas abnormally high in Na, e.g. nephelinites, and also usually silica-poor, seem to require for their generation the presence of CO_2 in their mantle source (see Chapter 4). If allowed to undergo fractional crystallisation, because phase relations have a habit of leading to converging residual liquid compositions (see earlier in this chapter) such magmas can also lead to phonolitic end-products.

The rare alkalic rocks enriched in K rather than Na are partly primitive magmas derived from the mantle and partly fractionation products from such parents. Potassium, as has already been mentioned, is an incompatible element in the main phases of the mantle, This means it is extremely susceptible to mobilisation by very low degrees of partial melting, and because of its chemical properties it is also relatively soluble in aqueous fluids. The mobility of K in the mantle means that it can be scavenged from some parts, and, correspondingly, deposited in others. The melting of K-enriched mantle domains can lead to the generation of small volumes of highly potassic magmas such as kimberlites, leucitites and absarokites. Leucitophyres, shoshonites, monzonites, and other evolved K-rich rock types can be generated by subsequent crystal fractionation.

To generalise: the fan of compositions displayed on the alkali-silica diagram can be seen as the consequence of crystal fractionation processes, or their equivalent partial melting processes, acting on basaltic, low-silica, magmas or rocks, which already have some degree of spread in their alkali contents. Crystal fractionation and partial melting processes generally lead to enrichment of the residual liquids or partial melts in both silica and alkalis, irrespective of the alkali-content of the parental basaltic liquids or source rocks. In the case of fractional crystallisation, extracts (the fractionating mineral assemblages) are usually silica-poor and alkali-poor relative to their liquids. In the case of partial melting, the refractory residues have the same characteristics. This is a very useful generalisation which, broadly speaking, gives an explanation of what we observe.

It is, finally, most important, however, to stress that a useful generalisation is not the same as a 'law of nature'. There are plenty of exceptions, which should warn us not to take things for granted. Figure 6.11 shows a number of specific cases, each of which is well documented and each of which shows some degree of exception to the general behaviour just described.

The trend A–A' shows the evolution of a relatively normal low-alkali basalt as it fractionates towards the rhyolitic end-point. It is the first part of the path which is exceptional, where fractionation of olivine, clinopyroxene and plagioclase leads to alkali enrichment *without silica enrichment*. The sharp bend in the path towards silica enrichment signals the entry of magnetite into the fractionating assemblage.

The second case, B–B', illustrates the fractionation path of a slight more alkalic basalt. Towards the silica-rich end of the trend, K-feldspar begins to dominate the fractionation process. The bulk composition of the extract at this stage is still silica-poor relative to the liquid, but it is alkali-rich. The path turns downward towards *alkali depletion*.

The third case, C–C', represents the fractionation trend of a yet more alkali-rich basalt. In this case, when the liquid is quite silica-rich, and very rich in alkalis, alkali feldspar again begins to fractionate in large quantities. Because of the high alkali content, alkali feldspar however begins to crystallise when silica is still quite low compared with the previous case. This time the extract is silica-rich and alkali-poor compared with the liquid. The subsequent fractionation path, towards the phonolite end-point, involves alkali enrichment as usual, but *silica depletion*.

Summary

Compositional variation in magmatic rocks is generated by fractional crystallisation of parental magmas and by variations in partial melting of source rocks both in the mantle and the crust. This concept can be verified by experiments which document the way magmas or simple synthetic analogues crystallise in the laboratory, and by the study of rock and mineral compositions. Many primary magmas originate in the mantle as basaltic or picritic liquids, and give rise by fractional crystallisation to more evolved magmas such as andesites, rhyolites, trachytes and phonolites. However, partial melting of existing crustal rocks in many instances also gives rise to rhyolitic (granitic) magmas, because, as a broad approximation, low degrees of partial melting give rise to products similar to those of high degrees of fractional crystallisation. As a result, in some cases it is difficult to be certain which process was dominant. The single unifying factor which controls all these processes is essentially the form of the appropriate phase diagram. The latter may be viewed as recording the relative solubilities of different minerals, which dictate the sequences in which they crystallise or melt. The compositions of the minerals which crystallise during cooling, and are removed from the liquid during fractional crystallisation, are the controlling factors on the compositional evolution of residual liquids. During melting, the compositional evolution of liquids is similarly controlled by the compositions of those minerals which contribute most to the melt, as opposed as those which are retained in the unmelted residue.

Further reading

Cox, K. G., Bell, J. D. & Pankhurst, R. J. (1979) *The Interpretation of Igneous Rocks*, Allen & Unwin, London, 445 pp.

CHAPTER

7

Chris J. Hawkesworth and Peter van Calsteren
Open University

GEOLOGICAL TIME

Geological Time stands as one of the Earth sciences' major contributions to the human understanding of the natural world. The realisation that successions of solid rock could be used to interpret something as ephemeral as time started a revolution much more profound than that experienced recently in the development of plate tectonic theory. Traditional concepts of a young Earth created in a short period of time, and perhaps for us to live on, were gradually replaced with an acceptance that human habitation has occupied a fraction of one per cent of Earth history. We have had to come to terms with a self-image of being the product of a gradual evolutionary chain, rather than chosen beings thoughtfully placed on a purpose-built planet.

Time remains one of those concepts which are instinctively understood until we try to describe them. It ebbs and flows between mathematical descriptions in the halls of science, and the wider freedoms of poetic imagination, and so it has different meanings in different contexts. We cannot be in more than one place at once, and information cannot be transmitted faster than the speed of light. Thus measurements of time are described relative to an observer, and a common example is that of a clock in a spacecraft travelling away from an observer close to the speed of light. As viewed by someone on Earth, that clock appears to run slower than his or her clock, even though it is inferred that both clocks would appear synchronous to any very distant universal observer.

Then there is what we might term psychological time: measurements of time were to Longfellow 'but the arbitrary and outward signs', for 'time is the life of the soul'. We perceive that time moves forward, that we cannot put the clocks back. Yet it is not time which passes, but we who move through the dimension of time. History is littered with ideas which placed man, or our Earth, at the centre of things, and perhaps time stands as one of the last of such anthropocentric notions. Thus there is time in the sense of a record of successive events, or **chronology**, which is how historians and geologists tend to perceive time. Historically, time was measured by counting the convenient cycles of night and day, or the seasons; however there exists another method of measuring time which is not based on cycles, and which is irreversible. It is based on the decay of **radioactive isotopes**, where the nuclei of one isotope transform into another at a specific rate. Such clocks are read by counting the nuclei produced by **radioactive decay**, and because the rate of decay is constant in space and time the readings are often referred to as **'absolute time'**. Thus, our understanding of geological time has been built up from two components: first, the recognition of **relative time** in the ordering of geological events and, second, the measurement of absolute time. Together these have provided fundamental information on the rates of change brought about by geological and evolutionary processes. At issue in many discussions has been the interplay between the directional and cyclical characteristics of time, and in particular the extent to which, in James Hutton's view 'the present is the key to the past'. More formally this is termed the **principle of uniformitarianism**.

Relative ages

The first logical step in determining the relative ages of rocks is the so-called **'law of superposition'**, which simply states that in a given sequence the bottom layers were laid down first and the top layers later, thus being younger. This law was formulated as early as 1669 by the Danish naturalist and theologian Nicholas Steno. It has become customary to date the beginning of geochronology from Steno's relevant treatise, but to do so is historically inaccurate. Even though the law of superposition is the first logical step, its enunciation did not represent the first historical step in the construction of the geological time scale. After all, Steno's law had no effect on contemporary calculations of the age of the Earth. During the 17th and much of the 18th centuries, when Earth history and human history were reckoned to be virtually coeval, the age of the Earth was calculated using historical documents, calendar systems and astronomical observations. Much of this represented scholarship of the very highest order, and some of the finest minds of modern times, ranging from James Justus Scaliger to Isaac Newton, quite apart from James Ussher, were deeply involved in this type of chronology.

The idea that rocks and fossils rather than just scrolls and books carry chronological information is of a much later date than Steno's law of superposition. The first step in the development of the modern time scale of Earth history took place when the basic premises of chronology changed. This happened during the period 1780–1830, when it was increasingly recognised that Earth history stretches back to long before the dawn of human history, and that the Earth's crust is an archive of its own historical development. This insight was brought about by the discovery that certain organic fossils represent extinct forms of life which preceded the earliest evidence of human habitation. At the same time, efforts were made to classify rock formations lithologically. The two great figures of early stratigraphy were the German mineralogist Abraham Gottlob Werner, whose version of a **stratigraphic column**, based on lithology, was of seminal importance, and the French founder of vertebrate palaeontology, George Cuvier, who recognised that certain assemblages of fossils are characteristic of particular formations. At about the same time, the Englishman William Smith also demonstrated that strata could be correlated using their fossil content.

These critical breakthroughs were followed by the flowering period of **stratigraphy** when such systems as the Silurian, the Jurassic and the Quaternary were described and defined. In Britain these developments were associated with the names of William Buckland, Adam Sedgwick, Henry de la Beche and Roderick Murchison, who defined formations regionally and correlated them globally in producing the modern relative time scale of Earth history (Figure 7.1). In doing so they became aware of the enormous, though finite, duration of geological time (Buckland spoke of 'millions of millions' of years), and of the progressive nature of the fossil succession, and thus the directional nature of Earth history. Opposition to this vision of the past came from two sides. Biblical literalists objected to the notion of an old Earth, whereas James Hutton and his followers, who believed in an indeterminate, cyclical Earth history ('no vestige of a beginning, no prospect of an end'), denied its directional nature. Charles Lyell and other Huttonians clung to aspects of this anti-progressionist view until as late as after the publication of Darwin's *Origin of Species* in 1859. At the same time, however, their uniformitarian view of an indeterminate vastness of geological time contributed to the idea of a very old Earth, and provided Darwin with the enormous length of time which his theory of evolution by natural selection required.

In the development of a stratigraphic column rock units were grouped into **'systems'**, and each of these was taken to represent a particular stretch of time or **'period'**. While such stratigraphic columns were being constructed, efforts were made to quantify the relative time scale of Earth history. Some calculations were based on the rate of sedimentation, others on the rate of denudation and the increase of salinity in the oceans. Such methods involved a variety of assumptions and they could be stretched to produce a wide variety of results. A seemingly more precise, and therefore authoritative dating technique came from the application of physics to the study of the Earth, and was developed by Lord Kelvin. Lord Kelvin was an eminent physicist, and in 1897 he calculated the age of the Earth on the assumption that it had cooled from an initial molten state. As the co-discoverer of the second law of thermodynamics, Lord Kelvin had strong views on the evolution of planetary bodies. If the Earth functions as a heat machine, as is now widely believed, then it must constantly be losing energy. Thus the Earth was hotter in the past than it is now, and he envisaged that at least as a habitable planet it was in the process of evolving from a fiery beginning to an icy end. Lord Kelvin made several estimates of

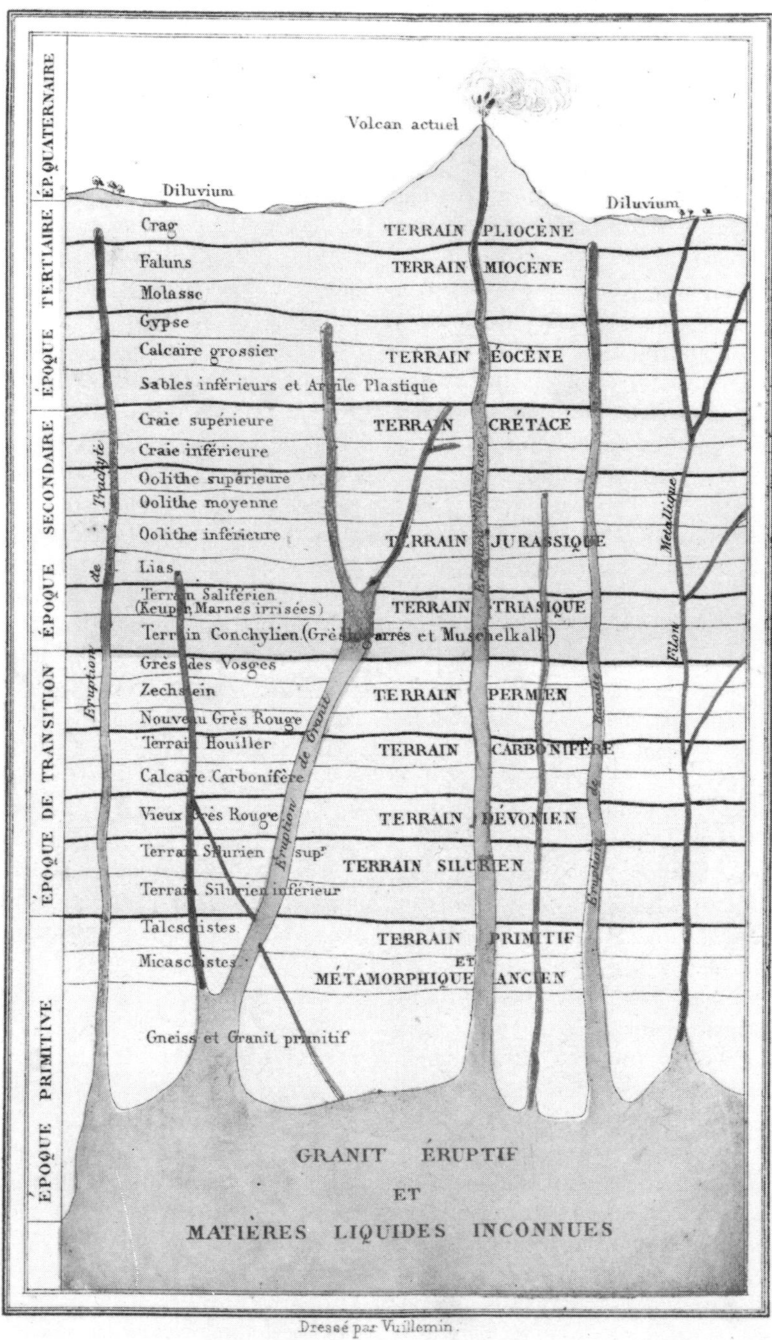

Figure 7.1 An early stratigraphic column, from L. Figuier, *La terre avant le Déluge*, published in 1864.

the age of the Earth, and as more data became available on the rate of increase of temperature with depth in the Earth's crust, and on the melting temperatures of rocks, he finally proclaimed an age of between 20 and 40 million years (Ma). So strong was the impact of these calculations, and so great the influence of precisely determined mathematical figures arrived at by making seemingly valid assumptions, that attempts were made to compress the complexities of the Earth's history into this very short period of time. For a while, the mathematical models of Kelvin the physicist held sway.

But Lord Kelvin was wrong. He had not taken into account that most of the heat movement within the Earth is by convection, rather than conduction (see Chapter 3), and he did not know that the isotopes of certain elements were unstable and, with time, decay to form other isotopes. This process of radioactive decay involves the liberation of energy in the form of heat, and so the thermal history of the Earth is that of a body with its own internal heat supply which can therefore stay warm for very much longer than Kelvin could have predicted. However, the second key point about radioactive decay is that it can be used to date rocks and minerals, and so it provided the first opportunity of determining absolute ages.

Radiometric ages

It was realised early in the 20th century by Rutherford and Holmes in England, and by Boltwood in America, that the decay of unstable isotopes to produce radiogenic isotopes could be used to date the minerals and rocks of the Earth's surface. However, at that time, analytical techniques were nowhere near precise enough to isolate and to determine the minute quantities of the radiogenic isotopes present in rocks. So it was not until the 1950s, when better mass-spectrometers were developed, that rock dating became commonplace.

Different isotopes of the same element have the same atomic number, but different atomic weights. Isotopes that are the product of radioactivity are then called radiogenic, and radiometric ages are those calculated from the accumulation of radiogenic isotopes. In many cases **radiometric ages** may be regarded as absolute ages in that they provide precise estimates of when minerals or whole rocks formed. However, geological processes may conspire to disturb the elements involved in a decay scheme and so radiometric ages need to be interpreted with caution. The subject of the dating of rocks and minerals is known as **geochronology**.

Radioactivity is a property of the nucleus and is independent of physical or chemical states; thus it is not affected by temperature, pressure or gravity or by the minerals or rocks that contain the radioactive isotope. The probability of a radioactive disintegration taking place depends only on the character of the nucleus and it is specific for each radioactive isotope. The number of atoms which disintegrate in a specific period of time just depends on the number of radio-active atoms which are present and, because no new radioactive isotopes are produced, the original quantity is reduced exponentially. More formally, the **law of radioactive decay** states that the number of atoms disintegrating per unit time ($-dN/dt$) is proportional to the total number of radioactive atoms (N) which are present. Thus $-dN/dt = \lambda N$, where λ is the decay constant and has a characteristic value for each radioactive isotope. This may also be expressed by the simpler concept of a half life $T_{\frac{1}{2}}$, the time required for half a given number of atoms of a radioactive isotope to decay.

The relationship between the **decay constant** and **half life**, and the way in which ages are calculated for the widely used decay scheme of ^{87}Rb (parent isotope) to ^{87}Sr (daughter isotope), are developed in more detail in Box 7.1 and Table 7.1.

Table 7.1 also summarises some of the decay schemes which are widely used in the dating of rocks and minerals. Isotope compositions are expressed as **isotope ratios**, $^{87}Sr/^{86}Sr$, $^{207}Pb/^{204}Pb$ etc., and the key point is that to calculate an age we need to know the difference between such isotope ratios at the time the sample was formed and the present day. However, as we clearly do not have samples of any old rocks at the time they formed, in practice the equations of radioactive decay are reorganised so that ages can be

Table 7.1. *Summary of the most widely used decay schemes. Half life can be calculated from* $T_{\frac{1}{2}} = \ln 2/\lambda$.

Parent	Daughter	Decay constant	Half life
^{14}C	^{14}N	1.21×10^{-4}	5730 a
^{87}Rb	^{87}Sr	1.42×10^{-10}	4.88 Ga
^{40}K	^{40}Ca	4.962×10^{-10}	1.40 Ga
^{40}K	^{40}Ar	5.81×10^{-9}	110 Ma
^{138}La	^{138}Ce	6.54×10^{-12}	106 Ga
^{147}Sm	^{143}Nd	6.42×10^{-12}	108 Ga
^{176}Lu	^{176}Hf	1.96×10^{-11}	35.3 Ga
^{187}Re	^{187}Os	1.52×10^{-11}	45.6 Ga
^{230}Th	^{226}Ra	9.217×10^{-6}	75.2 ka
^{232}Th	^{208}Pb	4.9475×10^{-11}	14 Ga
^{234}U	^{230}Th	2.794×10^{-6}	248 ka
^{235}U	^{207}Pb	9.8485×10^{-10}	704 Ma
^{238}U	^{206}Pb	1.55125×10^{-10}	4.468 Ga

BOX 7.1

The law of radioactive decay states that the number of atoms disintegrating per unit time ($-\mathrm{d}N/\mathrm{d}t$, with the negative sign indicating that N is decreasing with time) is proportional to the total number of radioactive atoms (N) which are present. Thus $-\mathrm{d}N/\mathrm{d}t = \lambda N$, where λ is the decay constant. On integration this becomes $N_o = Ne^{\lambda t}$, where N_o is the number of radioactive atoms at the formation of the sample, N is the number present today, and e is the base of natural logarithms. The number of radiogenic isotopes N_R, plus the number of radioactive parent isotopes N still present today, is equal to the number of radioactive parent isotopes present at the time of formation N_o, i.e. $N_o = N_R + N$. Substituting this into the previous equation gives:

$$N_R = N(e^{\lambda t}-1). \quad (1)$$

Two further points may then be illustrated using the decay scheme ^{87}Rb to ^{87}Sr. ^{87}Sr is the radiogenic isotope, and so $^{87}\mathrm{Sr} = {}^{87}\mathrm{Rb}\,(e^{\lambda t}-1)$. However, mass-spectrometers are more simply used to measure ratios of isotopes rather than their absolute abundance, and it is therefore convenient to rewrite that equation in the form $^{87}\mathrm{Sr}/^{86}\mathrm{Sr} = (^{87}\mathrm{Rb}/^{86}\mathrm{Sr})\,(e^{\lambda t}-1)$: ^{86}Sr being chosen because it is neither radioactive nor radiogenic, and so its overall abundance is assumed to be constant. The second point is that in the case of strontium, some of the ^{87}Sr is not radiogenic (i.e. it was produced during nuclear synthesis, see Chapter 2), and even some of the radiogenic ^{87}Sr will have been generated before the formation of the sample, at time t, which is being dated. Thus, to calculate the age of a sample it is first necessary to subtract the ^{87}Sr which was already present at time of formation t from that which is measured today. This allows us to estimate the amount of ^{87}Sr actually produced by the radioactive decay of ^{87}Rb since time t. So,

$$\frac{^{87}\mathrm{Sr}}{^{86}\mathrm{Sr}} - \left(\frac{^{87}\mathrm{Sr}}{^{86}\mathrm{Sr}}\right)_o = \left(\frac{^{87}\mathrm{Rb}}{^{86}\mathrm{Sr}}\right)(e^{\lambda t}-1),$$

or,

$$\frac{^{87}\mathrm{Sr}}{^{87}\mathrm{Sr}} = \left(\frac{^{87}\mathrm{Rb}}{^{86}\mathrm{Sr}}\right)(e^{\lambda t}-1) + \left(\frac{^{87}\mathrm{Sr}}{^{86}\mathrm{Sr}}\right)_o, \quad (2)$$

where $(^{87}\mathrm{Sr}/^{86}\mathrm{Sr})_o$ is the strontium isotope ratio which was present at the time of formation of the sample, t million years ago, and it is referred to as the **initial strontium isotope ratio.** The initial isotope ratio will itself vary depending on the history of the source rocks from which the rocks being dated were derived. For example, the Earth's upper mantle has low Rb/Sr ratios, so it has low Sr isotope ratios, and any rock derived from the upper mantle will therefore have low initial Sr isotope ratios. In contrast, the continental crust is characterised by high Rb/Sr, and so provided it is old enough it will have high $^{87}\mathrm{Sr}/^{86}\mathrm{Sr}$ and rocks derived from older segments of the continental crust will have high initial Sr isotope ratios.

If the $(^{87}\mathrm{Sr}/^{86}\mathrm{Sr})_o$ is assumed, an age may be calculated, and that is called a **model age** in that it clearly depends on the choice of $(^{87}\mathrm{Sr}/^{86}\mathrm{Sr})_o$. However, if the sample has a high $(^{87}\mathrm{Rb}/^{86}\mathrm{Sr})$, as for example in micas, and it is not too young, the calculated model age is not very sensitive to the choice of $(^{87}\mathrm{Sr}/^{86}\mathrm{Sr})_o$. Model ages are much used in the treatment of Nd isotope variations and these are discussed in more detail in Chapter 8.

Equation (2) is in the form of a straight line ($y = mx+c$), and in a diagram of $^{87}\mathrm{Sr}/^{86}\mathrm{Sr}$ against $^{87}\mathrm{Rb}/^{86}\mathrm{Sr}$ it defines a straight line with a slope of $(e^{\lambda t}-1)$, and an intercept on the $^{87}\mathrm{Sr}/^{86}\mathrm{Sr}$ axis corresponding to $(^{87}\mathrm{Sr}/^{86}\mathrm{Sr})_o$ (Figure 7.2a). Such a line is called an **isochron,** since it connects points of equal age. Clearly at least two samples must be analysed for an isochron line to be drawn, and in practice the more samples that are analysed and the greater the spread in isotope ratios, the more precisely will the age t be determined.

It further follows from equation (2) that a diagram of $^{87}\mathrm{Sr}/^{86}\mathrm{Sr}$ against time t will also yield straight lines which illustrate the changing strontium isotope ratios of different samples, or segments of the Earth, with time (Figure 7.2b). These diagrams are called **isotope evolution diagrams**, and the slopes of the straight lines are equal to $^{87}\mathrm{Rb}/^{86}\mathrm{Sr}$. Similar isochron and isotope evolution diagrams can be constructed for all the decay schemes listed in Table 7.1, although because isotopes of both U and Th decay to Pb, Pb isotopes can be used in slightly different ways.

A major requirement of age calculations, as illustrated in Figure 7.2, is that the parent/daughter element ratios should not have been disturbed since the event which is being dated. Unfortunately, U is highly mobile in near-surface environments and so the U/Pb ratios of many whole rock samples have been recently altered. However, by combining the equations for the two U–Pb decay schemes (decay constants λ_1 and λ_2) it is possible to calculate ages without knowing the unaltered U contents.

7.1 continued

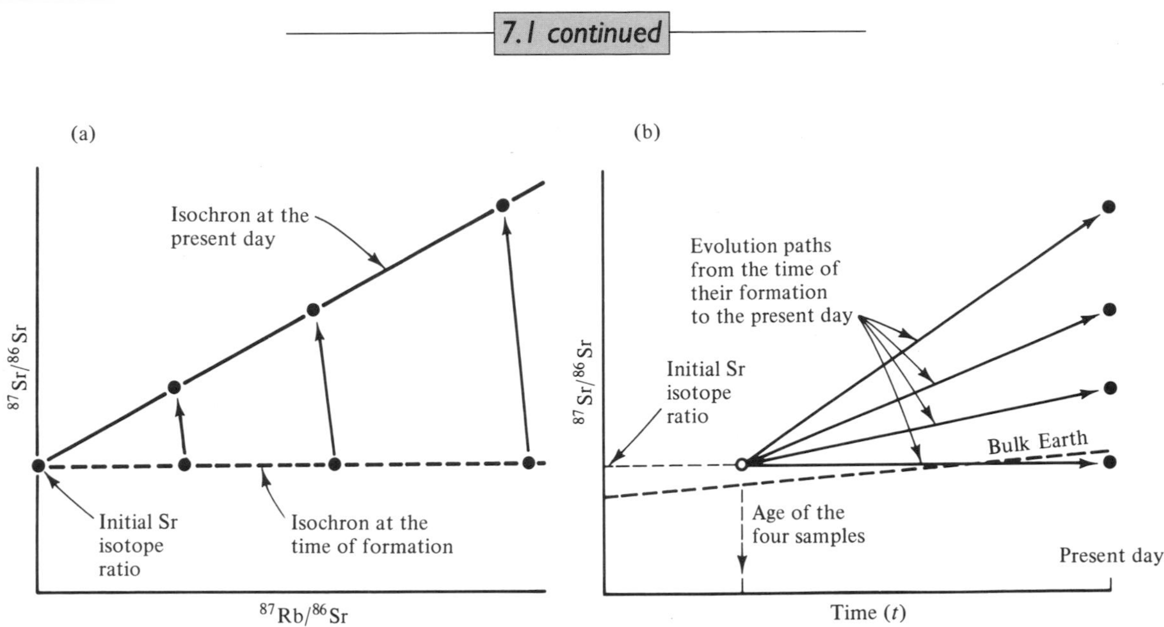

Figure 7.2 (a) An Rb–Sr isochron diagram illustrating the change in Sr isotope ratios in four samples with time from their formation to the present day. At the time of formation the samples had the same $^{87}Sr/^{86}Sr$ but different $^{87}Rb/^{86}Sr$, and since then their $^{87}Sr/^{86}Sr$ ratios have increased in response to the decay of ^{87}Rb to ^{87}Sr. The size of the increase in $^{87}Sr/^{86}Sr$ depends on the ratio of ^{87}Rb to ^{86}Sr, and in a sample with no Rb clearly $^{87}Sr/^{86}Sr$ does not change and thus it still records the initial Sr isotope ratio of the four samples. The slope of the isochron at the present day = $(e^{\lambda t}-1)$, and so once the slope has been determined by plotting the four points measured using a mass spectrometer, the age t can be calculated. (b) A Sr isotope evolution diagram illustrating the changes in Sr isotope ratios with time in the four samples plotted on the isochron diagram in (a). The slope of the evolution paths is proportional to the Rb/Sr ratio of the individual samples. The dashed line represents the evolution of the bulk Earth and, because the initial Sr isotope ratio of the four samples plots above that evolution line, it indicates that the source of those rocks already had a higher Rb/Sr ratio than the Earth. Since, in general, the rocks of the continental crust have higher Rb/Sr ratios than those of the upper mantle, it further suggests that these four samples were derived from source rocks within the continental crust. □

For the decay of ^{238}U to ^{206}Pb,

$$\frac{^{206}Pb}{^{204}Pb} = \frac{^{238}U}{^{204}Pb}(e^{\lambda_1 t} - 1) + \left(\frac{^{206}Pb}{^{204}Pb}\right)_o , \quad (3)$$

and for $^{235}U \rightarrow {^{207}Pb}$,

$$\frac{^{207}Pb}{^{204}Pb} = \frac{^{235}U}{^{204}Pb}(e^{\lambda_2 t} - 1) + \left(\frac{^{207}Pb}{^{204}Pb}\right)_o . \quad (4)$$

As before, the subscript $_o$ signifies the initial Pb isotope ratio, i.e. at time t. Combining these two equations,

$$\frac{\left(\frac{^{207}Pb}{^{204}Pb}\right) - \left(\frac{^{207}Pb}{^{204}Pb}\right)_o}{\left(\frac{^{206}Pb}{^{204}Pb}\right) - \left(\frac{^{206}Pb}{^{204}Pb}\right)_o} = \frac{^{235}U}{^{238}U}\,\frac{e^{\lambda_2 t} - 1}{e^{\lambda_1 t} - 1} \quad (5)$$

The ratio $^{235}U/^{238}U$ is a constant equal to 1/137.8 for all uranium of normal composition in the Earth, Moon and meteorites at the present time. Note that ^{235}U and ^{238}U have different decay constants (Table 7.1), and so the $^{235}U/^{238}U$ ratio has changed with time. Thus the rate at which $^{207}Pb/^{204}Pb$ increased with time relative to $^{206}Pb/^{204}Pb$ has also changed, and that is why the Pb isotope evolution lines in Figure 7.3 are curved. Note also that on a diagram of $^{207}Pb/^{204}Pb$–$^{206}Pb/^{204}Pb$ a suite of related rocks and minerals (i.e. of the same age and with the same initial Pb isotope ratios) should plot on a straight line of slope m:

$$m = \frac{1}{137.8}\,\frac{e^{\lambda_2 t} - 1}{e^{\lambda_1 t} - 1} . \quad (6)$$

7.1 continued

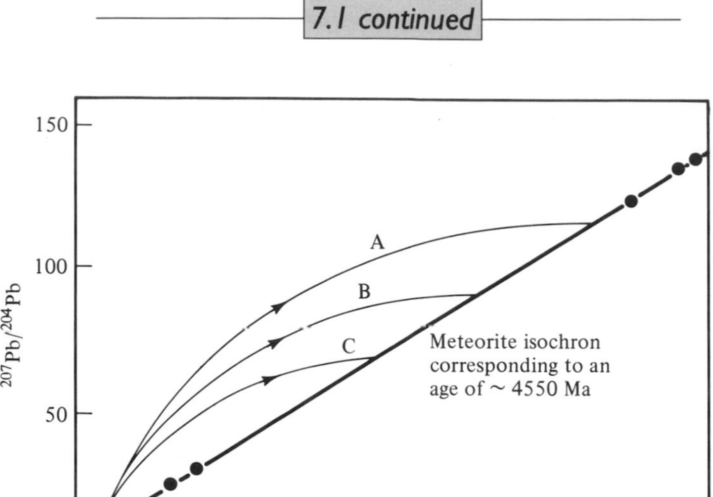

Figure 7.3 Lead (Pb) isotope ratios from meteorites plotted on a diagram of $^{207}Pb/^{204}Pb$ versus $^{206}Pb/^{204}Pb$. The slope of the straight line defined by the data points corresponds to an age of ~ 4550 Ma, which is interpreted as the age of the meteorites and of the Earth. The mineral troilite contains no uranium and so its Pb isotope ratios are the same now as when it was formed, giving the best available estimate of the initial Pb isotope ratios of meteorites and the Earth. Curves A, B and C are schematic evolution paths illustrating how the Pb isotope ratios of three samples with different U/Pb ratios changed with time. The U/Pb ratio is highest in A and lowest in C, and the shape of the curves reflects changes in the $^{235}U/^{238}U$ ratio with time. Early in Earth history there was relatively more ^{235}U so that more ^{207}Pb was generated, and the curves were steeper. Later on, when most of the ^{235}U had decayed, most of the radiogenic Pb produced was ^{206}Pb from ^{238}U, and so the curves, or evolution paths, flatten out. □

Thus a straight line on these Pb–Pb diagrams may be interpreted as an isochron, in which case the time t can be calculated from its slope. This technique made an important contribution in that it was one of the first to yield reliable estimates of the age of the Earth (see Figure 7.3 and its caption). However, it must be emphasised that whether or not straight lines on such diagrams are isochrons is still a matter of interpretation. In addition, the Pb–Pb isochron diagrams such as Figure 7.3 differ from those for Rb–Sr (Figure 7.2a) in that the initial Pb isotope ratios plot within the diagram, rather than on the y axis.

calculated from two or more samples of the same geological unit which have different isotope ratios at the present day (see Box 7.1). And different isotope ratios require samples with different parent/daughter element ratios, i.e. Rb/Sr, in the case of the ^{87}Rb to ^{87}Sr decay scheme.

In addition to finding samples with different parent/daughter element ratios, two other conditions must be fulfilled before geologically meaningful radiometric ages can be calculated. The first is that subsequent processes of alteration, weathering or metamorphism should not have affected the trace element or the isotope ratios involved in the decay scheme since the event which is being dated. Such processes are likely to result in a scatter of analyses, and so one test is how well the results define a straight line on an isochron diagram (Figure 7.2a). Second, all samples which are contributing to the age calculation must have had the same (initial) isotope ratio at the time of their formation. Experience shows that for rock samples this is much more likely in igneous rocks which may have crystallised from the same magma, and it is sometimes true for metamorphic rocks if metamorphism was sufficiently pervasive to homogenise the isotope ratios. However, the same initial isotope ratio is much less likely to occur among sediments with various amounts of detritus from different sources. The age of sedimentation is therefore difficult to determine directly by radiometric methods, and since the relative (stratigraphic) time scale was clearly developed from sediments, other less direct ways had to be found to establish absolute ages within the stratigraphic column, and thereby to establish its calibration in absolute terms. In practice, the most common is to use igneous rocks as time markers, either because they occur as volcanic rocks within sedimentary sequences, or because they intrude sediments and so provide minimum age limits (see Figure 7.4). More recently it has been demonstrated that under favourable conditions chemical sediments, such as limestones, can be dated directly by Pb isotopes, and by comparing their Sr isotope ratios with the changing value of Sr isotopes in sea water, particularly over the last 400 Ma.

The preceding discussion has been largely concerned with the ages of rock samples, but much geochronology has involved the dating of minerals rather than rocks. Initially minerals were chosen because some had very high parent/daughter element ratios (i.e. high Rb/Sr in Figure 7.2), so their isotope ratios changed relatively rapidly with time, and they were therefore easier to analyse and they yielded more precise ages. That remains a powerful reason and, for example, biotites have very high Rb/Sr ratios and so tend to yield significantly more precise Rb–Sr ages than their host rocks, whereas garnets have high Sm/Nd ratios and thus yield more precise Sm–Nd ages than their host rocks. However, there is a second reason, namely, that in some circumstances the minerals in a rock may preserve a geologically useful age which is different from the age of the rock itself.

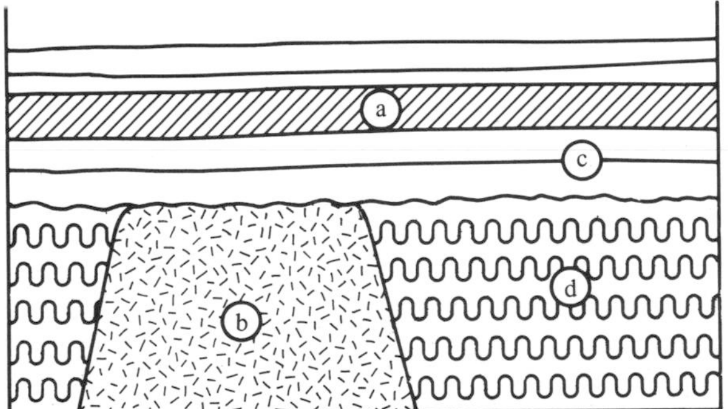

Figure 7.4 A geological cross-section illustrating a lava (a) within a sedimentary sequence (c), and an intrusive igneous rock (b) in a sequence of older sediments (d). The ages of the igneous rocks can be determined by isotopic methods, and so the age of the lava (a) indicates when the sediments (c) were being deposited, whereas the age of the intrusion (b) offers a maximum age for the sediments (c), and a minimum age for the older sediments (d). □

In general, if different decay schemes, or even separated minerals and their host rocks analysed for the same decay scheme, yield the same ages, the ages are termed **concordant**, and if the ages are different, they are termed **discordant**. Concordant ages indicate that the rocks have had a relatively simple geological history, whereas discordant ages are both evidence for a complex geological history and extremely useful when it comes to studying an area where the effects of more than one geological event have been superimposed on one another.

For example, clastic sediments are made up of detrital minerals eroded from pre-existing rocks. Thus if the minerals are sufficiently robust to survive weathering and erosion, they may preserve an isotopic record of their age from the pre-existing rocks, and in some areas this is one of the few ways of assessing the ages of sediment source terrains (see the discussion of Nd isotopes in Chapter 8). Good examples of minerals which preserve *older* age information through cycles of erosion and deposition are zircons, which may be dated by U–Pb, and garnets which may be dated by Sm–Nd. More common are minerals which are *younger* than their host rocks, either because different minerals remain open to the diffusion of elements, and hence isotopes, under different conditions, or simply because the rocks were metamorphosed, and so the minerals grew, in an event subsequent to the original formation of the rock.

If the temperature is high enough minerals will remain open to diffusion on the relatively long timescales which are available in geological systems. Then, as the system cools, it reaches a critical temperature, known as the **blocking temperature**, at which diffusion effectively ceases. The blocking temperature is different for different minerals, and for different elements, and the age obtained from a mineral analysis is the time at which that mineral cooled through the blocking temperature for the particular decay scheme. Thus detailed studies of the ages of different minerals within a single rock can unravel complex cooling histories from as high as 900 °C down to 100 °C.

If a rock is heated during a subsequent metamorphic event, the distances over which isotopes can move by diffusion in a given time increase. Provided that there is no fluid present, the isotopes of elements such as Rb and Sr are unlikely to move more than a few centimetres. Thus if the rock samples analysed are bigger than the distances moved by diffusion, their isotope and parent/daughter ratios will not have been disturbed by the metamorphic event. However, minerals

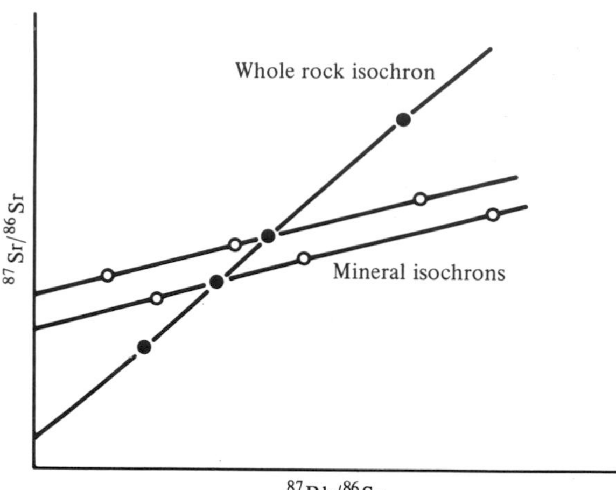

Figure 7.5 Whole rock and mineral Rb–Sr isochrons for a metamorphosed granite. Whole rocks are defined as rock samples (solid symbols) big enough not to have had their isotope and parent/daughter ratios disturbed during metamorphism. In this example analyses have been undertaken on minerals (open symbols) from two rock samples, and since the minerals are of the same age they plot on two parallel lines, with the same slope. The slope of the whole rock isochron is proportional to the age of metamorphism. □

separated from within a rock sample are likely to have re-equilibrated during metamorphism, and so their isotope and parent/daughter ratios will yield a new age, the **age of metamorphism** (Figure 7.5).

Earth history

The early development of stratigraphic columns relied on the presence of fossils, and the underlying rocks which contained no obvious fossils were simply labelled 'basement'. Inevitably the fossiliferous rocks were regarded as the major part of the geological record, and it came as a major surprise when absolute dating techniques demonstrated that most common fossils occur in rocks which are less than 500 Ma old. The oldest remains of hominoid species, by comparison, have been dated at just 1.7 Ma. A great deal of effort has gone into the establishment of a **Geological Time Scale**, and its major features are summarised in Figure 7.6.

A major scientific goal has been the age of the Earth, and so it is worth pausing to consider the events which were happening at about the time of the Earth's formation, and which might be amenable to dating. As indicated above, the age of formation of an igneous rock can be determined by analysing different rock

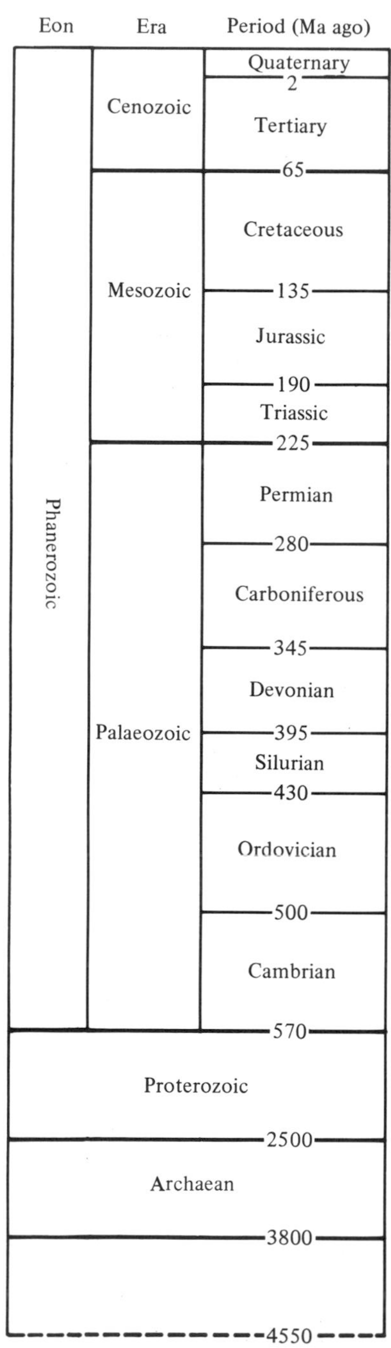

Figure 7.6 Geological time scale; the dates are based on the recommendations of the IUGS stratigraphy commissions. □

samples from the same body. Since those samples crystallised from the same magma, they are likely to have had the same initial isotope ratio, and the age obtained is that when the magma had crystallised sufficiently for the isotopes no longer to be able to diffuse outside the size of the samples analysed. The Earth is most unlikely ever to have been homogeneous in a way that might be analogous to a crystallising magma. Instead it formed in a number of stages which appear to have taken place within a sufficiently short space of time that their ages cannot be distinguished using presently available techniques. Isotope studies effectively date the last time when different portions of individual planetary bodies had similar (initial) isotope ratios, and when they developed significant variations in parent/daughter element ratios.

It is commonly assumed that the Solar System formed by the gravitational collapse of a cloud of interstellar gas and dust spiked with some material produced in a nearby supernova (Chapter 2). The supernova material contained some radioactive isotopes such as ^{26}Al, ^{129}I and ^{107}Pd with very short half lives (0.72 Ma, 17 Ma and 6.5 Ma respectively), so that they are now extinct, but for which evidence has been found in meteorites. From these data it has been argued that condensation to millimetre-sized particles in the solar nebula took place ~ 3 Ma after the supernova explosion, that 100 km bodies with differentiated cores were formed after another 40 Ma, and that it took a further 100 Ma for the 100 km bodies to coalesce into planets. Such times are simply relative, and the evidence for when these events took place had to be obtained from radioactive isotope systems which are still active at the present day. However, there are no samples of the Earth which have remained unchanged since its formation, so that estimates of when the Earth formed must be inferred from direct measurements on fragments of meteorites which have similar compositions to the bulk Earth, and from model ages of the Earth's crust (see below).

Figure 7.7 summarises the Rb–Sr isotope whole rock results on a number of chondritic meteorites, and they yield an age of 4520 ± 50 Ma (4.52 ± 0.05 Ga). The ages of minerals separated from these meteorites are indistinguishable from the ages of the whole rocks, and so they have experienced a relatively simple 'geological' history with no significant metamorphism since the time of their formation. Such chondrites are similar in composition to the Earth, and the age is taken to reflect the time that bodies of ~ 100–150 km diameter were formed, i.e. before they, in turn, coalesced into planets.

Model ages are ages calculated between a sample and some inferred original source material, such as between a fragment of the Earth's crust and the upper mantle from which it was derived (see also Chapter 8). It is not easy to assess the errors on any model age calculation, but Pb model ages provide one estimate for the age of the Earth. Some meteorites contain the iron sulphide mineral troilite, and since this has relatively large Pb contents and virtually no U, the measured Pb isotope ratios have not been changed by the radioactive decay of U since the troilite crystallised (see also Figure 7.3). Thus the measured Pb isotope ratios of troilite are the same as when the host mineral formed, and analyses of several Fe-rich meteorites have shown that their Pb isotope ratios at the time of formation were remarkably uniform. It is concluded

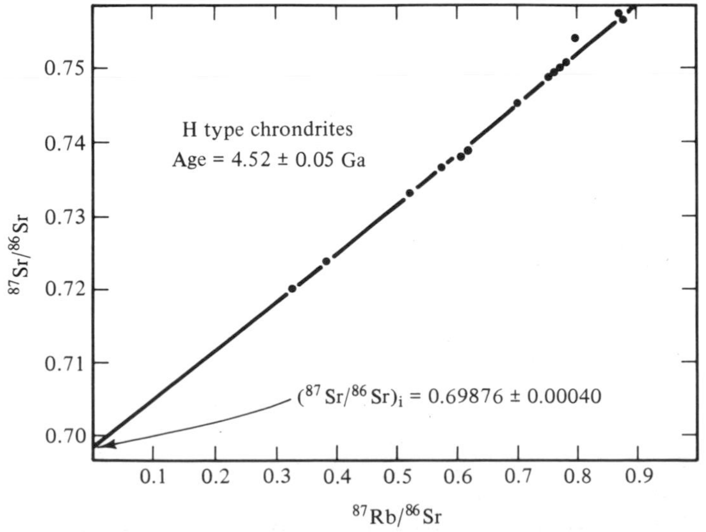

Figure 7.7 A Sr isochron diagram illustrating the $^{87}Sr/^{86}Sr$ and $^{87}Rb/^{86}Sr$ analyses on whole rock fragments of selected chondritic meteorites. They plot in a linear array, and the slope of the best-fit straight line yields an age of 4520 ± 50 Ma. □

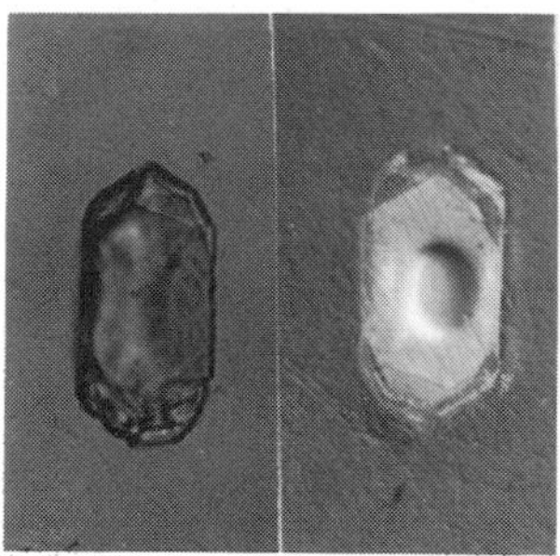
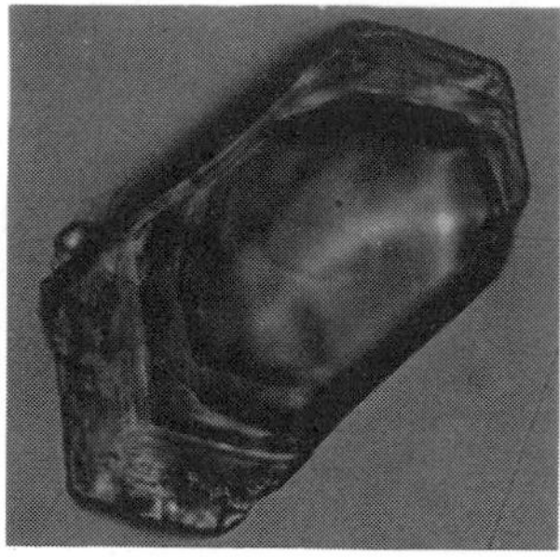
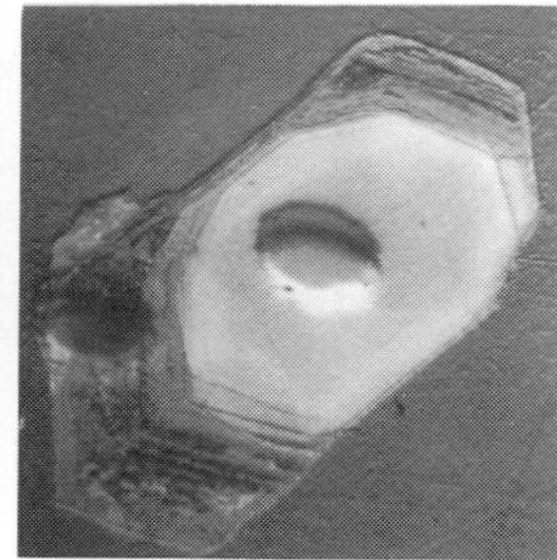

Figure 7.8 Photographs of two zircon crystals of which ages have been determined without chemical processing by using an ion beam to ablate the mineral slowly. On the left are transmitted light views and on the right are the reflected light pictures which clearly show the ~30 μm-wide/2 μm-deep ablation crators. The sputtering produces enough U and Pb ions to be analysed in the double focussing mass-spectrometer of this instrument known as the Ion Probe. □

that the Solar System was initially homogeneous with respect to Pb isotopes, and that the Pb isotope ratios of troilite are therefore similar to those of the bulk Earth at the time of its formation. On Earth the Pb isotope ratios have become more radiogenic with time, from the decay of U, and so if the present-day Pb isotope ratio of the bulk Earth can be estimated, its age can be calculated by reference to its inferred initial Pb isotope ratio. The best estimate of the present-day Pb isotope ratio of the Earth is probably that of oceanic sediments, and assuming that the initial Pb isotope ratio of the Earth is the same as troilite from meteorites, this yields an age for the Earth of 4550 Ma.

Although both the model Pb calculations and the available data from meteorites indicate that the Earth is 4550 Ma old, any direct record of the earliest rocks formed on Earth has been obliterated. The causes are a matter for speculation, but they include the effects of meteorite bombardment, and the much higher temperatures that were present early in Earth history and which caused rapid crustal reprocessing. The oldest known rocks are a suite of metasediments, volcanic rocks and associated tonalitic gneisses, the Amitsoq Gneisses, in Western Greenland which have been dated at 3800 Ma by several isotope techniques. Moreover, even though these rocks include sediments, which must in turn have been derived from some pre-existing source terrain, there is no isotope evidence to suggest that that source terrain was detectably older than 3800 Ma. Elsewhere, in Western Australia, a few zircon grains have been analysed which yield ages as old as 4200 Ma, even though the rocks in which they sit are only 3500 Ma old. Thus at present the geological record on Earth starts inauspiciously with just one or two minute grains of the mineral zircon (Figure 7.8), which were robust enough to survive for 4200 Ma (these grains are discussed further in Chapter 8).

But what is the present age distribution of rocks on the Earth's surface, and what may be inferred from the distribution of ages? Some 60 per cent of the Earth's surface consists of oceanic crust, and none of that is older than 200 Ma since it is continually being created at ocean ridges and destroyed at subduction zones. The remaining 40 per cent is continental crust and although ~30 per cent of it is submerged along continental margins there is increasing agreement on the relative proportions of continental crust of different ages. Figure 7.9 summarises the proportions (by volume) of continental crust of different ages, determined simply by estimating the areas of rocks of different ages and assuming that the ages do not vary with depth in the continental crust. The ages are those of the last major tectonic and thermal event in any area, and clearly the age distribution is biased towards the younger events and not many segments of old continental crust have survived until the present day.

It is a matter of long-standing debate whether, as perhaps suggested by the age distribution in Figure 7.9, the volume of continental crust has grown from 3200 Ma to the present day, or whether the volume has remained constant and most of the older crustal material has simply been reworked by the younger tectonic and thermal events. Such questions are difficult to answer geochemically, but it is now possible to investigate the relative amounts of new crust generated in different periods of Earth history. Nd isotopes are discussed in more detail in the next chapter, but they provide crust formation ages and the dashed lines in Figure 7.9 illustrate how the percentage of new crust generated has changed with time. Significantly,

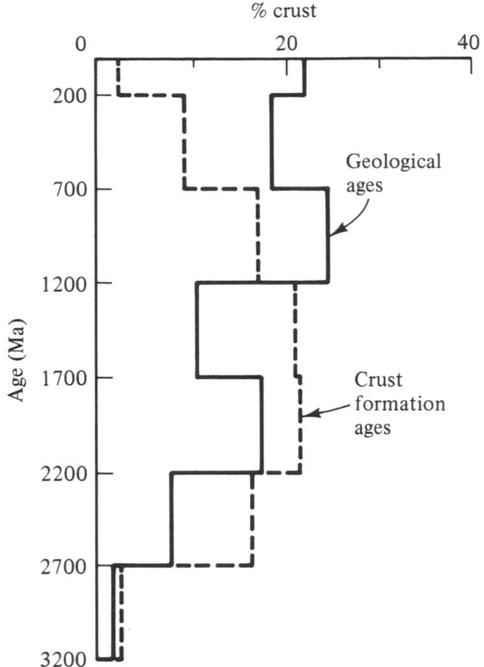

Figure 7.9 A summary of the volumes of continental crust of different ages, compared with the minimum proportion of new continental crust (by mass) generated at different times. The latter are estimated from the variations in Nd isotopes in continental sediments and granitoids discussed in Chapter 8. □

although approximately two-thirds of the continental crust yields ages which are less than 1200 Ma, it is estimated that only one-third of the crust was actually generated in that time. Overall the history of the continental crust is one of a sequence of tectonic and thermal events which have changed in tectonic style, in the composition of some of the associated magmatic rocks, and in the proportion of new crust generated in any event. Its evolution has been cyclical in that the crust has been affected by repeated episodes of tectonic activity, but because the Earth continues to cool down, the nature of those events has changed and so evolution has also been directional. Thus the evolution of the continental crust mirrors those discussions which highlight the directional and cyclical nature of time itself.

Thermal histories

Rocks conduct heat slowly, and both the thermal conductivities of common rock types and the major sources of heat within the Earth are reasonably well known. Thus a widespread application of geochronology is to provide the age framework for estimates of the thermal histories of particular areas, since these in turn constrain models for their tectonic evolution. The simplest example is when igneous rocks of different compositions are generated at different times in the development of an orogenic belt. Thus the tectonic models for the Himalayas are very dependent on the ages of the igneous rocks generated (i) in the Tethyan Ocean prior to continental collision, (ii) above the subduction zone as that ocean basin was destroyed, and (iii) by the partial melting of the continental crust in response to the increase in temperatures which resulted from the thickening of the continental crust after plate collision (see Chapters 12 and 14).

In more detail, analyses of different isotopes in different minerals from a single rock can chart the rate at which it cooled. Cooling may be linked to the cessation of igneous activity, since one obvious source of additional heat to the Earth's crust is the upward movement of hot magmas. However, cooling is also linked to uplift and erosion, and the amount and rate of uplift depends on the amount of prior crustal thickening, and hence on the regional tectonics (Chapter 14).

One technique which is now much used in the study of thermal histories at relatively low temperatures, is that of **fission tracks**. When heavy charged particles travel through a solid they leave a trail of damage, and such tracks are produced by the products of radioactive decay of isotopes such as ^{238}U. The host mineral is polished and etched so that the tracks can be more readily seen and counted, and then the number of tracks in a given area is proportional to the U content of the mineral (which can be measured), and to the time since it cooled through its blocking temperature for ^{238}U. For minerals such as apatite the blocking temperature is as low as 100 °C, and so it is applicable to the study of sedimentary basins which might, for example, contain significant oil or gas reserves, as well as to the cooling histories of igneous and metamorphic rocks.

Figure 7.10 summarises the ages from a number of radioactive decay schemes determined on minerals separated from an intrusive igneous rock in Scotland, and compares them with ages determined on minerals from rocks in the Himalayas. The calculated ages are plotted against the blocking temperatures for the different decay schemes in the different minerals, and although some of the estimated blocking temperatures should be viewed with caution, some useful patterns can be detected. **Cooling curves** which are concave-up are diagnostic of rapid cooling, as follows the emplace-

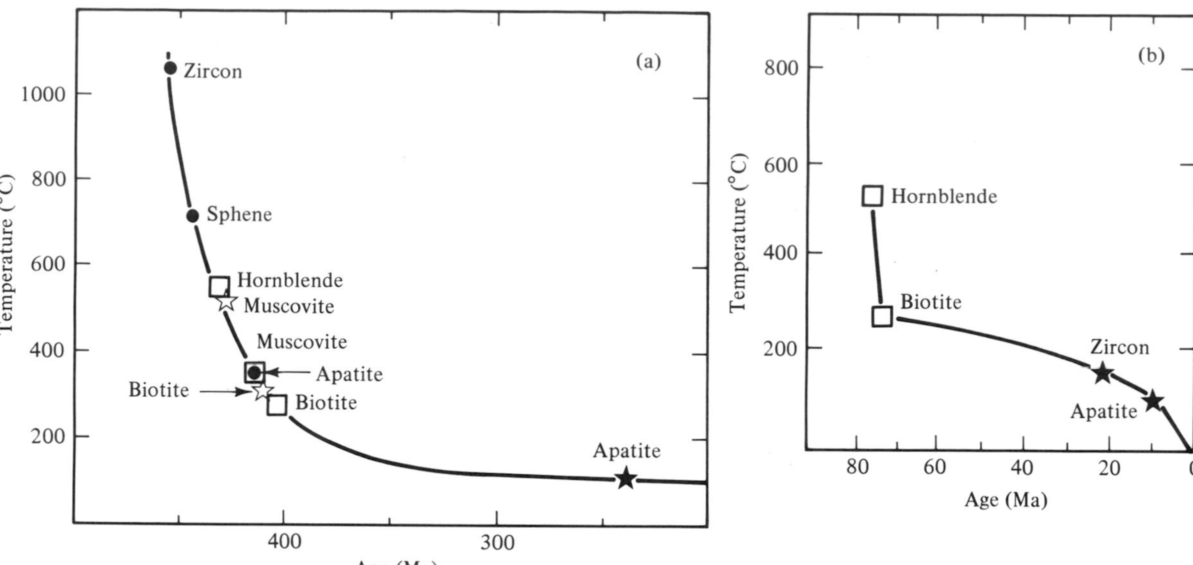

Figure 7.10 Mineral cooling ages from (a) the Glen Dessary Syenite in Scotland, and (b) a section within the Kohistan Arc in the Himalayas. Closed circles represent U–Pb ages, filled stars are Rb–Sr, open squares are K–Ar and open stars are fission tracks. □

ment of a hot magma, whereas curves that are convex-up are diagnostic of cooling resulting from the uplift and erosion of tectonically thickened crust (Chapter 12). Thus, although the Scottish example simply reflects cooling of the igneous intrusion, the Himalayan rocks appear to have cooled quickly after igneous activity ~ 80 Ma ago, and then more slowly in response to uplift and erosion.

Concluding comments

Radiometric dating has established the age of the Earth, and of the major changes in its history, and in so doing it has revolutionised our understanding. Microbial colonies have left fossil records (stromatolites) of life since 3500 Ma, the first vertebrates appeared ~ 530 Ma ago, the first land fossils occur at the end of the Silurian, and hominoid species first appeared a mere 1.7 Ma ago. We may not readily comprehend the vastness of the lengths of time involved, but it is no longer questioned that evolutionary change is extremely slow relative to our own life spans, and that we are simply the current state of an evolutionary chain which should continue unless we contrive to break it. Such acceptance may be the legacy of any successful revolution, but even now it makes for sharp contrast with the sense of wonder and excitement expressed just over thirty years ago at the end of a short book entitled '*How Old is the Earth?*'.

> How majestic are these broad reaches of time! Looking into an abyss, one senses the gigantic form of the void only in comparison to one's own minute stature. It is almost incomprehensible that only a few billion years ago our galaxy was born in a giant bomb-flash of nuclear energy. What an inspiring picture of the process of creation! But awesome and inspiring as it is to contemplate this mighty spectacle, the true reward is not to be found in whether our calculations are correct, give or take a few million years: it lies in the discoveries, in the advancement of human knowledge and philosophy that are the inevitable products of scientific search for law in nature.
>
> (Hurley, 1959)

Further reading

Faure, G. (1986) *Principles of Isotope Geology*, John Wiley & Sons, 589 pp.
The basic text on the various radioactive decay schemes commonly used in the Earth sciences.

Hurley, P. M. (1959) *How Old is the Earth?* Heinemann, London, 160 pp.
A pleasing, easy to read short book addressing the question of radioactivity and how it changed our understanding of the age of the Earth.

Rupke, N. A. (1983) *The Great Chain of History*, Oxford University Press, 322 pp.
A stimulating description of the early 19th century British contributions to the revolutionary discovery that the Earth had existed long before man.

CHAPTER 8

THE CONTINENTS

Keith O'Nions
University of Cambridge

Introduction

The Earth, like the other terrestrial planets, formed close to 4.5 Ga (4500 Ma) ago and became chemically differentiated soon after. Within the first 100 Ma the bulk of the Fe–Ni core had formed and the interior became substantially degassed (see Chapters 2 and 3 for details). With the exception of the very lightest gases, such as helium, much of this early evolved gas has been retained by the Earth's gravitational field and forms part of the present-day atmosphere. No direct information is available to us about the crust during the earliest part of the Earth's development, simply because none has survived. However, the geological record that is present in the continents extends back clearly to about 3.8 Ga, and only the first 700 Ma or so is unrepresented.

Continental crust covers approximately 30 per cent of the Earth's surface and the oceanic crust makes up the remainder; continental and oceanic crust differ in a number of important ways. Oceanic crust is basaltic in composition, about 6 km thick, and is formed predominantly at spreading ridge axes. It reaches a maximum age of about 200 Ma in the Pacific and has a mean age of about 100 Ma. The continental crust, on the other hand, is about 40 km thick on average and more silica-rich, with a composition close to that of a tonalite. It has a maximum recorded age of ~ 3.8 Ga and a mean age of about 2.2 ± 0.3 Ga. Although the continental crust has less than 1 per cent of the mass of the entire mantle, it has concentrated a sizeable portion of the **incompatible trace elements** – those elements which show a strong preference for the liquid phase upon mantle melting. The heat-producing elements potassium, uranium and thorium are among the most important of these and are enriched in the continents by a factor of 200–400 over typical upper mantle values (see Chapters 4 and 12 for implications in terms of geotherms and melting conditions). A number of estimates of Earth composition place somewhere between 25 and 50 per cent of the Earth's total K, U and Th in the continents. The upper mantle sampled by volcanism at the spreading ridges shows a complementary depletion in the more incompatible elements – their present abundance being down to 10–20 per cent of the initial complement.

It is perhaps surprising that the formation of oceanic crust is better understood than that of continental crust, despite its relative inaccessibility. This is because the oceanic crust is relatively simple geologically and may be examined at all stages of its development from is formation from basalt melts at ridges through its spreading history and finally to subduction. Continental crust, on the other hand, being much older, has been subjected to repeated tectonism which has erased much of the key evidence as to its formation. Only very recently have all the sites where melts are added to the continents and contribute to their growth started to become clear.

The long-held view has been that continents are

formed only through the accretion onto existing continent of the volcanic products produced at subduction zones. Originally this view arose because of the similar chemical compositions of some subduction related volcanics and the continental crust. There is little doubt that contributions to the continents do occur from this source, but island arc volcanics produced away from continental margins are both too basaltic and too poor in incompatible elements to be direct sources for continental crust. At some stage in their development, melts much more enriched in incompatible elements must be added to the continents, in addition to island arc type basaltic material. The mechanism and timing of this process is the subject of great current interest.

These introductory remarks serve to raise several basic questions about the Earth's evolution. For example, have the continents grown and developed over most of Earth history or did they form in a single episode? How much of the mantle has been processed and depleted in producing continental crust? Is growth occurring at the present day and if so where? These and associated questions will be discussed in the remainder of this chapter.

What are continents?

The continental and ocean crust are physically part of the Earth's lithosphere (Figure 8.1). **Oceanic lithosphere** consists of a crustal part produced by basalt melt extraction at spreading ridges and averages about 6 km in thickness, and a mantle part composed of depleted harzburgite containing olivine and orthopyroxene, and peridotite containing in addition clinopyroxene and either a spinel or garnet depending on the depth. The thickness of the crustal part remains more or less constant with age but the mantle part thickens as the lithosphere cools. The lithosphere may be viewed as a **boundary layer** divided into a **mechanical** and a **thermal** part. The cooler mechanical portion is separated from convecting asthenosphere by a thermal boundary layer within

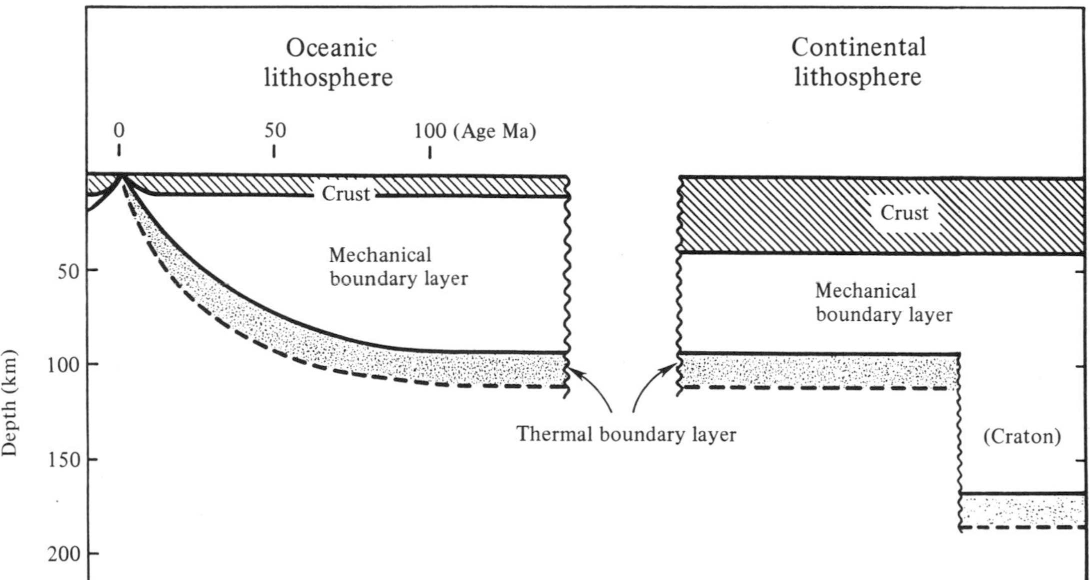

Figure 8.1 Comparison of idealised oceanic and continental lithosphere. The oceanic lithosphere consists of a basaltic crust averaging about 6 km in thickness which is produced by volcanism at spreading ridge axes. The thickness of the crust remains fairly constant as the seafloor ages. The mechanical boundary layer, separated from the convective upper mantle by a thermal boundary layer, thickens with time to a maximum of around 100 km after 100 Ma of cooling for the plate model (which defines the plate in terms of a rigid outer layer of the Earth).

Continental lithosphere is produced in quite a different manner, as discussed in the text, but also consists of a crustal portion and a mechanical boundary layer. Again the lithosphere is separated from the convecting mantle by a thermal boundary layer. The crustal portion has an average SiO_2 content of 65 per cent compared to 45–50 per cent for basaltic oceanic crust (Table 8.1) and has an average thickness of 35–40 km.

The thickness of the mechanical boundary layer in stable parts of the continents is very similar to the maximum thickness attained by oceanic lithosphere. However, there is good seismic and mineralogical evidence for a thickness around 200 km beneath the oldest parts of the Precambrian cratons. In these instances the composition of the mantle part of the lithosphere is probably of lower density than normal mantle peridotite. □

which the thermal gradient changes from conductive in the mechanical boundary layer to adiabatic in the freely convecting region beneath. Because the newly formed oceanic lithosphere cools as it ages, the mantle part increases in thickness as material becomes too viscous to remain part of the convective flow and passes through the thermal boundary layer to become mechanically part of the lithosphere. The contrasting situation with **continental lithosphere** is shown in Figure 8.1. In this case the crustal part is thicker than oceanic crust and averages about 40 km in tectonically stable regions, but may be locally much thicker, such as in areas of continental collision. Unlike the oceanic case, there is no straightforward relationship between mantle temperature, melting and continental crustal thickness, but some overall similarity of oceanic and continental lithospheric thickness is expected.

If the **potential temperature** of the mantle (the temperature the adiabatic gradient has if extrapolated to the surface) were the same everywhere beneath continents and oceans, then stable continents and old ocean crust would be expected to have similar lithospheric thicknesses (Figure 8.1). This appears to be the case for post-Archaean continental lithosphere, but the lithosphere beneath Archaean (> 2.5 Ga old) cratons is substantially thicker, between about 150 and 200 km. Evidence for this greater thickness comes from two sources. Firstly, seismic velocities are fast relative to the oceans down to depths of around 200 km or more, suggesting lower temperatures to greater depths beneath continents. Secondly, fragments of mantle peridotite brought to the surface from beneath the old cratons in kimberlite pipes and other volcanic eruptions are associated with and sometimes contain diamonds. The existence of diamond, given the known pressure at which the graphite to diamond transition occurs, requires that the mechanical boundary layer extends to more than 150 km in these regions. In the case of the Kimberley diamond pipes in South Africa, dating of garnet inclusions within the diamonds themselves has shown that this thickened lithosphere has existed from about 3.5 Ga ago. The reason why the lithosphere is overthickened beneath the old stable cratonic regions is not fully understood. One possibility is that the mantle part of old continental lithosphere has a lower density than normal mantle and is therefore relatively buoyant. There is some support for this idea from peridotite xenoliths which show a depletion in Fe, and a lower Fe/Mg ratio and therefore mean atomic number than normal peridotite, probably due to an earlier phase of basalt melt loss. An interesting feature of the low Fe/Mg ratio xenoliths is that they often show a high degree of trace element enrichment.

We turn now to the continental crust itself and its chemical composition. Some basic data about the continents are presented in Table 8.1. The chemical composition of exposed continental crust, which includes high-grade metamorphic rocks exhumed from

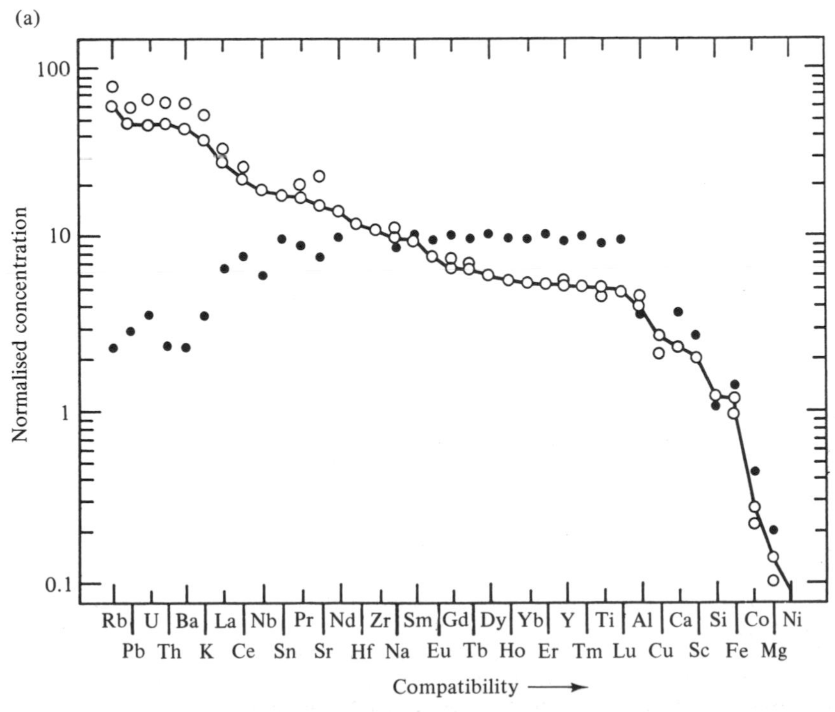

Figure 8.2 (a) Concentration of elements in the continental crust (open circles), and ocean crust basalt (filled circles) generated at normal spreading ridges, normalised to estimated values for bulk Earth. The strong enrichment of the most incompatible elements into the continents, which includes the heat-producing elements K, U and Th, is mirrored by their relative depletion in oceanic basalt. For some elements two different estimates for the chemical composition of the continental crust are shown. □

Table 8.1. *Upper and bulk continental crust and oceanic crust (oxides mass percentages)*

	Continental upper crust	Continental bulk crust	Oceanic crust
SiO_2	66.0	57.3	49.5
TiO_2	0.5	0.9	1.5
Al_2O_3	15.2	15.9	16.0
FeO	4.5	9.1	10.5
MgO	2.2	5.3	7.7
CaO	4.2	7.4	11.3
Na_2O	3.9	3.1	2.8
K_2O	3.4	1.1	0.15

depths up to 20–30 km as well as unmetamorphosed rocks, is close to that of tonalite with about 65 per cent SiO_2. However, samples of lowermost continental crust occasionally brought to the surface as xenoliths are most frequently basic granulites pointing to a higher proportion of basalt in the lowermost crust. Included in Table 8.1 is an estimate of the composition for the total continental crust making allowance for more basaltic compositions in the lower part.

An outstanding feature of continent chemistry is the great enrichment of the incompatible elements – those showing a greater affinity for liquid rather than solid phase upon mantle melting. The degree of enrichment increases with increasing incompatibility as shown in Figure 8.2. The abundances of a number of trace elements in the continental and oceanic crust are compared to their estimated abundances in the whole Earth (Figure 8.2a) and the trends are modelled using bulk distribution coefficients (Figure 8.2b). Note that the important heat-producing elements, K, U and Th, are among the most enriched of the incompatible elements in the continental crust.

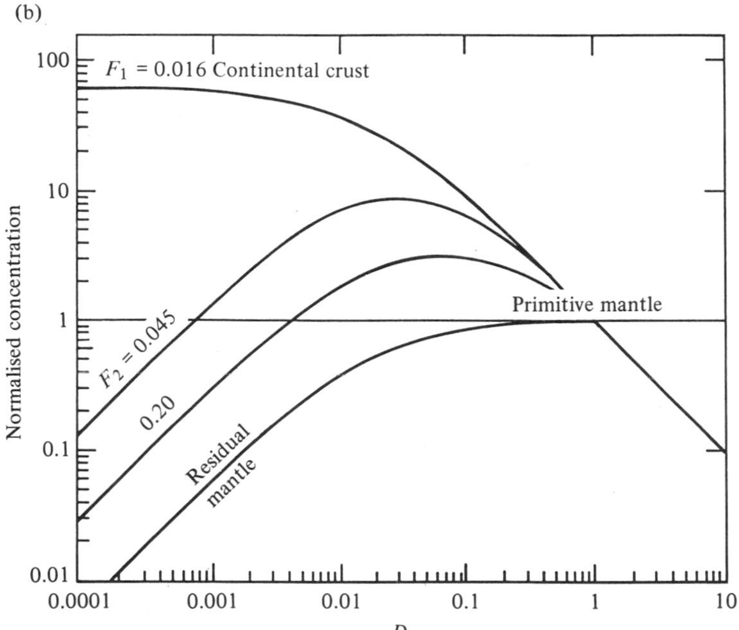

Figure 8.2 (b) The curve labelled $F_1 = 0.016$ is the calculated normalised concentration for elements of different bulk distribution coefficients, *D* values, resulting from a 1.6 per cent melting of primitive mantle. The residue remaining from this melting is labelled residual mantle, and F_2 melts generated by 4.5 and 20 per cent partial melting of this residuum are shown.

The broad similarity between the form of F_1 and the distribution of elements in the continental crust (all normalised concentrations ‡ 10 for elements with $D \leq 0.1$), and the form of F_2 and the distribution of elements in oceanic basalts (normalised concentration ≤ 10 for $D \leq 0.1$) is evident. Although the generation of continental and oceanic crust are not in fact related by such simple melting relationships, they do emphasise that small volume mantle melts (~ 1 per cent) are needed to generate the incompatible element enrichment in the continents. The depleted and 'residual' nature of the present-day upper mantle which melts to form oceanic crust is also evident.

Note: the bulk distribution coefficient is the concentration of a particular element in the solid ratioed to that in the melt. Increasing values of *D* reflect increasing compatibility (Figure 8.2a). The normalised concentration for the F_1 melt is related by the melt fraction F_1 and bulk distribution coefficient, *D*, by:

Normalised concentration = $1/[D_1+F_1(1-D_1)]$. □

We have seen in this section the importance of considering the continental crust within the framework of the whole continental lithosphere, which differs in some important ways from the oceanic lithosphere. The greater thickness, age and more silica-rich composition of continental crust distinguishes it from oceanic crust, as does the great enrichment of incompatible trace elements into less than 1 per cent of the Earth's mass.

How old are the continents?

The geological record contained in the continental crust spans almost 90 per cent of the time that the Solar System has been in existence. Stable cratonic areas in the shields of North America, Greenland, southern Africa, northern Europe, and Western Australia, for example, all have tracts around 2.5 Ga old, and in some cases these extend back to 3.5–3.8 Ga. The identification of these oldest remnants of continental crust has proceeded from the development of Rb–Sr, U–Pb and Sm–Nd **radiometric dating** techniques (see Chapter 7). However, at present the oldest reliable age reported on terrestrial material is 4.2 Ga for zircons from a sedimentary quartzite at Mt Narreyer, Western Australia, measured using an ion microprobe technique at the Australian National University.

Now that it is established beyond doubt that the continents are secondary features of tectonic and geochemical processing, the basic question is how they have grown and developed with time. An answer to this question cannot be derived simply from the distribution of radiometric ages obtained for crustal rocks, although this has been attempted at various times in the past. This is because radiometric ages may be reset by tectonic processes that post-date a rock's formation

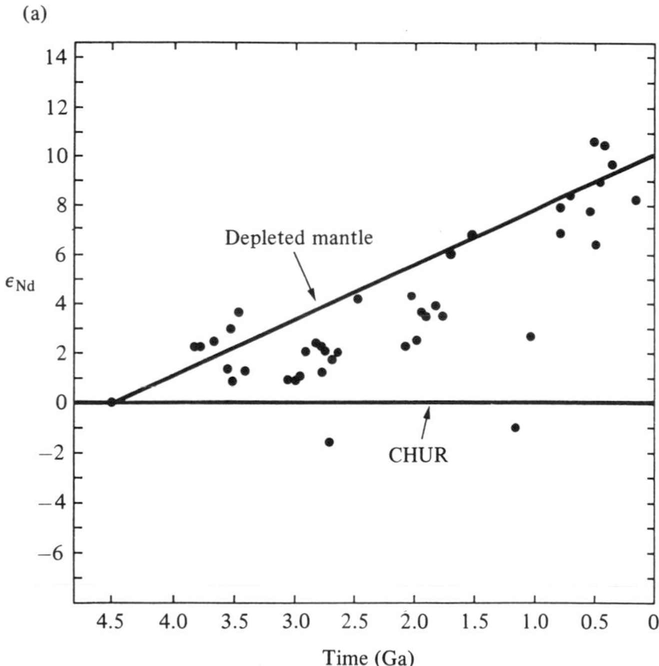

Figure 8.3 (a) Evolution of Nd isotopes in the Earth's mantle.

The $^{143}Nd/^{144}Nd$ ratio of natural Nd has increased with time from a uniform initial value through the α-decay of radioactive ^{147}Sm to ^{143}Nd. Although the half-life of ^{147}Sm is 110 Ga, differences in the $^{143}Nd/^{144}Nd$ ratio of different geological samples are easily resolved. The $^{143}Nd/^{144}Nd$ ratio of natural Nd is usually expressed in terms of ε_{Nd} as a deviation (in parts per 10^4) from the bulk Earth value, taken to be equal to that in chondritic meteorites (CHUR). The data shown are estimates for the isotopic composition of Nd in the mantle (95 per cent confidence limits) at various times in the past, and are derived from samples surviving in the continents.

Most of the samples ranging in age from 3.8 Ga to the present day have values that are positive, demonstrating that the mantle has had a Sm/Nd ratio greater than that in the bulk Earth, for almost the entire history of the Earth. This has arisen because Nd has become preferentially enriched in the continents over Sm, leaving depleted mantle with a higher Sm/Nd ratio. The solid black line is the adopted evolution of Nd in the depleted mantle.

The continental crust, because of its approximately 40 per cent lower Sm/Nd ratio, evolves to negative ε_{Nd} values and is complementary to the depleted mantle. □

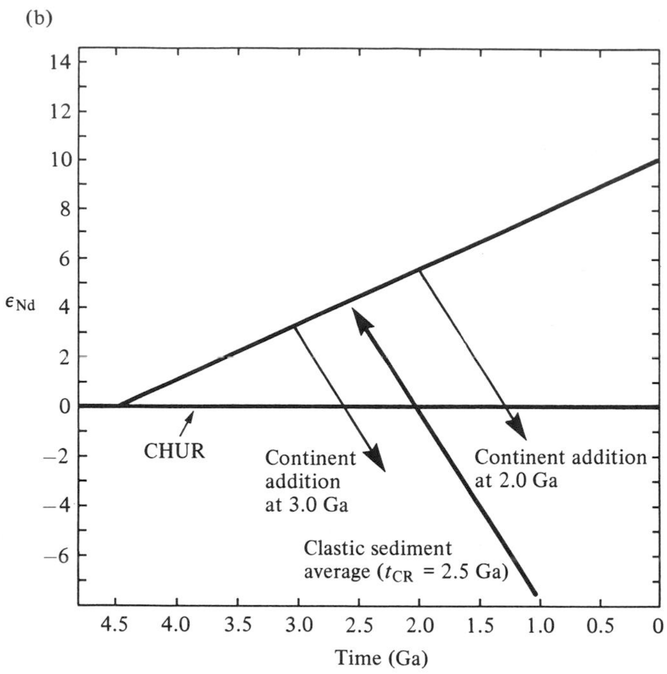

Figure 8.3 (b) Illustration of the method used to estimate crustal residence ages for samples of continental crust. Consider the production of a new segment of continental crust at 3.0 Ga. Because it will have an approximate 40 per cent lower Sm/Nd ratio than its parent depleted mantle it evolves in the direction of the arrow to negative values of ε_{Nd}. A remarkable feature of continental processes of erosion, clastic sedimentation and metamorphism is their inability to change significantly the original Sm/Nd ratio. Therefore the 3.0 Ga continent evolves as shown even if it goes through sedimentary recycling.

Similarly a younger continental segment formed at 2.0 Ga, for example, will evolve to negative ε_{Nd} as shown. In both cases the time of initial continent formation at 3.0 Ga and 2.0 Ga may be deduced from the analysis at the present day of these segments or clastic sediments derived from them. This estimate is called the crustal residence age and may be thought of as the length of time for which the Sm and Nd components have been part of the continental crust rather than the mantle.

In general, however, any fine-grained clastic sediment will contain contributions from a number of sources. The crustal residence age in this case is a weighted average of the various contributing sources and will not refer to a specific geological event. This is illustrated here (thick arrow) for an average which is made up from equal contributions from the 2.0 Ga and 3.0 Ga sources. □

– this may occur at temperatues as low as a few hundred degrees in the case of the K–Ar system or may even require melting to reset the Sm–Nd system on a whole rock scale. Radiometric ages therefore more often reflect the thermal history of the continental crust rather than its generation or growth. The problem may only be addressed through the determination of the abundances of natural radiogenic **isotopic tracers** in the crust at particular times in the past, but, most important, on length scales greater than the diffusive length scales accompanying tectonic reheating events. Such an approach usually involves determining the abundances of the tracer radiogenic isotopes such as ^{87}Sr, ^{143}Nd and ^{208,207,206}Pb, produced from radioactive decay of parent ^{87}Rb, ^{147}Sm, ^{232}Th and ^{235,238}U, which have half-lives between about 10^9 and 10^{11} a, and are therefore useful on the geological time-scale.

These principles are illustrated in Figure 8.3 for the ^{143}Nd tracer, which provides the basis of the most successful approach so far to these problems. The evolution is monitored through changes in the ^{143}Nd/^{144}Nd ratio, but in terms of the ε_{Nd} units, which are deviations from the bulk Earth ^{143}Nd/^{144}Nd ratio as described in the caption to Figure 8.2. Reference of the isotope ratio of a particular sample to the **bulk Earth evolution line** (CHUR) is made in order to identify deviations from a predicted value for the Earth system, had it not been subject to modification by the melting and chemical differentiation accompanying continent generation. Changes in ε_{Nd} in Figure 8.3 take place as a function both of time and Sm/Nd ratio. The idealised evolution for continental crust and the depleted mantle complement shown in Figure 8.3 makes the essential point – continental crust forms with a lower Sm/

Nd ratio than the mantle and evolves to negative values of ε_{Nd} along a characteristic vector. The residual, depleted mantle complement evolves to increasingly positive values for ε_{Nd}. Continental crust formed at 3.0 Ga, for example, should be immediately distinguishable from crust formed, say, 2.0 Ga ago. As shown in Figure 8.3b, the older 3.0 Ga crust will have a negative ε_{Nd} value by 2.0 Ga, and tectonic reworking of this material should leave the ε_{Nd} values unaffected. This older 'reworked' crust will be easily distinguished from a 'new' melt derived from the mantle at 2.0 Ga which will, as it forms, acquire the positive ε_{Nd} value of its mantle source, only evolving subsequently to lower ε_{Nd}. The distinction between 'new' and 'reworked' continental crust becomes straightforward. This approach has been applied widely and has led to greatly improved understanding of crustal evolution in a variety of geological terrains.

However, it would require a prohibitively large number of measurements of this type in order to specify the proportions of 'new' or 'reworked' continental crust that existed globally at any point in the past. The requirement is for estimates of the average isotope tracer composition of the entire continental crust as it existed at all times in the past as well as at the present day. Fortunately, a large degree of averaging has occurred naturally for many elements during the development and recycling of clastic sedimentary rocks. This statement is particularly pertinent to the rare-earth elements including Sm and Nd, whose relative abundances show very little variation in fine-grained sedimentary rocks. This contrasts with Rb and Sr, for example, which range widely, in part due to separation of Rb and Sr between carbonate and silicate rocks. The rare-earths are so strongly partitioned onto clay mineral surfaces that they have exceedingly low abundances in river and seawater, and fine-grained clastic sediments have Sm/Nd ratios that faithfully record their source rocks and are little modified during recycling and cannibalism of sediments. This coherence of Sm and Nd through erosion, sedimentation and sediment recycling is both a remarkable and helpful property, which does not extend to Rb–Sr and U–Pb, and uniquely qualifies them for investigation of **continent evolution** and generation. Returning to Figure 8.3, which portrays the idealised evolution of ε_{Nd} in the continents and mantle, the clastic sediment produced from a 3.0 Ga old crustal segment should have the identical ε_{Nd} evolution to its precursor. Thus the Nd–**isotopic memory** of a fine-

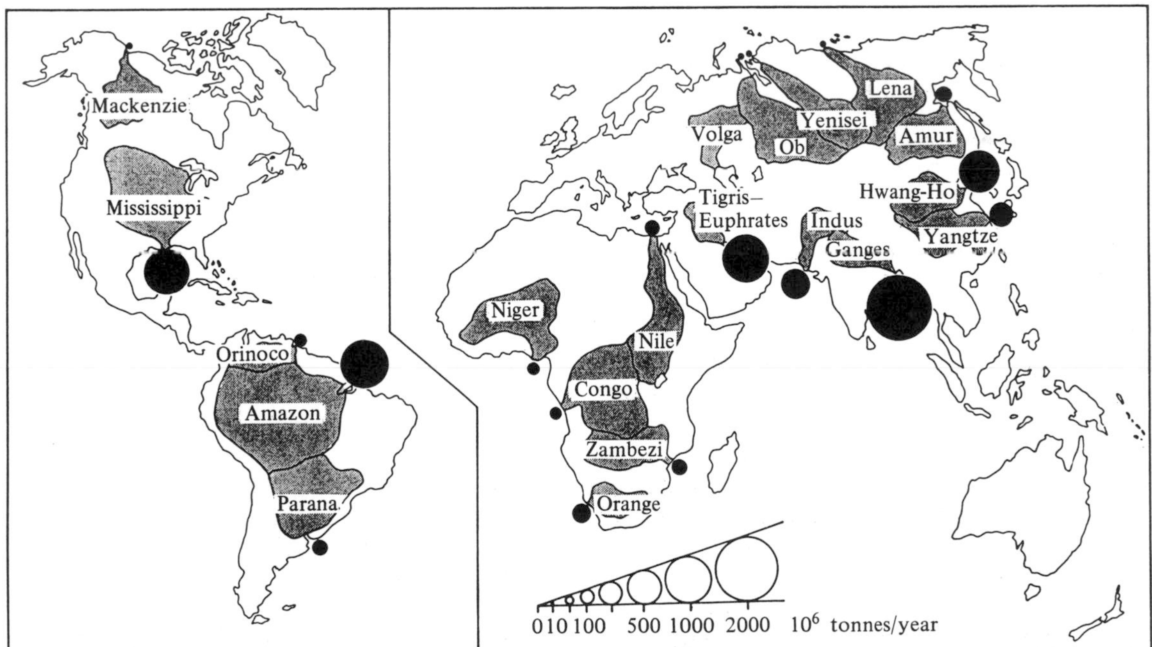

Figure 8.4 World map showing the distribution of the major drainage basins. The solid circles at the river mouths indicate the relative sizes of particulate loads discharged. It should be noted that a comparatively small number of rivers carry the bulk of the world's particulate load, and together their drainage basins cover a sizeable portion of the exposed continental crust.

The crustal residence ages calculated for the river particulates, and atmospheric (aeolian) dusts blown off the continents, are shown in Figure 8.5. Together they provide the most direct source of information about the average crustal residence age of the continental crust. □

grained sediment relates to the time of a melting event in the formation of continents rather than a sedimentary event. In reality, of course, any sediment will be a mixture of contributions from various sources and will record the average time for which Nd has been part of the continents. This average time will not refer in general to a specific geological event, but will be weighted in proportion to the amount of Nd component contributed from each source. Armed with this useful isotopic approach, the question of continent evolution may now be tackled.

As with the many efforts to understand geological processes, it is appropriate to start with the situation as it exists at the present day. Erosion and transport of clastic particulates occurs within the major drainage basins of the world and, as shown in Figure 8.4, these encompass both stable shields, platforms and active tectonic areas, and vary greatly in the particulate loads that they finally deliver to the oceans. The Nd budgets of these systems are divided between the dissolved and the particulate load, but in all cases the particulate load is overwhelmingly dominant, accounting for about 98 per cent of the discharge. Model Sm–Nd isotopic ages of the particulates, calculated as described in Figure 8.3, are now known for most of the major rivers. They are surprisingly uniform given the geological diversity of the terrains sampled and, as illustrated in histogram form in Figure 8.5, average 1.7 Ga. Expressed in a different way this result means that the Sm and Nd components of the contemporary global particulate load have resided as part of the continental system for an average period of 1.7 Ga. For this reason this model age is termed the **crustal residence age**, but it is emphasised once again that generally this will not correspond to a single event of continental crust formation, and because of the thorough mixing of contributions from different sources is an *average* period of residence weighted according to the amounts of Sm and Nd contributed.

The well-mixed nature of the river particulate load, reflected in the uniformity of crustal residence age estimates, is expected to extend to sedimentary ma-

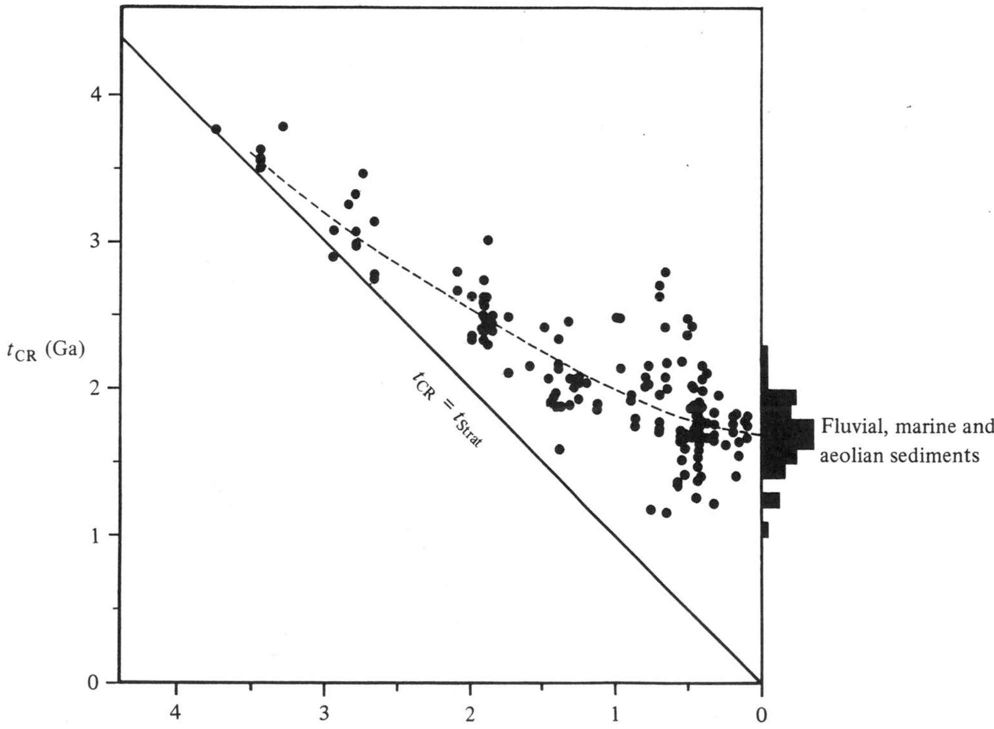

Figure 8.5 Comparison of stratigraphic age (t_{Strat}) and crustal residence age (t_{CR}) for modern river, marine and aeolian particulates (histogram), and for fine-grained sediments of various ages (dots). The crustal residence ages are calculated from Sm–Nd isotopic measurements following the procedure described in Figure 8.3b. Note the small range of crustal residence ages for modern river and aeolian particulates, averaging 1.7 Ga. This demonstrates how effectively materials are homogenised within the large drainage basins. Ancient fine-grained clastic sediments have similar values for t_{CR} and t_{Strat}, but these become more dissimilar towards the present. The average crustal residence age of the entire sedimentary mass is estimated at 2.0 ± 0.2 Ga from these data (see also Figure 8.6).

terial deposited earlier in the geological record. In principle it should provide estimates for the mean crustal residence age of exposed continental crust back through time. A compilation of crustal residence ages for fine-grained clastic sediments is shown as a function of stratigraphic age in Figure 8.5. Several interesting points emerge. The oldest sediments with stratigraphic ages of 3.5 Ga or more have almost identical crustal residence ages. This immediately demonstrates that their Sm and Nd was resident in the continental crust system for only a short time, probably of the order of 10^8 a, before the sediments were formed. Had they been derived from very much older crust, say 4.2 Ga in age for example, then their crustal residence ages would also be 4.2 Ga. In fact, Figure 8.5 shows that none of the sediments have crustal residence ages exceeding 3.8 Ga – this does not exclude contributions from some earlier continental crust, but it does require that it is relatively minor. The Proterozoic and Phanerozoic sediments in Figure 8.5, deposited after about 2.5 Ga ago, have crustal residence ages greater than their stratigraphic ages, demonstrating that material derived from older continental crust has made an increasingly important contribution. These data on sediments together provide the best insight currently available into the chronology of the continents as they existed in the past. Armed with these data, and information about the sedimentary mass and stratigraphic age relationships, as shown in Figure 8.6 as a cumulative distribution, it is straightforward to estimate a mean age for the total sedimentary mass that exists today. Because the amounts of sediment surviving from the early periods of Earth history (Figure 8.6) are very small relative to the most recently deposited sediments, the mean crustal residence age is strongly weighted to the values for Phanerozoic sediments. The mean age for the entire sedimentary mass obtained in this way is 2.0 ± 0.2 Ga. We may be confident that this estimate is valid for the sedimentary mass which itself is 10 per cent by mass of the entire continental crust. It is also reasonable to assert that this estimate is applicable to at least the upper part of the continental crust, and is currently the best estimate available for the continents as a whole.

In summary, it has been shown in this section how some unexpected properties of the clastic sedimentary mass provide the best insight available into the mean age of the continental crust. The mean crustal resi-

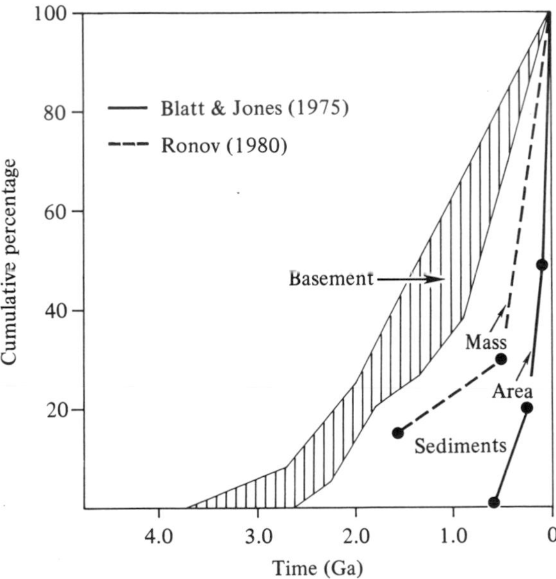

Figure 8.6 Age distribution of sediments by area and by mass compared to two estimates for the distribution of basement ages.

Sediments comprise 10 per cent of the continental crust by mass, and as shown have a half-mass age of approximately 500 Ma (i.e. half the mass of the sediments, that now exist had formed by 500 Ma ago). The young half-mass age reflects the more efficient recycling of sediments at the surface compared to the basement.

An estimate of the mean crustal residence age of the entire sedimentary mass may be obtained by combining the cumulative distribution curve for sediments with crustal residence age data in Figure 8.5. Because early Precambrian sediments make a small contribution to the total mass of sediments, their older crustal residence ages (Figure 8.5) make a relatively small contribution to the mean crustal residence age. The weighted mean age is 2.0 ± 0.2 Ga. □

dence age of the sedimentary mass is 2.0 ± 0.2 Ga, but this does not imply formation in a single event, of course, because it contains a spectrum of ages. Evidence for continental crust much older than 3.8 Ga has not been found in the sedimentary record so far. Older crust is not excluded by the data, but if it did exist then it made only relatively minor contributions to younger Archaean sediments. Therefore the occurrence of 4.2 Ga zircons in Western Australia does not contradict this statement.

Recycling and growth

The continents, like the atmosphere, are secondary geological features. As shown in the previous section there is no firm evidence for the existence of continental crust as we know it prior to about 3.8 Ga, and with the exception of the Mt Narreyer zircons there is very little preserved record until about 700 Ma after the time the Earth formed. From 3.8 Ga onwards there is a more or less continuous chronological record in the continents.

The significance of the approximately 2.0 Ga mean age estimate for the continents may be viewed in two quite different ways. On the one hand the continents may have formed very soon after 4.5 Ga with a mass more or less equal to the present-day mass (Figure 8.7). In this case the crustal residence age of the entire continental crust at the present day is expected to be 4.5 Ga. The much lower age of 2.0 Ga must then arise from the return of continental material to the mantle and its replacement by new younger *additions* which both maintain the mass of the continents constant and also reduce its mean age. In this case the record found in the continents is a **continental survival record**. An alternative view is that whereas continental material may be recycled through sedimentary systems, and even reprocessed during melting within the continents, recycling into the mantle does not occur, and in this case the record preserved in the continents is one of **continental growth**.

Assume first that recycling of continental crust conserves mass and does not return material to the mantle (the growth model). If the mean age of sediments at any time in the past is taken to equal the mean age of continents, then an apparent growth curve for Sm and Nd in the continental crust may be calculated (Figure 8.7) from the body of crustal residence age data for sediments. The result obtained implies a somewhat decreasing rate of continental growth towards the present day. If any return of continental material to the mantle did in fact occur in the past then this apparent growth curve would be a minimum growth curve with the real growth curve lying somewhere above it. At the present day there is no good evidence for the large-scale recycling of continental material to the mantle, but some *addition* can be shown to take place. It seems likely therefore that the apparent growth curve shown in

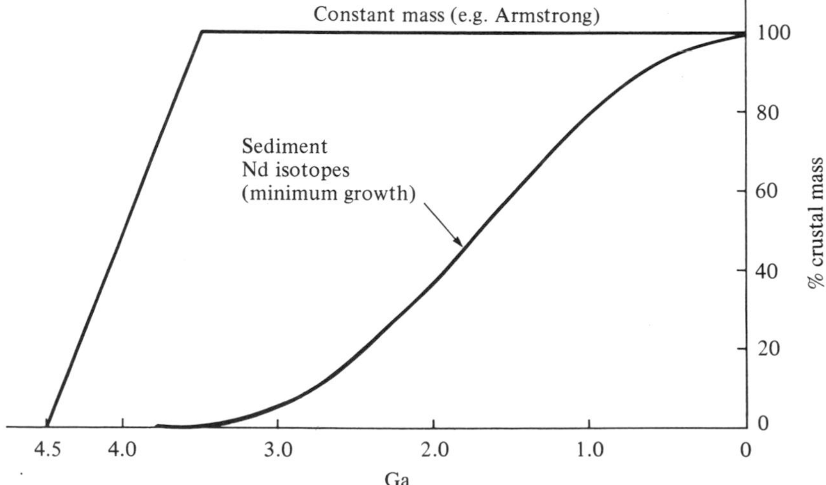

Figure 8.7 Minimum crustal mass curve with time estimated from sediment crustal residence time data, such as shown in Figure 8.5. In constructing this growth curve it is assumed that the average crustal residence age of clastic sediments shown in Figure 8.5 at any point in time is valid for the entire continental crust. Furthermore, in equating the Sm–Nd system with crustal mass it is assumed that the mean ages of the major elements, e.g. Si, O, and Al, are the same as Sm and Nd.

The apparent growth curve is a minimum curve for the actual crust. If any continental crust has been recycled into the mantle then the curve describes survival rather than growth. The result is compared to an extremum model of continental development. In this model the mass remains constant from 3.8 Ga to present, and material recycled to the mantle is exactly balanced by new additions. □

Figure 8.7 is not far removed from the real growth behaviour. So far the discussion on additions and growth has only referred explicitly to the two trace elements Sm and Nd. Only by implication does this refer to the major element components of the continents such as Si, Al and O.

The data available for clastic sediments also shed some useful light on the importance of erosion and recycling of older continental material into younger sediments or igneous rocks. The sum total of these processes is often simply referred to as intra-crustal recycling, implying internal conservation of mass. The relationship between crustal residence ages and stratigraphic and intrusive ages of sedimentary and granitic rocks, respectively, is shown in Figure 8.8. Recycling of pre-existing material would maintain the crustal residence age (as measured today) at a fixed value. In contrast, additions of new material from the mantle take place to reduce the crustal residence age of the total system as shown. Comparison between the two vectors describing recycling and addition with the sediment data shows (Figure 8.8a) the increasing dominance of recycling on the sedimentary system towards the present day. There is an important distinction to be drawn between clastic sediments and granitic rocks. Whereas both systems demonstrate the roles of recycling and new addition, the efficiency of sediment recycling makes average properties of the system easier to extract than is possible with the granitic data. The great value of the granitic data, however, is that they clearly demonstrate that some addition has occurred within the last 500 Ma to 1 Ga, because some samples have identical crustal residence and intrusive ages in this time interval. The sediment data however reveal that this has become relatively minor towards the present day.

It is quite remarkable just how much information about the fundamental processes is 'there for the taking' in the geological record.

Melting and continent generation

The point has already been made that the relationship between mantle melting and formation of oceanic crust is generally better understood than that for continental crust. It is clear that whereas mantle melting directly produces oceanic basalt, the more silica-rich tonalite composition of continental crust (Table 8.1) cannot be produced directly by melting of mantle peridotite. The formation of tonalitic crust must involve further processing of basaltic melts. The trace element abundances in continental crust are quite different from those in oceanic crust and mantle as seen in Figure 8.2a, particularly in the relative enrichment of the large ionic radius elements such as Th, U and Rb. Forgetting for the moment that tonalite melts cannot be produced directly from the mantle, the question of how much melting is required to produce the incompatible element abundances that they host can be addressed. This is a straightforward exercise because the **bulk distribution coefficients** – the ratios of trace element concentrations in the residual solid mantle divided by those in the melt – are well known for many of the elements shown in Figure 8.2a (see Chapter 6). The calculated pattern of enrichment in Figure 8.2 is given as a function of increasing distribution coefficient which follows the element sequence shown. It is evident from this distribution that the continental pattern of trace element enrichment hosted by tonalite corresponds quite closely to a mantle melt fraction between $F_1 = 0.01$ and 0.02. This result at first sight appears at odds with the **experimental petrology** which has been carried out on tonalite compositions (see Chapter 4). Not only does this show that tonalite cannot be produced directly by melting of mantle peridotite but also that 10 per cent or more melting of a basaltic parent is required for its formation. This problem of the large amounts of melt required to produce a tonalitic composition yet very small melts for the trace element abundances has only been addressed quite recently. In the past, models of continent generation have simply assigned the processes responsible to the subduction environment. However, island arcs do not themselves contain anything like the amount of trace element enrichment required – unless the subduction occurs against a continental margin where andesites are produced.

There is now sufficient understanding of the physics of **melt separation** from partially molten regions to pursue the above problem further. The basic equations that govern melt separation from a partially molten layer with a free upper boundary but a fixed lower boundary against which compaction occurs are known. Although this is a considerable simplification for the real Earth, a consideration of melt extraction from such a system provides some valuable insights. It is possible to derive estimates for the time-scale of melt extraction from such a system where the melts differ in both amount and composition. Intuitively, differences might be expected in the rates of melt separation depending upon melt viscosity. For example, a dry granitic melt is exceedingly viscous at around 10^{11} Pa s, and although this is reduced dramatically to about 10^4 Pa s when the melt is water-saturated, it is still

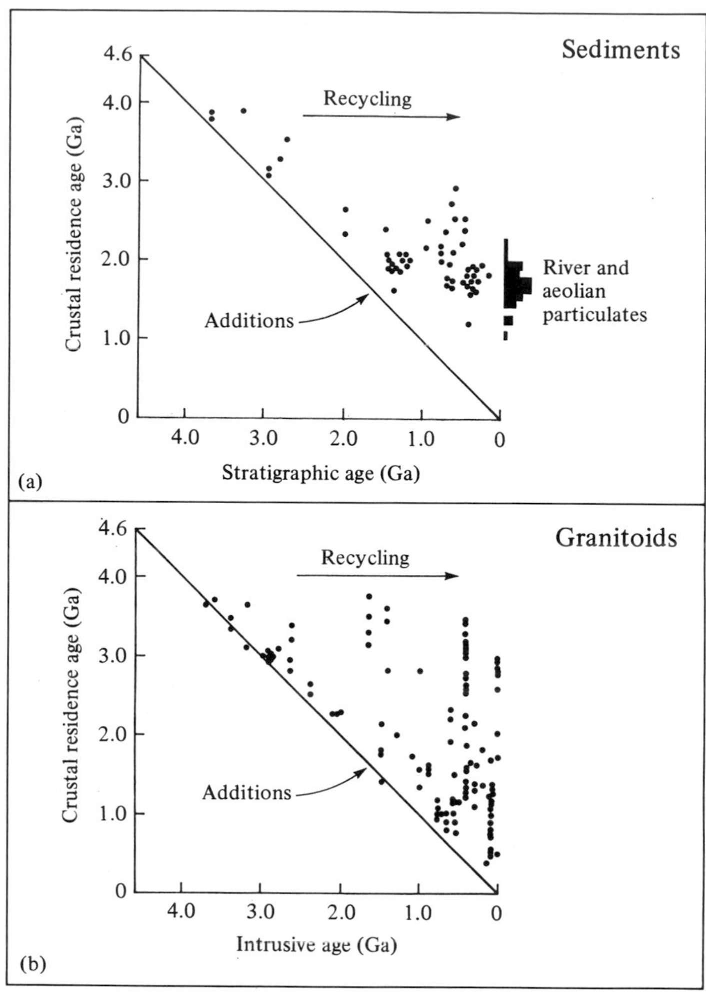

Figure 8.8 Relative roles of intracrustal recycling and additions of new material from the mantle in the evolution of the continental crust. The crustal residence age derived from Sm–Nd isotope measurement of the individual sediment particulate (a) or granitoid (b) sample shown is compared with the relevant stratigraphic or intrusive age.

In (b) additions of Nd to the continental crust at a particular time will, at that time, have a crustal residence age – intrusive age difference of zero and will occur as shown. In (a) the average value for the continental crust as a whole will then be shifted towards the diagonal line $t_{Strat} = t_{CR}$ in proportion to the amount added. Recycling of pre-existing materials within the continental crust causes no change of crustal residence age and is readily distinguished from additions as shown.

Comparison of the data for sediments and granitoids immediately demonstrates the more homogeneous and representative nature of the sedimentary mass. Granitoids are generated with highly variable amounts of new additions, and pre-existing continental crust and average properties are much harder to deduce than for the sediments. The data for sediments show the greater importance of recycling in continental evolution in the more recent geological past. Additions appear to have been relatively minor over the last 1 Ga or so. However, the granitoid data do show evidence for some additions, given that some samples plot very close to the diagonal line $t_{CR} = t_{Int}$. □

much greater than a typical basaltic melt of 1 Pa s (see Figure 4.5).

A crucial aspect of melt separation in partially molten silicate systems is the requirement for the **dihedral angles** between melt and grains to be less than 60°, at which point the partial melt is fully interconnected and can separate (see Chapters 4 and 5). The problem then may be reduced conveniently to one of examining the **melt separation velocity** from a partially molten region as a function of the porosity or melt fraction (Figure 8.9). The separation velocity scales to a time-scale for melt separation which is actually the

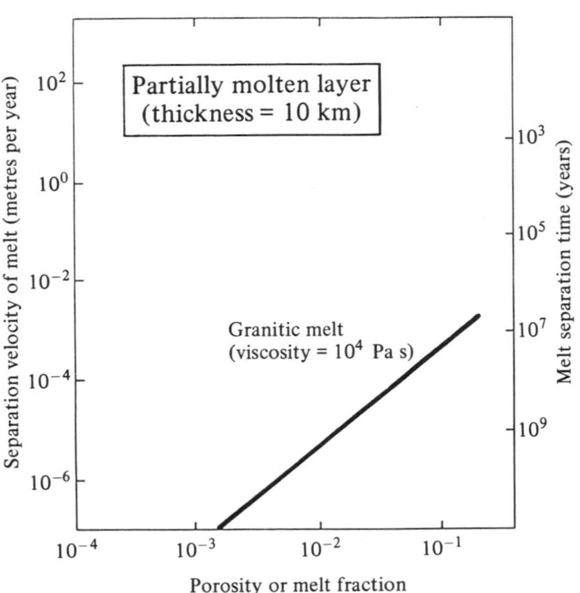

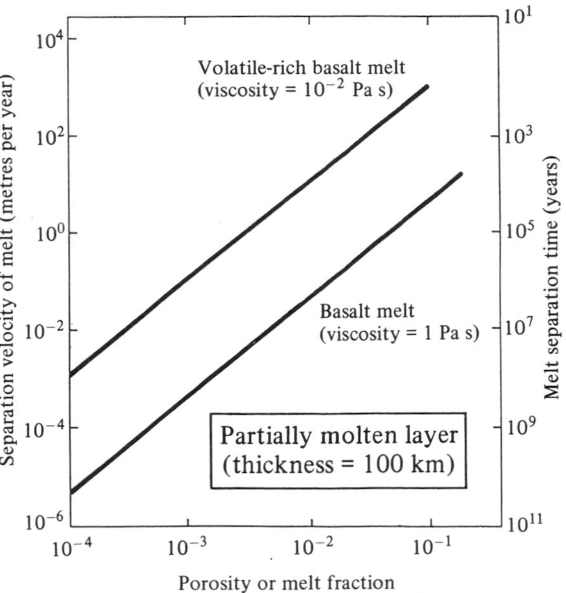

Figure 8.9 The relationship between melt fraction (or porosity) in a partially molten layer and the time taken for melt to separate. The relationships should be applicable to melting processes in the Earth's mantle and continental crust.

(a) This corresponds to a 10 km thick partially molten region in the continental crust, where the melt is a water-saturated granite, with a viscosity of 10^4 Pa s. The separation velocity is exceedingly low at small melt fractions, around 10^{-6} metres per year for a 1 per cent melt, and only reaches 10^{-3} per year when 20 per cent or more melt exists. The corresponding time-scale for melt separation is actually the time taken to reduce the porosity or melt fraction by 1/e, and is greater than 10^9 years for a 1 per cent melt. It is concluded that such small-volume melts will not separate on a geologically reasonable time-scale. Only when melt fractions exceed 10 per cent does the time-scale reach 10^7 a and become geologically plausible. Note that if the granitic melt were dry the viscosity would be around 10^{11} Pa s and the relevant time-scales *increased* by a factor of 10^7.

(b) This is an analogous situation corresponding to a 100 km thick partially molten region in the mantle. In this case the relationship between the separation time-scale and melt fraction is illustrated for basaltic melts with viscosities of 1 and 10^{-2} Pa s. The low-viscosity basaltic melt would be alkali and volatile rich. In this case the melts move with much greater velocities at small melt fractions – between 10^{-2} and 10 metres per year for 1 per cent melts. The corresponding time-scales are very short and approach 10^3 years for the low viscosity basaltic melt. In contrast to granitic melts produced in the continental crust, small degree basaltic melts separate readily from mantle regions.

The trace element fractionation summarised in Figure 8.2a will only be produced where small-volume melts separate, and this is most easily accomplished in the mantle. The generation of the more granitic crust must occur with little change in the trace element pattern in its source which is dominated by small-volume, trace element enriched melts. □

time taken for compaction to reduce the porosity or melt fraction to 1/e of its initial value. In Figure 8.9a a 10 km thick layer is considered in which the melt is a water-saturated granite with a viscosity of 10^4 Pa s. It is apparent immediately that this melt is very difficult to separate on relevant geological time-scales of 10^6 to 10^7 a. Only when the system melts to 10 or 20 per cent does the melt move quickly enough to get out on this time-scale. At smaller melt fractions the times become comparable to the age of the Earth and separation of very small amounts of melt between 0.1 and 1.0 per cent is therefore unlikely. Note that if the granite melt were dry with a viscosity of 10^{11} Pa s, then all the time-scales in Figure 8.9 would be increased by seven orders of magnitude. These considerations of granitic melt accord well with the results of experimental petrology which also indicate large melt fractions for the production of rocks of this composition. Unfortunately, when melt fractions become as large as 20 per cent there is very little scope for fractionating the incompatible elements. For example, consider two trace elements of different incompatibilities where their bulk distribution coefficients are $D = 0.001$ and $D = 0.1$ respectively. The highly incompatible element ($D = 0.001$) is only enriched in the final melt by a factor of 1.4 over the incompatible element ($D = 0.1$). (Note that this calculation follows from the equation in the caption to Figure 8.2b.)

The situation is quite different when the separation of basaltic melts from a 100 km thick partially molten mantle region is considered (Figure 8.9b). Two situations are considered here: one is a typical basaltic melt and the other is an **alkali** and volatile rich **basalt**, and these have viscosities of 1 and 10^{-2} Pa s respectively. It is immediately apparent that the time-scale for separation of these melts from a 100 km thick region is very much shorter than for the granitic melts. Reference to Figure 8.9 shows that melts of 1 per cent or even 0.1 per cent are able to separate on the 10^6 a to 10^7 a time-scale. Separation of such small degree melts has a much more profound effect on the incompatible trace element enrichment pattern. Considering again the same example as above with two trace elements having bulk distribution coefficients of 0.1 and 0.001, in this case for a 0.1 per cent partial melt, the highly incompatible element with the smaller distribution coefficient is some fifty times more enriched in the melt than the incompatible element.

In summary, silica-rich melts (granites and tonalites) which are in effect the basic building blocks of the continents, form as **large-volume partial melts** that have little or no capacity to produce a significant fractionation of the incompatible trace elements. Quite simply they are only able to inherit the trace element pattern that has already been established in their more basic parental source. In essence the question of trace element abundances in the continents reduces to how the source material that melted to form the granitic and tonalitic melts acquired the trace element signature of **small-volume** basaltic **partial melts**. The answer appears inescapable – this source is dominated by small-degree basaltic melt fractions, which (as seen in Figure 8.9b) are able to escape rapidly from the mantle. If they are very small then they will carry very little heat and freeze into the lithosphere whenever they encounter it – but also be very easily remobilised by any subsequent thermal perturbation of the lithosphere. The influence of these melts is seen as 'metasomatic products' in peridotite xenoliths from both the mantle and lower crustal part of the continental lithosphere. A last important point to make here is that the trace elements may enter the continental crust at different times and in different places from the major elements. All that has been said so far about the age of continental crust and its growth pattern is based on the trace elements. These conclusions must only be extended to include the major elements with caution.

Thus the high levels of incompatible trace elements in continental tonalite and granite are inherited from more basaltic precursors, which must include very-low-volume melts. The signature imparted by these small-volume melts stays with continental material even when it passes through a clastic sedimentary cycle.

Where do continents grow?

A geologist would consider continental growth to describe the increase in mass and volume of the continents and therefore imply additions of the major oxides of crustal rocks such as SiO_2 and Al_2O_3. However, as already noted, it is the trace elements which together make up less than 1 per cent of the crust that provide information on the time-scale of continent generation. Only if the major elements are added to the continental crust at the same time and in the same place as the trace elements do these results relate directly to continental growth. As shown in the previous section, this cannot be taken as granted, although it has been generally assumed to be the case in the past. In this section the problem of where continental growth occurs will be considered in two parts: firstly the sites for trace element addition and secondly the sites of major element addition. Although the discussion will be strongly influenced by the present-day tectonic situation, basic principles arrived at are probably relevant to the earlier continental crust also – there is no strong evidence for a fundamental change in the melting processes that produce the continents.

The small-volume melt fractions

It has been argued above that the trace element budget of the continents is determined by the addition of small, around 1 per cent or less, basaltic melt fractions. Basaltic melts are known to be generated at several sites. For example, the greatest proportion of basalt is generated at spreading ridges, some 20–25 km^3 a^{-1}, but these are large-volume melts produced by the extraction in total of 10 to 15 per cent of melt from the mantle. Overall, the fractionation of incompatible elements between oceanic crust and the mantle is relatively small as seen in Figure 8.2a, where the oceanic crust is shown to possess a depleted incompatible element pattern complementary to the enriched pattern of continental crust. Similarly island arc tholeiites do not exhibit marked incompatible element enrichment; these and spreading ridge volcanics together comprise the greatest volumes of basalt production on Earth, but do not provide the required fractionation. There are several possibilities (Figure

(a) Extension of lithosphere

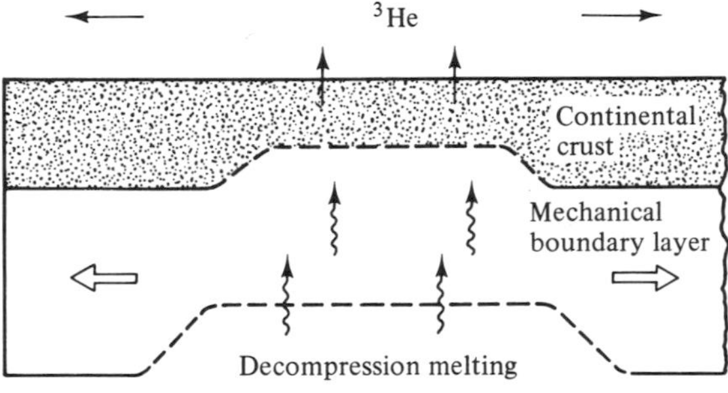

(b) Plume beneath lithosphere

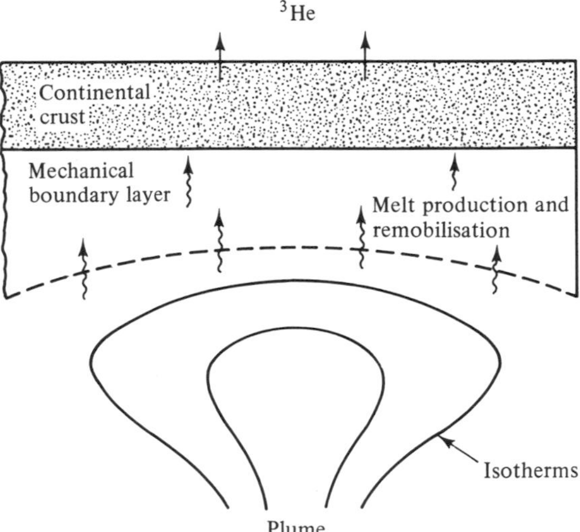

(c) Dehydration and melting beneath arcs

(d) Extension over a plume

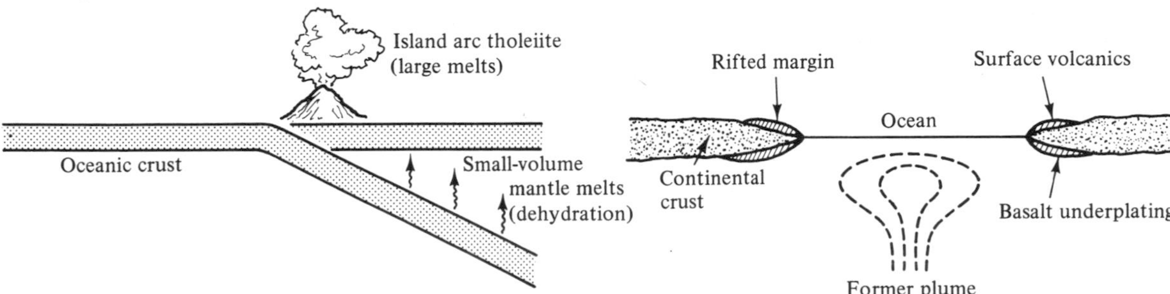

Figure 8.10 Schematic illustration of situations where the small and large-volume melt fractions may be added to the continental lithosphere (see Figure 8.9).

The small-volume basaltic melts will carry the incompatible trace elements as described in Figures 8.2a and 8.9. These may be formed where extension of the lithosphere gives rise to a small amount of decompression melting (a), where a plume causes local increases in lithospheric temperatures (b), and where dehydration in subduction zones reduces the solidus temperature of the mantle (c). In each of these cases the small-volume melt fractions carry little heat and may freeze upon encountering cooler conditions as they move into the lithospheric mantle, or lower part of the crust. They would occur there as 'metasomatic' products. Eventually they will be remobilised and their trace elements inherited by more silica-rich crustal melts.

The addition of major elements to the continental crust is likely to include large-volume melt fractions produced as island arc tholeiites, and where extension occurs over a plume (c and d). In the case of the north west Atlantic rifted margin, for example, the amount of basalt produced may be very large. □

8.10), as follows:

(a) *Lithospheric extension sites*: Small basalt melt fractions are generated where limited extension of the lithosphere results in thinning and decompression of the underlying asthenosphere. The amount of melt generated in this way will depend upon the amount of extension and the potential temperature of the mantle. In continental areas such extension occurs in the Basin and Range province, for example, and often leads to the formation of sedimentary basins such as the North Sea and Pannonian Basins. In each of these situations volcanics do occur at shallow levels or at the surface, but small-degree melts are likely to have frozen in at greater depths.

(b) *Plume sites*: In principle an elevation of temperature in the mantle over and above the ambient, if sufficient, will produce melts. In some situations these melts may be very small, if the temperature excess above the solidus is also small. This may occur where hotter plumes impinge on the lithosphere – but again small-volume melts produced in this way may freeze into the cooled lithosphere and not reach the surface. Such a plume is known to exist, for example, beneath the Yellowstone area at present and it is reasonable to assume that similar phenomena will have occurred in the past.

(c) *Subduction sites*: Although the major island arc tholeiite volcanics do not show the required enrichment, it seems likely that small-volume basalt fractions will be generated in the mantle wedge above the subduction zone. This will happen at normal mantle potential temperatures (about 1280 °C) if dehydration fluids (mainly H_2O) are released into the wedge from subducted ocean crust and lower the solidus temperature of the mantle. These melts will be emplaced into the base of the overlying lithosphere whether it be oceanic or continental (see earlier section), but may not appear as surface volcanics.

The balance of contributions to the continental crust of small basaltic melt fractions produced at these sites is difficult to assess at the present day and of course harder still for the geological past. It seems likely, however, that these will have been the principal mechanisms of trace element additions to the continents over most of the Earth's history.

The large-volume melt fractions

The absence of a major element with a convenient natural radiogenic isotope to serve as a tracer for major element addition to the continents makes the relationship between the addition of small and large-volume melts to the continents difficult to assess. At the present day, large amounts of melt are generated above subduction zones both along continental margins and in island arcs remote from them. The island arc tholeiites produced away from the margins clearly do not usually involve much recycling of pre-existing continental crust and there is evidence that some of these arcs do accrete to the existing continents. We have already seen that they will require further processing to produce the more tonalitic composition of the continents and must also join with small basaltic melt fractions on the way to produce the right end product. However, this is certainly one way in which major elements are added.

A second, less widely recognised, but important site is where continental extension occurs to produce sizeable volumes of melt. This is particularly the case where extension occurs above a plume, a region of higher than average mantle potential temperatures. Very large volumes of basaltic melt were added around 60 Ma ago along the **Atlantic margins** off west Greenland and west Scotland and Rockall – the estimated total volume is 5×10^6 km^3. Basaltic melts are likely to have been added at both of these principal sites (island arcs and continental extension zones) repeatedly over geological time. As with the small basaltic melt fractions the balance between major element additions at these two principal sites is difficult to establish.

Monitors of melt additions

Because the small basaltic melt fractions described above will carry little heat they will generally freeze into the deeper parts of the continental lithosphere (Figure 8.10). The absence of alkali basaltic melts at the surface in an area of extension therefore does not mean that small-volume melts highly enriched in incompatible trace elements do not occur; they may freeze into the lower crust or mantle parts of the lithosphere beneath. Fortunately, not all of the incompatible elements will freeze together with their basaltic host – this is particularly true in the case of the rare gases. Of these helium is currently the most valuable because it appears to escape from the continental crust to the surface even faster than heat.

Helium has two isotopes, ^{3}He and ^{4}He. Of these ^{4}He is by far the most abundant in the Earth and has been largely produced by the radioactive decay of ^{235}U, ^{238}U and ^{232}Th. A very small amount of ^{3}He is produced from the interaction between naturally produced neutrons in rocks and ^{6}Li – this is of subordinate importance and the helium produced by these reactions has a **helium isotope ratio** ^{3}He/^{4}He ~ 10^{-8}. This is the isotopic composition of helium produced in the continental crust. However, the composition of helium trapped deep in the Earth at the time of its formation was quite different. By analogy with the composition of helium retained in some meteorites, this initially trapped helium should have had ^{3}He/^{4}He ~ 10^{-4}. Some of this initially trapped helium is still present in the mantle today, and the addition of radiogenic helium since the Earth formed has only reduced the ^{3}He/^{4}He to ~ 10^{-5}. Therefore there is a three orders of magnitude difference in ratio between present-day mantle helium and the radiogenic helium (10^{-8}) generated in continental crust. The great value of this difference when coupled with the high diffusivity of helium is that small helium-rich melts in the lithosphere will, upon cooling, release helium with a characteristic isotope composition identifiable at the surface.

The correspondence between ^{3}He loss from the

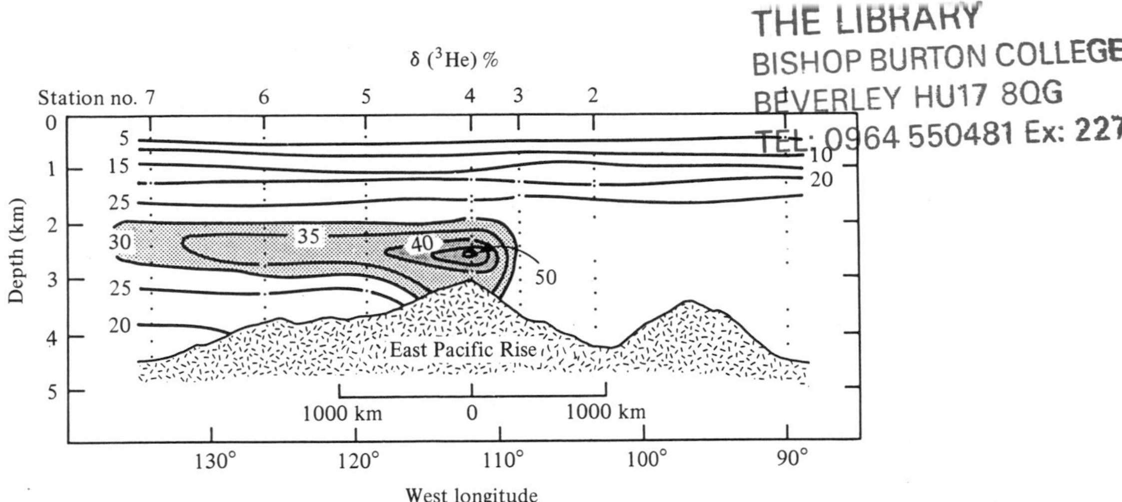

Figure 8.11 Primordial ^3He, trapped in the Earth at the time of its formation some 4.5 Ga ago, is still escaping at measurable rates from the East Pacific Rise. The principal mechanism is made evident by the excesses of ^3He in the deep oceans which clearly emanate from the spreading ridge axes associated with the production of new basaltic oceanic crust. Helium is partitioned into basaltic melts produced in the mantle, emplaced into newly formed crust, and separates during its cooling mainly by hydrothermal processes.

The excess ^3He present in the Pacific Ocean, above that which can be accounted for through equilibration with the atmosphere, is shown on this diagram as a percentage increase. □

Earth's interior and the sites of basalt melt production in the oceanic crust and at sites of island arc volcanism is now well known. In Figure 8.11 the escape of ^3He is clearly seen to be associated with the spreading ridge axis in the East Pacific. In this situation the rate of loss of ^3He may be calculated, as may the rate of basalt melt production. The helium appears to escape quickly from the most recently formed crust, with the loss centred upon the ridge axis itself.

Turning now to the continental environment and the use of ^3He as a tracer of mantle melting, the situation is best documented in western Europe. The results of a major survey of the helium isotope composition in the near-surface fluids (hydrocarbons, groundwaters, geothermal fluids) are summarised in Figure 8.12. The results are shown very simply, and a distinction is made between sites where mantle helium can and cannot be identified in the near-surface.

The distribution of the data shown in Figure 8.12 makes some interesting points. The first is that the occurrence of mantle-derived ^3He at the surface is much more widespread in Europe than are Recent volcanic products. Immediately this suggests that something is happening deeper in the lithosphere over a surprisingly large area. Those areas where no mantle helium is identifiable at the surface include tectonically stable areas such as the United Kingdom mainland and continental shelf, and parts of west Germany and France, and tectonically active areas such as sedimentary basins formed by loading (e.g. the Molasse Basin on the north side of the Alps). Mantle helium appears to be present everywhere where there is recent volcanic activity, such as the Eifel region of Germany and the Massif Central of France, but is distributed more extensively in these areas than are the surface volcanics themselves. In addition there is a very clear association between areas of active extension and the mantle helium. These areas include the Pannonian Basin of Hungary, the southern Rhine Graben, the Egergraben in Czechoslovakia, western Turkey and the Aegean, even in places where Recent volcanics do not exist at the surface at all. The amounts of basaltic melt emplaced into the lithosphere beneath these sites of extension need only be very small to supply ^3He and probably freeze quickly at depth. The distribution of mantle helium at the surface is the most effective monitor of melt and fluid addition to the lithosphere that we have available to us at present.

Because helium is an incompatible element, its presence in the lithosphere may record and monitor melt input but provides little information about the volume of melt introduced. Even in places where the flux of helium is known, such as at ridges and through some continental regions such as the Pannonian Basin, the identification of melt fractions is difficult. However, the global distribution of ^3He clearly identifies the ridges and island arc regions as sites of mantle helium loss and melt input, and to these we can add various parts of the continental lithosphere, particularly those places in active extension. In the latter case the melt fractions must be very small, yet incompatible elements are emplaced into the continents at these sites. Volcanic arcs are also one of the important sites of major element addition (Figure 8.10).

Summary and conclusions

This chapter has emphasised the differences between continental crust and oceanic crust in terms of their thickness, chemical composition and age structure. The formation of oceanic crust and development of the oceanic lithosphere is basically simpler and better understood than the continental lithosphere, despite the wealth of geological observations made on the continents.

Over a development period from 3.8 Ga to the present day the continents have concentrated some 25–50 per cent of the Earth's heat-producing elements K, U, Th, and other incompatible elements, which suggests that 1 per cent or less melt effectively controls their extraction from the mantle and their introduction into the continents. The identification of these small-volume melt fractions as mainly basaltic rather than tonalitic or granitic is of fundamental importance, because the addition of trace and major elements may occur both in different situations and at different times. Small-volume basaltic melts should enter the lithosphere at sites of subduction in response to dehydration, in areas of lithospheric extension, and above plume structures which will produce and remobilise small-volume melts. The major elements, which are in effect the continental volume, will be

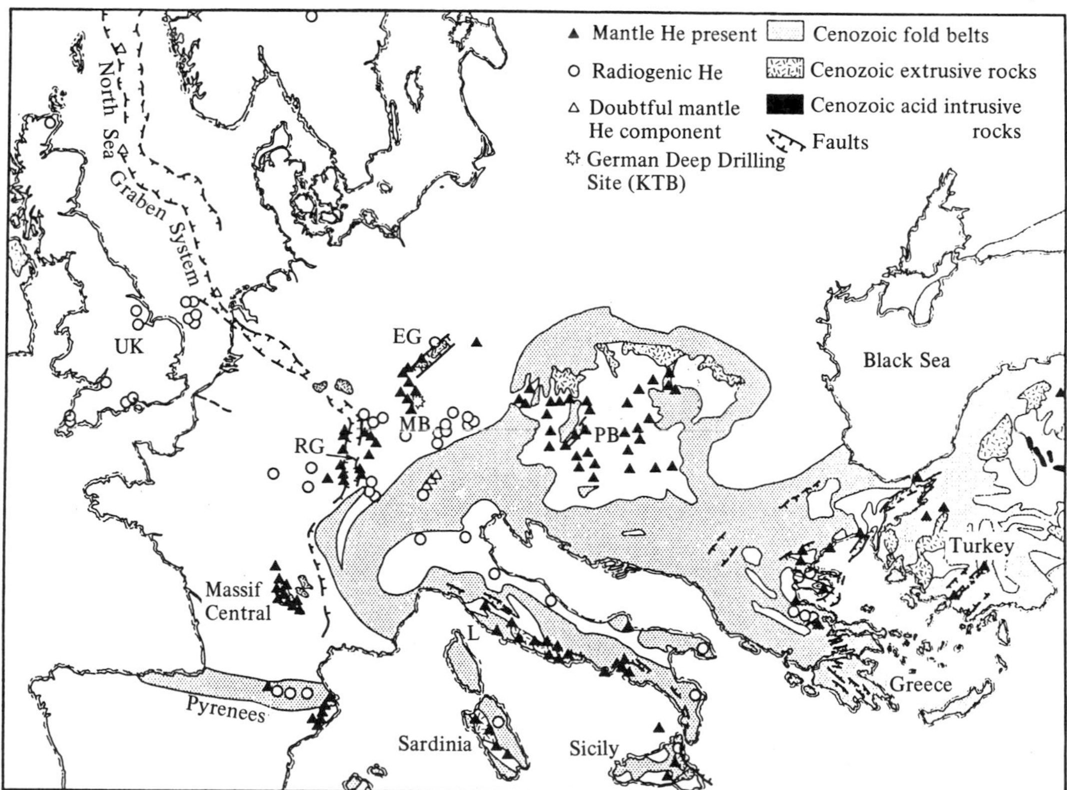

Figure 8.12 Distribution of mantle-derived helium in western Europe. The map is based upon the analysis of $^3He/^4He$ ratios in spring waters, groundwater and geothermal fluids from boreholes and hydrocarbon deposits, but only distinguishes between locations where a mantle-derived helium component is recognised or not.

The helium found in relatively stable areas, such as the United Kingdom mainland and shelf area, and in tectonically active areas where the crust is undergoing loading, such as in the Molasse Basin (MB), is predominantly crustally produced and radiogenic in origin. In areas undergoing active extension, however, mantle helium is present in near-surface fluids. This includes the southern part of the Rhine Graben (RG), the Pannonian Basin (PB), western Turkey and the Aegean. L = Larderello, EG = Egergraben, KTB = Kontinentales Tiefbohrprogramm.

To a first approximation the mantle He distribution in western Europe is considered to provide a map of mantle melts entering the continental crust. In many instances, the amounts of melt are too small to produce any volcanic expression at the surface (see Figure 8.10). □

added from island arc basalts and at places where substantial extension occurs above a plume. Helium isotope tracers provide an excellent picture of where these sites are, and their extent is surprising, but the relative importance of the various sites for both major and trace element additions is still difficult to assess fully. The rate of addition to the continents has been derived from incompatible elements, and suggests that the continents have grown throughout geological time but that the rate has decreased towards the present day.

What for the future? A major challenge must be to quantify the amounts of small-volume basaltic melts added to the crust in various environments in the geologically recent past. This will require further understanding of small-volume melt generation and new ways of monitoring its addition to the lithosphere. An additional requirement is to resolve the relationship between incompatible element and major element addition. The ideal approach would be a major element which had a radiogenic isotope which underwent sufficient change in abundance to serve as a tracer. The candidate is yet to appear.

Further reading

Taylor, S. R. & McLennan, S. M. (1985) *The Continental Crust: Its Composition and Evolution*, Blackwells, Oxford, 312 pp.

A highly informative and accessible text covering the subject of this chapter.

INTRODUCTION TO CHAPTERS 9–12

Since the late 18th century geology has been a clearly recognised and often influential subject. The study of rocks and fossils showed that the history of the Earth extended back far beyond the arrival of man, and that the nature of the historical succession had been progressive through to the present day. Much important work was done, but it is only in the last three decades with the development of plate tectonic theory that it has been possible to link the results of separate studies into an overall global context. Already it is difficult to recall what the Earth sciences were like before plate tectonics, and so in Chapter 9 Tjeerd van Andel gently reminds us of the history of the revolution which we may now too often take for granted, while at the same time summarising the main geological features which are so elegantly linked within plate tectonic theory. Many branches of the Earth sciences have traditionally been qualitative, and largely observational in their approach, and one of the other major contributions of plate tectonics has been to provide a framework for quantitative, predictive science. Alan Smith has been one of the leading geologists to explore the implications of a more quantitative approach, and in Chapter 10 he explains how the ability to apply a precise, mathematical description of the movement of rigid bodies on a sphere to the motion of plates on the surface of the Earth continues to revolutionise the interpretation of the geological record.

The acceptance of plate tectonic theory has resulted in a new understanding of many well-documented geological observations and, in turn, these have spawned new techniques for examining the evolution of plate boundaries. In Chapter 11, Michael Russell illustrates how the development of different types of orebodies can now be simply related to their tectonic setting, and evaluates the features which are significant for designing a strategy for mineral exploration. Many orebodies form because elements which occur naturally in tiny amounts in most rocks and minerals have been scavenged and concentrated into economically viable deposits by the movement of fluid. The movement of fluids is also likely to modify the minerals which are present, and this therefore links to the much broader subject of 'Metamorphism and Fluids' discussed in Chapter 12.

Metamorphism involves the modification of rock mineralogy and texture in the solid state under the influence of changing temperature, pressure, hence deformation, and fluid composition. Not surprisingly, therefore, metamorphic rocks are particularly common in crustal zones that have been subject to collision, burial and subsequent uplift, such as the Alps, Himalayas and many older so-called orogenic belts that mark the trace of ancient continental plate sutures. Clearly, the variation of temperature with depth, in the form of continental geotherms, is a critical determinant of metamorphism, and we strongly recommend the three boxed sections of text in Chapter 12 to readers unfamiliar with such concepts. After describing the way that rocks respond to different metamorphic conditions, Alan Thompson explains how data from metamorphic terrains can be used to examine the past processes affecting those terrains. Indeed, Alan has contributed extensively to the way that metamorphic petrologists are now able to read pressure–temperature–time histories from the rocks themselves.

CHAPTER

9

Tjeerd H. van Andel
University of Cambridge

SEAFLOOR SPREADING AND PLATE TECTONICS

The birth of continental drift and plate tectonics

Alfred Wegener and continental drift

This chapter was drafted in a year when a hot mudflow from a Colombian volcano took thousands of lives, and there were two devastating earthquakes within two months in the USSR. In California also, a major earthquake was expected soon. I am revising it not long after this Californian earthquake indeed came, destroying among many other things the building in which I worked for many years before moving to Cambridge. A mere three decades ago, these events would have been understood, but not in their global context. Now a major revolution in geological thought has allowed us to relate all three events to the fundamental dynamics of the Earth.

This revolution had a forerunner whose fate is instructive as well as unfortunate. In 1912 Alfred Wegener, a German known mainly for his work in meteorology, proposed an entirely new geological concept, one that conflicted sharply with the then dominant view of a static Earth where the continents, forever fixed in shape and position, were warped slowly up and down by contractions resulting from the cooling of the Earth. It was argued that the same force caused the crust to wrinkle and form mountain ranges, parts of continents to collapse into ocean basins, and portions of the ocean floor to be raised to form **land bridges**. Such land bridges were especially popular among palaeontologists to explain similar fossil assemblages on continents separated by wide oceans.

Geophysicists, armed with the principle of **isostasy** that had been established half a century earlier, realised that continents cannot sink to ocean depths nor oceanic crust be raised to form a continent, but geologists and palaeontologists alike ignored them. Wegener, however, recognised the fundamental validity of the argument and sought other explanations. Starting with the striking parallelism between the opposing coasts of the Atlantic Ocean, he proposed a single supercontinent **Pangaea** that would have existed in the later Palaeozoic and subsequently broke up. Since the Mesozoic the fragments, our continents, would have drifted slowly to their present positions. His main work, *Die Entstehung der Kontinente und Ozeane* (1915), marshalled an impressive array of geological and palaeontological matches between opposing continental shores and produced a configuration for Pangaea that differs little from what we accept today; some additional evidence excepted, later editions of his book required little revision.

Among British and Continental geologists Wegener's ideas attracted only modest interest but no strong opposition. On the other side of the Atlantic, however, the reception was decidedly hostile. In 1926, a major meeting of geologists in Atlantic City virtually to a man (and passionately) rejected **continental drift**, and the idea disappeared from the list of respectable geological concepts for more than thirty years. Only in the southern hemisphere, especially in southern Africa where the evidence for a former supercontinent is particularly impressive, did Wegener's ideas survive as an acceptable hypothesis.

The opposition made much of the fact that Wegener had not presented a convincing mechanism for continental breakup and drift, but the emotional nature of the rejection shows that it derived less from an insufficiency of arguments in favour than from an unwillingness to regard the conventional wisdom as flawed. As late as the 1950s it remained unwise for young geologists intrigued with the idea of continental drift to express this interest openly. In science as in politics revolutions are successful only when dissatisfaction with the current state of affairs is very widespread.

During that same decade, however, entirely new evidence emerged that forced those who obtained it to reconsider continental drift seriously. Many igneous rocks, especially **basalt**, contain enough iron to acquire a measurable magnetic field when they cool and crystallise, so preserving the magnetic field of the Earth at that time. Magnetic measurements on such rocks, if suitably oriented, thus yield the direction to the **palaeomagnetic pole** (see Chapter 10).

The palaeomagnetic pole positions proved most informative. If the continents were permanently fixed into place, the pole positions determined from their rocks should cluster around the present poles. This, however, they refused to do; instead, they traced separate **polar paths** for each continent that only slowly converged to meet at the present pole at the present time. On the other hand, if one fitted the continents together as suggested by Wegener and then allowed them to drift to their present positions, their polar paths did not diverge at all. A strong argument for continental drift had thus emerged and many a geophysicist was impressed, but still geologists found endless reasons to quarrel with the quality or relevance of the data or they ignored the issue altogether. More was obviously needed before the discussion of continental drift could be reopened. The impetus came next from the growing knowledge of the relief, geophysics and geology of the ocean floor which had been but little known in Wegener's time.

Oceanic crust and seafloor spreading

The great upsurge of oceanographic research that began in the 1940s revealed a number of curious facts. The relief of the ocean floor was dominated by three kinds of major topographic features (Figure 9.1): the **mid-ocean ridges** (not always located in mid-ocean), very long escarpments (fracture zones) that seemed to offset the ridges laterally, and **trenches**, narrow elongate deeps located either within ocean basins and then paralleled by island arcs, or along some continental margins, where the islands arcs were replaced by volcano-studded mountain ranges.

A typical traverse of an ocean basin contains a more or less centrally located mid-ocean ridge rising about 3 km above the deep ocean floor. The ridge crest

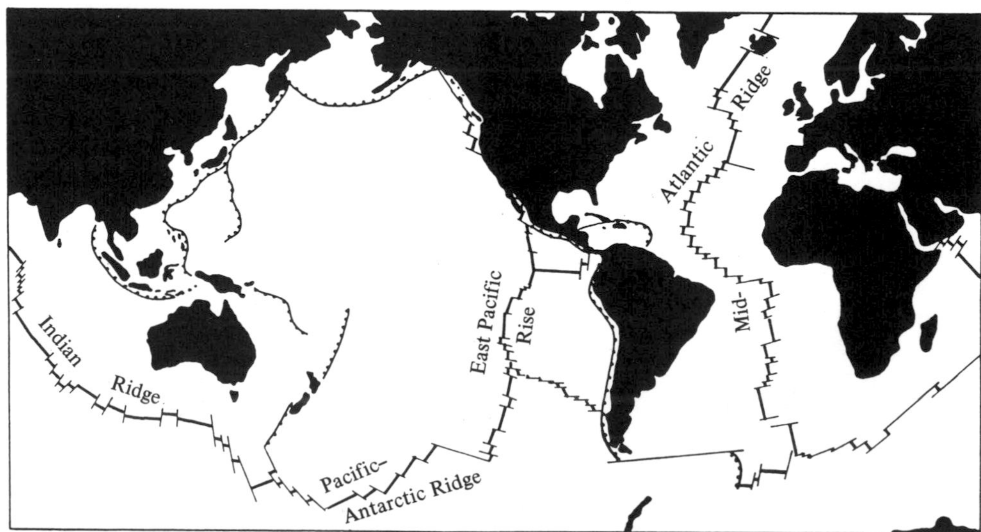

Figure 9.1 The major topographic features of the ocean basins are broad elevations called mid-ocean ridges and elongate, narrow and deep depressions, the trenches. The latter are shown on the figure by lines with small ticks to one side. Mid-ocean ridges are frequently offset by scarps called fracture zones.

is often marked by a deep longitudinal cleft, later recognised as a rift valley, by numerous shallow extensional earthquakes with foci to a depth of 20–30 km, and by active volcanism dominated by basaltic lavas. **Heat flow** tends to be high, and a thin, patchy, sediment cover suggests a young age. To each side the flanks fall away, steeply at first, then more gently, the heat flow diminishes, the blanket of sediments steadily thickens and seismic activity ceases. Only scattered volcanism continues away from the ridge crest, forming seamounts.

The trenches, in contrast, are marked by thick sediments and by an anomalously low heat flow and a virtual absence of volcanic activity. Most striking is an abundance of deep compressive earthquakes, their foci aligned along an inclined plane, the **Benioff zone**, which descends away from the trench at an angle of up to 60° to a depth of at least 200–300 km and in many places to 600–700 km. Fewer and smaller tensional earthquakes occur on the side of the trench opposite the Benioff zone. Trenches situated away from continents are paralleled on the side of the Benioff zone by two island arcs, one sedimentary, the other volcanic in origin.

This curious arrangement recalled Sir Arthur Holmes' suggestion in the 1920s that convection might occur in the mantle; the mid-ocean ridge crest would be located over the rising column, whereas the tenches were shaped by the sinking limbs (Figure 9.2). H. H. Hess of Princeton University, taking Holmes' idea a step further in a paper widely circulated in the late 1950s but not published until 1962, postulated that the oceanic crust was constantly being torn apart by the ascending flow which filled the crack with new crust. The old crust would be carried away passively on the flow and transferred towards the trenches for recycling in the mantle. Thermal expansion of the oceanic crust and mantle would be responsible for the relief, volcanism and seismic activity of the mid-ocean ridges, and downward drag for the topography, the kind and intensity of seismicity, and the low heat flow of the trenches.

The **Mohorovicic Discontinuity** or **Moho** for short, long known but just then the subject of much speculation regarding its nature, was the logical boundary of the thin skin of oceanic crust travelling from ridge crest to trench. In this scheme, the fracture zones offsetting the ridge crests might be the response to discontinuities between adjacent convection cells in the mantle underneath.

Some evidence, for example the increasing age of volcanic islands away from the Mid-Atlantic Ridge, indicated that the oceanic crust was indeed youngest at the ridge crest and grew progressively older toward the trenches, and supported Hess's proposition. Hess's ideas became renamed as the **seafloor spreading hypothesis**, and in the 1960s catalysed a revolution in the Earth sciences which ultimately rehabilitated Wegener and continental drift.

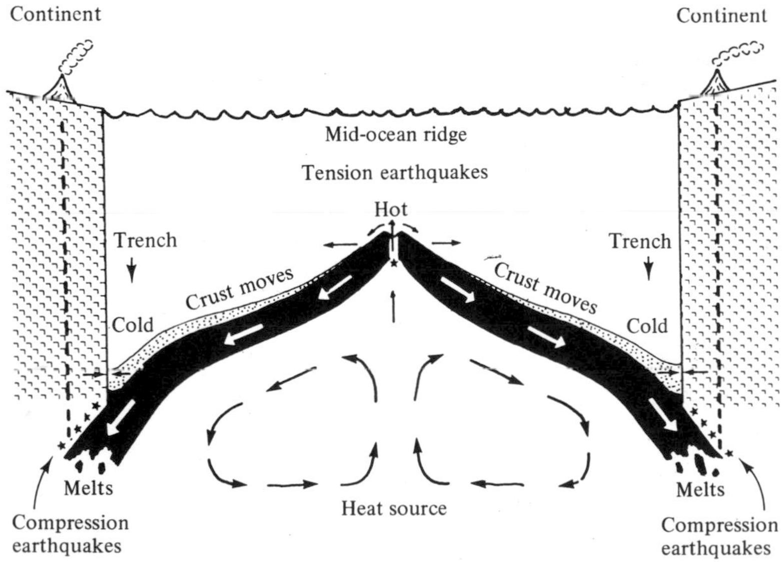

Figure 9.2 The seafloor spreading model of H. H. Hess. New oceanic crust is generated at the crest of a mid-ocean ridge by a rising convective plume in the mantle. Older crust is laterally moved to trenches where it is recycled by descending convective limbs. □

Tests of the seafloor spreading hypothesis

Except among marine geologists, Hess's seafloor spreading hypothesis attracted little attention, until much more striking confirmation began to emerge.

The hypothesis had three consequences amenable to testing: (1) the increase in the age of the oceanic crust away from the ridge crests, (2) the lateral motion of the crust, and (3) the limitation of crustal movement to the oceanic crust. Continental crust was not involved in the process, however, and continental drift not a required result.

A large part of the oceanic crust exhibits an odd pattern of positive and negative **magnetic anomalies** that resembles the skin of a zebra because of our habit of depicting positive anomalies in black and negative ones in white (Figure 9.3). The stripes appear to be offset along fracture zones and disappear at or under the continental margins. Although these anomalies had been widely discussed since their discovery in the late 1950s, no convincing explanation was forthcoming until F. J. Vine and D. H. Matthews of the University of Cambridge interpreted them as imprints of reversals in the polarity of the Earth's magnetic field on the crust produced by seafloor spreading. A similar proposition was simultaneously put forward by the Canadian L. Morley, but published much later.

Episodic reversals of the magnetic field of the Earth had been known since the beginning of this century; once in a while magnetic north becomes magnetic south or vice versa. In two papers in 1963 and 1964 Vine and Matthews noted that the ridge crest was always marked by a large positive anomaly which they explained as the result of the superposition of the Earth's magnetic field on a field of the same polarity induced in the new basalt crust. Assuming that the seafloor spreading hypothesis is correct, the central anomaly will eventually split and the halves drift away, narrower then but retaining their positive signature. A segment of crust intruded during a period of reversed magnetic polarity, on the other hand, would be magnetised opposite to the present Earth's field and be read as a negative anomaly (Figure 9.3).

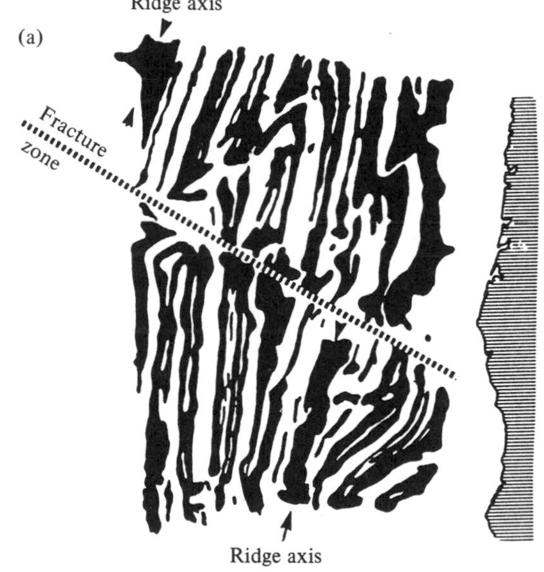

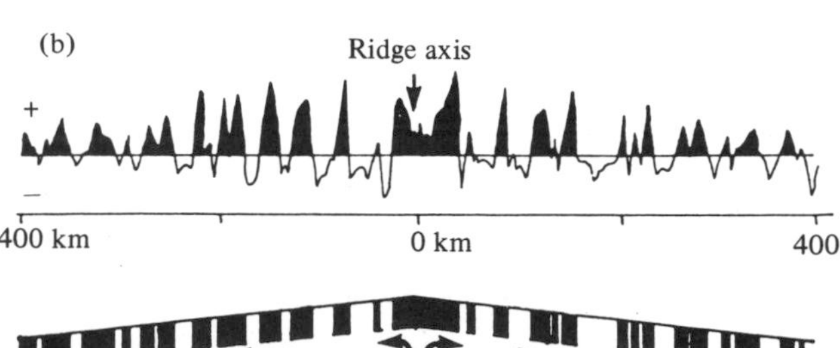

Figure 9.3 (a) An example of a magnetic anomaly pattern of the ocean crust; note the wide anomaly at the ridge crest and a fracture zone offset. Black = positive, white = negative anomalies. (b) A magnetic traverse of a mid-ocean ridge with positive (black) and negative (white) anomalies above the explanation according to the model of Vine, Matthews and Morley. □

The magnetic anomaly patterns on the flanks of a spreading mid-ocean ridge thus ought to be symmetrical. The widths of the alternatingly positive and negative anomalies, each half as wide as the original intrusions, should be proportional to the time elapsed between polarity reversals. By sheer good fortune, the chronology of several million years of magnetic polarity reversals had just been established from sequences of basalt lava flows on land, and in 1964 Vine produced a simple plot of anomaly widths against reversal dates that furnished strong support for the seafloor spreading hypothesis.

At about the same time, J. Tuzo Wilson of the University of Toronto presented an explanation of the mysterious fracture zones which also involved seafloor spreading. These long escarpments which offset mid-ocean ridge crests at approximately right angles had been regarded as **transcurrent faults** (Figure 9.4), so implying that the distance between offset crest segments was gradually increasing. Wilson, in contrast, postulated that the offsets had been there to begin with. If the seafloor spreading hypothesis was correct, crustal blocks formed on adjacent crest segments would move in directions opposite to each other on the part of the fracture zone between the crests. Their sense of motion would thus be the reverse of that expected from a transcurrent fault. Beyond the crest-to-crest segment opposing blocks would move in the same direction and with about the same speed until they were consumed in a trench. Seismic activity should therefore be restricted to the crest–crest part of the fracture zone. Wilson called this new class of faults **transform faults**, and a study of the distribution of earthquakes and their first motions on transform faults shortly afterwards confirmed his hypothesis.

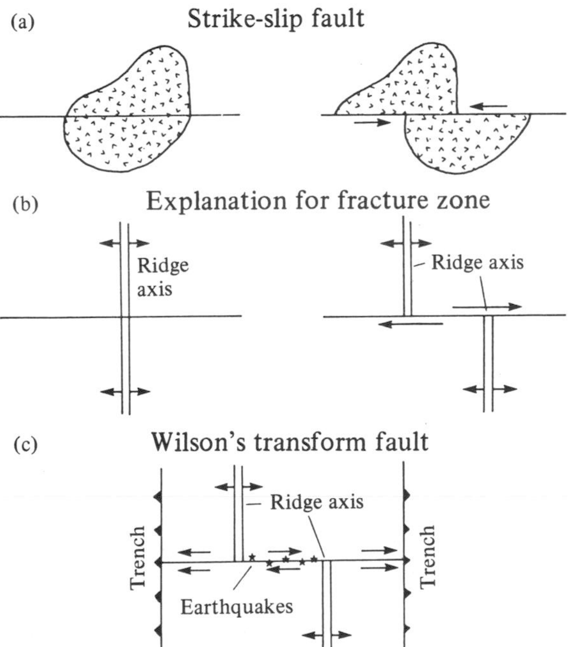

Figure 9.4 (a) The sense of motion on a strike-slip or transcurrent fault using a granite batholith as a marker. (b) The original explanation of fracture zones as transcurrent faults, regarding ridge crests as markers equivalent to the batholith above. (c) Wilson's model for displacement on a transform fault. Crustal blocks generated at an offset mid-ocean ridge crest move in opposite directions between crests but in the same direction beyond. Seismic activity is restricted to the crest-to-crest section of the fault. □

Unsolved questions

The third testable consequence of the seafloor spreading hypothesis, that only oceanic crust need be involved, could not be confirmed. Neither the Atlantic nor the Indian Ocean, for example, has bordering trenches where new crust generated on their mid-ocean ridges can be disposed of, and the fate of that new crust remained unclear.

Furthermore, although the seafloor spreading hypothesis elegantly explained many of the tectonic and volcanic features of the oceanic crust, it failed to account for so many others, small and large, that scepticism remained justified. The transform faults, for example, were assumed to be boundaries between **mantle convection cells** that were themselves also offset. However, as new surveys increased the number and decreased the spacing of transform faults, the individual convection cells began to seem strangely long and narrow. It also seemed odd that the oceanic crust, only 5–6 km thick, could be moved over thousands of kilometres and shoved back down into the mantle without any visible signs of compressive deformation.

Seafloor spreading was born in a time when much attention was given to the Moho which, separating the crust from the mantle, was then regarded as the most important physical boundary within the upper part of the Earth. It was only natural to assume that most tectonic activity, and especially lateral motion, should take place at or above this boundary. Moreover, above the Moho the difference between continental and oceanic crust loomed large as a global tectonic feature. Consequently, it was difficult to couple the thin oceanic crust to the much thicker continental blocks, and rendered it likely that the effects of seafloor spreading were confined within the ocean basins. This, of course, precluded the incorporation of continental drift.

Synopsis of plate tectonics

These problems were swept away by a fresh look by some young scholars free of Moho bias, that caused them to establish where on Earth the main zones of tectonic (seismic and volcanic) activity were located. These zones of activity proved so narrow and so well-defined (Figure 9.5) that they had to be regarded as primary boundaries between segments of crust (plates) moving differentially with respect to one another. It is true that seismic and volcanic activity, sometimes of substantial magnitude, occurs within plates, but the aggregate expenditure of energy there is small compared to that of the plate boundaries.

The pattern so revealed was not related to the geography of continents and ocean basins, and therefore required that the units forming the outer shell of the Earth possessed a lower boundary located well below the Moho. A suitable boundary, separating an upper lithosphere from an underlying weaker asthenosphere at a depth of approximately 100 km, had long been known to exist but had attracted little attention.

The recognition that the Earth's outer skin consisted of a relatively small number of segments or plates, along whose borders most of the tectonic activity of the Earth took place, produced the concept of **plate tectonics**, proposed almost simultaneously in 1968 by Dan McKenzie at Cambridge and Jason Morgan at Princeton.

Above the **asthenosphere**, the outer zone of the solid Earth is thought to consist of plates of **lithosphere**, initially thought to bc about tcn in number (Figure 9.5), but now known to be much more numerous. These plates move across the Earth driven by forces that are still unresolved. Because the carapace of plates is complete, the movement of each plate is severely constrained by those of all the others, and a major change in the direction or rate of motion at any point usually causes the pattern of motion to be globally rearranged. Several such reorientations have occurred in the past 100 million years (Ma).

The plates themselves were originally regarded as rigid and all major deformations seen to occur at plate boundaries. The idea that the plates are not internally

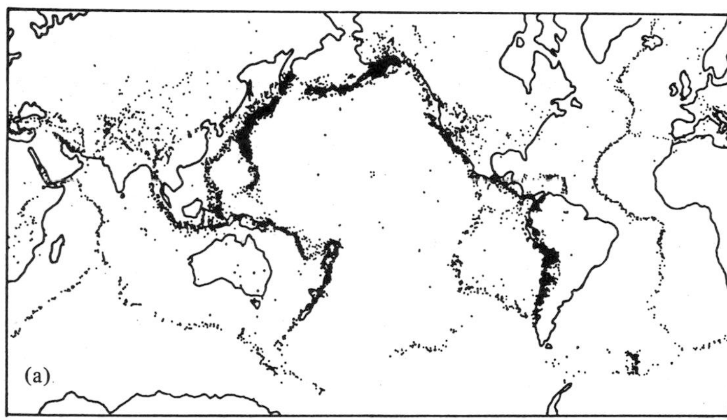

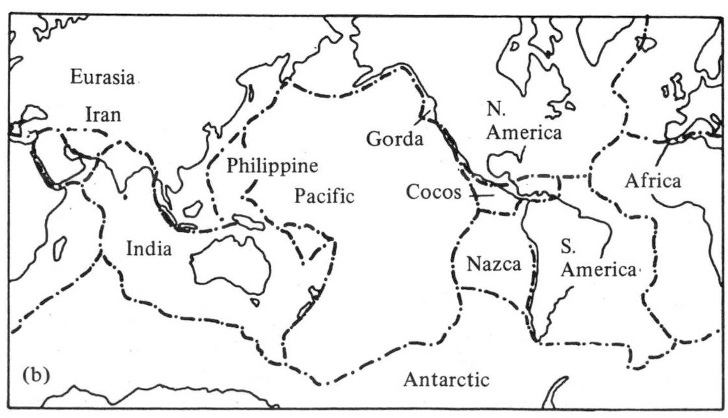

Figure 9.5 (a) Principal zones of global seismic activity between 1961 and 1967. Volcanic activity (not shown) largely coincides with the seismicity. (b) Narrow, well-defined active zones mark the boundaries between tectonic plates composing the outer shell (lithosphere) of the Earth. Note that the active zones are not usually located at continent/oceanic crust boundaries.

deformed is a powerful one because it permits the application of geometric rules to the analysis of plate motion but it is an artificial construct. Since the plate tectonics model was first proposed, evidence for the internal deformation of plates has increased, and more plates, mostly small ones, have been added as the original boundaries proved insufficient to accommodate all observed movements.

There are three fundamental ways in which a plate may move relative to another (Figure 9.6). Two plates may separate across a divergent or spreading boundary, almost always represented by a mid-ocean ridge such as the East Pacific Rise. The gap is continuously filled with newly created crust. The plates may converge instead of diverging; here old lithosphere attached to one or the other plate is returned to the mantle. Convergent boundaries, also known as **subduction zones**, are found between two oceanic plates and are then represented by a trench–arc system of the kind common in the western Pacific, or between a continental and an oceanic plate, with the South American Andes as an example. Convergence (collision) between two plates with continental edges, on the other hand, takes quite a different form. Subduction does not occur; instead, the boundary is marked by major compressional mountain-building. The Himalayan range is a striking consequence of a continent–continent collision.

The movements of plates need not be perpendicular to their boundaries; they may converge or diverge at angles of less than 90°. In the extreme case, the vector representing the relative motion of two plates may be parallel to their boundary so that the plates slip past each other without, at least in theory, any harm; this is a transform boundary.

The geometry of plate motions and the rules that govern the reconstruction of past plate positions are simple, but they do require practice. For more detail the reader is referred to Chapter 10, and to Cox & Hart (see Further reading). Suffice it to say here that, provided enough oceanic crust is preserved to determine spreading rates and transform fault trends reasonably accurately, a remarkably reliable reconstruction of the pattern of oceans, continents, mid-ocean ridges and subduction zones of the past can be obtained.

Given the finite surface area of the Earth, the sum of all divergent and convergent movements must be zero or, in other words, the total amount of new crust created at mid-ocean ridges must be balanced by an equal amount recycled by subduction. It is not, however, necessary that this balance should be achieved within individual ocean basins: for example, the crust formed on the Mid-Atlantic Ridge is balanced mainly by the subduction of Pacific plates under the western edge of the Americas. This property of the plate tectonics model removes one of the principal difficulties of the seafloor spreading hypothesis.

Processes at plate boundaries

At the boundaries between plates the major tectonic events of the Earth take place; new crust is generated at divergent boundaries, and it is there also that continents break apart and new ocean basins are created. In subduction zones old crust is recycled into the mantle or accreted against continental margins, thereby increasing the size of the continent. The process is accompanied by mountain-building (**orogeny**). Major orogeny also occurs when two continental plates collide, and old zones of convergence and collision, called **sutures**, contain key records of Earth history. Transform faults, although less spectacular in appearance, are significant for their use in deriving plate motions, because plate fragments can travel long distances along them. They are also one of the main sources of devastating seismicity, such as on the San

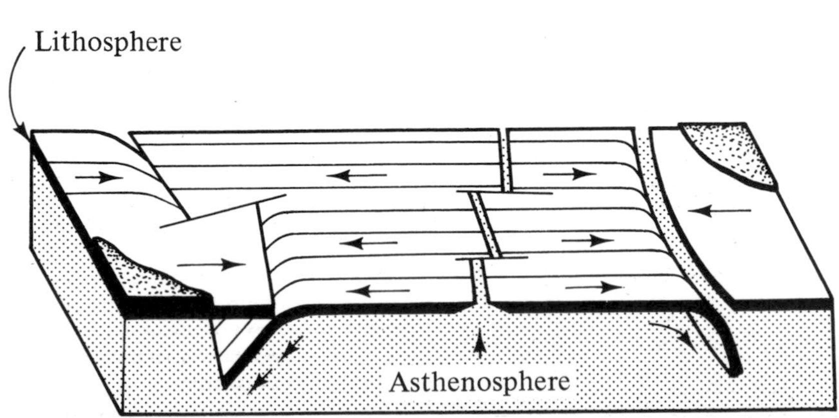

Figure 9.6 The plate tectonics model. The surface of the Earth is composed of c. 100 km thick lithospheric plates (black) floating on a weaker asthenosphere (grey). At their boundaries the plates may diverge (on mid-ocean ridges), converge (in trenches) or slip past each other (along transform faults). The plates may carry continents, ocean basins, or both. □

Andreas fault system in California or on the northern Anatolian faults in Turkey.

Divergent boundaries

Divergent plate boundaries may originate in continents as well as in ocean basins but, because even the former eventually produce spreading oceanic crust, mature divergent boundaries are found only in ocean basins.

The processes and events that create a divergent boundary within a continent must be inferred from the geological record of continents and continental margins and our knowledge of them is still to some degree speculative. The best modern example of the early stages of **continental breakup** is the East African rift zone; although parts of this system are at least 20 Ma old, the rifts have so far opened up only by normal faulting; the formation of oceanic crust and continental drift have not yet begun. For the subsequent stage, when drifting has followed rifting, we may turn to those embryonic oceans, the Gulf of California and the Red Sea/Gulf of Aden, where drifting started, respectively, 5 and 10 Ma ago. The end result is a new ocean basin bordered by a set of so-called passive continental margins, once the site of the breakup. Active margins occur where a transform or subduction boundary lies along a continental margin.

The process of continental breakup starts with a thermal expansion of the crust accompanied by high heat flow, due to a heat source located deep in or below the lithosphere (Figure 9.7). As the brittle upper crust swells to form a broad dome a few kilometres high and up to 1000 km wide, normal faults accommodate the stretching and allow keystone blocks to sink to relieve stress and provide the increase in surface area. The small amount of extension accompanying this rifting process foreshadows the eventual separation of two plates, but a true plate boundary has not yet emerged. Continents may remain in this rifting stage for millions of years and sometimes never evolve beyond it.

The most efficient way to relieve the strain in such a dome is the formation of three rifts that come together in the centre. Two of those may eventually extend and join rifts in adjacent domes to form a continuous rift valley, the future plate boundary. The third atrophies and eventually fails, but because the early marine environments in continental rifts are suited to the generation and accumulation of oil and gas, such failed rifts (aulacogens) which are not buried deeply under the thick silts and clays of a mature continental margin contain a significant part of the world's oil reserves.

Further heating will cause thinning of the lithosphere and the intrusion of mantle-derived basaltic lavas. The density of the crust increases, enhanced subsidence to below sea level is the result, and early marine deposits appear. Ultimately, the continuing intrusion of basaltic magma begins to create oceanic crust, real drifting starts, and an embryonic ocean basin forms.

As the edges of the rift drift away from the heat

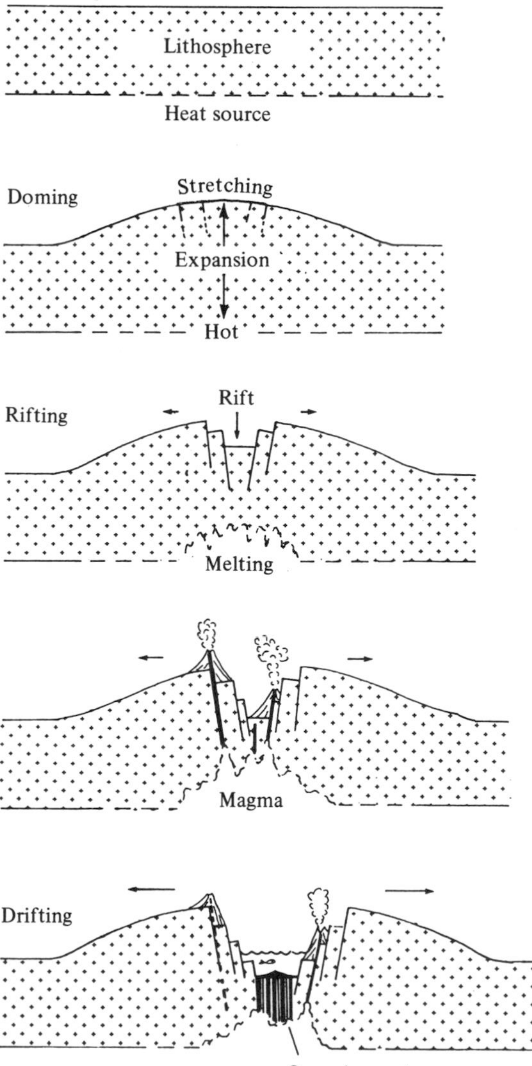

Figure 9.7 Continental breakup initiated by heating from beneath the lithosphere causes thermal expansion which domes the surface. The surface responds by normal faulting and the formation of a rift. Volcanoes form on rift edges and, as melting and intrusion of dense basaltic magma in the crust continue, the rift floor sinks below sea level and drifting replaces extension by rifting. □

source, now located under an embryonic mid-ocean ridge, they cool, the crust shrinks and the slope of the continental surface reverses. The regional drainage reorients towards the new ocean, and river-borne sediments bury the rift margin under a thick shelf prism.

On the mid-ocean ridge, the processes that generate new crust have been clarified in recent years by the use of new technology such as deep-diving research submarines. The plate boundary itself, where volcanic intrusions and extrusions are converted into oceanic crust, is in the rift valley, a fault-bound depression in the crest of the mid-ocean ridge. It is most strikingly developed on very slow-spreading ridges such as the Mid-Atlantic Ridge, but can also be seen at much higher spreading rates.

The floor of the rift valley is occupied by one or more rows of small elongate volcanoes (Figure 9.8). Along a mid-ocean rift axis, different segments are active at different times. The erupting row is paralleled by other, older ones, their age increasing away from the active row. The volcano bodies nearest the foot of the rift valley walls are episodically raised in their entirety to the bordering crests where they begin their way down the slope of the mid-ocean ridge. The axial zone within the rift valley is thus a plate tectonics no-man's land; its crust belongs to neither plate and, depending on where across the width of the valley the next active chain appears, the older chain may become attached to either side.

At slow opening rates (2–3 cm/year) the rift valley is deep and volcanic eruptions are followed by ten or more millennia of stagnation, faulting and uplift. At intermediate opening rates (6–7 cm/year), active phases occur much closer together in time and the newly formed crust is alternatively attached to one or the other rift wall. At highest rates (up to 15 cm/year) the rift is so shallow that much of the time it is buried entirely under profuse lava flows that flood the landscape and give the ridge crest a rounded rather than a cleft appearance.

The volcanic events at the surface reflect the behaviour of a magma chamber at depth. The episodicity of the alternating volcanic and tectonic activity at the crest thus implies episodic events at the base of the plates as well, but this conflicts with the sensible assumption that the plates, because of their great momentum, must move continuously.

On the ridge flanks volcanic activity becomes insignificant, but deformation continues as the plate subsides, old faults are reduced in height by reverse slip, and new ones form due to the cooling and shrinking of the crust.

Transform boundaries

The mid-ocean ridges are cut by numerous transform faults ranging in size from small offsets of a few tens of metres to large offsets of a few to a few hundred kilometres and fracture scarp lengths of several hun-

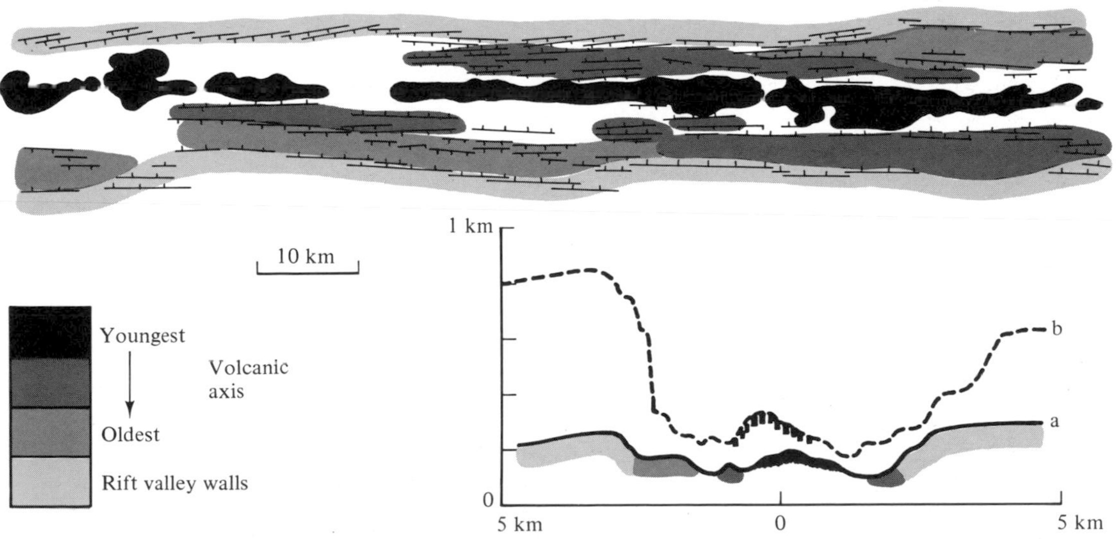

Figure 9.8 Volcanism and deformation in the rift valley of a mid-ocean ridge with a moderate spreading rate (Galapagos rift). Normal faults are straight lines hachured on the down-thrown side. The active volcanic ridge is black. Lower right: transverse profiles of this rift valley (a) and of a slow-spreading ridge such as the Mid-Atlantic Ridge (b); the active volcanic ridge is shown in black. □

dred kilometres. With the increase in offset comes an increase in spacing and a large decrease in abundance. These transforms always connect ridge crests and have lifespans that range from very short to the age of the ocean basin in which they occur. Most of them, and especially the smaller ones, should probably be regarded as part of the process of generation of new crust rather than as true fundamental plate boundaries.

At the high end of the range and not very common are the great fracture zones that extend many thousands of kilometres across the ocean basins and offset ridge crests by hundreds of kilometres. They include the equatorial fracture zones in the Atlantic, the 90° East Ridge in the Indian Ocean, and the Mendocino, Clipperton and other large fracture zones in the eastern Pacific. These are long-lived and seem to be related either to an initial continental breakup or to as yet poorly understood regional tectonic processes occurring entirely within an ocean basin.

True transform plate boundaries may also connect ridges to trenches and are common on land as well; examples are the fault system that extends from the Gulf of California to about 45° N and forms the boundary between the North American and Pacific plates, or the large fault systems of western China, Tibet, Afghanistan and Iran that are the result of the collision of India and the Asian continent. The structure of these transforms is better known than that of their oceanic counterparts; they consist of numerous major and minor subparallel **strike-slip faults** which often have vertical components of motion as well. Slip occurs now on one, then on another of these anastomosing branches. Just a cursory look at the geology map of one of these transform boundaries will convince one that even this simplest of the three plate boundaries is broad, complex, and often very diffuse.

Convergent or subduction boundaries

There are three kinds of convergent plate boundaries: ocean–ocean convergence, ocean–continent convergence, and continent–continent collision. The simplest one is the convergence between oceanic plates (Figure 9.9); trench–arc systems in the western and southwestern Pacific fall in this category.

In this case, either plate may be subducted under the other, as both have about the same density and thickness. Approaching the subduction zone from the subducting plate, one traverses a broad rise stepping down into the trench with a series of blocks bound by normal faults. Upwarping and down-faulting are the result of the bending and hence stretching of the crust as the plate begins to subduct. The trench itself is often

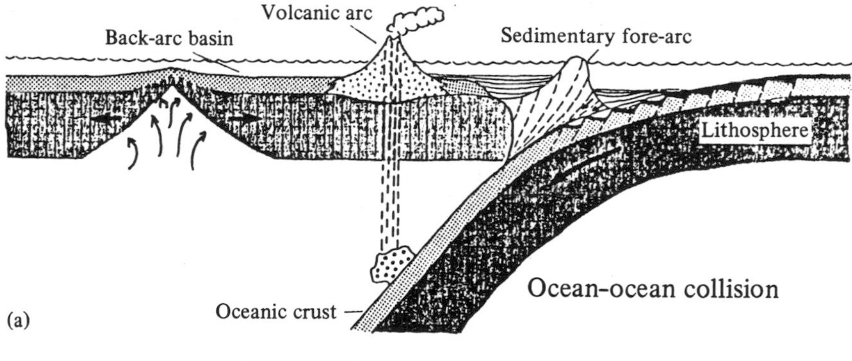

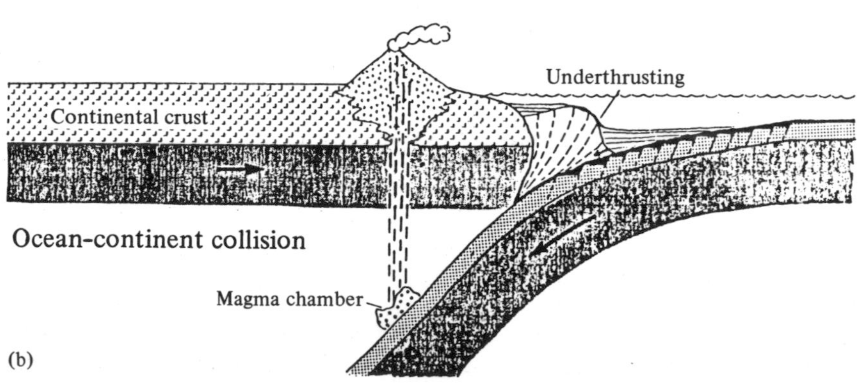

Figure 9.9 Tectonics of subduction zones. (a) Convergence of two oceanic plate edges. (b) Convergence of a continental and an oceanic plate edge. For discussion see text.

partly filled with stratified sediments laid down by **turbidity currents** (Chapter 16). The thickness of the fill varies as a function of the sediment supply but may, in extreme cases, conceal the existence of the trench. On the side of the overriding plate a deformed sedimentary wedge forms the **fore-arc**. The fore-arc builds from slices of pelagic and trench sediment and sometimes basaltic upper oceanic crust that are sheared off and thrust under the already existing wedge, the oldest slice ending up on top. The entire pile may be raised above sea level to form an **island arc**. Behind this arc lies the fore-arc basin filled with sediments washed in from both sides. At the far side andesitic volcanoes constitute a volcanic arc located where the subducted slab reaches a depth of *c*. 200–300 km. Beyond the volcanic arc a **back-arc basin** with a young crust and high heat flow may itself be formed by localised extension within the overriding plate.

Because the continental crust is of lower density and much thicker than oceanic crust, the convergence of an oceanic and a continent-bearing plate always involves subduction of the oceanic one. In this case the fore-arc is accreted to the continental margin and may cause substantial seaward progradation of the continent, while the fore-arc basin forms the shelf and the andesitic volcanic arc a major mountain chain on the continent beyond it (Figure 9.9). The classic example is the South American Andes, part of the eastern margin of the Pacific which almost entirely falls in this class.

Continental accretion does not always occur; often part or all of the pelagic and trench sediment is subducted together with the oceanic plate. In that case the high content in volatiles (carbon dioxide, chlorine, water etc.) of the sediment layer and the upper crust, and the iron and manganese oxide ores rich in copper, zinc and molybdenum that are often found between the basalt and the pelagic sediment sequence, may be mobilised by heat at a depth of 200–300 km and released through the volcanic arc to form major ore deposits (see Chapter 11).

A special case exists when the plate boundary lies some distance seaward of the continental margin in an oceanic slab attached to the continent. Either plate edge may then be subducted, but if it happens to be the slab attached to the continent, the direction of subduction must reverse when the ocean crust has been entirely consumed. There is evidence for such reversals but the mechanism remains obscure.

Subduction on one or both sides of an ocean basin may end by consuming all of the oceanic crust. Continued convergence will then produce a collision between two continental plate edges. The best known example of complete consumption of an ocean basin and the subsequent collision of two continents is the collision in the Eocene between India and central Asia.

The suture of this collision is not, as one might surmise, at the Himalayas but lies well north of them (Figure 9.10). The momentum of the Indian plate was so great that now, 40 Ma after the first contact, India, having moved another 2000 km farther north, has not yet stopped. Half of this distance can be accounted for by compression of the Indian plate to form the Himalayas, the other half is due to penetration into the Asian block in a manner analogous to hammering a chisel into a block of metal. The Asian crust is getting out of the way along a set of subparallel shear zones extending from western China to Iran that behave like strike-slip faults and form, in a sense, a complex transform plate boundary.

The India–Asia collision has its puzzling aspects. Granites have been intruded copiously into the Himalayas, heat flow is high there and the crust underneath appears to have been greatly thickened. None of this fits the model for converging plates and it has not been fully explained. One interesting hypothesis says that the lower lithosphere, below the Moho, being relatively cool and hence dense, would sink into the asthenosphere if not kept afloat by the continental crust. The subducted slab, having ceased to function and now detached, might have dragged down with it this lower lithosphere. This would clear the way for heat to rise and melt continental crust to form granite intrusions, and would provide the space for the continental crust of one plate to slide under that of the other (Figure 9.10).

Orogeny, the building of mountain ranges, is a prominent feature of plate convergence, raising some mountain ranges by underthrusting and uplift, piling up others by means of volcanic eruptions, and constructing the largest of all through compression and associated processes in continent-to-continent collisions. Plate tectonics has provided a simple set of basic models for this, one of the most important subjects in tectonics, hence its earliest name: 'The New Global Tectonics'.

Exotic terranes

The sutures of old plate convergences and continental collisions are identified by the highly deformed and metamorphosed remains of subduction zones and orogens, often so eroded that deep parts of the former plate boundaries are exposed. Close examination of sutures has shown, however, that, even if we ignore continent–continent collisions, there is more to the

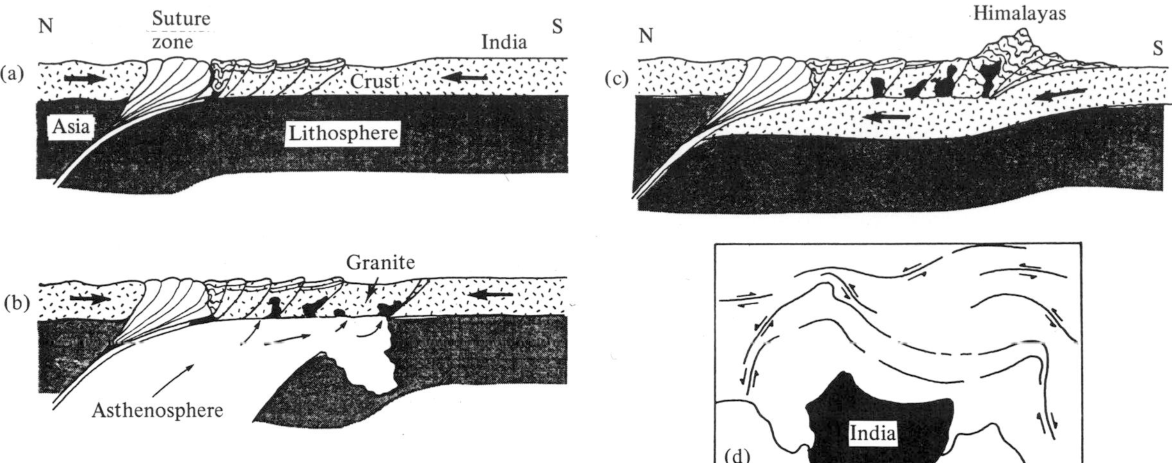

Figure 9.10 Collision between continental plate edges of India and Asia. (a) The suture zone marks the initial collision during the Eocene; (b) The deactivated subducted slab tears away the lower lithosphere, allowing access of heat and the intrusion of granite; (c) Underthrusting from the south doubles the lithosphere layer and the Himalayas are formed by compression; (d) The Indian wedge, driven northward into Asia, produces a set of strike-slip faults along which Asian crust gets out of the way. □

concept of converging plates than is indicated by the simple models discussed above. Many subduction zones have turned out to be places where pieces of oceanic crust from distant sources on one plate have been swept together with fragments torn loose from the other and have moved laterally along the plate boundary. They have produced complex *melanges* of crustal blocks of many compositions, sizes, ages and provenances, now known as **exotic terranes** (Figure 9.11a).

This crustal flotsam and jetsam so gathered in subduction zones consists of distinct tectonic–stratigraphic units, each with its own history different from that of the adjacent units from which it is separated by faults.

The terranes have a wide variety of origins. Most were once pieces of anomalously thick oceanic crust within ocean basins. Abandoned parts of mid-ocean ridges, former volcanic arcs, fracture zone ridges, large, thick volcanic plateaus such as Iceland, or seamounts and seamount complexes have all made their contributions. A few may even be of continental origin. Even a sizeable seamount descending a trench can for a while bring subduction at that spot to a halt, and very large blocks of thick crust may end subduction altogether, forcing a seaward jump of the blocked trench segment (Figure 9.11b).

Palaeomagnetic studies, sediment properties and fossil assemblages indicate that some terranes have come from distant palaeolatitudes; a few of those assembled in northwestern North America may have had their origin in the far southern or western Pacific. These far-travelled segments of ocean crust often constitute a large part of the volume of a continent–ocean subduction zone and may have contributed a major share to the growth of continental blocks.

The discovery and at times somewhat too eager application of the terrane concept has been disconcerting to those who liked the simple elegance of the original plate tectonics. And indeed, if the history of convergences is really as complex and regionally variable as the exotic terranes imply, the globally unifying power of plate tectonics is substantially diminished.

Thermal processes associated with plate tectonics

Cooling and subsidence of the oceanic crust

The width, height and cross-section of mid-ocean ridges are a function of heating at the ridge axes and subsequent cooling as the new crust is displaced laterally. As a result, there is little variation in the elevation of a ridge above the adjacent deep-seafloor and in its crestal depth. The width of a ridge, on the other hand, depends on the spreading rate, because the

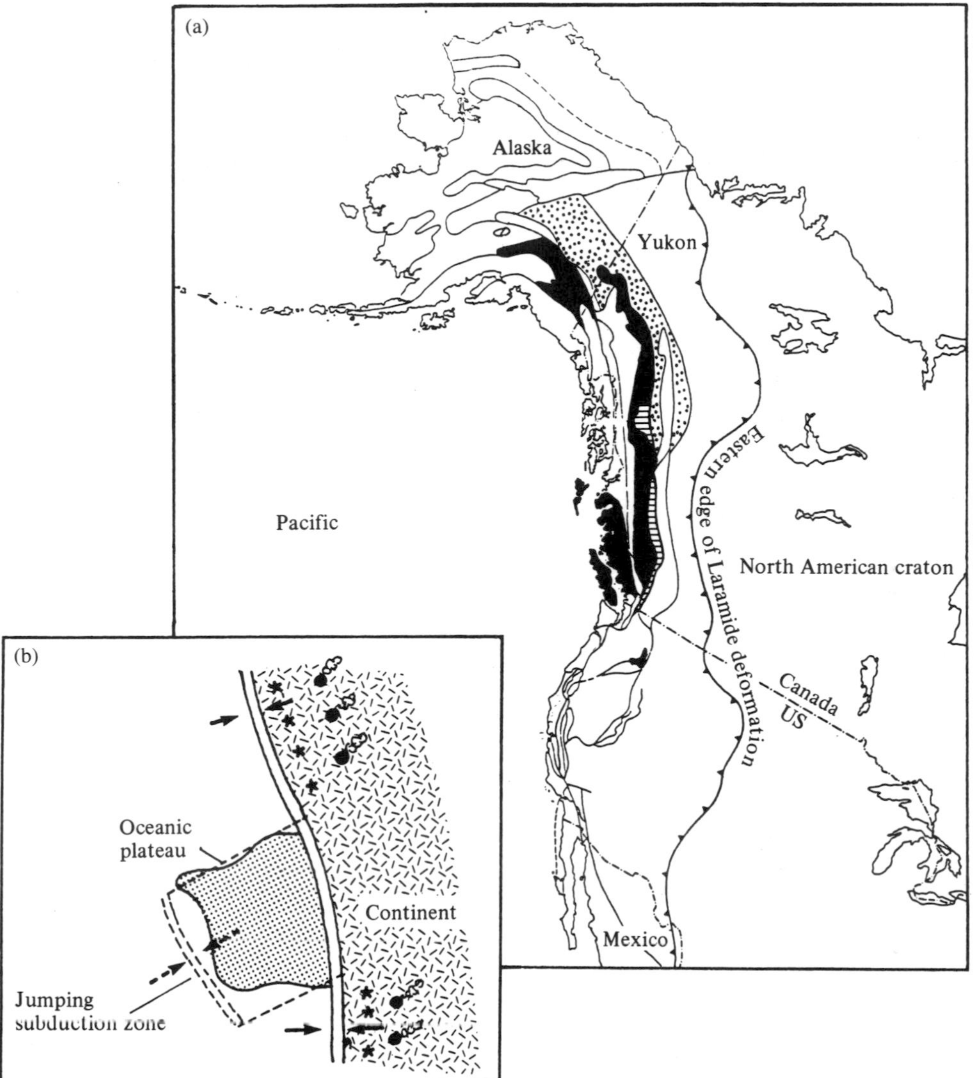

Figure 9.11 Exotic terranes. (a) The flotsam and jetsam that forms a large part of the continental accretion of western North America. Black: terranes of volcanic island arc origin. Striped area: terranes originally formed as oceanic plateaus. Dotted shading: metamorphic terranes. (b) An anomalously thick oceanic plateau segment forces the subduction zone to jump seaward. Stars are earthquakes, smoking circles volcanoes. □

cooling of the crust is only a function of time, and therefore a fast-spreading ridge subsides more gradually with distance from its axis than a slow-spreading one.

The high crestal heat flow of up to 7 or 8 microcal/cm²/s decays exponentially with increasing distance from the axis to one or less in the subduction zone (Figure 9.12, curve A). Consequently, the depth profile of a ridge is a function of its age and of the initial depth at its crest (which tends to be quite uniform at 2500 ± 300 m). The subsidence resulting from the cooling of a 100 km thick slab of lithosphere can be modelled quite accurately, being proportional to the square root of the age of the crust:

$$\text{depth of basalt} = 2500\ (\text{crestal depth}) + 350\ (\text{age})^{\frac{1}{2}}.$$

Because the upward flow of heat from the underlying asthenosphere becomes progressively more important as the new lithosphere cools, the equation needs to be slightly modified for crust older than 60 or 70 Ma.

The theoretical curve agrees well with observational

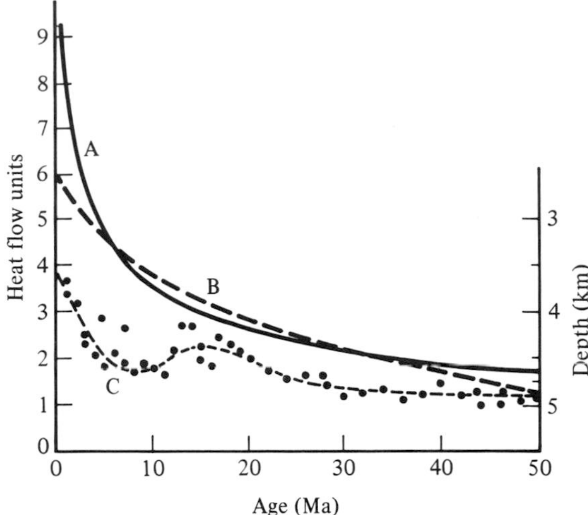

Figure 9.12 Crustal cooling, heat flow and the subsidence of a mid-ocean ridge. Line A shows the theoretical heat flow curve (in heat flow units = microcal/cm²/s). Line B depicts the subsidence of the new crust as a function of the square root of its age. Line C is a best fit line for a sample of actual measurements of conductive heat flow showing the deficit compared to curve A (Galapagos ridge). □

data (Figure 9.12, curve B), and on most mid-ocean ridges the water depth increases from *c.* 2500 ± 300 m at the crest to 3 km at 2 Ma, 4 km at 20 Ma, and 5 km at 50 Ma. This relation between depth and age enables us to reconstruct the palaeobathymetry of the ocean basins, provided a correction is applied for the additional subsidence resulting from isostatic compensation for the increasing sediment load. Inversely, one can estimate the approximate age of the ocean crust from its regional depth.

Because the depth of the ridge flanks depends only on age, a fast-spreading ridge will have a broader, more gentle profile than a slow-spreading one, and a larger cross-section (Figure 9.13a). If the spreading rate should increase, the volume of the ocean basins would be reduced by the increase in volume of the expanded mid-ocean ridge and sea level will rise. Conversely, a decrease in the average global spreading rate will produce a lowering of sea level (Figure 9.13b). The same result is, of course, achieved by increasing the aggregate length of mid-ocean ridges without changing the rate of spreading.

An increase in the spreading rate from 2 to 6 cm/year for a ridge system 10 000 km long, for example, will raise sea level by *c.* 110 m. Such a spreading rate change is by no means unusual and variations in sea level of several hundred metres can be attributed to the

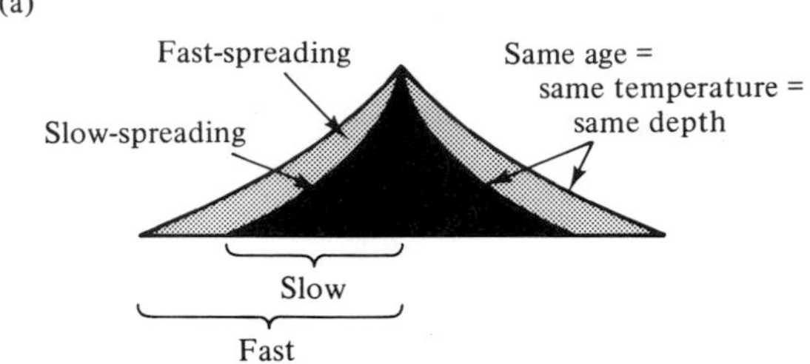

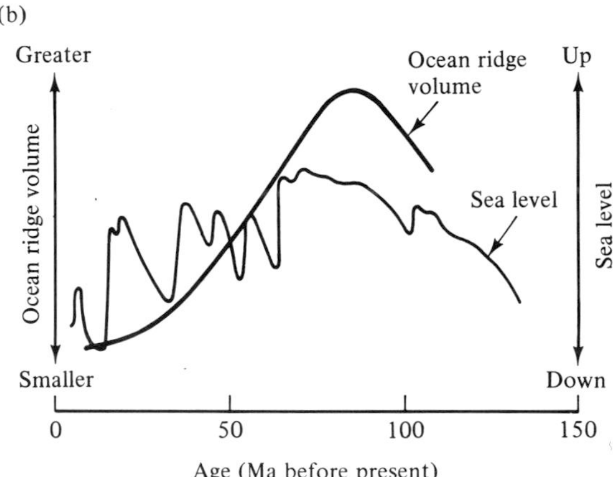

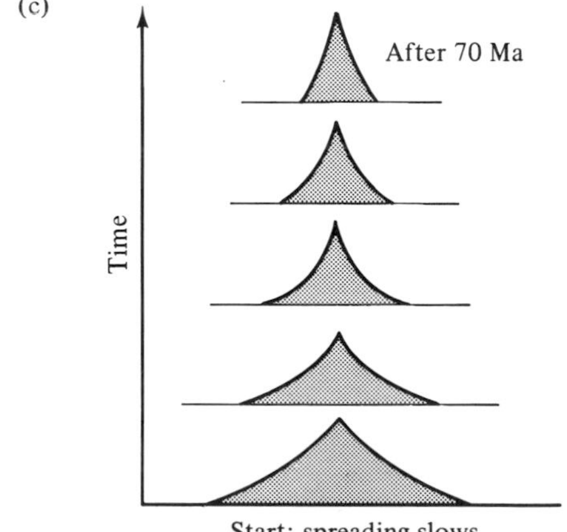

Figure 9.13 Volume change of a mid-ocean ridge as a result of a change in spreading rate. (a) The difference in cross-section between a slow-spreading and a fast-spreading ridge. (b) Comparison between the estimated change in ridge volume and sea-level changes since the Cretaceous. Gradual change in ridge volume after a sudden decrease in spreading rate; the new equilibrium profile takes 70 Ma. □

thermal consequences of plate dynamics, but because a change in spreading rate does not achieve an equilibrium profile for 60–70 Ma (Figure 9.13c) the explanation applies only to sea-level changes on that time-scale. For the very common sea-level changes of shorter duration (see Chapter 20) this mechanism is not appropriate.

Hydrothermal processes on mid-ocean ridges

Theoretical models of conductive heat flow due to cooling of the crust are simple and on the flanks of mid-ocean ridges they correspond closely to reality (Figure 9.12). At the ridge axis, however, they predict a much higher conductive heat flow than is actually observed, a discrepancy that persists even after account has been taken of various factors capable of distorting the observations. The shortfall, almost twice the observed value, is much larger than all possible errors of measurement. To resolve this contradiction, it was proposed more than two decades ago that the conductive heat flow represents only part of the cooling process, the rest being accomplished by the circulation of seawater through fissures in the young hot crust. Cold water would enter the new crust, descend to considerable depth while picking up heat, and return to the surface when hot. There it would exit in the form of submarine **hydrothermal springs** (see Chapter 11). The process would continue until, some considerable distance out on the ridge flanks, the ageing crust was sealed by a sediment blanket.

This hypothesis was confirmed in the late 1970's when hydrothermal springs were discovered on mid-ocean ridges in the eastern equatorial Pacific; they are now known to be ubiquitous. Spring water temperatures vary from a few degrees to 350 °C. Various chemical properties of the entering and exiting waters and theoretical considerations suggest that the heat released by hydrothermal flow is indeed approximately equivalent to the estimated total shortfall of axial heat flow of 10^{19} kcal/year.

Hydrothermal circulation of deep ocean water through hot, new oceanic crust is thus an important process; under present circumstances a volume of water roughly equal to that of the whole ocean passes through the upper oceanic crust every 8 to 10 Ma.

During its passage through hot basalt, seawater is greatly altered chemically; magnesium and sulphate, abundant in ocean water, are lost wholly to the hot basalt, while potassium, calcium and silica are extracted and brought to the surface. The springs also contribute great quantities of dissolved carbon dioxide and gases of mantle origin such as the helium isotope ^{3}He to the deep water. Furthermore, they appear to be the main source of manganese and barium in the oceans, and contribute about one-fourth as much calcium and silica as the combined rivers of the world and a larger quantity of ferrous iron. Upon entering the cold bottom water, many of these elements precipitate in the form of iron–manganese hydroxide deposits that are sometimes rich in valuable metals such as copper, silver, zinc or lead (see Chapter 11).

Apart from its role in disposing of the excess heat of the oceanic crust, this so recently discovered major Earth process is also potentially capable of altering the chemistry of ocean water on a timescale of 10 Ma or so. If, for instance, the average spreading rate were to double, as may have been the case during the Cretaceous, the carbon dioxide, calcium carbonate and silica contents in the oceans would be raised significantly while much of the dissolved oxygen would be consumed by the increased supply of reduced iron. It is possible, although still controversial, that a more vigorous hydrothermal circulation would also reduce the sodium and chlorine ion contents of the ocean. Significant changes in salinity and in the composition of seawater might thus result from variations in the seafloor spreading rate, with potentially a considerable effect on marine life. As there is every reason to expect that hydrothermal processes on spreading plate boundaries have been active from very early on in the history of the Earth, then probably functioning at a higher rate, their role in the evolution of life may have been considerable; this interesting issue remains to be thoroughly explored.

Reconstruction of past plate configurations

Magnetic anomaly patterns and transform faults provide the means for determining poles of rotation and relative velocities of pairs of plates. From them we may derive plate migration paths and past global plate configurations, but the results become progressively less certain as the amount of surviving crust decreases, and beyond 50–80 Ma other techniques based mainly on palaeomagnetic data must be used (see Chapter 10). Even for more recent intervals, however, the calculations are often beset with uncertainties due to lack of precision of the data, to plate boundary complexities, and to the breakdown of the axiom that plates remain internally undeformed throughout time.

Consequently, the best reconstructions are those of the new oceans, the North and South Atlantic, the

Indian and the Southern Ocean, where little or no ocean crust has yet been lost. The South Atlantic provides an instructive example (Figure 9.14).

Even before any oceanic crust had been emplaced, a gradual scissors-like opening beginning in the south about 125 Ma ago can be inferred from African and South American palaeomagnetic data. The narrow embryonic ocean evolved into two basins separated from each other and from the North Atlantic and Southern Ocean by shallow barriers. At times these barriers may have been above sea level so that behind them Early Cretaceous evaporites could form from overflow of the northern or southern oceans. These barriers gradually subsided as the ocean widened and the flanks of the Mid-Atlantic Ridge cooled, until around 65 Ma ago a direct deep-water connection existed between the Southern Ocean and the North Atlantic. Since then the South Atlantic has steadily widened and deepened.

This approach yields a pattern of plates and their relative motions that, although internally consistent, is not fixed with respect to an Earth frame of reference such as the equator. Over time, the lithosphere shell itself might have slipped relative to the equator and the north and south rotation (spin) poles of the Earth. Are there geological or geophysical markers that allow us to relate plate trajectories to the earth's rotational frame of reference?

The answer is affirmative and at least three approaches have been used. The first is only approximate, but informative and independent of any plate tectonic assumptions. The equatorial current system traverses the Pacific Ocean from east to west as a relatively narrow (about 4° latitude) zone of high fertility and biological productivity. Because some dominant planktonic microfossil groups produce abundant calcareous and siliceous shells, the path of the equatorial current system is marked on the ocean floor

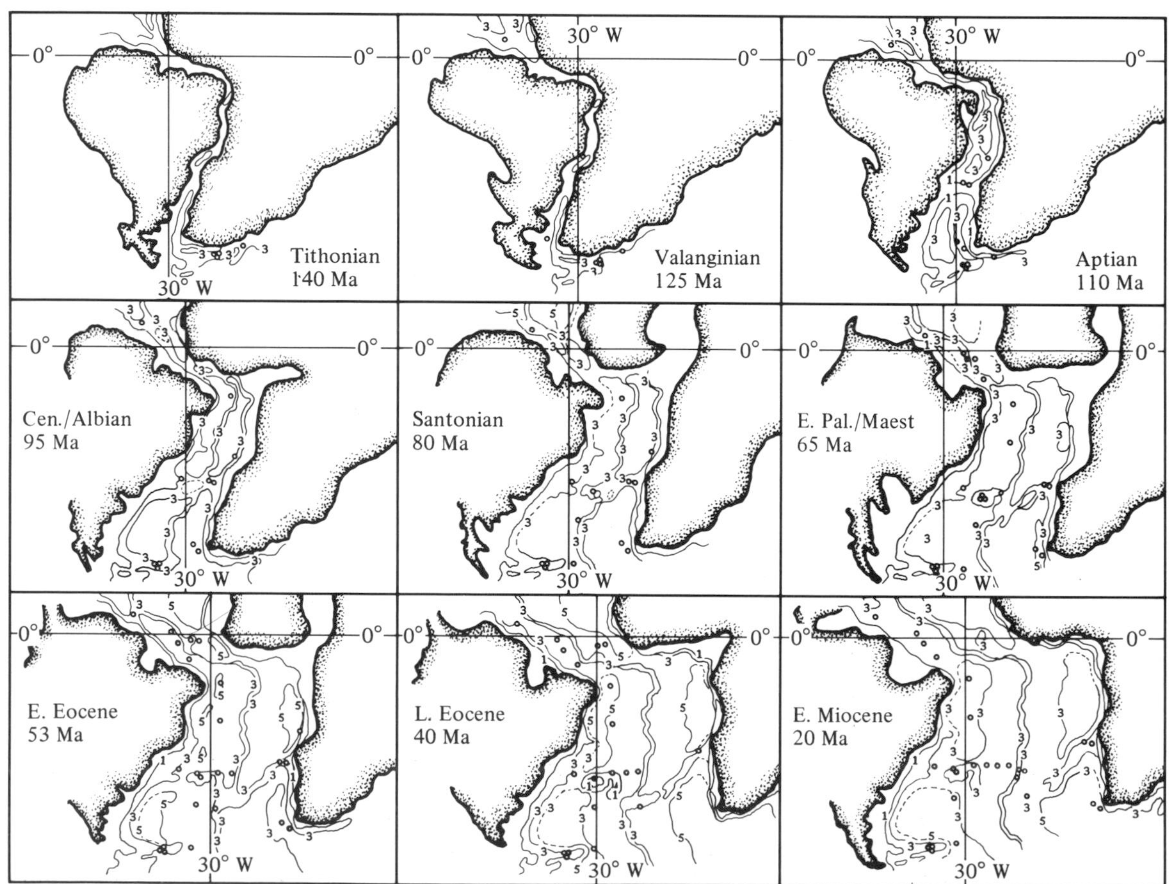

Figure 9.14 Reconstruction of the history of the South Atlantic with seafloor spreading data. The palaeobathymetric contours (in km) were derived from the age–depth relation of the oceanic crust. Shorelines were adjusted for the transgressions and regressions of the last 150 Ma. □

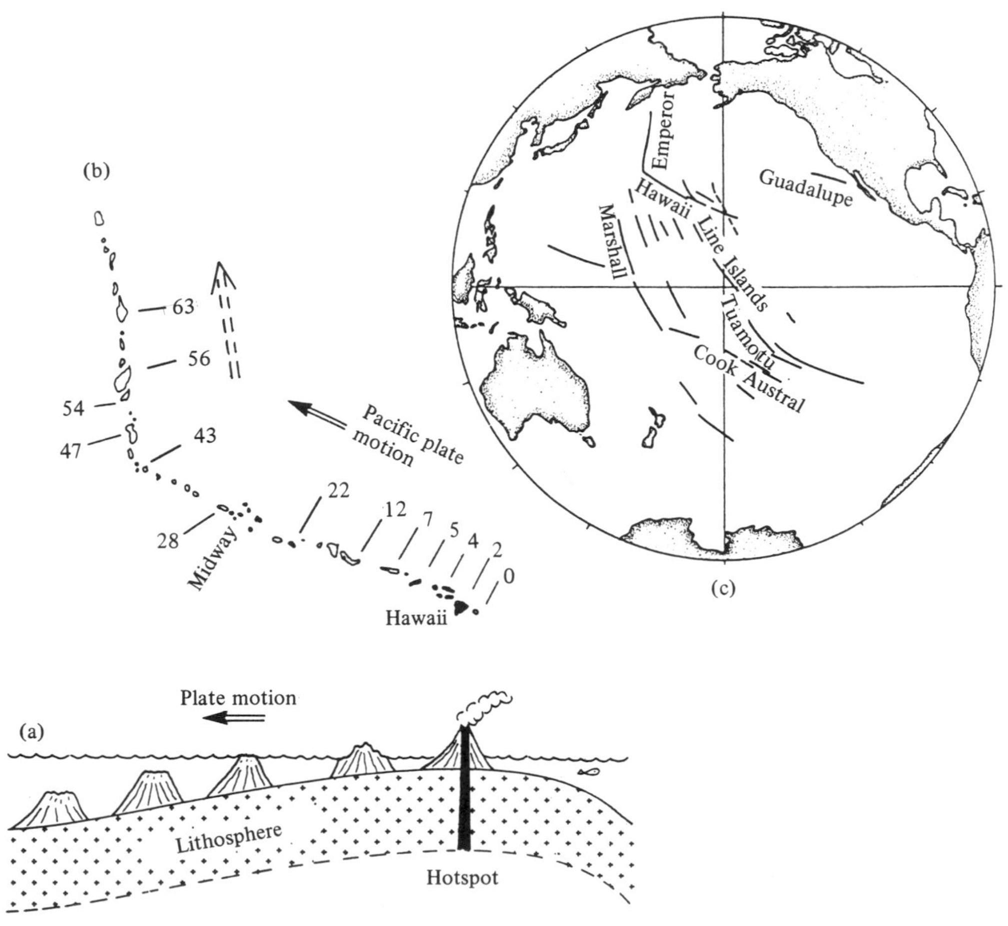

Figure 9.15 Reconstructing plate history in a whole-Earth frame of reference. (a) Formation of a volcanic chain over a hotspot. (b) The Hawaiian chain in the Pacific with volcano ages in Ma. (c) Linear volcanic chains related to hotspots trace the movement of the Pacific plate. □

by a belt of thick siliceous–calcareous oozes, a chalk line so to speak. Plates migrating across the equator bear subparallel belts of calcareous ooze increasing in age in the direction of movement. The orientation and velocity of the plate in the rotational frame of reference of the Earth are recorded in the azimuths and spacings of these equatorial chalk lines.

Another method rests on hotspots, long-lasting, stationary sources of magmatic activity in the mantle deep below the lithosphere. These hotspots give rise to volcanic eruptions on the seafloor which over time construct large volcanic complexes. As the plate moves over the hotspot, however, the volcano, detached from its source, becomes extinct, is eroded and sinks below the level of the sea while a new one forms behind it (Figure 9.15a). In the long run a chain of volcanoes ranging from active at one end to long dead at the other is created. The alignment gives the motion vector of the plate in the rotational frame of reference of the Earth; the absolute velocity can be estimated from the age progression in the queue. The Hawaiian–Emperor island chain in the Pacific is a good example (Figure 9.15b); the sharp bend marks a change in plate direction around 45 Ma ago, accompanied by a change in velocity from 6 to 10 cm/year.

The Earth has and had many hotspots; if they lie on moving plates they create linear volcano chains (Figure 9.15c). Whether the hotspots themselves are entirely stationary has been lengthily but inconclusively debated; the weight of the evidence tends towards very slow drift for the most important ones.

Finally, plate motions across the latitudes can be obtained from palaeomagnetic data from seamounts and continents but, as noted before, the lack of a means of obtaining palaeolongitudes hampers accurate reconstruction.

The forces that drive the plates

Wegener was criticised most severely for his inability to present a plausible force capable of causing the continents to drift. It is therefore ironical that even now the forces that drive the plates remain elusive, although this has not hindered acceptance of the plate tectonics theory. The subject, vigorously discussed during the early years following the revolution, has received little attention recently; the problem seems intractable at this time and must await a better understanding of the rheology of the mantle.

There is no lack of suggestions (Figure 9.16), however, all resting on the plausible assumption that the mantle is a hot solid capable of flowing at a slow rate of the order of 1 cm/year. The principal hypothesis, that the plates passively ride on the convective flow of the mantle like ice floes on a river, derives from Holmes' proposed mechanism for Wegener's continental drift, a proposition made as early as 1929. The directions of mantle flow, however, cannot be inferred from the motion vectors of the plates because the tight cover constrains their freedom of motion. Unfortunately, too little is known yet about the physical properties of the mantle to model the convective flow itself except that it is likely to be restricted to the upper part of the asthenosphere and unlikely to be turbulent.

Alternatively, it is possible that the plates are driven by the pull of the long, cold and therefore dense slabs suspended deep into the hotter and lighter mantle at subduction zones. Equally permissible is the suggestion that magma welling up at mid-ocean ridges exerts a lateral push that drives the plates. Present concepts of fluid behaviour in the mantle further suggest that under the pull of gravity some plates might simply slide down slight slopes created thermally at the base of the lithosphere.

Attempts have been made to test especially the latter possibilities with the directions and velocities of plates in cases where the subducting slab is absent or particularly large, or where push by upwelling can be ignored. Some interesting patterns have emerged from this search; plate velocities, for example, are inversely proportional to the amount of continental area on the plate; this suggests that a deep continental root exerts drag. The mainly continental Eurasian and Antarctic plates, for example, move far slower than oceanic plates do. Plates with long subducting boundaries tend to move faster than those not so endowed, but no correlations have been found between the length of divergent boundaries and plate velocities or with the length or angle of a subducted slab. Unfortunately, these attempts have become mired in that classic scientific swamp where the significant differences one seeks are uncomfortably close to the probable errors of the estimates.

One would not be surprised, however, if the driving force ultimately turns out to be a composite one, the plates being driven primarily by convective flow induced by a hot plume rising from great depth, but significantly aided and modified by ridge crest push and subduction zone pull.

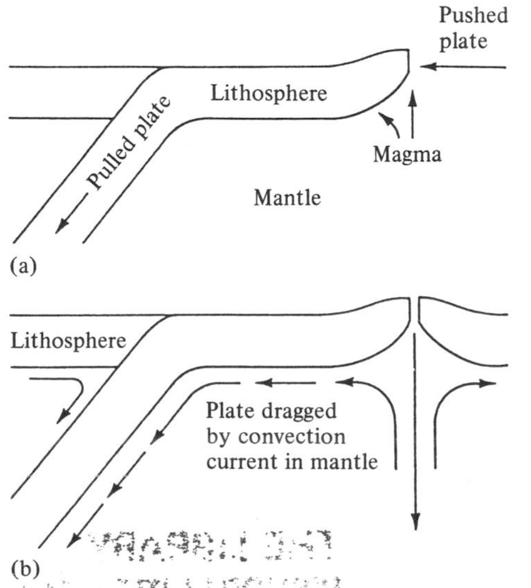

Figure 9.16 Driving forces of plate tectonics. (a) Push at the divergent edge and pull by a subducting slab. (b) A plate is dragged by convection. □

Looking back on the revolution

Any attempt to see how well the plate tectonics revolution has served us can begin with two clear statements. First, it has brought us the recognition that continental drift is a reality. Whatever may happen to this and future ruling theories of the Earth, we are not likely to return to a static world. Secondly, the model has greatly clarified our understanding of the ocean basins, of the formation of oceanic crust, of the genesis of ocean floor relief and deformation, and also of the history of ocean circulation and environment, sedimentation, and marine life. Palaeogeographic reconstructions, however, although very useful have required more *ad hoc* assumptions than is comfortable, because, as is so often the case when we try to impose strict rules on the Earth, purely geometric methods have proved to be more limited than we originally anticipated.

The theory has been less successful in dealing with the geological evolution of the continents, and with events and processes at converging plate edges involving continents and continental growth. Here the New Global Tectonics has been subjected to so many adjustments and special additions that it has been robbed of some of its universal explanatory power. Exotic terranes and highly specialised hypotheses to account for high heat flow in collision orogens, to name but a few, make us suspect that the basic model is either too simple overall or lacks some essential component.

Our continuing inability to accommodate long-sustained vertical movements of whole continents as easily as we can deal with long-distance lateral motion also suggests defects in the theory. Whether we consider global sea-level changes or local, regional and continental (and perhaps even world-wide) vertical motions of the Earth surface, in each case plate tectonics appears either not to demand them or to accommodate them only with difficulty and at the price of more *ad hoc* assumptions. It is not surprising that geologists in the USSR, a country distant from most active plate boundaries but with a geological history rich in events of regional uplift and subsidence, came rather late to the acceptance of the new theory of the Earth.

Few sciences go for long without revolutions and it seems certain that the plate tectonics revolution will not be our last one either. Still, the next one may take some time in coming because, notwithstanding the ever more complex adjustments needed, the current theory remains capable of much new insight, especially in the history of the physical and chemical environment at the Earth's surface and the evolution of life.

Summary

The ruling plate tectonics theory states that the outer shell of the Earth, the *c.* 100 km thick lithosphere, is divided into closely fitting plates, some large, some small, some including continents, some not. Floating on the weaker asthenosphere, the plates drift across the surface of the Earth at rates from one to more than ten centimetres per year.

Almost all tectonic and volcanic activity on the Earth's surface takes place at boundaries where plates diverge, converge or slide harmlessly past one another.

At divergent boundaries new crust is generated, while old crust is destroyed at convergent ones by accretion to the overriding plate and/or by subduction and recycling in the mantle. At boundaries where plates move parallel to each other, there may be much seismic activity, but crust is not created nor destroyed there to any great degree.

When two oceanic plate edges collide, either may subduct, but a continental edge will always override an oceanic one because the continental crust is the more buoyant one. On the overriding plate a deformed sedimentary fore-arc and a volcanic arc form.

The collision of two continental plates does not result in subduction but in compression and interpenetration of the two lithosphere slabs and in the formation of compressive mountain ranges. Orogeny is a general term for mountain-building where plates converge.

At subduction zones exotic terranes, fragments mainly of oceanic, more rarely continental crust from distant sources, gather and are incorporated into the overriding plate margin.

In the oceanic crust magnetic anomalies record rates of spreading at divergent boundaries, while transform faults indicate the direction of spreading. These features allow us to track the relative motions of pairs of plates and to reconstruct past plate configurations provided enough old oceanic crust is preserved. The movement of plates relative to the frame of reference of the Earth can be determined by means of palaeomagnetic measurements of latitude and the azimuth of the palaeomagnetic pole, and from linear volcanic chains formed as a plate moves across a hotspot fixed deep in the mantle.

The forces that drive the plates remain controversial, but it seems fairly certain that convection below the lithosphere plays a significant and perhaps the dominant role.

Further reading

On the revolution in the Earth sciences:

Hallam, A. (1973) *A Revolution in the Earth Sciences*, Oxford University Press, Oxford.

Oreskes, N. (1988) The rejection of continental drift, *Historical Studies in Physical Sciences,* **18** (2), 311–48.

A set of classic papers:

Cox, A. (ed.) (1973) *Plate Tectonics and Geomagnetic Reversals*, W. H. Freeman, San Francisco.

Full treatment of plate tectonics:

Cox, A. & Hart, R. B. (1986) *Plate Tectonics – How It Works*, Blackwell Scientific Publications, Oxford, 392 pp.

Uyeda, S. (1978) *The New View of the Earth*, W. H. Freeman, San Francisco.

On exotic terranes:

Howell, D. G. (1989) *Tectonics of Suspect Terranes*, Chapman & Hall, London.

CHAPTER

10

Alan G. Smith
University of Cambridge

PLATE TECTONICS AND CONTINENTAL DRIFT

Introduction

Continental drift is the name given to the notion that continents move slowly – or *drift* – across the surface of the Earth (see also Chapter 9). The idea originated largely with Alfred Wegener, who showed that the 300 million year (Ma) old ice-age deposits of the southern continents had a rational distribution when plotted on a geometric reconstruction of them known as **Gondwanaland** (Figure 10.1a), whereas their present-day distribution is inexplicable in terms of fixed continents (Figure 10.1b). How could an ice age exist in India if India had had its present geographic position? How could all the continents in the southern hemisphere have been affected by an ice age when most of the northern continents were basking in the tropical warmth of the Carboniferous coal forests or the sweltering heat of enormous Permian deserts and salt pans? From a climatic viewpoint, the more one thinks about it, the more nonsensical is the present geographic distribution of these 300 Ma rocks.

With the advent of **plate tectonics**, the relative motions between continents implied by continental drift can be cast into a precise form. Continental drift is simply the motion of the continental parts of rigid tectonic plates moving across a deeper layer of rock in the Earth's mantle that can flow, but is still solid, as opposed to the motion of continental rafts smashing their way through the ocean-floor at the surface. Quantitative predictions can now be made about the motions of the major continents for several million years to come, though, assuming the human race survives that long, these will be verifiable only by geologists who are many tens of thousands of generations away in the future. But inferences can also be made about their behaviour in the past which can be verified by entirely independent evidence that is available to the field geologist. **Tectonics**, that is, the study of the large-scale structure of the Earth and its causes, is now a science.

Plates

The distribution and nature of the present-day **earthquakes**, particularly those at shallow depths, is the key to plate tectonics (see Chapter 9). Earthquakes are found mostly in three distinct areas: firstly, along the oceanic ridge-and-transform systems; secondly, near deep ocean trenches; and thirdly, diffused through some continental areas typified by high mountains such as the Alpine–Himalayan belt (see Figure 9.5, p. 172). Since earthquakes represent the sudden release of stored elastic energy by rupture along fractures in the Earth, then areas of present-day seismicity are zones where such energy is being released at the present time. Today shallow earthquakes form a continuous network (see Figure 9.5(a), p. 172). They divide the surface of the Earth into areas without seismicity, which are assumed to be rigid and not to be undergoing deformation, from areas of active defor-

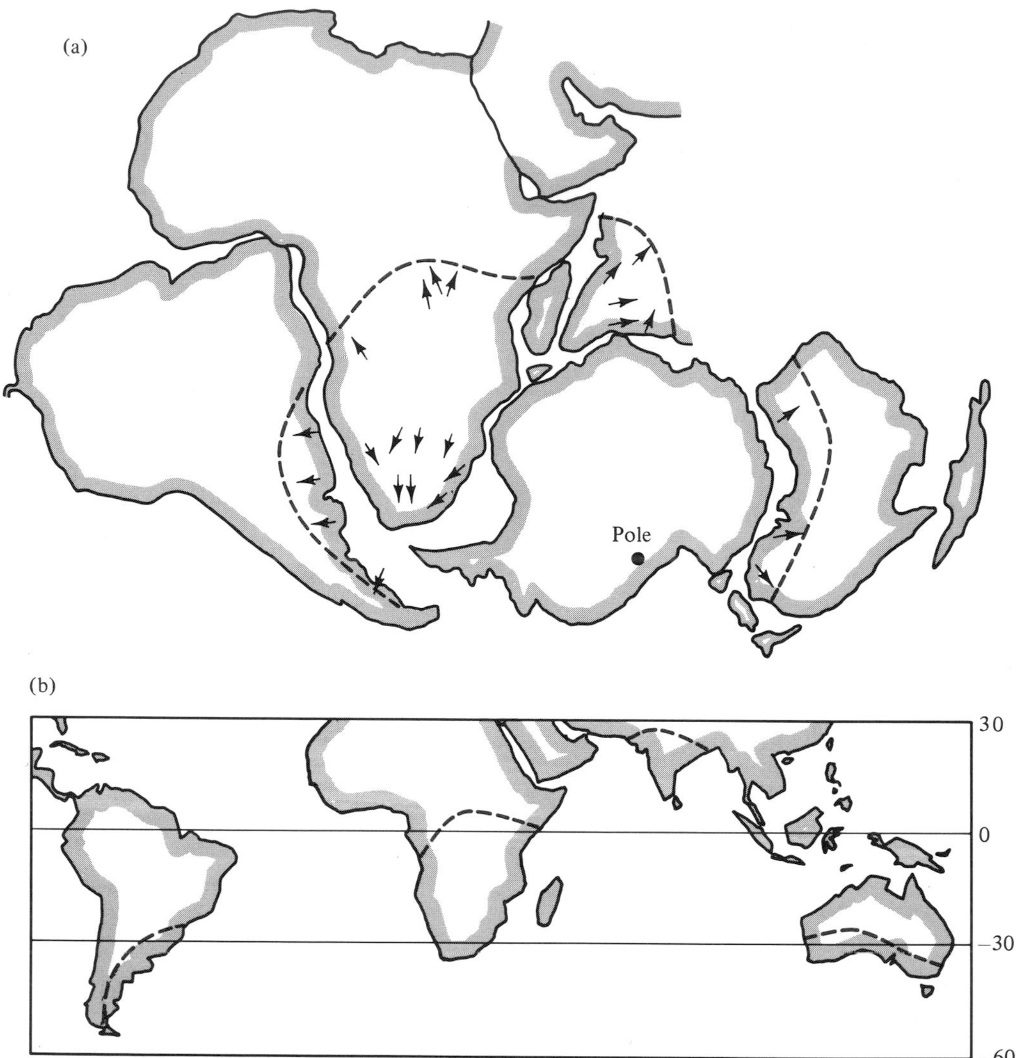

Figure 10.1 (a) Ice-flow directions, palaeomagnetic pole position and extent of Late Carboniferous glaciations. All three features have a rational distribution on the continental reassembly. (b) The nonsensical distribution of Late Carboniferous glaciations (within the area bounded by the dashed lines) plotted on a present-day map. □

mation. The rigid areas are known as **lithospheric plates**. Plates are 100 to 150 km or so thick and lie on the solid but mobile **asthenosphere**. The actively deforming areas are of two types: continental areas in which the seismicity is irregularly distributed over broad areas, and narrow belts found in continents and oceans. The narrow belts are **plate margins**; the broad belts are **actively deforming continental areas** (see Chapter 14 for further details).

Detailed surveys of the mid-ocean ridge systems show that the earthquakes coincide with the segments of the **oceanic ridge-and-transform systems**. The relative motions taking place can be inferred from the earthquakes themselves (see Chapter 9). At a ridge segment the motion is a divergence, i.e. a relative separation of one part of the ridge relative to the other. Along a transform segment the motion is a translation, i.e. a sliding motion, in which the lithosphere on one side of the transform fault slides past the lithosphere on the opposite side (see Figure 9.4, p. 171). By contrast, the earthquakes near deep **ocean trenches** show that the ocean-floor is generally moving toward the adjacent continent or island arc, i.e. the plates are converging, at trenches. The relative motions in actively deforming continental areas are complex (see Chapter 14).

Euler's fixed point theorem

How can we describe plate motions? To do this we use a theorem of spherical geometry known as **Euler's** (pronounced as Oiler) fixed point **theorem**. This theorem states that any movement of rigid material on the surface of a sphere can be described in terms of rotation about two diametrically opposite fixed points, or poles.

At first sight this result seems to have no relevance to tectonics, but a consequence of the theorem is that it is possible to describe the motion between two rigid bodies on a sphere as a rotation about an axis through the centre of the sphere (Figure 10.2).

Because plates are rigid bodies, their relative motions can be described as rotations about axes through the Earth's centre. We describe these rotations as Euler rotations that take place about points where each axis cuts the Earth's surface. We shall refer to such points as **Euler poles**. If we take the geographic position of the Euler pole, P, closest to a pair of plates, its position can be described by its latitude and longitude. The amount of the rotation about an Euler pole is referred to as the Euler angle, θ, and the rotation itself is referred to as an Euler rotation. The displacement between the two plates is maximum at 90° from the Euler poles and falls to zero at the poles themselves.

The relative motion on a plate boundary depends on the trend of the boundary relative to the Euler pole and on the sense of relative motion. Imagine that the Euler axis is the N–S geographic axis, the spin axis, of the Earth (Figure 10.3). Consider a plate margin between two plates A and B. Clearly, the displacement at a plate margin depends on the trend of the margin relative to the latitude and longitude lines. If the margin is parallel to a latitude line, i.e. E–W, then the two plates slide past each other. Separation or overlap do not take place. The motion is *pure translation*; the margin is a **transform fault**. All parts of the margin that are latitude lines are transform faults. They form arcs of 'small circles' centred on the N and S geographic poles.

If the margin is parallel to a longitude line, i.e. N–S, then the two plates separate or approach one another at right angles to the margin (Figure 10.3). If the plates are moving apart then the margin is divergent and shows *pure extension*; if they are moving towards each other, the margin is convergent and displays *pure convergence*, which will usually be compression. In general plate margins will be parallel to neither a latitude nor a longitude line: they will be oblique to both.

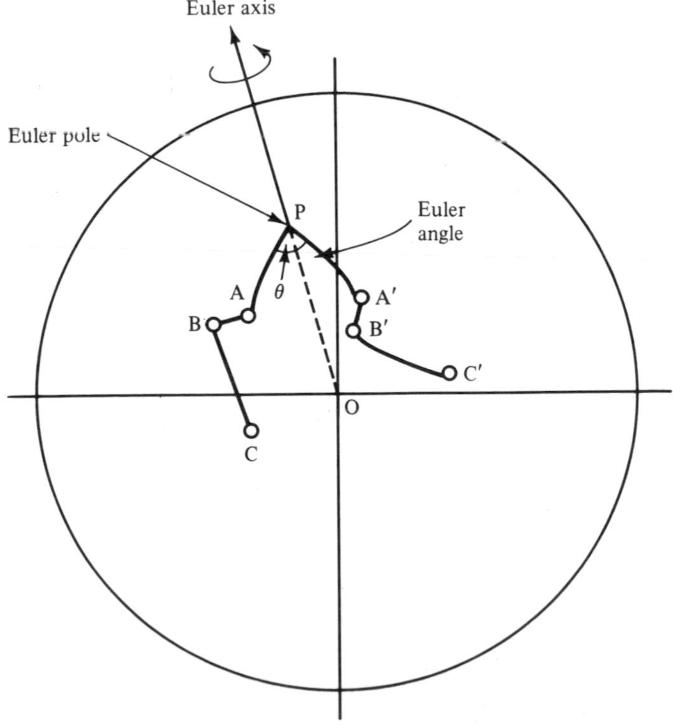

Figure 10.2 O is at the Earth's centre. P, A, B, C, A', B' and C' are on the Earth's surface. Rotation of the line ABC about the Euler axis OP and by the Euler angle θ brings it to A'B'C'. During this rotation P does not move. A moves less than B, which moves less than C. Movement is maximum at an angular distance of 90° from P. □

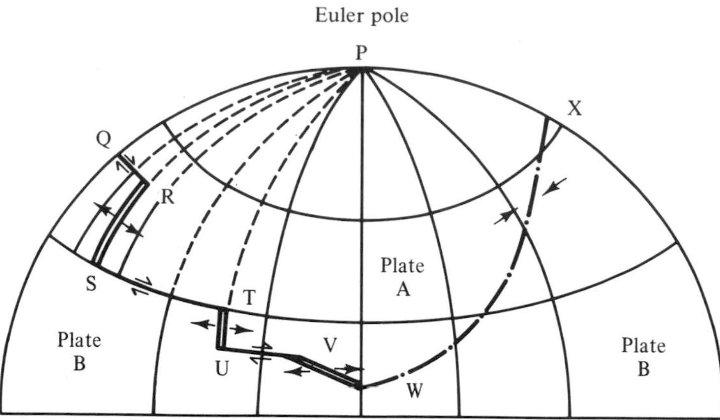

Figure 10.3 P is the Euler pole. A and B are plates. The boundary between A and B is irregular. RS and TU are ridge segments spreading perpendicularly to the ridge axis; VW is a ridge segment spreading obliquely to the ridge axis; QR and ST are transform faults along which the plates slide past one another; WX is a zone of oblique convergence. □

The present-day plate boundary between Africa and the Americas

To apply these results we must know the geometry of a plate margin and where its Euler poles are. For an oceanic margin the Euler poles can be found by drawing **great circles** at right angles to each transform segment (Figure 10.4). The margin is the trace of the active ridge-and-transform as marked out by present-day seismicity.

The mid-Atlantic ridge between the Americas and Africa was the first plate margin to be analysed in this way. The Euler poles lie at the intersections of the great circles drawn perpendicular to the transform segments. Because some transform segments have small extensional or small compressional components that are difficult to detect, the great circles do not intersect at a point but rather at two clusters of intersections, one cluster near 60° N, 30° W and the other diametrically opposite at 60° S, 150° E. For convenience we deal only with the Euler pole closest to the area of interest. In this case we would use the Euler pole at 60° N, 30° W. This pole is the Euler pole for the average trends of the transform segments formed in the particular time interval. The transform faults have not had a fixed orientation throughout this interval, otherwise the intersections of the great circles would lie in a much smaller area than they do.

Motion at a triple junction

A **triple junction** is simply a place where three plate margins meet one another. Consider first the triple junction near the Azores (Figure 10.5). The three plates that meet here are the North American, Eurasian and African plates, which we shall abbreviate to NAM, EUR and AFR respectively.

The Euler poles for the motion of NAM and EUR and of NAM and AFR can be found from the transform faults north and south of the triple junction. The angle that NAM has moved through relative to AFR in the past few million years is known from the spacing of the ocean-floor magnetic anomalies south of the triple junction. Similarly, the angle that NAM has moved through in the same time interval relative to EUR is also known from the anomalies north of the triple junction. From these known motions we can find the Euler rotation for AFR to EUR in the same time interval. If we know the angle and the time it took to move through that angle we can calculate the rates of motion. They are shown in Figure 10.5.

Plate theory predicts that along the western part of the Eurasian/African margin the motion is one of oblique spreading at about 4 mm per year; along the Gloria Fault it is pure translation at about the same rate; and to the east is oblique convergence, with rates increasing to 7 mm per year near Sicily. This change in tectonics along the margin is an excellent example of how quite different tectonic effects along a plate margin result from the relationship between the margin and its Euler pole and how these effects can be predicted from plate theory.

Before the advent of plate tectonics it was impossible to make such predictions because there was no quantitative model that could connect the tectonics of one part of an oceanic ridge with that of another. In fact, prior to the late 1960s, few geologists envisaged the ocean ridges to be tectonically active. The questions we ask nowadays would have been meaningless then.

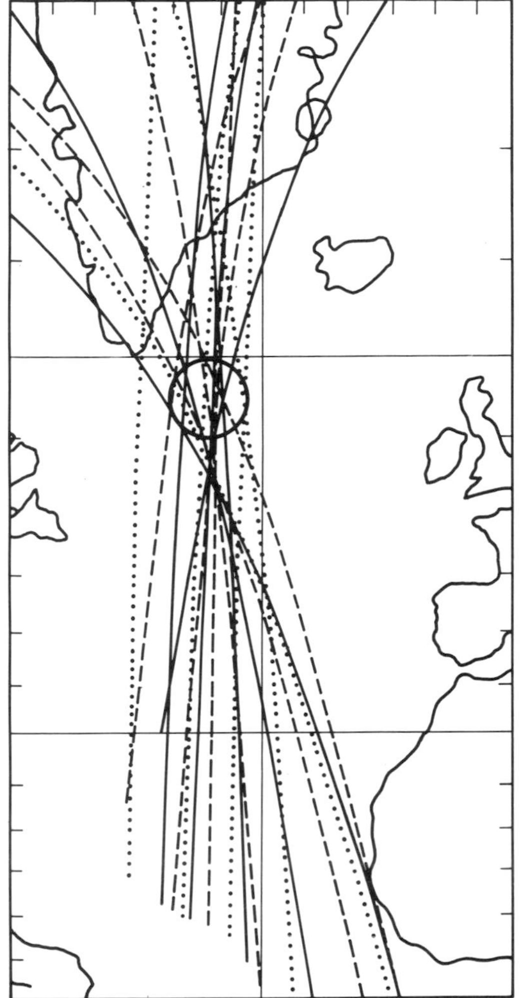

Figure 10.4 The full, dashed and dotted lines are great circles drawn perpendicular to short transform segments in the central Atlantic. Most of them intersect near or in the circle whose centre was the first estimate of the Euler pole for the African and American plates. □

The present-day plate model for the Earth

The past 3 Ma

The most consistent present-day plate model divides the Earth's lithosphere into twelve plates (Figure 10.6). For a full description of the relative motion, each plate margin needs an Euler pole and a rate of rotation about the Euler pole. The Euler poles have been determined from great circles drawn perpendicular to the trends of transform faults or to the direction of motion determined by studying earthquakes originating at the plate boundary. The rate of motion can be found directly only along those margins where oceanic crust has been formed. The model assumes that the present-day rate is the average of the ocean-floor spreading rate for the past 3 Ma.

The data used for the model are shown as symbols on the various plate boundaries. For example, in the Indian Ocean there is a triple junction formed where the Antarctica/Africa, Africa/Australia and Australia/Antarctica plate margins meet. The squares show that the motion on all three plate margins is known from ocean-floor spreading and seismic data. However, at the Azores triple junction, discussed above, only the direction of motion of Africa relative to Eurasia is known from studying earthquakes, as shown by the triangles on that boundary. In the case of the Pacific/Australia boundary there are no data points. The motion on this boundary has to be found by adding together the motions on the Pacific/Antarctica and Antarctica/Australia boundaries.

There is enough data to calculate the motions on all present-day boundaries either directly or indirectly by adding motions at triple junctions.

Direct measurements

The motions in the model discussed above are averaged over a 3 Ma time-span. We do not actually measure these motions, but we infer them from the ocean-floor anomaly record and other data. The results may be compared with estimates of some plate motions obtained from direct measurements. Surveying techniques have advanced so much in the past decade that they can now detect differences of a few hundred millimetres in the distance between two points that are thousands of kilometres apart. Though plates move at rates that are as slow as those at which one's fingernails grow, the total growth over a period of several years can amount to a few hundred millimetres.

For example, one can use a global **satellite positioning system** to measure the location of a site, say, in London, and of a second site, say, in Washington. If the positions are measured after an interval of ten years, spreading on the mid-Atlantic ridge will have moved them 200 mm or so further apart. Alternatively, one can use laser beams to measure the distance directly between two points and detect the motions.

The initial results are entirely consistent with the predictions from the various plate models that have been proposed. For the first time geologists not only

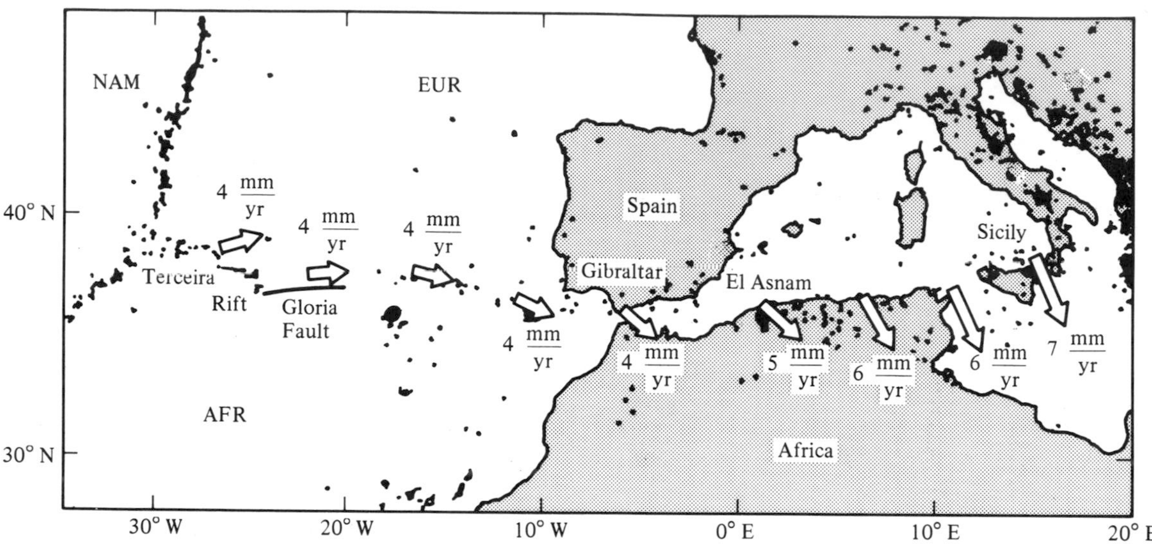

Figure 10.5 The black dots are earthquake locations. In the west they form three well-defined lines marking the present-day plate boundaries between the North American plate (NAM), the Eurasian plate (EUR) and the African plate (AFR). The lines meet at a triple junction. The ocean-floor anomalies and transform faults on the NAM–EUR and NAM–AFR margins give the Euler poles and rates of rotations for these margins. From them the Euler pole and rate of rotation for the AFR–EUR margin can be calculated. The displacements caused by this rotation rate are shown by the arrows, with the displacements referring to the tails of the arrows. They show how Eurasia is moving with respect to a fixed African plate. □

have direct evidence of the reality of 'continental drift', but in the next century they will have measurements whose errors will be small enough for them to be able to distinguish between different plate models.

Reference frames

Plates as reference frames

The present-day plate tectonic model gives the relative motions that take place along present-day plate margins. The relative motion of any one plate to any other plate can be found by adding up the relevant motions. For example, the relative motion of the Indian plate to the American plate can be found by adding the motion of the Indian plate relative to the Eurasian plate to the motion of the Eurasian plate relative to the North American plate. Any plate can be used as a **'reference frame'**, i.e. it can be considered to be fixed with all other plates moving relative to it. No one plate provides a fundamental reference frame.

Palaeomagnetism and the palaeomagnetic reference frame

Palaeomagnetism means 'old magnetism', and it is the name given to the study of ancient magnetism preserved in rocks. The most obvious example of fossil magnetism is the magnetism preserved in the magnetic anomalies formed by ocean-floor spreading (see Figure 9.3, p. 170). Here the lava extruded at an oceanic ridge crest cools down and is weakly magnetised by the Earth's field existing at the time. When the field reverses, the lavas are magnetised in the opposite sense. The juxtaposition of normally and reversely magnetised lavas gives the ocean-floor its characteristic magnetic striping.

In addition to field reversals in time, at any instant the *direction* of the magnetic field varies over the Earth's surface in a systematic way. At any given spot the lines of magnetic force emerge or enter the Earth with a fixed attitude. This attitude can be described by two numbers: the **inclination** (I) and the **declination** (D). The inclination of the magnetic field is analogous

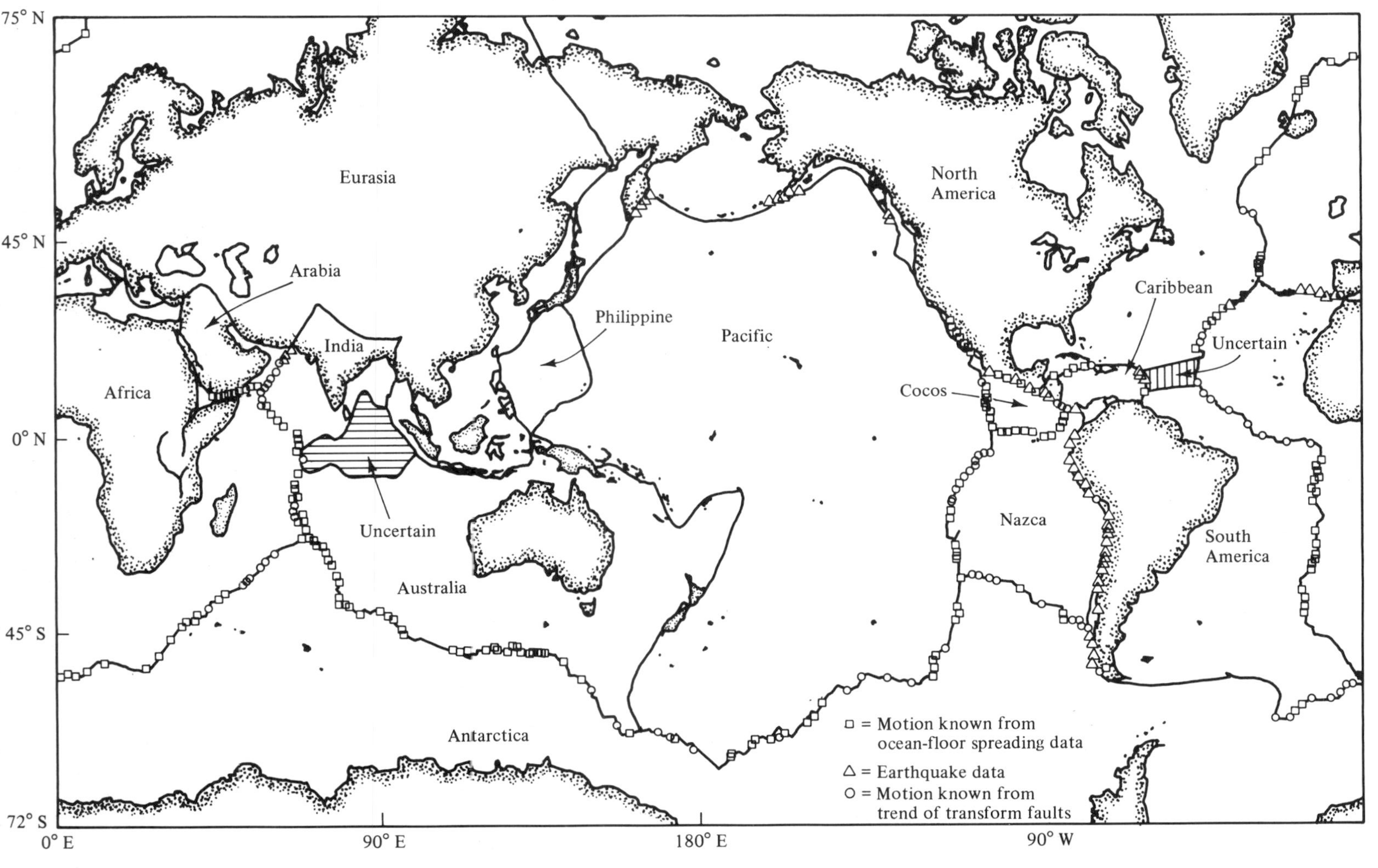

Figure 10.6 Present-day boundaries of the major plates based mostly on the distribution of earthquake activity. □

to the dip of sedimentary rocks, i.e. it is defined as the angle that the lines of force make with the Earth's surface (Figure 10.7). The declination is the bearing of the field (Figure 10.7), i.e. the angle between the geographic north and the direction of the field at a point. It is analogous to the dip direction of bedding in sediments. If the lines of force are horizontal the inclination is 0°; if they are vertical the inclination is 90°. If the lines of force point due north, the declination is 0°, and so on.

Although the Earth's field varies significantly over periods of tens of years, when averaged over time periods of the order of 1 Ma, it closely resembles that of a bar magnet at the Earth's centre with its axis coincident with the spin axis, the geographic axis, of the Earth (Figure 10.8). For such a field there is a simple relationship between the inclination, I, at any point and the geographic latitude, L, given by:

$$\tan(I) = 2\tan(L).$$

For this ideal field, the inclination at the equator is 0° and at the poles it is 90°. Except at the equator and the poles, the inclination is always greater than the latitude. For example, at a latitude of 45° the inclination is $\tan^{-1}(2)$ or 63.4°. The declination, D, gives the direction in which the pole must lie. Its angular distance from the point at which the magnetic measurement is made is (90–latitude).

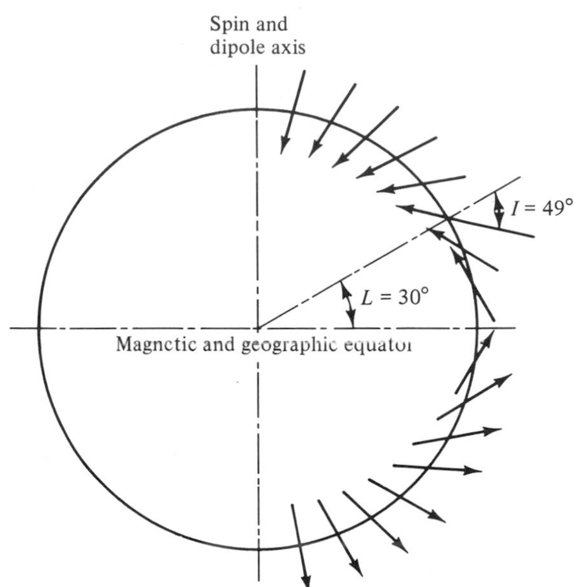

Figure 10.8 A cross-section through the Earth showing the relation between the inclination (*I*) and the latitude (*L*) for a centred dipole, or 'bar magnet', – whose axis parallels the Earth's spin (= geographic) axis.

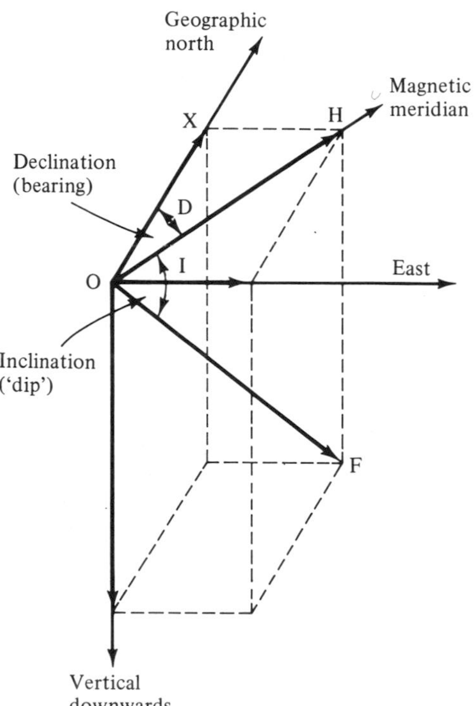

Figure 10.7 OF is parallel to the magnetic field. *D* is the declination, or angle between geographic north (the line OX) and magnetic north (the line OH). *I* is the inclination, or the 'dip' of the field. It is the angle between the horizontal line pointing to magnetic north (OH) and the direction of the magnetic field (OF).

By measuring the declination and the inclination in ancient rocks of a known age, the position of the magnetic pole can be found. The best data come from lava flows that can be directly dated by radiometric means. Almost as good magnetically are data from 'red beds', usually red siltstones, but these are often difficult to date. They neither contain minerals with suitable radiogenic isotope ratios, nor are they very fossiliferous. By making enough measurements of the declination and the inclination and averaging them, the positions of the geographic poles at the time can be estimated. Thus a map of the time can be drawn showing where the area on which the palaeomagnetic measurements have been made was in relation to the geographic poles. Thus palaeomagnetism provides us with a **palaeolatitude reference frame**. To make a map of some time in the past, all we need to know are the positions of the continents relative to one another and the location of the geographic pole relative to one of these continents.

The reliability of this method can be tested by plotting the pole positions of igneous rocks that are no more than 20 Ma old. During this time period the major continents have not moved much relative to the uncertainties in the palaeomagnetic pole positions. What would we expect? We obviously expect that the palaeomagnetic pole positions should coincide with the geographic pole. Figure 10.9 shows that *on average* there is an excellent correspondence between the two, although individual poles may lie up to 65° from the pole. Such extreme departures could be due to aberrant data, or to catching the field during a reversal, and so need not affect this argument.

Provided two continents have merely 'drifted apart' without any of the new ocean-floor between them being destroyed, they can be reassembled at any time between break-up and their present-day positions by mentally removing the ocean-floor between them. All that is required is a knowledge of the Euler rotations describing how they have moved apart in the relevant time interval. For example, North America can be repositioned relative to Africa at any time since their initial separation in Jurassic time by matching anomalies of the same age on opposite sides of the mid-Atlantic ridge (Figure 10.10). In practice, we can reassemble all the major continents around the Atlantic and Indian Oceans by mathematically winding up the old ocean-floor between the continents bordering these oceans.

Palaeomagnetic measurements then allow us to place this continental reassembly into its proper geographic orientation (Figure 10.11). This procedure works well for most of Mesozoic and Cenozoic time, the past 250 Ma or so, but, as we shall see, it cannot

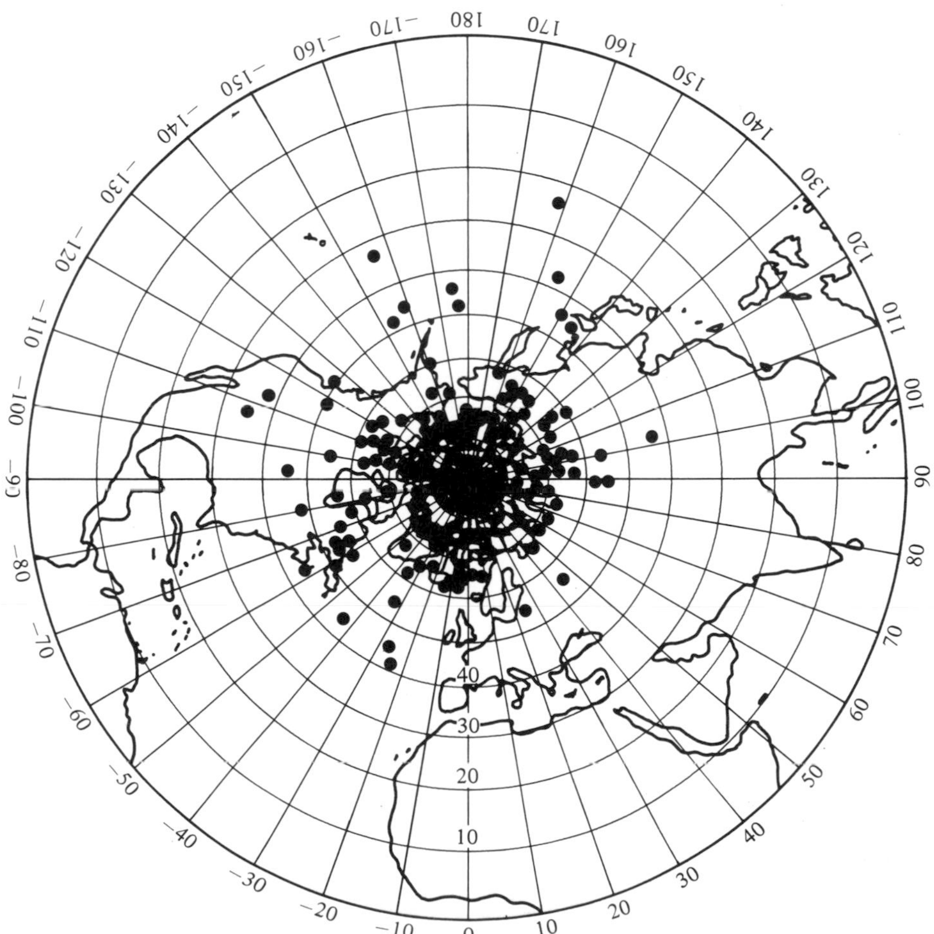

Figure 10.9 The dots represent palaeomagnetic pole positions from igneous rocks that are up to 20 Ma old. They cluster around the Earth's spin axis, i.e. the geographic pole, clearly showing that the average palaeomagnetic pole is close to the geographic pole, at least for this time interval. □

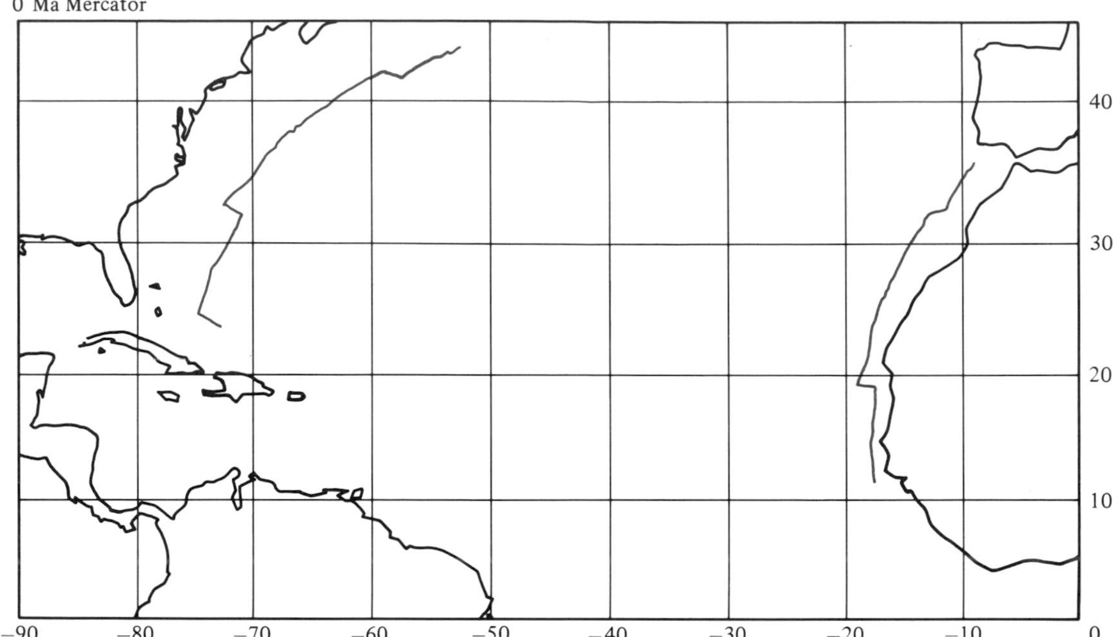

Figure 10.10 A present-day map of the central Atlantic Ocean showing the approximate positions of Mesozoic ocean-floor anomaly M29 off eastern North America and West Africa, dated at about 165 Ma. □

be used to make pre-Mesozoic maps.

The resulting maps are not palaeogeographic maps because they do not show old geographic features such as coastlines, but they provide base maps on which such features can be plotted. Nor are they palaeo-plate tectonic maps because they do not show where any of the plate margins of the past lay, though again they provide base maps on which old plate margins can be plotted. So what are they? They are **palaeocontinental maps**: they show the past positions of the continental crust.

Some continental areas cannot be repositioned in past time. These are the areas involved in orogenic activity, i.e. areas that have been deformed by continental collisions or by being near compressional plate boundaries. For example, we do not know how to unscramble the *melange* of microcontinents, ocean-floor remnants and island arcs that make up the present-day Alpine–Himalayan mountain belt. We can move much of India back in time until it joins onto Antarctica, but we are uncertain what to do with those parts of northern India now in the Himalayas. Should they remain attached to northern India? If so, do they need unscrambling in some way? If not, where was the ocean between northern India and the Himalayas? Was there just one ocean, or several? How can we tell? Such questions cannot be answered precisely at the present time.

Hot-spots and the hot-spots reference frame

Right in the middle of the Pacific plate lies the Hawaiian island chain. It is almost entirely basaltic. The largest island, Hawaii, known as the 'Big Island' to Hawaiians, is also the site of the most active volcanism. Northwest of Hawaii the islands become smaller and smaller and the volcanism older and older. A map of the ocean-floor shows that there are submerged islands still further to the northwest. Those that have been dredged and dated show a remarkably uniform increase in age and depth of water above them (Figure 10.12, inset). Some 3000 km from Hawaii, the submerged island chain makes an abrupt bend, the Hawaiian–Emperor bend, forming a still older chain known as the Emperor Seamounts. These eventually disappear in the subduction zone at the western end of the Aleutian Islands.

Apart from subtle differences in mineralogy and chemistry, there are no obvious differences between the basalts of the Hawaiian–Emperor chain and those of the ocean-floor. Seismic measurements show that the chain is characterised by a thicker-than-normal crust. The crust and mantle under Hawaii itself seem to be hotter than in adjacent areas: Hawaii is a veritable 'hot-spot'. The Hawaiian–Emperor chain seems to be the trace of a long-lived hot-spot that has

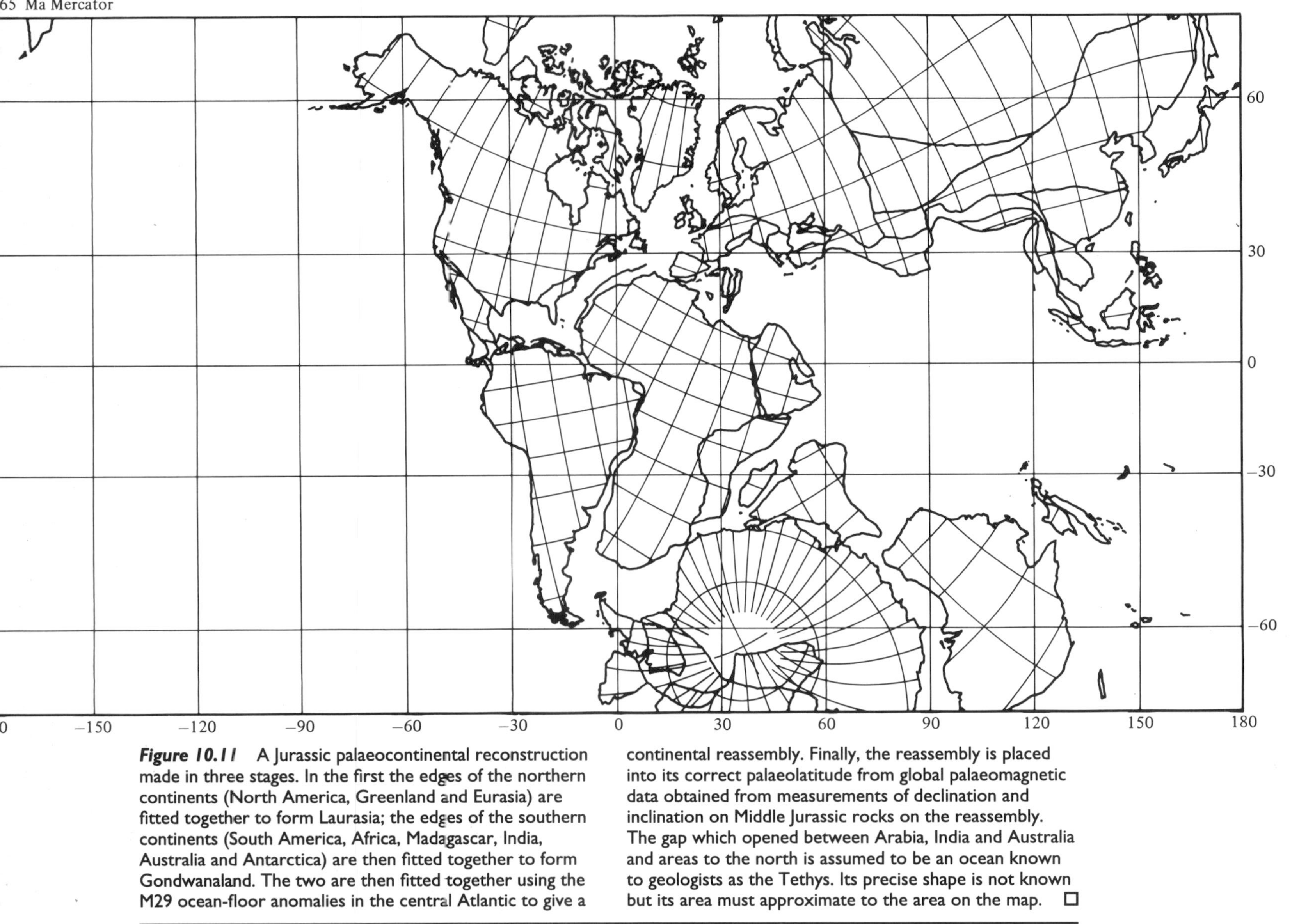

Figure 10.11 A Jurassic palaeocontinental reconstruction made in three stages. In the first the edges of the northern continents (North America, Greenland and Eurasia) are fitted together to form Laurasia; the edges of the southern continents (South America, Africa, Madagascar, India, Australia and Antarctica) are then fitted together to form Gondwanaland. The two are then fitted together using the M29 ocean-floor anomalies in the central Atlantic to give a continental reassembly. Finally, the reassembly is placed into its correct palaeolatitude from global palaeomagnetic data obtained from measurements of declination and inclination on Middle Jurassic rocks on the reassembly. The gap which opened between Arabia, India and Australia and areas to the north is assumed to be an ocean known to geologists as the Tethys. Its precise shape is not known but its area must approximate to the area on the map. □

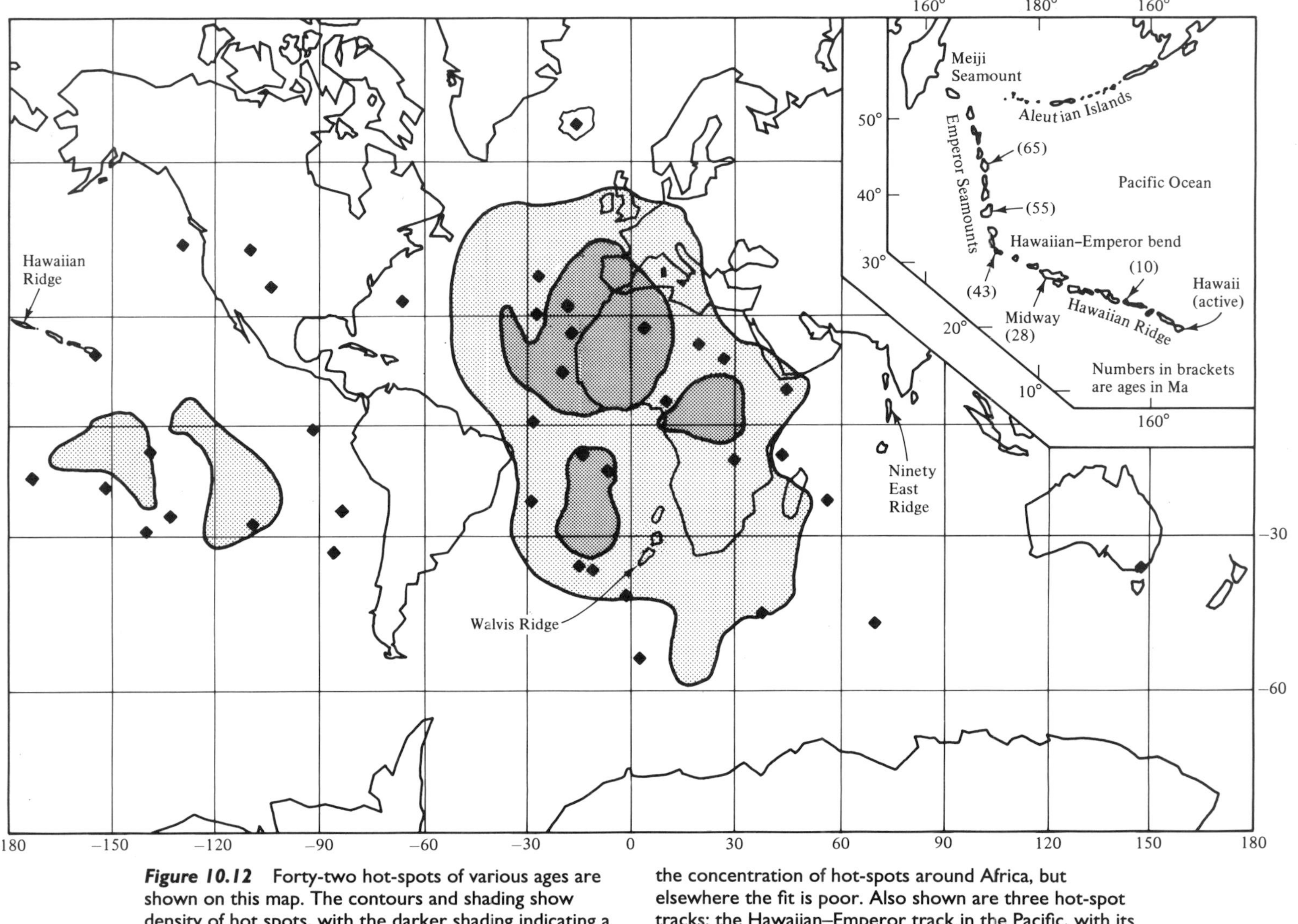

Figure 10.12 Forty-two hot-spots of various ages are shown on this map. The contours and shading show density of hot spots, with the darker shading indicating a greater density of hot-spots. The contours are a mathematical attempt to fit a smooth surface to a very scattered distribution. The surface is a reasonable fit to the concentration of hot-spots around Africa, but elsewhere the fit is poor. Also shown are three hot-spot tracks: the Hawaiian–Emperor track in the Pacific, with its bend dated at about 43 Ma (see also inset); the Walvis Ridge in the South Atlantic; and the Ninety East Ridge in the Indian Ocean.

migrated from the Aleutian Islands to Hawaii in the past 75 Ma or so, with a distinct bend in it at about 43 Ma.

There are similar areas of concentrated volcanism in other parts of the Pacific and in other oceans. Iceland is the best example from the Atlantic Ocean, but Iceland lies astride the mid-Atlantic ridge, rather than lying in the middle of a plate. There are also other long-lived island chains elsewhere on the Pacific floor. In addition there are higher-than-normal submarine ridges that seem to bear no relation to oceanic ridge systems, old or young, in all the major oceans. Two of the best known are the Walvis Ridge in the South Atlantic and the Ninety East Ridge in the Indian Ocean. All of them may be hot-spots or hot-spot traces (Figure 10.12).

The remarkable thing about hot-spots and their traces is that they move much more slowly relative to one another than do the plates. It is as if the plates were independently running around on the Earth's surface while a series of fixed, deep-seated hot fountains, the hot-spots, played on them, occasionally spraying out at the surface as hot-spot volcanoes.

Exactly which volcanic chains are hot-spot traces and which are not is a matter of debate. The number of Mesozoic and Cenozoic hot-spots ranges from about 30 to 120, depending on which geologist analyses the data. A 42 hot-spot distribution is shown on Figure 10.12. Whichever distribution is accepted it is clear that hot-spots are irregularly distributed. Continental hot-spots are uncommon except under Africa (Figure 10.12). Higher-than-average hot-spot concentrations are associated with higher-than-normal values of the geoid (see Chapter 14). The significance of this association is not yet clear.

Because the hot-spots move much more slowly than the plates, hot-spots provide a frame of reference in which to analyse plate motions. Such a frame has been described as an 'absolute' frame of reference in the sense that it enables the changes of latitude *and* longitude to be estimated, rather than just the latitude changes that the palaeomagnetic reference frame provides. For example, the 'absolute motions' of the major plates about 60 Ma ago are shown in Figure 10.13.

A remarkable feature of the hot-spots and palaeomagnetic reference frames for the past 60 Ma is that the latitudinal changes predicted by both reference frames, which are completely independent, differ by no more than 7°, which is to say they are essentially identical within the limits of error.

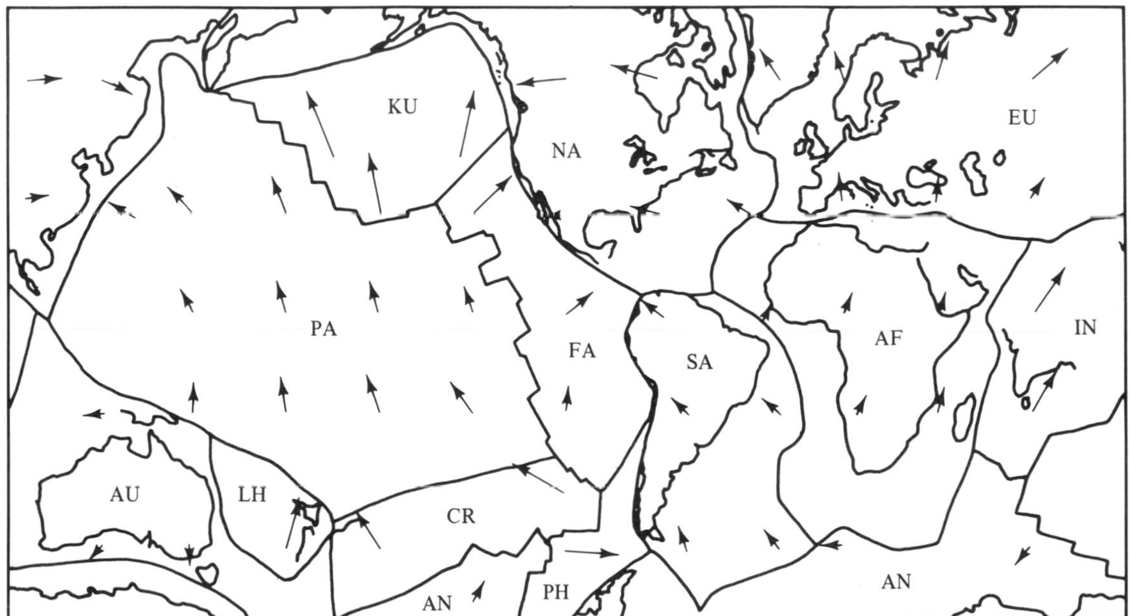

Figure 10.13 The map shows a global tectonic reconstruction for Palaeocene time, about 60 Ma ago. Several oceanic plates are present that are not shown on Figure 10.6: Kula (KU), Farallon (FA), Phoenix (PH), Chatham Rise (CR), and Lord Howe (LH) plates. They have either been subducted or are much smaller today. The arrows show the relative motions of these plates in the hot-spots reference frame of 60 Ma ago; the longer arrows show faster relative velocities. □

Hot-spots are unlikely to originate in the asthenosphere (see Chapter 3) because this is moving as fast as the plates, which are moving much faster than the hot-spots themselves. They probably come from a part of the deeper mantle which is moving much more slowly than both the overlying plates and the asthenosphere. Thus the apparent motion of the Hawaiian hot-spot from the northern end of the Emperor Seamount chain to its present-day position (Figure 10.12) can be much more usefully thought of as reflecting the trace of a *fixed* (*or slowly moving*) *hot-spot* on the Pacific plate as the Pacific plate has moved over it in time, rather than the trace of a moving hot-spot on a fixed Pacific plate.

All Palaeozoic ocean-floor has been subducted (or emplaced as ophiolitic shreds). Thus there is no record of oceanic Palaeozoic hot-spots and, with the possible exception of some volcanics in parts of Soviet Asia, there are no agreed continental Palaeozoic hot-spots either. The Precambrian hot-spot record, a time-span representing well over 80 per cent of geological time, is virtually non-existent. Whether this is due to differences between the way geologists study the Precambrian and the Phanerozoic or to a genuine scarcity of Precambrian hot-spots is not yet known.

Plate margins and continental drift

Let us now re-examine some of the kinds of evidence that were originally put forward in support of continental drift. One argument was the similarity of shape of the continental margins of continents on opposite sides of an ocean. The classic example is the similarity of western Africa to eastern South America. In plate tectonic terms, the preservation of this similarity of shape for well over 100 Ma reflects the fact that plates are rigid. Only where geological processes have significantly modified the margin is the fit poor. For example, overlaps of the present-day margins occur where a hot-spot trace intersects one of the margins, as in the case of Walvis Ridge, with the result that the southwest African margin now protrudes well beyond its original position at the time of break-up. Protrusion may also be caused by the outgrowth of large deltas, as in the case of the Niger delta.

A tectonic argument for drift was the similarity of the history of orogenic belts that now appear to be thousands of kilometres apart. For example, the evolution of parts of the Appalachian orogenic belt of Newfoundland closely resembles the evolution of the Caledonian orogenic belt of Britain. This similarity does indeed reflect the fact that they were once joined together. Such similarities are to be expected for those parts of the orogenic belt that once lay on the same continental margin. In general, a continental margin evolves from a rift margin to a passive continental margin to an active continental margin. Its evolution should be broadly synchronous along its entire length, with corresponding similarities in sedimentary facies, faunas, structures and igneous and metamorphic rocks. Thus it should no longer surprise us that some episodes of orogenic deformation appear to be synchronous over distances of thousands of kilometres; rather it should surprise us if this was not at times true.

The Palaeozoic problem

How to infer rates and directions of continental drift in Palaeozoic time (roughly 570–250 Ma) is an unsolved problem. The difficulties are due to several problems. First, the boundaries of the Palaeozoic continents have to be determined. On a gross scale, the boundaries are found relatively easily. One starts in the middle of an area that is known to have been stable throughout Phanerozoic time, i.e. since the beginning of Cambrian time, and moves outward until one encounters a Phanerozoic orogenic belt or a passive continental margin. A stable Phanerozoic area surrounded by Phanerozoic orogenic belts or passive margins is part of a Phanerozoic plate, and therefore of a Palaeozoic plate.

For example, North America is surrounded on its Pacific and Atlantic margins by Phanerozoic orogenic belts. Its Arctic boundaries are either Phanerozoic orogenic belts or passive margins. Therefore much of North America is part of a Palaeozoic plate.

There are no agreed Palaeozoic hot-spots in North America. However, there are abundant Palaeozoic palaeomagnetic data and the continent can therefore be oriented into its former palaeolatitude. What cannot readily be reconstructed are the positions of those parts of the North American continent that have been caught up in an orogenic belt that is of the same age or younger than the age of the reconstruction: we do not know precisely how to unravel orogenic belts.

We can start on a global scale. First we outline the stable areas of the time on Pangea. These areas are then broken away from other stable areas along somewhat arbitrary boundaries lying in the orogenic belts separating the stable areas from one another. The areas involved in orogeny are attached to the stable areas, but their positions will not be known. Next, we select the palaeomagnetic poles that are in the time range and use them as estimates of the geographic pole positions. Each continent is then drawn onto a map in its correct palaeolatitude and orientation. Because there is no Palaeozoic ocean-floor, we cannot determine the longitude separation of these stable continents.

The resulting composite map, i.e. a map made up of projecting several continents independently onto a single map frame, has to satisfy some general conditions. The first is fairly precise, except in areas that have been stretched at a later period, overlaps of continental areas are not allowed. The other two conditions are rather vague. One is that areas on different continents with similar faunas should not lie too far from one another. The other is that the composite map should be capable of evolving into younger composites or palaeocontinental maps without the need to propose unrealistically large or complex displacements.

Treating the Mesozoic and Cenozoic as a Palaeozoic problem

We can obtain some idea of the plausibility of Palaeozoic reconstructions by treating the present-day world as if it were Pangea, i.e. as if we know where the continents are, but do not know anything about present-day oceans. As we go back in time, the areas of Cenozoic and Mesozoic deformation increase, in turn decreasing the continental areas that can be reconstructed from palaeomagnetic measurements. We illustrate these methods by making a palaeomagnetic reconstruction at 65 Ma, roughly the same age as the map on Figure 10.13.

For example, to make a 65 Ma palaeomagnetic composite, we take the present-day world and look for the Cenozoic orogenic belts. The Alpine–Himalayan chain is the only major belt that separates two or more stable Cenozoic continents. We draw a line along it, attaching some of the orogenic belt to southern Eurasia, and the rest to northern Africa, Arabia and India. Each continent is then oriented independently by its own palaeomagnetic poles (Figure 10.14). The result is quite close to Figure 10.13 based on ocean-floor and hot-spots data.

Nevertheless, as far as palaeomagnetism is concerned, we could slide each continent along a latitude line, as if the latitude lines were wires on an abacus from which the continents were hanging, and still have as good an agreement with the palaeomagnetic data. To improve a palaeomagnetic composite, we need to use some geology that gives estimates of the likely separation among the continents at this time.

In particular, we could look at the ages of the continental margins around each continent. For example, the South American/African margins are passive margins. Their formation can be dated from the geology of the margin as probably of Early Cretaceous age. Since that time they have been passive margins. Thus, since Early Cretaceous time, the two continents have either been stationary relative to one another or have been moving apart from one another. They cannot have been moving toward one another. We know their present relative positions. By assuming that the two continents were joined together in Early Cretaceous time and finding the Euler rotation to bring them back together then, we could guess that the 65 Ma position would be the appropriate fraction of the Euler rotation.

By using such evidence, composites can be improved to give pictures quite similar to Mesozoic or Cenozoic palaeocontinental maps based on ocean-floor and palaeomagnetic data. We can therefore anticipate that, when better Palaeozoic palaeomagnetic data become available, the resulting composites will probably be quite close to Palaeozoic reality. Unfortunately there is as yet no independent way of testing such anticipations.

Summary

The Earth's outermost rocks, forming a layer known as the lithosphere, behave mechanically like a moving eggshell that is broken into several large fragments. The broken lithosphere forms irregular, sub-spherical fragments known as tectonic plates, and the movements of the plates at the present day are expressed physically by earthquakes. Earthquake distributions show that, to a good approximation, plates are rigid. A theorem in spherical geometry, known as Euler's theorem, states in effect that the relative motion between two plates may be described as a rotation of one plate relative to the other about an axis through the Earth's centre.

The ability to apply this precise, mathematical description of how rigid bodies move on a sphere to the motion of lithospheric plates has turned the qualitative, largely observational, study of 'continental drift' into an important part of the quantitative, predictive science known as global tectonics. The necessary data are provided by estimates of the directions of motions on active faults inferred from earthquakes ('fault-plane solutions'), the trends of transform faults, the rates of motion inferred from ocean-floor magnetic anomalies, and the latitudinal changes of plates in-

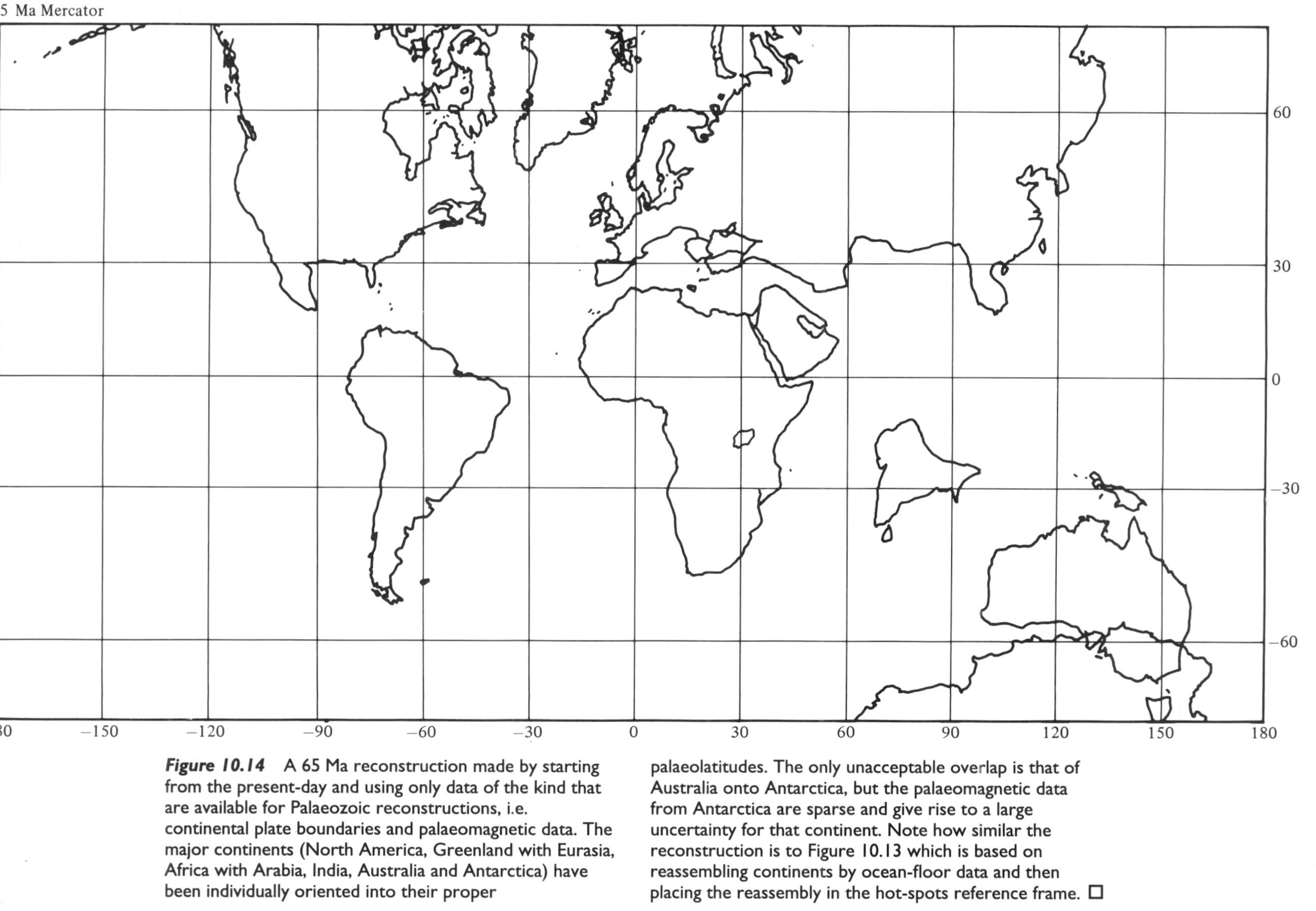

Figure 10.14 A 65 Ma reconstruction made by starting from the present-day and using only data of the kind that are available for Palaeozoic reconstructions, i.e. continental plate boundaries and palaeomagnetic data. The major continents (North America, Greenland with Eurasia, Africa with Arabia, India, Australia and Antarctica) have been individually oriented into their proper palaeolatitudes. The only unacceptable overlap is that of Australia onto Antarctica, but the palaeomagnetic data from Antarctica are sparse and give rise to a large uncertainty for that continent. Note how similar the reconstruction is to Figure 10.13 which is based on reassembling continents by ocean-floor data and then placing the reassembly in the hot-spots reference frame. □

ferred from the old magnetic field preserved in rocks, known as palaeomagnetism. Direct measurements of 'continental drift' are now possible using satellite and high-precision studies.

Further reading

Cox, A. & Hart, R. B. (1986) *Plate Tectonics – How It Works*, Blackwell Scientific Publications, Oxford, 392 pp.

Fowler, C. M. R. (1990) *The Solid Earth*, Cambridge University Press, 472 pp.

Keary, P. & Vine, F. J. (1990) *Global Tectonics*, Blackwell Scientific Publications, Oxford, 302 pp.

CHAPTER

11

Michael J. Russell
University of Glasgow

PLATE TECTONICS AND HYDROTHERMAL ORE DEPOSITS

Basaltic magmatism along mid-ocean ridges remains the site of major heat loss from the Earth's mantle. The amount of heat escaping is less than that predicted by plate tectonic theory, but that is because much of it is dissipated by water convecting through the upper kilometre or so of the oceanic ridges. For example, where the Mid-Atlantic Ridge surfaces in Iceland, groundwater circulates and escapes in pulses along with steam, notably at the **hot springs** at Geysir. Further south, at latitude 26° N on the Mid-Atlantic Ridge, water up to 366 °C issues into the ocean 2.5 km below sea level. In this case it is seawater that is convecting, and tens of litres a second of hydrothermal solution continuously escape through **hydrothermal chimneys** (Figure 11.1) composed of metal sulphides and anhydrite ($CaSO_4$). Boiling is prevented by the confining pressure at such depths. But when the hot water, which has become acidic and contains dissolved metals and sulphur leached from the oceanic crust, meets the cold, rather alkaline seawater on the floor of the ocean, a black cloud of metallic sulphides is the immediate result. Such springs are called 'Black Smokers' (Figure 11.1).

Black Smokers consist of suspensions of sulphide minerals such as sphalerite (ZnS), pyrrhotite (FeS), pyrite (FeS_2), chalcopyrite ($CuFeS_2$) and minor molybdenite (MoS_2) as well as oxides such as MnO_2. In certain circumstances, precipitation at, around and just beneath these chimneys can produce substantial quantities of zinc and copper sulphides. Indeed it is possible that on the Juan de Fúca Oceanic Ridge in the northeast Pacific, west of Oregon, such a deposit may be mined one day for zinc and copper.

Any metalliferous mineral deposit which is worth mining for its metal content is termed an **orebody**. For it to be exploited as ore a copper deposit must have a copper content of at least half to one per cent, depending on accessibility, depth, size, accompanying minerals, current metal price, and environmental and political considerations. Zinc must grade five to ten per cent before it can be considered as an ore.

The first discovery of a potential ore deposit associated with an oceanic ridge was in the Atlantis II Deep in the centre of the Red Sea, where up to 25 m of hydrothermal mud has been deposited in a brine pool 2000 m below the sea surface. This mud comprises iron oxides as well as zinc and copper iron sulphides and may be mined by dredging or suction techniques. Oxidation of the sulphide layers is prevented by the pickling effect of the stagnant brine, which has a salinity about seven times that of normal seawater. The brine is probably derived from dissolution of **evaporites** precipitated earlier in the history of the Red Sea, when it was an enclosed evaporating basin.

Similar mineral deposits to that discovered on the Juan de Fúca Ridge were generated in a comparable tectonic setting 80 million years (Ma) ago in the ophiolite complex on Cyprus. In this chapter, by combining the evidence from existing hot spring deposits at oceanic ridges with the ancient Cypriot

Figure 11.1 Photograph of Black Smoker issuing at 350 °C from the East Pacific Rise at 21° N. The chimney, which has been broken by the Alvin Submarine, is about 30 cm across, cf. Figure 11.11. Oxidation of the massive sulphides at the TAG hydrothermal field on the Mid-Atlantic Ridge has led to enrichment of the copper grades, and more importantly the gold, from 2 ppm to 11 ppm, because the nobler elements are not leached and dissolved away by the sulphuric acid which is produced on oxidation of the sulphide, and so are concentrated. □

examples, we first consider a theory which has been developed to explain the origin of ore deposits hosted by volcanic rocks extruded into the sea at spreading centres (**Cyprus Type**). We then go on to examine orebodies generated at submarine hot springs during the volcanic episodes accompanying the first stages of plate separation where seawater leaches ore constituents from continental rather than oceanic crust (**Rio Tinto Type** in Spain). Rather similar are the orebodies formed at embryonic oceanic ridges developed in back arc settings (**Kuroko Type** in Japan). A complete suite of orebodies are formed above subduction zones in Andean type settings where magmatic fluids are also implicated in the deposition of disseminated copper, molybdenum, gold or tin minerals (**Porphyry Type**). Magmatic fluids escaping from this environment go on to precipitate ores of lead, zinc, copper, tin, tungsten, silver and gold (**Epithermal Type**). As Cordilleran and Alpine mountain belts rise, so groundwaters may migrate towards the continental interior dissolving metals and sulphur in their path. Zinc and lead sulphide ± barite ($BaSO_4$) are precipitated from these hot, and chemically modified solutions, in caves, faults and breccias (**Mississippi Valley Type, MVT**). Finally, we trace the plate tectonic cycle through to the earliest rifting stage where giant base-metal deposits are generated around hot springs on the sea floor from convection of seawater driven by heat in the upper crust. Zinc and lead sulphides ± barite are the main components of these deposits (**Sedimentary–Exhalative or SEDEX Type**). The rifts or extension zones that host SEDEX deposits do not necessarily develop into oceanic rift margins. Although understanding the tectonic setting is important in developing a strategy for mineral exploration, we shall also remark on features significant in the prospecting for each type of orebody.

Orebodies at constructive plate margins (Cyprus Type)

Copper takes its name from the island of Cyprus where the metal has been mined since the Bronze Age. Here it occurs in chalcopyrite ($CuFeS_2$) in **ore lenses** up to 50 m thick and ~ 500 m across (Figure 11.2). The lenses are in fault-controlled depressions within a sequence of pillow lavas erupted on the floor of an ocean basin in the late Cretaceous. The ore lenses consist of massive **pyrite** with between 1 and 4 per cent copper and up to a few per cent of zinc. The top of the pyritic ores were being oxidised to ochre or **goethite** (FeO(OH)) in the late Cretaceous Sea when extrusion of submarine lavas had the effect of protecting them from further dissolution. Moreover, because these deposits were formed at the same time as the host rocks, they are called **syngenetic** deposits.

The orebodies occur along the northern margin of the **Troodos Massif**, an obducted slab of Cretaceous ocean floor probably generated above an intra-oceanic

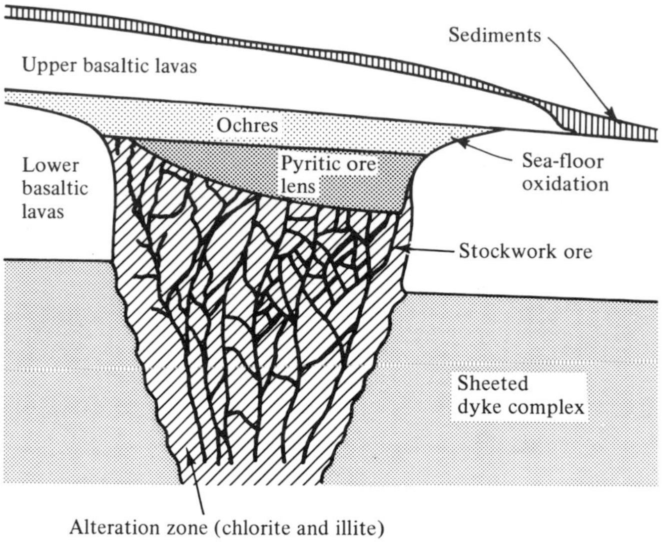

Figure 11.2 **Cross-section of a Cyprus Type copper orebody and its host rocks in a fragment of oceanic crust, based on the work of Hutchinson and Searle. Stockwork is the name given to a chaotic network of mineralised veins produced by hydraulic and chemical brecciation.** □

subduction zone. About twelve deposits of significant size occur at intervals along a 50 km stretch of the margin. The two largest, Mavrovouni (15 million tonnes of ore) and Skouriotissa (6 million tonnes) are about 5 km apart, and were formed at or very close to a ridge axis.

Let us build a mental picture or model for the genesis of these deposits. As the two oceanic plates diverge along a spreading ridge, basaltic magma at about 1200 °C rises to about a kilometre beneath the ridge crest (see Figure 5.1, p. 92). Stresses associated with this rifting cause vertical fractures to develop, especially near transform faults, down which seawater gravitates into the hot oceanic crust bearing sodium, calcium and magnesium cations and chlorine and sulphate anions. Because anhydrite is more soluble in cold water than hot, it is precipitated in cracks in the oceanic crust as the seawater heats up. The most important chemical reaction is the loss of magnesium above 200 °C to make chlorite and/or serpentinite from feldspar and mafic minerals. This has the effect of turning rather alkaline seawater into an aggressive acid, at the same time producing magnesium hydroxide, or brucite as its is commonly known, a component of hydrous magnesian silicates:

$$Mg^{2+} + 2H_2O \rightarrow Mg(OH)_2 + 2H^+. \qquad (1)$$

The hot, acidic fluid then attacks the basaltic rocks of the ocean crust, partially converting them into a mineralogy typical of greenschist facies metamorphism, i.e. producing minerals such as albite, chlorite, epidote and serpentine. Sulphur, iron and trace elements, particularly copper, zinc and manganese, as well as major elements such as calcium, are released to the solution as it approaches within a few metres of the magma chamber (now preserved as gabbro). The solutions can get this close because of contraction cracks in the country rock, and because water at such high temperature has a low viscosity.

Once the solution exceeds the **critical point of seawater** (~ 400 °C) the volumetric expansion is so great that buoyancy drives it back up to the sea floor in a matter of hours. As the hydrothermal solution exhales in a turbulent plume it entrains cold, rather alkaline seawater and a 'Black Smoke' of fine sulphide, is produced. Because calcium sulphate is less soluble in hot seawater than cold, a sleeve or chimney of anhydrite forms around the hot plume from a few centimetres to tens of centimetres across. Sulphides precipitated from the hydrothermal solution replace the anhydrite precipitated from the heated ocean water, to produce large chimneys similar to that depicted in Figure 11.1. Chimneys may grow to a height of 1 to 10 m before becoming unstable and toppling over, other chimneys rising to take their place. A sulphide mound is formed from the debris. Residual anhydrite re-dissolves on meeting cool seawater again and so the deposits become richer in sulphides. Fragments of such sulphide chimneys have been found in the Cyprus deposits. Precipitation also takes place as veins below the seafloor. The threat to the survival of these deposits lies in their vulnerability to subsequent oxidation and dispersion prior to burial.

The pathways that the acid solutions take can be mapped out within the sheeted dyke complex of the Troodos Massif where certain dykes have been converted to albite ($NaAlSi_3O_8$) and finally to the green mineral epidote ($Ca_2Fe^{3+}Al_2Si_3O_{12}(OH)$). They have concentrations of copper and zinc which are an order of magnitude less than those normally found in unaltered basalts which typically average around 100 ppm for both metals.

In the case discussed here we have seen that the temperature of the exhaling hydrothermal fluid is controlled by the high buoyancy of salt water above its critical point (~ 400 °C) at moderate pressure (400 bars, which is equivalent to an hydrostatic head of about 4 km, i.e. 2.5 km of seawater + 1.5 km from the sea bottom down to the magma chamber). As the expansion of aqueous fluids is strongly suppressed above the critical point at pressures in excess of about 500 bars (equivalent to 5 km or more total head of seawater), the fluids could reach even higher temperatures if the magma chamber were deeper.

Critical aspects of ore genesis

The four critical aspects of ore formation have been identified as: the source of the ore constituents, the transporting agent, the energy, and deposition. These are well illustrated with reference to the hydrothermal activity at mid-ocean ridges.

(i) The *source* of the ore constituents is mainly from disseminated sulphides in the basaltic dykes. Such sulphides will contain some copper, either as a trace element in pyrite and pyrrhotite or as discrete particles of chalcopyrite. Zinc may be present in small concentrations of sphalerite. There are also minor amounts of cobalt and molybdenum that may be leached during this open-system hydrothermal metamorphism. A smaller proportion of metal is probably scavenged from the silicates and aluminosilicates.

(ii) The *transporting agent* in this case is seawater, modified to an acidic solution, which dissolves and then carries the ore constituents towards the focussing point at the top of a convective updraught.

(iii) The *energy* driving this geothermal system is provided by the heat of the magma and the latent heat of its crystallisation.

(iv) *Deposition* occurs because metallic sulphides are relatively insoluble at low temperature and so they precipitate when the acid hydrothermal solution mixes with rather alkaline, cold seawater. The availability of reduced sulphur is usually the limiting factor in the deposition of base-metals.

Prospecting for 'Black Smoker' and Cyprus Type mineralisation

To prospect for deposits associated with Black Smokers, pollutants from plumes of the 'spent' mineralising solution may first be identified within the oceanic column and traced back to source. Such plumes are found at water depths of about 2000 m, and are detected by their slightly elevated temperatures, fine particles of manganese and iron oxides, and, more subtly, by their methane and helium isotope contents. The plumes may be followed for thousands of kilometres and they provide a continual supply of manganese, iron and, at lower concentrations, nickel, copper, cobalt and molybdenum to **manganese iron nodules** which concentrate in some places on to the ocean floors to such an extent that they also may be worth mining. A particularly prospective area lies south of Hawaii, where manganese nodules are probably sourced from hydrothermal solutions related to the Hawaiian hot-spot rather than ocean ridge activity. We may expect a myriad of discoveries of hot springs along the oceanic ridges which have a combined length of over 55 000 km.

Black Smoker related mineralisation is revealed by physical and chemical signals, much as the whereabouts of game may be betrayed to a hunter or predator. In the case of Cyprus Type deposits many of the signals have been obliterated by time but instead we can sometimes follow the **alteration zones** that betray the pathways the hydrothermal fluids took on their way to the site of ore deposition, much as a hunter tracks 'quarry'. This is certainly possible on Cyprus where also, more subtly, it has been shown that the ratio of **oxygen isotopes** $^{18}O/^{16}O$ is low along these pathways relative to the unaltered basalt. Oxygen isotope exchange during hydrothermal interaction between host rock and circulating aqueous fluid generally results in a lowering of altered host rock $^{18}O/^{16}O$.

Chemical considerations

Orebodies generated at oceanic ridges have a restricted mineralogy and metal content. This is because basalt contains negligible concentrations of other ore-forming elements such as tin, tungsten, lead, barium and silver. But where submarine volcanic rifts occur within continents then these latter metals may be represented in orebodies that otherwise were generated in a similar way to the Cyprus Type.

Why is this? We can answer this question by referring to the geochemical behaviour of elements as controlled by their ionic charge and size. Although predictions based on these rules, as formulated by the Norwegian geochemist **Goldschmidt**, are not infallible, the **rules** summarise general geochemical behaviour. They state that, in an aluminosilicate structure, ions of similar charge and size may substitute for one another if the bonds do not differ markedly in covalent character. If the ions are broadly similar in ionic radius (i.e. they do not differ by ≥ 15 per cent) then the higher charged ion enters the structure preferentially. If the charge is the same, the slightly smaller ion normally enters the lattice first.

Although strictly speaking the bonding is intermediate between ionic and covalent, common minerals may be considered as ionic structures. Because of the overwhelming abundance of Si, Al, Fe, Ca, Na, Mg and K in the Earth, these ions control the mineralogy of most rocks. During igneous differentiation, Fe–Mg silicates such as olivine and pyroxenes crystallise before aluminosilicates; and nickel, for example, can substitute for magnesium in the olivine structure. Calcic plagioclase precedes sodic plagioclase, and in K-feldspar lead and barium are enriched in the crystal structures.

The ore-forming elements are in rather short supply within the Earth and do not control the structures of the common rock-forming minerals. Because of this they tend to be enriched in the final fractions of the igneous differentiation series. Referring to the plot in Figure 11.3, we can see that apart from Cr, Ni, Cu, Zn and Mg (the ores of which are often associated with mafic rocks) the ionic radii of most of the other ore metals referred to are too large to sit easily in silicate or aluminosilicate structures. These elements are enriched in the 'granitic' continental masses. But even here they are not evenly distributed and certain elements are enriched in certain areas or '**metallogenic provinces**' of the Earth's continental crust so that during successive mineralising events the same metals characterise the resulting orebodies. In the next section we examine ore genesis in continental crust.

Ores associated with submarine rifting of continents (Rio Tinto Type)

Where continents experience submarine rifting and associated volcanism then hydrothermal convection

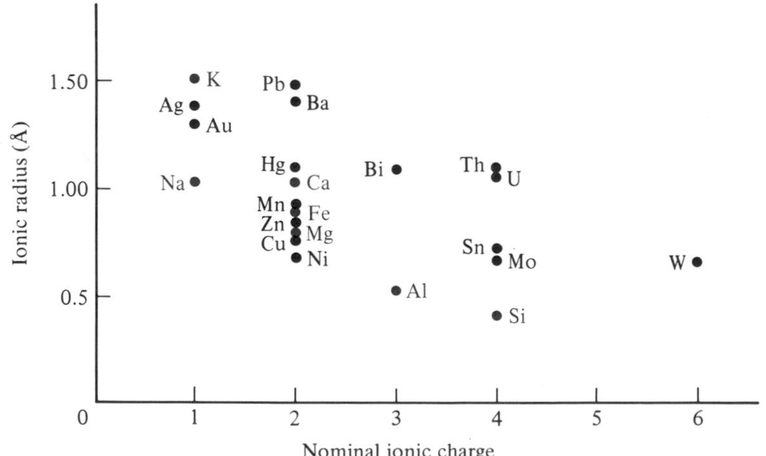

Figure 11.3 Plot of ionic radius against nominal ionic charge. Small, highly charged cations contribute early to crystallisation sequences. Therefore in minerals comprising the major elements (red), silicates generally precede aluminosilicates, and in these minerals Mg^{2+}, with an ionic radius of 0.80 Å, usually precedes Fe^{2+} (0.86 Å), and Ca^{2+} precedes Na^{+} and K^{+}. Minor and trace elements (black) contributing to ore minerals, such as W, Sn, Mo, Th, U, Bi, Hg, Ba, Pb, Ag and Au are usually *incompatible* with silicates and aluminosilicates because of (i) their ionic radii which contrast with those of the major elements; (ii) they are chemically inert; e.g. Au or (iii) their affinity for sulphur when available (e.g. Fe, Zn, Cu, Ni, Co). As with the major elements there is a tendency for the smaller highly charged incompatible minor elements to contribute to the earlier ore minerals (e.g. W, Sn, Mo). □

systems develop in just the same way and at similar temperatures to those at the oceanic ridges. Such rifts may or may not develop into fully fledged oceans, but because the source rocks have different chemical compositions so do the resulting ore deposits. Generally the main difference is the addition of substantial lead and barium as galena (PbS) and barite ($BaSO_4$) respectively. This is because lead and barium occur in K-feldspar and micas in the upper continental crust and are released (during albitisation and/or kaolinisation) to the corrosive, hot, acid, saline solutions involved in convection cells driven by a shallow magmatic intrusion. The magma is generally more felsic than its oceanic counterpart because of the greater time available for magmatic differentiation as it struggles upwards through the thick continental crust. A feature of these ores is their association with felsic breccias which include sulphide clasts, an important characteristic useful in mineral exploration. These **breccias** are fragmentary rocks produced as the magma meets seawater in the uppermost crust, and the seawater boils at low pressure causing phreatic explosions.

In Portugal and Spain giant pyritic bodies are associated with early Carboniferous felsic volcanism. The name of the most famous deposit, Rio Tinto (Red River), indicates how the oxidising orebody was detected. The deposits are mined for copper, zinc and lead. Iberia is generally enriched in mercury, too, and the largest mercury deposit in the world occurs in Silurian sandstones at Almaden in Spain. It is no surprise, then, to find this metal represented in some of the Rio Tinto Type ores along with copper, zinc and lead. This repetition of the same ore metal or metals in time is known as **metallogenic inheritance**. Finally, the development of bimodal volcanism at the time of this mineralisation in Iberia, has encouraged suggestions that the plate tectonic setting was one of aborted rifting associated with Hercynian subduction.

Ores associated with rifting of island arcs (Kuroko Type)

Rather similar hydrothermal systems to those described above develop in extensional regimes in island arcs above subduction zones, and about 13 Ma ago submarine volcanism led to the generation of what are now regarded as the classic Kuroko deposits in Japan. These are exhalative as evidenced by clear sedimentary bedding and the recognition of fossil hydrothermal chimneys, although much of the copper mineralisation was formed as a replacement at the base of the sulphidic sediments (Figure 11.4). The deposits occur in clusters, often within a collapsed submarine caldera. As in the case of the Iberian massive sulphides, the volcanism is bimodal, with tholeiite flows and dacitic and rhyolitic plugs and tuffs, and it probably formed along a rift in the volcanic chain which was aborted during an attempt to create a new marginal sea above a subduction zone.

The deposits themselves comprise a pyritic lens with a chalcopyrite-rich base (yellow ore) overlain by a black mixture of fine-grained sphalerite and galena (Kuroko ore *sensu stricto*) with marginal gypsum ($CaSO_4 2H_2O$) and mudstone (Figure 11.4). Overlying the Kuroko ore is a lens of barite topped by a ferruginous chert. These ores are also associated with felsic tuff breccias, in this case developed at the top of a white kaolinised rhyolite dome. The orebodies are up to 500 m long and 50 m thick (similar to the Cyprus ores) and exploration is facilitated by the associated

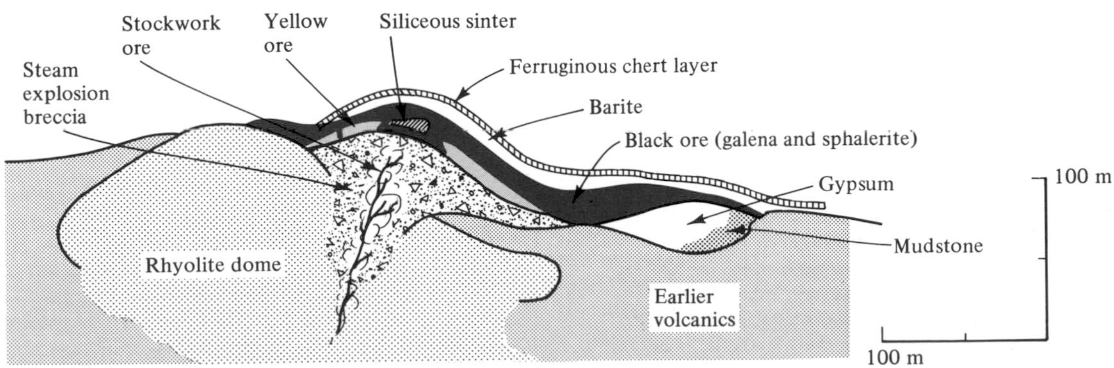

Figure 11.4 **Diagrammatic cross-section of a Kuroko deposit based on the work of Sato.**

alteration haloes comprising chlorite and illite which extend laterally 1 km beyond the ore and 300 m upwards into the hanging wall. Significantly, the ratios between the metal contents in individual orebodies is broadly similar to the ratios of the same elements present in trace concentrations in the crustal rocks beneath. The largest Kuroko deposits occur over slate belts in which zinc averages 80 ppm, copper ~ 27 ppm and lead ~ 17 ppm. These contrasting but generally high concentrations are reflected in the ore metal ratios of Zn:Cu:Pb ~ 3:2:1. Over the granitic rocks, on the other hand, the orebodies are less rich, and the copper to lead ratio is reversed so that Zn:Cu:Pb is 4:1:2, reflecting the relative concentrations of zinc, copper and lead in the granites. Calcium sulphate is a major constituent of these latter ores.

Judging from hydrothermal fluids trapped in **mineral inclusions**, as well as the alteration of the volcanic arc rocks to green tuff (i.e. greenschist facies minerals as in the ocean ridge and Cyprus examples), the hydrothermal systems operated at similar temperatures (350–400 °C) to those of the other ore-generating systems associated with submarine volcanism. A genetic model for the Kuroko deposits follows.

A rhyolite melt differentiates from tholeiitic magma, to rise buoyantly in the crust until it meets seawater in joints associated with marginal faults to the caldera. Hydrothermal convection ensues, but because it takes time for the magma to heat the water and 'cook' the crust the early exhalations of moderate temperature only deposit the sulphides of zinc and lead, sphalerite and galena (Figure 11.5). As the rhyolite magma continues ascending, the base of the hydrothermal system reaches a peak temperature of ~ 400 °C and copper and iron are dissolved from the country rocks along with appreciable hydrogen sulphide. When the rising fluid plume meets the beds of sphalerite and galena, then pyrite and chalcopyrite are precipitated by replacement of the zinc and lead sulphides. The rhyolite plug continues to rise and may invade and disrupt the very mineral deposit it spawned. Finally its top may surface at the sea floor and catastrophic cooling cause hydraulic brecciation. Seawater coursing through this newly emerged dome converts the rhyolite to the white kaolinite so typical of this type of deposit. This and other indications of hydrothermal alteration mentioned above are key indicators in the exploration for these ores.

Subduction and orebodies associated with subaerial volcanism (Porphyry and Epithermal Type)

The great cordilleras of the western Americas are host to an extraordinary variety of rich or vast ore deposits containing all of the ore metals described as incompatible in Figure 11.3. Although broadly related to granodiorite batholiths, and more especially **stocks**, associated with the easterly subduction of the Pacific

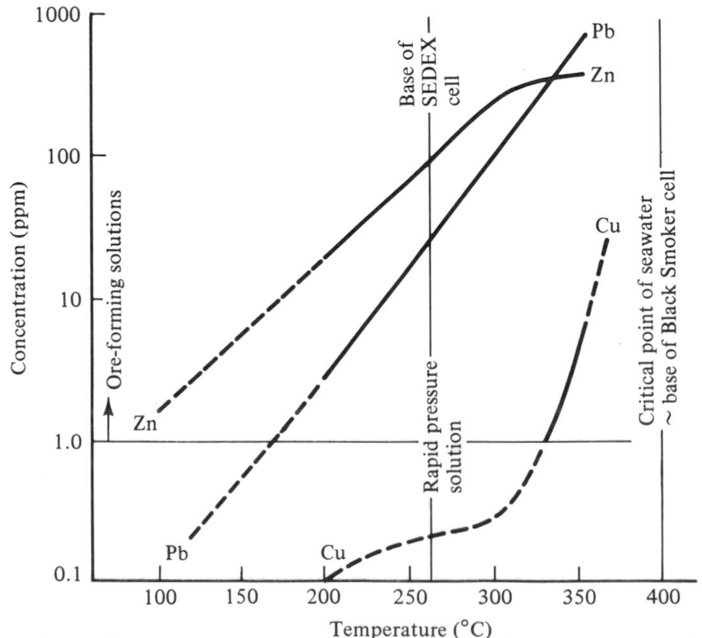

Figure 11.5 Solubilities of lead, zinc and copper in saline acid solution in the presence of hydrogen sulphide and pyrite. This plot explains why SEDEX deposits generally contain little copper and why an early phase of Zn + Pb precedes copper mineralisation in Kuroko deposits as the convecting seawater heats up to 350–400 °C. Based on data from Ohmoto and Walshe. In the absence of hydrogen sulphide and pyrite, copper is appreciably soluble in low-temperature solutions. □

oceanic lithosphere over the last 100 Ma, Dick Sillitoe has shown that andesitic volcanoes are the key. They provided the focus or hearth for the late-stage magmatic aqueous hydrothermal solutions which fed the orebodies 2 to 3 Ma after the main volcanic edifice was built (Figure 11.6a). The transporting agents were the felsic magmas and ultimately the highly saline residual magmatic fluids.

We will attempt to make sense of these mineral deposits using an evolutionary approach developed in studies of the Central Andes of Chile and Argentina, which among other important metals, contain nearly a third of the world's copper resources (Figure 11.6b). We may imagine a late stage batch of felsic magma rising into the base of a volcano where the outside freezes. As this magma slowly crystallises, a highly saline aqueous fraction separates and the buoyant fluid is trapped near the top of the magma chamber by the fine frozen skin. At this stage, and depending on its initial chemistry, metals with strong affinity for sulphur (such as iron, copper and molybdenum) or oxygen (such as tin) precipitate as disseminations of chalcopyrite, molybdenite or cassiterite (SnO_2) between the silicates. The silicates, growing slowly in a well-fluxed magma, develop as phenocrysts giving the final intrusion its **porphyritic** character (Figure 11.6a) and thus the rocks are called **porphyries**. The build-up of volatile pressure eventually fractures the carapace to the intrusion causing hydraulic brecciation, and boiling of the solution on pressure loss. Further precipitation of ore metals takes place in fine fractures. Depending on the metals deposited in this late-stage magmatic process, such orebodies are known as porphyry copper, porphyry molybdenum or porphyry tin deposits.

As the stock continues to crystallise, part of the late-stage magmatic fluid may escape along fractures and faults. These rising, buoyant fluids then boil again as the external pressure drops and a proportion of the surviving solutes, containing metals inclined to partition into saline fluids as chloride complexes, are deposited in relatively low-temperature (**epithermal**) **veins** of Pb, Zn, Cu, Ag and Au. This is because sulphides and gold are not appreciably soluble in steam and they are even less soluble in cool water. Some copper and tin may well be enriched in the veins close to the igneous source, whereas nearer the volcanic crater gold, silver and then manganese oxide veins form as disseminations in pyroclastics, and native sulphur sublimes in the crater itself. It may be that a proportion of the lead and silver in such epithermal deposits was introduced later, leached from country rocks by meteoric fluids circulating at lower temperatures (~ 200 °C), the convective system driven by the cooling stocks. Nevertheless, the zoning is broadly what we would expect from Goldschmidt's rules (see Figure 11.3).

The copper porphyries (Figure 11.6a) are often significantly enriched in molybdenite (MoS_2) (the Cu:Mo ratio generally ranges from 20:1 to 50:1) and more rarely gold (< 0.7 g per tonne Au), and they are particularly economic where there has been supergene enrichment. **Supergene enrichment** occurs where the ore has been exposed to weathering. The copper is dissolved by acidified rainwater in the oxidising zone and is re-precipitated as sulphide just beneath this zone at the top of the water-table because it is less soluble as a sulphide than baser metals such as iron and zinc. Iron dissolves to be re-precipitated as an insoluble iron oxide cap, or **gossan**, above the ore.

Although copper is chemically enriched during supergene alteration, this is not the case for tin which occurs as the oxide cassiterite (SnO_2) and is therefore chemically stable in the near-surface oxidising environment. Nevertheless, this chemical stability allows tin to be enriched physically during weathering, in **placers** in low-energy environments of streams, where it concentrates by virtue of its high specific gravity. Tungsten, as wolfram ($(Fe, Mn)WO_4$), may accompany the cassiterite.

Iron invariably features in this type of mineralisation either as a component of the ore minerals or, particularly in the case of copper porphyries, in pyrite, the major sulphide in such deposits. These orebodies are mined by open-cast methods, and pits up to 2 km across and 500 m deep are developed to mine ore which is low grade (e.g. 0.8 per cent Cu and 0.4 g/tonne Au) but high tonnage (up to a thousand million tonnes or more). The largest deposit in Chile is the Chuquicamata orebody with ~ 9 gigatonnes of ore grading ~ 0.46 per cent copper.

In the Andes the zones of Cu(Mo) mineralisation have migrated eastwards approximately 400 km from Chile to Argentina over the last 90 Ma, and the last period of mineralisation (5–15 Ma) is largely confined to Argentina. It has been widely argued that these deposits are associated with subduction processes, and three key points should be borne in mind as we explore this idea further. First, this mountainous province, which is a little shorter than an average island arc, contains an estimated 30 per cent of the world's copper resources. Second, since the ratio of copper to iron in the orebodies is similar to that in the surrounding granodiorites, perhaps the copper porphyry deposits should be primarily considered as large sulphur anomalies. Third, the tin porphyries occur to the north of the

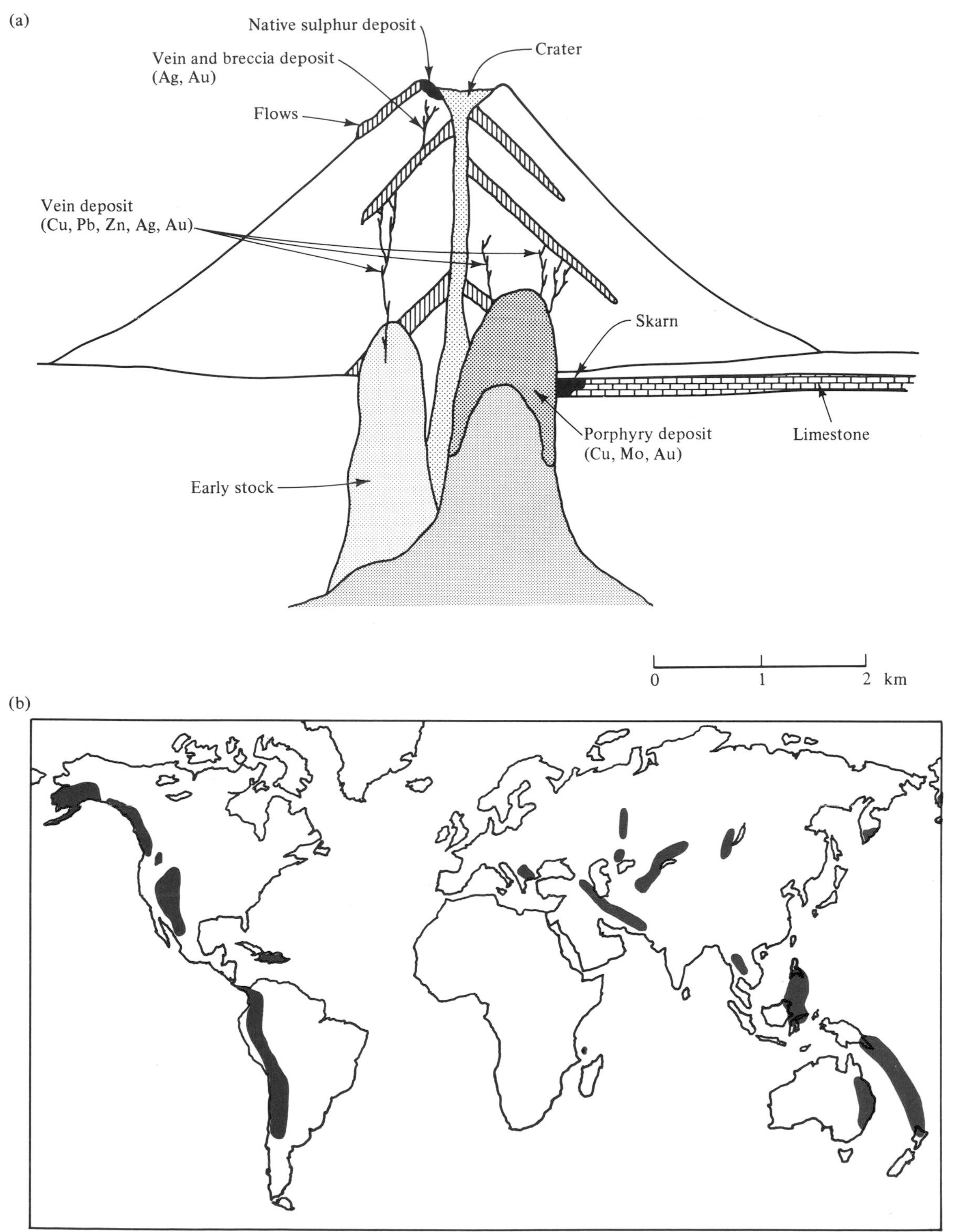

Figure 11.6 (a) Diagrammatic cross-section showing how particular types of ore relate to a stratovolcano. Where a stock intersects a carbonate horizon a richer ore or 'skarn' is often formed by replacement. (b) Distribution of the main porphyry copper provinces (based on a diagram by Titley and Beane). Note how they relate to Tertiary and Mesozoic subduction zones (see Chapter 9). □

Argentinian copper deposits, in Bolivia.

We have already seen how iron sulphides with chalcopyrite, sphalerite and minor molybdenite are produced at Black Smokers on ocean ridges. It appears that these deposits and their sedimentary host rocks are not scraped off the subducting oceanic plate at the Peru–Chile trench, and so they are likely to be subducted beneath the site of Andean magmatism. Thus, most observers regard the subducted oceanic crust as the principal source of the sulphur which is the key ingredient of the Cu(Mo) mineralisation (Figure 11.7). What is less clear is why the mineralisation has migrated eastwards over the last 90 Ma. Magmatism has also migrated towards the east over the same period, and it is clearly the main source of the necessary heat and sulphur.

So far so good. However, tin and lead are not well represented in ocean floor sediments, and so the subducted ocean crust is an unlikely source of the associated lead–zinc–silver mineralisation and, in particular, of the tin porphyries in Bolivia. Instead, we appeal to the concepts of metallogenic provinces and metallogenic inheritance touched on in the discussion of the Rio Tinto orebodies.

In western Brazil there is a Precambrian belt of tin-bearing granites which extends beneath the Andes, and outcrops on the Pacific coast (Figure 11.7). Subsequent Silurian sediments in the eastern Andes are strongly enriched in cassiterite (up to 1 per cent Sn), presumably derived from the Precambrian granites. The preferred model is that these old placer deposits were involved in partial melting, especially in the Tertiary when felsic magmas intruded the Silurian metasediments and gave rise to hybrid magmas rich in tin. Vein and porphyry tin deposits resulted as the hydrothermal fluids escaped and boiled, but overall the reason for the tin-rich province is that Andean magmatism took place in an area which already had relatively high tin abundances.

Where two continents converge as a result of subduction, crustal thickening depresses crustal rocks to such a depth that melting can occur as the isotherms rise. The magmas produced will tend to be drier in this environment, and any mineralisation associated with near-surface volcanoes and stocks is often generated by the circulation of meteoric and ground waters. Again, metals are leached from the crust to be precipitated generally in veins at the top of the updraught of such magma-driven hydrothermal systems. A case in point here is the copper, tin and tungsten mineralisation found in Cornwall, England.

Prospecting for Porphyry and Epithermal Type deposits

Surface manifestation of porphyry mineralisation may be as gossan (red iron oxides), the concentration of K-feldspar, and the sericitic and argillaceous alteration zones (muscovite and clay zones respectively) which generally extend well beyond the disseminated ores. As much of this alteration has been superimposed on the country rock by later circulation of **meteoric water** in convection cells which can be 40–50 km in diameter, it is only useful as a guide to disseminated mineralisation. It remains important to map out the volcanic stratigraphy and thus the present orientation of the 'palaeo-volcano', but the final search for epithermal deposits is the expensive one of diamond drilling often at an angle to the vertical. Even where ore is not intersected, the drill-core is invaluable for the information it provides on the volcanic history and the pattern of hydrothermal alteration. Porphyry copper mineralisation is generally related to the calcalkaline magmatism associated with compressional regimes above subduction zones. Because of the effects of erosion, the most prospective terrains are the late Mesozoic and Tertiary cordilleras and compressive arcs (Figure 11.6b).

Subduction, uplift and Mississippi Valley Type (MVT) deposits

There is a region in northern France called Artois where, if one sinks a well through impermeable rocks into aquifers, water springs to the surface without pumping, whence the term artesian. The water issues naturally from the wells because the aquifer occurs on the higher ground. So, when tapped at depth water is driven through the permeable aquifers by a hydraulic head and escapes upwards to any man-made well through the overlying impermeable rocks (the aquiclude).

The early French economic geologists, such as Daubrée, realised the potential of this hydrodynamic driving mechanism in explaining certain hydrothermal ore deposits. If we think of an artesian system on a much larger scale, we can imagine that as a mountain chain is formed, so a large hydraulic head may be developed with respect to the same permeable sedimentary horizons well out in the foreland (Figure 11.8). In the Mississippi Valley of the United States

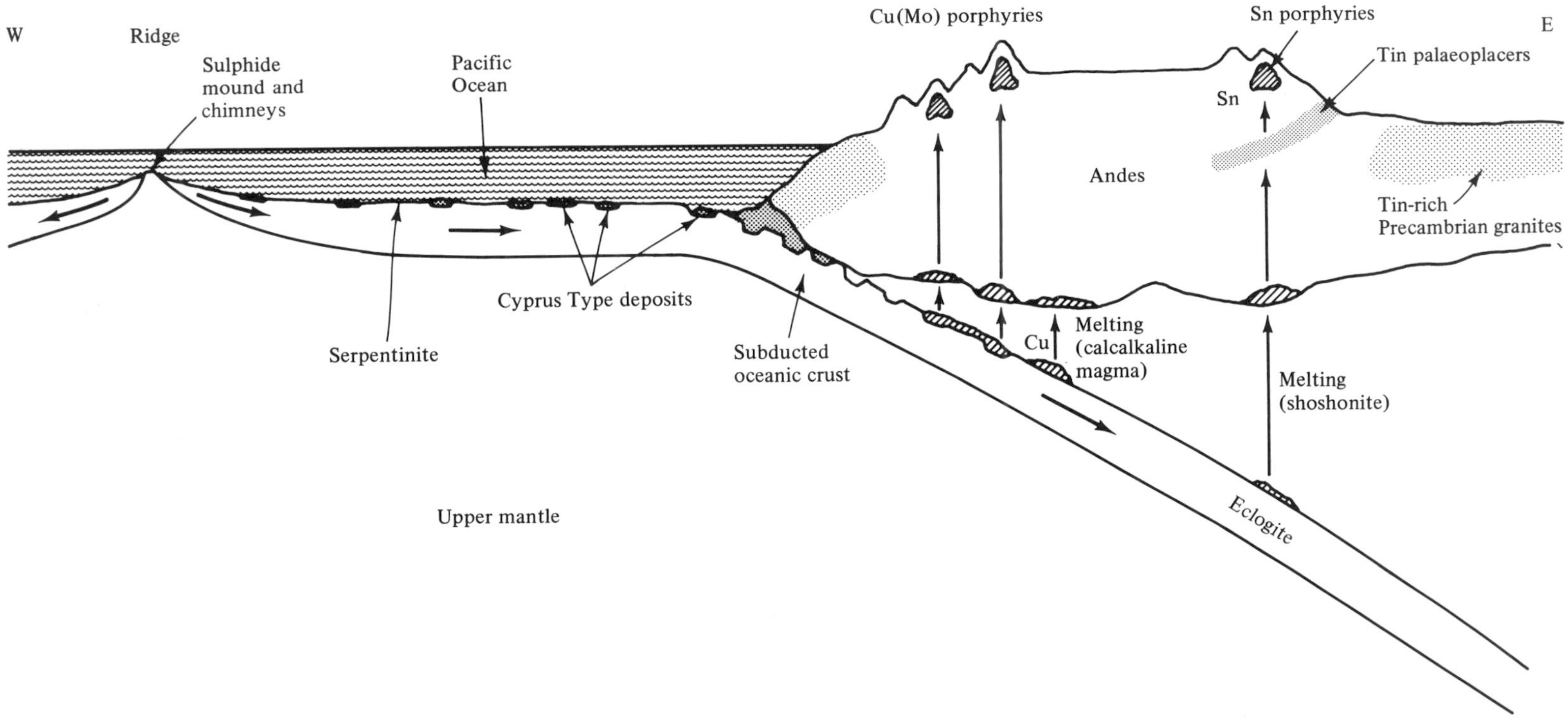

Figure 11.7 Diagram showing how ocean floor sediments with Cyprus Type copper sulphide deposits overlying serpentinised oceanic crust begin to melt as they are carried down to high-temperature zones in the mantle at a depth of 100–300 km. The subsequent melts contribute to sulphur and copper-bearing calcalkaline magmas. At greater depths alkaline magmas are produced which rise through the mantle and then the crust to digest Silurian quartzites bearing cassiterite within the orogenic prism. Tin as well as copper porphyries crystallise in the hearths of volcanoes. Not to scale.

□

there are many very large lead + zinc deposits in limestones that may have been produced by Artesian flow, in which the lateral migration of meteoric and ground water from mountain chains such as the Appalachians and the Ouachitas, may have been hundreds of kilometres (Figure 11.8). Similar deposits developed in Canada and the west central United States as the Rocky Mountains began to rise during the Laramide orogeny concomitant upon the onset of subduction of the East Pacific Plate. Meteoric water with a hydraulic head may effectively flush a foreland basin (and underlying basement) of its formation water, which at a depth of 3 or 4 km may have temperatures of between 80 and 150 °C. Where evaporites occur in the succession such waters become highly saline and are well able to dissolve lead and zinc, and in red bed sequences even some copper. These fluids may then be guided in an aquifer towards a basin margin, where they may rise and react with hydrogen sulphide formed by the reduction of sulphate in groundwater or, in evaporite horizons, by exothermic reaction with hydrocarbons as shown below:

$$CaSO_4 + CH_3CO_2H \rightarrow CaCO_3 + H_2S + H_2O + CO_2\uparrow. \quad (2)$$

The hydrogen sulphide reacts with the metal ions to produce the sulphides:

$$ZnCl_2 + H_2S \rightarrow ZnS + 2H^+ + 2Cl^-. \quad (3)$$

The hydrochloric acid produced in the reaction dissolves the limestone host, so preparing space for further mineralisation.

One of the best examples of the Mississippi Valley Type of orebody occurs outside of the Mississippi itself, in the North West Territories of Canada (Figure 11.8). This is the Pine Point zinc + lead district and, as with other Mississippi Valley Type deposits, although there is strong evidence for the former presence of hydrocarbons, there are no economic oilfields in the immediate vicinity of the ore district. This is presumably because the oil has been oxidised to carbonate or carbon dioxide and water by the exothermic reaction with sulphate which produces the reduced sulphur to precipitate the lead and zinc (equation 3). Some 10^{18} or

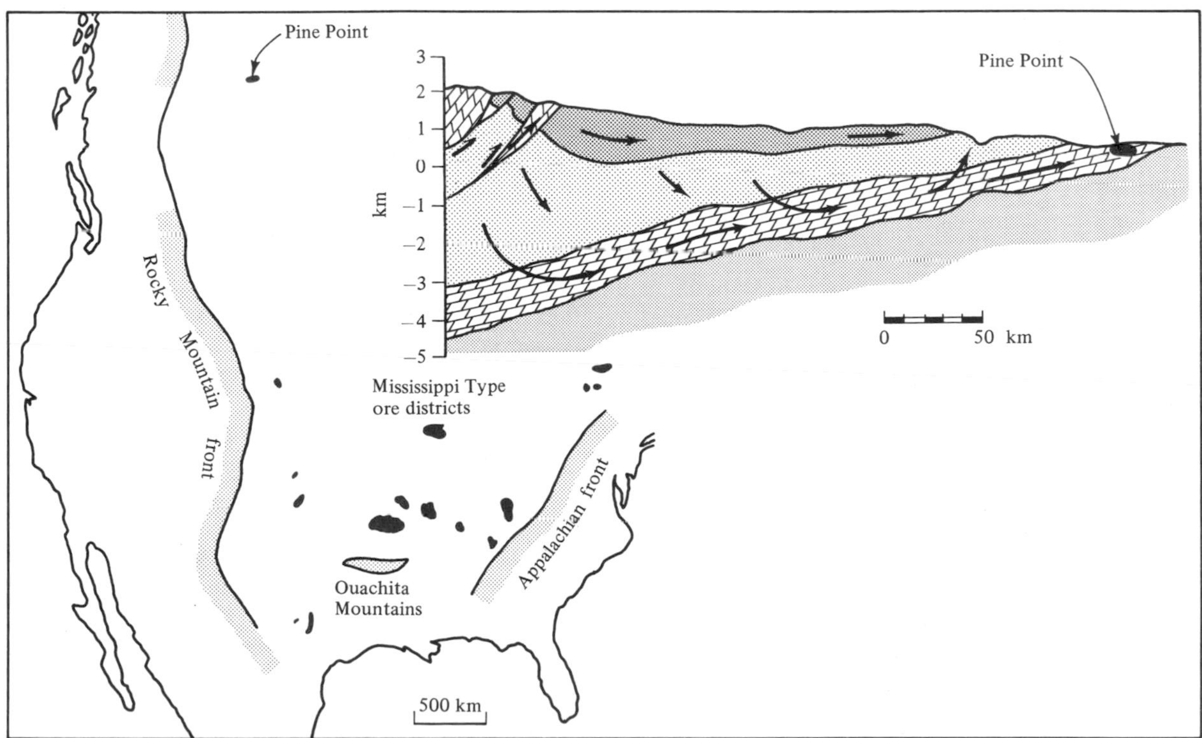

Figure 11.8 Map showing the relationship between Mississippi Valley Type ore districts (red) and mountain chains. The inset shows the artesian drive model used by Garven to explain the Pine Point zinc + lead orebodies. □

10^{19} g (> 1000 km^3) of water are required to feed a deposit of 10 million tonnes or more of metal, and this clearly requires an open-system with forced convection. These Mississippi Valley Type deposits stand in strong contrast to the Irish SEDEX Type discussed in the final section, in that not only are the MVTs entirely later than the host rock (i.e. **epigenetic**), but they are also much more extensive, and are referred to as mineral districts, generally measured in hundreds of square kilometres. More puzzling are those MVT deposits which also contain millions of tonnes of fluorite, often in a central zone. The North Pennines of England are a fine example, but opinion is divided as to whether the fluorine is of magmatic origin, or, as is generally agreed for the other ore components, derived from source rocks by leaching.

Mississippi Valley Type deposits generally occur in carbonate rocks (often dolomitised) at the margins of basins, abutting or over basement highs. Genetic models may be used to search for hidden ('blind') deposits where drilling is the only certain test.

Earliest rifting and Sedimentary–Exhalative (SEDEX) Type ores

We close the cycle of plate tectonics with a discussion of SEDEX deposits. SEDEX orebodies include in their number the largest base-metal accumulations on Earth (up to 50 million tonnes of zinc + lead as sulphide at grades of 10–20 per cent). Barite may accompany the base-metal, the quantity often totalling many millions of tonnes. Although these deposits are quite rare, where they do occur, several ore beds occupy a particular stratigraphic horizon separated from orebodies of the same age by distances of at least 20 km. Unlike the Rio Tinto Type, their genesis is not directly related to the igneous activity typical of a well-developed rift, but it may be broadly linked to the onset of extensional stress conditions. Famous examples are Mount Isa in Australia, Sullivan in Canada, and deposits in the Irish Carboniferous Limestone. We will examine these last in some detail because of their undeformed state and because they are relatively well understood.

The first orebody to be discovered in recent times in Ireland was at **Tynagh** in 1961. Struck by the similarities of terrain between the famous mining camps in the Archaean Superior Province of Canada (broadly comparable to the Kuroko ores), two highway engineers had returned to their native Ireland to search for orebodies beneath the poorly exposed bog-covered Central Plain. Asking advice of the Director of the Geological Survey of Ireland they were directed to the Tynagh area on the basis of red sandstone boulders in glacial till which were copper stained and apparently derived from an Old Red Standstone inlier at Tynagh. First analysis of soils to the north of this inlier revealed extraordinary concentrations of lead and zinc with important copper and silver values! After open pit extraction of loose zinc and lead ore, produced by the dissolution of the Lower Carboniferous Limestone matrix on oxidation of sulphide to sulphuric acid, drilling revealed an orebody within the limestone itself (Figure 11.9). It was a time when most such ores were thought to have had a sedimentary origin, and so this hypothesis formed the basis of the mining strategy at Tynagh. With so much of the evidence and structure of orebodies being hidden prior to mining, the temptation is to use models developed for other deposits. However, mining at Tynagh quickly ran into difficulties because the ore proved to be discontinuous within sedimentary units, and so the hypothesis of a sedimentary origin was thrown out in favour of one that assumed that the ore formed by replacement of the limestone some 100 Ma after its deposition. Nevertheless, a striking spatial association between the sulphide orebody and a beautifully layered but bioturbated haematite–magnetite–chert deposit (Figure 11.9) was reminiscent of similar associations found elsewhere in the world, and so perhaps a compromise was in order. Perhaps the sulphide orebody was formed mainly by the filling of voids in, and replacement of, the Carboniferous Limestone, but very shortly after its deposition.

One prediction, stemming from a theory of early emplacement of the Tynagh orebody and its association with the exhalative, sedimentary iron formation, is that there must have been submarine hot springs in the general area of Tynagh, operating in Early Carboniferous times. Thus we should expect to see the effects of 'spent' hydrothermal solutions to be present as geochemical additions, or anomalies, within the surrounding limestone of the same age (cf. Black Smokers). In practise, extensive manganese enrichment is observed, and this decreases to background levels at about 8 km from Tynagh (Figure 11.10a). This discovery gave strength to the view that the mineralisation was indeed of Early Carboniferous age (~ 360 Ma) and that it was associated with the iron formation.

Once Tynagh was discovered, a mineral rush took place and the ground within 15 km or so of Tynagh was thoroughly searched but to no avail. In fact, the next orebody discovered was 45 km to the south at

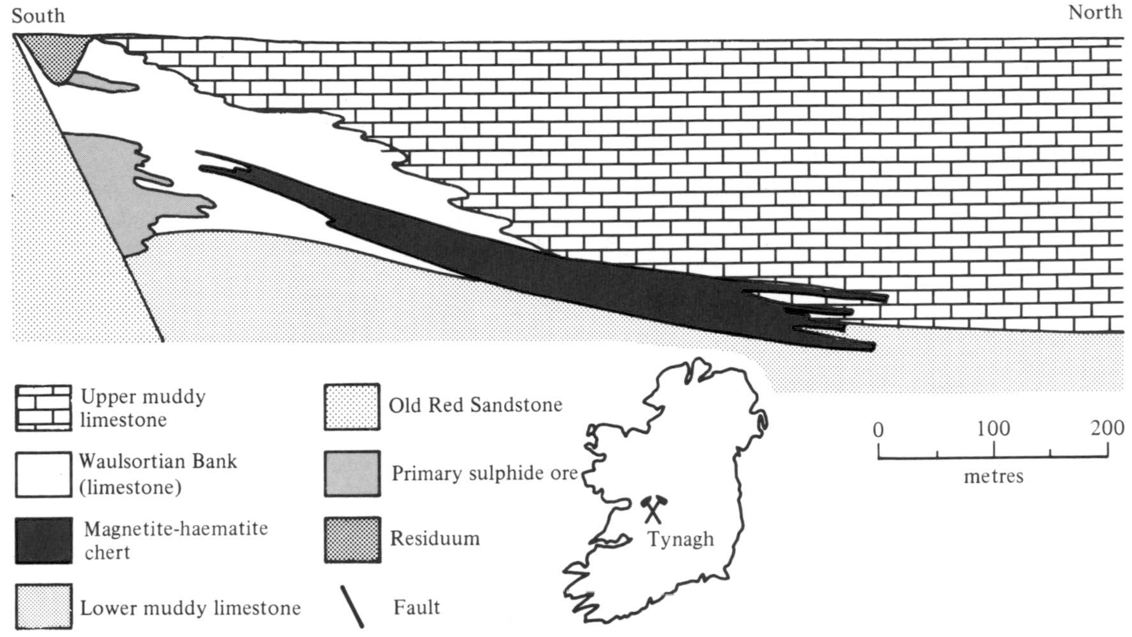

Figure 11.9 **Diagrammatic cross-section showing the relationship between sulphide ore which is dominantly epigenetic and the sedimentary iron formation at Tynagh (location shown on map by crossed hammers).** □

Silvermines. For previous centuries silver and lead had been mined intermittently from veins but the new deposit consisted of an extensive layer of pyrite with galena and sphalerite which passed laterally into a large barite orebody. These ores were clearly deposited directly onto the Carboniferous sea floor, that is they were of the Sedimentary–Exhalative (SEDEX) Type. Given the exhalative theory for both deposits and armed with the knowledge that hot springs below the sea floor build up chimney-like edifices, the presence of chimneys could be anticipated. Again these expectations were borne out (Figure 11.11), but the chimneys both at Tynagh and Silvermines were an order of magnitude smaller than those found at the East Pacific Rise and in the North Atlantic. This is partly because the hydrothermal solutions were of lower temperature than those associated with magmatism at the oceanic crests. We know this from the study of inclusions of fluids trapped within minerals in veins which were the feeders of conduits for the hydrothermal solutions at Silvermines. By taking thin slivers of quartz and sphalerite, and studying their **fluid inclusions** during heating on a special microscope stage, we can determine the temperature at which the fluid fills the inclusion, allowing for a small pressure correction. The highest temperatures recorded in these inclusions are around 220 °C, or 150 °C lower than Black Smoker temperatures. Slightly higher temperatures (up to 240 °C) are indicated for Tynagh.

It is clear that the mineralising fluids rose up structures closely associated with active faults. These faults partly defined small basins on the sea floor, and earthquakes periodically dislodged materials from the upthrow side of the fault down into the basin to form submarine debris flows. In the case of Silvermines the basin waters were saline and highly reducing. The iron as well as the lead and zinc were precipitated as sulphide and protected from oxidation by a brine as revealed by fluid inclusion studies of the barite that grew on the floor of the brine pool. The sulphide at Silvermines was produced by bacterial reduction of sulphate ions in seawater (cf. equation 2). At Tynagh most of the base metal that might otherwise have been precipitated as sulphide directly on the seafloor was oxidised and dispersed, and all the iron was precipitated in the oxide form. Pyrite chimneys have also been found at Tynagh along with fossil worms similar in morphology to worms found at the Black Smoker site at 21° N on the East Pacific Rise.

What turned out to be the largest zinc–lead deposit to be mined in Europe was found ten years after Tynagh at Navan, about 120 km to the north-north-east. The Navan deposit also occurs in Carboniferous Limestone. A small proportion of it was precipitated on the sea floor but a good part of it precipitated just beneath. It is the same age as the deposits of Tynagh and Silvermines but it occurs in four or five superposed lenses, occupying about 100 m of the Lower Carbon-

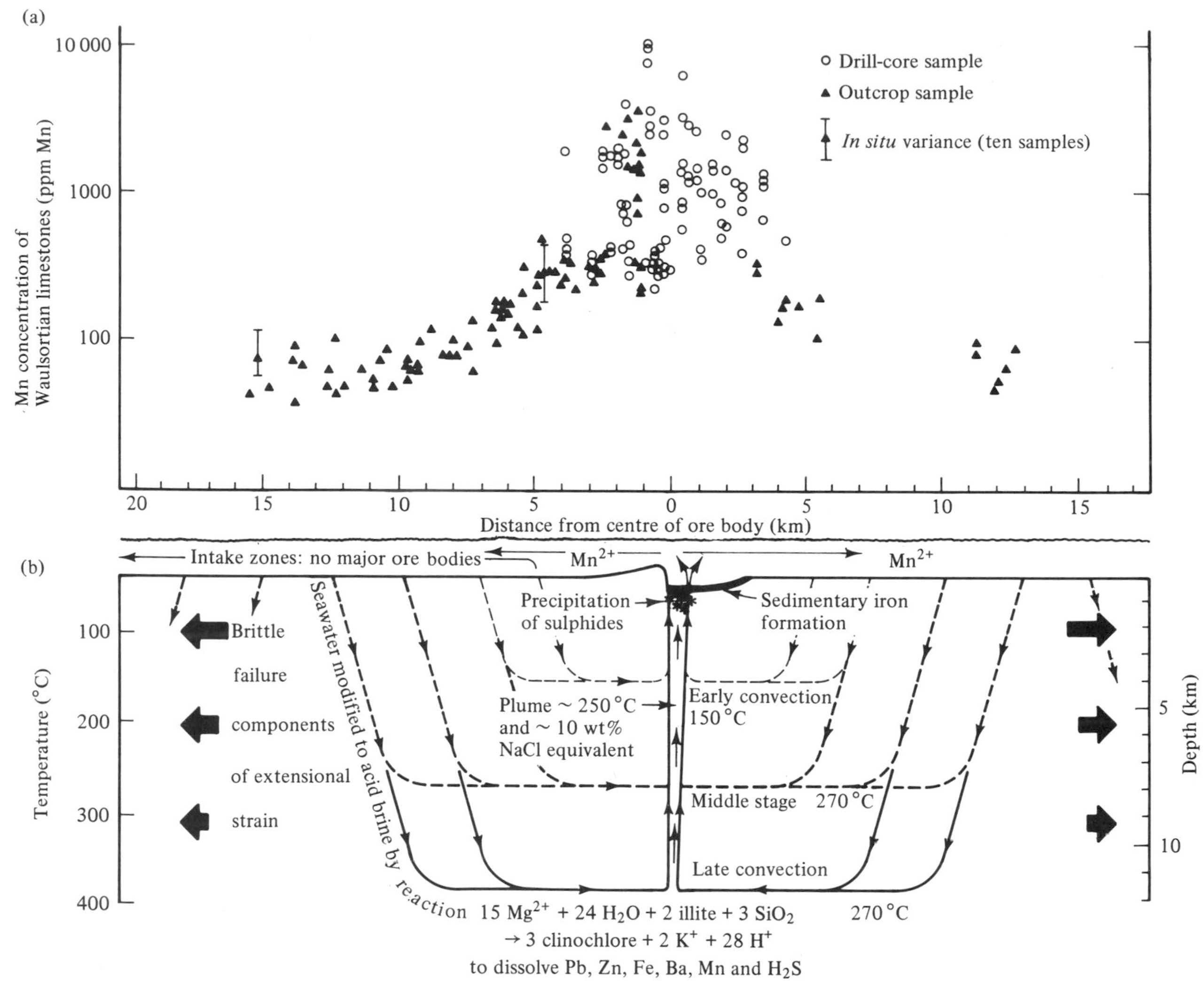

Figure 11.10 (a) Diagram to demonstrate the manganese contents of pure limestone around Tynagh zinc + lead orebody (note the logarithmic scale for manganese concentration in ppm). The manganese was exhaled into the Carboniferous Sea from hot springs associated with the orebody, became absorbed onto calcium carbonate mud and so gravitated to the bottom where it lodged in the lithifying sediment. (b) Model for the genesis of the Tynagh orebody by convection of modified seawater in a cell that expands both downwards (tracking the 270 °C isotherm) and laterally with time (both diagrams are to the same scale). □

iferous Limestone of shallow water facies (Figure 11.12). After deposition there was strong seismic activity in the region and continued differential foundering, and half the deposit seems to have slid down a slump scar to the southeast. The later Carboniferous Limestones overlying this slump were laid down in deep water, but there is evidence of continued mineralisation. We may assume that the mineralising process forming the orebody took about one million years, but that exhalations continued afterwards for perhaps another few million years as judged from enrichments of zinc and manganese in overlying limestones.

Given the method outlined in the section on critical aspects of ore genesis we may ask why there is a dearth of copper in these deposits? Why are they so large? Why are they all the same age, and yet so separated in space? Let us first consider the possible *source* of the metals. Given such a large tonnage, it seems likely that the metals were somehow leached by convecting

Figure 11.11 Pyrite chimney with an outer diameter of 8 mm, from the 360 Ma Silvermines orebody, Ireland. Contrast with Figure 11.1.

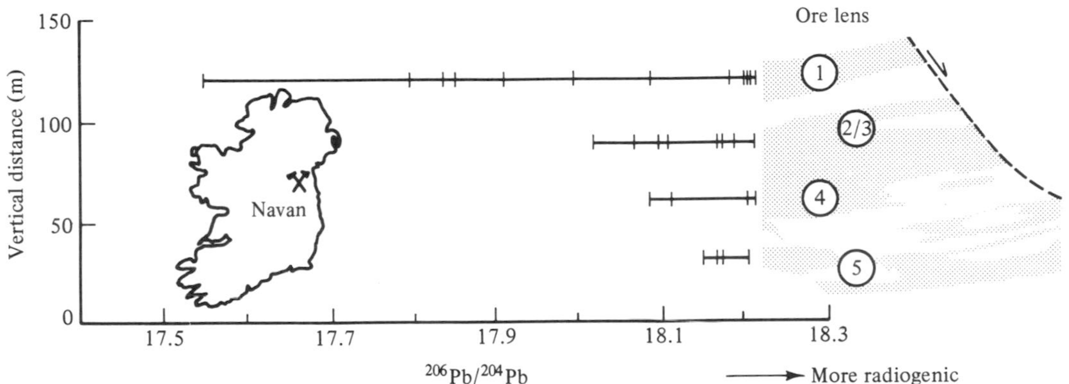

Figure 11.12 Plot of $^{206}Pb/^{204}Pb$ variation in Navan galenas in relation to ore lens position, showing how there is a tendency for a proportion of the samples to become less radiogenic (i.e. lower $^{206}Pb/^{204}Pb$) with time in accordance with expectations of cell deepening during mineralisation, cf. Figure 11.10b. The isotope ^{206}Pb is derived by radioactive decay of ^{238}U whereas ^{204}Pb is primeval, in that it has no known radioactive progenitor.

seawater from the metamorphic rocks beneath the Carboniferous. These consist of folded Lower Palaeozoic metasediments and metavolcanics intruded by Caledonian granodiorites. From evidence gleaned from xenoliths in Carboniferous volcanoes, as well as from geophysical studies, it appears that the Palaeozoic metasediments are underlain by rocks of higher metamorphic grade, probably originally derived from the ancient crystalline rocks of the Lewisian to the northwest. The Lower Palaeozoic metasediments average about 100 ppm zinc, 20 ppm lead, 40 ppm copper and 700 ppm barium. Clearly, it is possible that these metasediments are the source of the metals (remember the largest Kuroko orebodies overlie similar rocks). Below we develop a model that seeks to explain the relationships between ore genesis and the tectonic setting.

In the early Carboniferous, seawater gravitated down faults at a time of tectonic instability, was heated and converted to an acid solution (see equation 1). This solution became involved in large-scale convection cells, the updraughts of which focussed on permeable zones at the intersections of faults. As there is no evidence of significant magmatic activity at this same time, we must assume that the convection cells were driven merely by the heat residing in the crust (Figure 11.10b). Why then did the temperatures of the rising solutions only reach ~ 250 °C? Well it may be that they had lost some of their heat during adiabatic expansion, but nevertheless it seems that the peak temperatures of the mineralising solutions must have been around about 250 to 270 °C. At these temperatures, and in the presence of a certain amount of reduced sulphur such as hydrogen sulphide, lead and zinc are known to be soluble but copper is not appreciably soluble until the temperatures approach 300 °C (Figure 11.5). Why, then, did not the temperature rise above 250 °C? We saw in the case of the Cyprus deposits that the high-temperature fluids there were a feature of the sudden volumetric expansion of water at around 400 °C. No such expansion exists for pressurised water at about 250 °C, so we must look for another mechanism. It is likely that the mechanism is related to pressure solution of silicates which becomes very rapid at temperatures of around 250–270 °C at high pressures, sealing fractures and lowering permeability. This brings us to another aspect, and that is the peculiar tectonics operating at the time of mineralisation. We known from deep-drilling studies from geothermal energy and also from reservoir-induced earthquakes that if the crust is in horizontal tension in one direction and in compression in another, then vertical fractures are rejuvenated. Voids are produced, propped open by the uneven surfaces of the walls of these joints and faults. Clearly the seawater could gravitate to depth in such a tectonic situation, but we know that, where in contact, the walls are likely to dissolve away rapidly above 270 °C, so that the floor of a convection cell would be held at this temperature (Figure 11.10b). Convecting saline solutions dissolve lead, zinc,.barium and some sulphur from the country rocks and deliver them to the focussed updraught of the cell near, or at, the sea floor. Here precipitation takes place by cooling, an increase in pH, and also by reaction with hydrogen sulphide produced from sulphate by bacteria living off organic debris (cf. equation 2). Such precipitation of sulphides tends to increase the hydrogen ion content, that is, to decrease the pH (equation 3). The acid dissolves limestone away to give room for further precipitation below the seafloor although in many circumstances the mineralising fluid will escape through to the seafloor depositing sulphides thereon.

We can consider the geothermal process described above as open-system hydrothermal metamorphism accompanied by hydro-cooling of the crust, so that as the heat is lost from the base of the convection cell, so fractures may propagate downwards to follow the retreating 270 °C isotherm with time (Figure 11.10b). We know from laboratory studies, as well as mathematical analysis, that convection within porous media tends to have a reproducible **aspect ratio**, such that the height of a convection cell is normally about one-third of its diameter. Given that the expected geothermal gradient was about 35–40 °C per kilometre, it follows that initially the height of the convecting limbs in the crust will be about 7 km and the diameter of the system about 20 km. Thus, we would expect the nearest neighbouring updraught to be about 20 km away, which is about the shortest observed distance between giant orebodies of this type. But as the crust cools in this way, and the cell deepens to plumb the entire 15 km of the upper crust, it will widen and capture neighbouring cells. Neighbouring updraughts will then be forty or so kilometres apart. Given that the solubility of lead and zinc in hot saline solution is between 10 and 100 ppm in the presence of some hydrogen sulphide (Figure 11.5), we may calculate that to produce an orebody the size of Navan (~ 10 million tonnes of base-metal) requires about 10^{18} to 10^{19} g of water assuming an efficiency of deposition of about 10 per cent (i.e. 90 per cent of the metal ions escape and disperse in the sea). Only a few per cent of the metal occurring in trace amounts within the crust would be required, and only about 10 per cent of the heat would be used to drive the system, so it is feasible, but is the hypothesis testable?

One possible approach is to look at the **isotopic composition** in the **lead** ores throughout the five lenses at Navan. In the above hypothesis we expect the earliest, or lowest lens to be produced from the shallowest stage of hydrothermal convection of modified seawater, i.e. ~ 7 km. Lead and zinc in the succeeding lenses are then derived from ever greater depths with the top lens produced as the cell bottoms at about 12–15 km. We have already seen that the rocks nearer the surface are Silurian to Ordovician metasediments. Such sediments generally have a few ppm of the radioactive elements uranium (238) and thorium, and these decay radiogenically to give ^{206}Pb and ^{208}Pb, respectively. So we might expect the earliest orebody to reflect this so-called radiogenic lead content of the uppermost crust. The crystalline basement beneath is generally depleted in the large highly charged 'incompatible' elements, and uranium and thorium would have been driven off during earlier hydrothermal activities (see Figure 11.3). We might, therefore, expect lead leached from this greater depth to have a smaller proportion of the radiogenic isotopes of lead and more of the primeval ^{204}Pb, and in fact, such is broadly the case. Although radiogenic lead of the same ratio is found in all the four sampled lenses, the least radiogenic values do become significantly less in each succeeding lens (Figure 11.12) as expected. We assume the continued appearance of radiogenic lead is due to contamination by the later hydrothermal fluids recycling lead from the lowest, largest lens in the Navan orebody.

Prospecting for SEDEX deposits

How should we prospect for SEDEX deposits? First the favourable horizon must be identified. Such a horizon is often betrayed by evidence of a strong syntectonic influence on sedimentation, especially of submarine debris flows, and other indications of differential subsidence. Lithochemical haloes may also be expected; Mn, Fe and Mg in limestones; Ba, K, Na, Zn and B in pelites. Evidence of the former presence of evaporating brines are important both as mineralising solutions, and as stagnant covers, which prevent oxidation of sea-floor sulphide. In other words, to explore in low palaeolatitudes is an advantage. Lastly, tempting though it is to be one more bee around a known honey pot, one should have courage and look for new deposits at palaeo-distances of 20 or even 40 km or more from established SEDEX orebodies.

Summary

We have seen that the ore minerals which occur as sulphides or oxides are generally incompatible elements which are usually excluded from silicate and aluminosilicate structures throughout geochemical cycles. Hence, they tend to find their final resting place in marine sediments either in mineral deposits or as disseminations. Even here these disseminated metals are often remobilised during diagenesis but finally become fixed as sulphides or become adsorbed on to clays. The ore metal oxides such as cassiterite, being chemically stable, are often trapped in placer deposits in rivers, or alternatively manage to reach shallow marine shelves where they are often reworked by currents. Sedimentary rocks enriched to some extent in a variety of these minerals may give up their metals to aggressive ore-forming solutions which are driven towards a depositional site by (i) magma-driven thermal convection (e.g. Kuroko deposits), (ii) forced artesian flow (e.g. Mississippi Valley Type deposits) or (iii) by natural thermal convection (SEDEX deposits). Alternatively, the sedimentary source may be melted and the various metals concentrated in the aqueous phase of an intrusion, to be precipitated as disseminations (Porphyry deposits), replacements and veins (Epithermal deposits) near the surface. In contrast, ores derived from oceanic crust and ophiolites have a hydrothermal metallogeny restricted to copper, zinc and minor gold (Cyprus Type deposits).

Further reading

Edwards, R. & Atchinson, K. (1986) *Ore Deposit Geology*, Chapman & Hall.

Evans, A. M. (1991) *An Introduction to Ore Geology*, Blackwell Scientific Publications.

Sawkins, F. J (1990) *Metal Deposits in Relation to Plate Tectonics*, 2nd edition, Springer-Verlag, 461pp.

CHAPTER

12

Alan B. Thompson
Swiss Federal Institute of Technology

METAMORPHISM AND FLUIDS

Metamorphism inside the Earth

Most of the Earth consists of metamorphosed rock. We are not usually aware of this because the oceans and a thin veneer of sediment often cover the crystalline rocks of the crust.

Rock **metamorphism** is the process by which the mineralogy and textures of rocks are modified at depth in the Earth in response mainly to changes in temperature. Pressure, fluid composition and deformation can also influence the reactions that transform one **assemblage of minerals** into a more stable one reflecting the new environmental conditions of temperature, pressure etc.

Laboratory experiments on natural rocks and minerals at high pressure and temperature, have shown that specific assemblages of minerals are characteristic of distinct ranges of pressure, temperature and fluid composition in nature. Hence, the finding of specific groups of minerals across an exposed **metamorphic terrain** can be used to decipher particular pressure–temperature conditions during metamorphism. Rearrangement of lithospheric fragments by tectonic processes can result in temperature changes at depth, and allow the introduction of additional **heat sources for metamorphism**, such as crystallising magma. In this way, metamorphic petrology has contributed to the new and exciting developments in our understanding of the thermal evolution of the lithosphere, and the tectonic processes that are able to bury rocks and exhume them from depth.

Contact metamorphism

Sometimes the heat source causing metamorphism is obvious. **Contact metamorphism** produces an **aureole** of new mineral assemblages in the adjacent country rocks consistent with increasing temperature towards a hot magma intrusion contact. For example, at the northern margin of the 8-km diameter granodiorite–tonalite pluton of Ardara in Donegal in the Irish Republic (Figure 12.1a), pelitic schists in the contact metamorphic aureole show obvious mineralogical and textural reconstitution. The metamorphic zones are distinctly curved around the intrusive contact and found up to 1.5 km away. Prior to the contact metamorphism, the pelites were intensely deformed schists containing the fine-grained hydrous minerals muscovite ($KAl_2(AlSi_3)O_{10}(OH)_2$) and chlorite ($(Mg,Fe)_5(AlSi_3)O_{10}(OH)_8$), with some quartz and feldspar. At about 1.5 to 1 km away from the intrusion, biotite ($K(Mg,Fe)_3(AlSi_3)O_{10}(OH)_2$) has grown at the expense of chlorite, but the old schistosity can still be recognised. Within 550 m of the contact the pelitic schists have become reconstituted into **hornfels** (hard rocks without any macroscopic internal fabric). New growth of anhydrous aluminosilicate minerals (Al_2SiO_5), sillimanite within 140 m of the contact and andalusite further out, mark distinct mineralogical changes in

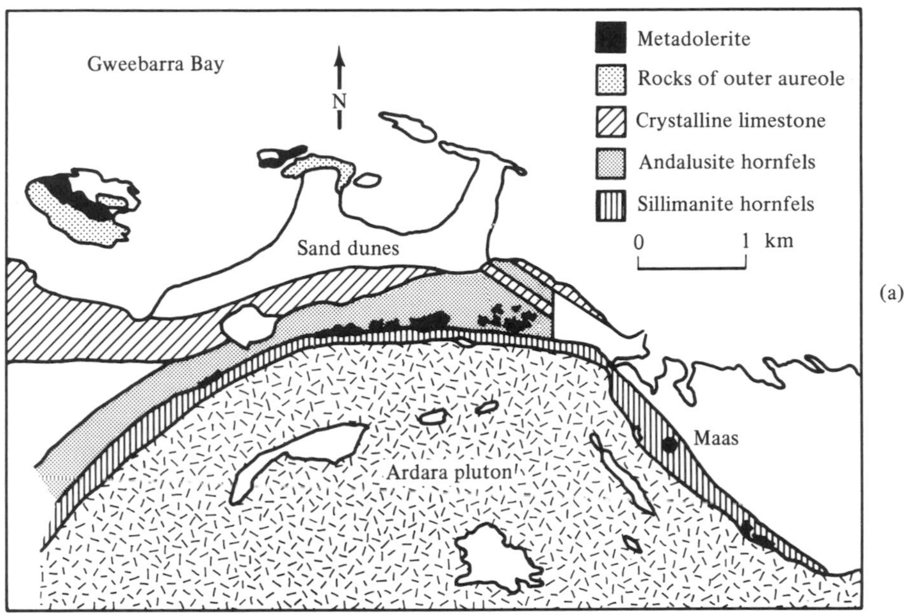

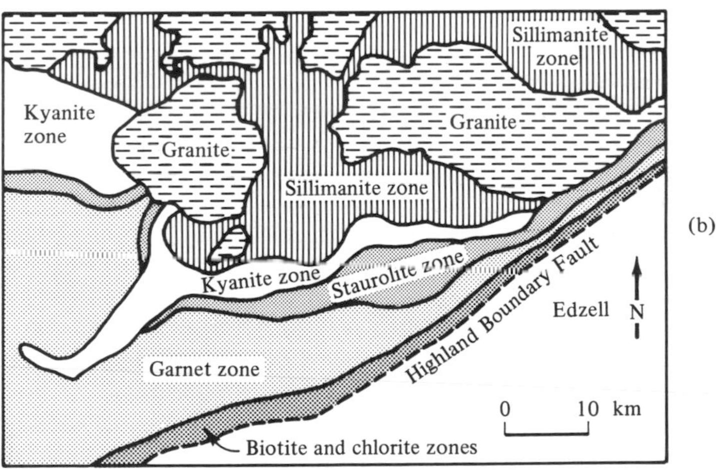

Figure 12.1 (a) Contact metamorphic mineral zones at the northern contact of the Ardara pluton, Donegal granitic complex. Note that the contact aureole follows closely the intrusive contact. Sillimanite occurs in the hornfels closer to the magmatic contact, consistent with its higher thermal stability compared to andalusite.
(b) Regional metamorphic mineral zones in the Dalradian rocks near Edzell at the southern border of the Scottish Highlands. Increasing metamorphic grade is indicated by the sequence of index minerals recognisable in hand specimens: chlorite, biotite, garnet, staurolite, kyanite, sillimanite (see Table 12.1). Younger, unmetamorphosed sediments and volcanics are found south of the Highland Boundary Fault. The intrusive granites post-date the regional metamorphism. They were intruded later during the Caledonian Orogeny that earlier caused the regional metamorphism. □

response to the heat released from the cooling igneous intrusion.

In a given rock unit, the most hydrous mineral assemblages (lowest **metamorphic grade**) show progressive dehydration towards the intrusive contact with the formation of anhydrous minerals at the highest metamorphic grade. Similar behaviour can be seen in carbonate rocks in some contact aureoles, where carbon dioxide, CO_2, has been progressively released through reactions involving calcite ($CaCO_3$) and dolomite ($CaMg(CO_3)_2$).

Regional metamorphism

Many **metamorphic** terrains show a **regional** development of new mineral assemblages indicative of increasing metamorphic grade in particular directions, but without visible evidence of the source of heat. For example, late in the 19th century Barrow (1893) was able to map metamorphic **zones** (now known as Barrovian zones) of metamorphic minerals in the exposed Dalradian rocks of Scotland (Figure 12.1b). The metamorphic zones are most spectacularly developed in former clay-rich (argillaceous, or pelitic) sediments. Each zone is characterised by the growth of a new **index mineral.** These tend to occur in the form of **porphyroblasts** with large grain size in hand specimens. The rock fabrics, as well as the mineral assemblages, also progressively change with increasing grade of regional metamorphism (Table 12.1). The chlorite zone metapelitic rocks are **slates** and **phyllites**, whercas the biotite, garnet staurolite, kyanite and sillimanite rocks are **schists** also containing muscovite and quartz. At the highest grade, sillimanite and K-feldspar (i.e. potassium feldspar) occur in **gneisses**, with quartz, garnet and biotite, but no longer with muscovite.

In regional metamorphism, as in contact metamorphism, minerals containing greater amounts of structurally bound volatile components (e.g. water, H_2O, as (OH) in mica, chlorite, amphibole; CO_2 in carbonates; nitrogen, N, as (NH_4) in mica) are stable at lower temperatures. Because an increase in temperature induces **devolatilisation** and leaves progressively more anhydrous minerals in rocks at successively higher metamorphic grade, the regional direction of increasing temperature can be deduced even if the heat source cannot be recognised.

Since the pioneering work of Barrow, many studies have revealed similar systematic distributions of mineral assemblages as zonal arrangements in **metamorphic belts**, which are usually deformed. These occur in many continental areas and in rocks of different metamorphic age (from Archaean to Alpine). Metamorphism is therefore a continuing, but sporadic, feature of lithosphere evolution.

Metamorphism in orogenic belts

Recent detailed structural mapping in deformed regional metamorphic terrains has distinguished lithospheric fragments somewhat chaotically juxtaposed, and of much smaller dimension (20–50 km) than the whole orogenic belt (> 100 km wide, several hundred kilometres long).

The traces of fossils in some metamorphosed sedimentary rock units show that the rocks originated at the surface. Since sedimentation they have experienced

Table 12.1. *Some typical mineral assemblages developed in metapelitic rocks of Barrovian zones*

Rock fabric	Metamorphic zone	Mineral assemblages	Metamorphic facies	
Slate | phyllite	Chlorite	Quartz–muscovite–**CHLORITE**–albite		Increasing metamorphic grade ↓
Schist	Biotite	Quartz–muscovite–chlorite–**BIOTITE**–albite	Greenschist	
	Garnet (almandine)	Quartz–muscovite–chlorite–biotite–**GARNET**–albite or oligoclase*		
	Staurolite	Quartz–muscovite–biotite–garnet–oligoclase–**STAUROLITE**		
	Kyanite	Quartz–muscovite–biotite–garnet–oligoclase–**KYANITE**	Amphibolite	
	Sillimanite	Quartz–muscovite–biotite–garnet–oligoclase–**SILLIMANITE**		
Gneiss	K-feldspar + sillimanite	Quartz–**K-FELDSPAR**–biotite–garnet–**SILLIMANITE**–plagioclase	Granulite	

* Oligoclase is a sodic plagioclase containing 10–30 per cent of the anorthite ($CaAl_2Si_2O_8$) end-member.

a complex history of burial, transformation at depth and elevated temperature (hence the name metamorphism), and, of course, a later uplift history. Metasedimentary rock units are often found adjacent to metamorphosed lower crustal magmatic rocks or discrete fragments of upper mantle (e.g. Alpine peridotites).

Frequently, metamorphic rocks record a much earlier set of tectonic events than those which led to their differential exhumation. Modern geochronology permits distinction between isotopic ages reflecting metamorphism deep within the crust and subsequent events during uplift.

Metamorphic fluids

The fluid phase is extremely important in some metamorphic rocks as it acts as a catalyst for many mineral transformations that otherwise would not take place, even on geological time scales. Fluids can facilitate deformation in rocks under stress, and enhance the transport of soluble mineral components.

Fluids released by metamorphism at deeper crustal levels appear to escape by being focused into distinct channelways. **Retrograde metamorphism** sponges up these fluids and transforms higher grade into lower grade assemblages. Luckily, this happens patchily and not pervasively, so that high-grade metamorphic mineral assemblages are frequently preserved even in lower crustal rocks exhumed to the surface.

Under certain conditions the metamorphic fluid phase, when grossly out of chemical equilibrium with rocks through which it passes, can induce significant chemical changes (**metasomatism**). For example, seawater circulation through ocean-floor basalt frequently produces marked **fluid–rock interaction**. But fluids derived within the crust can be important for many other geological processes. For example, influx of fluids of certain chemistries can cause metamorphic reactions to occur at much lower temperatures. Water-rich fluids can induce many lower crustal rocks to melt at temperatures hundreds of degrees below melting of an anhydrous equivalent rock assemblage. These lower crustal melts are almost granitic in composition and often can rise to shallower crustal levels. Fluids released from crystallising granitic magmas may produce economically important pegmatites and also, though often following interaction with meteoric waters, hydrothermal ore deposits (see Chapter 11).

Metamorphic rocks as assemblages or minerals

Metamorphism changes the mineralogy and texture of sedimentary, igneous (magmatic) and pre-existing metamorphic rock. Actually, many physical and chemical features of metamorphic rocks are inherited from pre-metamorphic processes. For example, weathering causes enrichment of the less soluble elements in the residue as the more soluble elements are removed to be precipitated elsewhere. During sedimentation, compositional layering can form, and clasts or nodules can grow. Explosive volcanism, stratified flow of lava and crystal settling in magma chambers can also produce layering that is retained despite subsequent metamorphism. Progressive metamorphism generally preserves the local chemical composition of **domains**. Metamorphism is largely **isochemical**, apart from the progressive loss of mineralogically bound volatile components (mainly H_2O and CO_2) which are expelled as metamorphic fluid.

The use of metamorphic mineral assemblages and reactions in the deduction of pressure – temperature conditions of metamorphism

Careful petrographic studies of contact and regional metamorphic sequences worldwide have revealed systematic changes in mineral assemblages for a wide range of rock compositions. Table 12.2 shows which minerals are commonly found in metamorphic rocks of different bulk chemical composition and parentage. In order to use the observed different metamorphic mineral assemblages to tell us about how temperature (T) has varied with pressure (P) during metamorphism, we need to be able to anticipate the normal (average) temperature–depth distribution **(geotherm)** in the lithosphere and to known how pressure relates to depth (Box 12.1) We also need to know something about how metamorphic minerals react.

Metamorphic mineral reactions can be viewed as three types, namely, devolatilisation, solid–solid and melting reactions. In general, increasing temperature favours devolatilisation and eventually rock melting, whereas increasing pressure favours solid–solid reactions producing denser mineral assemblages.

The various curves shown in Figure 12.2 were all

Table 12.2. *A simplified classificiation of metamorphic rocks**

Bulk composition	Former rock types	Common minerals	With oriented fabric	No oriented fabric
Pelitic	clay-rich (argillaceous) sediments – some extremely weathered volcanic (laterite) and plutonic rocks	quartz, muscovite, biotite, chlorite, feldspars, garnet, staurolite aluminium silicates	slate, phyllite, schist, gneiss	pelitic hornfels pelitic granulite
Mafic (basic)	basaltic volcanics, gabbroic intrusives	chlorite, epidote, amphibole, plagioclase, pyroxene, garnet	greenschist, amphibolite	mafic hornfels, mafic granulite
Quartzo-feldspathic	sandstones, arkoses, granitic syenite intrusives, rhyolites, felsic ashes	quartz, feldspars mica	schist, gneiss	hornfels felsic granulite
Quartzose	quartzites, cherts	quartz, (mica)	schist, gneiss	quartzite
Calcareous	mainly limestones and dolomites, rarely carbonatites and kimberlites	calcite, dolomite, talc, Ca-garnet, tremolite, Ca-pyroxene, wollastonite	marble, calc-silicate rock	marble, calc-silicate hornfels
Ultramafic	peridotite, serpentinite	talc, serpentine, (Ca) Mg-amphibole, (Ca) Mg-pyroxene, Mg-olivine	schist	ultramafic hornfels

* In classifying metamorphic rocks the descriptive fabric type is usually preceded by the names of the distinctive porphyroblast minerals, e.g. garnet–biotite–muscovite schist (as in Table 12.1).

obtained from careful laboratory experiments that have explored the *P–T* stability ranges of many common metamorphic minerals. In the *P–T* diagram the **dehydration reactions** exhibit distinct curvature especially at low pressures, whereas the solid–solid reactions plot as almost straight lines. Together, the mineral reactions define a grid, which enables distinct *P–T* fields to be assigned to specific assemblages of minerals. Actually, laboratory experiments have been performed on many more mineral reactions than shown in Figure 12.2, so that the *P–T* stability of most common metamorphic assemblages can be defined in great detail.

Despite the simplifications, we can use some aspects of Figure 12.2 to deduce approximate *P–T* conditions of metamorphism in the contact aureole of the Ardara pluton (Figure 12.1a) and in the Dalradian regional metamorphism (Figure 12.1b). Aluminosilicate (Al_2SiO_5) occurs as three polymorphs – andalusite (And) at low pressures, kyanite (Kya) at higher pressures, and sillimanite (Sil) at high temperatures. Being a one-component system (Al_2SiO_5), the stability fields of the polymorphs should define a triple point in the *P–T* diagram. In the outer part of the contact aureole at Ardara, kyanite formed during a previous metamorphism has been replaced by andalusite and, close to the hot magmatic intrusion, sillimanite has replaced andalusite. Thus the metamorphism took place at depths shallower than the pressure of the Al_2SiO_5 triple point. A temperature increase from at least 400 to 600 °C was achieved within the contact aureole, as indicated by the dashed line (labelled ARDARA) in Figure 12.2.

The higher grade part of the Dalradian regional metamorphic sequence (Figure 12.1b) shows kyanite followed by sillimanite zones. The absence of andalusite here indicates metamorphism at pressures greater than the Al_2SiO_5 triple point (dashed line labelled DALRADIAN in Figure 12.2). The implied temperature range (400 to ~ 700 °C for the Dalradian regional

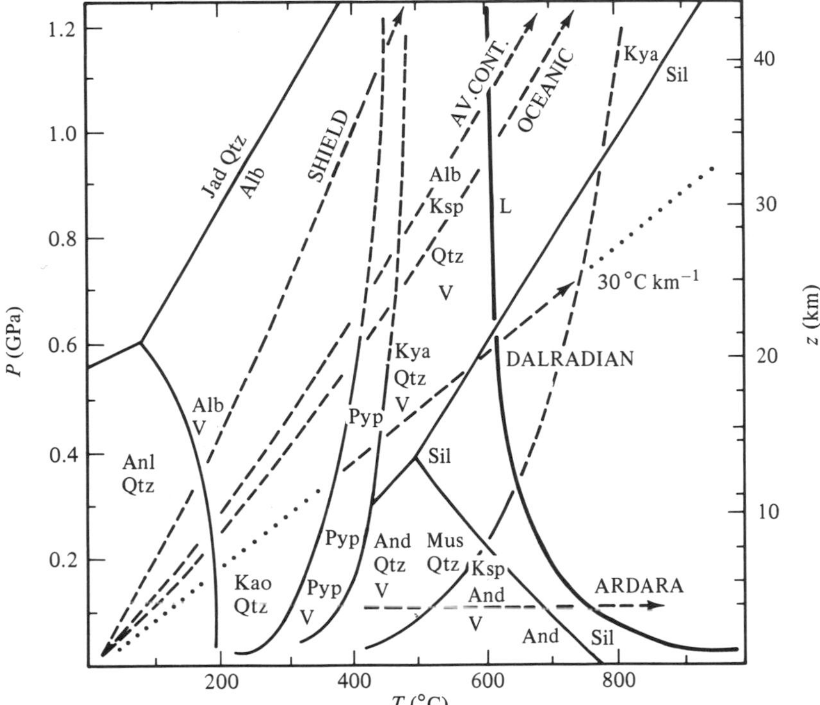

Figure 12.2 Pressure (*P* in GPa) against temperature (*T* in °C) diagram showing some important metamorphic mineral reactions located by laboratory experiments. The correspondence between pressure and depth is outlined in Box 12.1 The average geotherms for stable (ancient) continental shields, within average continental crust and beneath oceanic lithosphere are obtained as outlined in Box 12.3. The heavy arrows show consistent metamorphic gradients for the contact aureole at Ardara (Figure 12.1a) and the regional Dalradian metamorphism (Figure 12.1b). These metamorphic gradients are clearly much hotter than the average long-term gradients beneath continents. Examples of solid–solid, dehydration and melting reactions of metamorphic mineral assemblages, are shown. The extensions of these reaction boundaries into *P–T* space where other (mineral or liquid) phases would be more likely to occur instead are indicated with dashed lines. The abbreviations correspond to the following minerals: Alb (albite) $NaAlSi_3O_8$; Kao (kaolinite) $Al_2Si_2O_5(OH)_4$; Jad (jadeite) $NaAlSi_2O_6$; Pyp (pyrophyllite) $Al_2Si_4O_{10}(OH)_2$; Ksp (potassium feldspar) $KAlSi_3O_8$; Anl (analcite) $NaAlSi_2O_6 \cdot H_2O$; Qtz (quartz) SiO_2; Mus (muscovite) $KAl_2(AlSi_3)O_{10}(OH)_8$; And (andalusite), Kya (kyanite), and Sil (sillimanite) Al_2SiO_5. V = vapour, fluid, H_2O and L = liquid, melt, 'pseudo-granite' composition.

metamorphism) is comparable to that for the Ardara contact aureole but is spread over a much larger area (see the scale bars in Figures 12.1a and b) and has occurred at noticeably higher pressures. Geochronology has established that the granitic intrusions in the Dalradian post-date the regional metamorphism, thus they cannot have produced contact metamorphism on a regional scale. It is quite likely, however, that the same thermal processes that caused the regional metamorphism could also have been important in the production and emplacement of the granites.

The range of *P–T* conditions recorded by metamorphic rocks

Metamorphism occurs at pressure–temperature conditions between sedimentation and rock melting (Figure 12.3). Sedimentary **diagenesis**, which results in the change of mineral assemblages not too far removed from the surface conditions, overlaps the process of very-low-grade metamorphism. Because metamorphic grade is mainly a function of temperature, low, medium and high grades of metamorphism cover temperature ranges of approximately 250 to 400 °C, 400 to 600 °C and 600 to 800 °C respectively, although the quoted ranges strongly overlap (Figure 12.3). Thus the Ardara aureole and the Dalradian zones represent medium to high-grade contact and regional metamorphism respectively.

High-grade metamorphism often coincides, in the field, with partial melting of some rock types (**anatexis**), a process which produces layers, sometimes injected, of **migmatite** (mixed magmatic rock). Pressures of metamorphism are harder to estimate, but low, me-

BOX 12.1

Pressure – depth relations in the lithosphere

The pressure at any depth is determined from the weight (i.e. the force = mass × acceleration) of the overlying column of rock in the solid Earth, or of water in the oceans. A dimensional analysis of the relation

$$\underset{(\mathrm{kg\,m^{-1}\,s^{-2}})}{P} = \underset{(\mathrm{kg\,m^{-3}})(\mathrm{m\,s^{-2}})(\mathrm{m})}{\rho \times g \times z}$$

for an average rock density (ρ) of 2800 kg m^{-3} (2.8 g cm^{-3}) at a depth (z) of 10 km (10^4 m), for g (acceleration due to gravity) = 9.81 m s^{-2} gives

$$P = 2800 \times 9.81 \times 10^4 = 2.747 \times 10^8 \text{ pascals.}$$

Now because 1 bar = 10^5 Pa = 10^{-1} MPa (megapascals),

or 1 kbar = 10^8 Pa = 10^{-1} GPa (gigapascals),

the equivalent pressure to the rock depth (z) of 10 km, with ρ = 2800 kg m^{-3}, is

$$P = 2.747 \text{ kbar.}$$

Table 12.3 shows the pressure–depth relations for lithosphere consisting of quartz + feldspar (ρ = 2800 kg m^{-3}, continental crust) and for lithosphere consisting of olivine + pyroxene (ρ = 3200 kg m^{-3}, oceanic crust and upper mantle).

Table 12.3

Continental crust			Oceanic lithosphere		
z(km)	P(GPa)	P(kbar)	z(km)	P(GPa)	P(kbar)
1	0.027	0.27	1	0.031	0.31
10	0.275	2.75	10	0.314	3.14
20	0.549	5.49	20	0.627	6.27
30	0.824	8.24	30	0.941	9.41
40	1.098	10.98	40	1.255	12.55
50	1.373	13.73	50	1.569	15.69
60	1.648	16.48	60	1.883	18.83
70	1.923	19.23	70	2.197	21.97

dium and high-pressure ranges may be considered as about 0.5 to 3 kbar (0.05 to 0.3 GPa), 3 to 8 kbar (0.3 to 0.8 GPa) and greater than 8 kbar (0.8 GPa). Average continental crust has a thickness of about 35 km (about 0.95 GPa, Box 12.1) so that metamorphism could occur by increasing the heat supply without changing crustal thickness.

Some rare metamorphosed rocks that formed as sediments at the Earth's surface, contain the SiO_2 polymorph, coesite, and other diagnostic high-pressure assemblages. These appear to indicate metamorphic pressures of about 25 kbar (2.5 GPa > 90 km, see Box 12.1). This implied burial depth is even greater than the depth to Moho beneath recently tectonically thickened regions, such as the Himalaya (about 70 km). The exhumation of such higher pressure rocks found, for example, in the French Western Alps and the Norwegian Caledonides, indicates the removal of material from above, principally by erosion and gravitational collapse of the mountains (for details see Chapter 14).

Metamorphic mineral stability and the metamorphic mineral facies

For a given rock composition, certain ranges of T and P are characterised by distinct mineral assemblages (Box 12.2) This notion is embodied in the principle of **metamorphic mineral facies** proposed by the Finnish geologist, P. Eskola, in 1920. The original facies names, and those from later modifications, were formulated for particular mineral assemblages observed in **metamorphosed basalt** (Figure 12.5). However, as shown in Table 12.1 and 12.2, the metamorphic mineral assemblages in any rock composition are approximately diagnostic of metamorphic P–T conditions. Higher temperature minerals facies are characterised by less hydrous mineral assemblages (e.g. Al_2SiO_5 plus quartz instead of pyrophyllite) and higher pressure ones by denser minerals (e.g. jadeite plus quartz instead of albite) as shown in Figure 12.2. Some general correspondence can be seen between P–T calibration of the selected mineral reactions in Figure 12.2 and the metamorphic facies regions in Figure 12.5 Much current research in metamorphic petrology is

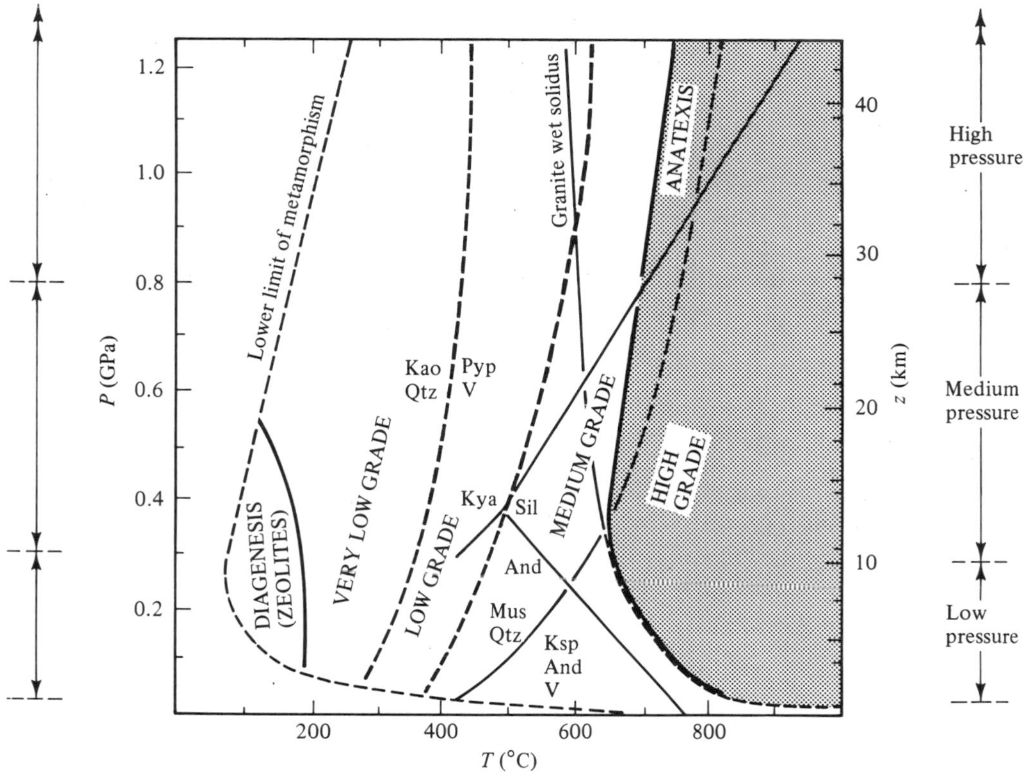

Figure 12.3 The approximate conditions for the various grades of metamorphism. As illustrated, metamorphic grade is primarily a function of temperature and culminates in partial rock-melting (anatexis). Approximate pressure ranges of metamorphism are shown by the vertical arrows labelled low, medium and high pressure. The correspondence between observed mineral assemblages and assigned metamorphic grade can be seen by comparing Figures 12.2 and 12.3. The abbreviations are the same as those used in Figure 12.2. □

focussing on how observed mineral assemblages can be used as fossil barometers (**geobarometry**) and thermometers (**geothermometry**) to track the *P–T* path of a given rock during its tectonic history.

Metamorphic mineral reactions and geothermometry and geobarometry

All of the metamorphic facies boundaries shown in Figure 12.5 are transitional because of the ways in which crystalline (solid) solutions in minerals and variations in the composition of the metamorphic fluid can influence the *P–T* location of specific mineral reactions.

Many of the common metamorphic minerals show extensive crystalline solution. Replacement of Mg^{2+} by Fe^{2+}, Al^{3+} by Fe^{3+}, $Ca^{2+}Al^{3+}$ by $Na^{1+}Si^{4+}$, Mg^{2+} by Ca^{2+} etc. in many minerals means that their breakdown reactions are 'smeared out' over sometimes quite large ranges of pressure and temperature instead of occurring at the well-defined *P–T* curves of the simple reactions in Figure 12.2. Thus in some cases the same mineral assemblage may be stable over a 5 to 10 kbar (0.5 to 1.0 GPa) pressure range and up to a 250 °C range in temperature. Fortunately, the individual minerals in specific assemblages change composition sympathetically and in a predictable fashion.

Recent experimental and theoretical work has enabled much more precise evaluation of these complex multicomponent mineral assemblages for use as geobarometers and geothermometers. It is now possible to predict how the *P–T* conditions of reaction boundaries (Figure 12.2) are displaced by the introduction of chemical impurities which form extensive crystalline solution. For example, the *P–T* location of the reaction.

albite (feldspar) → jadeite (pyroxene) + quartz

is shown in Figure 12.2 for the end-member components

$NaAlSi_3O_8 \rightarrow NaAlSi_2O_6 + SiO_2$.

BOX 12.2

Mineralogy and reactions in metamorphic rocks

The main types of metamorphic rocks, with their common minerals, are presented in Table 12.2. The mineral assemblages observed can be distinctly related to pressure (P) and temperature (T) of metamorphism. At any particular grade of metamorphism, the mineralogy can be diagnostic of the P–T conditions. To illustrate the changing assemblages with increasing grade, examples are given in Figure 12.4 for commonly occurring rock compositions in the most widespread kyanite–sillimanite metamorphic facies series (Figure 12.5). The principal mineral assemblages (Table 12.2) for metabasaltic, metapelitic and metacarbonate rocks are now shown (Figure 12.4) by triangular diagrams that represent their chemistry reasonably well (remember that we are visually limited in our perception of more complex geometrical figures). The compositions of the minerals occurring in these rock types are listed in Table 12.4 as an introduction to the graphical representations. Because many minerals exhibit extensive crystalline solution (see p. 229) simplified representations of their chemistries have been used.

The rows in Figure 12.4 approximate the metamorphic facies (Figure 12.5): (a), (b) – lower and upper greenschist, (c), (d) – lower and upper amphibole, (e) – granulite.

For metapelites (column II) metamorphic reactions represented by changing tie-lines link the prograde mineral assemblages listed in Table 12.1.

For metabasaltic rocks (column I), carbonate + silicate reactions first produce actinolite (a), and epidote + aluminous silicate react to produce plagioclase (b). The transition from greenschist (b) to amphibolite (c) facies permits the widespread coex-

Table 12.4

Metabasalt	Metapelite	Metacarbonate
Epidote (Epi) $Ca_2Al_3Si_3O_{12}$ (OH)	Muscovite (Mus) $KAl_2(AlSi_3)O_{10}(OH)_2$	Calcite (Cal) $CaCO_3$
Actinolite (Act) Ca_2 $(Mg, Fe)_5Si_8O_{22}(OH)_2$	Chlorite (Chl) $(Mg,Fe)_5Al(AlSi_3)O_{10}(OH)_8$	Dolomite (Dol) $CaMg(CO_3)_2$
Plagioclase (Plg) $NaAlSi_3O_8$–$CaAl_2Si_2O_8$	Biotite (Bio) $K(Mg,Fe)_3(AlSi_3)O_{10}(OH)_2$	Talc (Tal) $Mg_3Si_4O_{10}(OH)_2$
Hornblende (Hbl) $NaCa_2$ $(Mg,Fe)_4Al(Al_2Si_6)O_{22}(OH)_2$	Garnet (Gar) $(Fe,Mg,Ca,Mn)_3Al_2Si_3O_{12}$	Tremolite (Tre) $Ca_2Mg_5Si_8O_{22}(OH)_2$
Clinopyroxene (Cpx) $Ca(Mg,Fe,Al)(Al,Si)_2O_6$	Chloritoid (Ctd) (Fe,Mg) $Al_2SiO_5(OH)_2$	Diopside (Dio) $CaMgSi_2O_6$
Orthopyroxene (Opx) $(Mg,Fe,Al)(Al,Si)_2O_6$	Staurolite (Sta) $(Fe,Mg)_2Al_9Si_4O_{23}(OH)$	Forsterite (For) Mg_2SiO_4
Garnet (Gar) (Fe,Mg, $Ca,Mn)_3Al_2Si_3O_{12}$	Kyanite (Kya) Al_2SiO_5	Quartz (Qtz) SiO_2
Calcite (Cal) $CaCO_3$	Sillimanite (Sil) Al_2SiO_5	
Dolomite (Dol) $CaMg(CO_3)_2$	Andalusite (And) Al_2SiO_5	
Chloritoid (Ctd) (Fe,Mg) $Al_2SiO_5(OH)_2$	Cordierite (Crd) $(Mg,Fe)_2Al_3(AlSi_5)O_{18}$	
Chlorite (Chl) $(Mg,Fe)_5Al$ $(AlSi_3)O_{10}(OH)_8$	Potassium feldspar (Ksp) $KAlSi_3O_8$	
Staurolite (Sta) $(Fe,Mg)_2$ $Al_9Si_4O_{23}(OH)$		
Sillimanite (Sil) Al_2SiO_5		
Kyanite (Kya) Al_2SiO_5		
Major chemical components (apart from SiO_2, H_2O and CO_2):		
CaO, Na_2O, Al_2O_3, (FeO+MgO)	K_2O, Al_2O_3, (FeO+MgO)	CaO, MgO

12.2 continued

istence of amphibole (Act to Hbl) with plagioclase at the expense of epidote + chlorite, which finally disappear in the amphibolite facies (d). The transition from amphibolite (d) to granulite (e) is marked by the breakdown of the remaining hydrous mineral hornblende to leave only anhydrous minerals (Cpx, Opx, or Gar with Plg).

In metacarbonates (column III) the reaction sequence and mineral assemblage at different grades depends strongly upon the H_2O–CO_2 composition of a coexisting metamorphic fluid. Sometimes the assemblage (e) can be observed in the amphibolite facies.

Crystalline solution of, say, $CaMgSi_2O_6$ (diopside) in pyroxene will enlarge the pyroxene (plus quartz) stability field by displacing the reaction boundary to lower pressures and higher temperatures. Conversely, crystalline solution of $CaAl_2Si_2O_8$ (anorthite) in feldspar will displace the reaction boundary to higher pressures and lower temperatures. Thus, generally speaking, the mineral containing the greater amount of crystalline solution has its *P–T* stability field enlarged.

The *P–T* conditions for dehydration reactions are also displaced if the metamorphic fluid contains components other than H_2O. For example, CO_2 mixes extensively with H_2O but does not enter the structure of most silicate minerals. Because increasing temperature generally favours dehydration (Figure 12.2), the equilibrium *P–T* conditions of reactions such as

$$\text{muscovite} + \text{quartz} \rightarrow \text{K-feldspar} + Al_2SiO_5 + H_2O$$

will be displaced to lower temperatures and higher pressures in the presence of a mixed $H_2O + CO_2$ fluid phase, than with H_2O alone.

Recent developments in metamorphic petrology permit evaluation of *P–T* conditions of metamorphism independently of the possible role of metamorphic fluid composition in catalysing mineral reactions and enhancing rock deformation or melting.

Metamorphic facies series and the petrologic distinction of tectonic regimes

In 1960, the Japanese geologist A. Miyashiro noted that distinct successions of metamorphic facies were mapable in certain terrains. He referred to these as **metamorphic facies series**, linking together facies in the sense of Eskola, and identified low, medium and high-pressure types. Subsequent workers also identified facies series in the intermediate pressure ranges. Miyashiro's facies series have also been referred to, respectively, as **andalusite–sillimanite** (low *P*), **kyanite–sillimanite** (medium *P*) and **glaucophane–jadeite** or **blueschist–eclogite** (high *P*) types on the basis of commonly observable minerals diagnostic of different pressures (see Figures 12.2 and 12.5). Jadeite and glaucophane are, respectively, Na-bearing pyroxene and amphibole. As can be seen in Figure 12.2, jadeite ($NaAlSi_2O_6$) with quartz (SiO_2) occurs instead of albite ($NaAlSi_3O_8$) at high pressure (> 12 kbar at 400 °C). Glaucophane ($Na_2Mg_3Al_2Si_8O_{22}(OH)_2$) has likewise been shown by experiments to have a high-pressure stability – glaucophanitic amphibole is one of the diagnostic minerals of the 'blueschist facies', on account of its colour.

Some workers have considered regional metamorphism as regionally distributed contact metamorphism, induced by mantle magma at various depths of intrusion. Thus it is important to consider whether the different pressure ranges implied by the metamorphic facies series of Miyashiro could actually depict metamorphism as the introduction of a single type of heat source at different crustal depths.

However, detailed studies of contact aureoles have revealed incomplete mineral transformation and the common occurrence of metastable mineral assemblages – reflecting the geologically rapid cooling of magmatic intrusions. Most regional metamorphic sequences usually preserve mineral assemblages in chemical equilibrium – consistent with thermodynamic prediction and reflecting thermal equilibration with geologically long-lived heat sources. Actually, implicit in Miyashiro's work and that of subsequent workers is that the different facies series constitute really quite different metamorphic environments. It is instructive to attempt to decipher what kinds of temperature evolution would be expected at different lithosphere depths following various kinds of tectonic processes. But first we need to establish the normal types of temperature–depth distribution expected in the crust – i.e. geotherms.

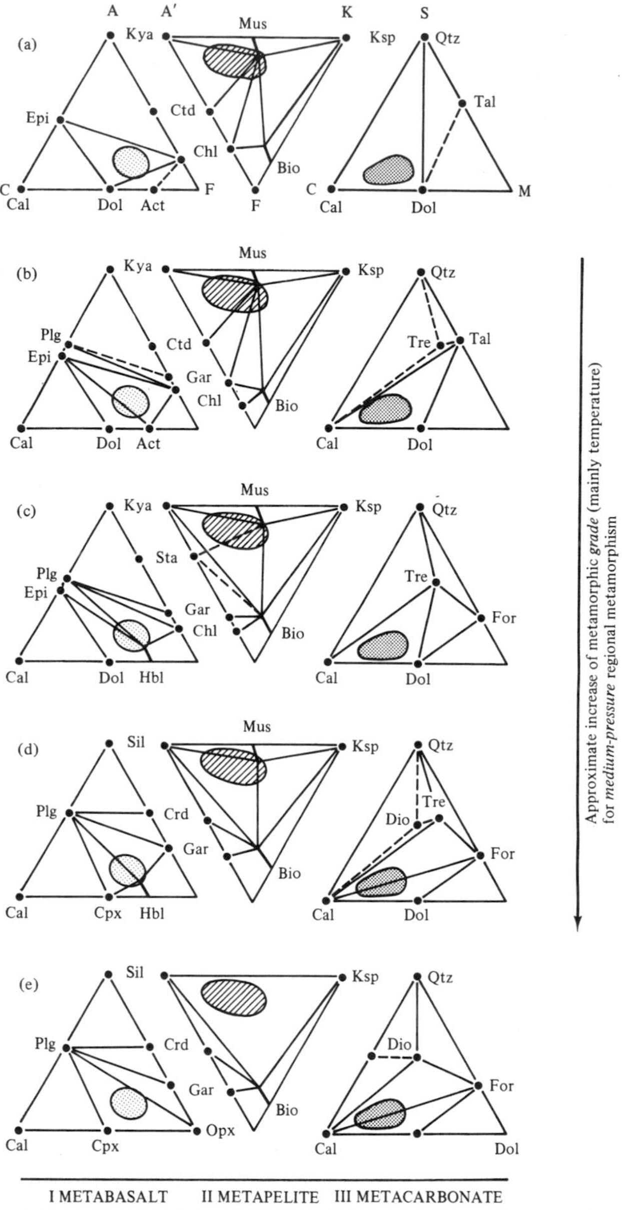

Figure 12.4 Composition diagrams for commonly observed mineral assemblages with increasing grade for medium pressure (kyanite–sillimanite type) regional metamorphism. The three columns represent simplified prograde changes for (i) metabasalt, (ii) metapelite and (iii) metacarbonate. In each diagram the mineral chemistry has been simplified into the following groupings of mol. % oxides: (i) ACF = (Al_2O_3–Na_2O), CaO, (FeO+MgO), (ii) A'KF = (Al_2O_3–(Na_2O+K_2O), (K_2O + Na_2O), (FeO + MgO), (iii) CMS = CaO, (MgO + FeO), SiO_2. Because quartz is frequently present among the minerals in metabasalt (i) and metapelite (ii), SiO_2 is regarded as an excess component. The A component groupings allow for Na_2O and K_2O in excess of that in feldspar (e.g. albite = Na_2O+Al_2O_3+6SiO_2). SiO_2 is included as an *actual* component in the CMS diagrams for (iii), metacarbonates, to show how the observed mineral assemblages in these rocks depend upon initial quartz content. In all cases H_2O or CO_2 are assumed to be in excess. The tie-lines in each diagram connect minerals in common assemblages. In all diagrams, the probable initial rock composition (shaded regions) strongly determines which metamorphic assemblage will be observed, in addition to the effect of increasing (a–e) metamorphic grade. Unique compositions are indicated by a dot, whereas the effect of solid solution on a particular mineral is indicated by a short, thicker line. □

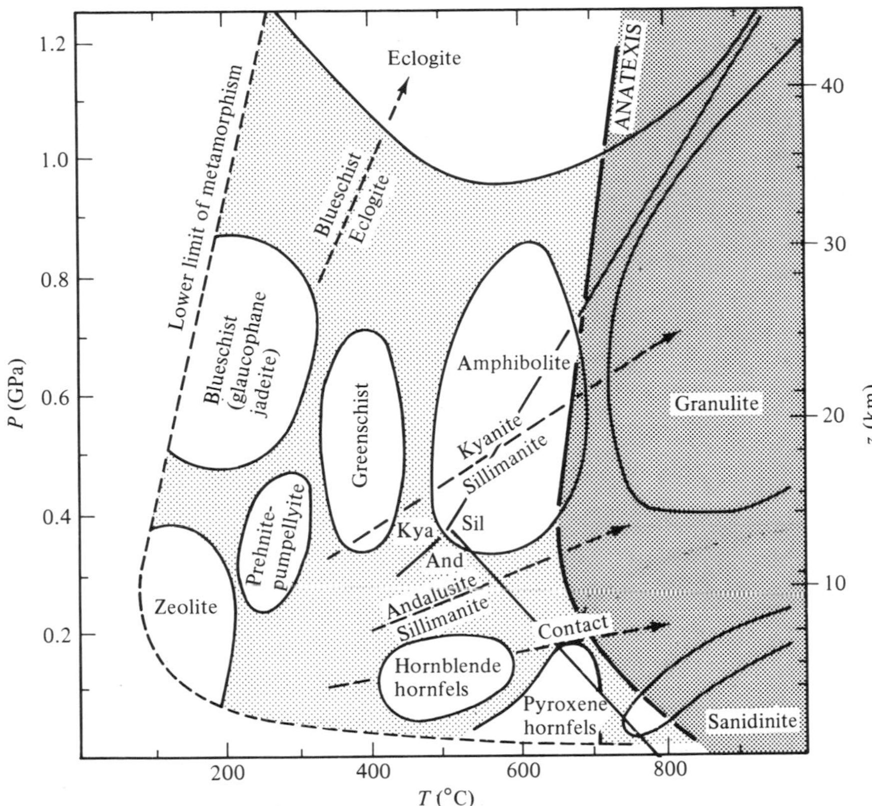

Figure 12.5 The approximate locations of the metamorphic mineral facies. The facies names are assigned for the assemblages observed in metamorphosed basaltic compositions at the various metamorphic grades and pressures (Table 12.2). Note that all transitions between the mineral facies are gradations, due to continuous ('smeared out') reactions involving crystalline solutions or impure fluids. The approximate *P–T* ranges corresponding to the principal metamorphic facies series are superimposed (see page 227). □

Metamorphic temperature gradients and geotherms

By comparing Figures 12.5 and 12.6 we see that the principal metamorphic facies series are approximately coincident with the simplified **linear geotherms** in the *P–T* diagram. The most commonly observed regional metamorphic sequence is the kyanite–sillimanite facies series (not only in the Scottish Dalradian, but worldwide). This facies series appears to correspond to an approximate linear geotherm of about 30 °C km^{-1}. It can be seen immediately from Figures 12.2 and 12.6 that such a linear geotherm would result in temperatures in excess of 900 °C at the base of continental crust of average thickness (about 35 km). Such temperatures would result in widespread crustal melting (anatexis, Figure 12.3). Seismic observations and thermal arguments (see Chapters 5, 13 and 14) indicate that the crust is not normally partially molten.

The rarer lower pressure andalusite–sillimanite metamorphic facies series would require linear gradients closer to 50 °C km^{-1} and result in temperatures greater than 900 °C shallower than 20 km. Conversely, the higher pressure blueschist (glaucophane) – eclogite (jadeite) facies series would correspond to a linear geotherm of about 10 °C km^{-1}.

Even though the three major facies series appear to correspond to distinctly different linear temperature gradients through the crust (Figures 12.5 and 12.6), it is certainly a reasonable question to ask what temperatures should be expected, on geophysical arguments, at the base of a continental crust of present average thickness (around 35 km to Moho depth). Once a normal 'steady state' geotherm can be established, then we need to consider how such geotherms can be perturbed by the variety of tectonic processes that can affect the lithosphere and lead to the different observed metamorphic facies series.

As presented in Box 12.3, expected temperatures in the continental crust should follow a mildly quadratic, rather than linear, distribution with depth. This reflects radiogenic heat produced in the crust. Obviously, in ancient (Archaean) continental terrains the contribution of radiogenic heat to measured surface heat flow will be much less than for more recent terrains, because of radioactive decay.

Surface heat flow measurements in ancient continental shields have been used to construct the **continental SHIELD geotherm** shown in Figures 12.2 and 12.6. This average geotherm is close to the 10 °C km^{-1} linear thermal gradient considered above to be characteristic of the glaucophane (blueschist) – jadeite (eclogite)

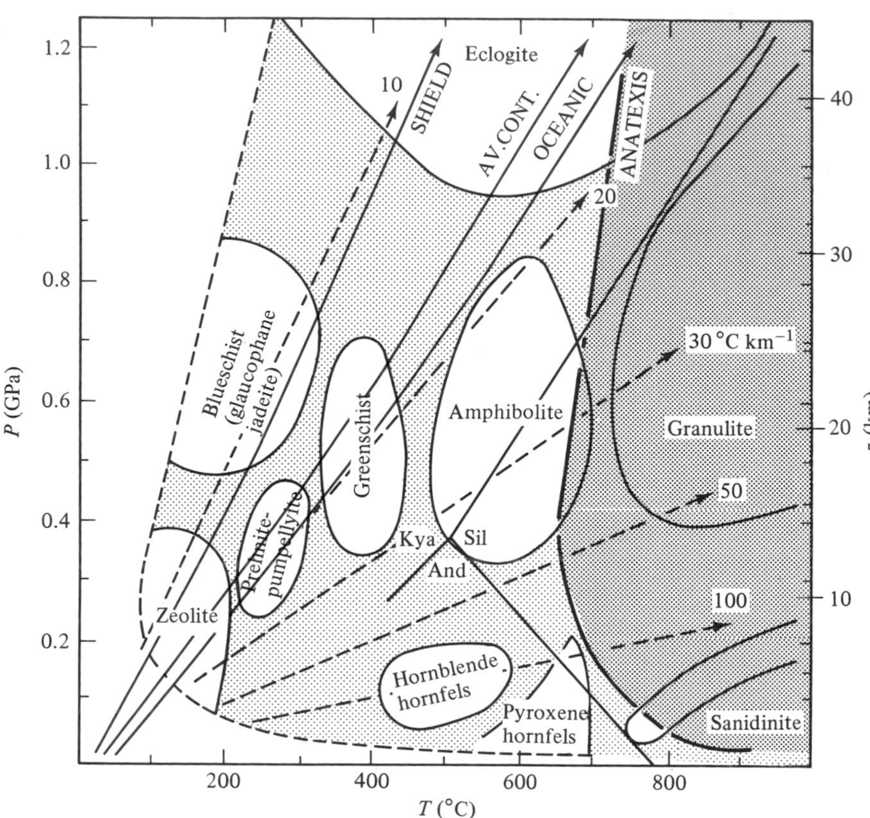

Figure 12.6 The *P–T* conditions of the metamorphic mineral facies in relation to steady-state geotherms and linear gradients of metamorphism (Box 12.3). Linear geothermal gradients appropriate to the various metamorphic grades and pressure ranges are shown by dashed arrows radial to a surface temperature of 20 °C. (AV.CONT. is the average geotherm for continential crust – see Box 12.3.) □

facies series. The shield geotherm is clearly much lower than the 30 to 50 °C km^{-1} gradients apparently necessary to achieve regional metamorphism of the kyanite–sillimanite or andalusite–sillimanite types. We can thus see that most metamorphic rocks are recording in their mineral assemblages, rather unusual temperature conditions in the continental crust.

Continental geotherms calculated from heat flow measurements in tectonically young terrains (250 Ma or younger) can be as high as 20 °C km^{-1} (**AVerage CONTinental geotherm** in Figure 12.6). The highest surface heat flow is found above presently active terrains, often reflecting the operation of extensional processes, and/or the presence of cooling magma bodies close to the surface; moreover, in some cases, greater amounts of radioactive decay occurs in young continental crust, also augmenting heat flow.

The average **OCEANIC geotherm**, relevant to the crystalline oceanic crust (Figure 12.6), corresponds to a linear geotherm of about 20 °C km^{-1}. This reflects the thinner oceanic lithosphere and lesser depth to the asthenosphere, the upper bound of which is nearly isothermal beneath the oceanic lithosphere. Amphibolites, sampled from the oceanic crust, indicate temperatures noticeably increased above the steady-state oceanic geotherm.

It is usually assumed that the commonest radioactive heat-producing isotopes of uranium, thorium and potassium are mostly concentrated in the upper 10 to 15 km of the crystalline continental crust. This means that the 'boost' to the continental geotherm will occur in this upper layer. Thus the underlying geotherm could be much steeper (smaller dT/dP). Before attempting to correlate grade and gradient of continental metamorphism primarily with areas of concentrated radiogenic heat, we should perhaps examine various ways in which tectonic processes and the introduction of heat sources can also influence regional metamorphism.

Heat, metamorphism and tectonic processes

Continental crust of about 35 km thickness is in isostatic equilibrium with the underlying mantle. According to the average continental geotherms (Figures 12.6 and 12.8) temperatures of around 400 °C should be typical at this depth under tectonically inactive areas, and around 550 °C under more recently active continental crust. The characteristic metamorphic rocks would be in the blueschist, high-pressure

BOX 12.3

Conduction temperature profiles for the crust – ingredients of steady-state geotherms

We could define the temperature distribution with depth in the crust (the geotherm) by **Newton's Law of Heat Conduction,** if indeed conduction was the mode of heat transfer, the crust was in a steady-state thermal condition, and no heat sources or heat sinks were present in the crust. Thus the simple equation

$$T_z = T_s + \frac{qz}{K}$$

would define the temperature at depth $z(T_z)$ with reference to the surface temperature (T_s), as functions of the surface heat flux (q, Wm^{-2}), and thermal conductivity (K, $Wm^{-1} K^{-1}$). The plotting of this linear geotherm in a temperature–pressure (or depth, Box 12.1) diagram; e.g. Figure 12.7, can be done by choosing values of the heat flux measured in boreholes near the surface (q_s), or the deduced heat flux entering the crust from the mantle (q_m). In the steady state, and if none of the mantle heat is consumed or no heat generated within the crust, then q_m should equal q_s. The linear geothermal gradients passing through the metamorphic facies diagram in Figure 12.6 are examples of this and can be reformulated to determine the heat flux needed to support the conduction thermal gradient.

Metamorphic processes within the crust obviously influence the geotherm. Dehydration and melting reactions are endothermic and use heat that would otherwise be available to raise crustal temperatures. Radioactive decay of uranium, thorium and the potassium-40 isotope contributes to crustal temperatures. On geological time scales, radiogenic heat production boosts the geotherm, the amount depending on the concentration of the radioactive isotopes of these elements and how they are distributed. The geotherm which includes radiogenic heat production is no longer a straight line and can be represented by

$$T = T_s + \frac{q_m}{K} z + \frac{\rho H}{2K} z^2.$$

where q_m is the mantle heat flux ($W m^{-2}$), H is the radiogenic heat production rate per unit mass of

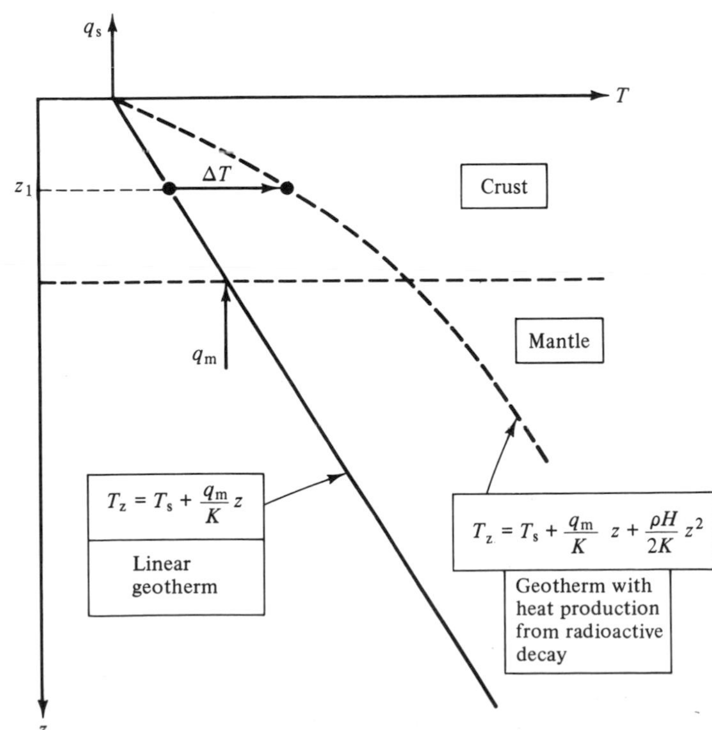

Figure 12.7 Schematic temperature–depth (*T–z*) diagram showing how a linear steady-state geotherm is modified by the presence of radioactive heat sources in an upper crustal layer. In the steady state the heat flux at the surface (q_s) should be the same as the heat flux from the mantle (q_m). □

12.3 continued

crust (W kg^{-1}) and ρ is the density (kg m^{-3}). The new term in the equation provides the quadratic shape to geotherms and shows how radiogenic heat modifies the linear temperature–depth distribution.

We can assess the magnitude of the effect of radioactive heat production on boosting geotherms for two extreme cases. In Table 12.5 we consider a crust made entirely from a quite radioactive rock – granite, or a very unradioactive rock – basalt, then assess the deviation from a linear geotherm (ΔT) at various depths (z_i = 10, 20, 35 km) by evaluating the quadratic term in the geotherm equation.

Clearly, the actual crustal radiogenic heat contribution to the geotherm will lie between the extremes. Conductive linear geotherms are appropriate if the crust is made entirely of basalt (e.g. under the ocean floor). Continental crust will generally have some radiogenic heat and thus non-linear geotherms.

Table 12.5

	Granite	Basalt
ρ (kg m^{-3})	2800	3200
H (W kg^{-1})	9.6×10^{-10}	6.2×10^{-12}
K (W m^{-1} K^{-1})	3.3	4
10-km thick layer	$\Delta T \approx$ 40 °C	$\approx$ 0.25 °C
20-km thick layer	160 °C	1 °C
35-km thick layer	500 °C	3 °C

Based upon surface heat-flow measurements and estimated radioactivity in an underlying crustal column, several workers have computed average geotherms for stable continental shields, for average continental crust, and under the ocean floor (labelled as SHIELD, AVerage CONTinental and OCEANIC geotherms in Figures 12.2 and 12.8). It can be seen that conventional linear geotherms approximate these calculated average geotherms reasonably well under steady-state conditions.

Complex transient geotherms result from tectonic processes such as thrusting and subduction. Likewise consideration of additional heat sources (cooling magmas, frictional heating, exothermic reactions) or heat sinks (endothermic reactions) will introduce further perturbations.

It will be noted by comparing Figures 12.7 and 12.8 that conventional portrayals of depth–temperature (z–T) and pressure–temperature (P–T) diagrams show opposite orientation of the depth and pressure axes.

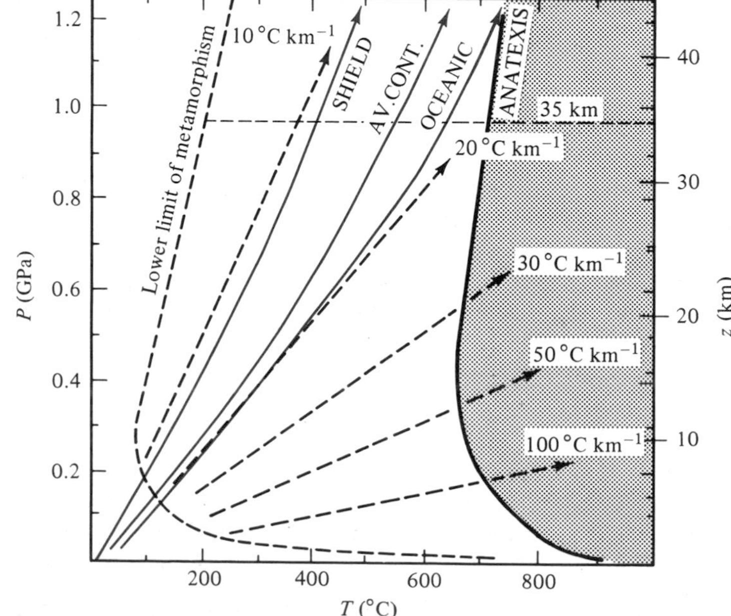

Figure 12.8 Average geothermal gradients in the steady state, long after any tectonic or magmatic perturbations have disappeared, are shown for ancient continental SHIELD terrains, for AVerage CONTinental crust and for sub-OCEANIC regions. Linear geothermal gradients approximate reasonably well these steady-state geotherms (see Figure 12.6) – but not those transient geotherms resulting from tectonic or magmatic perturbations. The average thickness of present continental crust (~ 35 km) and the field for rock melting (anatexis) are emphasised. □

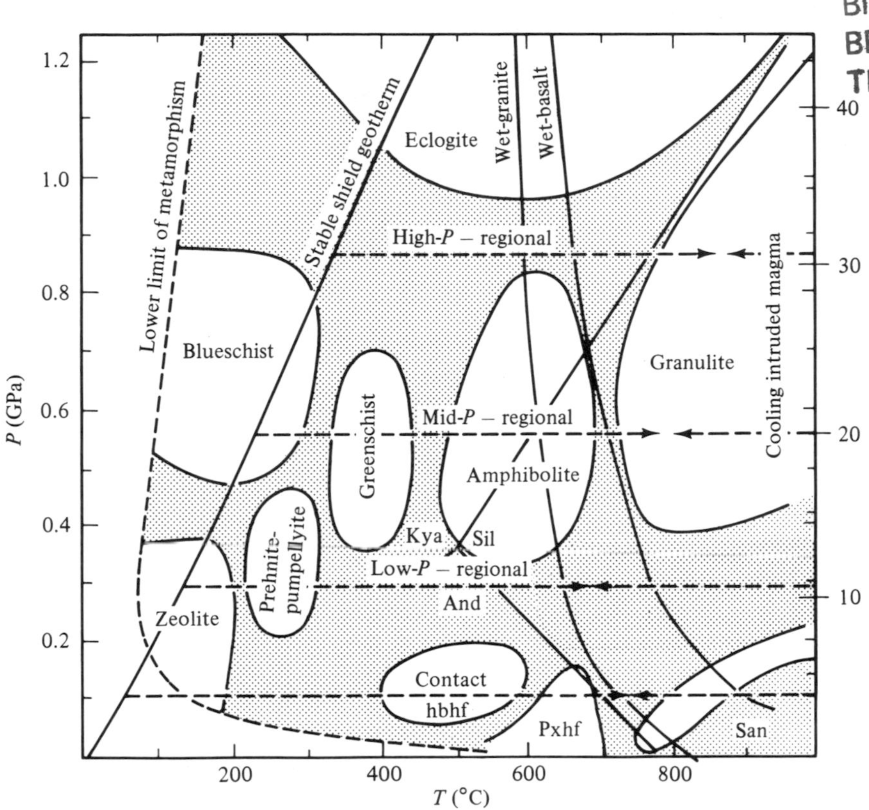

Figure 12.9 Regional metamorphism considered as being due to regional-scale contact metamorphism. Here cooling magmatic intrusives at various crustal depths, correspond to low, medium and high-pressure types of metamorphism. Note that crustal deformation would need to accommodate large volumes of intrusive magma to achieve the implied temperatures of metamorphism. Unless magmatic crustal thickening is extensive, there is no impetus for the regionally metamorphosed rock to return to the surface through erosion. □

greenschist or perhaps eclogite facies (Figures 12.6 and 12.9). However, many metamorphic rocks preserve much higher grade mineral assemblages, because they have not suffered retrograde metamorphism by fluid influx during later deformation.

Regional metamorphism is the response of buried rocks to thermal perturbations. These often result from tectonic processes that thicken the crust and affect the distribution of **radioactive heat sources**. Such metamorphism is often associated with continental collision and rarely shows the involvement of mantle magma as a heat source. However, during lithosphere extension, mantle magma may intrude at various crustal depths. Cooling magma, at different depths in the continental crust, could therefore be an important heat source for some types of regional metamorphism.

Regional metamorphism caused by cooling of basaltic intrusions

One simple model for metamorphism is shown in Figure 12.9. The upper crust is buried to varying depths by different (unspecified) processes and a heat source is introduced at this depth. This could be mantle magma (see Chapter 5). Such a model interprets regional metamorphism as being caused by regional-scale invasion of the crust by magma generated in the mantle.

From simple considerations of a thermal energy balance, it is possible to calculate how much magma would need to be intruded to raise the temperature of specific crustal volumes by certain amounts. Let us say that basalt is intruded with a temperature (T_b) of 1250 °C and during cooling through its solidus (say at 1150 °C) it releases an amount of latent heat (L_b). This heat, plus that released on cooling the solidified basalt to its final temperature (T_f), is available to heat crust (Figure 12.10). The amount of heat needed to change the temperature of the crust (ΔH_c) from its initial temperature (T_c) to the final temperature (T_f), when the crust has gained all the available heat from the freezing and cooling basalt is:

$$\Delta H_c = m_c C_c (T_f - T_c),$$

where m_c is the mass of crust, C_c is the heat capacity of the crust and $\Delta T = (T_f - T_c)$ is the temperature change achieved. This heat is gained from the basalt by freezing ($m_b L_b$) and cooling [$m_b C_b (T_b - T_f)$], together:

$$\Delta H_b = m_b L_b + m_b C_b (T_b - T_f) \text{ or } m_b [L_b + C_b (T_b - T_f)].$$

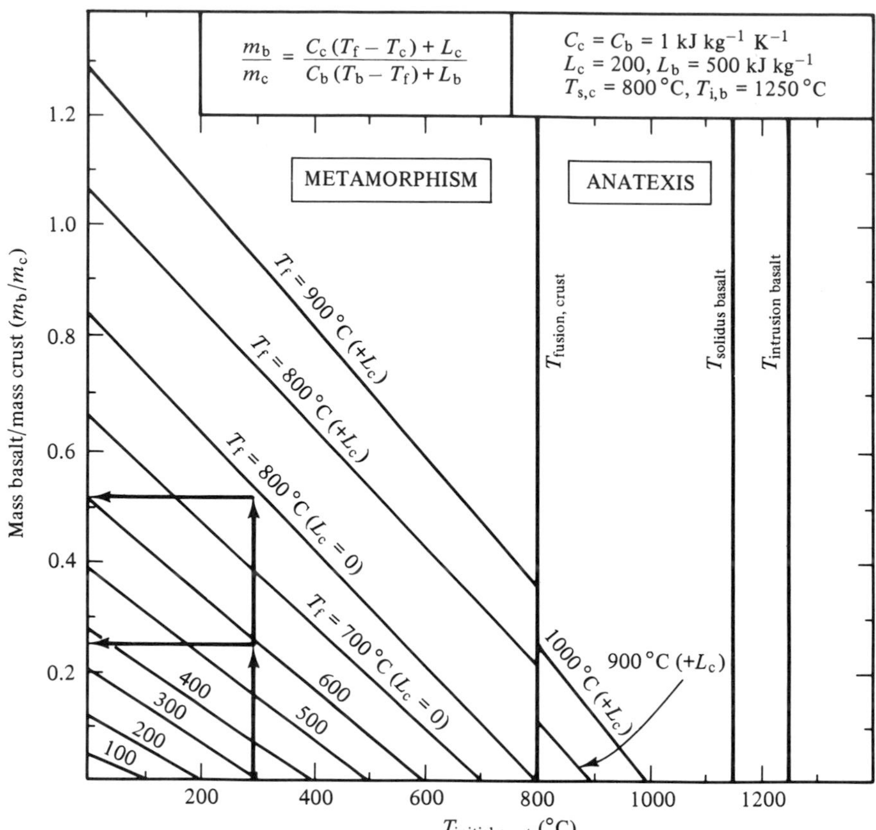

Figure 12.10 Heat balance for metamorphism caused by cooling and freezing of molten basaltic intrusions. The amount (mass) of basalt is ratioed to the mass of crust that becomes metamorphosed, assuming complete heat transfer. A ratio $m_b/m_c = 1.0$ on the ordinate axis means that the equivalent masses are equal. The abscissa shows the initial temperatures of the crust before intrusion. The linear contours show the final temperatures (T_f) reached by both bodies after exchange, with or without the inclusion of L_c (the latent heat of the crust, applicable during anatexis). The basalt intrusion temperature $T_{i,b}$ is assumed to be 1250 °C and the melting temperature of the crust ($T_{s,c}$) 800 °C. Latent heats for basalt $L_b = 500$ and for crust $L_c = 200$, kJ kg^{-1} were used. Heat capacities C_b, C_c were both assumed to be 1 kJ kg^{-1} K^{-1}. The linear relationships result from solving the mass–heat balance $m_bC_b(T_b{-}T_f){+}m_bL_b = m_cC_c(T_f{-}T_c) + m_cL_c$. □

Because, in our simplified model, all the heat is transferred completely from magma to metamorphism, it is easy to calculate the relative masses, or volumes, of crust and magma involved:

$$\frac{m_b}{m_c} = \frac{C_c(T_f - T_c)}{L_b + C_b(T_b - T_f)}.$$

For likely values of latent heat of basalt crystallisation and heat capacities of crustal rocks and basalt, the graph in Figure 12.10 shows the amount of basalt needed to heat crust to a final common temperature (T_f) as a function of the average initial crustal temperature (T_c). For example, to heat a volume (or mass) of crust from an initial temperature (T_c = 300 °C) to a final temperature (T_f = 600 °C) needs 25 per cent of basaltic magma; and for T_f = 800 °C needs 50 per cent of magma (see heavy arrows in Figure 12.10).

To induce crustal melting by basalt intrusion would require injection of even more mantle magma to overcome the latent heat of crustal fusion (L_c), hence the version of the equation given in Figure 12.10. The conduction cooling time for single magma bodies is quite short (a 10-km-thick sill will be solid in about 10^5 years). Obviously repeated intrusion by mantle magma will become a more efficient process for thermal metamorphism, as the cooling of each successive magma intrusion will prewarm the crust. Nevertheless, the quantities of mantle magma required to induce regional-scale contact (thermal) metamorphism are formidable and, in general, far in excess of the amounts of derivative amphibolite and mafic granulite observed in exhumed lower crustal terrains. Also, because average crust has a density of about 2800 kg m^{-3} and basalt about 3000 kg m^{-3}, mantle magma

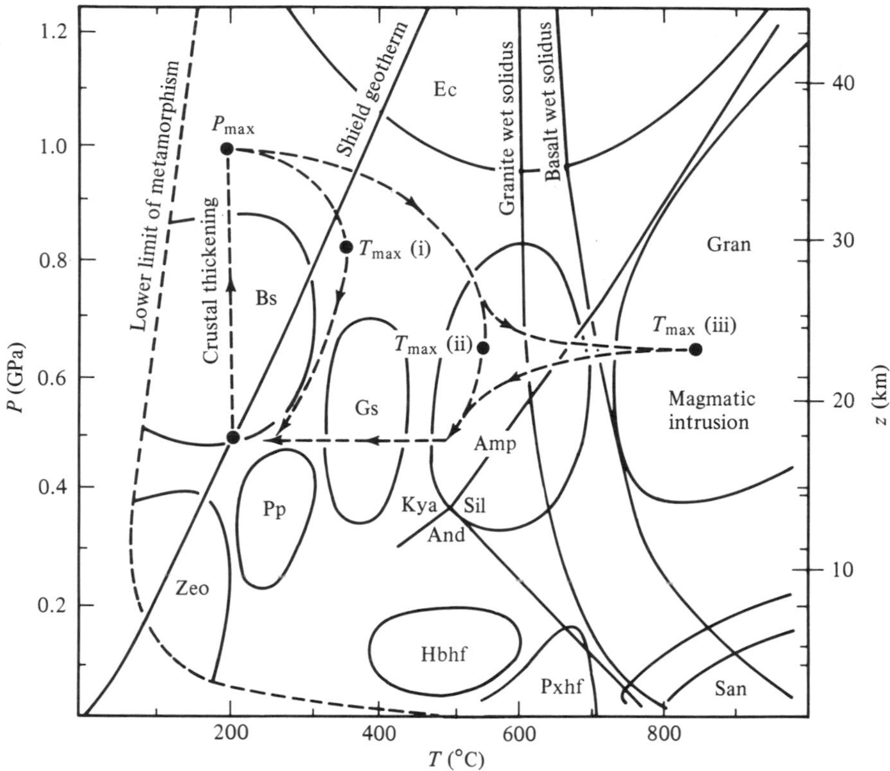

Figure 12.11 Regional-scale metamorphism viewed as a process of continental thickening followed by erosional controlled uplift. This will ultimately result in heating because of the average higher temperature at depth. Increasing crustal thickness provides the impetus for erosion of the thickened amount. The lower part of the crust will remain buried unless tectonic processes induce exhumation. Path (i) shows that rapid uplift may preserve assemblages of low metamorphic grade. Path (ii) shows that slow uplift may enable higher grades of metamorphism to be encountered along the metamorphic *P–T* path. Path (iii) illustrates the metamorphic effects of a magma intrusion at some time along path (ii).

If metamorphic mineral assemblages preserve *P–T* conditions close to T_{max}, then the depths of maximum burial (P_{max}) are not easily inferred. □

should pond at the Moho and only penetrate the crust when under excess pressure. Magmatic underplating beneath continental crust could nevertheless deliver heat enough to produce some lower crustal granulite facies rocks.

The above scenario may be a successful explanation for regional metamorphism in terrains undergoing active continental extension (see Chapters 5 and 14). Widespread intrusion of mantle magma can be accommodated by laterally extending lithosphere and high grades of metamorphism could be achieved regionally at quite low pressures (andalusite–sillimanite facies series). Here crustal anatexis could be extensive. The exhumation of such high-temperature–low-pressure metamorphic terrains, reflecting crustal thinning by extension, could occur by simple erosion only if the thinned crust were subsequently thickened by a significant amount. Otherwise, later (probably compressional) tectonic processes would be required to exhume them. We now consider thermal perturbations that might occur during crustal thickening following continental collision.

Regional metamorphism as a response to crustal thickening by tectonic processes

Another concept for regional metamorphism is sketched in Figure 12.11. Simply stated, when rocks are buried to crustal depths where the surroundings are hotter, they will be metamorphosed. Collision of two continents will cause crustal thickening through underthrusting or by homogeneous deformation of ductile lower crust. This causes the building of mountains, as well as the formation of a root that intrudes downward perturbing the mantle and temporarily deepening the Moho (see Chapter 14). Thickened continental crust is obviously no longer in isostatic equilibrium with the underlying mantle, and there is thus an impetus for erosional exhumation. The P–T paths followed by such rocks clearly depend upon their rate of uplift. Rapid uplift might follow a path such as (i) shown in Figure 12.11. Slower uplift would permit the buried rocks to remain in the deep crust for longer periods of time and experience further prograde metamorphism. This reflects long-term radioactive heating in rocks which are good thermal insulators. Such rocks will follow different P–T paths (such as (ii) in Figure 12.11) and will reach their maximum temperatures (T_{max}) much later and at shallower levels than their maximum burial depth (P_{max}).

If we assume that metamorphic rocks record, in mineral assemblages, the conditions of metamorphism close to their maximum temperature (T_{max}), then an earlier blueschist–eclogite facies mineralogy could be obliterated (hopefully only incompletely) by an amphibolite facies mineralogy (path (ii) in Figure 12.11). This amphibolite facies mineralogy would be retained by the rocks provided they escaped subsequent deformation and hydration, which otherwise could destroy higher grade metamorphic facies by retrograde metamorphism. Magmatic intrusion late in the uplift history could produce an additional episode of localised prograde metamorphism (path (iii) in Figure 12.11).

In any case, thickened continental crust can only eventually erode back to its pre-thickened depth, because of isostatic constraints. Thus the metamorphic rocks in the lower part of the crust will not reach the Earth's surface, unless exhumed by later tectonic events.

Metamorphism related to subduction zones

Subduction transports oceanic crust into the mantle. Continental crust is, in general, too buoyant to be subducted, but the transformation of abundant plagioclase feldspar to high-pressure denser minerals (pyroxene, garnet), obviously aids its subduction through decrease of the density contrast. The remarkable finding of coesite (a high-pressure polymorph of quartz, see above) in some rocks in the French Alps and Norway, means that some continental materials can be subducted to more than 100 km. More remarkable, perhaps, is the fact that these tectonic blocks, at least 10 km across, have been returned back to the surface from such depths.

In Figure 12.12, the path labelled (i) could represent burial and metamorphism within a subduction zone. Glaucophane–jadeite (or blueschist–eclogite) facies series metamorphism clearly indicates metamorphic conditions of high pressure and low temperature. Such rocks would need to be rapidly exhumed after their burial (path (i) in Figure 12.12) to preserve such P–T conditions. With time the material in the subduction zone will be heated by the surrounding mantle, and earlier high-pressure–low-temperature metamorphic assemblages will become partially or completely obliterated.

Even when subduction has ceased, the subducted material has little impetus to return back to the surface. The exhumation of regional-scale eclogite facies terrains of continental crust (such as the Sesia Lanzo Zone of the Italian Alps) appears to be entirely through tectonic processes. The later Alpine continental collision appears to have reactivated older fault systems or other weak zones in the lithosphere, resulting in the exhumation of extensive eclogite facies terrains in the Western European Alps.

Magma generated in the vicinity of the subduction zone often penetrates the overlying mantle wedge and crust, and erupts at volcanoes. If the magma stops rising, then metamorphism can be induced at depth. (The application of experimental data to the interpretation of high-pressure–low-temperature continental metamorphic rocks is also discussed in Chapter 4.)

Metamorphic facies series and tectonic processes

One important current aim of metamorphic petrology is to reconstruct tectonic processes in the distant past from observations on metamorphic rocks at the present surface. We have already seen that the tectonic processes that bury surface rocks are distinct from those that exhume their metamorphosed equivalents.

As outlined above, it does seem that the various regional metamorphic facies series can be usefully interpreted as being diagnostic of distinct previous tectonic regimes. The decipherment of many metamorphic histories is often complicated because most metamorphic rocks remain at depth within the crust and consequently suffer subsequent or **polymetamorphism**. This frequently destroys mineralogical evidence of the previous history. Sometimes tectonic exhumation is sufficiently rapid following the burial event, that tectonic reconstructions can be made using diagnostic mineral assemblages.

In ideal cases, the high-pressure facies series (glaucophane–jadeite, blueschist–eclogite) can be diagnostic of previous subduction zone metamorphism (Figures 12.5 and 12.12). The medium-pressure facies

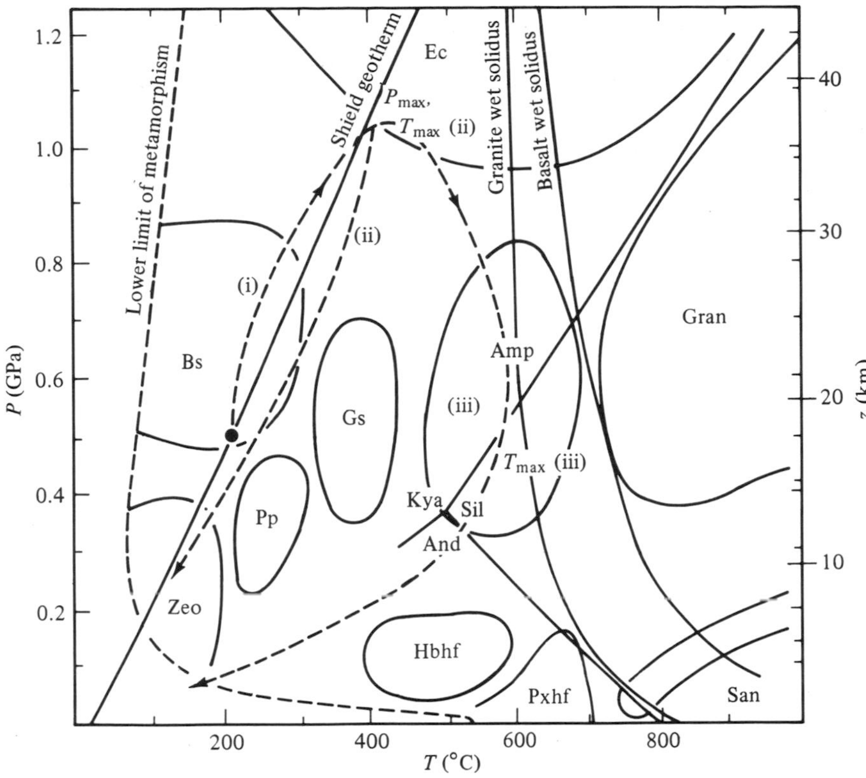

Figure 12.12 Metamorphism in and near a subduction zone. Burial (path i) followed by rapid uplift (path ii) will preserve high-pressure – low-temperature blueschist–eclogite facies. Slow uplift (path iii) will cause attainment of higher metamorphic grade and the partial, or complete, obliteration of the earlier metamorphic history. Extreme tectonic exhumation is required to excavate subducted material to the surface. □

series (kyanite–sillimanite) can indicate fossil zones of continental thickening by collision. The low-pressure regional metamorphic facies series (andalusite–sillimanite) may indeed by diagnostic of massive invasion of the crust by mantle magma. This *P–T* regime is more readily achieved if the crust fails during extension with the formation of a lithospheric rift system, and mantle magma is able to intrude, inducing regional-scale contact metamorphism.

Actually, the metamorphic *P–T* conditions recorded by rocks probably tells us more about when the process of mineral transformation was *interrupted*, in a continuous series of tectonic events, rather than indicating the greatest burial depth achieved (P_{max}) or the highest temperatures experienced (T_{max}).

Until recently it has been difficult to reconstruct regional tectonic histories from detailed observations made on metamorphic rocks from specific localities within a metamorphic terrain. New approaches in metamorphic petrology, using the chemistry of zoned minerals and their solid inclusions, permit the evaluation of parts of a metamorphic rocks' *P–T* history. Coupled with geochronology, the timing of distinct metamorphic events can now be calibrated with reasonable resolution. Recent developments in palaeomagnetism and seismology complement metamorphic petrology in the reconstruction of orogenic histories of juxtaposed lithospheric fragments (see Chapters 9 and 14). It is possible to distinguish between metamorphic terrains exhumed by tectonic processes within continental areas, from those accreted onto continental margins during seafloor spreading. Exactly how the large-scale tectonic forces and the small-scale rheological properties of lower crustal rocks interact to exhume pieces of the deeper lithosphere to make a metamorphic terrain is one of the most exciting current problems for metamorphic petrologists and structural geologists.

Fluids in the Earth's crust

Much geochemical and isotopic evidence indicates that the interior of the Earth is still degassing (see Chapter 8). Constituents such as H_2O, CO_2, CH_4, NH_4 etc. are probably stored in minerals in the mantle. Helium, being a noble gas, will not easily be bonded in a mineral structure.

Mantle melting to generate basalts, or higher temperature magmas such as picrites and komatiites, occurs relatively close to the surface, as the mantle solidus is crossed by ascending mantle rock (see Figures 4.5 and 4.9, pp. 75 and 80). If hydrous mantle minerals (such as amphibole, phlogopitic mica, humites, or unknown minerals) are also melted, the H_2O is transferred to the magma (see Chapter 4). If melting also involves mantle carbonates or elemental carbon (hopefully graphite, not diamond) then some CO_2 might be dissolved in the magma at high pressure; the remainder would form a coexisting fluid (vapour) phase.

If the mantle magma containing dissolved volatiles rises rapidly through fractures in the overlying lithosphere, boiling occurs close to the surface, resulting in explosive volcanism (see Chapter 5). Active geothermal systems on the ocean floor as well as on land, are often related to magma boiling. If the mantle magma crystallises at depth, the release and ascent of dissolved H_2O can induce crustal melting, or may form some ore deposits (see Chapter 11). Any dissolved CO_2 exsolved from mantle magma is hardly soluble in the 'granitic' liquids produced by fusion of continental crust, and remains in a fluid phase available for subsequent metamorphic processes. For the most part, volatile phases from the mantle are escaping via magmatic events and probably contribute only occasionally to the metamorphic fluid phase.

Geochemical evidence, mainly from stable isotope measurements on metamorphic minerals, indicates that many 'metamorphic' fluids ultimately have had a near-surface origin. That is, the fluids were incorporated in the minerals in the diagenetic regime, and were subsequently released by metamorphic reactions deeper in the crust.

In considering the role of fluids in metamorphic processes, we need to treat separately the shallower and deeper parts of the crust.

Fluids in the shallower parts of the Earth's crust

Pore waters are squeezed out during induration and diagenesis of fresh sediments. These are sampled as 'formation water' during drilling of bore-holes. Consolidation and compaction drives out pore water as a clay-rich sediment becomes a shale. Brittle deformation introduces a cleavage during slate formation. Recrystallisation of an existing clay mineralogy reduces pore space until eventually the porosity becomes discontinuous. The motion of near-surface fluids can be through flow in more porous layers. If the permeability of a porous layer is high enough, then pervasive flow occurs.

The flow network also depends upon whether fractures are available, or can be made by the fluids cracking the rock (**hydrofracturing**). Much of the near-surface fluid-flow of geological consequence (i.e. resulting in concentration of elements into exploitable mineral deposits in veins), occurs in fracture systems, reflecting brittle behaviour of the wall-rocks (see Chapter 11). The flow can be convective, driven by temperature–density gradients in the fluid (e.g. in geothermal systems), as well as by pressure gradients.

Thus, depending upon rock type and rock strength, fluid migration in the outer crust will be mainly through flow along cracks. These are frequently open to the surface in geothermal fields. Boiling of natural fluids in the near surface expels dissolved gases and is accompanied by focussed deposition of ore minerals. Fluid flow through the higher permeability layers of a porous medium is particularly important for water supplies and for hydrocarbon migration.

As has been observed in many well-holes drilled through newly consolidated sediment, the pore-fluid pressure (P_{fluid}) changes from being '**hydrostatic**' (P_{fluid} ~ 1/3 P_{rock}, because the density of water is about 1/3 of that of the rock), to being near **lithostatic** (P_{fluid} ~ P_{rock}). This change in fluid pressure occurs at 4 to 6 km in recently consolidated sediments (Figure 12.13). In mature crystalline lithosphere the transition zone lies deeper (5–12 km), depending upon crustal strength (Figure 12.13) – as witnessed by the influx of water at depths of at least 11 km in the deep-continental drilling through the ancient rocks of the Kola peninsula, USSR.

Metamorphic fluids and their migration in the deeper crust

In deeper crustal regions, volatile species remain structurally bound in minerals until released by metamorphic reactions. Sometimes metamorphic fluids are trapped as **fluid inclusions** within minerals, or adsorbed at certain locations along the grain boundary network of an annealed metamorphic rock. The high temperatures of metamorphism, as in the manufacture of ceramics, tend to anneal grain boundaries and diminish pore space. Ductile deformation in the deeper crust does not permit open (air-filled) pores. They are completely filled by metamorphic fluids or gases, or do not exist at all.

Generally, metamorphic fluid inclusion compositions fall into two distinct chemical types: (i) brines, often with daughter crystals of alkali-halide salts (mostly NaCl, KCl); and (ii) H_2O–CO_2 mixtures with high calcium concentrations. Sometimes other gaseous species, such as H_2, H_2S, CH_4 and NH_4, are present. Metamorphic reactions involving silicates, carbonates or sulphide minerals can provide most of these volatile elements, with the obvious exception of the high quantities of chlorine. Here we must appeal to buried evaporites (a former sea-water source), primary magmatic fluid, or a concentration mechanism for metamorphic brines.

Often it can be demonstrated that fluid inclusions have had their compositions modified, or that they have grown late in the metamorphic history. Thus the chemistry of metamorphic fluids is, in general, hard to characterise.

Sometimes rising metamorphic fluids can modify the chemistry of rocks through which they pass (metasomatism). This **fluid–rock interaction** consequently also modifies the fluid composition by chemical exchange. Metamorphic fluids need to be out of chemical equilibrium with the adjacent rocks, be in sufficient supply, and be in contact long enough for the diffusional processes responsible for metasomatism to be efficient. The best examples of metasomatism are in the aureoles of some magmatic intrusions, where ore minerals frequently constitute a large proportion of the **skarn** mineralogy. Chemical exchange can also be seen adjacent to some mineralised veins and along some faults and shear zones. Vein formation, although commonly able to be demonstrated to be a feature of the shallow crust, even occurs at the high pressures of the eclogite facies. In the high-temperature metamorphic facies (amphibolite–granulite, Figures 12.5 and

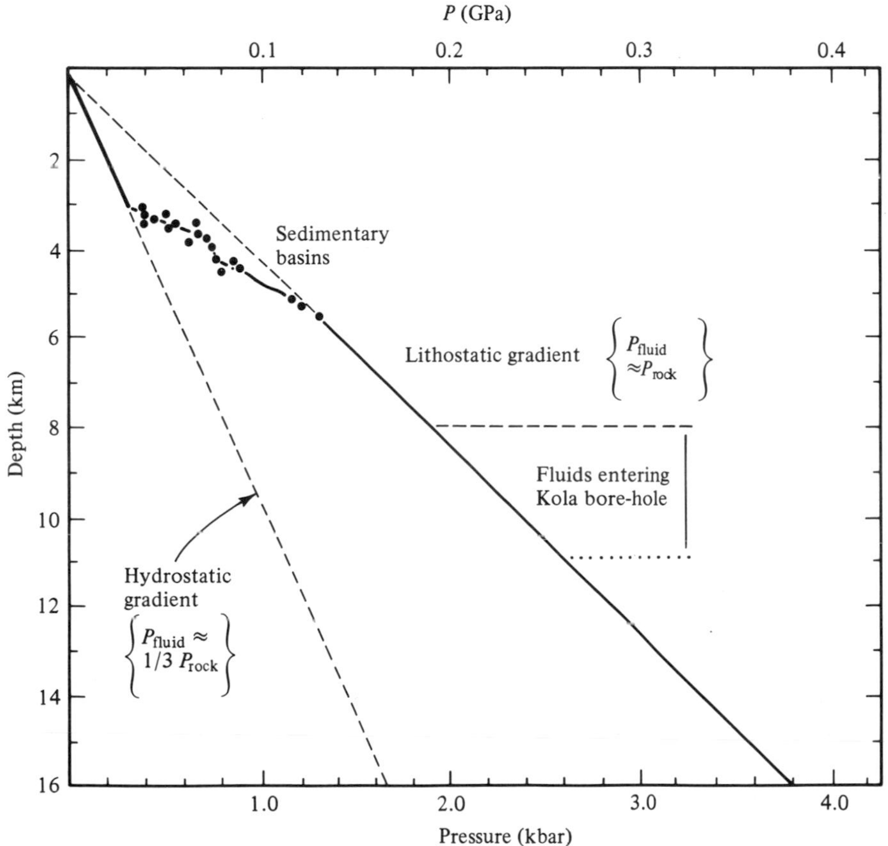

Figure 12.13 Pressure–depth diagram showing hydrostatic and lithostatic gradients for the Earth's crust. The hydrostatic gradient refers to the outer (more brittle) part of the crust where fluids can exist in cracks, or a connected porosity. Because the density of natural fluids ρ_{fluid} ($\approx$ 1000 kg m^{-3}) is much less than that of rock ρ_{rock} ($\approx$ 2800 kg m^{-3}), the fluid pressure is approximately one-third of rock pressure ($P_{fluid} \approx 1/3\ P_{rock}$). The lithostatic gradient pertains in the lower part of the crust, where disconnected pockets of fluid are compressed to pressures equalling the rock pressure ($P_{fluid} \approx P_{rock}$).

The depth and width of the transition zone depends upon the type of material and its rheological properties. Pressures measured in bore-holes, especially in recently compacted sediment, reveal the transition zone to be much shallower than in zones where the lithosphere is old and cold (e.g. Kola peninsula).

The term 'hydrostatic' here refers to a column of H_2O. In the geological literature the term 'non-hydrostatic' is often intended to distinguish systems where the three resolved orthogonal principal stresses are *not* equal. In that sense 'hydrostatic' would mean equal orthogonal stresses, and should be distinguished from the present usage. □

12.6), hydrous rock-melts often appear as injected veins (migmatites).

The mode of metamorphic fluid motion in the middle and lower crust is very much governed by rock rheology. Quartzo-feldspathic rocks are quite weak at lower crustal temperatures, even along the relatively low-temperature shield geotherm in Figures 12.2 and 12.6. Weak rocks can deform relatively easily by ductile deformation, even at slow geological strain rates. Ductile deformation will reduce porosity and pressurise pore fluid to near lithostatic pressure (see Figure 12.13).

Even in a 'tight' metamorphic rock there is a minute flow network, involving tubules connecting three mineral contacts at corners. An analogy can be drawn here with migration of melt (see Figures 4.6, 5.7 and 8.9, pp. 76, 98 and 157) or of hydrocarbons (see Chapter 18). There will always be some fluid flow, depending upon rock permeability. D'Arcy's Law ($q = [K/\mu]\ dP/dz$) defines the vertical fluid flow (q, kg $m^{-3}\,s^{-1}$; or kg $m^{-1}\,s^{-1}$, crossing an area m^2) due to the pressure gradient (dP/dz); K is rock permeability and μ is fluid viscosity. It is to be expected that in such 'tight' deep crustal metamorphic rocks, the fluid flow will be upwards. This occurs because the density of metamorphic fluid is always much less than that of rock. This strong vertical pressure gradient ($dP/dz = \Delta\rho g$) means that fluid downflow will not normally occur and thus the lower crust will not contain significant fluid convection systems.

Because of the strong dependence of rock rheology on mineralogy, some rock types, e.g. amphibolite, may behave in a brittle fashion. This can occur when adjacent quartzo-feldspathic or carbonate rocks exhibit ductile behaviour and take up the strain. Buildup of metamorphic fluid pressure often induces hydrofracture in brittle rocks. In the lower crust, short-term brittle behaviour may be widespread and be manifest in syntectonic veins. This can happen even when on longer time scales the rocks could deform in a ductile fashion. Hydrofracturing may thus cause transient high-permeability channelways for metamorphic fluid, in a similar way that magma fracturing rapidly propagates and channels hot mantle magma to the surface (see Chapter 5).

If rising metamorphic fluid encounters an inclined layer of higher permeability, then the lower crust will be more rapidly drained. Certain horizons, for example shear zones in the lower crust, could be regions of enhanced flow of metamorphic fluid. Rapid drainage effectively removes all available fluid, until devolatilisation reactions can release more. Thus metamorphic fluid production is essentially controlled by the heat supply that induces the metamorphism. Metamorphic fluid migration is mainly controlled by the ways in which temperatures influence the rheology of the lithosphere.

Metamorphism and rock melting

On average, both the Earth's crust and mantle should be regarded as metamorphic rocks – in the sense that the relevant geotherms normally lie beneath those of the appropriate **solidus curves** at depth. These solidii, which define the beginning of melting, are quite different for crustal and mantle materials, because the average mineralogies of the appropriate rock types are quite distinct. As discussed in Chapters 3 and 4, mantle melting is dominated by the mafic minerals olivine, pyroxene, spinel and garnet, whereas crustal melting is dominated by the felsic minerals – feldspar and quartz. In both cases, the solidus temperature at a particular depth is very much influenced by the presence of water and other volatiles. To assess how important crustal melting has been in the evolution of the lithosphere, we are primarily interested in knowing the solidus temperatures of lower crustal rocks at depth, the amount of melt that can be formed and the compositions of the melts produced.

Water, temperature and crustal melting

If the crust and mantle were truly *dry* (i.e. no free water along grain boundaries and no water structurally bound in minerals such as mica and amphibole), then melting of both crust and mantle would occur at very high temperatures. The curves labelled GDS (Granite Dry Solidus) and BDS (Basalt Dry Solidus) in Figure 12.14a are located by experiments, as described in Chapter 4. For a reference pressure P = 1.0 GPa (~38 km) they lie at about 1080 °C and 1140 °C respectively. However, if enough free water is available at this reference depth, then the rock-melting temperature for granite is lowered from 1080 °C to about 630 °C (Granite Wet Solidus, GWS), and for basalt from 1140 °C also to about 630 °C (Basalt Wet Solidus, BWS), as shown in Figure 12.14a.

The most abundant high-grade metamorphic rocks at the surface are metasediments and quartzo-feldspathic types, as well as amphibolites and granulites of mafic composition. Seismic velocity data and the kinds of xenoliths brought up by mantle magmas suggest that these rock types are typical of the lower crust as

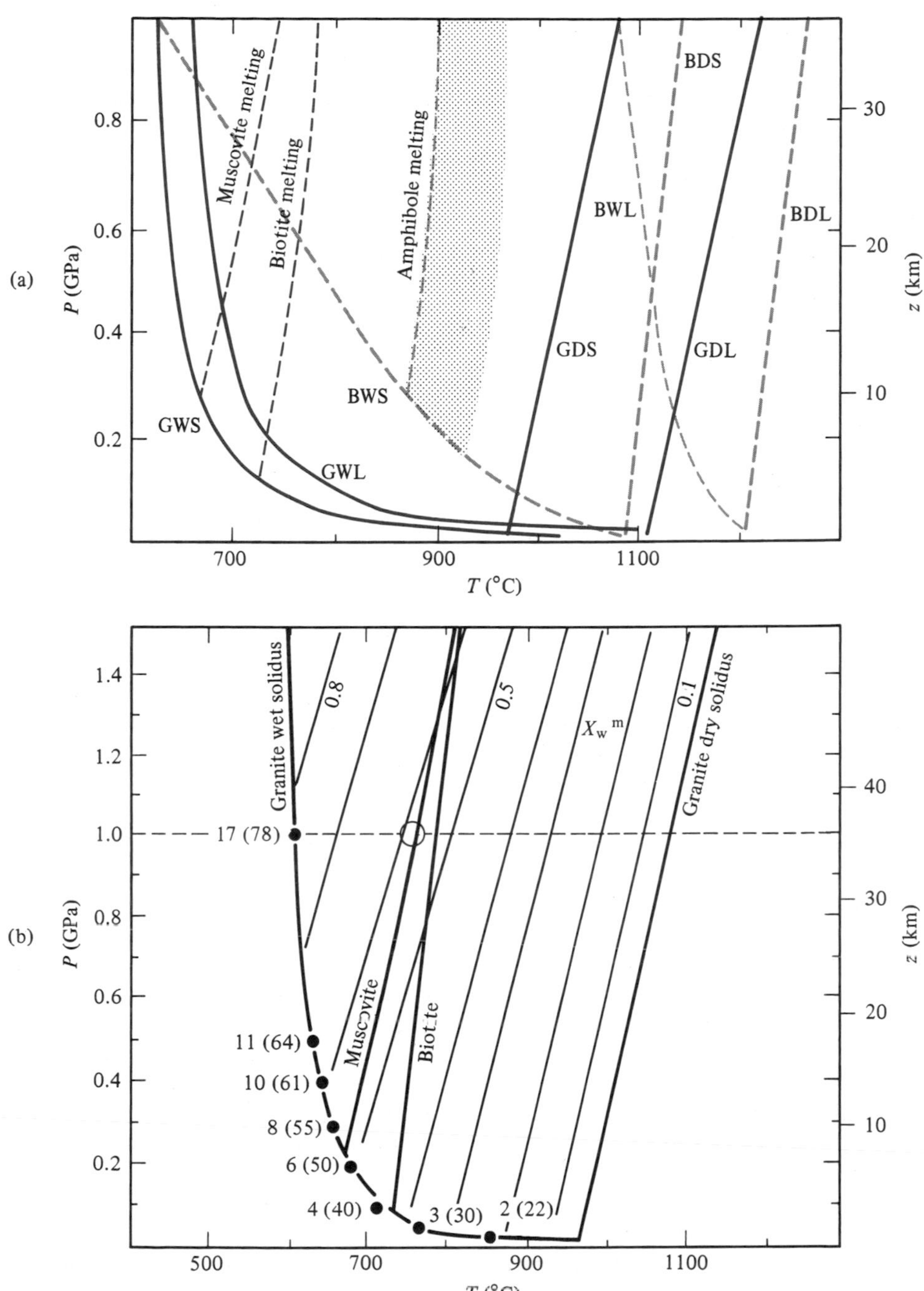

Figure 12.14 (a) The influence of water on the temperatures of melting of granite (G) and basalt (B). The solidus (S) marking the beginning of melting is considerably lowered in temperature for water-saturated (wet, W) conditions compared to dry (D). The width of the temperature interval between solidus (S) and liquidus (L) is a measure of the departure of rock composition from an eutectic. The beginning of dehydration-melting of muscovite and biotite in felsic rocks (e.g. granite) and of an amphibole in mafic rocks (e.g. basalt) are shown by the dotted lines emanating from GWS and BWS respectively (see also Figure 4.10, p. 82). (b) The solubility of water in granite melts. Along the wet solidus, the numbers given are the wt % (mol %) of water required to produce a 100 per cent water saturated granite melt at the given *P–T* conditions for each point on the GWS curve. These are *experimentally* determined values and, as can be seen from the figure, they correspond well with the thermodynamically calculated molecular fractions of H_2O (0.1 is 10% etc.), in a granite melt given by the sloping lines. □

well. As all of these abundant rock types contain quartz and various feldspars, the melting of the lower crust can be crudely approximated by the melting of granitic rocks. In fact, experiments have shown that the initial anatectic liquids produced from this whole range of rock types, including amphibolite, are 'pseudo-granitic' in composition.

The amount of temperature lowering of the granite dry solidus is directly proportional to the amount of available water near to a potential melting site. The quantity of melt formed depends on the width of the temperature interval between **solidus** (where melting begins) and **liquidus** (where the rock is wholly molten) for a given rock composition at a particular water content and pressure. Experiments have shown that with large excess of water the melting interval for granite is less than 50 °C (see Figure 12.14a). This occurs because the modal proportions of quartz and feldspar in granite are close to those of the **eutectic** (see Chapter 4). It is possible to examine the effects of temperature, pressure and quantity of available water on the amount of melt produced, with the help of experimental results on the 'pseudo-granite' system Alb($NaAlSi_3O_8$) – Ksp($KAlSi_3O_8$) – Qtz(SiO_2) – vapour (v, H_2O), shown in Figure 12.14b. The experimentally determined water contents of the eutectic liquids are shown adjacent to the solid circles along the **water-saturated** granite **solidus**. In an eutectic system this is the same as the liquidus for the special rock composition at the eutectic.

Taking a pressure of 1.0 GPa (~ 38 km) to represent the depth of average continental Moho, then from Figure 12.14b we can see that about 17 weight per cent (wt %) H_2O is needed to produce 100 per cent granite melt at 630 °C. If this amount of water were required to be available as metamorphic fluid just before melting, we can calculate the equivalent volume percentage of fluid-filled porosity in the metamorphic rock. With ρ_{fluid} = 1000 kg m^{-3} and ρ_{rock} = 2800 kg m^{-3}, 17 wt % water corresponds to about 36 volume per cent of water in the rocks at the melting site. Fluid-filled porosities of this magnitude are not expected in the lower crust. For a possible lower crustal porosity of 1 per cent of free water, the amount of granitic melt produced would be at most 3 per cent at 630 °C, 1.0 GPa.

As can be seen in Figure 12.14b, the degree of water solubility in granitic melts, and amount of lowering of the melting temperature compared to dry conditions, is a strong function of pressure (depth). Thus for the same quantity (1 vol. %) of free water, more granite liquid (~ 7 per cent at P = 0.2 GPa, where 6 wt % = 15 vol. % is needed to saturate the melt) would be generated at shallower depths by eutectic melting, because of the decreased water solubility at lower pressures. However, to attain the temperature of the solidus at low pressures, much greater perturbation of the geotherm is required, compared to deeper in the crust.

If only trivial amounts of free metamorphic fluid are normally present in the lower crust, what are other possible sources of water to induce crustal melting at accessible temperatures? An obvious possibility is from hydrous minerals in the metamorphic rocks themselves.

Hydrous minerals and crustal melting

The common hydrous silicates, muscovite, biotite and amphibole, are found not only in high-grade metamorphic rocks of the appropriate composition, but also in many magmatic rocks. Experiments have shown that the dehydration breakdown of these minerals overlaps the region of melting (see Figure 12.14). For example, muscovite undergoes **dehydration melting** near 750 °C at 1.0 GPa to produce a 'pseudo-granitic' liquid. Muscovite contains about 4 wt % water, so that a rock containing 25 per cent muscovite would release about 1 wt % water near 750 °C at 1.0 GPa. At these P–T conditions, 100 per cent granitic melt contains about 7 wt % water (about 57 mol. % water, X^m_w open circle at 750 °C, 1 GPa in Figure 12.14b) which is considerably less than the 17 wt % water needed to produce the 100 per cent granitic melt at 630 °C, 1.0 GPa. As the 1 wt % water is completely transferred from mica to melt near 750 °C, 1.0 GPa, the amount of melt produced is about (1/7) or 14 per cent. Because the lines defining the dissolved water contents in granitic melt are very steep in the P–T diagram (labelled X^m_w in Figure 12.14b), very similar melt quantities will be produced at intermediate crustal depths (0.4 to 0.6 GPa). Such small quantities of granitic melt are not likely to be able to migrate far from their source regions in view of the only slightly lower melt density compared to the residual rock and the relatively high viscosity of granitic liquids. Such granitic melts probably remain close to their source regions as migmatites.

The amount of melt produced by dehydration melting clearly depends upon the amount of hydrous minerals and the temperature of the melting reactions. Of course, higher temperatures result in greater amounts of melt production because of the lesser amount of water required to cause melting (shown by the lines labelled X^m_w in Figure 12.14b). But temperatures greater than about 850 °C are difficult to attain in the crust without addition of significant extra heat sources, for example, ascent of mantle-derived basalt magmas (see Figure 12.10). That such temperatures can be achieved in some environments is indicated by the occurrence of felsic magmas as volcanic rocks. We should perhaps expect some correlations between the geological setting, the types of magma and their ascent mechanism, as also discussed from another viewpoint in Chapter 5.

Crustal melting in various tectonic settings: granulites and granites

The P –T field for crustal anatexis (Figures 12.3 and 12.14) overlaps that for the **granulite facies** of metamorphism (Figure 12.6). However, if granitic intrusives are found in granulite facies terrains they are usually either much older or much younger than the particular metamorphic episode. In fact, in some cases, granulitic rocks are considered to be residues after granitic melts have been removed by partial melting (anatexis). Because crustal melting occurs to an extent governed by heat supply and the availability of water, environments with insufficient water will not melt, but produce directly granulite facies metamorphic assemblages. Flushing of the lower crust with carbon dioxide from the mantle could produce granulite facies assemblages, by causing the hydrous minerals to break down at lower temperatures but without partial melting.

The thermal evolution of the lower crust in continental collision zones will rarely achieve temperatures in excess of 850 °C (see Chapter 14). Fertile metasedimentary material in the lower crust will release water for dehydration melting, but the small amounts of melt produced will remain *in situ* as migmatites. Sometimes granitic bodies can be observed with chemical affinities to melted metasediments, but these have apparently not migrated far from their source regions. Amphibolites will probably not melt even in the lower parts of continental collision zones because the temperatures do not normally become high enough.

In zones of continental extension, massive invasion of the continental crust by mantle magma can lead to widespread crustal melting (see Chapter 5). The increased temperatures in these zones could cause amphibolites to partially melt producing tonalite–granodiorite magma, as well as extensive melting of metasediments to produce granites. Because high temperatures favour ductile rather than brittle behaviour for a wide range of rock types, the rheological behaviour of the crust exerts strong controls on magmatic and metamorphic evolution with time.

The voluminous felsic plutonics and volcanics in major mountain belts appear to be related to subduction zones and processes in the overlying lithospheric wedge. Mixing of mantle magmas with a whole range of crustal anatectic melts leads to great magmatic diversity in these regions. All of these features reflect the strongly coupled interactions of metamorphism, magmatism and deformation in the active tectonic regimes of the Earth's lithosphere.

Summary

Most of the Earth's crust and mantle consists of metamorphic rock. Fragments of the lower continental crust and mantle and parts of the oceanic crust have been exhumed on a regional scale to the continental surface through erosion and tectonic processes. These regions often constitute distinct linear metamorphic belts which can be correlated with large-scale orogenic processes. Such belts are diagnostic of dynamic lithosphere evolution throughout geological time (Archaean to Alpine).

Metamorphic mineral assemblages similarly are diagnostic of particular ranges of pressure (P) and temperature (T). Most metamorphic rocks show P–T conditions much hotter than steady-state geotherms at any depth. Metamorphism therefore reflects abnormal thermal conditions in the lithosphere. Sometimes contact metamorphism may be directly correlated with proximity to hot magmatic intrusions. Different types of regional metamorphism appear to be correlated with fossil continental collision zones (kyanite–sillimanite type), with subduction zones (blueschist–eclogite type) or with zones of continental extension (andalusite–sillimanite type). Metamorphic rocks therefore constitute vital marker horizons in the thermal and tectonic evolution of the lithosphere.

Fluids in metamorphic rocks are mostly derived from devolatilisation reactions deep in the crust occurring in response to enhanced heat supply. Usually the volatile components were originally derived from the surface, being structurally bound in minerals to be released later in the metamorphic history. Occasionally, mantle fluids contribute to the metamorphic fluids, when released through the boiling of mantle magma. Fluids are vital to many metamorphic mineral transformations which they catalyse. Fluid influx and rock deformation are often related. High-grade metamorphic rocks only retain their high P–T mineralogy when they have escaped later penetrative deformation and rehydration. Fluids expelled from metamorphic rocks are particularly important in the formation of ore deposits and in enhancing deformation, and perhaps exhumation by erosion, of lower crustal and mantle metamorphic rocks.

Further reading

Best, M. (1984) *Igneous and Metamorphic Petrology*, W.H. Freeman, 630 pp.
A comprehensive descriptive overview of rocks, minerals and processes.

Clark, I. F. & Cook, B. J. (1983) *Perspectives of the Earth*, Australian Academy of Science.
An excellent, well-balanced overview of Earth sciences.

Fyfe, W. S., Price, N. J. & Thompson, A. B. (1978) *Fluids in the Earth's Crust*. Volume 1 of Developments in Geochemistry, Amsterdam, Elsevier, 383 pp.
An integrated text on the role of fluids in metamorphic, magmatic and deformational processes.

Miyashiro, A. (1973) *Metamorphism and Metamorphic Belts*, Halstead Press, 492 pp.
Metamorphic minerals and rocks viewed in the context of their geographic location and relation to tectonic processes.

Wyllie, P. J. (1971) *The Dynamic Earth*, Wiley & Sons, New York, 416 pp.
A fascinating early integration of the various disciplines of Earth science into a comprehensive overview of the physical and chemical workings of our planet.

Yardley, B. W. D. (1989) *An Introduction to Metamorphic Petrology*, Longmans, 264 pp.
An up-to-date and very readable view of metamorphic rocks.

INTRODUCTION TO CHAPTERS 13–15

Following the acceptance of plate tectonics, new insights have been gained into the structure and evolution of plate boundaries and the intervening slabs of relatively stable lithosphere. This trilogy of chapters shows how geophysical observations and modelling have been integrated to provide a better understanding of lithospheric processes.

A critical contribution to our understanding of the nature of the lithosphere in the post-plate tectonic era has been the identification of horizontal and gently sloping boundaries within it through modern techniques of reflection seismology. High-resolution images can now be produced showing structure right through the crust and well into the sub-continental lithospheric mantle. Deep seismic data are extremely expensive to acquire, but have been obtained solely for academic purposes and through joint projects with oil companies who depend on similar data for hydrocarbon exploration. Most data have been obtained in North America and western Europe. Simon Klemperer, who was a member of the pioneering group in the USA, joins with Carolyn Peddy to explain with elegant simplicity how deep lithosphere reflection seismology works and what sort of tectonic problems have been addressed. Although Chapter 13 focusses principally on the evidence for continental extension and shortening on a variety of scales, it provides seismic analysis of other tectonic boundaries and so builds bridges to later chapters.

In the second part of this trilogy, Philip England, a geophysicist whose research is directed at resolving relative motions across modern plate boundaries, examines continental deformation from a physical standpoint. Much of the available evidence for continental dynamics stems from analysis of gravity data over topographic anomalies, and from earthquake focal mechanisms that tell us about the sense of natural movements in regions of tectonic activity. One of the most elegant results is that mountains produced in convergence zones tend to collapse under their own weight, so being characterised by shallow extensional faulting. The mechanical contrasts between oceanic and continental crust which helps us to appreciate their different response to convergence are reviewed; and an introduction is given to continental rifting and the formation of sedimentary basins. Finally we establish links with Chapter 12 by relating the physical properties and vertical movements of continental crust to the state of metamorphism reflected in the rocks.

The formation of sedimentary basins is the subject of Chapter 15. Tony Watts picks up the theme of Chapter 12 and, in the context of sedimentary basins, examines further the likely causes of *vertical* movements of the crust; these cannot be accounted for directly by plate tectonics which stresses the role of horizontal movements. Vertical movements produced by crustal thinning during extension, and crustal loading by the emplacement of thrust sheets during compression are explained. These basin models are of economic importance, for they help to explain the distribution of sediment types that may form reservoirs for oil and gas, and also the burial history of petroleum source rocks, enabling the time to be determined when they reached depths at which temperatures were high enough to 'cook' them to produce oil and gas (see Chapter 18). Chapter 15 also discusses the likely causes of sea-level change, both local and global, and its importance in modelling the nature of the sedimentary fills of basins. The theme of global sea-level change is revisited in Chapters 16, 17, 20, 21 and 24.

CHAPTER

13

Simon L. Klemperer and Carolyn Peddy
University of Cambridge

SEISMIC REFLECTION PROFILING AND THE STRUCTURE OF THE CONTINENTAL LITHOSPHERE

If we want to learn about the interior of the Earth, we can choose between several methods of study. By mapping rocks which formed at deeper levels of the crust, but which are now at the surface, we can learn about the composition of the Earth's interior. By constructing cross-sections from geological maps, we can develop hypotheses about structures at depth, though these extrapolations become progressively more uncertain at greater depths. In order to test such cross-sections, we can drill into the Earth – but it is as yet impossible to drill deeper than about 15 km (because of the increased temperature at depth), and in any case such holes would be very expensive, costing more than £100 million.

The best alternative is **seismic imaging**: using sound waves to locate geological structures by remote sensing. In principle, the study of natural earthquakes can tell us a lot about the deep structures of the Earth. However, we cannot predict when and where earthquakes will occur, so we cannot deploy enough recording instruments in the right place in order to build up a highly detailed picture of the lithosphere. Instead, in controlled-source seismology, the seismologist creates shock waves by artificial means and deploys geophones (microphones) in the best positions to collect the sound waves after they have travelled through the Earth.

There are two sorts of sound waves that propagate through the interior of the Earth: **P-waves** and **S-waves**. P-waves (primary waves or compressional waves) are faster and involve particle motion (alternating compression and rarefaction) in the direction of propagation, just like sound waves in air. S-waves (secondary waves or shear waves) are slower and are propagated by a shearing motion that involves oscillation perpendicular to the direction of propagation, rather like waves running along a rope shaken vigorously at one end. P-waves are easier to produce, easier to record, and easier to interpret than S-waves, so in this chapter all the references to seismic waves, velocities, data etc. are to P-waves, P-wave velocities, P-wave data etc.

As in geometrical optics, sound waves – P-waves – in the Earth show two principal sorts of behaviour (Figure 13.1). They refract, or bend, as they pass between rock layers which have different velocities because of their variation in density and elastic properties. They are also reflected from the boundaries between such layers. In **refraction seismology** and in **wide-angle reflection seismology** sound waves may travel hundreds of kilometres through the Earth's crust or mantle, with raypaths that become horizontal at their greatest depth of penetration, as the seismic waves are refracted and turned back towards the surface by the increasing seismic velocities at greater and greater depths (Figure 13.1). The sound waves feel the effects of all the rock types through which they travel, so it is only possible to build up a picture of the average properties of the crust by this means. The travel-time delays of such refracted rays allow seismologists to measure the seismic velocity of different

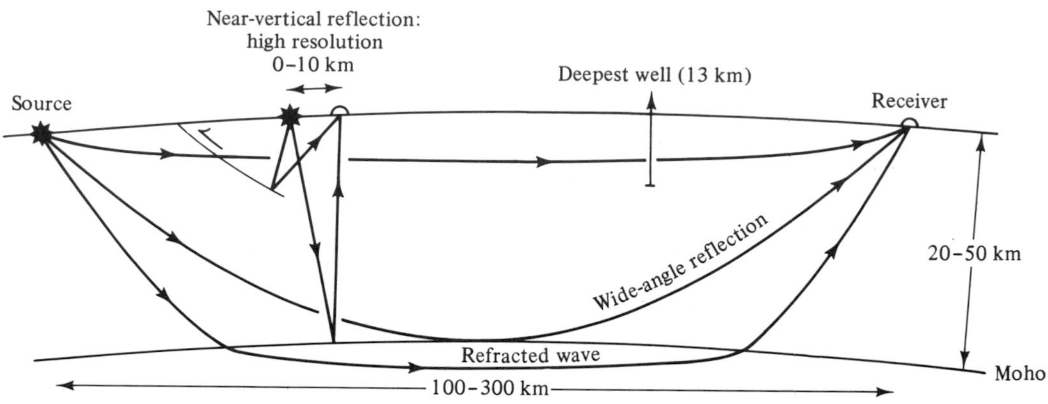

Figure 13.1 Schematic comparison of seismic raypaths for near-vertical reflection profiling, and for refraction/wide-angle reflection seismic studies. Also shown for scale is the deepest drill-hole in the world.

parts of the crust and upper mantle. For example, the **Moho** (the **Mohorovicic Discontinuity**, the base of the crust) can be located as the boundary between the crust with seismic velocity generally less than about 7 km s^{-1} and the mantle with velocity typically about 8 km s^{-1}. In contrast, in near-vertical **reflection seismology** (normally called just reflection seismology), the sound waves travel nearly vertically down to and back from reflecting interfaces (Figure 13.1). Because such rays only travel a short distance laterally, they can be used to locate reflecting boundaries with high precision, and to construct detailed cross-sections of the whole crust.

The first seismic-reflection recordings were made in the 1920s. Since then, reflection seismology has become a business worth thousands of millions of dollars a year, because it is the prime means of exploring sedimentary basins for geological structures that are potential hydrocarbon reservoirs. The technique of using seismic reflections to map geological structures received major boosts from the advent of magnetic recording and digital computer processing in the 1950s and 1960s. As seismologists look deeper into the Earth, computer processing becomes more and more essential in order to pick out increasingly weak reflections and distinguish them from background noise. By the mid-1970s, technology was sufficiently far advanced that university geologists could apply it to the study of the whole crust, looking far beneath the top few kilometres where oilfields are found, to study structures and tectonics on a continental scale. In 1975 the pioneering **COCORP (Consortium for Continental Reflection Profiling)** group in the USA began recording the first large-scale deep reflection profiles to image the continental basement, working on land. Inspired by the success of COCORP, a British group, **BIRPS (British Institutions' Reflection Profiling Syndicate)**, began profiling at sea in 1981, and other groups around the world, most notably in Canada, France and Germany, also began large-scale programmes. But before we look at what has been discovered, it is appropriate to go into a little more detail about how the technique works.

How the technique works

Reflection seismology requires a sound source, a sound detector, a means of recording the signals detected, and a way of processing and displaying the data recorded. A range of sound sources is used. On land, shock waves may be produced by dynamite blasts or by large **seismic vibrators** (heavy trucks which jack themselves up on a solid wooden pad and bounce up and down on top of it) shaking the ground; at sea, the most common technique is to release a bubble of highly compressed air which expands in the water, generating a pressure wave. The crust beneath the shallow water of continental shelves or large lakes is continental crust, just as beneath the continents themselves, so it is possible to study continental structure and tectonics in this way. On board a seismic reflection profiling ship, large air-compressors charge arrays of **airguns** which are towed behind the ship, 5 to 10 m deep in the water. The guns are fired electonically, to release a bubble of high-pressure air (typically 0.1 m^3 at about 200 atmospheres pressure) into the sea. The expansion of the bubble generates a sound wave which penetrates deep into the Earth, and is partially reflected back to the surface at each change in rock type (Figure 13.2).

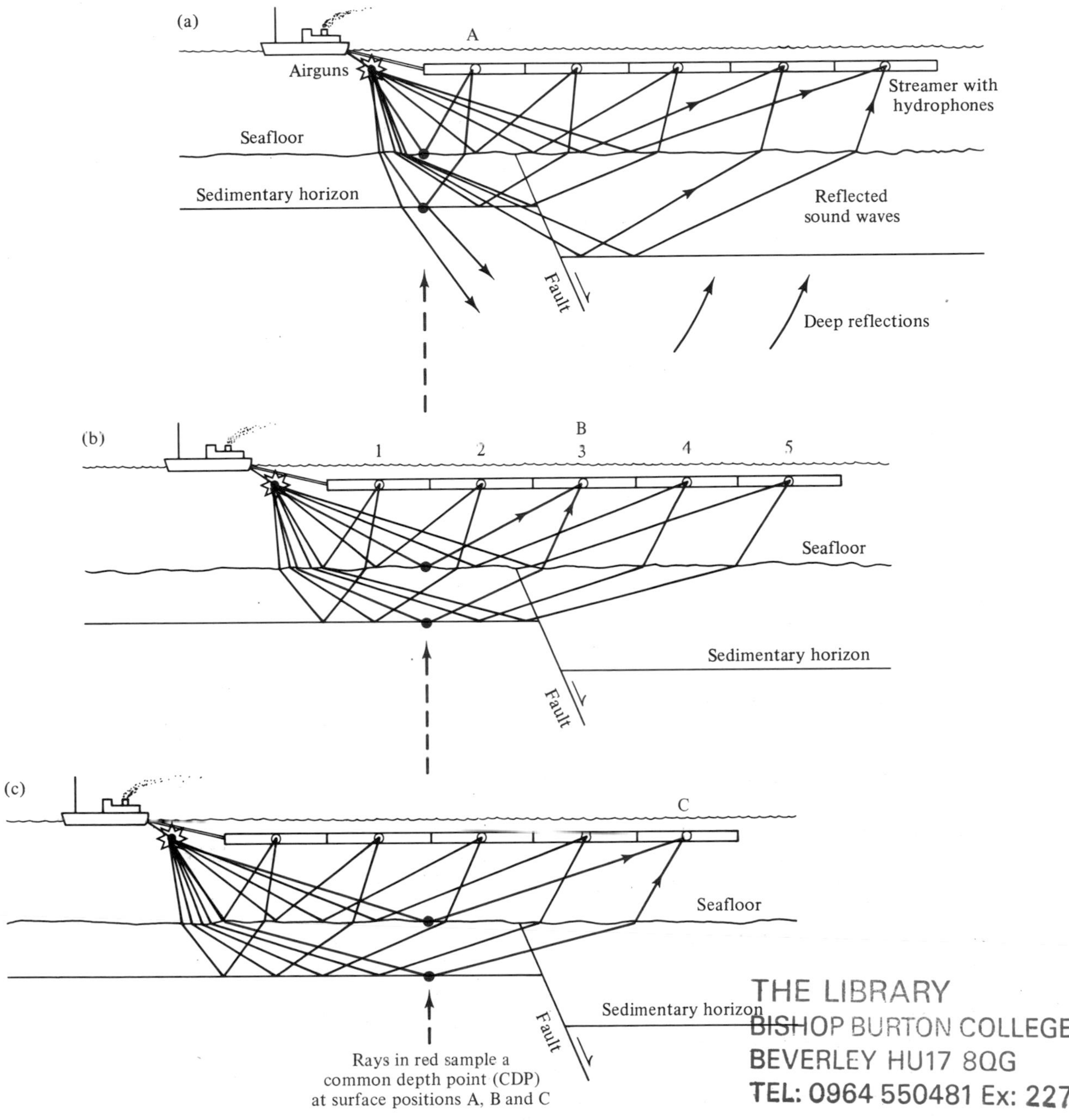

Figure 13.2 Schematic cartoon of marine seismic acquisition, illustrating the geometry of the airgun seismic source and the hydrophone streamer, and raypaths to subsurface reflectors for successive airgun shots (a, b and c). Each shot is fired after the ship has moved forward a distance, typically 25 to 100 m, equal to the receiver spacing (or sometimes double the receiver spacing). Rays in red reflect from the same subsurface point (the common depth-point or CDP) and will be added together in the process called stacking, which is shown in Figure 13.3g. The common depth-points on the seafloor and the sedimentary horizon are aligned at the midpoint between the source and receiver for each shot-point. □

The returning sound wave is recorded, in land experiments, usually by 60 to 240 *groups* each of 6 to 24 **geophones** (microphones) stuck into the ground in a line that may extend between 6 and 20 km from the seismic source. When working at sea, **hydrophones** (microphones designed to work underwater) are towed behind the ship in a tube or **streamer**. The streamer may contain 50 or 100 groups of hydrophones, and extend for 3 or 4 km behind the ship. The signals picked up by each group of hydrophones are separately recorded on magnetic tape aboard ship. Because the reflected signals may be more than one million times weaker than the initial shock wave, complicated data processing is required to extract these tiny but real reflected signals from all the noise produced by wave action or the ship's propellers etc. The other reason for employing computer processing is the amount of data recorded. In a typical deep profiling experiment, the airguns may be fired every 50 m along a 250 km profile; each shot is recorded by 60 groups of hydrophones; and the signal at each hydrophone group is recorded on magnetic tape 250 times a second for 15 seconds. These thousand million data samples require a large computer for their analysis.

In the processing operation, seismic signals recorded from rays incident at different angles upon a particular subsurface point (the **common depth point or CDP**) (Figure 13.2) are appropriately combined to improve the signal-to-noise ratio by summing, and to cancel noise signals by destructive interference. This process is called **CDP stacking** (Figure 13.3g). Though stacking is the single most important step in seismic data processing, many other computer processes are commonly used to remove noise (e.g. Figure 13.3e) or to increase resolution of the seismic data (e.g. Figure 13.3h).

After most of the data processing is completed, the **seismic section**, known as a **stack** from the process of CDP stacking, may be passed through another process called **migration**. Migration is used to place dipping reflections in their correct spatial location. Figure 13.4 shows why migration is necessary. Dipping planes such as fault planes reflect seismic energy sideways (Figure 13.4a) rather than vertically, as is the case with flat-lying layers. At any one geophone the interpreter has no way of knowing from which direction each reflection arrives, so all reflections are plotted directly beneath the surface point at which they are recorded (Figure 13.4b). This results in horizontal layers being plotted in the correct location, but dipping layers being plotted to the side of the correct location. Migration moves the fault-plane reflection (Figure 13.4b) back to the real location of the fault, and also causes dipping reflections to steepen (Figure 13.4c). The amount that a reflection must be moved during migration depends on the dip of the reflection and on the seismic velocity above the reflection. Only dipping reflections have been moved on a migrated seismic section – migration has no effect on flat-lying reflections. Be sure to note whether the example seismic sections which follow have been migrated.

The end product of data processing is a seismic section (e.g. Figures 13.5, 13.6, 13.8, 13.9, 13.10, 13.13 & 13.14) composed of many hundreds of wiggly traces (as in Figure 13.3g) placed side-by-side (as in Figure 13.3i). Each trace in detail is a complex waveform in which peaks (coloured black) and troughs (white) correspond to compressions and dilatations in the returning sound wave. This section looks like a cross-section through the crust, but it is very important to note that the vertical scale is not in depth but in seconds, and corresponds to the echo-time, or **two-way travel-time**, for a sound wave to reach a reflector and return to the surface. To convert the travel-time to

Figure 13.3 Simplified seismic processing sequence. The five traces in parts (a), (b) and (c) are those shown in Figure 13.2b, while the three traces in (d), (e) and (f) are the three red traces in Figure 13.2, one each from Figures 13.2a, b and c. The traces displayed in part (a) result from a single shot (Figure 13.2b) and were each recorded by a single receiver. The direct wave (refracted wave) appears at a time proportional to the receiver distance from the shot, while the reflections lie on a hyperbolic travel-time curve. Any bad or noisy traces are deleted (Figure 13.3b). The seismic amplitudes recorded decrease with increasing travel-time because the later reflectors are further away, so these weaker amplitudes are boosted (Figure 13.3c). In a very important step, the traces are re-sorted so that all traces with an identical source–receiver midpoint are **gathered** together (one trace from each of Figures 13.2a, b & c) in Figure 13.3d. The direct wave is removed (shaded area in Figure 13.3e), and then each trace is separately corrected for the time-delay appropriate to its source–receiver offset (known as the **normal move-out** or **NMO correction**) (Figure 13.3f). The wavelets from each reflection are now lined up, as if each trace had been recorded with coincident source and receiver (i.e. without source–receiver–offset dependent delays), and all the traces in the gather can be summed (stacked) to give a single trace (Figure 13.3g) with a higher signal-to-noise (S/N) ratio. If the wavelet is rather reverberatory, or extends for a long time, then its resolution is correspondingly poor, but it can be shortened by a digital filtering technique called **deconvolution** (Figure 13.3h). Finally, all the traces are displayed as a seismic section (Figure 13.3i), and interpreted (Figure 13.3j), just as in Figures 13.5, 13.8, 13.9, 13.10 etc. □

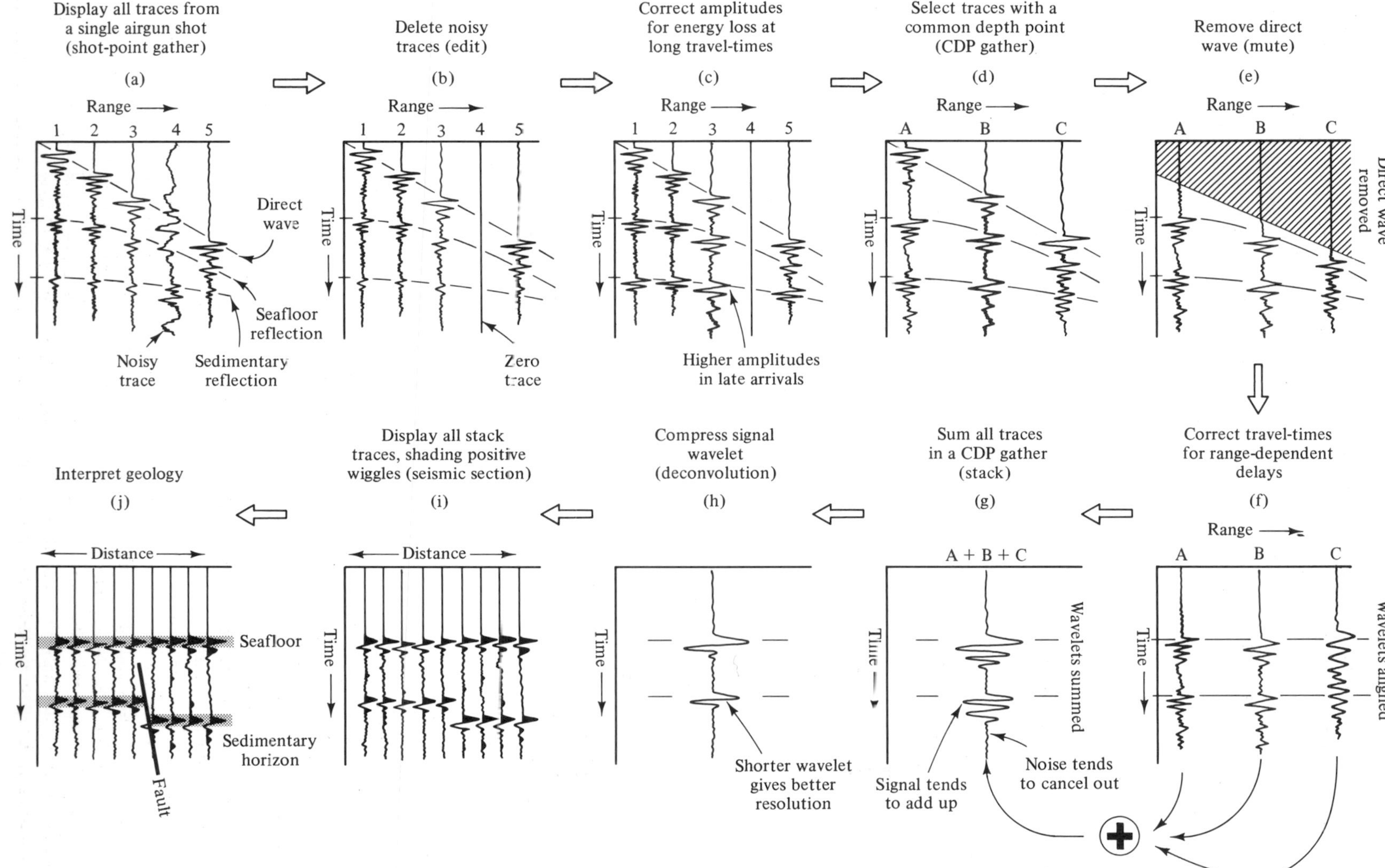
Display all traces from a single airgun shot (shot-point gather)
(a)
Range
1 2 3 4 5
Time
Direct wave
Seafloor reflection
Sedimentary reflection
Noisy trace
Delete noisy traces (edit)
(b)
Range
1 2 3 4 5
Time
Zero trace
Correct amplitudes for energy loss at long travel-times
(c)
Range
1 2 3 4 5
Time
Higher amplitudes in late arrivals
Select traces with a common depth point (CDP gather)
(d)
Range
A B C
Time
Remove direct wave (mute)
(e)
Range
A B C
Time
Direct wave removed
Correct travel-times for range-dependent delays
(f)
Range
A B C
Time
Wavelets aligned
Sum all traces in a CDP gather (stack)
(g)
A + B + C
Time
Wavelets summed
Noise tends to cancel out
Signal tends to add up
Compress signal wavelet (deconvolution)
(h)
Time
Shorter wavelet gives better resolution
Display all stack traces, shading positive wiggles (seismic section)
(i)
Distance
Time
Interpret geology
(j)
Distance
Time
Seafloor
Sedimentary horizon
Fault

depth, one must multiply the one-way travel-time by the average seismic (sound) velocity to the depth – say, by 2 to 6 km s^{-1} in sedimentary rocks (though this is highly dependent on age and lithology), and by about 5 to 7 km s^{-1} in the crystalline crust. If the rock types, and hence the seismic velocities, vary laterally across a seismic section, then the stack section can provide a distorted geometric view of the crust.

Although reflection seismology has a much greater spatial resolution than other geophysical techniques, this resolution is still limited by the physics of the experiment. The smallest vertical separation between two objects that can be measured is about a quarter of a wavelength, thus the higher the frequency content of a seismic signal, the greater the vertical resolution, since, for a constant velocity, frequency is inversely proportional to wavelength. The frequency bandwidth of sound waves used in deep reflection experiments is around 10 to 40 Hz (that is 10 to 40 black peaks, and the same number of white troughs, per second on an average trace). Vertical resolution decreases with depth because the high-frequency component of the seismic signal is preferentially absorbed by frictional energy loss caused by small-scale motions of grains and crystals as the seismic wave passes through the Earth. The dominant bandwidth of reflections from the middle to lower crust is often decreased to about 20 Hz. For a mean crustal velocity of ~ 6 km s^{-1} this means that layers thinner than about 75 m (one-quarter wavelength) cannot be individually distinguished. Horizontal resolution also decreases with depth, but for a different reason. Seismic waves spread in the Earth in three dimensions in much the same way as a ripple spreads in two dimensions from a pebble dropped in a pond. As the diameter of the ripple gets bigger, a longer arc of the ripple is approximately straight. The portion of the wave that is approximately straight to within an error of one-quarter wavelength of the ripple is known as the **Fresnel zone**, which may be familiar from the study of optics. The seismic response from the entire area contained within the Fresnel zone adds together to give a single response; any heterogeneities within the Fresnel zone are smoothed out. The Fresnel zone is smallest – hence resolution is best – close to the seismic source, but it increases to a couple of kilometres in the middle crust. Two objects closer than this resolution limit cannot be separately imaged, while an object smaller than this limit cannot be seen clearly and appears on the seismic section as a characteristic **diffraction hyperbola**, a convex-upward hyperbola with its apex at the object (see e.g. Figures 13.5 and 13.12). At the base of the crust, an object must be at least the size of a thousand football fields in order to be accurately resolved; an object smaller than this would appear only as a diffraction.

In addition to problems associated with the finite resolution of the technique, the interpreter of deep reflection data is faced with a more fundamental difficulty, that of assigning physical significance to each reflector on a deep profile. If a well is drilled on a seismic profile, then the cause of the reflections on the seismic section can be identified, as, for example, the change from sandstone to shale. However, wells typically provide direct correspondence between rocks and seismic reflections only to depths of no more than about 5 km, or 2 to 4 s two-way time. Below that, the interpreter has no direct information about the rocks which give rise to reflections. One is, in a very real sense, 'interpreting' the pattern of reflected energy, using available regional geological and geophysical information to provide constraints for interpretations of the reflection profiles.

Given the above limitations, what can the seismic reflection technique 'see' within the crust? How do Earth scientists use the technique to study the deep crust? In the following sections, some of the major results of deep reflection profiling are discussed. At the end of the chapter is a list of publications which give more detailed information about the technique of seismic reflection profiling as a whole, and about specific deep reflection surveys.

Compressional structures: thrusts and sutures

Geological mapping of the exposed levels of many of the world's mountain chains has been largely completed in the 20th century. The next step, mapping the subsurface levels of mountain chains using deep seismic reflection profiling, is only just beginning. Geologists have many questions about mountain belts which cannot be answered by surface mapping. For example, what lies beneath the often chaotic assemblages of **thrust sheets** mapped at the surface of a mountain belt? What happens to thrust faults deep within the crust? Do they form flat-lying or shallow-dipping décollement structures, or do they become vertical? These are the sorts of questions which deep seismic reflection profiling can address (see also discussion of thrust tectonics in Chapter 14).

One of the first deep reflection surveys across a mountain range was carried out by the COCORP group in the Rocky Mountains of Wyoming, western USA. The Wind River mountain range is one of a

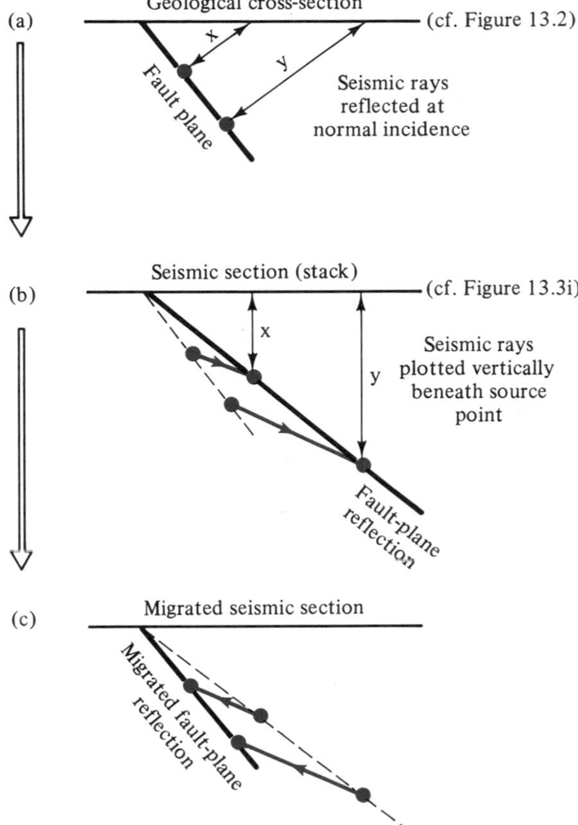

Figure 13.4 Schematic illustration of the process of migration, which moves seismic reflections recorded from dipping structures into their correct location on a seismic section. The amount of movement required depends on the dip of the structure and on the seismic velocity. Flat horizons (as in Figures 13.2 & 13.3) do not move when migrated. □

group of Precambrian basement blocks uplifted during the early Tertiary during the Laramide orogeny. Several attempts had been made to explain the formation of the Wind River mountains. According to one hypothesis, the uplift took place by essentially vertical motion along a very steep bounding fault. Alternatively, thrusting along low-angle faults involving mainly horizontal motion had also been suggested as the cause of the Wind River uplift. COCORP collected the deep reflection data shown in Figure 13.5 to see if deep seismic reflection profiling could solve the question. The vertical scale of Figure 13.5 (and of all the seismic sections in this chapter) is in two-way travel time. To obtain very approximate depths, multiply the two-way time by half the average crustal velocity, by say 3 km s^{-1}; thus 10 s two-way time typically corresponds to about 30 km depth. The mapped location of the Wind River thrust is labelled WRT at the top of the seismic section. From the surface outcrop of the fault an east-dipping reflection extends across the figure to 6 s and then beyond the figure down to about 8 s two-way time, or about 25 km depth in the Earth. The reflection is from the thrust fault along which the Precambrian basement was uplifted. Southwest (left) of the thrust are bright layered reflections from Palaeozoic and Mesozoic sedimentary rocks, which also extend beneath the thrust fault. Northeast of the Wind River thrust is Precambrian basement, which is not layered like the sedimentary rocks but is transparent (non-reflective) near the surface. At greater depths, beneath the sedimentary rocks, the Precambrian basement is characterised by small-scale discontinuities or random lithological changes, as shown by the diffractions seen in this part of the seismic section. At the extreme northeast of the section, Palaeozoic strata can be seen, carried on the back of the thrust sheet and tilted due to rotation by movement along the thrust fault. The fault-plane reflection is fairly planar to a depth of about 15 km, and would have a true geologic dip of about 30° after migration. The reflection from the Wind River thrust provides unequivocal evidence that the Wind River mountains formed by moderate-angle thrusting due to dominantly horizontal stresses from the northeast, and not purely by vertical uplift. The type of thrusting observed in the Wind Rivers is called **thick-skinned thrusting** because essentially the whole thickness of the crust was involved in the thrust motion.

Fault zones such as the Wind River thrust are often observed directly as reflections on deep seismic data sets (see Figures 13.8, 13.9 & 13.13 for other examples), though the precise reason for their reflective character is unknown. Mylonitic fabrics, fluids and metasomatic mineral replacement within fault zones are some of the hypothesised causes of fault reflectivity. In order to be imaged at all, the fault zone, where it juxtaposes Precambrian basement against Precambrian basement, must be at least a quarter wavelength thick – around a hundred metres. Whatever its cause, fault-plane reflectivity has contributed a great deal to the efficacy of deep seismic reflection profiling in solving tectonic problems.

The next deep seismic reflection profile is chosen to show the variability in tectonic style that can occur during compression. The profile is from the Appalachian orogen which extends the entire length of the east coast of North America. The Appalachians are the result of a protracted series of orogenic events extending in time from the Ordovician to the Permian, and correspond loosely to the Caledonian and

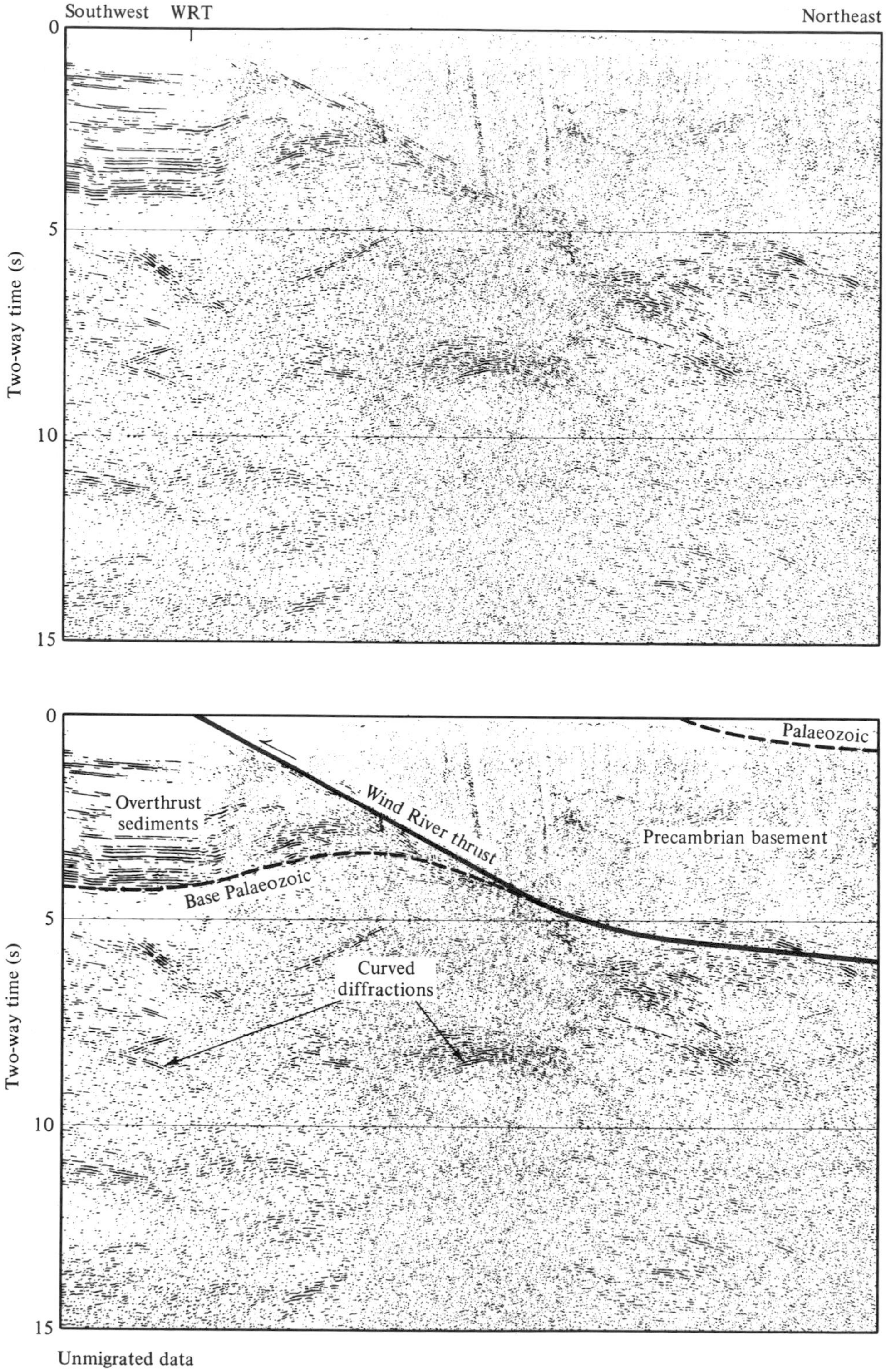
Southwest
WRT
Northeast
Two-way time (s)
0
5
10
15
Two-way time (s)
0
5
10
15
Overthrust sediments
Base Palaeozoic
Wind River thrust
Palaeozoic
Precambrian basement
Curved diffractions
Unmigrated data

Figure 13.5 COCORP profile across the Wind River thrust, uninterpreted and interpreted data. Look at the interpretation, then try to find the interpreted reflections on the uninterpreted data. The reflection from the Wind River thrust extends to 6 s two-way travel-time or about 18 km, showing a 'thick-skinned' style of thrusting. □

Hercynian orogens in Europe. In the late 1970s COCORP began a programme of deep profiling in the southern Appalachians in order to study the deep parts of the orogen. Other deep seismic reflection data have also been collected in the same area by the Appalachian Ultra-Deep Core Hole (ADCOH) study group, in preparation for a deep drill-site in the Appalachians. The surface features of the Appalachians are well-mapped, and consist of metasedimentary, crystalline and volcanic assemblages bounded by major thrust faults. The amount of movement on the thrust faults cannot be directly determined by surface mapping, but early workers theorised that large parts of the exposed rocks were part of an **allochthonous terrane** (i.e. had been moved large distances into their present position) and so predicted large amounts of thrust motion.

Figure 13.6 is an unmigrated ADCOH seismic section processed so that only continuous reflections are plotted, suppressing random noise. What does the seismic section tell us about the way the Appalachian orogen formed? The first observation from the seismic section is that there are many reflections in the upper 3 s two-way time and very few reflections below about 3 s. The boundary between the reflective upper 3 s and the nearly transparent lower part of this seismic section is an extremely continuous, high-amplitude reflection (A on Figure 13.6) which can be traced on ADCOH data and on other COCORP data for over 250 km beneath the surface thrust sheets. A well drilled through the rocks which cause this bright reflection (at a shallower level, off this seismic line) shows that the reflections are from lower Palaeozoic sediments deposited on a continental shelf. Thus, the crystalline rocks at the surface must have been emplaced by horizontal

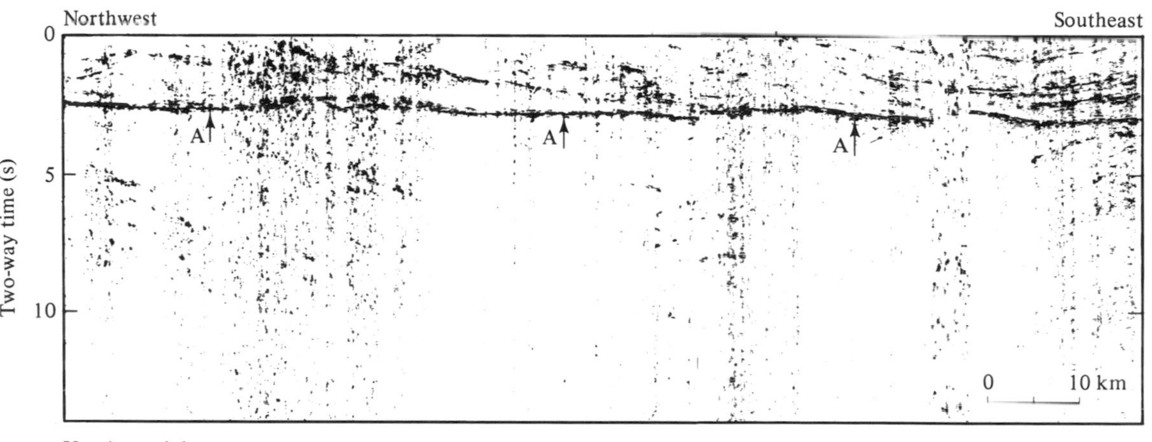

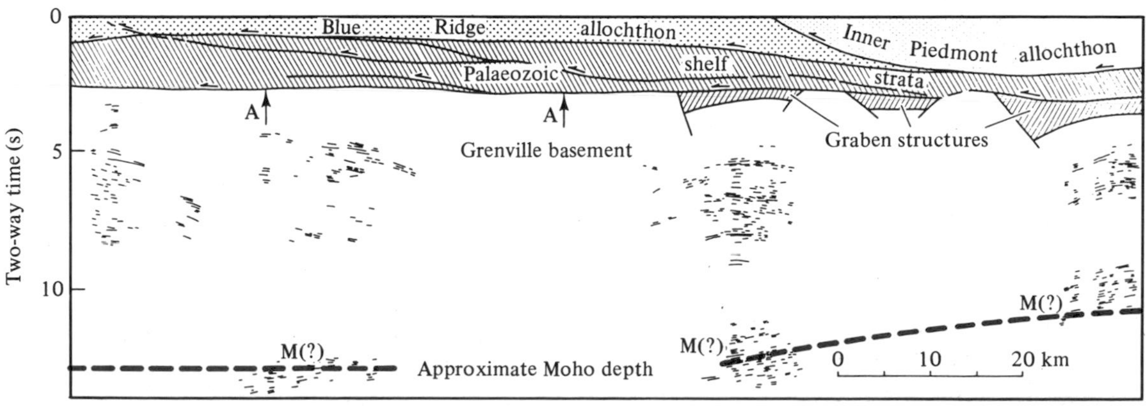

Figure 13.6 ADCOH profile across the southern Appalachians overthrust, uninterpreted data and interpreted line-drawing. A is the major thrust zone or décollement, and the crust beneath – 80 per cent of the whole thickness – is essentially undeformed. This type of faulting is known as thin-skinned. □

thrusting for at least the distance beneath which the reflective shelf strata subcrop. Thus this seismic profile shows that an ancient continental margin is now covered by thin thrust sheets which have travelled over 250 km to their present position. This interpretation is shown in the lower part of Figure 13.6, a **seismic line-drawing** in which each reflection in the lower part of the crust is represented by a single line. This area of the Appalachians is a classic example of **thin-skinned thrusting** which has placed thin flakes (< 10 km) of continental crust over older basement and sediments, but which has not deformed the whole thickness of the crust. The shelf sediments have undergone some internal thrust duplication but were otherwise remarkably undisturbed by the large-scale thrusting. The deep profiling in the Appalachians opened a new area for petroleum exploration by discovering this extensive area of underthrust sediments of unknown petroleum potential.

Compare the two styles of thrusting shown in Figures 13.5 and 13.6. In the thin-skinned Appalachians, thin slivers of crystalline crust (< 10 km) make up the overthrust sheets, and thrusting is at a very low angle. In contrast, in the thick-skinned Wind Rivers, a thrust fault extends from the surface to deep crustal levels, so that thrusting involves a block of crust rather than a sliver. Despite these differences, the thrust faulting in the Appalachians and in the Wind Rivers were both related to compressional forces acting on continental lithosphere at plate boundaries: for the Wind Rivers, subduction of oceanic lithosphere at the Pacific margin of North America, and for the Appalachians, subduction and island arc collision at the margin of the proto-Atlantic ocean. Whether thick-skinned or thin-skinned thrust belts develop in a particular area probably depends on the detailed rheology of the crust, and whether weak zones are present in the crust. Thin-skinned thrust belts (Figure 13.6) are often found in the external parts of orogens where the crust is thinnest and strongest; while thick-skinned belts (see Figure 13.7) are often found in the internal parts of compressional orogens where the crust is thicker and weaker (see Chapter 14 for further discussion of rheology in relation to tectonics).

The boundary between two collided continents or continental fragments is called a **suture zone**. Deep seismic reflection profiles have been collected in both Europe and in North America across the sutures formed by the closing of the Palaeozoic oceans. The provincialism of fossil faunae shows that, in the early Palaeozoic, before the modern Atlantic opened, part of the present-day southern Appalachians of North America was attached to the African continent, and hence that a suture between African crust and North American crust must exist somewhere within the Appalachian chain, southeast of the low-angle thrust shown in Figure 13.6. Though the location of the suture is known approximately from studies of fossil provinces, the suture zone is covered by younger sedimentary rocks so that it is not accessible to geological mapping. Figure 13.7 shows COCORP data from south-central Georgia, across the approximate location of the North American–African suture. This illustration is a line-drawing, which is an effective way of illustrating a large amount of data at a small-scale, and of excluding residual noise or side reflections from consideration. Beneath the thin veneer of flat-lying, mainly Tertiary, coastal plain sediments, and Mesozoic sedimentary rocks related to Atlantic rifting, is a wide, diffuse, south-dipping wedge of south-dipping reflections. The Moho is at 12 s two-way time, and the wedge of reflections extends from close to the surface to the base of the crust. This wedge of reflections has been identified as the suture zone by COCORP workers because of its location with respect to geological exposures of African and North American basement and because of its obvious importance as a crustal-scale feature. The dipping reflectors may be oceanic sedimentary rocks deposited between the African and North American continental blocks, and caught up in the suture zone during continental collision.

In Britain, deep reflection profiling carried out by the BIRPS group has also imaged dipping reflectors thought to be associated with the Palaeozoic closure of the proto-Atlantic, or **'Iapetus', ocean**. The seismic section shown in Figure 13.8 is part of a profile extending from the Southern Uplands, where the crust is known to be of North American provenance, to the Lake District, which contains European fossils. Thus, by definition, the profile must cross the 'Iapetus suture', though as in the example of Figure 13.7 the actual suture is covered by younger sedimentary strata. The major reflective zone that is believed to be the Iapetus suture is marked in Figure 13.8, which shows migrated seismic sections. The reflection labelled 'IS' probably marks a major discontinuity in crustal type – it may truly be the boundary between European and North American crust, which is the classical definition of the 'Iapetus suture', or it may represent a sliver of subducted oceanic crust or some other thrust-related structure. To the north (left) of IS is a group of very bright reflections interpreted to be from a subduction complex, that is, a collage of thrust slices of mixed oceanic and continental affinity, emplaced by underthrusting along the Iapetus subduction

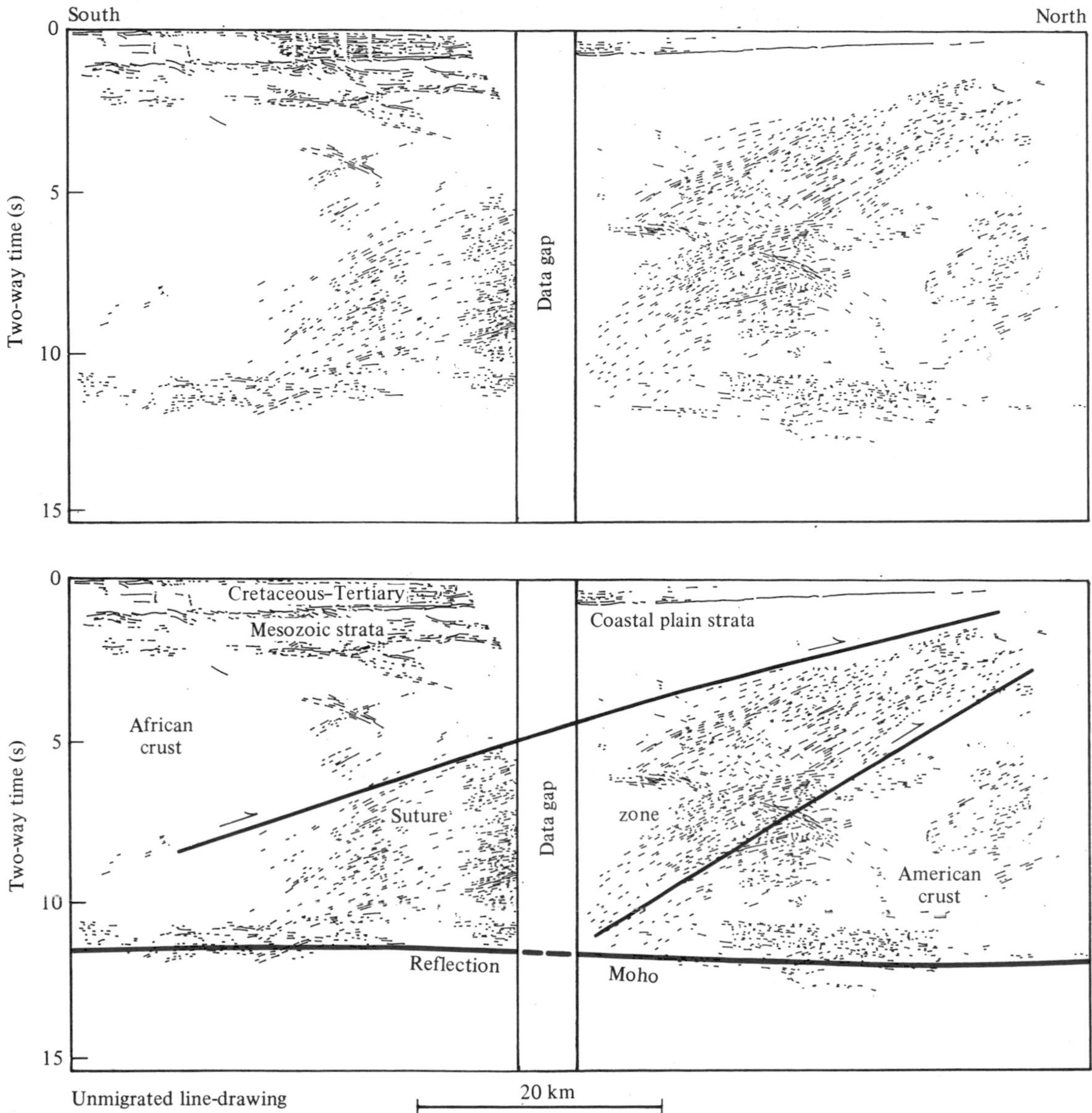

Figure 13.7 COCORP profile across the southern Appalachians suture zone, line-drawing and interpreted line-drawing. The dipping wedge of reflections may represent shelf and slope sedimentary rocks caught up in the collision between the two continental blocks. The deformation zone clearly extends down to the base of the crust, but the Moho truncates the dipping reflections and may be younger than the collisional suture. □

zone. This putative subduction complex is > 10 km thick, thicker than normal oceanic crust with its overlying sediments, suggesting that the oceanic crust may have been dismembered by faulting and underplated with complex internal geometries and much duplication. Notice that the reflection Moho runs straight across the seismic section at about 10 s two-way time (about 30 km depth), and truncates the IS reflection. At first sight this is surprising because if IS is a suture, then, during subduction 400 Ma ago, it should have continued beneath the Moho down into the mantle part of the lithosphere. However, there are at least two explanations for the lack of structure on the reflection Moho in the vicinity of the suture. The Moho may have acted as a **décollement horizon** for thrust motion during the Caledonian orogeny, that is, the thrust zone marked IS could have turned horizontally into the Moho and run along the Moho for some

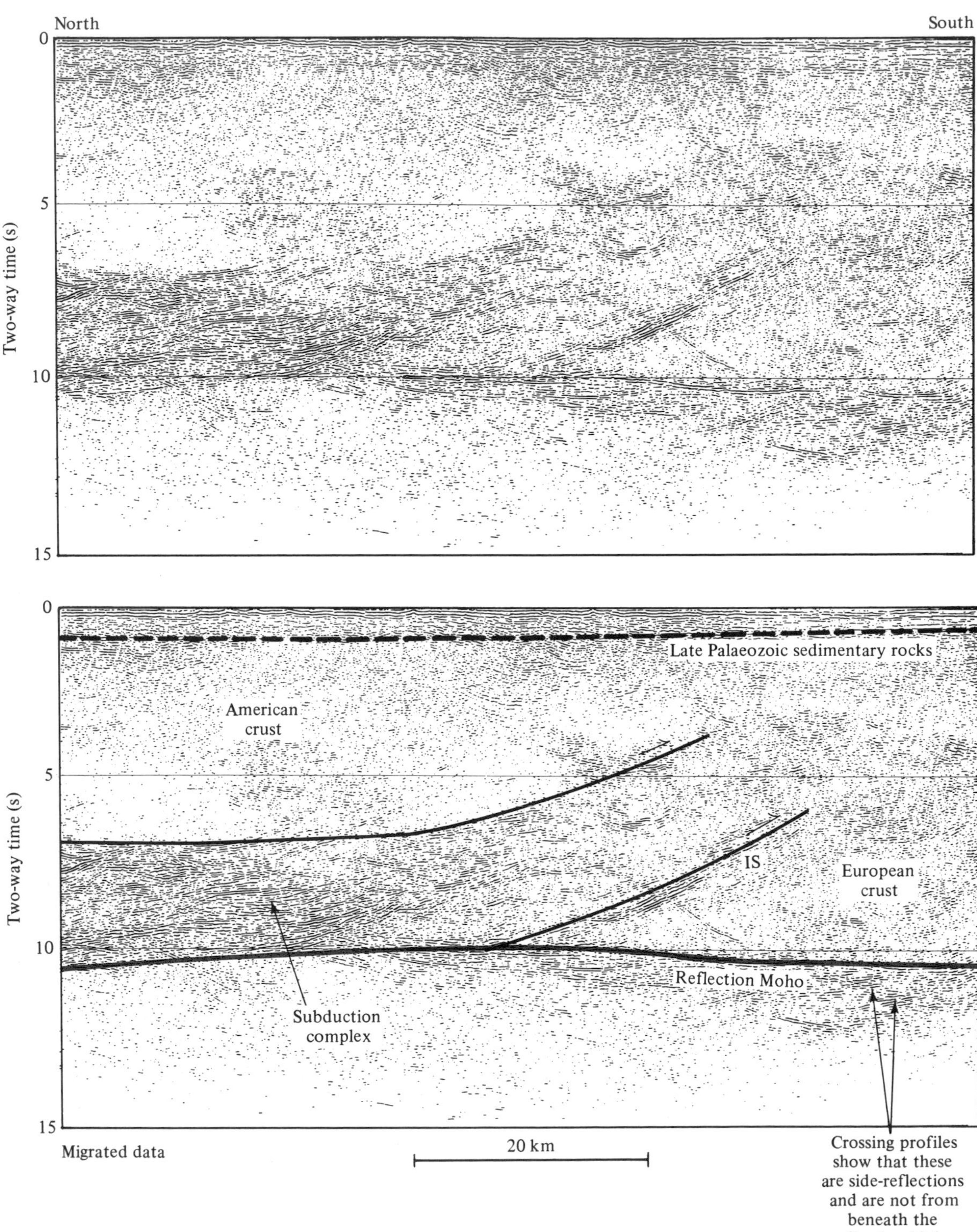

Figure 13.8 BIRPS profile across the Iapetus suture, around the Scotland/England border, uninterpreted and interpreted migrated data. The red line off the east coast on the map (opposite) indicates that part of the NEC (North-East Coast profile) BIRPS line that is represented in the seismic profiles. As in the southern Appalachian suture zone, the dipping reflections either merge into or are truncated by the reflection Moho. □

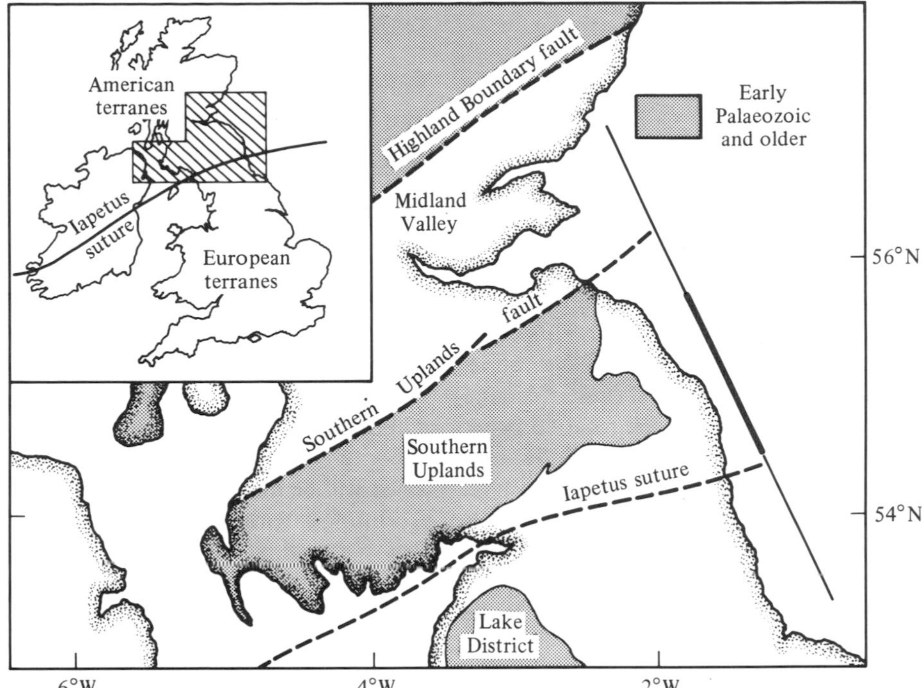

distance to the north. Alternatively, it is possible that the reflection Moho formed after subduction ended, due to isostatic re-equilibration of crustal thickness by ductile flow in the lower crust, so that the truncation of IS by the Moho represents an age relationship, a younger structure cutting an older one.

Extensional structures: normal faults and rifts

Just as thrust faults take many forms, so do **normal faults**. This can have great economic importance. On a small scale, oil is often trapped against faults in sedimentary basins, and the geometry of faulting controls the geometry of the oilfield. At the largest scale, the way in which the crust and mantle extend during rifting controls the way in which heat flows from the mantle into the crust and so controls the thermal history of a sedimentary basin and hence the relative distribution of oil and of gas in a petroleum province (see Chapters 15 and 18).

In order to understand how the crust extends, geologists study regions where extension is either very recent or is currently occurring. One of the most intensively studied extensional provinces is the Basin and Range province of the western USA. In the Basin and Range province, many normal faults cut the surface creating the characteristic topography which gives the area its name. At the surface, these faults may have dips of around 60°. What happens at depth? The answer varies from place to place. One possibility is seen in Utah, where a COCORP seismic profile (Figure 13.9) shows that the steep faults at the surface are cut off at depth by a much shallower **detachment**, or **extensional décollement**. In this instance, the Sevier Desert detachment (Figure 13.9) runs from the surface to 5 s (about 12 km) over a distance of 60 km, a regional dip of only 12°. The reflections from this detachment fault span quite a thick zone, up to 0.5 km, suggesting that a correspondingly wide zone of brecciation or hydrothermal alteration may exist. Above the detachment, Oligocene and Miocene sedimentary beds (dated from drill core) are highly rotated by movement along the detachment. The dipping sedimentary rocks were subsequently cut by steep normal faults which offset a Pliocene basalt flow but which do not cut the main detachment. Thus the detachment has been active at least from Miocene to Pliocene, and may still be active. Correlation of structures on either side of the detachment (not shown) indicates a total movement of 30–60 km, so this is quite a major fault. The thin-skinned, low-angle extensional faulting shown in Figure 13.9 is analogous to the thin-skinned thrusting shown in Figure 13.6, and it is possible that the Sevier Desert detachment is in fact a thrust fault that has been reactivated. Similarly, the structurally higher fault shown in Figure

was the overall thickness of the lithosphere. Because rocks expand and so decrease in density when they are heated, this heating causes uplift of the extending zone. This thermal uplift is partly why the Basin and Range province, an area of active extension and recent volcanism, is currently more than 1 km above sea-level. After extension ends, the lithosphere cools and subsides, allowing an undeformed **thermal-subsidence** or **sag basin** to form. It is this Cretaceous to Tertiary post-rifting basin that is seen above the tilted fault blocks in the North Sea in Figure 13.10. The Tertiary sedimentary horizons are clearly unfaulted.

If the stretching and heating are great enough and rapid enough, then igneous rocks may be intruded into the lower crust. If continental rifting proceeds sufficiently far, enough igneous melt is produced to break through the whole crust and an oceanic rift may form. In the North Sea the pre-Mesozoic crust was thinned by a factor of 1.8 from a typical thickness of about 32 km around Britain to about 18 km in Figure 13.10 (measured from base Triassic at about 4 s to the Moho at about 10 s). (It is possible that ductile flow since the Mesozoic extension has caused the crust to thicken slightly, moving back towards an equilibrium thickness. If so, then the estimate of thinning by a factor of 1.8 is a slight underestimate.) A continent–ocean transition is shown in Figure 13.11, a line-drawing (as for Figures 13.6 & 13.7) of a Canadian profile recorded offshore Newfoundland. Here, on the margin of the Atlantic, the crystalline crust is thinned by a factor of at least 3.5, from about 28 km to about 8 km. In fact, part of the 8 km-thin transitional crust is probably made of new igneous intrusions, so that the degree of thinning, or extension, is much greater than the apparent factor of 3.5. Figure 13.11 shows that most of the crustal thinning occurs within a very narrow lateral zone of about 50 km. At the left of the figure the continental reflection Moho is clearly visible at 10.5 s beneath the continental shelf, while on the right of the figure the rough surface of the oceanic crust can be seen beneath 4 s (3 km) of water and 1.5 s (3 km) of sedimentary rocks, with the oceanic Moho beneath at 8 s. It appears that the crust has been thinned by stretching, or necking, without obvious faults cutting through the whole crust.

Strike-slip faulting

The geometry of seismic reflection acquisition means that it is simplest to image flat-lying, or only gently-dipping, reflectors. This is because if a steep structure is to be imaged, the seismic wave must travel sideways from the source, rather than vertically downwards, in order to intersect the steep structure at nearly normal incidence and be reflected back to the recorder. It is rare to see reflections from faults steeper than about 45°, such as **strike-slip faults**. However, just as the normal faults in Figure 13.10 can be recognised by the offsets they produce on the sedimentary horizons, so can strike-slip faults be recognised by their truncation of other seismic reflectors. One important example is shown in Figure 13.12a. The Great Glen fault is a major intra-continental transform fault in Scotland on which palaeomagnetic evidence suggests there has been several hundred kilometres of motion. This fault may have played a similar role in the Lower Palaeozoic Caledonian orogeny as is played today by the San Andreas fault in California between the Pacific

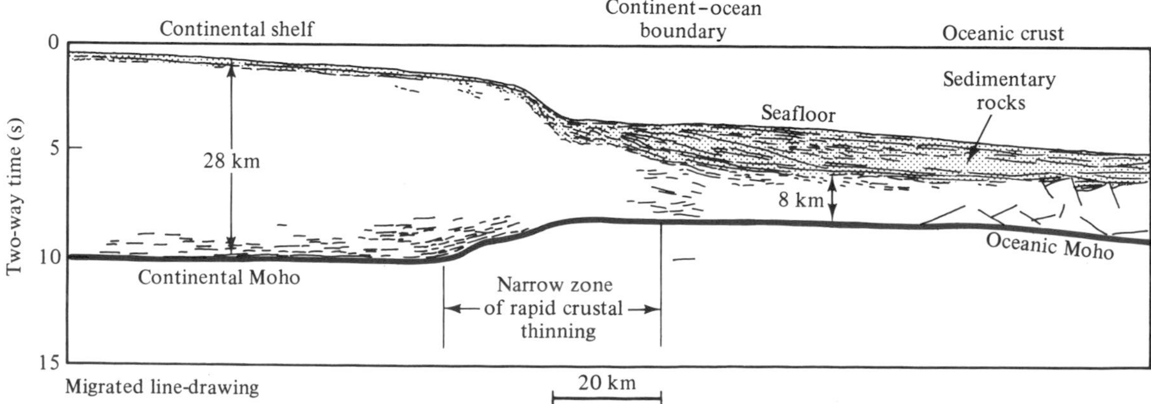

Figure 13.11 LITHOPROBE profile across the Newfoundland Atlantic continental margin, interpreted line-drawing. This is a line-drawing of migrated data, so all the dipping reflections are in their correct structural positions.

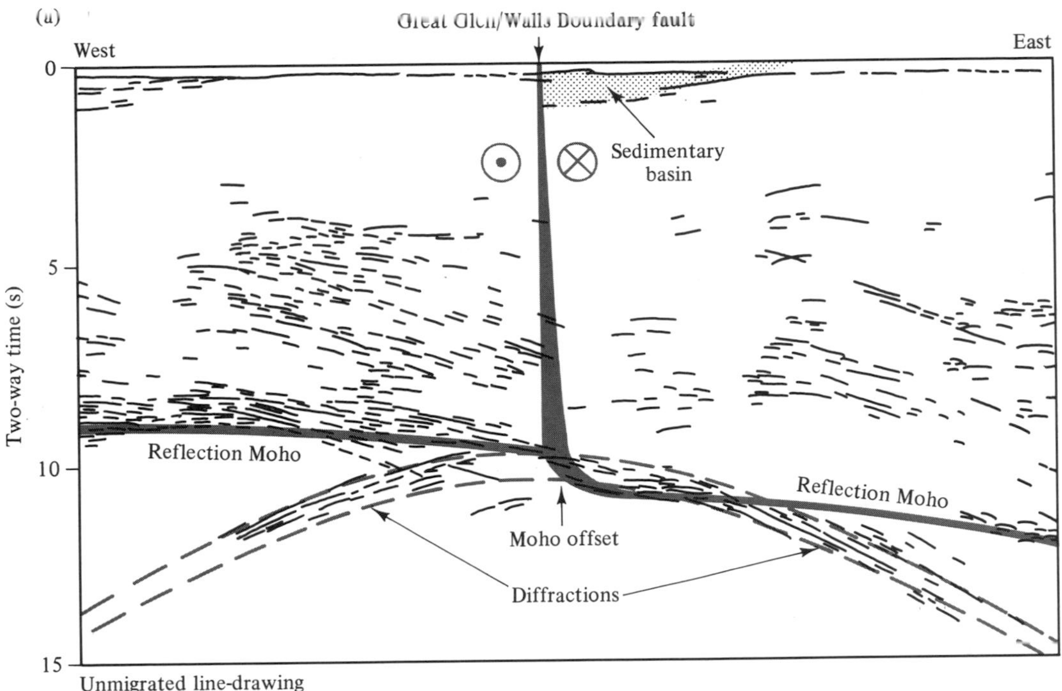

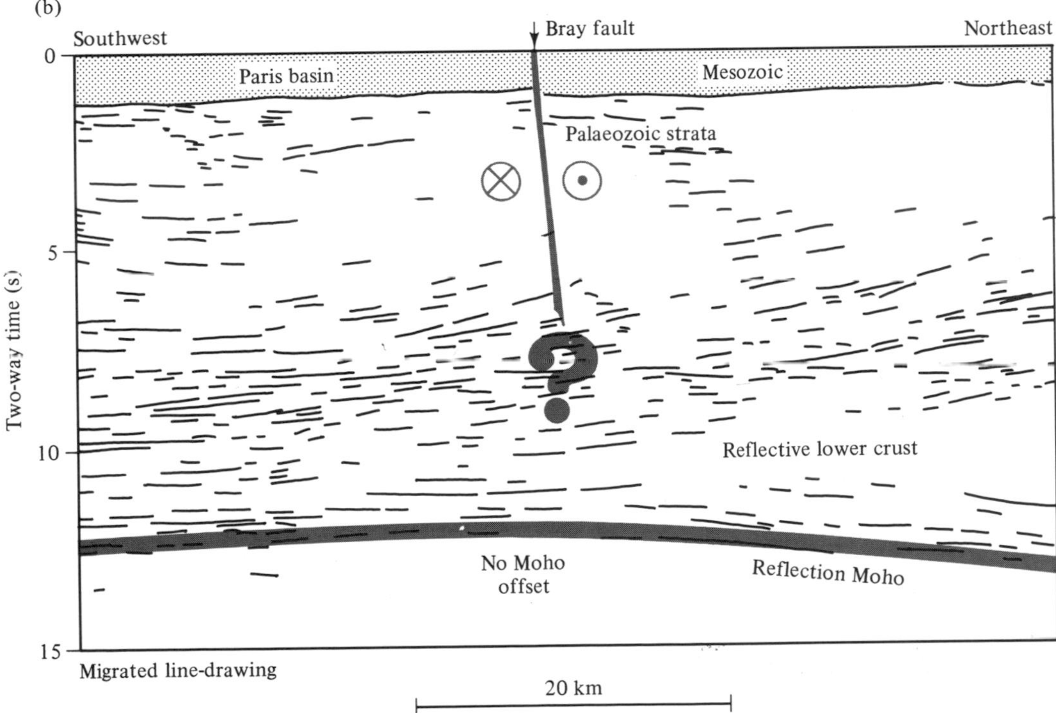

Figure 13.12 (a) BIRPS profile across the Great Glen fault, north of Scotland, interpreted line-drawing. This line-drawing is shown unmigrated; after migration the diffractions collapse to a single point (the Moho offset from which they are diffracted). (b) ECORS profile across the Bray fault, northern France, interpreted line-drawing, migrated data. This line-drawing is shown migrated, so that the reflections in the lower crust are in their correct position vertically below the surface trace of the Bray fault. (The symbols in circles indicate the probable sense of movement along these strike-slip faults. A cross means that the block has moved into the paper and a dot indicates motion out of the paper.) □

and North American plates. Figure 13.12a is a line-drawing of a BIRPS profile 200 km north of Scotland which crosses the Great Glen fault (known in this area as the Walls Boundary fault). The fault can be located in this area as the boundary to a small sedimentary basin, which is younger than the main episodes of strike-slip motion, and therefore formed by slight normal-fault reactivation of the transform fault rather than being half of a basin truncated by transcurrent faulting. The profile shows no reflections from the Great Glen fault itself because the fault is too steep, but it does show the Moho to be about 1 s two-way time (3 km) shallower to the west of the fault than it is east of the surface trace of the fault. Because the Moho has been abruptly truncated by the strike-slip fault, characteristic hyperbolic diffractions are formed. Diffractions originate at abrupt discontinuities due to the seismic waves 'seeing' the edge of a geological structure. The diffractions in Figure 13.12a may be thought of as side-reflections from the sharp corners on the crust–mantle boundary where it is cut by the fault; the reflection is plotted vertically beneath the surface point from which the sound wave travelled out and to which the reflection returned, rather than at the depth point at which the reflection originated – see Figure 13.4. This profile shows that the Great Glen fault has moved sufficiently far to juxtapose two crustal types of somewhat different thickness. Also, the Great Glen fault cuts vertically all the way through the crust as a rather narrow zone of disruption (no more than a few kilometres wide), and may continue through the entire lithosphere as an essentially vertical structure. Because no deeper structures were imaged, we do not know if the fault ends at the Moho (and the crustal blocks moved past each other using the Moho as a décollement surface), or whether the Great Glen fault continues down to the asthenosphere.

Though it might not seem surprising to find that strike-slip faults are narrow, vertical faults penetrating the whole crust, it is certainly not the only possibility. Geological field mapping cannot indicate whether strike-slip faults might instead broaden out into a very wide zone of ductile shear in the hot lower crust, or if a strike-slip fault might be confined to the upper crust and be riding above a detachment in the mid-crust which carries the transcurrent motion well away from the surface trace of the fault. An example of a strike-slip fault that does not seem to be a vertical fault through the whole crust is the Bray fault in northern France. This fault, along which there may have been 100 km of strike-slip motion during the Upper Palaeozoic Hercynian orogeny, has been studied by reflection profiling (Figure 13.12b). In this area there is very prominent lower-crust reflectivity, from 7 s (about 20 km) to the Moho at 12 s (35 km) (compare with reflective lower crust in Figure 13.10, and see also Figure 13.14). Reflections in the lower crust and at the Moho seem to pass unbroken beneath the surface trace of the Bray fault, so the fault does not seem to penetrate the whole crust at the present day. It is possible that the Bray fault never penetrated as a discrete structure into the lower crust, and that the transcurrent motion in the lower crust took place over a very wide area of shearing. However, the reflectors in the lower crust have never been drilled, and so it is uncertain whether they pre-date or post-date the strike-slip movements. If the reflections in the lower crust formed after the fault ceased to move (if, for example, the lower crustal reflections are related to the formation of the Mesozoic Paris basin seen at the top of the seismic section), then they can, of course, tell us nothing about the relationship between the fault and the base of the crust. Thus reflection profiling evidence can often be ambiguous unless enough supporting information is available to constrain the interpretation of the seismic sections.

The mantle – the new frontier

So far our discussion of the reflection profiles has emphasised the discoveries made by studying crustal structure down to the Moho. When deep profiling began in 1975 it was a cause for excitement that reflections could be obtained from the Moho, and that seismologists could begin to study how the Moho varies from place to place and how it evolves through time. However, on most profiles recorded around the world, no reflectors were found within the mantle, except at sites of recent continental collision (e.g. the Pyrenees) or active oceanic subduction (e.g. Vancouver Island) where structures probably related to the transport of crustal material into the mantle were imaged. Over most of North America, Europe and Australia (the only continents where extensive reflection profiling has been carried out), the crust is frequently strongly reflective, but the mantle is normally transparent. The exception to this rule is the continental shelf of northwest Europe. On many academic and industry deep seismic profiles around the British Isles, prominent reflectors, some dipping, some horizontal, are present in the mantle. We do not know if this region of the Earth is geologically special, or whether the thin crust in this area just makes it easier to get seismic waves down into the mantle than elsewhere in the world, so that the apparently transparent mantle elsewhere may just be a question of insufficiently energetic seismic sources.

One of these intra-mantle reflections is quite spectacular. Figure 13.13 shows the **Flannan reflector**, which has been traced in three dimensions, for over 100 km along-strike and down-dip, northwest of Scotland. Figure 13.13a is a west–east dip-line, showing the east dip of the Flannan reflector, while the north–south strike-line in Figure 13.13b shows the reflector to be broadly arched about an east–west axis. Figure 13.13c shows how it is possible to use these two profiles (and in fact a grid of over twenty deep seismic profiles) to map out the three-dimensional structure of the Flannan reflector. It runs from about 20 km depth within the lower crust down to at least 80 km total depth (further east than the data shown in Figure 13.13a or the map in Figure 13.13c) in the sub-crustal lithosphere, dipping at about 30° to the east. Although the Flannan reflector extends up from the mantle into the lower crust, it does not appear to offset the Moho: in Figure 13.13a the Moho is at essentially the same travel-time both west and east of the Flannan reflector. Nor has it been possible to trace the reflector up into the upper crust, no matter how far to the west the profiles are extended: the Flannan reflector is observed to roll over and flatten out in the lower crust.

In addition to mapping out the geometry of the Flannan reflector, we can also try to relate it to

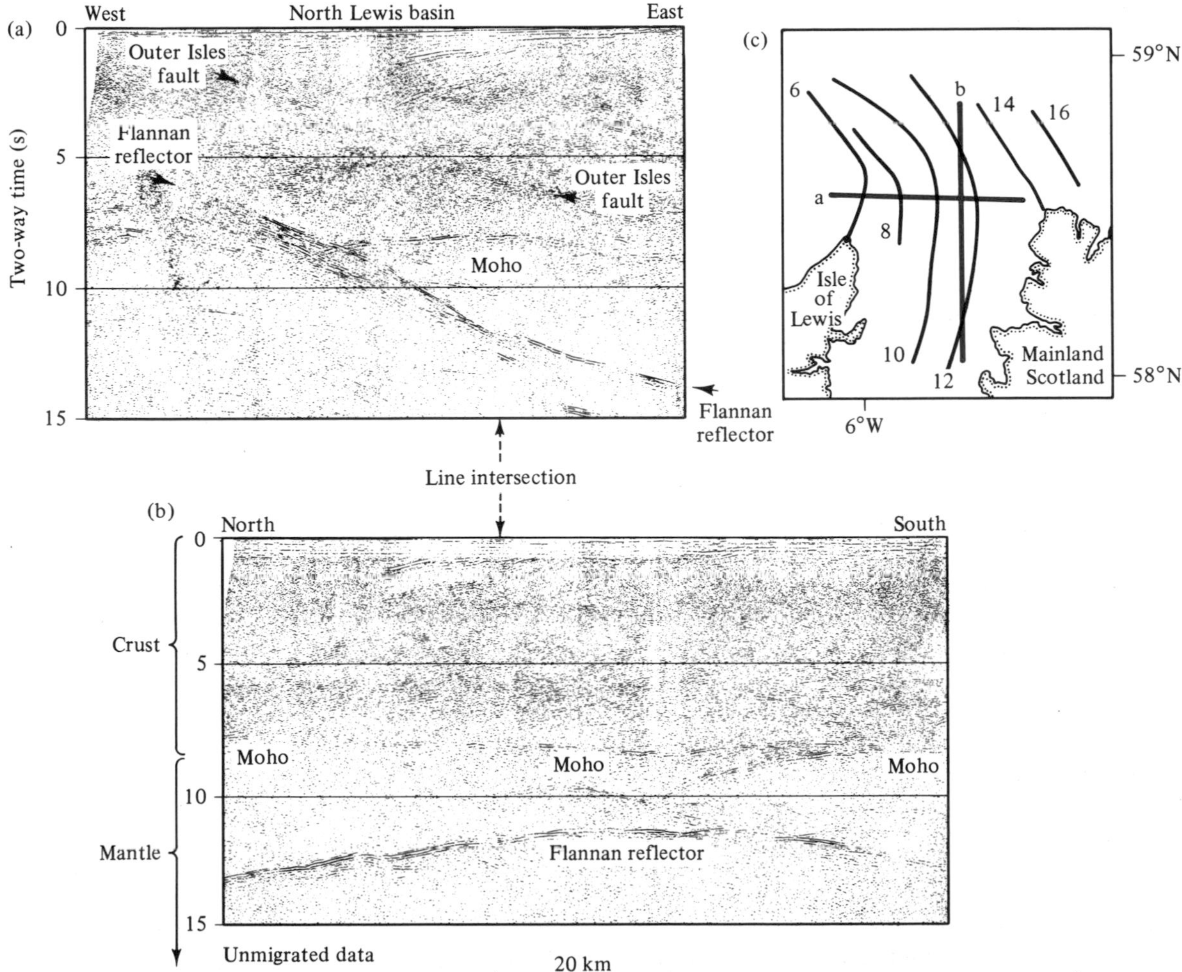

Figure 13.13 (a) BIRPS profile shot west–east across the Outer Isles fault and Flannan reflector, northwest of Scotland, unmigrated section. This is approximately a dip-line. (b) BIRPS profile shot north–south along the Flannan reflector, unmigrated section. This is approximately a strike-line. (c) Map locating reflection profiles in Figures 13.13a & b and showing contours of two-way time to Flannan reflector, which is an east-plunging antiform. Note that these contours are based on over twenty deep seismic profiles of which only two are shown. □

regional tectonics. This area of northwest Scotland has been subjected to west-directed Caledonian thrusting, followed by extension in the Mesozoic to form sedimentary basins in the offshore areas. The dip-line (Figure 13.13a) across the Flannan reflector shows that it is vertically beneath the Outer Isles fault. The Outer Isles fault is believed to have been a Caledonian thrust fault (compare its geometry to that of the Wind River thrust, Figure 13.5), subsequently reactivated as a Mesozoic normal fault (cf. the reactivated Pavant Range thrust, Figure 13.9) with a sedimentary basin formed above it. The throw on the Outer Isles normal fault was about 10 km, based on the deepest sedimentary rocks in the North Lewis basin (seen as shallow, west-dipping reflections in Figure 13.13a) which must have been deposited near sea-level but which are seen on the seismic section down to about 4 s two-way time. However, there cannot be nearly so much vertical displacement on the Moho where the Outer Isles fault reaches it – no more than 2 to 3 km – and other profiling shows that the Outer Isles fault does not extend through the Moho into the mantle. But when the crust extends, the lithospheric part of the mantle must do so also. It is probable that the Flannan reflector is related to the extensional faults in the crust, and represents a shear zone along which at least some extension in the lithospheric mantle took place. It is also possible that the Flannan reflector had an earlier history as a Caledonian compressional structure, just as the Outer Isles normal fault probably is a reactivated Caledonian thrust.

The Flannan reflector is currently the brightest, most laterally extensive intra-mantle reflection identified on any deep reflection data-set, anywhere in the world. However, there are now other dipping reflectors within the mantle on profiles recorded by the BIRPS group, albeit of a less spectacular nature than the Flannan reflector. These reflections are often vertically beneath upper-crustal faults (rather than continuous prolongations of the crustal faults), just like the Flannan reflector, suggesting that the intracrustal faults and the mantle reflectors are in some way related. The significance of these mantle 'faults' is a topic of considerable debate.

General results of deep seismic reflection profiling

Since 1975, when the first COCORP profile was collected, over 40 000 km of deep reflection data have been acquired worldwide; this rapid pace of acquisition shows no sign of slowing, and more countries are beginning deep reflection programmes every year. Our ideas about the nature of lithospheric processes and properties, such as its rheology, composition and growth, have changed dramatically as the result of deep reflection profiling. Thus this overview may soon become outdated, as newly collected profiles are incorporated into our changing picture of the Earth's lithosphere. Some general findings can be discussed, however.

Rheology

One of the most interesting results of deep reflection profiling concerns the mechanical, or rheological properties of the lithosphere. Predictions of the **rheology** of the lithosphere based on its temperature gradient and its mineralogical stratification suggest that there will be a change in behaviour of the crust at about 10–20 km depth, and another change at the Moho. The upper crust is cold and permits brittle faulting. At about 10–20 km (300–400 °C), the quartz and plagioclase-rich rocks making up the crust cease to behave in a brittle manner, and are warm enough for solid-state creep to occur, so that they flow rather than fracture. The depth at which this change in behaviour occurs, the **brittle–ductile transition**, varies from place to place depending particularly on heat flow and crustal lithology. At the Moho, another change in rheology is predicted because there is a change from the quartz–plagioclase–dominated lithologies of the crust to the olivine-dominated lithologies of the mantle. Olivine requires a much higher temperature than plagioclase to become ductile so, even though the mantle is hotter than the crust, the uppermost part of the mantle may show brittle rather than ductile behaviour, and may support localised faulting, if the stresses are great enough.

Does reflection seismology give any indication that these rheological predictions are true? Is the lithosphere a sandwich of brittle–ductile–brittle (upper crust–lower crust – uppermost mantle) rocks? Many of the results of reflection seismology support the sandwich theory. As we have seen, upper-crustal faults are not seen to continue throughout the crust on deep reflection profiles. Instead, they broaden out or disappear downwards into a zone of sub-horizontal reflections within the lower crust (the reflective lower crust in Figures 13.10, 13.12b and 13.13) which might perhaps represent ductile shear zones (see below). Dipping reflectors that may well be faults are again seen in the upper mantle (the Flannan reflector, Figure 13.13). It is certainly possible that this reflective pattern echoes the mechanical properties of the crust, and that the reflective lower crust can be equated with the ductile lower crust. Thus, deep seismic sections are exciting because they represent real data against which these types of speculation about lithospheric rheology can at last be tested.

Reflective lower crust

The observation that the lower part of the crust can differ radically from the upper part, on reflection records, has been made in many different parts of the world. Figure 13.14 is part of a BIRPS profile recorded southwest of Britain. At about 7 s two-way time (20 km depth), the almost transparent (unreflective) upper crust gives way to long, sub-horizontal reflections which correspond to geological layering or interfingering lamellations that extend 10 km or more horizontally but are less than 1 km thick. The reflections extend from 7 to 10 s two-way time (from about 20 to 30 km depth) and then terminate at the approximate depth of the Moho even more abruptly than they began. This reflectivity pattern, a seismically transparent upper crust, a reflective lower crust, and a transparent upper mantle (Figure 13.14), has been observed on data from Europe, Australia, the Basin and Range province of the western USA, and the east coast of both Canada and the USA. It has been only infrequently observed in cratonic areas (e.g. the Archaean Wyoming craton which is the Precambrian basement in Figure 13.5), and is not commonly present on profiles where compression has been the last tectonic event (e.g. the southern Appalachians, Figure 13.6). Because reflective lower crust occurs in so many different regions of the Earth, it must represent an important crustal process, the nature of which is still controversial.

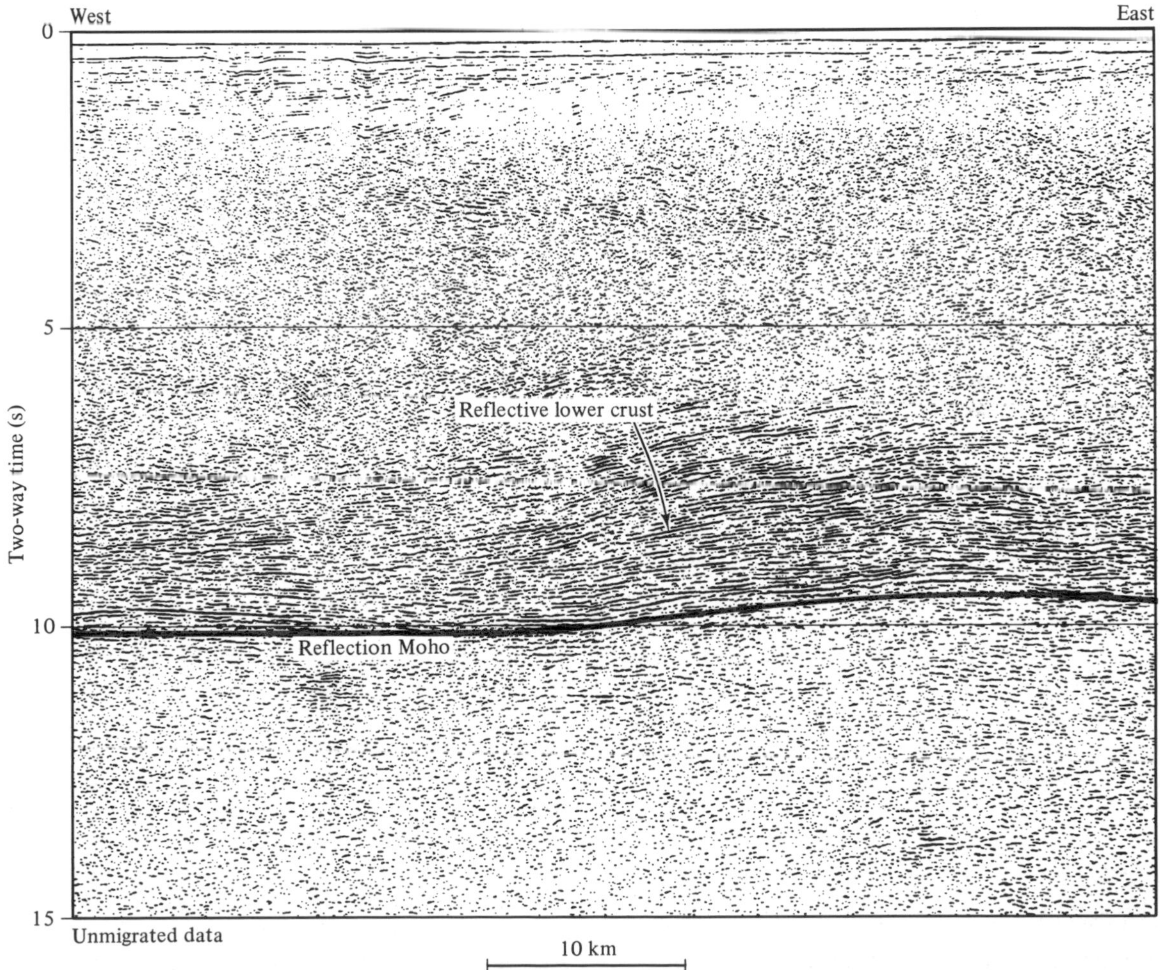

Figure 13.14 BIRPS profile southwest of Britain, unmigrated data. Note the great length (> 10 km) of some of the lower-crustal reflections.

Why is the lower crust reflective? There are several competing hypotheses. One hypothesis is that the reflective properties of the lower crust are caused by ductile shear zones within the lower crust. We have seen that brittle faulting does not seem to occur within the lower crust, but rather that extension is accomplished within wide zones of sub-horizontal ductile shear. Alignment of metamorphic minerals, and segregation of minerals into felsic/mafic layered gneisses on a large scale might produce the layered reflections seen on so many deep profiles. Because the reflections are most frequently observed in areas where the crust has been stretched as the last major tectonic event, some scientists believe that the reflections are due to fabrics produced in the crust during extension. Reflection profiles across areas where gneissic layering produced at mid-crustal depths has been brought to the surface show that rocks of this type can produce strong reflections. This hypothesis, equating the lower-crustal reflectors with the results of ductile shearing, provides a natural link between the rheology and the reflectivity of the crust that was discussed previously.

An alternative hypothesis is that the reflectors are primarily magmatic contacts between mafic sills and country rock, or within and between differentiated intrusions. The high proportion of mafic rocks among **xenoliths** brought to the surface from the lower crust indicates that much of the lower crust must be mafic. However, seismic velocities of mafic rocks do not match the results of refraction profiling, which indicate that the lower crust is, on the average, of intermediate composition. Intrusions of mafic partial melts from the upper mantle into a granodioritic lower crust would produce a lower crust with the correct seismic velocity, seismic layering, and also provide the source of mafic xenoliths.

A third hypothesis is that the cause of the reflectivity is related to the cause of large electrical conductivity anomalies often observed in the lower crust. The conductivity of the lower crust in many areas is so high that only free water (i.e. not bound in minerals) and some unusual materials such as graphite, are sufficiently conductive to provide an explanation. It has been argued that if even small amounts of free water under very high pressure are present in the middle and lower crust, this would lower seismic velocities sufficiently to produce the bright reflections observed on deep reflection profiles. All three possibilities – shear zones, mafic intrusions and high-pressure fluids – are under intense investigation at the present time.

Moho

The base of the crust, or Moho, has long been a topic of interest to Earth scientists. The Moho is defined as a jump in seismic velocities, and so can only be located by refraction data (which measure velocities) and not, strictly speaking, by near-vertical reflection data which image reflections without measuring detailed seismic velocities. What does the Moho look like on deep reflection profiles? The answer varies tremendously from place to place. On some profiles, typically those from cratonic areas, there are few distinctive reflections from the expected depth of the Moho (e.g. Figure 13.6). In other areas, there is a very bright, continuous reflection or band of reflections from depths corresponding to the refraction Moho (e.g. Figures 13.13a & b). A third common reflection pattern is for the base of the reflective lower crust to coincide approximately with the refraction Moho, without there being any unusually bright reflection from the Moho itself (e.g. Figures 13.10 and 13.14). Unless there are refraction data (Figure 13.1) to constrain the true position of the Moho at the same place as a seismic reflection profile, one cannot tell exactly what the reflection patterns correspond to in relation to the refraction Moho. Therefore a new term, **reflection Moho**, is used to describe the deepest sub-horizontal reflection observed on deep reflection profiles. The different patterns of reflection Moho just described probably correspond to different sorts of transition from crust to mantle. If the transition is very gradual, spread out over several kilometres, then there may be no distinctive reflections from the Moho (e.g. Figure 13.10). However, if the crust–mantle boundary is very sharp, localised perhaps along a décollement, or represented by a single intrusive boundary, then a distinctive, continuous, reflection Moho may be observed (e.g. Figure 13.8).

Is the Moho formed just once, when a piece of continental crust is created, and does it remain as a permanent marker ever afterwards? Or can the Moho be re-formed during tectonic events such as rifting and collision? What is the age of the Moho? Reflections cannot be directly dated, of course, but inferences concerning the age of the reflection Moho can be made in some areas. For example, in the Basin and Range province the reflection Moho is a fairly flat, continuous reflector which continues essentially unbroken beneath the rift. Because the area of the present-day Basin and Range province has a complex history of both rifting and subduction-related thrusting which stretches back to the Proterozoic, it is unlikely that a flat, continuous reflector formed at an early stage would have been preserved, and continue unbroken beneath collisional suture zones. Thus, the reflection Moho in the Basin and Range province seems to be a young feature, and is postulated to be of the same magmatic origin as the Tertiary volcanism that is widespread in the Basin and Range province. The relative youth of the reflection Moho in the Basin and Range province may be contrasted with the evidence from the Shetland Islands of north Scotland (Figure 13.12a). The Great Glen fault, on which the last significant strike-slip movement occurred in the late Palaeozoic, offsets the reflection Moho, as discussed above. Thus the Moho in this area must be at least Caledonian (Palaeozoic) in age, and possibly older.

Summary

Deep seismic reflection profiling has only become a widely available tool in the last decade or so. This technique has enormous potential to help develop our understanding of the Earth. And so data gathering and interpretation continue around the world. Improved acquisition and processing techniques are being developed, both by the seismic industry and by academic deep profiling groups, which will enable us to use deep seismic data more effectively to answer the questions above, and others as yet unthought of. Many of these new techniques rely on super-computers to carry out the billions of calculations which transform sound waves into a detailed picture of the unseen Earth.

Further reading

Popular articles and reviews with more examples of the use of deep seismic data, and pictures of the equipment used in the experiments:

Cook, F. A., Brown, L. D. & Oliver, J. E. (1980) The Southern Appalachians and the growth of the continents, *Scientific American*, **243** (4), 156–68.

De Voogd, B. & Keen, C. (1989) Deep seismic-reflection profiling, in *The Encyclopedia of Solid Earth Geophysics*, ed. D. E. James, Vol. 16 of The Encyclopedia of Earth Sciences, Van Nostrand Reinhold, NY, pp. 181–90.

Klemperer, S. L. & Fifield, R. (1988) Sound waves reflect Britain's deep geology, *New Scientist*, no. 1598 (4 February 1988), 73–8.

Matthews, D. H. (1989) Profiling the Earth's Interior, in *1990 Yearbook of Science and the Future*, ed. D. Calhoun, Encyclopaedia Britannica Inc., Chicago, pp. 178–97.

Mutter, J. C. (1986) Seismic images of plate boundaries, *Scientific American*, **254** (2), 66–75.

Textbooks on the seismic reflection technique: these cover in great depth the section 'How the technique works'. McQuillin et al. *is an undergraduate text; Sheriff & Geldart bridge the gap to graduate level.*

McQuillin, R., Bacon, M. & Barclay, W. (1984) *An Introduction to Seismic Interpretation: Reflection Seismics in Petroleum Exploration*, Graham & Trotman, London, 287 pp.

Sheriff, R. E. & Geldart, L. P. (1982). *Exploration Seismology: Volume 1: History, Theory and Data Acquisition*, Cambridge University Press, 253 pp.

Sheriff, R. E. & Geldart, L. P. (1983) *Exploration Seismology: Volume 2: Data-processing and Interpretation*, Cambridge University Press, 221 pp.

Textbook on related geophysics of the continental crust, which is a reference/source book for undergraduates, or a reader for graduates, on continental refraction, rheology, conductivity etc:

Meissner, R. (1986) *The Continental Crust: A Geophysical Approach*, Academic Press, London, 426 pp.

Conference volumes containing detailed technical articles about numerous reflection surveys and their geological results: these volumes contain peer-reviewed conference articles from the three international conferences to date and they show the great range of geological problems to which deep seismic reflection profiling has been applied:

Brown, L. & Barazangi, M. (eds.) (1986a) Reflection seismology: a global perspective, *American Geophysical Union Geodynamics Series*, **13**, 1–311.

Brown, L. & Barazangi, M. (eds.) (1986b) Reflection seismology: the continental crust, *American Geophysical Union Geodynamics Series*, **14**, 1–339.

Leven, J. H. *et al.* (eds.) (1990) Seismic probing of the continents and their margins, *Tectonophysics,* **172**, 1–641.

Matthews, D. & Smith, C. (eds.) (1987) Deep seismic reflection profiling of the continental lithosphere,

Geophysical Journal of the Royal Astronomical Society, **89** (1), 1–447.

Two monographs which present large numbers of deep seismic reflection profiles in high-quality, large-format displays, together with geologic interpretations. The first includes expanded reproductions and descriptions of the data displayed in this chapter as Figures 13.8, 13.10, 13.12, 13.13 and 13.14:

Klemperer, S. L. & Hobbs, R. W. (1992) *The BIRPS Atlas: deep seismic reflection profiles around the British Isles,* Cambridge University Press, 128 pp. +100 seismic sections.

Meissner, R. & Bortfeld, R. K. (eds.) (1990) DEKORP – *Atlas: Results of Deutsches Kontinentales Reflexionsseismisches Programm*, Springer-Verlag, Berlin, 18 pp. +80 seismic sections.

CHAPTER 14

Philip England
University of Oxford

DEFORMATION OF THE CONTINENTAL CRUST

Introduction

The theory of plate tectonics provides a unifying explanation for many of the motions that take place on the Earth's surface. The idea that oceanic lithosphere moves from the ocean ridges to the trenches without much internal deformation explains the major features of the ocean basins. In many ways, plate tectonics appears to account for the major features on the continents as well. Each of the three types of plate margin has its expression in deformation of the continental lithosphere. Mountain belts are formed where two plates converge: for example, in the Alps or the Himalaya. Continents can extend as, for example, in the East African Rift zone. Large zones of strike-slip deformation occur where plates move horizontally past each other as, for example, in the San Andreas Fault system of California.

We shall consider the different types of continental deformation and their relation to plate tectonics in more detail below. Before this, however, we must discuss several features common to all deforming continental regions which, on closer inspection, reveal that the continental and oceanic parts of the plates differ in their behaviour in quite important ways.

We begin by looking at the distributions of surface heights, earthquakes, and faulting, which show that deformation is much more diffuse on the continents than in the oceans. Furthermore, the style of deformation that is observed, at least in some parts of the continents, suggests that forces are acting in the continental lithosphere that cannot be explained by the theory of plate tectonics. Then we address the mechanical differences between continental and oceanic lithosphere that are responsible for the differences in their behaviour. The final part of the chapter describes the different types of continental margin, and the deformation and metamorphism that is associated with them.

Surface heights and isostasy

The topographic effects of **continental deformation** are well known, and may be seen upon opening any atlas. Each continent contains at least one major mountain range that is the result of Mesozoic or Tertiary **compressional deformation**. Figure 14.1 shows the distribution of regions where the surface height exceeds 1000 metres (m) above sea level. A large belt of mountains stretches almost continuously from the straits of Gibraltar as far as eastern China. The best-known of these mountain ranges are the Alps and the Himalaya, and the whole region is often referred to as **'The Alpine–Himalayan Belt'**. This name is rather misleading, as the Alps and the Himalaya in fact make up a rather small portion of the entire mountain belt. The Alps are barely visible on the scale of Figure 14.1, and although the average elevation of the Himalaya is about 5000 m, with several peaks exceeding 8000 m, even this mighty range is only about 250 kilometres

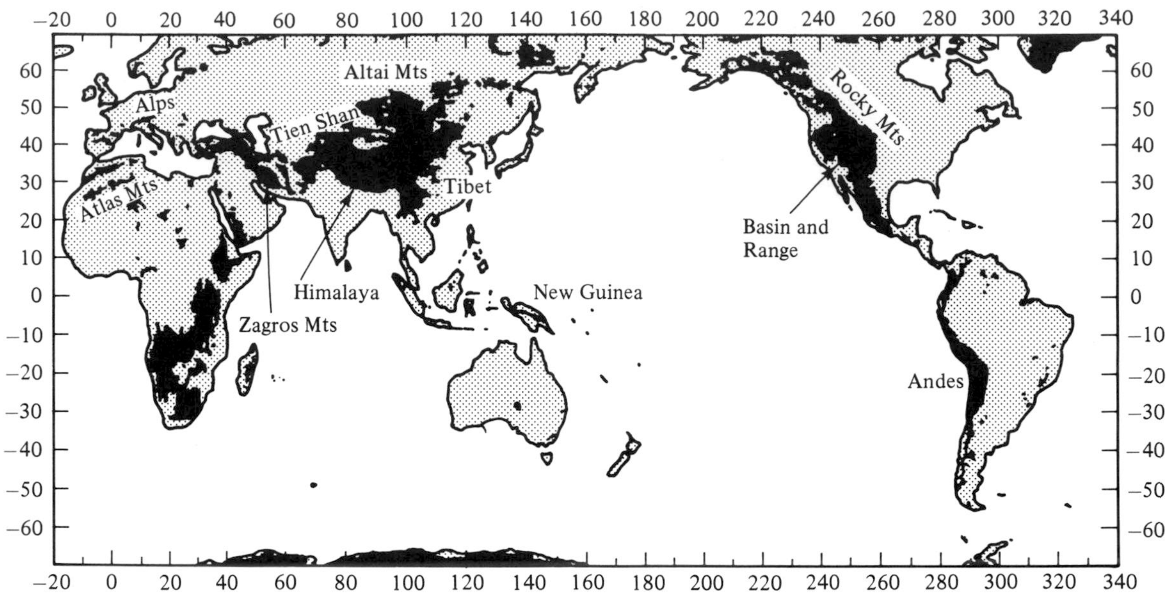

Figure 14.1 The continents, with regions of surface height greater than 1000 m shown in black. With the exception of Greenland, which is covered by a thick ice cap, and southern Africa, all the regions with surface height above 1000 m have been tectonically active in the Tertiary and most are still active now.

(km) wide. The topography of Eurasia is dominated by the three million square kilometres of the **Tibetan Plateau**, which has an average elevation of 5400 metres; subsidiary mountain ranges, such as the Tien Shan and Altai mountains, exceed 4000 m in elevation and are considerably larger than the Alpine chain of Europe.

The Alpine–Himalayan Belt is the site of the convergence between four different continental masses: Africa, Arabia and India are all moving approximately northwards with respect to Eurasia. The mountain ranges of Asia were formed during the penetration of the Indian plate into Eurasia. Much of the deformation in the Middle East (for example the Zagros Mountains of southern Iran, the Elburz Mountains of northern Iran and the Caucasus Mountains of the southern USSR, which can be seen in the eastern part of Figure 14.5, and in Figure 14.7) is caused by the convergence between Arabia and Eurasia. Africa and western Eurasia have been fairly close to each other for the last hundred million years or so. During this time there have been several episodes of compression that formed the mountain ranges surrounding the Mediterranean Sea. The most prominent of these ranges is the Alps of southern Europe which extend about 800 km from southeastern France to the east of Austria. The Atlas Mountains, on the southwestern coast of the Mediterranean Sea, exceed 4000 m in height, and there are mountain belts in Spain, Italy, Turkey and the Balkans, that were formed in the same time interval, though some of these – particularly in Greece and Turkey – are now regions of crustal thinning rather than thickening. We discuss the transition from thickening to thinning of the crust in mountain belts towards the end of this chapter.

The other major mountain ranges on Earth at present lie along the east coast of the Pacific ocean. The Andes of South America are nearly as high as the Tibetan Plateau, and stretch from 10° N to 55° S: a distance of over 7000 km. Much of western North America, too, is elevated, lying above 1500 m. Unlike the Alpine–Himalayan belt, which is the result of the horizontal compression of continents, the mountain ranges of North and South America were formed during the convergence of oceanic plates with the continents.

One of the most important conclusions that we can draw from observations of mountains is that large areas of the continents are underlain by crust that is thicker than average. This conclusion is based on observations of the **gravitational attraction of large mountain ranges**. Early investigators thought of mountain ranges as extra masses sitting on top of the Earth (Figure 14.2a). It is simple to calculate the gravitational attraction of such a mountain (a force that would point toward the mountain). The gravitational attraction of the rest of the Earth points towards the centre of the Earth, and in consequence the combined gravitational attraction of the mountain and the rest of the Earth is in a direction slightly away from the centre of the Earth (Figure 14.2a). This proposed **deflection of vertical** is small, but readily measurable using accurate astronomical observations.

In fact, when geodesists first investigated this proposal they did not find the expected deflections. Both Pierre Bouguer's survey, near the Andes between 1735 and 1745, and Sir George Everest's survey near the Himalaya a hundred years later, showed that the deflections of the vertical due to the presence of the mountain ranges are very much smaller than those calculated on the assumption that mountain ranges are simply extra masses on top of the Earth's surface (Figure 14.2b). The extra masses of the mountain ranges that are above the surface must be counterbalanced by lighter material beneath the mountain ranges.

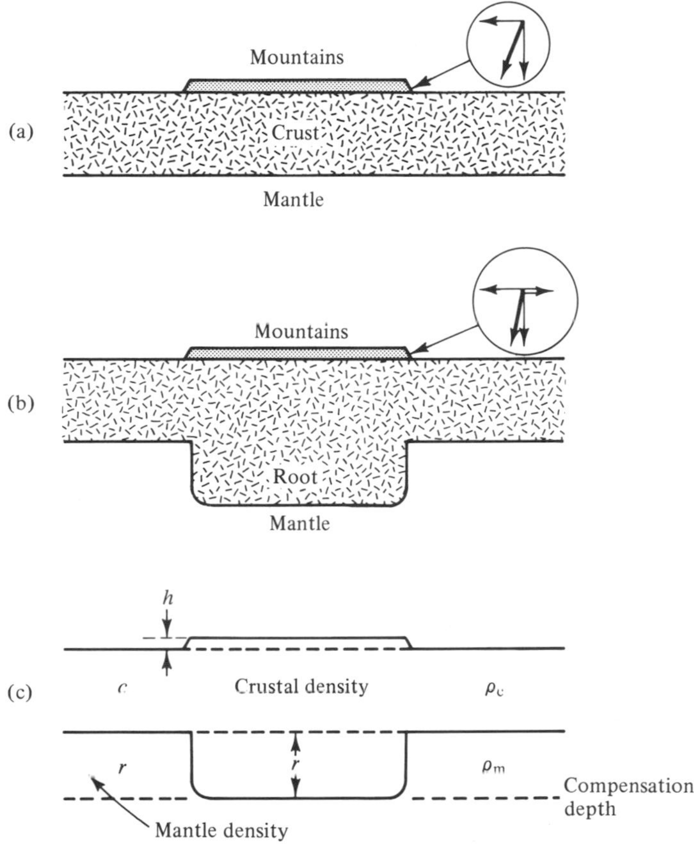

Figure 14.2 Sketch of isostatic compensation of mountain ranges. (a) A hypothetical mountain range sits on top of crust of constant thickness. The inset shows the result of a determination of gravity near such a mountain. The gravitation attraction of the Earth points towards the centre of the Earth, and the gravitational attraction of the mountain points towards the centre of the mountain (thin arrows). The combined gravitational attraction does not point precisely towards the centre of the Earth, so there is a local deflection of the vertical (thick arrow). The gravitational attraction of the mountain is greatly exaggerated in this sketch; actual deflections would be about a hundredth of a degree. (b) Observations of the deflection of the vertical near large mountain ranges show that the deflections are much smaller than expected for the case shown in (a). The attraction of the mountains above the surface is counterbalanced by the presence of a thick root of low density crust underlying the mountain. Because the root has less mass than the mantle it displaces, the gravitational attraction of the mountain range as a whole is reduced (open arrow) compared with the case illustrated in (a). (c) The theory of isostasy holds that the mass per unit area of each column of rock above a certain level (usually called the compensation depth) is the same. This condition is nothing more than Archimedes' Principle – applied to crust floating on the denser mantle, rather than to the more familiar examples of a ship or an iceberg floating on water. In this illustration there is no variation of density or pressure beneath the base of the crustal root. The extra mass represented by the mountain of height h and density ρ_c is counterbalanced by a root of thickness r, density ρ_c that displaces mantle of density ρ_m. It follows that $\rho_c h = (\rho_m - \rho_c)\, r$, and thus that $r = \rho_c h/(\rho_m - \rho_c)$. The total thickness of the crust is $t = h + c + r$, where c is the average thickness of the crust – 30–35 km. □

These observations gave rise to what is called the theory of **isostasy**. This states that the upper mantle behaves as a fluid on a geological timescale; therefore the continental crust, which is less dense than the mantle, floats upon it, as an iceberg does on water; the thicker the piece of crust, the higher is its surface. It is now generally accepted that mountain ranges are counterbalanced by their thickened crustal 'roots' (Figures 14.2b and c).

Two columns of lithosphere are said to be in isostatic balance if their masses (per unit area) above a given horizontal surface (usually called the **compensation depth**) are the same. It is, therefore, simple to test the theory of isostasy – in principle at least. If the masses per unit area of two columns of lithosphere are the same, then so are their gravitational attractions. So, provided that we correct for the change in gravitational acceleration as we change height, the acceleration due to gravity in the foothills of an isostatically compensated mountain range should be the same as at its top. In practice, it is difficult to make accurate determinations of gravity in mountain belts, and data are scarce. Nonetheless, it is now fairly well established that all large elevated regions are close to being in isostatic balance.

The density of the continental crust is about 2800 kg m^{-3}, while the density difference between crust and mantle is about one-fifth to one-seventh of this amount – about 400–500 kg m^{-3}. In other words, it requires 5 to 7 km of crustal root to counterbalance each kilometre of mountain range above sea level (Figure 14.2c). Thus the thickness of the crustal root in Tibet, where the average surface height is over 5 km, should be 25 to 35 km, and in the Alps, where the average surface height is about 2 km, the root should be 10 to 15 km thick. The average thickness of continental crust is 30 to 35 km, so the total crustal thickness in Tibet should be about 60 to 75 km, and in the Alps it should be about 40 to 50 km (see Figure 14.2 caption for calculation method). Where independent data – such as determination of Moho depth from seismic refraction studies (see Chapters 3 and 13) – are available, these conclusions are borne out. The crustal thickness in southern Tibet is between 55 and 70 km and the crustal thickness along the axis of the Alps is between 45 and 55 km.

Earthquakes

It is difficult, for logistical or political reasons, to work in the field in many of the largest regions of active continental deformation (for example, Iran, Tibet or Afghanistan at the time of writing). For this reason, much of the information on large-scale deformation of the continents that has accumulated during the 1970s and 1980s comes from earthquakes, which are easily studied from distant seismic stations.

What is an earthquake? **Earthquakes** may have more than one cause; indeed, the precise conditions that lead to earthquakes at depths of several hundred kilometres in subduction zones have still to be determined. Within the continental crust, however, earthquakes have a single cause: an earthquake occurs when the elastic strain of part of the Earth is *suddenly released* by slip on a fault. The sequence of events that involve the building up of elastic strain, its release in an earthquake and its building up again are known as the **seismic cycle** (Figure 14.3). The present view of earthquakes stems from a proposal by H. F. Reid to explain the 1906 San Francisco earthquake, which occurred on part of the San Andreas Fault system in California.

In order to determine the movement across the fault during a seismic cycle we need a marker, or markers. Let us imagine that someone builds three fences across the fault at one time (lines 1, 2 and 3 in Figure 14.3). These fences are markers that will move as movement takes place near the fault. Figure 14.3 illustrates the seismic cycle for a strike-slip fault – such as the San Andreas fault – across which all relative movements are in a horizontal direction; a similar picture can be drawn, though, for a fault of any orientation. As time progresses, the fences become bent (reflecting the bending – or strain of the ground near the fault) until, eventually, an earthquake releases some of the strain. Note that after the earthquake only some of the fault has broken: fences 1 and 3 are still deformed, and the strain near those parts of the fault will be released in later earthquakes on the same fault.

If we wish to measure the size of an earthquake we must determine the orientation of the fault, the direction of the slip, and the area of the fault that broke in the earthquake. The area of the fault that breaks, and the amount of slip, determine the amplitude of the seismic waves generated. It is a matter of routine, these days, to determine the size of an earthquake from the amplitude of its seismic waves. The commonly reported measure of size is the **magnitude** on the **Richter scale**. Table 14.1 lists the approximate ranges of fault areas and slips that apply for each magnitude on this scale.

The same routine determinations that identify the sizes of earthquakes also yield information on the orientations of the faults and of the movement upon them, for earthquakes of magnitude 5.5 or greater – that is for between 300 and 500 earthquakes a year, worldwide. Figures 14.7, 14.8 and 14.10 show the results of such study for the regions of Iran, the Aegean

Table 14.1. *The relationships between the magnitude of an earthquake, the area of a fault that broke in the earthquake, the amount of slip in an earthquake, and the amount of strain energy released in the earthquake. All numbers for fault dimensions, slip and energy released are approximate; any individual earthquake may depart from the numbers given in this table by a factor of three at least. Note that the energy released in an earthquake goes up by a factor of about thirty for each unit of the magnitude scale.*

Magnitude	Fault area (km^2)	Fault length (km)	Slip (m)	Energy (joules)	Number per year
3	0.15	0.4	0.008	2×10^9	100 000
4	1.5	1.2	0.025	6×10^{10}	15000
5	15	4	0.08	2×10^{12}	3000
6	150	12	0.25	6×10^{13}	100
7	1500	40	0.8	2×10^{15}	20
8	1.5×10^4	120	2.5	6×10^{16}	2
9	1.5×10^5	400	8	2×10^{18}	

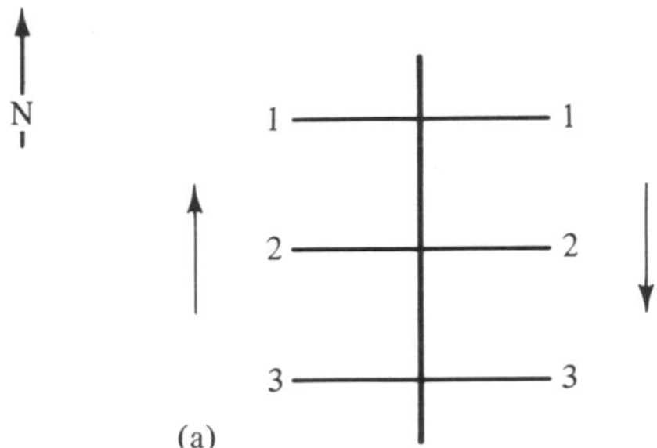

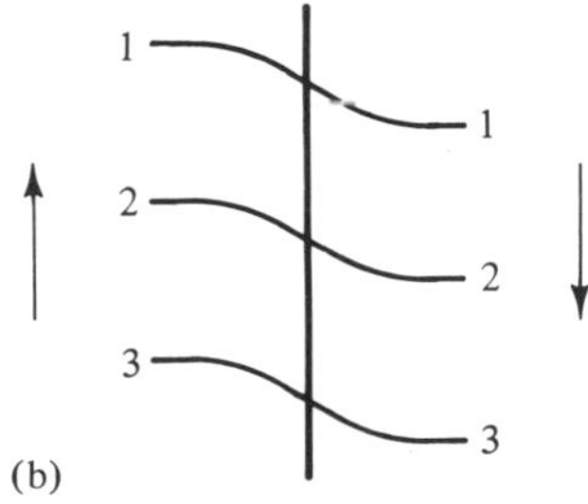

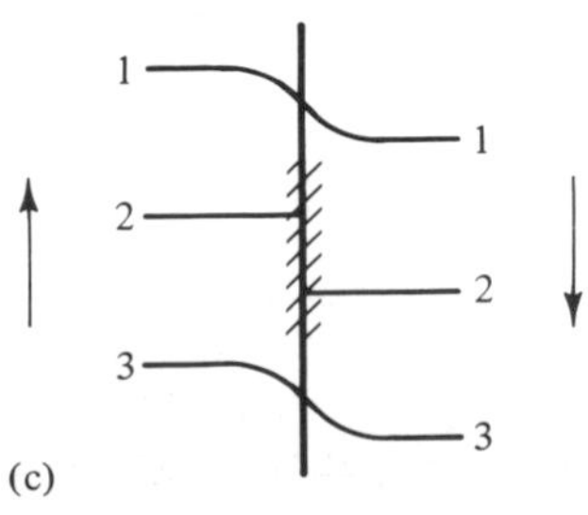

Figure 14.3 The seismic cycle for a strike-slip fault (one with horizontal relative motion only) such as the San Andreas Fault in California. The thick line running north–south represents a fault, on part of which an earthquake will occur in (c). (a) Three fences have just been built spanning the fault, they are shown by the lines 1, 2 and 3. The large arrows to the left and right of the figure indicate that the ground at the left of the picture is steadily moving northward with respect to the ground at the right. At the fault itself, no relative movement occurs. (b) The situation after an interval of several tens to hundreds of years; an earthquake is just about to occur. The ground has been strained as a result of the relative motions, and this strain is shown by distortion of the fences. (c) A few minutes after (b); the earthquake has just taken place, rupturing the section of the fault shown by the hatching. Fence 2 is now in two separate, but unstrained, pieces, and the elastic strain near this part of the fault has been released. The distance by which the two pieces of fence have moved apart is the slip in the earthquake. The strains illustrated in this figure are greatly exaggerated; the slip in an earthquake is usually a few ten-thousandths of the length of the fault break (see Table 14.1 for illustration of this relationship). □

Sea and Asia, but first we need to understand how to interpret these data.

Most modern techniques of studying an earthquake involve sophisticated analysis of all the vibrations that it excites, and are beyond the scope of this book. However, many of the earthquakes illustrated in this chapter were studied using an older technique that involves looking at only the first seismic waves to arrive from an earthquake – referred to as **first motions** – in order to derive a **focal mechanism** for the earthquake. This technique embodies the same principles of determining the movement in an earthquake from distant observations employed in modern sophisticated techniques, and is considerably simpler to illustrate (Figure 14.4).

Let us look again at the earthquake of Figure 14.3, and now imagine that we could not see what happened at the site of the earthquake, but could only determine what went on by looking at the seismic waves arriving at distant stations. (In fact, this is the usual state of affairs: most earthquake-related fault movements occur in inaccessible regions, or do not break the surface of the Earth, so they cannot be observed directly.)

All such waves that travel north and west (or south and east) will have their first motions away from the earthquake, and vice versa. Thus by observing at distant stations the pattern of first motions generated by an earthquake, we can determine what the pattern of first motions would have been near the earthquake. The representation of the seismic radiation from an earthquake, such as is shown in Figure 14.4, is known as the focal mechanism of the earthquake.

Unfortunately, the focal mechanism alone is not enough to enable us to determine completely the nature of the earthquake. As Figure 14.4b shows, the same pattern of first motions can be produced by movement on either of two different faults. Thus, once a pattern of first motions has been obtained, some detective work is still necessary to determine which of the two possible faults did in fact generate the earthquake. There is a variety of techniques available, and one of the most powerful for the study of deformation on land is the analysis of satellite photographs. These give detailed coverage of otherwise inaccessible regions. An example of what can be achieved by the combined study of earthquakes and satellite images is given later (Figures 14.10 and 14.11, which show the major active faults in Asia). Regardless of the ambiguity just discussed, the focal mechanisms of earthquakes give valuable information on the deformation that is occurring in a region. As Figure 14.4 shows, the focal mechanism of an earthquake tells us whether the fault was a normal fault, a strike-slip fault or a thrust fault, and thus whether the crust in the region was extending, shearing or compressing.

Earthquakes in the continents generally represent the deformation of near-surface rocks. The locations of earthquakes and the directions of relative motions that their first motions reveal play a vital role in understanding how plate tectonics works in the oceans, and they play an equally important role in showing how and where the continents deform. The distribution of earthquakes shows that the continental lithosphere deforms over much larger areas than the oceanic lithosphere does. We shall now look more closely at the nature of that deformation, as revealed by the distributions and the focal mechanisms of large earthquakes in the continents.

Distribution of earthquakes in continents

In the oceans shallow earthquakes occur in narrow bands around the edges of plates (see Chapter 9) and the absence of major earthquakes in the middle of plates gives us confidence that the plates are not deforming. Figures 14.5 and 14.6 show the distribution of earthquakes in two regions where the relative motions of the plates have brought continental masses together. Figure 14.5 shows the positions of earthquakes in the region of the Mediterranean Sea and eastern Atlantic Ocean; Figure 14.6 covers the large zone of deformation associated with the convergence of India with Eurasia.

The first important point to make about the distribution of earthquakes on the continents is that they are nearly all shallow. All the earthquakes in Figures 14.5 and 14.6 were certainly shallower than 35 km, and it seems probable that most of the major earthquakes within continental crust have focal depths shallower than 20 km. Large continental earthquakes break through the entire upper crust. In consequence, the study of major earthquakes in regions of active deformation yields a clear picture of the style of deformation in at least the upper 10 to 20 km of the lithosphere. In many regions (see for example Figures 14.6, 14.7 and 14.8) one style of deformation persists over a region several hundred kilometres on a side, and it is a reasonable assumption that the pattern of deformation seen in earthquakes in the top part of the lithosphere is also occurring at greater depth in the lithosphere, but (as we shall discuss at the end of the chapter) by other mechanisms, which do not cause earthquakes.

In some parts of Figures 14.5 and 14.6 the earthquakes result from the relative movement of plates as

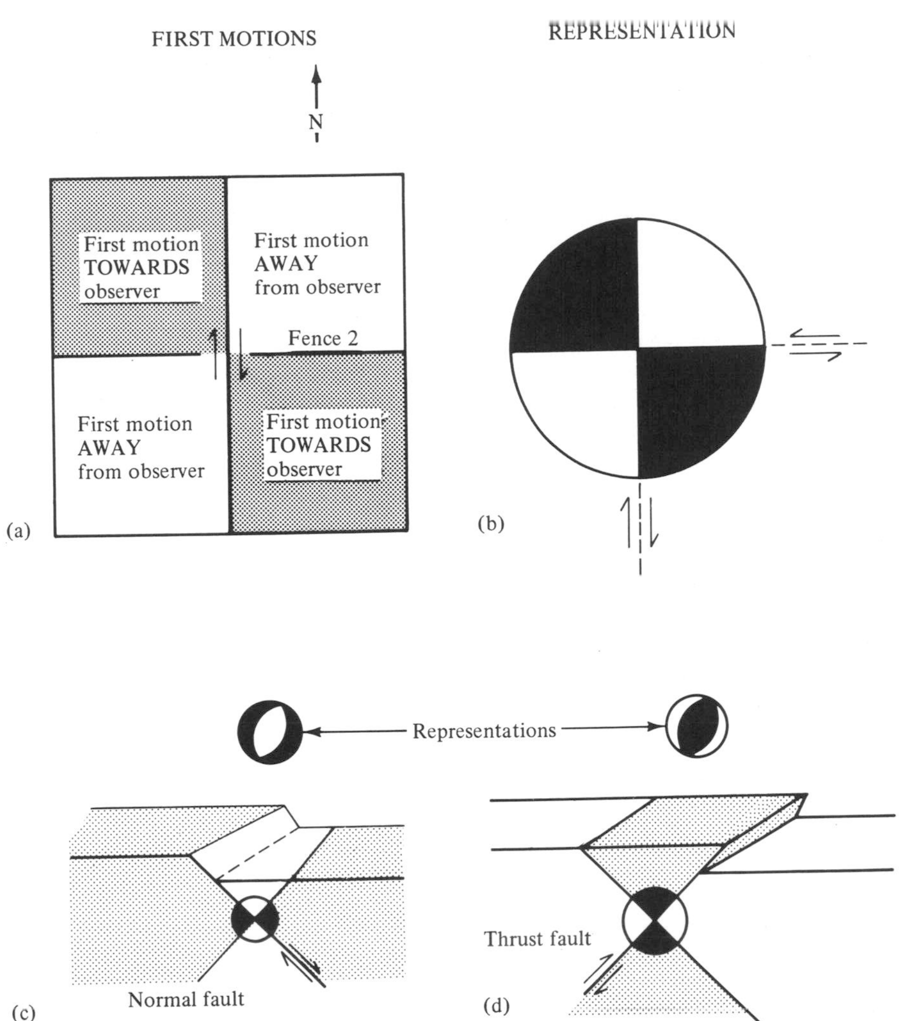

Figure 14.4 Focal mechanisms for earthquakes. Determination of the way in which a fault slipped in an earthquake is based on the fact that the nature of the observed seismic wave depends upon where (relative to the fault plane) the observation is made. (a) This illustrates a portion of the fault in Figure 14.3 in the neighbourhood of Fence 2 with the first reactions as they would be seen by an observer. The earthquake has just occurred, breaking the fence, and moving ground to the left of the fault rapidly northward, and ground to the right of the fault rapidly southward. The first movement of the ground experienced by any observer to the north and west of the earthquake will be away from the earthquake, towards the observer. Such *compressional first motion* is indicated by dark shading in this figure. There are two quadrants that experience compressional first motions, and two that experience *extensional first motions*: towards the earthquake and away from the observer. The fault illustrated here is a strike-slip fault. Faults can have any orientation and any direction of slip; observation of waves at distant seismic stations determines the orientation of the compressional and extensional quadrants, and so the nature of movement on a fault. These observations give us what is called a focal mechanism for the earthquake. (b) A standard representation of the focal mechanism in used: the quadrants are always illustrated *as if we were looking down on top of the earthquake*. Note that the same pattern of seismic waves would be observed if the earthquake had occurred by movement on a fault at right angles to the one we consider. The compressional and extensional quadrants are separated by a pair of lines at right-angles; these lines represent two vertical planes: the north–south plane on which the earthquake occurred, and an east–west plane, with the opposite sense of slip, which would have generated the same pattern of seismic waves. This ambiguity cannot be resolved unless independent evidence is available to tell us which is the fault plane. (c) and (d) The representations of the other two basic types of earthquakes – on *thrust and normal faults*. Note that, as viewed from above, the ground above the fault plane moves away from the observer (downward) in normal fault movement, producing extension and an 'open' fault plane solution (light centre). In contrast, compression in the case of the thrust fault yields movement toward the observer on the ground above the fault plane, and a fault plane solution with a dark centre. □

they are understood in plate tectonic theory. The narrow band of earthquakes in the western part of Figure 14.5, between 25° and 40° W, marks the position of the Atlantic ridge, where the separation of the North American and Eurasian plates is going on. Most of these earthquakes are small (magnitude 5 or less), and are associated with the extensional deformation; some are on transform faults – such as the cluster near 53° N on the Charlie Gibbs Fracture Zone. A similar band of shallow earthquakes may be seen in Figure 14.6 (10° N–10° S, 50–70° E), marking the position of the Carlsberg ridge. The subduction of oceanic material of the Indian plate beneath the Eurasian plate is marked by a band of earthquakes stretching from 90° E, 20° N to 120° E, 10° S in Figure 14.6. (The width of this seismic zone is exaggerated on the map because the subduction zone dips at a shallow angle, and so its projection onto the land surface is much broader than its actual thickness.)

In contrast to the narrow bands of seismicity in the oceans (cf. also, Chapter 9), the earthquakes in the continental parts of Figures 14.5 and 14.6 spread out over areas that are several hundred kilometres on a side. The present determinations of plate motions tell us the relative motions of the major plates in the region. For example, India is moving northwards with respect to Eurasia at about 50 mm yr^{-1}; the east–west dimension of the zone of earthquakes in Asia that result from this relative motion is approximately 2500 km, and the zone extends northwards for between 1200 and 2000 km from the northern edge of India. Similarly, the major east–west zone of earthquakes in Figure 14.5 results from the movements between three major plates: Africa is moving northwards with respect to Eurasia at about 10 mm yr^{-1}, Arabia is separating from Africa along the Red Sea and the Gulf of Aden and, as a consequence, is converging with Eurasia – along the Persian Gulf and the Zagros Mountains – at about 30 mm yr^{-1} in a northeasterly direction. The earthquakes in this region occur in bands up to 1000 km wide.

Figures 14.7 and 14.8 show the focal mechanisms (see Figure 14.4) for the largest earthquakes in two parts of the area covered by Figure 14.5, Iran and the Aegean Sea. The convergence between the Arabian and Eurasian continents in Iran takes place in a roughly northeasterly direction. Most of the earthquakes are on thrust faults running about NW–SE; if the earthquakes represent the deformation of the whole crust, we should conclude that the crust is

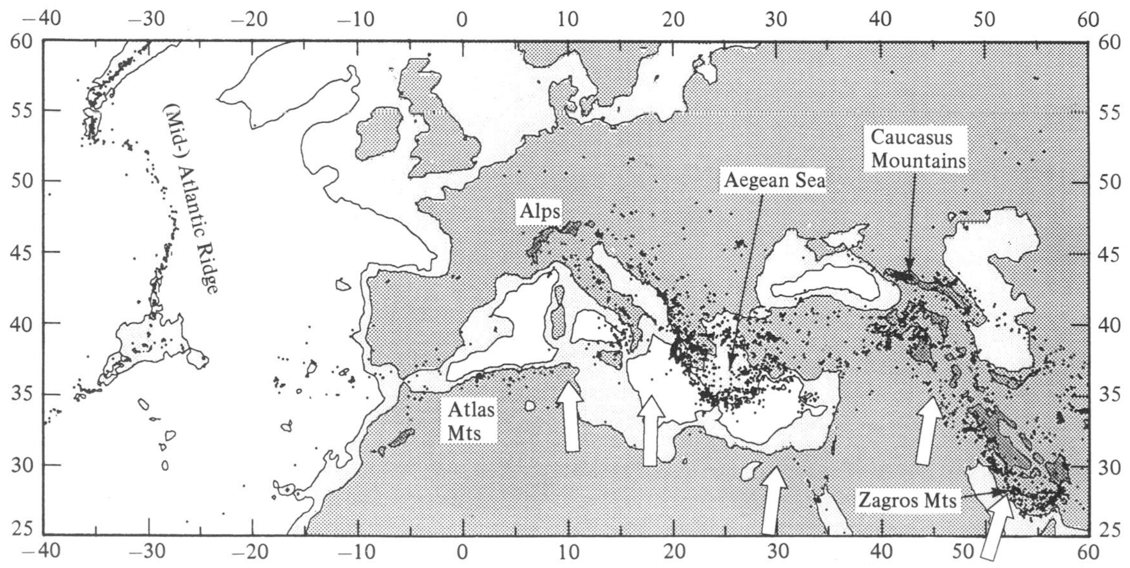

Figure 14.5 Distributions of surface height and earthquake activity in the Mediterranean and eastern Atlantic region. Surface height is indicated by shades of grey, with darkest shades corresponding to greater surface height. The three grey shades correspond to: light, areas with water depths 0–2000 m which are generally regions of continental crust; medium, land areas with surface elevations 0–2000 m; dark, land areas with surface elevations over 2000 m. Unshaded areas are covered by more than 2000 metres of water. Dots indicate the epicentres of earthquakes greater than magnitude 4 (see Table 14.1) in the interval 1961–84. Arrows show the direction of relative motions of the African and Arabian plates with respect to Eurasia. □

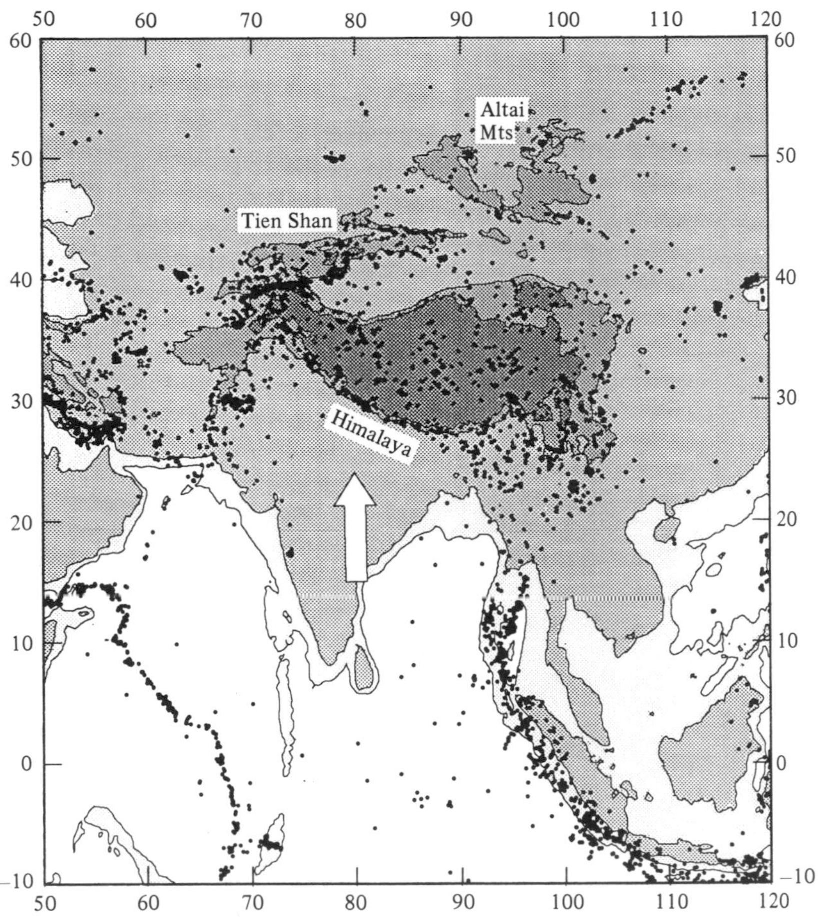

Figure 14.6 Distributions of surface height and earthquake activity in the Indian Ocean and central and southern Asia. Surface height is indicated as in Figure 14.5, but with the addition of a further, darker shading for land with elevations over 4000 m. Dots indicate the positions of earthquakes greater than magnitude 4 (see Table 14.1) in the interval 1961–84. Note that most of the earthquakes occurring within Asia are within regions where the surface height exceeds 1000 m above sea level. The open arrow shows the direction of movement of India relative to Asia. □

shortening in a NE–SW direction. None of the large earthquakes in the region occurs deeper than 15 to 20 km; thus there is no subduction zone, such as would be expected in an oceanic setting. The convergence of the Arabian and Eurasian continents causes the crust to *decrease in surface area and increase in thickness.*

The region of intense earthquake activity some 800 km by 800 km around the Aegean Sea (Figure 14.8) is one of the most rapidly deforming areas of the continents. The region is bounded to the north by the stable continent of the Eurasian plate and to the south by oceanic material of the African plate. Although the African plate is moving towards Eurasia, in this region at about 10 mm yr^{-1}, the Hellenic Trench is moving southwards with respect to Eurasia, resulting in overall extension of the region in a roughly north–south direction at about 70 mm yr^{-1}. In consequence, the net rate of underthrusting of the African plate in the Hellenic Trench is about 80 mm yr^{-1}. We shall discuss later (see Figure 14.15) the forces that may be responsible for these motions. For the moment, however, we concentrate on the motions alone. Apart from earthquakes associated with the subduction of the African plate, the earthquakes in this figure are mostly on normal faults. These faults run approximately east–west, and so accommodate the roughly north–south extension. As in Iran, the region is cut by many faults. There is no single spreading centre, such as there would be in an ocean. The crust of the Aegean region is, therefore, *increasing in surface area and decreasing in thickness.*

The features that we have discussed in detail for Iran and the Aegean Sea may also be observed in other regions of active deformation, and they exemplify a key point that emerges from a study of earthquakes on the continents. The upper continental crust in deforming regions appears to be broken completely by many faults over wide areas, in contrast to the oceanic plates, which are undeformed except in narrow bands round their edges. Although only the top part of the crust breaks in earthquakes, it seems likely that the rest of the crust also deforms, but we cannot observe this in regions that are deforming today. We discuss below the information derived from the study of rocks in

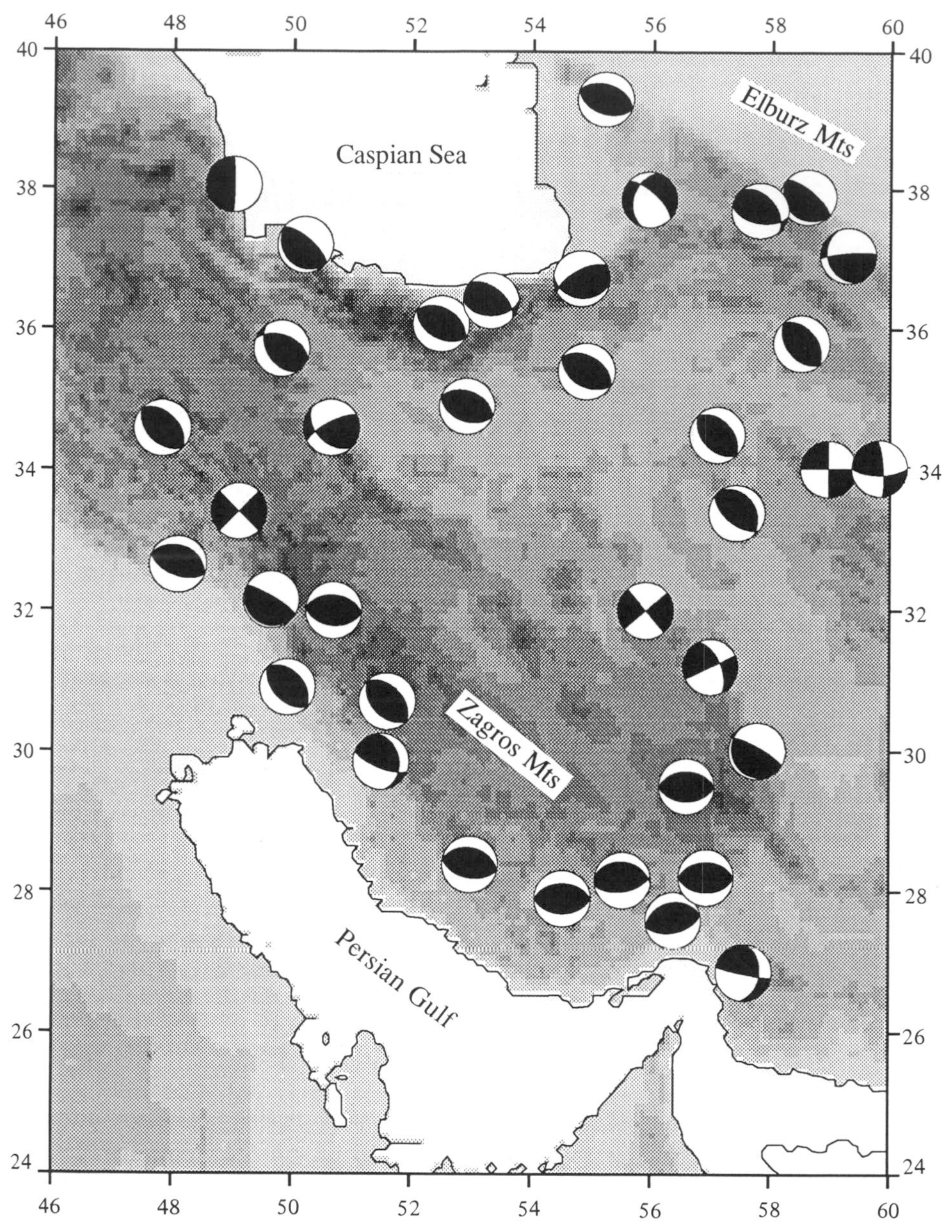

Figure 14.7 Focal mechanisms of earthquakes occurring in Iran between 1900 and 1987. The shading represents topography on a finer scale than in Figures 14.5 and 14.6. The predominant style of deformation is by thrust (compressional – see Figure 14.4) faulting in the south, and by a combination of thrust and strike-slip faulting in the north. The deformation results from the convergence of the continental masses of Eurasia and Arabia. □

older mountains ranges, where erosion has exposed rocks from deeper levels, which show ductile (fluid-like) deformation equivalent to the brittle deformation associated with earthquakes near the surface.

The deformation of Asia

The distribution of earthquakes just discussed tells us that the continental parts of lithospheric plates do not behave rigidly. Nonetheless, continental earthquakes are generally associated with the movement of the plates and it might seem that the only difference between continents and oceans lies in the *width* of the zones of deformation that result from the plate movement. The study of deformation within Asia has been central to demonstrating that this view is not correct, and that the continents are deforming in response to forces in addition to those resulting from plate motions.

The geological history of Asia since the end of the Mesozoic Era has been dominated by its convergence with India. Figure 14.9 shows the relative motion of India with respect to Asia since 70 million years ago. At first this relative motion was accommodated by the

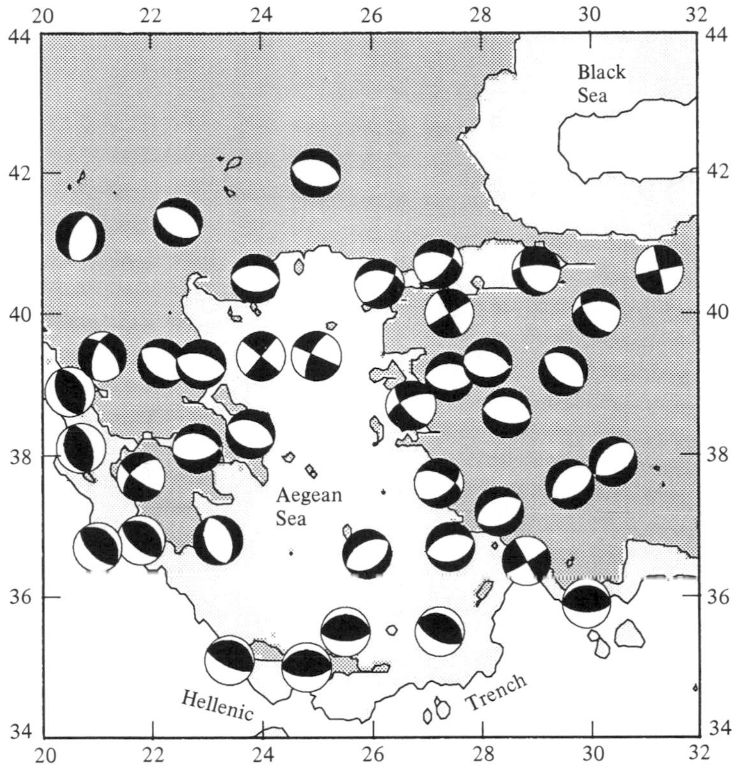

Figure 14.8 Focal mechanisms of earthquakes occurring in Greece, the Aegean Sea and western Turkey between 1900 and 1981. The predominant style of deformation is by normal (extensional – see Figure 14.4) faulting that results from the southward movement of the Hellenic Trench with respect to Europe. Focal mechanisms indicating thrust faulting in the south of the region represent the subduction of Africa beneath Eurasia in the Hellenic Trench; the remainder of the earthquakes represent mainly horizontal north–south extension within the continental crust of the Eurasian plate. (Surface heights are as indicated in Figure 14.5.) □

subduction of the ocean floor of the Tethys Ocean, but approximately 45 million years ago, the continents of Eurasia and India came into contact. Since that time, India has moved northward by over 2000 km with respect to Asia, at an average speed of about 50 mm yr^{-1}. The major result of this convergence has been to produce a region whose surface height exceeds 1 km, covering a large part of Asia (see Figures 14.1, 14.6 and 14.10). The most impressive feature in this region is the Tibetan Plateau (30° to 40° N, 80° to 100° E) which has an average height of over 5 km. Simple isostatic calculations, such as those at the beginning of this chapter, suggest that the crust of the plateau is as much as twice the normal thickness of continental crust. Much of the plateau is covered by marine limestone of Cretaceous age, which shows that it was below sea level before the convergence of India with Asia began. It seems, therefore, that much of this convergence was taken up by the kind of distributed crustal thickening that we discussed in the case of Iran at the present day (Figure 14.7).

Figure 14.10 displays the focal mechanisms of the major earthquakes in Asia during this century. The dominant movement on the scale of the plates in this Figure is the convergence of India with Asia in a roughly northerly direction, and this is reflected in thrust faulting on approximately east–west planes beneath the Himalaya, on the southern margin of the deforming zone, and beneath the Tien Shan at the northwest margin (40° N, 72° E). (Note that the focal mechanisms around 28° N, 85–100° E are low-angle thrust faults with shallow dips.) The deformation of the rest of the region, however, is not consistent with north–south shortening. Within the Tibetan Plateau the earthquakes show a combination of strike-slip and normal (horizontal extensional) faulting that indicate a large amount of east–west extension. On the eastern and western edges of the plateau, the earthquakes show thrust faulting on approximately north–south planes – consistent with *east–west* shortening. The transition between compressional deformation beneath the Himalaya and extension in the Tibetan Plateau, immediately to the north, takes place over a distance of less than 100 km.

Of course, earthquakes have only been studied seriously in the last hundred years or so, and a century is only a short time in geological terms. The use of high-resolution satellite photographs has, however, provided a valuable addition to the study of earthquakes. Strike-slip and normal faults show up on such photographs as clear lines cutting across country. Each of the earthquakes represented in Figure 14.10 broke only a small section of fault (recall Figure 14.3), but each fault may be traced for a considerable

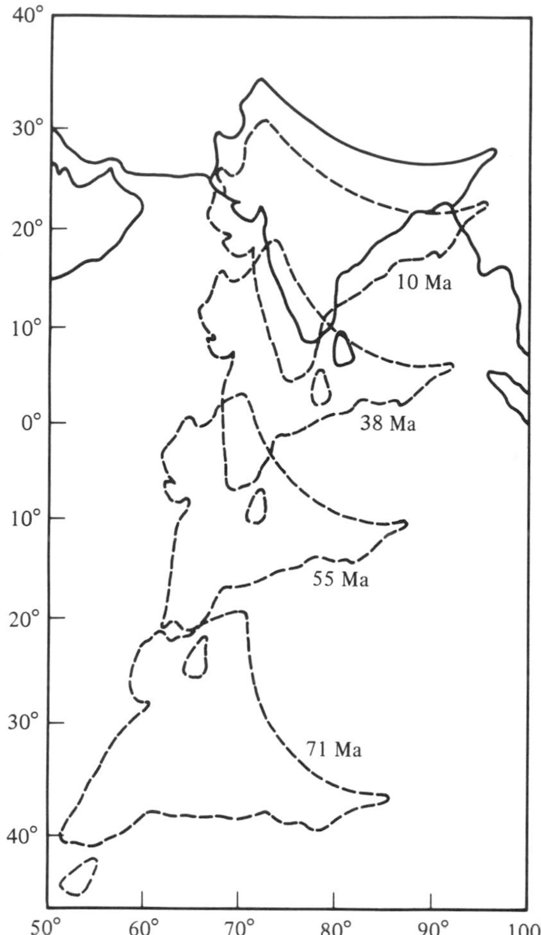

Figure 14.9 **The movement of India with respect to Asia over the last 70 million years. India collided with the coastline of Eurasia – which was about 2000 km further south than the site of the present Himalaya – about 45 million years ago.** □

distance. Most of the major active faults in Asia have been active throughout the Quaternary, and leave clear traces that may be seen from satellite images, such as that in Figure 14.11. The slip on these faults inferred from satellite images is consistent with what is seen in this century's earthquakes.

The pattern of deformation is rather surprising. We would expect, by analogy with subduction zones, to find major thrust faults at the site of the convergence between two continents and, indeed, they are found in the southernmost part of the region. Major thrust faults are active in the Himalaya, where the Indian subcontinent is moving beneath the Tibetan Plateau. This thrusting dies out, however, beneath the highest part of the Himalaya, less than 300 km north of the foothills of the Himalaya, and is replaced by the extensional and strike-slip faulting that covers the 1000-km-wide Tibetan Plateau. This extensional faulting in turn dies out northwards to be replaced by thrust faulting on the northern and eastern margins of the plateau. It seems clear that the forces acting within Asia today are not simply those due to the relative motion of the Indian and Eurasian plates. We shall discuss the origin of these forces below.

Summary of the nature of continental deformation

The observations discussed in the preceding sections demonstrate three key features of continental deformation.

First, the deformation is widely distributed: it is spread over regions hundreds to thousands of kilometres wide, and not concentrated on single faults or narrow fault systems, as is the deformation in the oceans.

Second, this deformation is different in character from the corresponding deformation in the oceans. A distributed zone of continental compression, for example, is not simply a collection of subduction zones in which lithosphere is pushed into the mantle, but is a region in which the crust is thickened. Equally, distributed continental extension does not take place by the creation of lithosphere at many spreading centres, but by the widespread thinning of continental crust and, probably, the lower lithosphere as well. In either case, the material of the continental crust is approximately conserved; it is redistributed by the deformation, but apparently is not substantially added to by plate creation, nor subtracted from by subduction.

Third, the forces acting within the continental lithosphere are not always those that would be predicted simply from plate tectonic considerations. Sometimes the style of deformation does appear to be consistent with the relative motions of two plates. For example, in Iran the convergence between Arabia and Eurasia is accommodated in a zone approximately 1000 km wide, whose deformation is mainly compressional and in the orientation that would be expected from plate tectonic considerations (Figure 14.7). In other cases the deformation cannot be predicted from the motion of the major plates. The African and Eurasian plates, which border the Mediterranean region, are converging in a roughly north–south direction at about 10 mm yr^{-1}. The most rapid rates of deformation in the region, however, are in the Aegean Sea (about 40° N, 25° E in Figure 14.5), which is experiencing roughly north–south *extension* (Figure

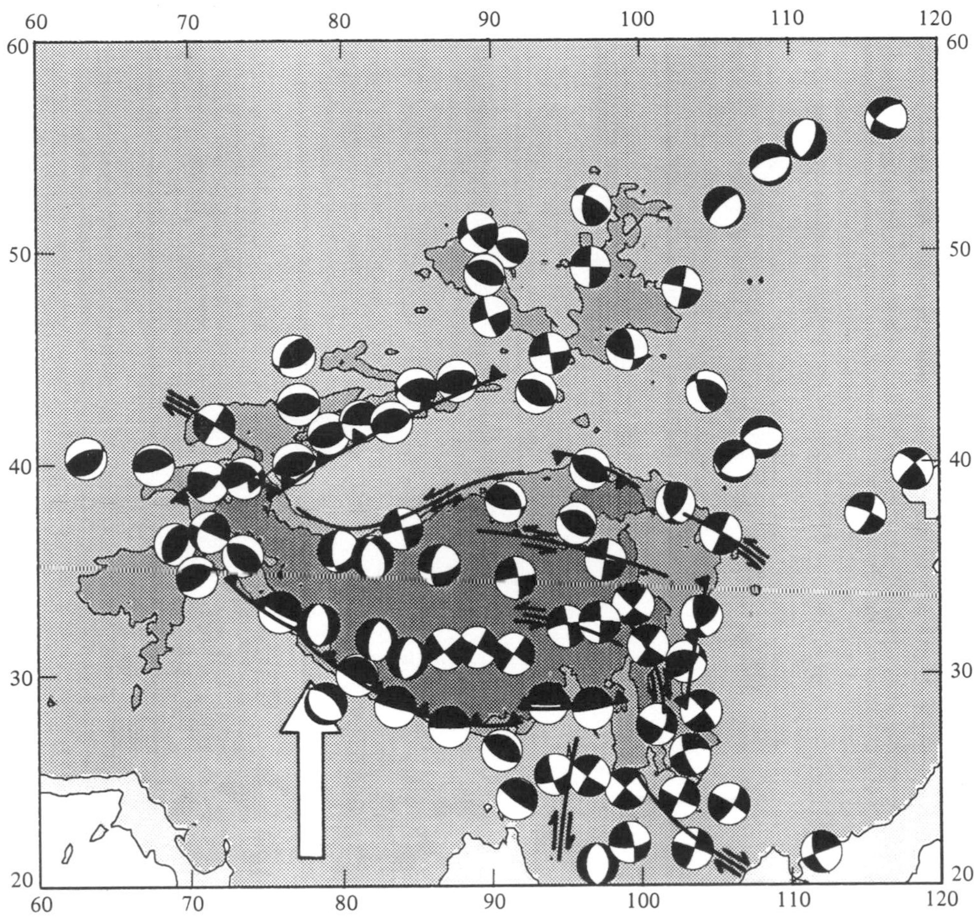

Figure 14.10 Focal mechanisms of earthquakes occurring in Asia between 1900 and 1987. The open arrow indicates the direction of relative movement of India with respect to Asia. The major faults associated with these earthquakes are shown as thick lines. Faults with teeth marks indicate convergence; faults with twin opposing arrows indicate strike-slip motion. Many of these faults have broken in earthquakes during this century along only short segments. Much of the information on these faults comes from the study of satellite images such as that displayed in Figure 14.11. Only the largest faults in the region are shown. Note that many earthquakes happen on faults too small to show at this scale. The highest part of the region, the Tibetan Plateau, is surrounded by thrust faults, but the active faults in its interior are all strike-slip or normal faults – in other words, the crust of the plateau is thinning. The state of affairs we see today has existed for probably only the last few million years. Recent geological expeditions to Tibet have found clear evidence there of the thrust faulting that must have taken place to produce the thick crust of the plateau. (Surface heights are as indicated in Figure 14.6.) □

14.8). Finally in Asia, the largest active zone of continental convergence, while the intense compressional deformation and active thrust faulting in the Himalaya seem entirely consistent with what one would expect on plate tectonic grounds, from the convergence between two plates, the whole of the Tibetan Plateau is undergoing crustal thinning at the present day (Figure 14.10). To find the explanation for this behaviour we need to examine some of the differences between the continental and oceanic lithosphere.

The continental lithosphere

The lithospheres beneath continents and oceans both participate in plate motions, and are presumably subjected to roughly the same forces, yet continental lithosphere deforms extensively while oceanic lithosphere does not. Clearly, there must be important mechanical differences between them. The principal mechanical differences are in their *densities* and *strengths*: the continental lithosphere is less dense

Figure 14.11 Landsat image of a portion of southern Tibet 180 km across. Mt Everest can be seen in the lowermost right-hand corner, its north face being the triangular region in shadow. The generally east–west fabric of the terrain in most of the image is due to folding and thrust faulting of Jurassic and Cretaceous sediments that were laid down on the northern continental margin of India, and were subsequently deformed and thickened following the collision of India with Asia. The Indus–Tsang Po thrusted suture at the junction of these two terrains runs east–west near the top of the image. The major active faults, however, are normal faults that strike approximately north–south (east–west extension); notably those bounding the Guzuo graben, which begins to the northwest of Mt Everest whence it can be traced to the centre of the image and beyond. □

than oceanic lithosphere, and it is weaker (Figure 14.12).

Explosion seismology shows us that the crust is 30 to 35 km thick beneath continents with a surface close to sea level. The crust generally has a layer of sediments on top, which may be as thin as a few metres, or as thick as 10 to 20 km. The sediments are underlain by a layer whose average composition is similar to that of granites, and this is, in turn, underlain by a layer that is more basic in composition (see Chapter 8 for crustal growth, layering etc.). The continental crust is highly variable, and any given section of continental crust probably violates this generalisation in at least one respect. In contrast, the oceanic crust is much more consistent in composition; it has a thin layer of sediment on top, which varies in thickness with age of the ocean floor and proximity to sediment sources, but underneath this there is usually a uniform layer of about 5 km of basalt and gabbro, which have a density of approximately 2960 kg m^{-3}. Old oceanic crust has its surface some 6 km below sea level, while continental crust that is 30 to 35 km thick has a surface height of a couple of hundred metres above sea level. This allows us to estimate that the average density of continental crust is about 2800 kg m^{-3} (see Figure 14.2).

One important consequence of the differing structures of continental and oceanic lithosphere is that old oceanic lithosphere is denser than its underlying upper mantle, because it is colder and has only a thin layer of light crust, whereas continental lithosphere is less dense than the upper mantle, because of its great thickness of lower density crust (Figure 14.12a). Thus old oceanic lithosphere has a tendency to sink into the upper mantle (cf. Chapter 9).

A second important difference between continental and oceanic lithosphere is that the continental crust contains larger quantities of the *heat producing radioisotopes* of uranium, thorium and potassium, than does the oceanic crust. The average *surface heat flow* from the continents is about 60 mW m^{-2}, of which

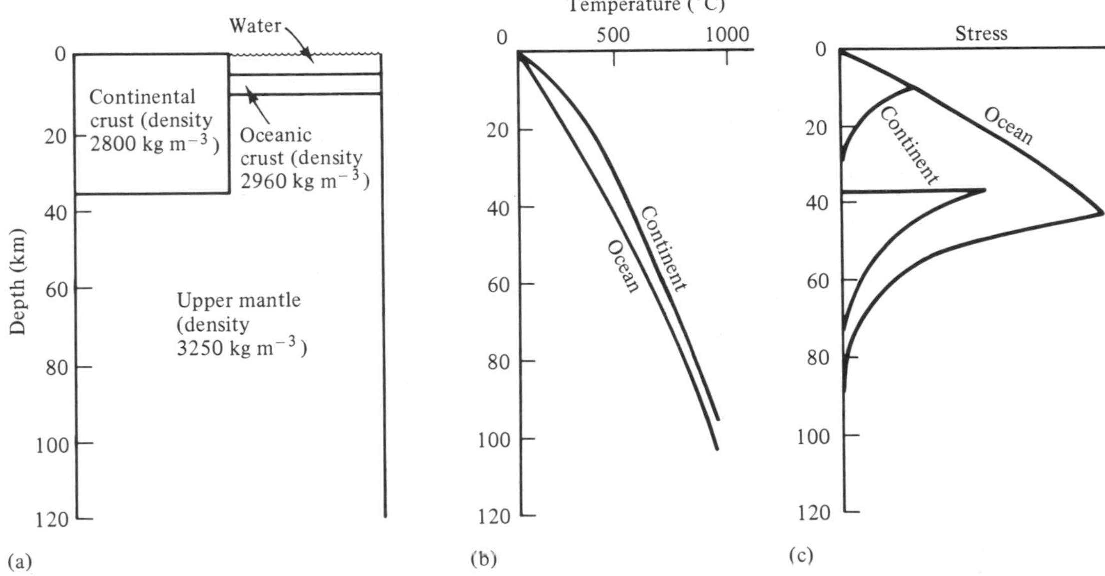

Figure 14.12 General cross-section through continental and oceanic lithosphere, with schematic profiles of the structures of density, temperature and strength in the two types of lithosphere. (a) The density structure of continental and oceanic lithosphere. Continental crust with a surface at sea level is approximately 35 km thick, whereas oceanic crust is usually 5 to 8 km thick. In each case the crust is underlain by denser upper mantle that makes up the rest of the lithosphere. The lithosphere rides on top of the asthenosphere, which has the same properties as the mantle portion of the lithosphere, except that it is hotter – and so less dense, and less strong. The density of the asthenosphere is about 3170 kg m^{-3}; in the situation illustrated here the average density of the oceanic lithosphere is 3240 kg m^{-3}, and that of the continental lithosphere is 3120 kg m^{-3}. (b) Approximate temperature profiles (geotherms) through continental and oceanic lithosphere. The temperatures shown are appropriate for 100-Ma-old ocean lithosphere, and for continental lithosphere having average surface heat flux. (c) Schematic profiles of the strengths of continental and oceanic lithosphere. The curves show the stresses that would be required to deform the two kinds of lithosphere at a reasonable geological strain rate; no absolute values are shown because they are uncertain, and what is of importance is the relative strengths of the columns. Continental lithosphere is weaker than oceanic lithosphere because it contains a thick layer of weak (quartz- and feldspar-rich) crust in the depth range where the oceanic lithosphere has its greatest strength, because it consists mainly of less deformable olivine and because temperatures, beneath the old oceans at least, are generally somewhat lower than beneath the continents. □

approximately 40 per cent is generated *within* the crust by the decay of these radioactive isotopes. Of course, the surface heat flow in young oceanic lithosphere can be much higher than this, but in oceanic lithosphere that is appreciably older than 60 Ma the surface heat flow is less than 60 mW m^{-2}. In consequence, temperatures in the upper mantle beneath the continents may be considerably higher than those beneath the oceans at the same depth (Figure 14.12b; see also Chapters 7 and 12 for discussions of radioactivity and heat flow).

The significance of this discussion of temperature profiles lies in the fact that the strength of rocks depends greatly upon their temperature. As we have seen in Chapter 9, the oceanic portions of the plates form part of the convection in the mantle. They behave rigidly because they are much colder and stronger than the ductile mantle beneath them. The strength of the plates lies not in the oceanic crust, which contains many fractures, and is probably quite weak, but in the olivine-rich upper mantle. The top of the oceanic upper mantle is at a depth of only 5 to 10 km, and at a temperature of about 100 °C. Under these conditions, olivine is very strong, and it would require forces far greater than those that do act on the plates for the upper mantle beneath the oceans to deform over geological timespans. In contrast, the crustal part of the continental plates is thicker; as we have seen, its upper 10 to 20 km is broken by many faults, and below this depth the continental crust is made up of minerals that are generally much weaker than olivine is under the same conditions of temperature and pressure (Figure 14.12c). As discussed below, the most common deformational features of rocks from the deeper parts of mountain belts are the folds formed by the flowing of crustal rocks. Furthermore, the mantle portion of the continental lithosphere may often be hotter – and so weaker – than the equivalent portion of the oceanic lithosphere (Figures 14.12b and c).

Thus continental lithosphere differs from oceanic lithosphere in two important ways. First, *it floats on the upper mantle, whereas oceanic lithosphere has a tendency to sink* as it ages because its density becomes greater than that of the underlying asthenosphere. It is easy, when two oceanic plates converge, for the convergence to take place by the subduction of one or other of the plates, but when two continental plates converge, subduction does not occur to any great extent. Second, *the continental lithosphere is much weaker than the oceanic lithosphere*; consequently convergence between continental plates involves deformation over very wide regions.

There is one crucial aspect of the mechanics of the continents that we have still not addressed. We still have to account for extensional deformation such as that seen in Asia, immediately in front of compressional plate boundaries (Figure 14.10). Let us return to the isostatically balanced mountain belt in Figure 14.2, where we assumed that the pressure everywhere was approximately equal to the weight per unit area of overlying rock. This assumption allows us to draw a simple sketch of the distribution of pressure beneath a mountain range, and to compare it with the distribution of pressure beneath lower land nearby (Figure 14.13). If isostatic balance holds, the pressure at the bottom of these two columns is the same. At the top of each column, the pressure is just the atmospheric pressure. The tops of the columns are at different heights, however, and the pressure at sea level in the mountainous column is equal to the weight per unit area of the mountains – which for mountains 5 km high, like Tibet or the Himalaya, is $9.81 \text{ m s}^{-2} \times 5000 \text{ m} \times 2800 \text{ kg m}^{-3} = 1.4 \times 10^{8} \text{ Nm}^{-2}$ = 1400 times atmospheric pressure. The pressure in the mountainous column stays greater down to the level of the base of the crust in the lowland column (M2 in Figure 14.13). Below this level, the pressure increases more rapidly with depth in the lowland column, because the mantle is more dense than the adjacent, submountain crust. Below the level of the base of the crust in the mountainous column (M1 in Figure 14.13), the pressures are the same (in this simple model that assumes a compensation depth at M1).

We can see, therefore, that the pressure in the crust beneath mountains is everywhere higher than at equivalent levels beneath lowlands, and that there must therefore be a tendency for mountains to spread under their own weight (Figure 14.13b). We can now look again at the tectonics of Tibet (Figure 14.10). The deformation is the result of the addition of two forces: one a compression resulting from the convergence of the two continental plates, the second an extension resulting from the tendency of the plateau of Tibet to spread under its own weight. Figure 14.14 illustrates how the sum of a north–south compression and a tendency of the plateau to spread in all directions can lead to compression directed radially outward around the edges of the plateau and to a combination of normal and strike-slip faulting within the plateau.

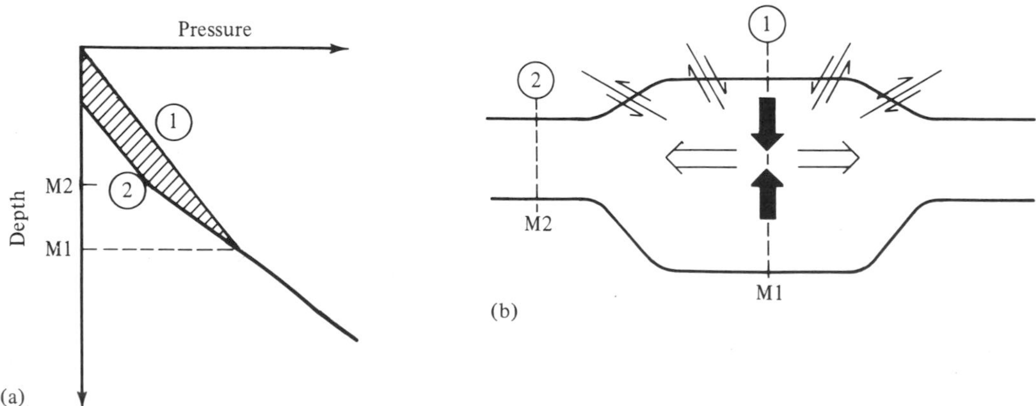

Figure 14.13 Sketch of pressures beneath mountains and lowlands. In (a) pressures are plotted against depth for two columns of lithosphere in isostatic balance at a compensation depth, MI, as shown in (b). Column I lies beneath a mountain range, column 2 lies beneath land whose surface is close to sea level. Pressure at any depth is assumed to be equal to the weight, per unit area, of overlying rock. (The excess pressure of column I over column 2 creates a sideways force, as shown by the large open arrows in part (b), that is in every way analogous to the 'ridge push' force that operates to either side of ocean ridges.) The faults illustrated schematically in (b) show that the high ground can be extending while the lower ground is shortened by thrust faulting. □

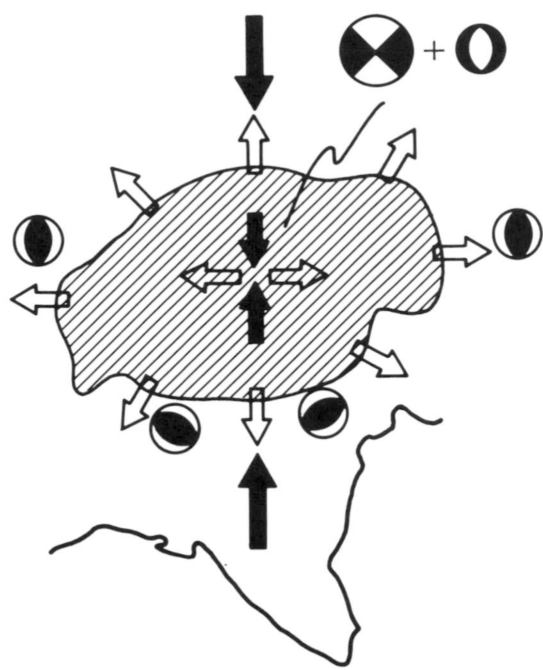

Figure 14.14 Schematic representation of horizontal stress components in the India–Asia collision zone. The coastline of the Indian subcontinent and the approximate outline of the Tibetan Plateau are shown. Compressional stresses (solid arrows) are approximately aligned with relative motion direction (Figure 14.10), while extensional stresses from topographic contrasts (open arrows) act outward from the Tibetan Plateau. Around the edges of the plateau there is thrust faulting with horizontal shortening roughly perpendicular to contours of surface height. The interior of the plateau is deforming by a combination of strike-slip and normal faulting. □

Plate motion and continental tectonics

The earlier sections have concentrated on the physical properties of the continents, and how these modify the nature of continental tectonics. We now consider how the boundary conditions applied to the continents by the relative motion of plates influence the style of deformation.

Ocean–continent convergence

When oceanic lithosphere converges with continental lithosphere, the oceanic lithosphere is subducted. The kind of deformation that the overlying continental lithosphere experiences depends less upon the relative movement of the two plates, and more on the movement of the plate boundary with respect to the interior of the continent. If we imagine that the continental interior is fixed, and that the oceanic lithosphere is moving towards it at a constant speed (Figure 14.15) we can see that what determines the type of deformation inside the continent is the direction of movement of the outer edge of the continent with respect to its interior. This edge forms the plate boundary (the trench); if the trench moves away from the continental interior, extension occurs, and if it moves towards the continental interior, the continent must shorten horizontally and thicken.

Andean continental margins

The largest region in which **ocean–continent convergence** is happening today is along western South America, where the Nazca plate is sinking below the South American plate. Over most, if not all, of this region, the trench appears to be moving towards the continent, and the continental margin is shortening in an east-west direction. The type of orogeny associated with this class of plate boundary is often called 'Andean', after the Andes, which lie along the whole length of the coast of South America. At their widest part, in the Altiplano–Puna plateau of Peru and Chile, the Andes are about 400 km wide with an average elevation of over 3000 m, and peak heights up to 6000 m. The crustal thickness in this region is about 70 km. Like the Tibetan Plateau, this greatly thickened crust has experienced extensional faulting in the geologically recent past. Unlike Tibet, however, there is no clear evidence of this from earthquakes, but there are many normal faults that appear to have been active in the last few thousand years.

The deformation also has much in common with that in island arcs at the edges of oceanic plates. There is a Benioff zone beneath the whole length of the Andes, and there are active volcanoes along most of the chain. Although there are gaps in volcanic activity, at present in northern Peru and central Chile, the whole region has experienced large amounts of intrusive and extrusive igneous activity since at least the Cretaceous. For example, the Peruvian batholith is made up of over a thousand plutons in a belt over 1000 km long and about 200 km wide that was emplaced over a timespan of 70 Ma between the Cretaceous and the Oligocene. The source of this igneous activity is probably partial melting in the mantle that overlies the subducting oceanic crust. The melts move upwards into the continental crust, which they heat up, and in turn partially melt. There are geochemical uncertainties in estimating what fraction of a particular pluton is derived from the mantle and what fraction is made up of re-worked continental crust, and there are uncertainties in estimating the make-up of the crust just from observations of the rocks that are at the surface. Consequently it is difficult to determine what fraction of the continental crust in such regions is newly derived from the mantle, but it seems likely that addition of material to the continental crust at Andean margins is an important contribution to the growth of continental crust (see Chapter 8).

It is certain that not all of the great elevation and 70 km crustal thickness of the Andes is the result of the addition of melt from the mantle. There is abundant earthquake activity within the South American continent that indicates thrust faulting – particularly on the eastern flank of the Andes (cf. Figure 14.15b)

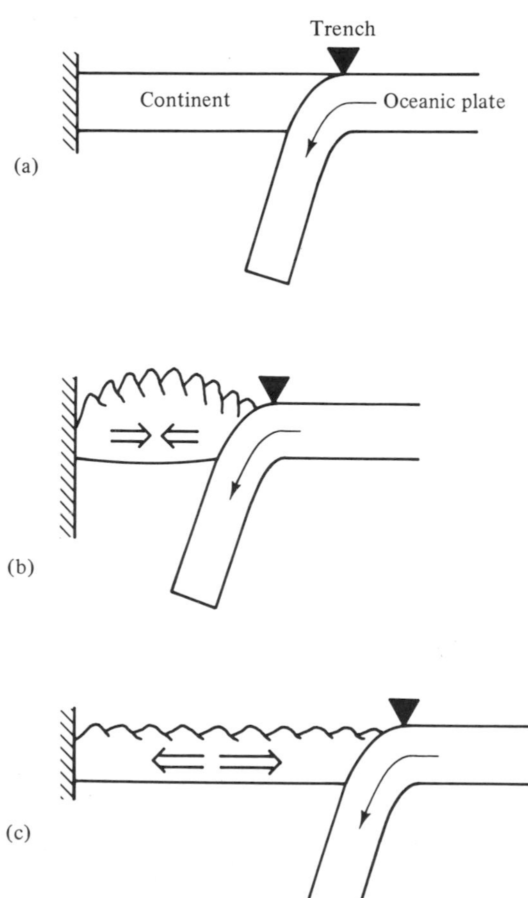

Figure 14.15 Sketch of movement of trench with respect to a continent. (a) The initial position of the trench is shown by an arrowhead, and the continent is taken as fixed. If the trench moves towards the continent (b), the continents will be compressed, as is probably occurring in the Andes at present. If the trench moves away from the continent (c), extension will occur, as is happening in the Aegean. □

Andean-type continental margins existed along the western coast of North America for much of the Mesozoic and early Tertiary, resulting in the emplacement of the major batholiths of the Sierra Nevada, the Coast Ranges and Idaho. Erosion has exposed some of these batholiths to depths of more than 30 km and the geological structures that are revealed indicate large-scale horizontal shortening and vertical thickening, showing that these regions, too, owed their increased crustal thickness to a combination of magmatic and tectonic thickening.

Accretion of oceanic crust

There are several large regions of the oceans that are underlain by crust of greater thickness than average for oceanic crust. The most obvious example at present is Iceland, where the oceanic crust of the Atlantic Ridge is so thick that its surface is above sea level. There are many other places – particularly in the Pacific Ocean (Figure 14.16) – where equivalently thick crust was produced in the past, and has now sunk below sea level as the plate cooled. In these regions, where the oceanic crust may approach, and even exceed 20 km in thickness, our arguments sketched in relation to Figure 14.12 may no longer hold. When oceanic lithosphere having such thick crust reaches a subduction zone at the edge of a continent it may produce compression in very much the same way that lithosphere carrying 20 or 30 km of continental crust would. If all such regions of thick crust were produced in a roughly symmetrical fashion at the ocean ridges, as Iceland is being produced at present, then these plateaus would have had their equivalents on the eastern side of the East Pacific Rise, which have presumably now collided with the continents of North and South America. There is clear evidence, at least in the Cordillera of western North America, that oceanic crust sometimes becomes attached to continental crust, a process known as **oceanic crust accretion** (cf. discussion of exotic terranes in Chapter 9).

Continental back-arc basins

In most regions where oceanic trenches are moving away from continental interiors, the divergence has resulted in **back-arc basins** in which spreading occurs within oceanic lithosphere in a fashion very similar to ocean ridges. The principal exception to this at present is the region of the Aegean Sea, Greece and western Turkey. The Aegean Sea is bounded to the south by the Hellenic Trench, where old oceanic lithosphere of the African plate is sinking into the upper mantle; the

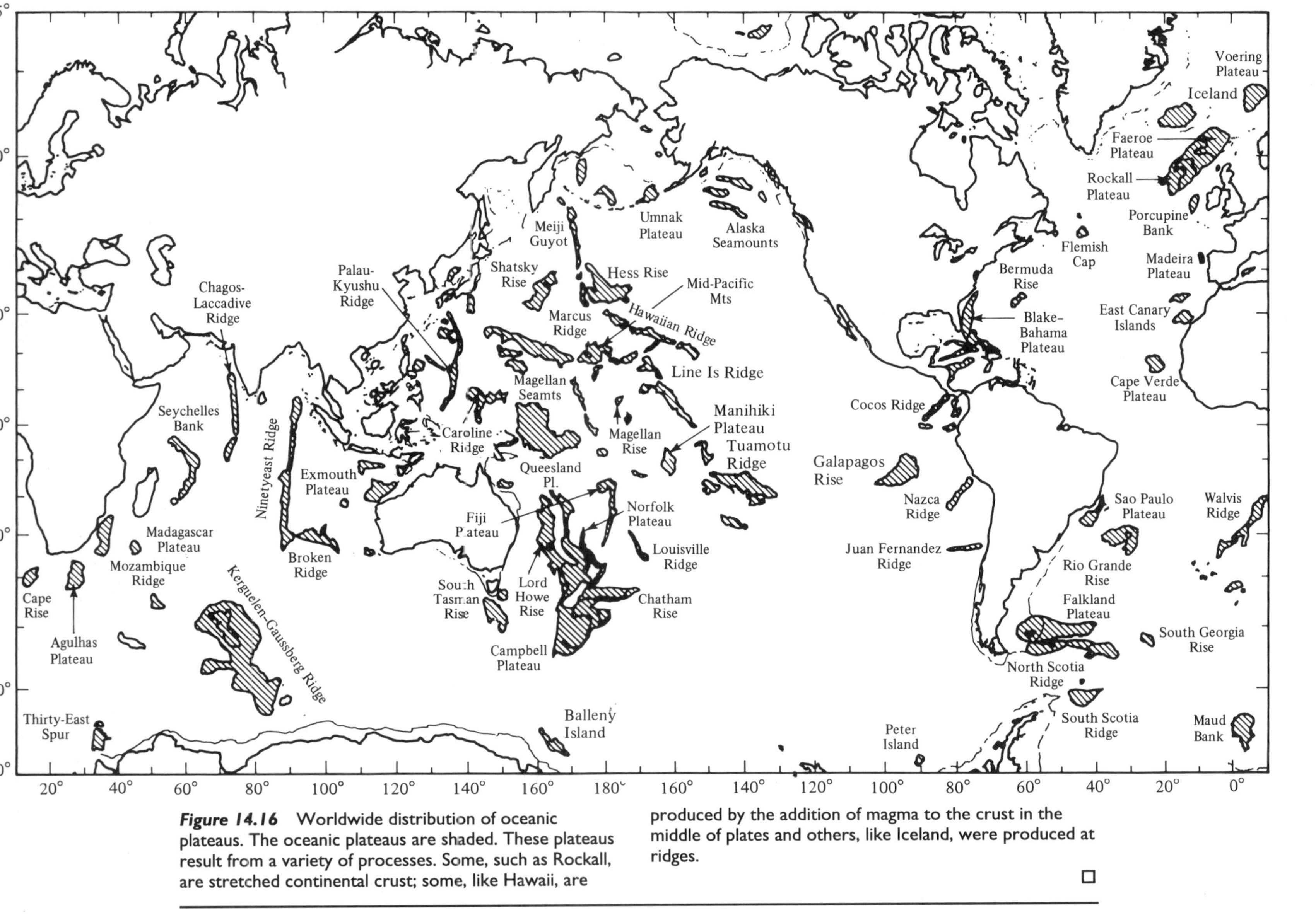

Figure 14.16 Worldwide distribution of oceanic plateaus. The oceanic plateaus are shaded. These plateaus result from a variety of processes. Some, such as Rockall, are stretched continental crust; some, like Hawaii, are produced by the addition of magma to the crust in the middle of plates and others, like Iceland, were produced at ridges. □

shallow seismic activity associated with this may be seen in Figure 14.8 as the arcuate set of thrust faults at about 35° N near the island of Crete. An area of equally intense seismicity covers much of Greece and western Turkey, immediately to the north of the Hellenic Trench. Whereas the earthquakes in the Hellenic Trench are on subduction-related thrust faults, the earthquakes in Greece and Turkey all occur on strike-slip and normal faults and show that the region is extending in an approximately north–south direction, and that the continental crust is thinning.

Continental rifting

The major rift systems of the world are along the ocean ridges (see Chapters 9 and 10), but occasionally, as in East Africa at present, long, linear **continental rift systems** develop on land. The rifts of East Africa form a chain that is over 2000 km long, and the individual rift valleys – like those in the oceans – are rarely much more than 50 km wide. The East African Rift runs northwards into the junction between the Gulf of Aden and the Red Sea, where active oceanic spreading is going on. Other examples of long continental rift zones include Lake Baikal in Central Asia, and the Rhine valley of western Europe. Although these systems bear a superficial resemblance to ocean ridges in that they are long, narrow valleys with edges formed by large normal faults, the amount of spreading that is occurring at present in continental rifts is a minute fraction of the spreading at the ocean ridges.

This is not to say that continental rift zones are unimportant; indeed, it seems probable that active continental rifting today represents a mild form of a process that is intermittently much more important. At times in the past – particularly during the break-up of the supercontinents of Pangaea and Gondwanaland – continental rifting must have been a dominant feature of continental tectonics. The opening of the Atlantic Ocean, for example, began with the stretching of continental crust, as in East Africa today. This stretching continued much further, though, until there was no continental crust left, and ocean floor was created along a narrow zone that is now the Atlantic Ridge.

Sedimentary basins

The principal feature common to continental back-arc basins and continental rift zones is the thinning of continental crust. As would be expected from considering isostasy (Figure 14.2), thinning of the continental crust results in subsidence and the development of **extensional basins**. For example, the Aegean Sea is underlain by continental crust that is between two-thirds and one-half the thickness it had 5 to 10 Ma ago. The water depth is now as much as 2000 m but – to judge by the elevation of the surrounding regions of Greece, Bulgaria and Turkey – the land surface was probably up to a kilometre above sea level before extension began. The active deformation extends on to the surrounding land, with several destructive earthquakes a year taking place in Greece and western Turkey; if extension continues, these areas, too, will sink below sea level.

A second important consequence of continental extension is that it thins the lithosphere without appreciably changing the temperatures of the rocks within and below it. In consequence the temperature gradient within the lithosphere increases as extension progresses (Figure 14.17) and the upper mantle below the lithosphere decompresses adiabatically. If extension goes on long enough the adiabatic decompression of the upper mantle leads to large amounts of partial melting, eruption of basalts, and finally sea-floor spreading. The continental margins of the Atlantic Ocean (and indeed of much of the Indian and Arctic Oceans) are the end products of this process (Figure 14.17).

Frequently, however, extension of the continental lithosphere stops before the ocean floor is produced. There are several regions in which the lithosphere appears to have been thinned by relatively small amounts, perhaps to between two-thirds and one-third of its original thickness, and then to have stopped deforming. At the end of this deformation, the temperature gradient in the lithosphere was higher than at the beginning of deformation, and consequently the lithosphere cooled off. Just as the surface of the oceanic lithosphere subsides as it cools, these stretched pieces of continental lithosphere, too, subside over several tens of millions of years after extension ceases.

The consequence of such extension and subsidence is to produce a stable region of continental lithosphere with a surface below sea level. Sediments gather in these basins, and trapped within these sediments are the organic remains whose alteration under the action of pressure and temperature lead to the generation of oil. Although oil is generated in other tectonic settings, the thick piles of sediments that accumulate on stretched continental lithosphere form the principal reservoirs of this rare commodity. Many important accumulations are on the margins of continents which stretched to form ocean basins, for example the western and eastern margins of the Atlantic Ocean, and small continental fragments isolated within the oceans, like the Rockall and Falklands Plateaus. Other important accumulations are found in the stretched

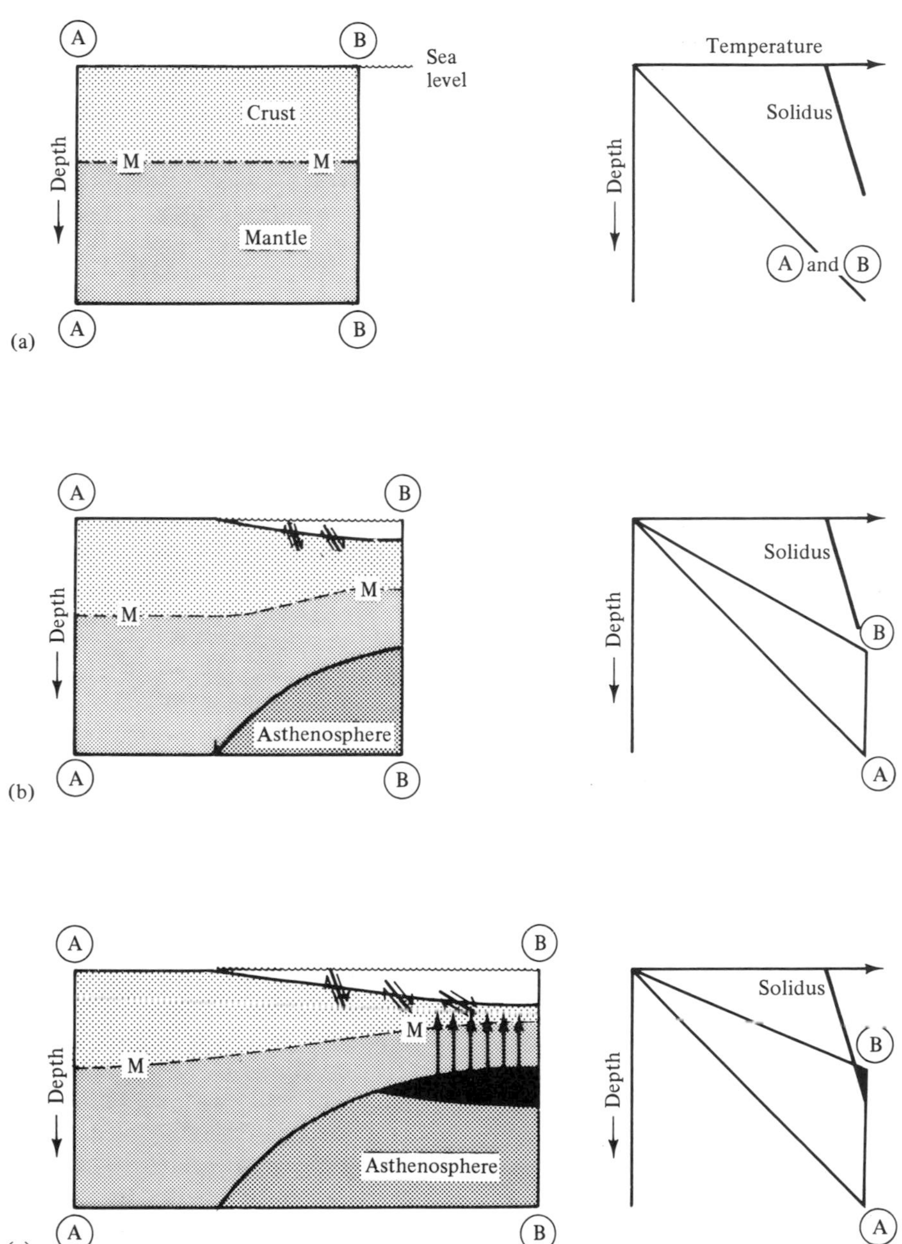

Figure 14.17 Sketch of the development of strain and temperatures in stretching continental lithosphere (a to c). The temperature gradients beneath the left and right of each cross-section are given by A and B respectively. The upper crust exhibits the 'domino' style of faulting that is commonly seen in actively extending regions. Faulting begins on steep faults, which rotate to shallower angles as the crust thins, causing earthquakes whose focal mechanisms resemble those in Figure 14.8. Below the upper crust, the lithosphere is shown as stretching in a ductile fashion; below the lithosphere, the asthenosphere rises passively in (b). While the deformation occurs, the rocks barely change in temperature, consequently the temperature gradients in the lithosphere steepen A to B, and rocks in the asthenosphere decompress adiabatically. As strain progresses the land surface sinks farther below sea level, temperature gradients become higher, and in (c) melting occurs in the asthenosphere (see also Chapters 4 and 8). At the end of extension the temperatures in the lithosphere are higher than their equilibrium values, so the lithosphere cools and subsides over the next few tens of millions of years. (M = Moho) □

interiors of the continents in places like the North Sea of northwest Europe. The topics of continental sedimentary basin formation and of hydrocarbon accumulation are developed further in Chapters 15 and 18.

Continent–continent convergence

We have already discussed extensively the active belts of continental convergence in Eurasia. In addition, western North America contains, in the Rockies and the mountains of the Basin and Range, the remnants of a large mountain belt assembled in late Mesozoic and early Tertiary time.

Continental convergence was equally important in the more distant past, and has been a cause of mountain belt formation almost continuously throughout Phanerozoic time. Several mountain belts formed in the early Mesozoic or in the Palaeozoic still have obvious topographic expressions: the Appalachians of eastern North America were a major mountain chain in the late Devonian and early Carboniferous (about 400 to 350 Ma ago). At approximately the same time much of east Greenland, Scotland and western Scandinavia, which have since been separated by sea-floor spreading, formed a mountain belt that was comparable in area to the present Tibetan Plateau and was probably also its equal in elevation.

The older orogenic belts have all been eroded considerably, so their elevations are much less than when they were active mountain chains. The depths to which they have been eroded may be inferred from the pressures recorded by minerals formed during the metamorphism that accompanied the mountain-building episode (see Chapter 12 for discussion of geothermometers and geobarometers). Depths of burial in excess of 30 km are commonly recorded by rocks now at the surface in old mountain belts. Even in young terrains, such as the Alps and the Himalaya, there are large areas in which the rocks appear to have come to the surface from depths of 30 km or more in the last 30 Ma. The rocks recording these high pressures are at – or well above – sea level and are underlain by continental crust of normal thickness (about 35 km) or greater. There is, therefore, a strong indication that the degree of crustal thickening in older belts was often a factor of two or more, entirely consistent with the degree of thickening strain observed today in active belts. Furthermore, it is beginning to be recognised that widespread extension may have marked the end of many episodes of mountain building. Traces of extensional strain at a late stage have now been found in the exhumed, deeper, portions of old mountain belts.

The deeply-eroded older mountain belts afford the opportunity to examine the processes taking place at depth, which we cannot observe directly in modern mountain ranges. As we have seen, the deformation at the surface of active mountain belts is characterised by the movement of faults and by earthquakes. In contrast, the rocks that are exposed in the deeper parts of old belts have usually deformed by folding or by flowing over large regions. It would not be true to say that one never sees faulting in rocks that have been deeply buried (any more than it would be true to say that one never sees folding in near-surface rocks). There is, nevertheless, a profound contrast in the types of deformation between rocks that have been deeply buried and rocks that experienced deformation in the upper 10–20 km of the continental crust. This contrast can be summarised by the statement that, above some level in the crust, rocks generally deform by **fracturing** and below that level they generally deform by **flowing**.

The fracturing of rocks involves the presence of cracks in the rocks, that can rapidly grow and move through the rocks if stresses become high enough. At high pressures – equivalent to a depth of burial of a few tens of kilometres – such cracks cannot remain open and it becomes easier to deform the rock by one of the mechanisms of ductile flow (see Figure 14.12c). The principal mechanisms by which this flow occurs are by the **creep** of atomic dislocations through the individual crystals in the rock, by the recrystallisaton of mineral grains and by the growth of new minerals (cf. Chapter 12). Each of the processes of ductile deformation involves the breaking and reforming of bonds in the crystal lattice, and so take place more readily the higher the temperature. Because temperatures and pressures both increase with depth in the Earth, brittle deformation becomes more difficult with depth while, at the same time, ductile deformation becomes more easy. The depth range over which rocks change from deforming by brittle fracturing to deforming by ductile flow is often referred to as the **brittle–ductile transition.** This transition may sometimes be recognised in crustal seismic reflection studies (see Chapter 13).

As stated earlier, most earthquakes in modern deforming continental regions take place in the upper 15 km or so of the crust. Examination of the rocks exhumed from old mountain belts show that the transition between the brittle fracturing and the ductile flow of rocks covers an interval of temperature between about 250 and 450 °C. This range of temperature is what one would expect to find in continental crust at depths of 10 to 20 km.

The lifetime of mountain ranges

As we have seen, most mountain ranges are produced

as a result of the motion of plates. Sometimes the pattern of plate motion that produces mountain ranges may persist for tens or hundreds of millions of years. For example, there have been subduction zones close to the western margins of North and South America for at least the last 200 Ma, and probably much longer; the collision of India with Asia has been going on for about 50 Ma. Alternatively, as in the Alps, there may be several sharp, but short-lived, phases of motion that produce mountains over a long interval. For these reasons, it is usually very hard to say how long a mountain belt took to form and, of course, while the mountains are forming, the forces that tear them down are also acting.

One of the main agents that reduces the heights of mountains is erosion. It is difficult to calculate rates of erosion over long time intervals. At the present day, rivers are carrying enough sediment into the oceans to account for the erosion of the active mountain belts at a rate of about 1 mm yr^{-1}. This is certainly an overestimate. Some of the sediment comes from elsewhere than the mountains but, more importantly, the sediment load at present is much greater than it would otherwise be because of the impacts of the recent Ice Age and of human activity upon the environment. Nonetheless, it seems from careful measurements made in small parts of present mountain ranges, that erosion rates may be as high as a few tenths of a millimetre per year. To reduce a mountain range 5 km high to sea level would require the removal of between 25 and 35 km from its top (Figure 14.2c); erosion at a few tenths of a millimetre per year could do this in 100 to 200 Ma.

Of course, the earlier text of this chapter demonstrates that erosion is not the only factor that greatly reduces the thickness of mountain ranges. We have seen that the Tibetan Plateau is extending by normal faulting (Figure 14.10), and Figure 14.13 shows that all mountains have a tendency to spread under their own weight. The extension in the Aegean Sea is occurring in continental crust that was a high mountain belt until recently. The Basin and Range province of western North America is a region that has been extended by a factor of two or so, yet still has crust that is about 25 km thick: clearly it, too, was a mountain range with a considerable thickness of crust before extension began. We may regard Tibet, and the Basin and Range, as the beginning and end of a thinning process that may occur in many mountain ranges. In the Basin and Range, this spreading has occupied the last 20 to 30 Ma and its last phases are still going on. The extension in Tibet began in the last few million years and it, too, might take only a few tens of millions of years – much less time than erosion.

Transcurrent boundaries

When the relative motion of two plates is parallel to the plate boundary, no wide-scale thickening or thinning of the lithosphere occurs, and the principal mode of deformation in the upper crust is strike-slip faulting. In oceans, this faulting occurs on discrete fracture zones, or transform faults (cf. Chapter 9). The largest example of a **transcurrent boundary** within continental lithosphere is the Pacific–North American plate boundary in California, which contains the San Andreas Fault, and which has been active for about the last 30 Ma. Though the San Andreas Fault has undoubtedly absorbed much of the relative motion of the Pacific and North American plates over the last 30 Ma, it is not a simple transform boundary such as we find in the oceans. Here, too, deformation is distributed on several faults, and recent very precise determinations of strain indicate that this deformation is spread out over a region 100 to 200 km wide away from the western edge of North America.

Metamorphism and continental deformation

The deformation of the continents involves the thickening of crust and burial of rocks, as well as the thinning of the crust and exhumation of rocks. These processes result in changes to the temperature gradient in the crust (see Figure 14.17) and, as a result of these changes, rocks may recrystallise existing minerals or grow new minerals. Such changes in the minerals in the rock are known collectively as metamorphism. Chapter 12 deals with the mineralogical changes in rocks during metamorphism. In this section we shall concentrate on the temperature and pressure changes that happen during tectonic activity, which result in the growth of new minerals in rocks.

Most metamorphic rocks that reach the surface of the Earth and that we can observe are the result of the processes of mountain building, which involve first thickening the crust and then thinning it by a combination of erosion and extensional faulting, to bring deeply buried rocks to the surface. Figure 14.18 illustrates four stages in the metamorphic development of an idealised mountain range. The left-hand column of sketches show cross-sections through the continental crust, with rough indications of the way temperature changes with depth. The right-hand column shows the temperature and pressure conditions experienced by three rocks that are caught up in the process of thickening the crust. In this simple example,

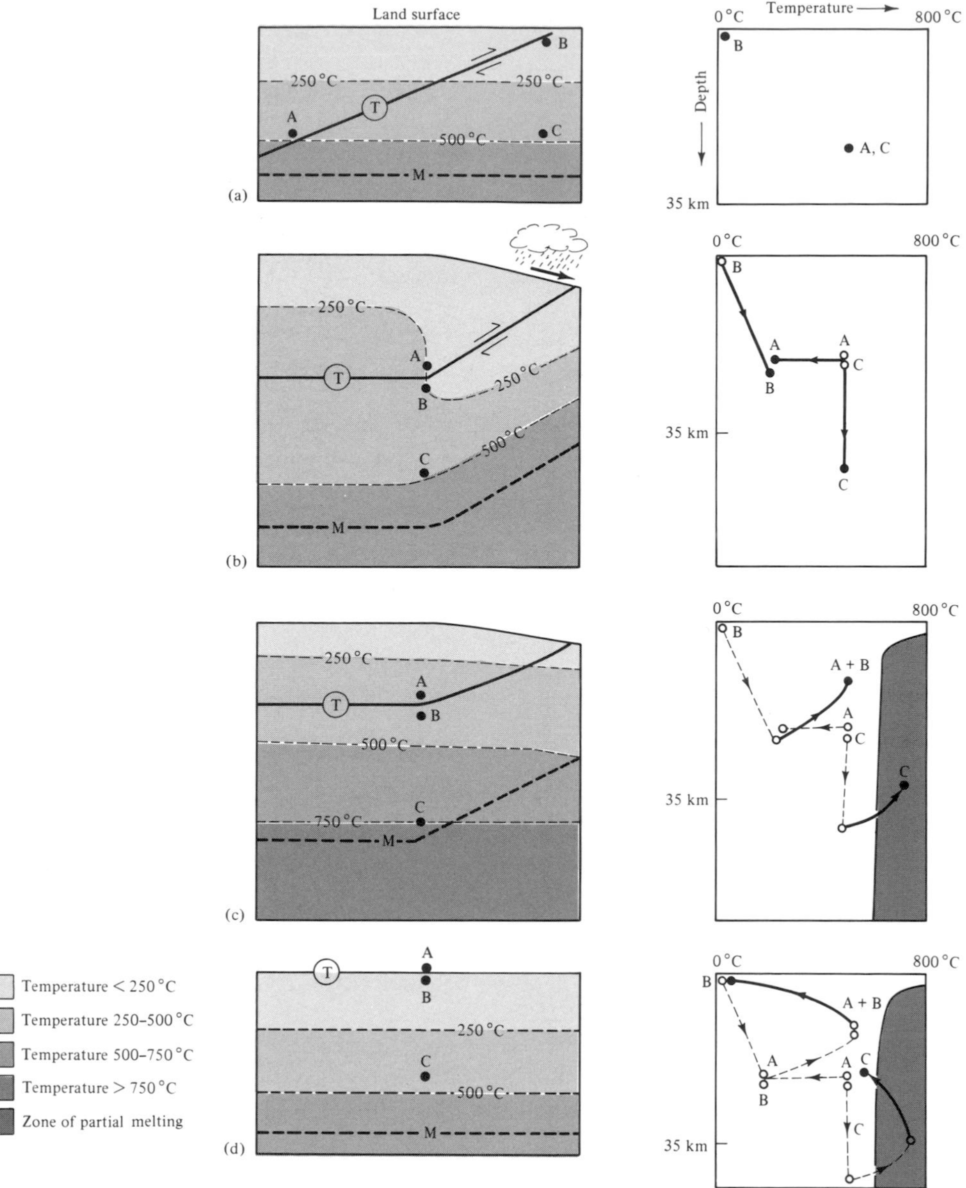

Figure 14.18 Sketch of the conditions leading to the metamorphism of three rocks A, B and C. (a) B lies initially close to the land surface, just beneath a large thrust fault. A lies to the left of B and at a depth of about 20 km, just above the fault, and C lies at the same depth as A, but below B. On the pressure–temperature diagram on the right-hand side, we see that the temperature and pressure of B are about 0 °C and 1 atmosphere. A and C are at about 500 °C and 7000 atmospheres (*c.* 20 km depth). (b) Movement on this thrust fault results in the block containing A being placed above the block containing B and C, and the crust is therefore thickened. A is at its original depth, but has cooled, because of contact with colder rock beneath it. B has been buried and heated. C has been buried, but has not changed in temperature. (c) The temperatures in (b) are much lower than normal, and the rocks heat up; at the same time, the mountain belt is being eroded, so that the rocks begin to return to the surface. A, B and C all heat up and decrease in pressure. The red shaded area to the right of the pressure–temperature diagram represents the zone of partial melting for wet crustal rocks. (d) After erosion of almost all the overthrust material, the temperatures of all three rocks have fallen; the crust is back to its original thickness, A and B are on the surface, and C is back to its original depth. (M is the Moho – the crust–mantle boundary, normally at 30–35 km depth.) □

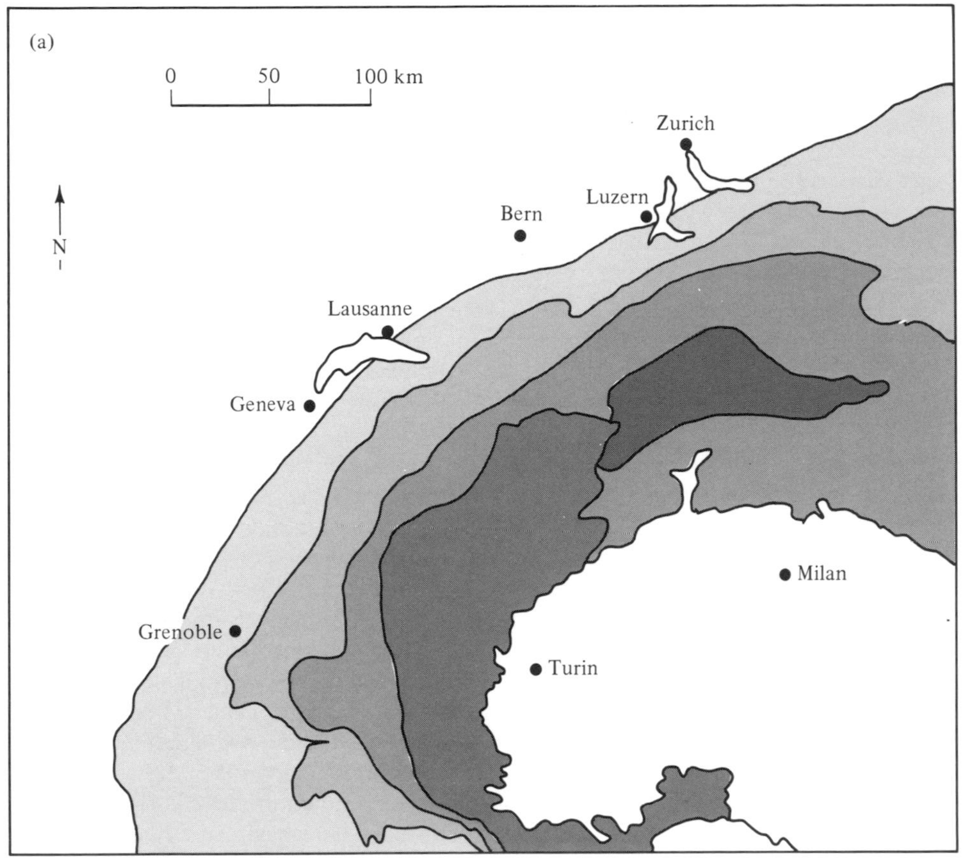

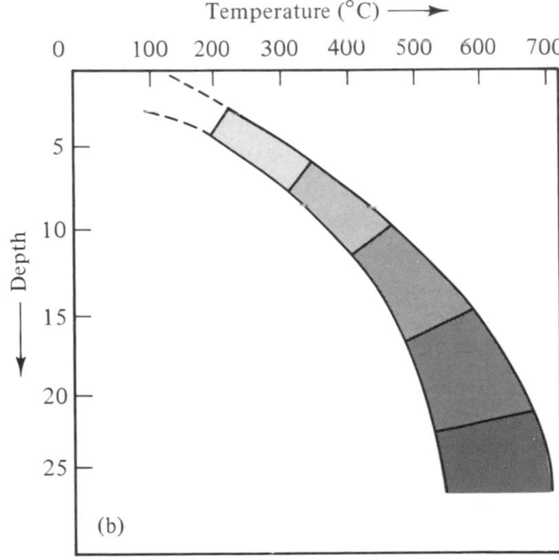

Figure 14.19 Metamorphic conditions recorded today in rocks at the surface of the western part of the Alpine Chain of Europe. Continental convergence between 60 and 30 Ma ago produced a wedge-shaped pile of thickened crust, with its thickest part in the south. Heating of the rocks buried in this pile (Figure 14.18) led to the metamorphic conditions shown. Erosion or exhumation by extension, brought to the surface rocks that were originally tens of kilometres down. Increasingly darker shades on the map correspond to increasingly higher grades (pressures and temperatures) of metamorphism recorded in the mineral assemblages preserved in rocks now at the surface. The approximate temperatures and pressures of formation of these mineral assemblages are indicated in the pressure–temperature plot to the right. □

the crust is thickened by a single thrust fault (T) that rapidly places one piece of crust on top of another; in reality, the thickening of the crust would involve many thrusts and folds in many possible configurations.

In Figure 14.18a, thrusting is about to begin on a plane that passes immediately on the left of B and immediately below A. When movement has finished (Figure 14.18b) B is immediately below A, and both are in the middle of crust that is now approximately twice as thick as it was originally. The rock C is at the base of the crust. Thickening of the crust has temporarily lowered temperature gradients; the radioactive decay of isotopes of uranium, thorium and potassium within the crust, and the flow of heat from below combine to raise temperature in the thickened crust, and all the rocks increase in temperature (Figure 14.18c). At the same time, the tops of the mountains are being eroded, so the crust is becoming thinner again, and the rocks are returning to the surface. After a long time (perhaps a couple of hundred million years), the crust is back to its original thickness, and rocks A and B are on the surface (Figure 14.18d). All the rocks have been heated. Rock B has experienced the most change; before the mountain belt was made, it lay on the surface – perhaps it was a piece of sediment. During the metamorphism it was buried perhaps as deeply as 30 km and its temperature was raised by as much as 500 °C. Rock B would experience many metamorphic reactions and would return to the surface as a crystalline rock containing minerals grown at great depth in the crust (Chapter 12). Rock A never experienced higher pressures than it had before the thrusting, but may have been heated somewhat above its original temperature (Figure 14.18c). Rock C was buried to at least 40 km, and then experienced a temperature increase great enough to cause it to melt partially, to form a granite magma. If this granite was mobile enough, it could move upwards through the overlying rock and intrude some higher-level, colder rocks, causing contact metamorphism.

The sketch of Figure 14.18 is highly idealised, and in any real mountain belt there are large variations in the depth of burial of rocks, their compositions, and the rate at which they generate heat, and in the rates at which they return to the surface. Nonetheless, the processes outlined in Figure 14.18 appear to be common to many mountain belts. Figure 14.19 shows the range of metamorphic conditions that are recorded in rocks of the Swiss and French Alps (see also discussion of the Dora–Maira massif near the end of Chapter 4). The crust in this region was thickened by approximately a factor of two in the interval 60 to 30 Ma ago. The rocks in the core of the Alps experienced metamorphism at temperatures above 600 °C and at pressures of more than 7000 atmospheres before being brought to the surface where we can now see them. The Alps are still over 1500 m high, on average, so the pattern of metamorphism that will be on the surface when the Alps are at sea level will show still higher temperatures and pressures.

Summary

Continental tectonics differs radically from the tectonics of oceanic lithosphere. The oceanic lithosphere moves as rigid plates (see Chapter 9), whereas the continents deform over wide areas. The difference between continental and oceanic lithosphere may be explained by the relative weakness and lower density of continental lithosphere. Differences in surface height lead to forces that make mountains spread under their own weight. The combination of these forces with the forces applied to the continents by the relative motion of the plates accounts for the rich variety of continental tectonics. Horizontal stretching of continental lithosphere produces regions of steepened thermal gradient which subside and collect large thicknesses of sediments. If continental stretching continues long enough, it can lead to melting of the upper mantle and production of new ocean floor. Horizontal compression of continental lithosphere produces thickened crust and mountain belts. Rocks buried in this thickened crust experience increases in temperature and pressure that lead to metamorphism. The final stage in the tectonics of many mountain belts may be their collapse by extension.

Further reading

Bolt, B. A. (1978) *Earthquakes*, W. H. Freeman, 241 pp.

Eminently readable book on the subject. Rich in anecdotes of disaster, but also covers the basics of earthquakes from a seismologist's point of view. Barely mentions the uses of earthquakes to determine tectonic motions.

Molnar, P. (1986) The structure of mountain ranges, *Scientific American*, **255** (July), 64–73.

Modern account of the tectonics of Asia and of the forces driving continental deformation.

Park, R. G. (1988) *Foundations of Structural Geology*, Blackie, 160 pp.

As its title implies, this book introduces the student to the many manifestations of deformation of rock. But it does more: the book provides an accessible account of the relations of these structures to the forces causing them, and of the relation of those forces to larger scale tectonic motions.

CHAPTER

15

Tony Watts
University of Oxford

THE FORMATION OF SEDIMENTARY BASINS

Introduction

Sedimentary basins are a characteristic feature of the Earth's crust and lithosphere and range in age from Archaean to the present day. While sedimentary basins have long been of interest to the exploration geologist, it is only in the last decade or so that their mode of formation has received much attention from Earth scientists in academia.

At the present day, the world's largest basins are found at the margins of the continents near the mouths of river systems. For example, up to 15 km of sediments have accumulated at the continental margin off the Amazon, Niger and Mississippi rivers. Other margins with large sediment thicknesses occur off the Ganges river in the Indian Ocean and the Colorado river in the Pacific Ocean. The largest sediment thicknesses in the geological record are also believed to have occurred at the margins of the continents. For example, several kilometres of sediments are known to have accumulated at the margins that flanked the Tethys Ocean prior to the collision of the African and Eurasian plates and formation of the Betic-Alps mountain ranges during the Tertiary.

Sedimentary basins are dominated during their evolution by *vertical movements*. They are therefore difficult to explain in terms of plate tectonics, since this theory emphasises the horizontal motions of the plates. On a large scale, an undeformed basin has a generally synclinal form that reflects the tectonic movements that modified it while it was being infilled by sediment. Basins may form close to plate boundaries, or in intra-plate settings. The continental margin basins described above are related to extension of the lithosphere that often leads to continental break-up and ocean opening. On the smaller scales of geological field mapping, the internal reflector geometry of a basin reflects not only tectonics, but the modifying effects of erosion, compaction, sea-level changes and sediment supply.

The past few years has seen a rapid increase in our understanding of sedimentary basin formation due in large part to the acquisition of high-resolution multichannel seismic reflection and refraction profile data and advances in our understanding of the thermal and mechanical properties of the Earth's lithospheric plates. In addition, sedimentary basins are the world's largest repository of oil and gas deposits (Figure 15.1) and so have been subject to intense study by the petroleum industry.

The different types of basins

If we define basins as depressions of the Earth's crust that contain more than one kilometre of sediment, then approximately 70 per cent of the Earth's surface is underlain by basins of one type or another. Such a definition excludes basins whose sedimentary infill is now incorporated in fold belts, but includes those in the stable continental interiors and flanking regions that have escaped the destructive effects of plate subduction and rifting.

Sedimentary basins are found in a variety of tectonic settings. Some of the world's largest basins occur within or on the stable continental interiors, such as the central USA (e.g. Michigan/Illinois), the Russian Platform, and Congo (e.g. Zaire) and are termed **intra-cratonic basins**. They are also associated with the flanks of orogenic belts as **foreland basins**, such as the Appalachians (e.g. Allegheny, eastern USA), Rockies (e.g. Denver, western USA), Alps (e.g. Molasse, Switzerland), Betic (e.g. Guadalquivir, Spain) and Pyrenees (e.g. Ebro, Spain) ranges. Other types of basins are associated with **passive continental margins** (e.g. Atlantic-type), **back-arc** (e.g. western Pacific) and **fore-arc** (e.g. Cook Inlet, Alaska) regions, and **strike-**

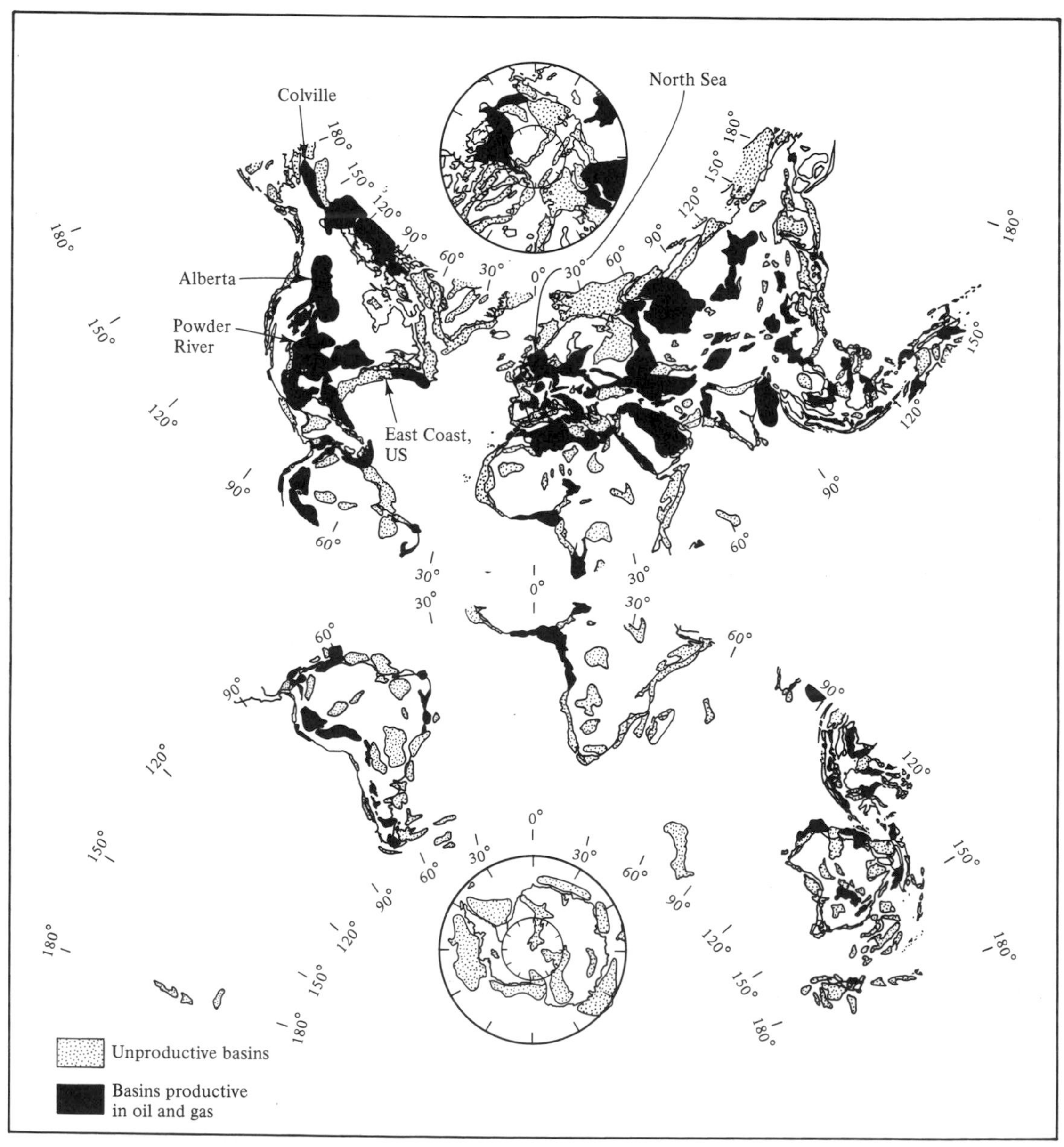

Figure 15.1 The global distribution of sedimentary basins. Heavily shaded regions correspond to basins that are productive in oil and gas. The basins referred to in the text are labelled. □

slip basins associated with faults showing lateral movement (e.g. Ventura, California).

Even though basins show considerable tectonic diversity, they have some common features that depend on their setting. For example, most passive continental margin and intra-plate basins are associated with rifting (i.e. extension of the crust along normal faults) of rigid continental lithosphere. Foreland basins are associated with compressional plate boundaries. By way of contrast, smaller basins of the fore-arc, back-arc and strike-slip type develop in response to an extensional, compressional or strike-slip stress field along a plate collision zone.

In some cases, a basin may change tectonic setting during its evolution. For example, it is quite common in the geological record to find a foreland basin located on or near a previous passive margin (e.g. Colville Trough of the North Slope, Alaska). During convergence, the steep plunge in basement at the edge of passive margin basins serves to localise thrusts and may cause them to preferentially 'telescope' onto the more gently dipping flanking coastal plain. Other examples of rift-type basins within compressional-type orogenic belts occur (e.g. Alboran Sea between the Betic and Rif mountains of southern Spain and north Morocco). The extension may, in this case, be the result of collapse following thickening of the crust and lithosphere during mountain building.

The deep structure of sedimentary basins

The shallow stratigraphy of many basins is now quite well known, but the underlying crystalline basement is often too deep to be penetrated by drilling or to be imaged by seismic reflection profiling techniques. Moreover, there have only been a few seismic refraction studies carried out to determine the velocity structure and depth to the Moho (see Chapter 3) beneath basins.

One example where both stratigraphic and deep seismic data are available is the North Sea basin (Figure 15.1). The Palaeozoic to Recent sediments that fill this basin reach maximum thicknesses of up to 9 km. There is good evidence in the basin of rifting in the form of graben formation, volcanism and flank uplift. Seismic refraction data indicate that the depth to the Moho (i.e. the seismic definition of the base of the crust) is about 18 to 25 km beneath the basin, and about 25 to 35 km beneath adjacent areas of Britain and Norway. The Moho therefore shallows locally beneath the North Sea basin.

While a shallow Moho depth is a feature of many basins, some notable exceptions exist. For example, deep seismic reflection profiling studies show that the Fastnet and Celtic basins in the continental shelf off Britain and France have a flat underlying Moho (see Chapter 13). Other examples appear to exist on deep seismic profiles of the Canadian margin.

An interesting feature of the deep structure of many basins is the presence of sub-horizontal reflectors in the lower part of the crust. The origin of these reflectors is not clear, but they have been attributed to either mechanical transformations of the crust during orogeny, or to some form of ductile flow in the crust during rifting. The **deep crustal reflectors** are quite common throughout Europe, but have only been clearly seen beneath the Basin and Range province in the USA.

Mechanisms of subsidence and uplift

A sedimentary basin will not form unless there is an initial depression for the sediments to fill in. Rift-type, compressional-type and strike-slip basins are often characterised by thick sequences of continental and shallow-water sediments and therefore require substantial tectonic driving forces in order to explain them.

Perhaps the best known of the basin-forming mechanisms is **thermal contraction of the oceanic lithosphere** as it cools away from a mid-ocean ridge crest. At a ridge crest, hot mantle material is accreted to the lithosphere which, as it cools, increases density and subsides (see Chapter 9). Observations of sea-floor depth show that the subsidence is exponential in form (Figure 15.2), increasing from depths of about 2500 m at the ridge crest (i.e. zero age) to greater than 6000 m for sea floor older than 150 million years (Ma).

At some continental margins, a considerable thickness of continental-derived material has built out onto the oceanic lithosphere. In the central Atlantic, for example, turbidity deposits have caused the Mesozoic age oceanic crust to be bent down at least 1 km more than would be expected on the basis of its age. Locally, at the mouth of the Mississippi and Niger rivers, continental-derived sediments have depressed the oceanic crust by as much as 4 to 5 km below expected depths.

As a rule, sediment thicknesses much in excess of 2 km rarely accumulate on the oceanic crust. The greatest thicknesses occur instead on the continental crust – at or near the present-day shelf break in slope. Off the East Coast US margin, for example, seismic reflection profiles reveal up to 10 to 15 km of Mesozoic–Tertiary sediments that overlie a crystalline basement of Palaeozoic or greater age.

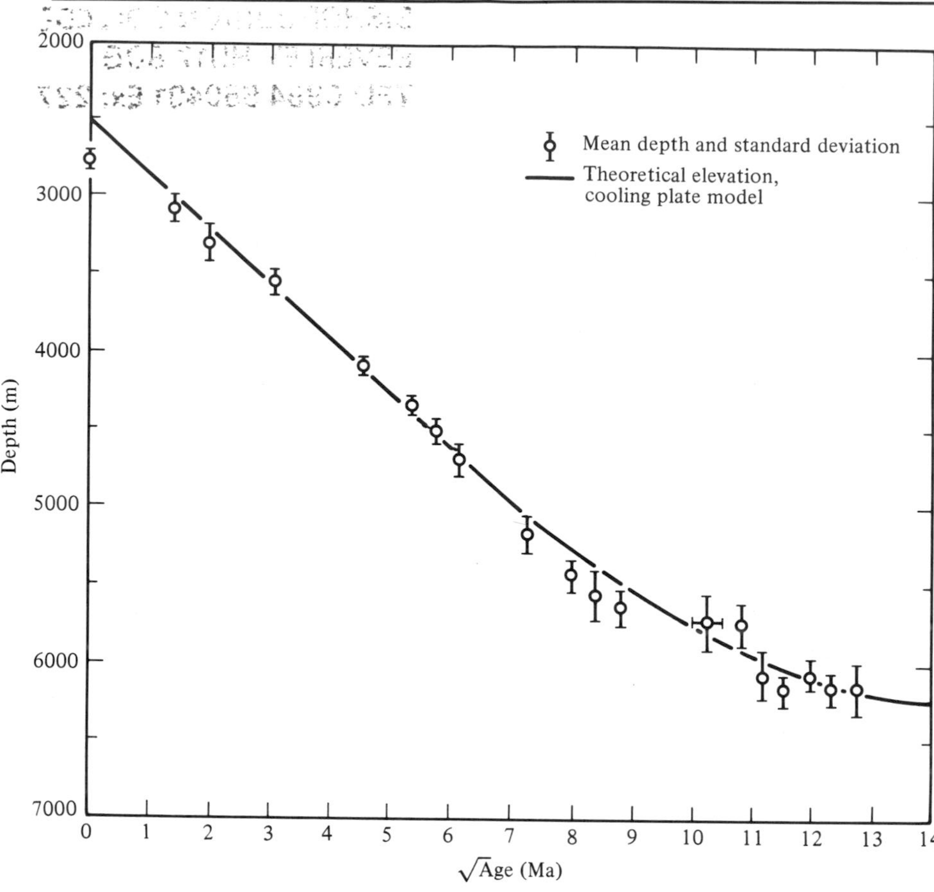

Figure 15.2 Subsidence of the North Pacific Ocean basin based on observations of the sea-floor depth and theoretical models based on a cooling plate that show an exponential decline in subsidence rates.

In the late 1970s, Dan McKenzie suggested that during rifting continental lithosphere is stretched, cools and subsides with age in a manner similar to that of the oceanic lithosphere. According to this model, **rift basins** are characterised by a relatively rapid initial subsidence, due to crustal and lithospheric thinning, followed by a slow thermal subsidence, as the crust and the lithosphere cool. This behaviour is commonly referred to as the **McKenzie model of basin formation**.

A **stretching model** is supported by the observation that rift structures are generally associated with thinner than usual crust. Off northwest Spain, for example, seismic reflection profiling (Figure. 15.3) and deep drilling data reveal a series of tilted fault blocks in the upper part of the continental crust. The crust in this region is 20 km thinner than it is beneath the adjacent shelf, and the geometry of the thinning suggests uniform extension. A gentle undulating reflector separates the faulted upper crust from a highly reflective lower crust below, and this reflector has been interpreted by some workers as a **'detachment'**, which during extension separated a region of brittle deformation above from one of pervasive ductile flow below.

The rift structures illustrated in Figure 15.3 are unusual in that they have only a thin sediment cover, despite their location in water depths of up to 4000 m. In most basins, the rift structures are obscured by substantial thicknesses of post-rift sediments. Since sediments displace water they represent a load on the crust and lithosphere which should sag under their weight. The extent of sediment loading depends, however, on the water depth that is available for sedimentation, the density of the sediments and mantle materials, and the strength of the crust and lithosphere.

Let us assume that due to tectonic or some other means a water depth, W_d, is made available for sedimentation. According to the **principle of isostasy** there will be some depth in the Earth, known as the **depth of compensation**, where the pressure (or force/unit area) on a loaded and unloaded column are equal. The pressure on an unloaded column is given by

$$W_d \rho_w g + T_c \rho_c g + r \rho_m g, \qquad (1)$$

where W_d = water depth of deposition, T_c = mean thickness of the crust, r = distance from the base of the crust to the depth of compensation, ρ_w = density of water, ρ_c = density of crust, ρ_m = density of mantle, g = average gravity, and

$$r = d - T_c - W_d,$$

where d = depth of compensation. The pressure on the base of a loaded column is given by

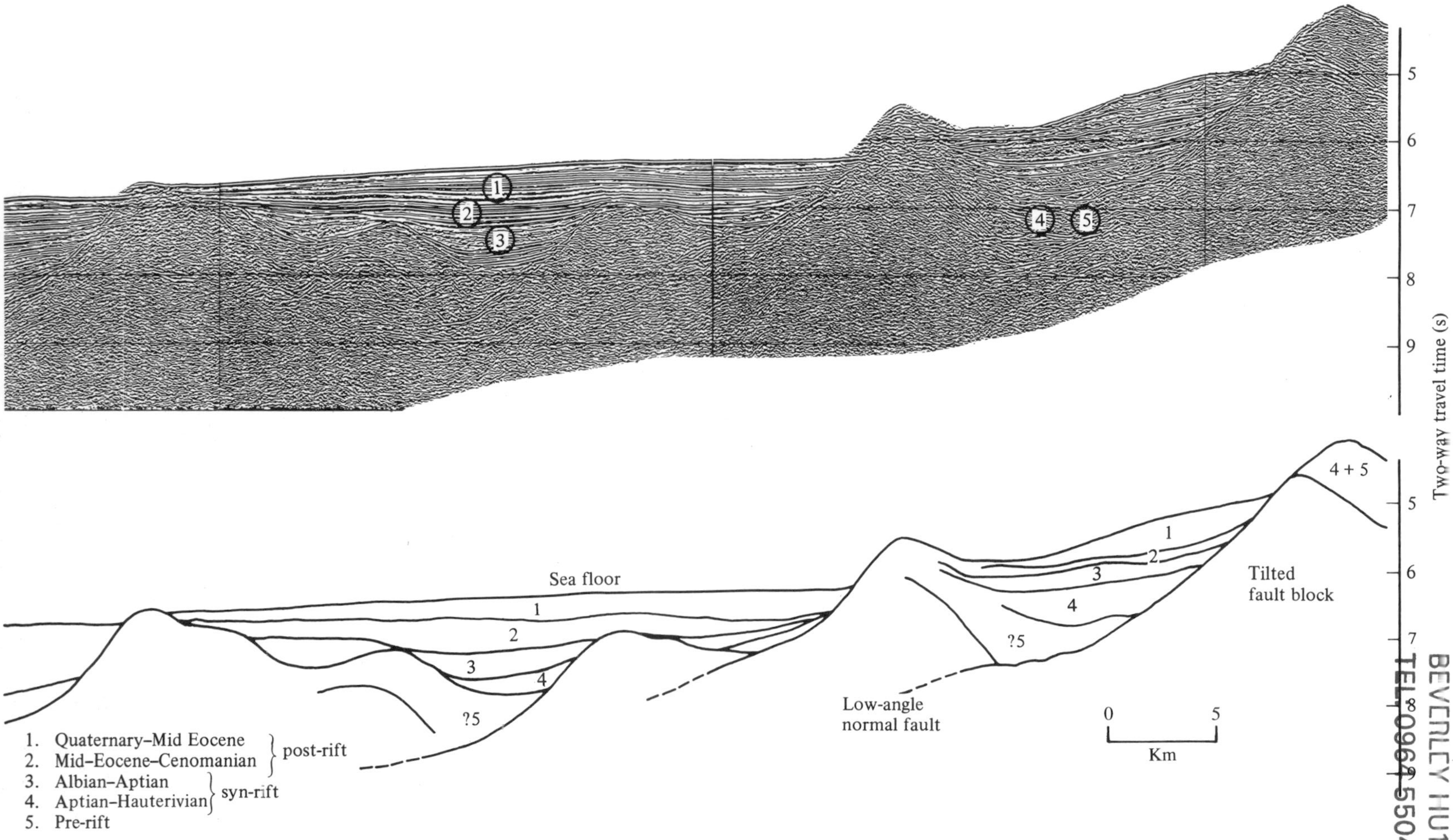

Figure 15.3 Seismic reflection profile of the Galicia Bank in the continental margin off northwest Spain. The profile, which was obtained by the Institut Français du Petrole (Paris, France), shows the tilted fault block structure associated with the initial rifting between Europe and North America. □

$$\rho_s S g + T_c \rho_c g, \quad (2)$$

where S = thickness of sediments and ρ_s = density of sediment. If the depth of compensation is at the base of the loaded crust then

$$d = S + T_c$$

or

$$r = S - W_d. \quad (3)$$

Combining equations 1–3 we then get (neglecting effects of sea-level and water depth changes):

$$S = W_d \frac{(\rho_m - \rho_w)}{(\rho_m - \rho_s)}. \quad (4)$$

With ρ_s = 2500 kg m^{-3}, ρ_m = 3300 kg m^{-3} and ρ_w = 1030 kg m^{-3}, equation (4) shows that only a thickness of up to about 2.5 times the water depth can form as a result of sediment loading.

In the case of a land-locked basin, the water is eventually replaced by sediment and no further deposition can occur. The situation is different in basins where there is a limited supply of sediment such that the basin is unable to fill completely. At a continental margin, for example, sediments may be able to build out, but they can rarely fill a growing ocean basin.

Let us consider the case of a 'hypothetical' margin in which sediment loads progressively build up and out from some initial shelf break in slope (Figure 15.4a). Oceanic flexure studies reveal that when the crust and lithosphere are loaded they respond by bending over a broad region. The pattern of deformation is similar to that which would be expected for an elastic plate overlying a weak fluid substratum (see Box 15.1 for a mathematical treatment of how an elastic plate deforms when loaded). Figure 15.4b shows that the flexure due to a wedge-shaped sediment load is characterised by a broad depression, a thinning of the stratigraphic unit in a landward and seaward direction, and a flanking region of uplift or bulge (often termed a **flexural bulge**).

As a margin builds out, the stratigraphy due to an initial load and its flexure is modified by subsequent loading. For example, a previous load and flexure that is located in the flexural depression of a new load will experience subsidence. Conversely, a previous load and flexure located on the bulge will be subject to uplift. The cumulative stratal geometry (Figure 15.4c,d) shows **onlap** (see Figure 20.4a, p. 394) in the direction of load migration and a basinward shift in the pattern of onlap, or **offlap**. Furthermore, each depositional surface is flexed to a sigmoidal shape. The calculated stratal patterns are similar to structures observed in river delta systems, especially those which are referred to as clino-forms, (see Figure 20.4b and c, p. 394) observed on seismic reflection profiles.

The calculated stratigraphy in Figure 15.4 is based on a model which assumes that during sediment loading there is little or no stress relaxation of the lithosphere. Oceanic flexure studies, however, show that the elastic thickness of the lithosphere (which is determined by the flexural rigidity) is some two to three times thinner than the seismic thickness, suggesting that on loading there must be some form of relaxation of the mechanically supportive part of the lithosphere from its short-term (? seismic) thickness to its long-term elastic thickness. Since oceanic island loads take at least about 1 to 2 Ma to form, an elastic model is probably a good approximation for the response of the lithosphere to sedimentary packages that are a few million years or longer in duration. Unfortunately, it is not currently known whether continental lithosphere behaves in a similar way to the oceanic lithosphere or, whether, because of compositional differences, it takes more or less time to complete its relaxation.

Irrespective of the relaxation time, Figure 15.4 shows that sediment loading is capable of forming basins of considerable thickness. The infilling of a water-filled depression 5 km deep (the maximum that could be produced by a thermal cooling model) by sediment would produce a basin up to 12.5 km thick. The sediment loading effect is large enough that it complicates the use of stratigraphic data to isolate the effects of those processes, such as thermal contraction, that are believed to be responsible for basin subsidence. The **'backstripping'** technique is one approach that has been adopted to attempt to correct the stratigraphic data for the effects of sediment loading, and is described in Box 15.2.

Backstripping of sediments from deep commercial wells shows that the post-rift subsidence of many basins is characterised by a simple exponential decrease – in accord with the predictions of the McKenzie model. At some basins, the agreement is remarkably close suggesting that thermal contraction may be a major contributor to basement tectonics. Others show irregularities on the smooth exponential decrease due to superimposed tectonic events. Off Sable Island, Nova Scotia, for example, an abrupt step in the tectonic subsidence at about 140 Ma has been attributed to piercement structures caused by the mobilisation of an underlying salt layer.

At some wells, the McKenzie model is unable to explain the tectonic subsidence deduced by backstripping. For example, wells in the East Coast US coastal plain show uplift, rather than subsidence, during the first few tens of million years of the post-

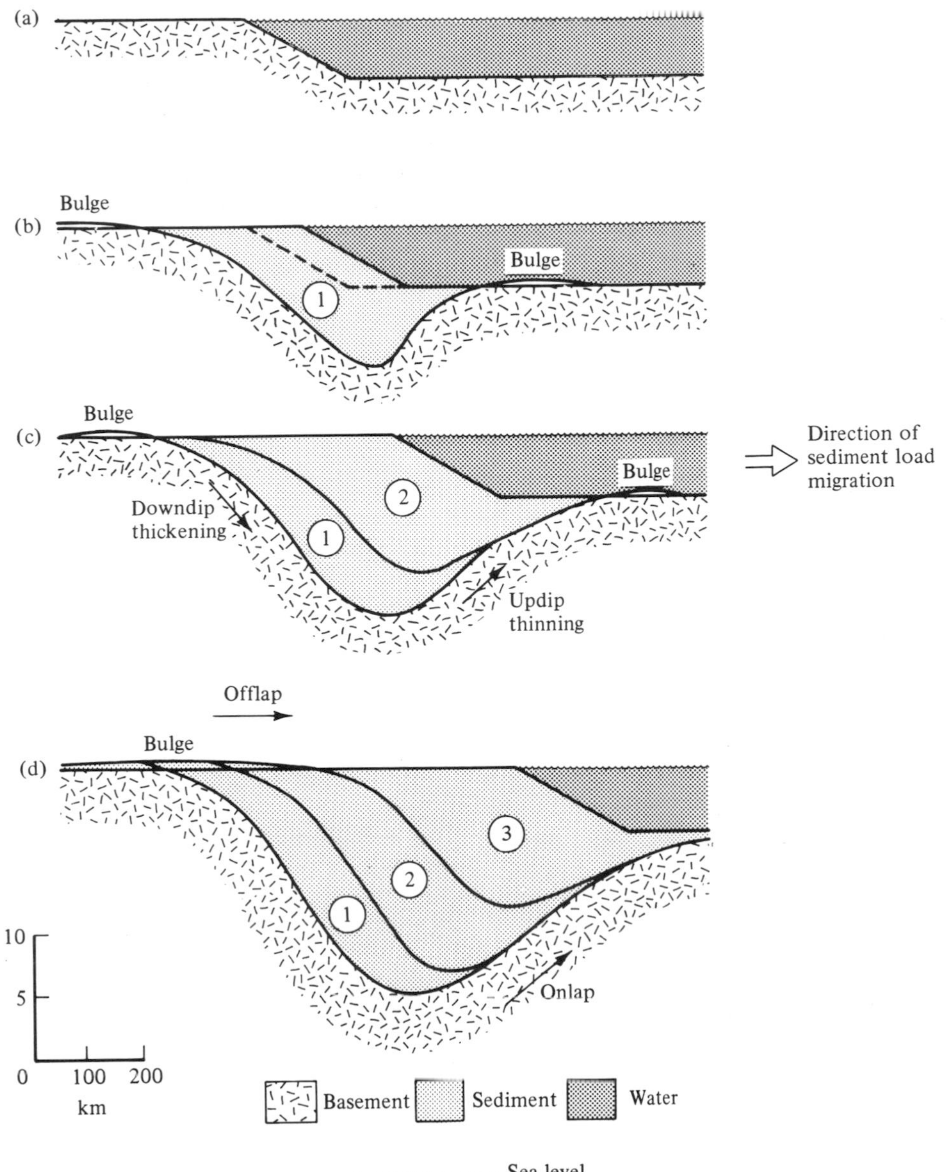

Figure 15.4 Simple model for the effects of sedimentary loading at a passive continental margin. (a) The initial state of margin with a slope and abyssal sea-floor depths of 5 km; (b) flexure due to a single wedge-shaped load; (c) the flexure after two loads have been applied; (d) the margin configuration after three loads. □

rift period. This uplift cannot be explained by a thinner than usual crust at the time of rifting (an initial crustal thickness of 18 km or less will cause uplift in the McKenzie model) since the edge of the margin eventually did subside forming a gently dipping coastal plain on to which Late Cretaceous to Recent sediments were deposited.

One way to explain the well data is by a **'two-layer' model** where the crust and mantle are extended by different amounts: the uplift being the result of mantle thinning beneath a region of little or no crustal thinning. The total amount of extension in the crust and mantle must be the same, of course, otherwise there would be a space problem. McKenzie and coworkers used simple Gaussian curves for the distribution of the extension in the crust and mantle. A model (e.g. Figure 15.5) in which the mantle extension occurs over a broader region than the crustal extension but the total amount of mantle and crustal extension are the same explains the absence of the Jurassic beneath

BOX 15.1

Loading of thin elastic plates

A useful way to calculate the flexure of the crust and lithosphere due to arbitrary shaped sediment loads is based on the response function technique. Consider, for example, the case of a load, $h(x)$, on the Earth's surface of the form:

$$h(x) = h \cos(kx)\, g\, (\rho_s - \rho_w), \tag{A.1}$$

where h = the maximum height of the load, g = average gravity, ρ_s = density of the sediment, ρ_w = density of water, k = wavenumber ($2\pi/\lambda$, where λ is the wavelength), and x = horizontal distance. The flexure of the crust and lithosphere, $z(x)$, is obtained by solving the fourth-order differential equation for the bending of a thin elastic plate overlying a weak fluid substratum and is given by:

$$z(x) = \frac{(\rho_s - \rho_w)\, h\, g \cos(kx)}{((\rho_m - \rho_s)\, g + Dk^4)}, \tag{A.2}$$

where ρ_m = density of the mantle and D is the flexural rigidity of the plate which is determined from the elastic thickness, T_e by:

$$D = \frac{E\, T_e^3}{12\,(1 - \sigma^2)}, \tag{A.3}$$

where E = Young's modulus and σ = Poisson's ratio. We see from equation (A.2) that when $k \to 0$ (i.e. $\lambda \to \infty$) then

$$z(x) \to \frac{(\rho_s - \rho_w)\, h}{(\rho_m - \rho_s)},$$

which if $\rho_s = \rho_c$ where ρ_c = density of the crust gives:

$$z(x) \to \frac{(\rho_c - \rho_w)\, h}{(\rho_m - \rho_c)}.$$

This is the well known 'Airy root' response to loading. If, on the other hand, $k \to \infty$ (i.e. $\lambda \to 0$) then

$$z(x) \to 0,$$

and there is no deflection. This can be considered as the Bouguer response to loading. We see from equation (A.2) that the crust and lithosphere is behaving as a sort of filter in the way that it responds to sediment loads. The term filter is used in the usual sense to denote a system with an input and output. In the application discussed here, the input is the load $h(x)$ and the output is the flexural response $z(x)$. The filter characteristics can be determined by defining a certain wavenumber parameter given by:

$$\phi_e = \frac{\text{input}}{\text{output}}.$$

For convenience, we will consider the Airy-type response as input and the flexural response as output. We are then able to write:

$$\text{Input} = \frac{(\rho_s - \rho_w)\, g\, h \cos(kx)}{(\rho_m - \rho_s)\, g},$$

$$\text{Output} = \frac{(\rho_s - \rho_w)\, g\, h \cos(kx)}{[(\rho_m - \rho_s)\, g + Dk^4]},$$

and

$$\phi_e = \left[\frac{D\, k^4}{(\rho_m - \rho_s)\, g} + 1\right]^{-1}. \tag{A.4}$$

This is the function that modifies the input (i.e. the Airy-type response) to produce the output (i.e. the flexure). Equation (A.4) shows that for long-wavelength loads, the crust and lithosphere do not modify the Airy-type response, and these loads are supported mainly by buoyancy of the substratum. For short-wavelength loads the crust and lithosphere greatly modify the Airy-type response. These loads are mainly supported by the rigidity of the plate.

The response function approach outlined here can be used to rapidly compute the flexure due to any shape of sediment load. Let us replace the load $h \cos(kx)$ by a function $\delta(x)$ such that $\delta(x) = 0$ when $x \neq 0$ and

$$\delta(x)\, \partial x = 1, \qquad x = 0.$$

Now, if $H(k)$ and $Z(k)$ are the discrete Fourier transforms of $h(x)$ and $z(x)$ respectively, then equation (A.2) reduces to

$$Z(k) = \phi_e(k)\, H(k)\, \frac{(\rho_s - \rho_w)}{(\rho_m - \rho_s)}.$$

To calculate the flexure, $z(x)$, due to an arbitrary shaped sediment load we simply take its transform, multiply by the wavenumber parameter, ϕ_e, and a density factor, and inverse transform the result.

BOX 15.2

Backstripping

Let us consider a so-called 'loaded' column that represents the sedimentary unit that accumulated during a certain interval of geological time and an 'unloaded' column that represents the position of the underlying 'basement' without the effects of the sediments (see figure). The pressure at the base of a loaded column is given by:

$$W_d \rho_w g + S^* \rho_s g + T_c \rho_c g, \qquad \text{(B.1)}$$

where W_d = water depth of deposition, T_c = mean thickness of the crust, S^* = the sediment thickness corrected for compaction (see discussion in later section), g = average gravity and ρ_w, ρ_s and ρ_c are the densities of the water, sediment and crust respectively. The pressure at the base of the unloaded column is given by:

$$Y \rho_w g + T_c \rho_c g + r \rho_m g, \qquad \text{(B.2)}$$

where Y is the corrected or tectonic subsidence, ρ_m is the density of the mantle, and r (see figure) is the distance from the base of the unloaded crust to the depth of compensation (assumed to be at the base of the loaded crust) and is given by:

$$r = S^* + W_d - \Delta_{SL} - Y, \qquad \text{(B.3)}$$

where Δ_{SL} = sea-level change as it would be viewed on the continent or 'freeboard'.

Combining equations (B.1, B.2 and B.3) we then get:

$$Y = S^* \frac{(\rho_m - \rho_s)}{(\rho_m - \rho_w)} + W_d - \Delta_{SL} \frac{\rho_m}{(\rho_m - \rho_w)}. \qquad \text{(B.4)}$$

Equation (B.4) corrects the observed stratigraphic record for the effects of sediment and water loading and changes in water depth – a technique that has come to be known as '**backstripping**'.

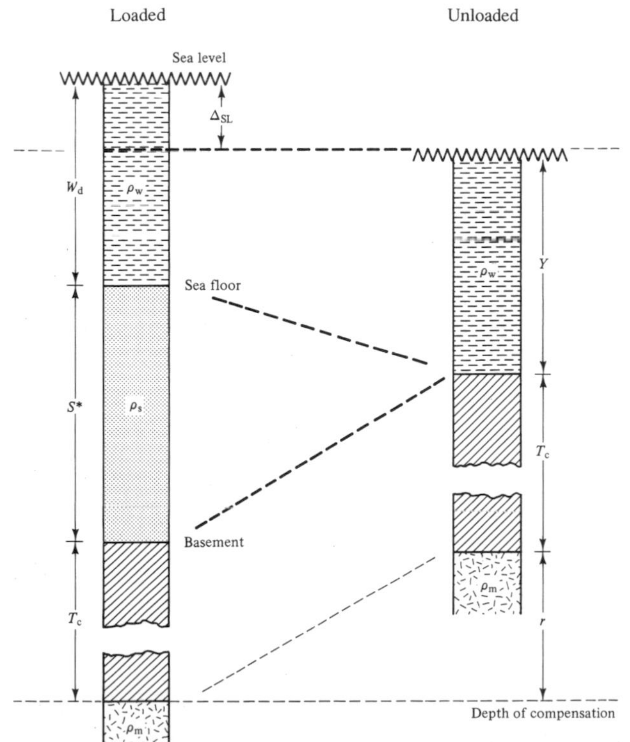

Figure B.1 Schematic diagram illustrating the backstripping technique.

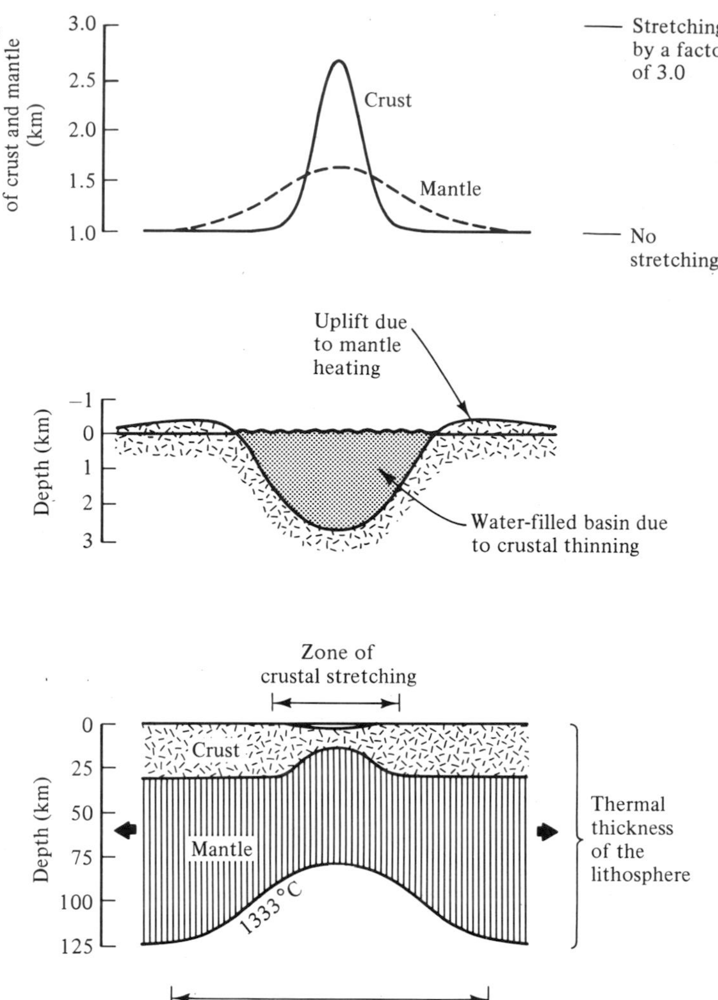

Figure 15.5 The initial water-filled subsidence for a rift-type basin formed by stretching of the crust and lithosphere. The thermal bulges are the result of a transfer of heat from the hot stretched region to the unstretched crust and lithosphere. The model assumes that the stretching is distributed unevenly with depth but that the total amount of extension in the crust and lithosphere is equal. □

the East Coast US coastal plain, even though the rift/drift transition (i.e. the change from rifting to sea-floor spreading) at this margin is believed to be of Late Triassic/Early Jurassic age.

The distribution of extension with depth will influence the shape of a water-filled basin (Figure 15.6). If the extension is accomplished by a simple necking of the crust and lithosphere (i.e. **pure shear**) then the maximum crustal and mantle extension will coincide and there will be an overall symmetry to the pattern of basin uplift and subsidence. If, on the other hand, the extension is by **simple shear** (as some recent geologic studies in the Basin and Range province, western USA suggest) then the crustal and mantle extension may be offset such that the syn- and pre-rift are displaced from the post-rift sediments and there is an overall asymmetry to the basin shape.

At present, it is not known whether the extension in rift-type basins occurs by pure or simple shear or by some combination of these mechanisms. A critical test is seismic reflection and refraction profile data since these have the potential to image deep layers in the crust as well as determine its detailed velocity structure. During the past decade several countries (e.g. USA, UK, France and Canada) have developed programmes to apply seismic techniques to the study of the deep structure of the continental crust (see Chapter 13).

A compilation of a deep seismic LITHOPROBE (Canada) line across the Canadian margin with a BIRPS and ECORS (UK, France) profile of the conjugate margin (i.e. one that was opposite to another prior to continental separation) off southwest Britain is shown in Figure 15.7. This profile shows a highly layered lower continental crust and several strong reflectors that may represent detachment surfaces within the crust. The most significant result of the seismic profiles is that they reveal that, while the

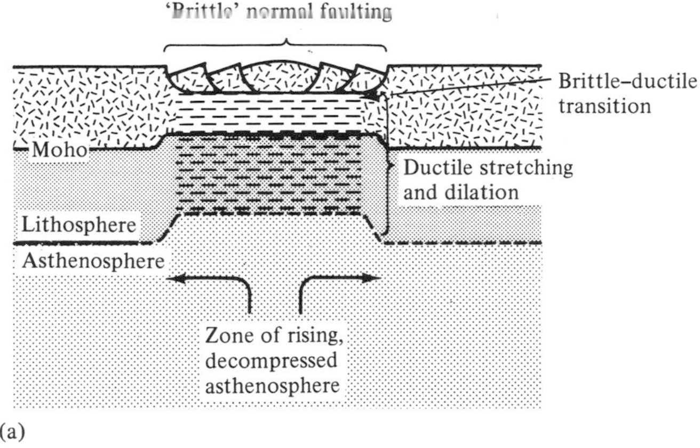

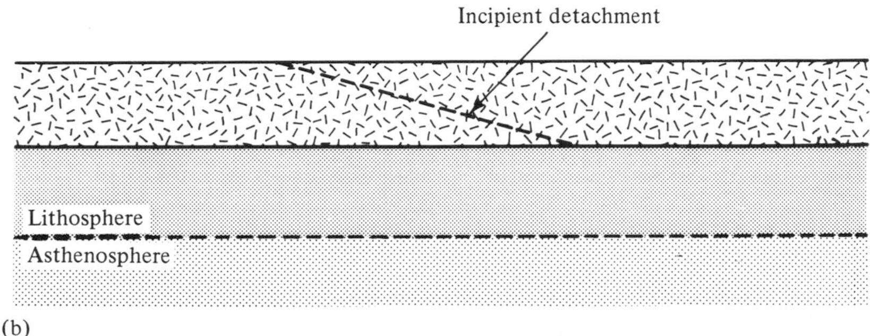

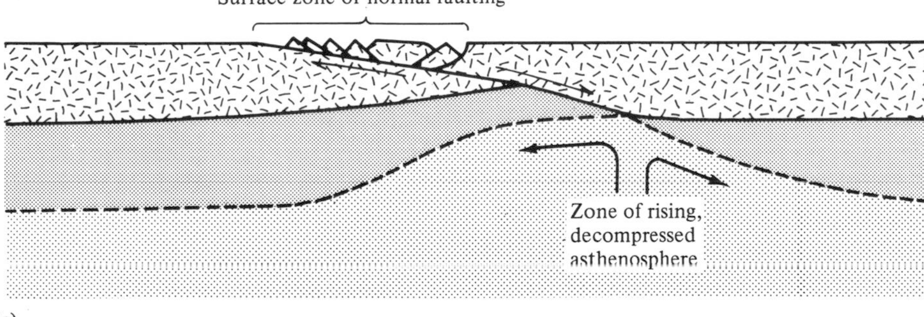

Figure 15.6 **Illustration of the two end-member models for lithospheric extension and basin formation. (a) Shows the pure shear model; deformation in both brittle and ductile layers is uniform. (b) and (c) are successive stages in the simple shear model, where brittle fault blocks are carried along a shear plane that propagates through the entire lithosphere; deformation in both brittle and ductile layers is asymmetrical.** □

geometry of faulting in the upper crust is asymmetric, the lower crust is more symmetric providing general support to the stretching model.

A number of basins fail to show any evidence of the exponential decrease of subsidence that is so characteristic of extensional basins formed by rifting and that is predicted by the McKenzie model. Foremost among these are the foreland basins that form in compressional settings in front of advancing thrust/fold belts. Although exceptions exist, many of these types of basins are characterised by an initially slow subsidence which progressively *increases* with time.

It is generally agreed that foreland basins are mechanically, rather than thermally, driven. Stratigraphic studies based on deep wells, seismic reflection profiles and gravity modelling show that basement beneath the Colville (Alaska), Powder River (western USA) and Alberta (western Canada) basins, for example, dips gently toward the mountain belt. Furthermore, these basins are bounded on their landward

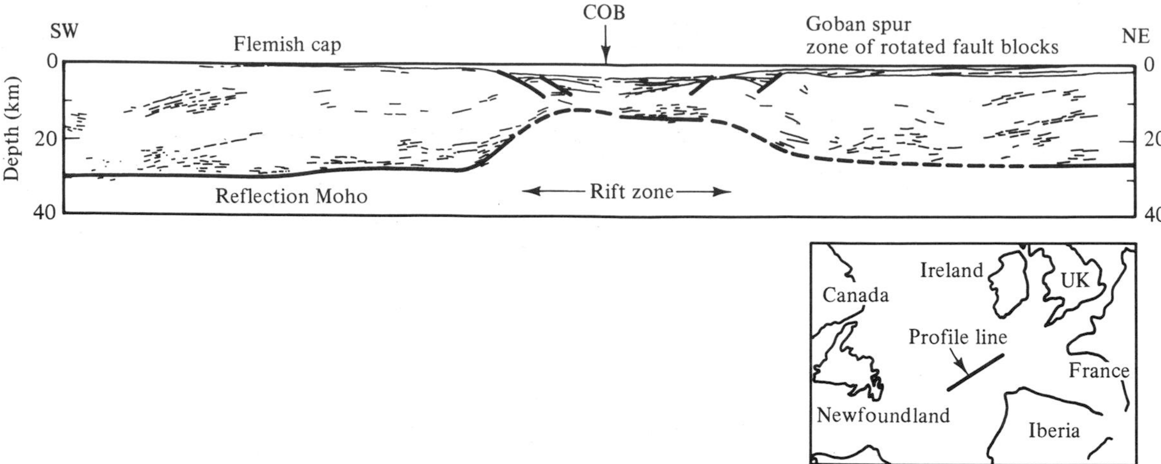

Figure 15.7 A compilation of deep seismic reflection profile data obtained over the eastern margin of the Canadian shelf and across the conjugate margin off southwest Britain. The compilation shows that, whereas the fault pattern in the upper crust is asymmetric, the thinning of the lower crust is symmetric, suggesting stretching occurred by pure shear (COB: continent–ocean boundary). □

side by a broad flexural arch which, in some cases, plays an important role in the supply of sediment to the trough. One of the most well-studied foreland basins is the Colville Trough in the North Slope, Alaska (Figure 15.8). The trough is bounded to the south by the Brooks Ranges and to the north by the Barrow Arch, which is one of the world's largest petroleum plays. Figure 15.8 shows that the basement underlying the Colville Trough has a flexural form and previous studies have shown that it can be explained by an elastic plate model which is loaded at one end.

If mountain belts comprise a series of thrust/fold loads then a flexure model predicts that as each load advances, the flexural depression and arch migrates away from the mountain belt. The tectonic movement at a particular locality in a basin would depend, however, on its location on the flexure profile. For example, a locality that is near the flexural node (i.e. a point of no subsidence or uplift) will not experience subsidence until the flexural depression of a new load reaches it. The tectonic subsidence, as obtained say by backstripping, would be expected to be initially slow and then progressively increase.

A difficulty with the modelling of foreland basins is that the sources of the loads responsible for the tectonic subsidence are not well known. In the case of the Himalaya, gravity anomaly studies suggest that the present-day topography is sufficiently large to account for the subsidence of the Australian–Indian plate beneath the Ganges basin. However, the Himalaya consists of a 'stack' of individual thrust/fold loads so that if we are to predict the stratigraphy of the flanking foreland, the geometry of each load needs to be reconstructed first. Unfortunately, this requires estimates of the amount of shortening, the height of the series of thrusts, and the extent of erosion, and these are difficult to access. Gravity anomaly studies of the Alps and Appalachians suggest, on the other hand, that the present-day topography is insufficient to cause flexure of the Swiss Molasse and Appalachian foreland. These mountain belts require **'buried' loads** within the crust. Again, buried loads pose a difficulty in stratigraphic modelling since their distribution with time is not known.

The observation in Precambrian shield areas of foreland basins *without* their accompanying mountain belt raises the possibility that buried rather than surface loads are the primary mechanism by which forelands are preserved in the geological record. A mountain belt is subject to erosion and, if it constitutes the main load, eventually the adjacent foreland basin would be removed as the mountain belt and its flank respond to the denudation by uplift. Buried loads, however, are 'protected' from erosion and could remain in the crust for long periods of geological time. Currently, there is considerable controversy regarding the origin of these loads. In the Alps and Appalachians, buried loads are associated with large-amplitude gravity anomalies and, in some cases, outcrops of ultra-basic rocks. They could be wedges of oceanic crust (and mantle?) material that were thrust into the thickened continental crust during convergence. Alternately, they may represent areas of formerly thin crust, formed in passive margin settings, the mantle 'anti-

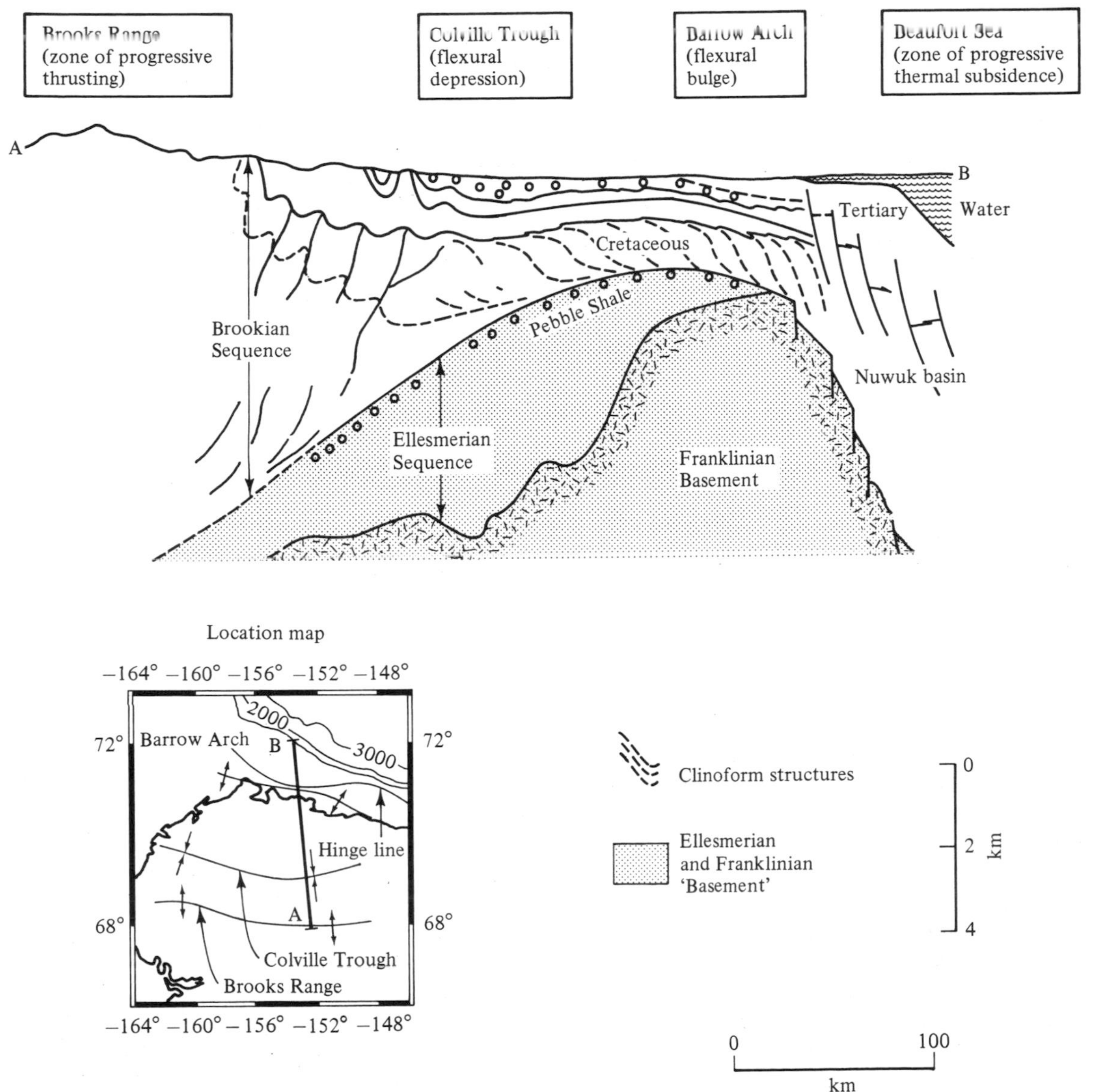

Figure 15.8 Schematic cross-section of the Brooks Range – Colville Trough – Barrow Arch system. The Pebble Shale marks the base of the fore-deep sequence and was probably flat prior to loading in the mountain belt. □

roots' of which later become loads on the base of the crust.

The mechanical response of the crust (and lithosphere) to loads also appears to play a role in the development of the small basins that develop in regions of localised extension and compression. These basins are distinguished from their rift-type and foreland counterparts by their narrowness, rapid subsidence, variable facies, and; in the case of marine basins, the inability of sedimentation rates to keep pace with subsidence. The stratigraphy of these basins suggests that they alternately experience compression and extension on very short time scales due to rotations of adjacent crustal blocks.

One of the best examples of the mechanically driven small basins are the **'pull-apart' basins** (e.g. Ventura, California) that form between overlapping strike-slip faults such as occur along the San Andreas fault system. Although, a McKenzie-type stretching model has been applied to explain subsidence in the Ventura basin, other basins in the fault system (e.g. Santa Barbara) show marked interruptions in their subsidence due apparently to compressional events and their associated flexural effects. Other examples of small basins occur in compressional and extensional settings. In compressional settings it is quite common to find 'piggy-back' basins perched above the main thrust front (e.g. Rocky Mountains). These basins occupy the sites of large pre-existing inter-montane depressions that probably formed in response to compression. Often these basins are characterised by sub-aerial, lacustrine or fluviatile sediments. The major control on the subsidence of these basins appears to be sediment loading. In extensional settings, small basins are found within rift systems, especially on the hanging walls of normal faults (i.e. the side of the fault that moved relatively downwards). Again, sediment loading is thought to be a major cause of subsidence in these basins.

Basin stratigraphy

The previous discussion suggests that it should be possible to construct synthetic stratigraphic cross-sections of basins simply by combining sediment loading with mechanisms such as thermal contraction and thrust/fold emplacement. This is difficult to do in practice because (a) the calculation of sediment load depends on a knowledge of the water depth through time and (b) there are certain feedback effects between sediment loading and thermal contraction and thrust/fold loading which need to be taken into account.

The problem is illustrated schematically in Figure 15.9. Let us assume a basin margin characterised by a wedge of sediment that formed in some depth of water. Within the wedge there will be a surface which will be the depth the underlying 'basement' would be without the sediment. This surface is the backstripped 'basement' depth, referred to here as the **base**. The concept of base is important since it allows the sediment thickness to be partitioned into two parts: a region above which defines the load and a region below which defines the flexure. Since we now have models for the movement of base (e.g. by thermal contraction) and the flexural response of the crust and lithosphere, stratigraphic modelling reduces to estimating the sediment load as increments through time. Unfortunately, estimation of the load depends also on the water depth of deposition and, for most basins, this is a poorly known parameter.

The water depth is a product of the combined processes of erosion and sedimentation, each of which will vary spatially and temporally. Coastal studies show that shorelines are continually being modified in such a way that they tend to acquire just the right slope to ensure that the incoming supplies of sediment can be removed at about the same rate as they are received. An adjusted profile of this type is referred to by geologists as an **'equilibrium profile'**. Unfortunately, because of the complexities of tides, storms, surges and climate variations it is unlikely that a long-lived approach to equilibrium can ever be achieved. The concept is a useful one, however, and most basin models now assume that sediments infill to either a horizontal, pre-deformation surface or to some constant sloping palaeo-bathymetric surface.

A further potential difficulty in stratigraphic modelling is that the mechanical and thermal properties of the crust and lithosphere are linked in such a way that the flexural strength depends on their thermal structure. For example, in a rift-type basin the lithosphere

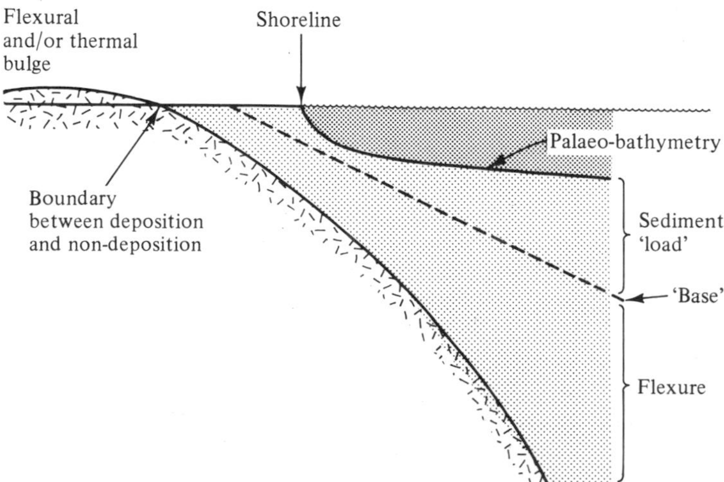

Figure 15.9 Schematic illustration of the 'beach problem' in basin modelling. For explanation, see text.

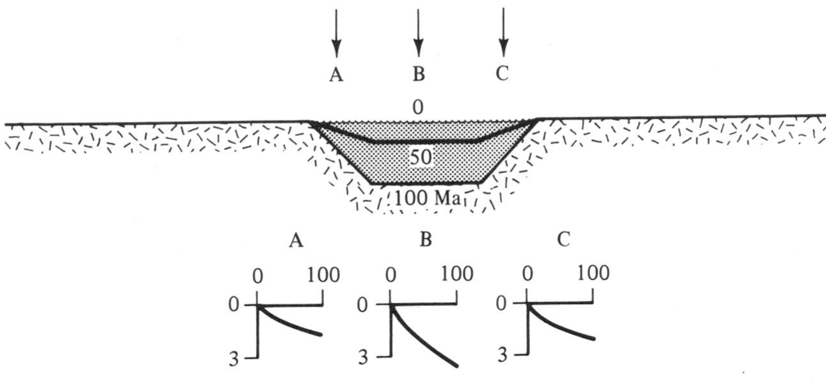

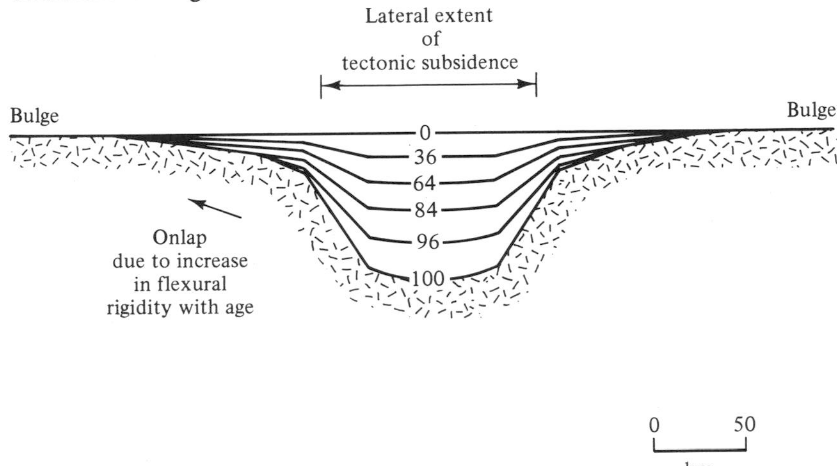

Figure 15.10 Simple model for the stratigraphic patterns that develop in a thermally subsiding basin formed by extension at the time of rifting. The onlap at the basin margin is produced by an assumption that as the lithosphere cools following rifting it becomes stronger, thereby spreading out the deformation due to individual sediment loads. □

is likely to be hot and weak early in basin evolution while later on it is likely to be cold and strong. This change in strength during basin evolution could greatly affect the basin architecture (i.e. the stratal patterns developed within the basin).

Oceanic flexure studies have shown that as the lithosphere cools, its flexural rigidity (which is a measure of its resistance to deformation) increases. The elastic thickness of the lithosphere, T_e, is determined by the flexural rigidity and is given approximately by:

$$T_e = Z_{450\,°\mathrm{C}}, \qquad (5)$$

where Z is the depth to the 450 °C oceanic isotherm. If the flexural rigidity is also coupled to the temperature as it is in the case of rift-type basins, then we would expect that T_e would increase from small (zero?) values for sediments deposited at the time of rifting to about 30 km for sediments formed 100 Ma after the basin-forming event.

Figure 15.10 shows a model for the stratigraphy of a rift-type basin based on the concept of an equilibrium profile and a T_e which increases with age as the crust and lithosphere cool. The tectonic subsidence is assumed to be exponential in form, and to increase from the basin edge to its centre. The predicted stratigraphy shows a progressive overstep, or onlap, of the basement at the basin edge. This onlap is the result of our assumption that the rigidity of the underlying lithosphere, like that of oceanic lithosphere, increases with age. The result after a number of stratigraphic units have been deposited is to produce a flexural basin with a characteristic sag shape, which is referred to by some as the **'steers head'**.

A number of rift-type basins show onlap at their edges especially during the thermal subsidence phase of basin development. Figure 15.11 shows an example which developed following the rifting and separation of Africa and North America during the Early Jurassic.

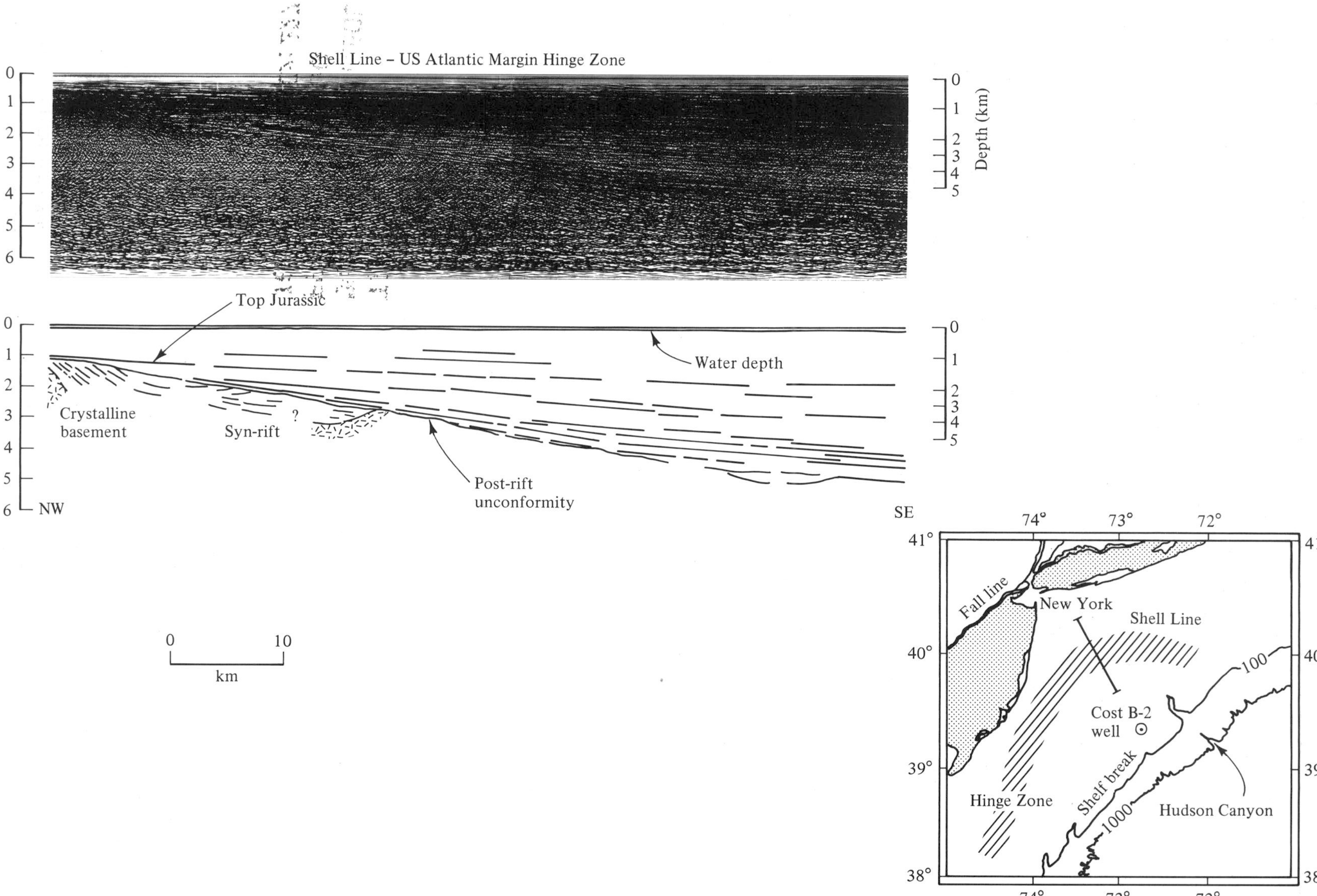

Figure 15.11 Seismic reflection profile obtained by the Shell Oil Company across a portion of the hinge zone of the Atlantic-type continental margin off New York. □

A flexure model with a T_e that increases with age predicts that a basin should widen with time and that young sediments progressively overstep basement rocks at the basin edge. The fact that some basins (e.g. Michigan, central USA) show the youngest sediments in the centre is therefore problematical. Initially, it was thought that the young sediments were the result of some form of viscous relaxation in the lithosphere. Inspection of Figure 15.10 shows, however, that an elastic plate model with a T_e that increases with age could explain the occurrence of young sediments at the basin centre, provided that its edges were subject at some time during their history to a widespread erosional event.

As pointed out earlier, the assumption that T_e depends on age is based mainly on oceanic flexure observations and not on any direct evidence for the behaviour of the continental lithosphere. Some researchers have recently suggested that the continental lithosphere may not follow such a simple T_e versus age dependence and that in extension it may be intrinsically weak. If this is the case can onlap patterns, such as observed in Figure 15.11, still be explained by a tectonic mechanism?

Several workers have pointed out that an Airy-type model (i.e. $T_e = 0$, see Box 15.1) could also explain the post rift onlap, provided it is combined with a thermal model in which extension varies with depth. The thermal model required is one in which the basin flanks are characterised by mantle heating with little or no crustal stretching. The mantle heating causes the flanks to rise while the lack of crustal stretching prevents subsidence. Eventually, the mantle cools, the basin flanks subside below sea level and sediments progressively overstep the basement.

Foreland basins apparently do not have a thermally induced driving component, although both the sediments (and basement) may heat up as subsidence proceeds. The movement of base (Figure 15.9) in this case is produced by flexure due either to thrust and fold loads emplaced on the surface of the crust or to 'buried loads' within the adjacent orogenic belt.

Figure 15.12 shows the stratigraphy that would be expected for a foreland basin formed by four separate thrust/fold loading events. Each load was emplaced on the plate assuming that sediment infilled the basin to a constant horizontal pre-deformation surface. As each load advances, the sedimentary infill progressively oversteps older units and the flexural bulge migrates outward of the thrust/fold front. The result

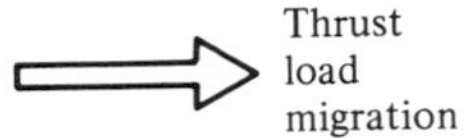

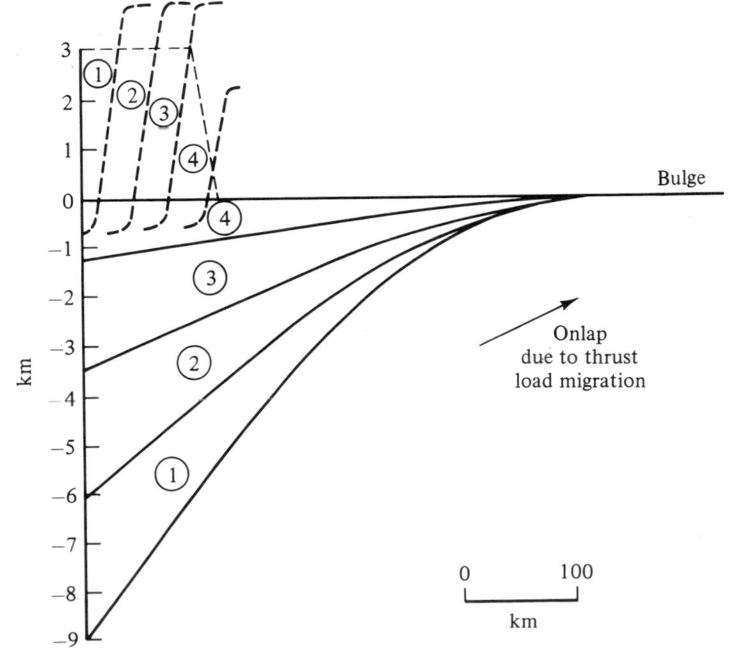

Figure 15.12 Simple model for the stratigraphy of a foreland basin formed by flexure due to successive advancing thrust/fold loads.

after four loads is to produce a flexural sag basin that is dominated by onlap at its edge. Onlap is a feature of all models: the widest pattern of onlap corresponding to large values and the narrowest pattern to small values of the flexural rigidity.

Foreland basins range in width from 20 to 300 km, suggesting that the rigidity of the underlying basement ranges from weak (e.g. Appenine, Ebro, Guadalquivir) to strong (e.g. Allegheny, Ganges). The cause of such large strength variations in the continental lithosphere is not clear, but it may be due to the different tectonic settings of these basins. For example, the Appenine, Ebro and Guadalquivir basins are located on lithosphere that was quite close to the site of passive margin development at the borders of the Tethys Ocean, while the Allegheny and Ganges basins developed on the cold, rigid, interiors of the continents.

Modulation of the stratigraphic record

The basin models discussed so far are based on only the simplest of thermal and mechanical concepts. They ignore the disturbing effects of other sedimentary factors such as erosion, variations in the supply of sediment, compaction and changes in sea level. While it is believed these factors do not contribute significantly to the overall geometry of basins, they nevertheless may interact with the primary controls in such a way as to exert a major control on the smaller-scale packaging of stratigraphic units.

Erosion

Basins formed by thermal contraction and flexure are associated with peripheral bulges which may be subject to erosion. Thermal bulges are a feature of the initial phases of rift-type basin evolution and can be produced by either the lateral flow of heat from the hot stretched region to the relatively cold unstretched continental lithosphere, or a region of mantle heating that extends over a wider region than the crustal stretching. Flexural bulges, on the other hand, are a consequence of the mechanical loading of an elastic plate overlying a weak fluid substratum and they can develop in strike-slip, extensional and compressional settings. The removal of material by erosion has two effects: one is to unload the crust and cause uplift and the other is to thin it and cause subsidence. At the edge of a basin, the interaction of these effects will have important consequences for the position of the shoreline.

Let us consider a hypothetical mountain range that initially is in isostatic equilibrium such that its elevation, h, is balanced by a deep crustal root, r. Erosion disturbs this balance and an uplift, w, results. The pressure at the base of a column in the mountain range is given by:

$$(h + r + T_c)\rho_c g,$$

where ρ_c = density of the crust, T_c = mean thickness of the crust and g = average gravity. The pressure at the base of the eroded column is:

$$(T_c + r)\rho_c g + \rho_m w g,$$

where ρ_m = density of the mantle. Assuming the columns before and after erosion are in isostatic equilibrium we then get

$$w = h\frac{\rho_c}{\rho_m}. \qquad (6)$$

Using typical values of ρ_m = 3330 kg m^{-3} and ρ_c = 2700 kg m^{-3} we see that an elevated region can be reduced to about 70 per cent of its height by erosion.

If the erosion is a prolonged event then this uplift will in turn be subject to erosion. Equilibrium is achieved when the uplift is reduced to zero which occurs as soon as the total amount of material that is removed, S_e, reaches $h + r$. The Airy root r is given by

$$r = \frac{h\rho_c}{(\rho_m - \rho_c)},$$

so the total amount that can be removed is given by

$$S_e = h\left[1 + \frac{\rho_c}{(\rho_m - \rho_c)}\right]. \qquad (7)$$

Using the same values of ρ_m and ρ_c as before, equation (7) shows that up to about 5.5 times the initial topography could be removed by erosion.

The estimates of the amount of erosion discussed so far are maximum likely ones since they ignore that the uplift may be limited by the flexural rigidity of the plates. Moreover, erosion may be inhibited by other factors such as lithology, vegetation and climate. Unfortunately the dependence of erosion rate on these variables is not well understood.

Studies of the catchment areas of large river systems have revealed that the rate of erosion, due to the combined effects of lithology, vegetation and climate, is approximately proportional to the average elevation. We can then write that

$$\frac{\partial h}{\partial t} = K\, w_{av}, \qquad (8)$$

where w_{av} = average elevation and K = the constant of proportionality. The rate of uplift (assuming an Airy-type crust) is obtained by differentiating equation (6) with respect to time:

$$\frac{\partial w}{\partial t} = \frac{\partial h}{\partial t} \frac{\rho_c}{\rho_m} .$$

The rate of change in the average elevation is given by the difference between the rate of uplift and the rate of erosion

$$\frac{\partial w_{av}}{\partial t} = \frac{\partial w}{\partial t} - \frac{\partial h}{\partial t} ,$$

which gives:

$$\frac{\partial w_{av}}{\partial t} = \frac{\partial h}{\partial t} \left[\frac{\rho_c}{\rho_m} - 1 \right]$$

or

$$\frac{\partial w_{av}}{\partial t} = -K \left[1 - \frac{\rho_c}{\rho_m} \right] w_{av} .$$

Hence,

$$w_{av(t)} = w_{av(t=0)} \left| e^{-(1-\frac{\rho_c}{\rho_m})Kt} \right| , \qquad (9)$$

where $w_{av(t=0)}$ is the initial elevation and $w_{av(t)}$ is the average elevation at subsequent times. Equation (9) shows that an initial uplift will decay exponentially with time. The parameter $1/K$ is defined as the **erosion time constant**.

The question for basin modelling is what are likely values for the erosion time constant? Modern estimates for the rate of erosion range from 50 cm per 1000 years for the Himalaya to 5 cm per 1000 years for the Appalachians. The average elevation of these features are 5 and 0.5 km respectively, so equation (8) suggests a time constant of about 10 Ma. At this rate, it would take about 60 Ma to reduce the Himalaya and Appalachian mountains to about a third of their present-day elevation.

Several workers have pointed out that since erosion continually modifies the landscape it is important to take into account the conservation of mass. One approach is to assume that erosion is a linear diffusion process and that a balance exists between the rate of erosion and the spatial variation in flux of material. The change in topography can then be written

$$w_{(t)} = w_{(t=0)} \left[e^{-(1-\frac{\rho_c}{\rho_m}\eta)Kt} \right] , \qquad (10)$$

where η is the slope of the topography and $1/K$ is the **diffusion time constant**.

A diffusion model is important to basin modelling since it allows the transport of sediment from source to sink to be specified, even in basins where the supply of sediment is not known. For example, the difference between the slope after diffusion and the original slope gives the amount of material removed by erosion and the amount that is added by sedimentation. Diffusion causes the area of the material that is removed and added to broaden with time. This effect has consequences for the development of stratigraphic patterns at the basin edge – causing offlap in the region of erosion and onlap in the more distal region of sedimentation.

Variations in the supply of sediment

It is generally assumed in basin modelling that the supply of sediment to a basin is sufficient to maintain a constant bathymetry profile through time. In fact, the amount of sediment supply is highly variable. Studies of bio-stratigraphic data from deep wells in the East Coast US outer continental shelf, for example, show that in Jurassic to Late Cretaceous time this margin was prograding and water depths were in middle shelf environments (i.e. water depths of about 20–100 m) but, by the Eocene, water depths had increased to slope environments (200–2000 m). In the Miocene there was a renewed period of deltaic progradation and a return to shallow-water depths of deposition.

One way to evaluate the effects of sediment supply is to use the backstripping equation discussed earlier. However, rather than consider the 'static' case, we will examine the relationships that exist between changes in the variables with time. Equation (B.4) can then be written

$$\dot{Y} = \dot{S} * \frac{(\rho_m - \rho_s)}{(\rho_m - \rho_w)} + \dot{W}_d - \dot{\Delta}_{SL} \frac{\rho_m}{(\rho_m - \rho_w)} , \qquad (11)$$

where the superscript dot refers to the rate of change of the variable. If we assume no sea-level change then $\dot{\Delta}_{SL} \to 0$ and

$$\dot{Y} \to \dot{S} * \frac{(\rho_m - \rho_s)}{(\rho_m - \rho_w)} + \dot{W}_d . \qquad (12)$$

Equation (12) shows that changes in water depth are determined by the relative difference between the rate of tectonic subsidence and sediment supply. If the sediment supply rate exceeds the tectonic subsidence, $\dot{W}_d < 0$ and water depths decrease. If, on the other hand, tectonic subsidence exceeds the sediment supply, $\dot{W}_d > 0$ and water depths increase. These considerations suggest that the sediment supply has the capacity to control the water depth or, as some workers have referred to it, the **accommodation space** (see Chapter 21) in a basin.

Usually, in the early evolution of a basin (e.g. during the initial subsidence phase) $\dot{Y}$ is at a maximum so there is an increased likelihood that $\dot{W}_d > 0$, and that water depths will increase. This may explain the gradual landward shift of transgressive sequences and onlap that dominates the early evolution of many rift-type basins. As a rift-type basin cools (e.g. during the thermal subsidence phase), $\dot{Y}$ becomes less significant and there is an increased likelihood that $\dot{W}_d < 0$ and water depths will decrease. This may result in a gradual seaward shift of regressive sequences and could explain the occurrence of offlap that tends to dominate the later stages in the evolution of many rift-type basins.

The problem is that the rate of supply of sediment to a basin is a poorly known quantity. This is because the supply of sediments from a source to a basin depocentre is frequently interrupted. Even if continental-derived sediments eventually make their way to a basin depocentre, erosion due to bottom counter currents could remove them and deposit them elsewhere as re-worked material.

As Figure 15.4 shows, the progradation of sediments in river delta systems is probably the simplest form of infill that can be taken into account in basin modelling. The geometry of progradation can be seen as 'clinoformal' structures on seismic reflection profiles. Other margins, however, during the same time period have mainly built up (i.e. have aggraded rather than prograded, see Chapter 20) and show only a limited development of such structures.

Some of the best-documented examples of prograding sediment wedges occur within a foreland basin setting – as clastics move out across a basin following deepening of a basin by thrust/fold loading. Figure 15.13 shows a model in which a prograding clastic wedge migrates out across a foreland basin from the thrust/fold tip towards the flexural bulge. The model includes a tectonic subsidence beyond the flexural bulge that causes a shift in the main depocentre and the migrating wedge to build out over the bulge, thereby limiting the amount of sediment that is deposited in the basin. Eventually, as the second thrust/fold load is emplaced, the main depocentre returns to its original position and the clastic wedge once again migrates out toward the bulge.

Compaction

One of the most important post-depositional processes that may considerably alter the 'architecture' of a sedimentary basin is compaction. The compaction may involve either a mechanical change that closes the pore spaces in a rock or a chemical alteration of the rock. Usually, sedimentary geologists restrict their definition of compaction to the physical removal of pore space, and ignore cementation. However, the latter process needs to be taken into account when modelling the burial history of sediments. Box 15.3 gives a mathematical treatment of mechanical compaction.

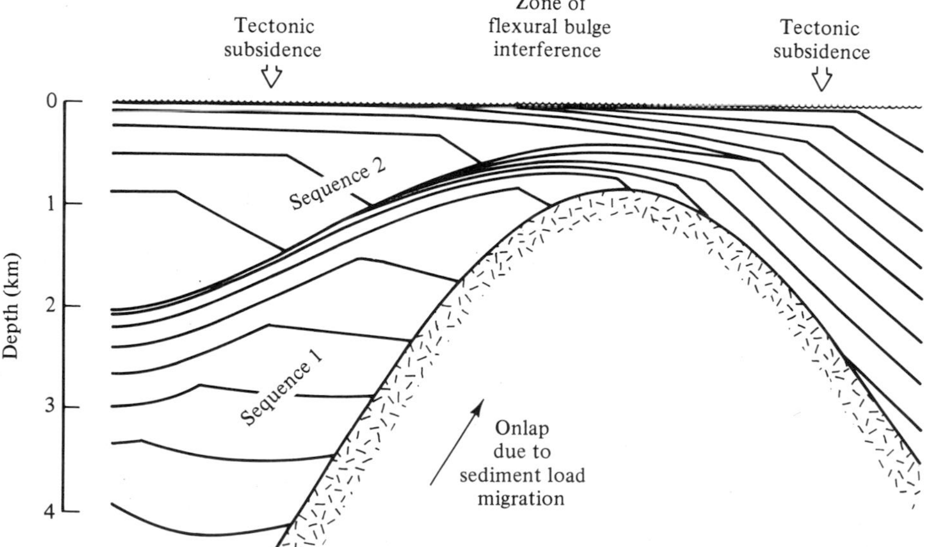

Figure 15.13 Stratigraphic model of a foreland basin assuming that it results from the superposition of two loads: one due to thrust/fold loading and the other due to thermal contraction of the lithosphere following rifting. The centres of the two loads are displaced by about 400 km, which causes uplift in the intervening region due to interference of two flexural bulges. The model assumes that in the foreland basin sediments prograde across the basin from the thrust/fold tip towards the bulge while in the rift-type basin sediments build up to a constant bathymetry profile. □

BOX 15.3

Compaction of sediments

The easiest process to take into account in modelling is a mechanical compaction. Let us consider a cylinder of rock such that its porosity (ϕ) is given by:

$$\phi = \frac{h_w}{h_t}, \tag{C.1}$$

where h_w is the height of the voids and h_t is the total height. Now

$$h_t = h_w + h_g, \tag{C.2}$$

where h_g is the height of the grains. So,

$$h_g = h_t\,(1 - \phi). \tag{C.3}$$

We can then write for a layer at shallow depth (say one that has not been very compacted) that

$$h_g = h_{ts}\,(1 - \phi_s),$$

where h_{ts} and ϕ_s are the thickness and porosity respectively at shallow depth. At great depth we have

$$h_g = h_{td}\,(1 - \phi_d),$$

where h_{td} and ϕ_d are the thickness and porosity respectively at great depth. If we assume that the grains are incompressible then h_g will be constant with depth and,

$$h_{ts} = \frac{h_{td}(1 - \phi_d)}{(1 - \phi_s)}. \tag{C.4}$$

Equation (C.4) can be used to evaluate the thickness any layer would originally have prior to it being buried from its present thickness, provided some information is available on the relationship between porosity and depth. For many normally pressured sediments, the variation of porosity with depth is thought to follow an exponential path.

In basins, there is usually a variation in lithology both horizontally and vertically that reflects changes in the environment of deposition. From the basin edge to its centre the sediment type may change from well-cemented quartz-rich sandstones, grade to siltstones and eventually shales. Since each rock type has a different porosity–depth curve, then the effects of compaction will vary greatly across a basin.

Consider, for example, a 100-m thick stratigraphic unit which changes in lithology across a basin. Towards the basin depocentre, the unit is a sandstone which has a porosity of 20 per cent at a depth of 3000 m. The same rock type at the surface would have a porosity of 56 per cent. Substituting these values in equation (C.4) gives an original thickness of 182 m, nearly twice the preserved thickness. Towards the basin margin, the unit is a well-cemented quartzite with a porosity of 10 per cent. The same rock has a porosity of 20 per cent at the surface. In this case, the original thickness is 120 m which is similar to the preserved thickness.

Simple considerations such as these show that compaction effects may significantly 'distort' the geometry of a basin with time. Since the porosity–depth profile can be expected to vary laterally, in sympathy with changes in sediment type, the distortion will not only be vertical, but also lateral in extent.

Sea-level changes

Sea-level changes modify the stratigraphic record in two different ways: by loading or unloading the basement and by causing a shift in the location of different types of sediment and limits of deposition. If the magnitudes of sea-level changes are known, then equation (B.4) can be used to incorporate their loading effects. Also, since the past few thousands of years were times of known sea-level changes due to waxing and waning of continental ice sheets, the sedimentary response and shoreline shifts that result from a sea-level change can be tested empirically using the geological record.

The main controversy at present is focussed not so much on the stratigraphic response to sea-level changes, but on questions concerning the causes, magnitude and period of these changes during the geological past. There are only a few generally accepted mechanisms for sea-level changes and these produce substantially different types of sea-level curve. One of the mechanisms that has already been mentioned is a change in ocean water volume due to the waxing and waning of continental glaciations (or **glacio-eustasy**). This mechanism tends to produce oscillatory sea-level changes that are of relatively short period with magnitudes of up to 100 m or more.

Another is the change in volume of the ocean basins due to variations in length and/or spreading rate along the world's ocean ridge system (**tectono-eustasy**). This mechanism produces 'smooth' sea-level changes that are of relatively long period. Although the individual contributions to sea-level changes of glaciation, changes in the volume of oceanic ridges and other plate tectonic mechanisms, such as mid-plate swells, volcanism or mountain building, can be estimated, it is not known to what extent these mechanisms contribute to intermediate-period sea-level changes in the past.

Most researchers are now of the view that long-term changes in sea-level (i.e. changes over periods of a few million years to a few hundred million years) are caused by changes in the volume of ocean ridge flanks caused by spreading rate changes. At times of fast spreading (e.g. during the Late Cretaceous) there is a relative increase in the amount of water displaced onto the continental edges, whereas at slow spreading ridges there is a decrease. For example, a ridge system 10^4 km in length that spread at 2 cm/year for 70 Ma and was then subject to a three-fold spreading rate increase of 6 cm/year would cause an increase in sea-level of about 127 m.

W. Pitman carried out a study of the cumulative contribution to sea level of spreading rate changes along the entire length of the world's ocean ridges as well as some ridges (e.g. Pacific/Kula) which have long since been consumed at the deep-sea trenches. The sea-level curve obtained by Pitman shows a maximum value in Late Cretaceous time of 350 m above present levels and then a steady decline to the present day.

The Pitman curve may be tested using stratigraphic data since times of sea-level rise should have left a record of transgressive and highstand deposits in the continental interiors. As sea level fell, these deposits would have been stranded as elevated outcrops providing, of course, that erosion did not completely remove them first. One group of rocks that apparently escaped significant erosion is the Late Cretaceous Coleraine formation of Minnesota in the central USA. This formation is at a present-day elevation of 390 m which, when corrected for local glacial loading/unloading effects, is close to the predicted value using the Pitman curve.

One problem with using continental flooding estimates as an indicator of past sea-level changes is that the continents themselves may have undergone large-scale tectonic movements subsequent to a sea-level change. There is therefore considerable debate at the present time about the suitability of certain continents to act as **stable reference frames**, although some workers have argued that the central USA and, perhaps, the Russian platform, have probably both been relatively stable since the Late Cretaceous.

Because their tectonic history can now be reasonably well described by thermal models, rift-type basins have become the main focus to test the effects of sea-level changes on the stratigraphic record. In equation (B.4), for example, if y is known and well data can be used to estimate W_d and S^* then Δ_{SL} could be isolated by backstripping. Several attempts have been made in the past to estimate Δ_{SL} from bio-stratigraphic data but, while the resulting curves agree with the timing of the Late Cretaceous maximum in Pitman's curve, they are about a factor of two smaller in magnitude. The reason for this difference is not clear, but it could be due to an inadequate global data set for the backstripped well data, errors in the calculation of ridge crest volume changes, or the competing effects of other tectonic factors.

Another test of sea-level changes is to examine the stratigraphic record itself. For example, if sea level fell by either 350 m or 150 m there should be an observable difference in the sediment accumulation, especially in the later stages of rift-type basin development when the tectonic subsidence is small. At the East Coast US margin, for example, the change in the thermal subsidence since 85 Ma is about 320 m. If on the shelf sea level fell by 350 m, then little sediment should have accumulated. If, on the other hand, sea level fell by 150 m, then as much as 375 m could accumulate. The post-Cretaceous sediment thickness at the East Coast US margin is 500 m or greater, suggesting the use of a low rather than a high sea-level curve.

To date, the most exhaustive study using the stratigraphic record to deduce sea level was the one carried out by P. R. Vail and colleagues at Exxon in the late 1970s. By correlating unconformity bounded stratigraphic sequences within and between basins, the Exxon group were able to document that certain unconformities are widespread in extent and therefore are most likely caused by sea-level changes. The resulting **'Vail sea-level curve'**, as it has come to be known, was an 'oscillatory' curve, with periods that vary from several hundreds of thousand years to a few tens of million years and amplitudes of up to 300 m, that was superimposed on the 'smooth' Pitman curve.

The method used by Vail and colleagues to estimate the magnitude of sea-level change was based primarily on the geometry of stratal relationships observed on seismic reflection profiles. Unfortunately, their procedure may have over-corrected for tectonic effects. The resulting sea-level curve was characterised by gentle sea-level rises, but very rapid falls. Later studies by the Vail group did apparently take into account tectonics, since the sea-level falls were less abrupt and, in some cases, as smooth as the sea-level rises.

Of particular interest in basin modelling is the stratigraphic response to sea-level changes – whether they be smooth of the type proposed by Pitman or oscillatory as suggested by Vail. It might be expected that a sea-level rise would cause a landward shift of sediment belts and onlap, while a fall would cause a seaward shift and offlap. The stratigraphic response to sea-level changes, however, is complex and depends on the relative rates of tectonic subsidence, sediment supply and water depth changes.

There has been considerable progress in the past few years in understanding the stratigraphic response to simple periodic changes in sea level. However, these studies are limited by the fact that we still do not know precisely enough the actual magnitude and period of sea levels in the past. The smooth Pitman curve is presently in dispute because of uncertainties in both the lengths and spreading rates of the ocean ridge systems and the geological time scales. According to one recent analysis, the amplitude of the Late Cretaceous sea-level highstand could range from –5 to 450 m. Similarly, the Vail curve is disputed because a sea-level rise and fall may give rise to quite similar depositional environments and onlap/offlap patterns, depending on the relative rates of tectonic subsidence and sediment supply. Thus, it may be difficult to use the stratigraphic record to infer sea-level changes, unless precise knowledge is obtained of the geometry (onlap and offlap) and depositional environment of individual sequences. One way to obtain this information for an offshore basin is by a combination of seismic reflection profiling and deep drilling data. While such a data set was apparently available to Vail and his colleagues for many of the world's sedimentary basins, most workers do not have access to it and it has proved difficult to verify the details of their sea-level curve.

As a result of these difficulties, some workers have taken a different approach, the most promising of which have been the derivation of sea-level curves from studies of the ^{18}O record in deep-sea sediments and basin modelling that incorporates tectonics and simple periodic waveforms for sea-level changes.

Conclusions

The various sedimentary factors discussed in the previous sections complicate our ability to construct more complex basin models than ones constructed on the basis of simple thermal and mechanical concepts alone. The problem is that while we understand how erosion, changes in the sediment supply, compaction and sea-level changes may influence the stratigraphy individually, it is not known how they combine with tectonics, or with each other, to control the architecture of basins. A single dynamical model for the formation of sedimentary basins is therefore still beyond reach at the present time.

A basin model which incorporates tectonics together with the sedimentary processes discussed in this chapter would have great predictive power. Of particular interest would be the ability to construct synthetic stratigraphic profiles for basins of different ages in different tectonic settings. By incorporating the effects of erosion, changes in sediment supply, and compaction it should be possible, for example, to quantitatively evaluate the effects of sea-level changes in the past. It has been suggested, on the basis of existing models, that sea-level changes may only be an important control on the stratigraphy during the *later* stages of basin formation when tectonic effects are subdued. During the *early* stages, tectonics may dominate. If subsequent modelling studies are able to verify a tectonic control, then this would have profound implications for correlative stratigraphy and our understanding of the nature of the control of the stratigraphic record.

One aspect of basin modelling that has not been discussed here is its application to oil and gas exploration. The thermal and mechanical parameters that best fit observed stratigraphic data can be used to compute the temperature history of the sediments and the maturation of any source beds that are thought to be present (see Chapter 18). Although most industry geologists consider actual measurements of maturity to be far more reliable, a predicted maturity may be useful in a frontier basin where there has been little or no drilling. Basin models may also be used to compute how the stratal relationships within a basin may vary with time and position. This is important for models which consider how fluids such as oil may flow from low to high points in a basin.

In addition to its importance to the oil and gas industry, basin modelling represents an unparalleled challenge to the academic community. The reason why individual basins have the shape they do is still enigmatic and remains one of the major outstanding questions to be addressed by the next generation of students in the Earth sciences.

Further reading

Allen, P. A. & Allen, J. A. (1990) *Basin Analysis: Principles and Applications*, Blackwell Scientific Publications, 451 pp.
A book that summarises the modern approach to basin

analysis. The book provides further reading on many of the concepts developed here, particularly the sedimentary factors of erosion, sediment supply, compaction, and sea-level changes.

Bally, A. W. (1981) *Geology of Passive Continental Margins: History, Structure and Sedimentologic Record* (with special emphasis on the Atlantic margin), American Association of Petroleum Geologists, Education Course Note Series No. 19, 350 pp.
An introductory text devoted to the structure, subsidence history and sedimentological evolution of passive continental margins. Emphasis is on the East Coast US Atlantic margin, but discussions of the Gulf Coast and Tethys margins are included.

Bally, A. W. (ed.) (1984) *Seismic Expressions of Structural Styles*, American Association of Petroleum Geologists, Studies in Geology Series No. 15.
An atlas showing the structure of sedimentary basins as revealed by multichannel seismic reflection profiling.

Biddle, K. T. & Christie-Blick, N. (eds.) (1985) *Strike-Slip Deformation, Basin Formation, and Sedimentation*, Society of Economic Paleontologists and Mineralogists, Tulsa, Oklahoma, 396 pp.
A collection of papers describing the origin of the small basins that form in strike-slip settings.

Kent, P., Bott, M. H. P., McKenzie, D. P. & Williams, C. A. (eds.) (1982) *The Evolution of Sedimentary Basins*, Philosophical Transactions of the Royal Society of London, Series 305A, 338 pp.
An authoritative account that summarises the state of knowledge in the early 1980s of the principal basin-forming mechanisms. Emphasis is on the rift-type and foreland basins.

St John, B. (1980) *Sedimentary Basins of the World*, American Association of Petroleum Geologists, Map and chart series at 1:40 000 000.
A useful map showing the distribution of the world's sedimentary basins.

INTRODUCTION TO CHAPTERS 16–19

These chapters explore the nature of the sedimentary record, the way in which it is interpreted in terms of past depositional environments, and, in one case, outline one aspect of the economic significance of sediments.

At the end of the 18th century, 'catastrophic' theories were used to account for the origins of rocks and fossils. These theories were a compromise between the scientific implications of observations of geological phenomena and the creation of our planet as told in the Bible. During the early part of the 19th century, a different view gradually prevailed, first described in meticulous detail by James Hutton, and later championed and developed by Charles Lyell. By observing the rocks and landscapes of his native Scotland, Hutton came to realise that the Earth 'is destroyed in one part and renewed in another'. He showed that debris from eroded mountain areas was transported to the sea, where it formed new sediments and, with time, new rocks. He stressed that these processes operated gradually, but that with sufficient *time*, they would produce the landscapes and rocks he saw around him. Thus the essence of Hutton's approach, and that later elaborated by Lyell in his book *The Principles of Geology* (1830), was that processes we can see today operating on the Earth, will, given sufficient time, account for all observable geological phenomena. This concept of *uniformitarianism* – the present being the key to the past – still guides Earth scientists today in interpreting the past as recorded in the rock record.

The uniformitarian approach is used in all four chapters, but the reader will find that there are dangers in employing it too literally. In Chapter 16 Roger Walker gives an example of how models for past depositional environments can be based on observations of analogous modern examples. He then goes on to show that the uniformitarian approach can, if applied too simplistically, lead to false conclusions, especially when no comparable modern analogues are available.

In Chapter 17, Maurice Tucker shows that changing chemistry and mineralogy of limestones through geological time indicates that atmospheric composition has changed, and that such changes may be linked to changing global sea-levels and climate. These conclusions do not contradict uniformitarianism, but they do, once again, show that caution must be exercised in using the approach.

Chapter 18 examines the depositional and post-depositional history of organic-rich sediments that comprise the source rocks for oil and gas. Rolf Littke and his colleagues again apply uniformitarian principles, which, as far as processes affecting these sediments during burial are concerned, must rely on the application of accepted physical and chemical laws to produce models that cannot be developed from direct observations at the present day.

When beginning to read Chapter 19, readers might gain the impression that its author, Adolf Seilacher, is advocating a return to catastrophism. He even uses the word to describe events which are sudden when judged in the perspective of our life spans. He does not deny uniformitarianism, but shows that most sedimentary processes are not as gradual as Lyell and Hutton envisaged, but are episodic in nature. For this reason, a dynamic approach to stratigraphy is advocated, enhancing the traditional approach of measurement and correlation of sequences by a consideration of the processes that lead to the formation of three-dimensional bodies of rock. Seilacher reviews this approach on a small scale; a global approach follows in Chapter 20.

CHAPTER

16

Roger G. Walker
McMaster University

CLASTIC SEDIMENTS

Introduction

Clastic sediments are derived from pre-existing rocks via the processes of weathering, transport and deposition. The topic is so broad that only a few aspects can be discussed here. I will first show how sediment is transported to the environments in which it is finally deposited and preserved, and how stratification is formed. I will then concentrate on the interplay of sedimentary processes, tectonic movements and relative sea-level fluctuations within sedimentary basins. One of the goals of sedimentology is to characterise or 'idealise' depositional environments by the construction of 'facies models'; such models serve as the basis for making predictions of economic importance in the mining and petroleum industries. I will use the barrier island environment as an example of how such modelling can be done, and the pitfalls that may be encountered. The petrographic characteristics of the rocks will not be discussed here. However, they do give important insights into the provenance of the sediment, and its diagenetic history during burial and lithification.

Transport of sediment

Sediment is **transported** away from its source, the weathering site, by many processes. Initially, these may include mass movements ranging from catastrophic rockfalls to gentle soil creep. However, for distances of transport in the range of tens to hundreds of kilometres, three processes dominate – dispersal by wind, and transport in ice or water. Although wind-blown and glacial sediments occur in the geological record, they are far less abundant than those transported by water, and water-lain deposits will be discussed exclusively in this chapter.

In **unidirectional flows**, without wave oscillations, sediment can move in continuous or intermittent contact with the bed, in suspension above the bed, or in solution. There is a **threshold flow velocity** necessary for the initiation of grain movement, which for **cohesionless grains** (coarser than very fine sand) is a function of size and sorting, immersed specific gravity of the grains, fluid viscosity, and flow velocity. For very-fine to very-coarse **sand sizes** (0.0625 to 2 mm), mean flow velocities required to initiate movement are roughly 10 to 40 cm per second, respectively; for 1 cm pebbles the velocity is about 100 cm per second. When the grains first begin to move, they roll or slide in permanent contact with the bed, as **bedload**. Movement as bedload is favoured by larger sizes and lower velocities. As the mean velocity increases, there is a tendency for grains to be lifted off the bed and transported briefly in suspension before falling back to the bed. This material is termed **bed-contact load**. For a given grain size, as the mean flow velocity (and hence turbulence) increases, so the time spent in suspension increases, until the clasts move in permanent suspension.

The **suspended load** is maintained in suspension by turbulent eddies in the flow, which are generated by **roughness elements** ('irregularities') on the bed. Roughness elements include (1) the clasts themselves projecting up into the flow, especially if there are some much larger clasts (pebbles, cobbles) projecting up above a generally sandy bed, and (2) bedforms, particularly ripples and dunes projecting up from the bed. As the mean velocity of the flow increases, so the turbulence increases, and coarser grains can be maintained in suspension. In an open channel such as a river, **clay sizes** (< 0.0039 mm) are normally in suspension all of the time. **Silt** (0.0039 to 0.0625 mm) moves in suspension when mean flow velocities are greater than a few centimetres per second, whereas coarse sand is only taken in permanent suspension when mean velocities are in the 2.5 to 3 m per second range. A range, rather than a single velocity, is given because the turbulence developed is both a function of flow velocity *and* the size of the roughness elements on the bed.

Bedforms

When cohensionless materials move as bedload or bed-contact load, the bed itself develops a characteristic morphology, or **bedform**. The nature of the bedform is a function of flow velocity, flow depth and grain size. A simplified version of a diagram relating bedforms to flow velocities and grain sizes for flow depths of about 50 cm is shown in Figure 16.1. The stability fields of the bedforms – ripples, dunes, and plane bed – have been determined from a large number of experiments, but the data points are omitted from Figure 16.1 for clarity. **Ripples** (Figure 16.2a) have a height of a few centimetres, and wavelengths of centimetres to a few tens of centimetres. **Dunes** are larger, with heights of tens of centimetres to a few metres, and wavelengths of many tens of centimetres to several metres. The distinction between straight crested (two-dimensional) and sinuous crested (three-dimensional) dunes has been omitted from Figure

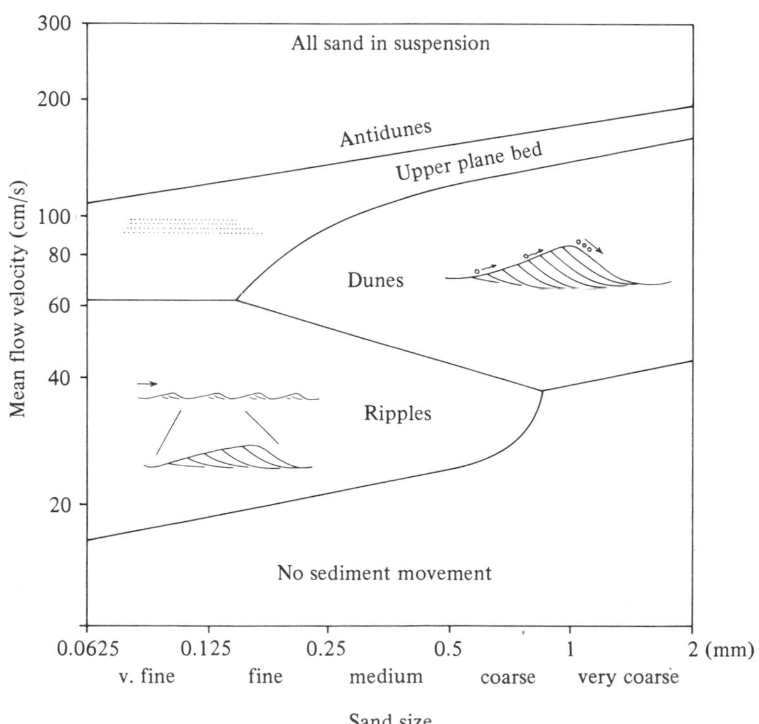

Figure 16.1 Stability fields of ripple, dune, upper plane bed and antidune bedforms, with respect to sand size and mean flow velocity for flows about 50 cm deep. Below 20–30 cm/s, no sediment moves on the bed; when velocities reach about 300 cm/s, all the sand moves in suspension and no bedforms aggrade. For dunes, the sand is shown rolling up the back of the bedform, and avalanching down the front – the avalanching produces preservable layers of cross-stratification that dip in the downstream direction. The same process operates for ripples, which are smaller than dunes. Aggradation of the upper plane bed produces horizontal stratification. Antidunes are very rarely preserved. □

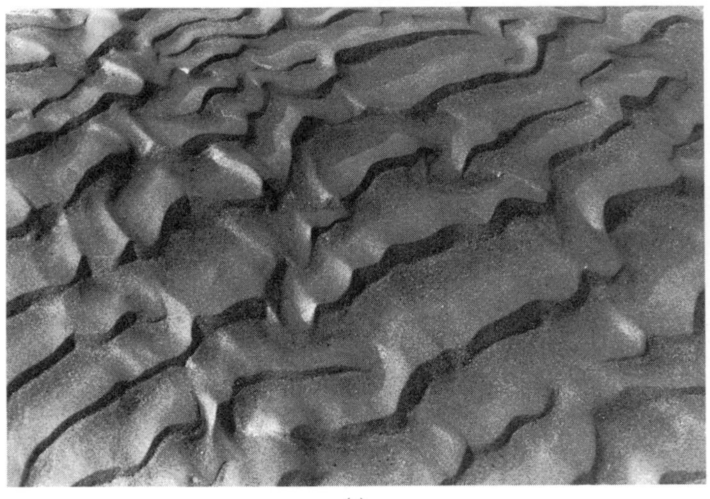

(a)

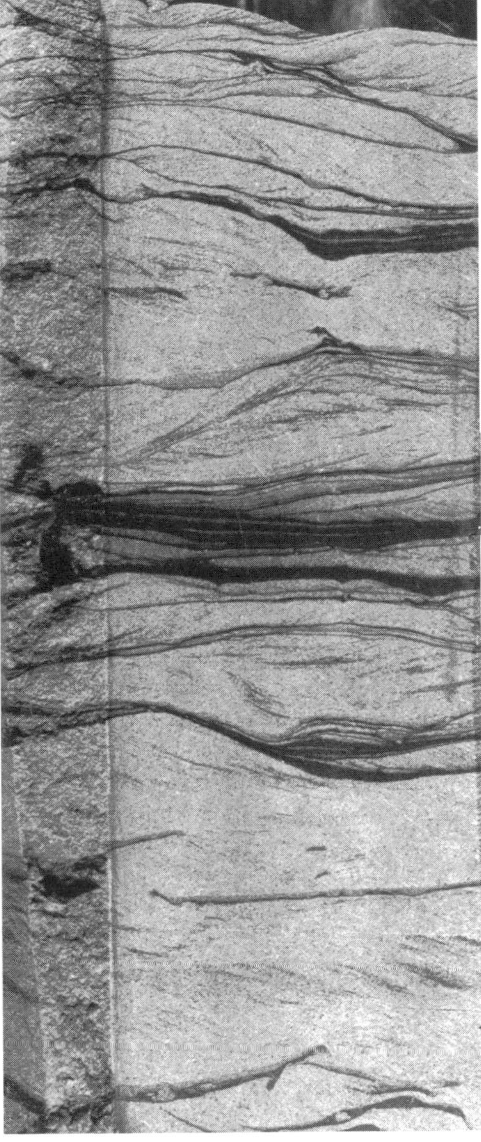

(b)

Figure 16.2 Current ripples and cross-lamination. (a) Current ripples in the South Saskatchewan river, Canada; the current flowed from top left to bottom right. The ripples are 2–3 cm high, and their wavelength is approximately 20 cm. Many of the ripples show straight crests, but the pattern is more sinuous in places. (b) Cross-lamination formed by current ripples. Note that cross-laminations dip both to right and left, and that in places the ripple profile is preserved, and is draped by mud. The bi-directional flow, and alternations between flow (sandy ripples) and stillstand (mud) suggest the influence of tidal currents. The core is about 7 cm wide. □

16.1. Both ripples and dunes have an asymmetrical profile; grains roll along the gentle upstream side, pile up at the crest, and intermittently avalanche down the lee face (Figure 16.1). The preserved layers are the lee-side layers, which dip at the angle of repose of wet sand – about 25°. The **sedimentary structure** produced by these accreting layers is termed **cross-stratification**. An example of ripple cross-stratification is shown in Figure 16.2b; note that the cross-laminations dip both to the right and left, implying reversing (possibly tidal) currents.

In the plane bed field, the topography made by ripples and dunes is no longer present, and the grains shear along an essentially flat, featureless bed, producing the sedimentary structure known variously as **planar, flat**, or **horizontal lamination** (Figure 16.1).

At the highest velocities, waves form on the water surface – these waves can migrate upstream, hence the term **antidune**. The waves can break (like waves on the beach) under certain conditions, and the turbulence developed is so intense that the bedform is temporarily destroyed and all of the sand thrown into suspension. Preserved antidune bedforms are very rare in the geological record because of their constant formation

(a)

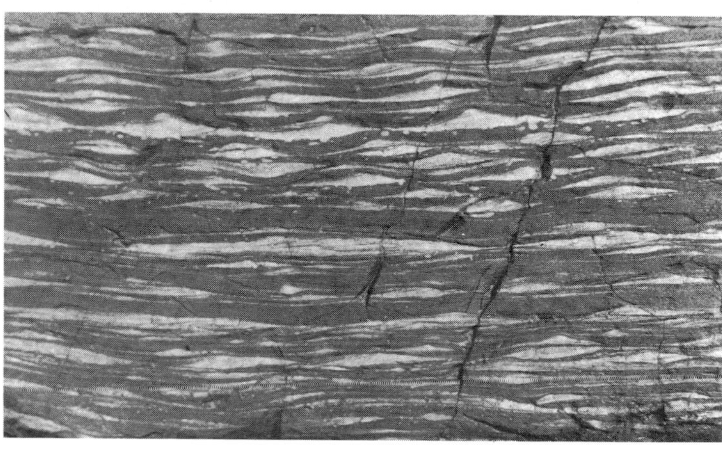

(b)

Figure 16.3 Structures produced by oscillatory currents. (a) Plan view of straight-crested symmetrical wave ripples; wavelength, crest to crest, is 7 cm. Upper Cretaceous Cardium Formation, Alberta, Canada. (b) Vertical cross-section through interbedded, isolated, symmetrical ripples and mudstone. Ripples are isolated because the area was starved of sand, with insufficient supply to form continuous beds. Note that the cross-stratification dips consistently to the left, indicating slight unidirectional flow superimposed on wave motions (to give combined flow). Each ripple is about 1 cm thick. Upper Carboniferous of North Devon, England. (c) (opposite) Interbedded, sharp-based sandstones and bioturbated mudstones. Note prominent gently curved stratification in centre bed, defining a hummock (foreground) and swale (in distance). Scale bar 15 cm long; Eocene Coaledo Formation, southwestern Oregon, USA. □

and destruction, and because, even if formed, they will be modified and reworked into plane bed, dunes or ripples as the rapid flow wanes.

If there is no net addition of sediment to the system, the bedforms will simply migrate across the bed with no net aggradation, and hence little chance of preservation in the geological record. To preserve beds of ripple or dune cross-stratification, or planar lamination, sediment must be supplied faster than it is transported through the system, causing the bed to aggrade.

Although Figure 16.1 has been simplified, it satisfactorily accounts for most of the sedimentary structures formed by unidirectional flow and preserved in sandstones. The major exception consists of massive or structureless sandstones in which there are *no*

(c)

sedimentary structures formed or preserved. The commonest way in which such beds are formed is by *very* rapid deposition directly from suspension onto the bed, such that equilibrium bedforms have no time to become established. Structureless sand is commonly found at the base of **turbidite beds**, where deposition is known to be directly from suspension.

Because of insufficient experimental work, bedform fields cannot be extrapolated to silt sizes, but ripple cross-lamination is probably the commonest sedimentary structure in silt. Experimentation is even more difficult in coarser grain sizes, but observations of natural systems suggest that cross-stratification and planar bedding are common in **granule** (2–4 mm) and **pebble** (4–64 mm) sized material.

The presence of one or more of the structures in Figure 16.1 can be used in the interpretation of an ancient sandstone. However, it must be emphasised that interpretations of depositional environments cannot be made directly from sedimentary structures. The structures provide a clue to flow depths and flow velocities for a given grain size, but specific combinations of size, depth and velocity may be present in many different environments. We shall see below that other data must be used in order to make environmental rather than hydrodynamic interpretations.

Effect of wave oscillations on bedforms

In rivers, unidirectional flows dominate, and waves can normally be disregarded. However, in marginal marine and shelf environments, the passage of wind-generated surface waves may produce an oscillatory water movement on the bed. Pure wave motions are uncommon – the oscillations are normally superimposed on weak unidirectional flows, producing **combined flows**. The surface waves are damped out downward through the water column, and are generally unable to move sand at depths greater than about a half of the wavelength of the surface wave. Under fairweather conditions on modern shelves, waves normally have wavelengths of a few centimetres to about 40 m, and hence affect the bed to depths of about 20 m. Thus the 20-m mark can be considered as a maximum value for **fairweather wave base** (the depth below which normal wave activity does not reach) – in most shallow seas, fairweather wave base is in the 5–15 m range. Under storm conditions, wavelengths may be much greater; the edges of the present-day continental shelves, with an average depth of about 130 m, are stirred by storm waves every few years.

Bedforms appear to exist at two main scales – small-scale symmetrical wave ripples, and large-scale hummocky cross-stratification (Figure 16.3). The symmetrical **wave ripples** (Figure 16.3a) have heights of a few centimetres and wavelengths up to a few tens of centimetres. Crestlines tend to be long and straight, and can be rounded or sharply peaked (Figure 16.3b). Although the back-and-forth water motion can give cross-lamination dipping in two directions, the effect of combined flow is normally to give a preferred transport direction superimposed upon the oscillatory motion, hence preserving cross-lamination dipping in only one direction (to the left in Figure 16.3b). Thus in vertical cross-sections without a plan view, wave ripples can sometimes be very difficult to distinguish from current ripples.

Hummocky cross-stratification (HCS; Figures 16.3c and 16.4) was first defined in 1975, and is now well known in the geological record. Because of the difficulty and danger of diving during storms, the formation of HCS has not been observed in modern shallow seas, nor has it been reproduced in full-scale experimental studies. Wavelengths range from about 1 to 5 m, and heights are a few tens of centimetres. The dips of laminae, from the tops of hummocks to the bottoms of intervening **swales** (troughs), are commonly less than 5° and rarely greater than 15°. In plan view, the hummocks tend to be circular to oval. The fact that HCS is normally unmodified by other currents suggests that it was formed by storm waves below fairweather wave base. As the hummocky and swaley bedforms shift slightly on the sea floor during storms, they produce a characteristic convex and concave upward stratification, commonly with low-angle curved truncations between sets of laminations (Figures 16.3c and 16.4).

Control of flow in different environments

Before examining depositional environments, it is necessary to look briefly at the processes that control water movements in various environments. The driving force for rivers on land is simply the hydrostatic head between the source of the river and 'base level', which is normally mean sea-level. Flow velocities in any reach of the river are given by the **Chezy equation**:

$$\bar{U} = C\sqrt{RS} \tag{1}$$

where $\bar{U}$ is the mean flow velocity in m/s, C is the dimensionless Chezy coefficient (which expresses the resistance to flow due mainly to varying bed roughness), R is the hydraulic radius (channel cross-sectional area divided by length of wetted perimeter; R has the dimension of length, and expresses the channel 'depth' in metres), and S is the slope, not of the bed but of the water surface (the energy gradeline).

In shallow seas, the driving forces of the flows are much more complicated. They are ultimately related to density and temperature variations in the oceans, to tides, and to storms. In the oceans, colder and/or more saline (thermohaline) flows originate in polar seas and are driven into deeper water by gravity, acting on the density difference between the polar water and the warmer, lower latitude ocean waters. The currents flow at a few tens of centimetres per second, and veer gradually to the right (in the northern hemisphere) due to Coriolis force. These **current flows** rarely spill up from the deep oceans onto continental shelves, but in the few places where they do, no new sediment is introduced onto the shelf.

Tidal flows are related to the tidal bulge made by the Moon, and to a lesser extent by the Sun. Bulges are formed both on the side of the Earth facing the Moon, and on the opposite side. In a simplified way, the tide can therefore be envisaged as a wave that circles around the Earth, with a wavelength of about 20 000 km (half the circumference of the Earth). In the deep oceans, the velocities close to the bed generated by the passage of this wave are too low to move sand. On the continental shelves, or in ancient epicontinental seas, sediment-moving flows are developed when a tidal wave propagates in from an adjacent deep ocean. On the shelf, the amplitude of the tidal wave and the associated flow velocities are magnified so that in enclosed areas, such as the southern North Sea, near-bottom velocities can reach about 100 cm/s. Dunes (Figure 16.1) are commonly observed in side-scan sonar pictures of shallow sea floors. On open, unobstructed shelves with smooth sea floors, the tidal current directions change continuously throughout the tidal cycle. In more enclosed areas, such as the southern North Sea or Georges Bank in the Gulf of Maine, there is commonly a series of linear sand ridges up to almost 40 m high on the sea floor. Tidal currents may be restricted such that the ebb flow hugs one side of the ridges and the flood flow hugs the other side. The sedimentary response through time may be more that of a quasi-unidirectional flow rather than that of a reversing tidal flow.

Storm flows affecting the sea floor are a response to surface winds piling water (commonly, but not exclusively) onshore, and giving rise to a storm surge. The elevated ocean surface results in pressure differences on the sea floor, setting up currents that flow down the pressure gradient. Thus if the water is elevated at the

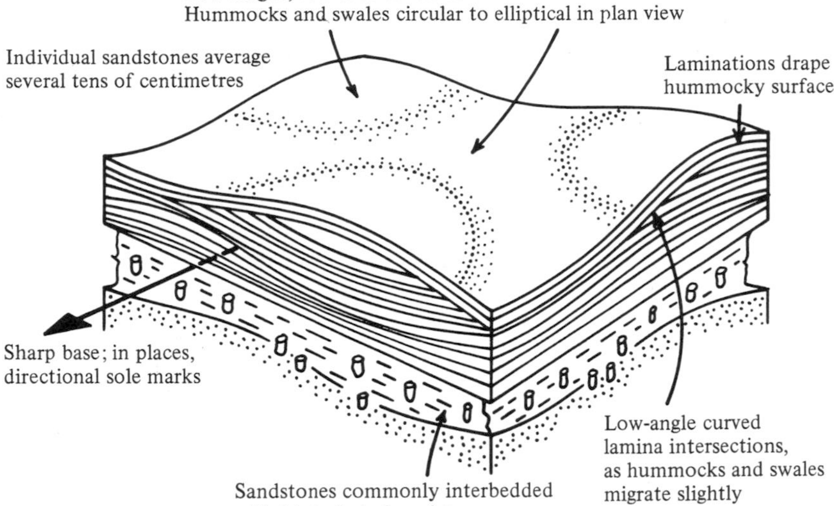

Figure 16.4 Block diagram illustrating the internal and external morphology of hummocky cross-stratification. Note long wavelengths and low heights, and the internal convex and concave stratification with low-angle curved lamina intersections. HCS sand beds commonly rest sharply on bioturbated mudstones, and may rarely have directional sole marks on their bases. □

shoreline, flows will be down the pressure gradient, or offshore. These flows will also be subject to Coriolis force, and hence will flow initially offshore, but will gradually veer to the right in the northern hemisphere to flow obliquely to shore. These are termed **geostrophic** ('influenced by the rotation of the Earth') flows. On the Atlantic shelf of North America, annual storms give near-bottom velocities of several tens of centimetres per second. In the Gulf of Mexico, tropical storms and hurricanes produce near-bottom velocities of over 100 cm/s. In both cases, these unusually high velocities can persist for times ranging from several hours up to a couple of days.

Storm flows in shallow seas can amplify the diurnal (once a day) or semi-diurnal (twice a day) tidal currents, and are powerful agents of sediment transport. It is important to note that even at long distances from the shoreline, storm winds steadily blowing in one direction for many hours can produce local relative elevations of the ocean surface. These in turn produce pressure differences on the ocean floor regardless of the water depth – thus storms can effect sediment transport in depths where day-by-day processes are too weak to allow anything but the quiet deposition of mud.

Sedimentary structures, facies and environmental interpretation

Individual sedimentary structures commonly allow interpretation of the flows which formed them – unidirectional, oscillatory, or combined. In the case of unidirectional flows, it may be possible to estimate a range of flow velocities for a given grain size if reasonable assumptions are made concerning flow depth (Figure 16.1). However, individual structures do *not* allow interpretations of the original depositional environments – other techniques and different data are necessary.

One of the more important environmental indicators is the fossil content of the rocks, including both the preserved **body fossils**, and the **traces** left in the sediment as organisms move around, find shelter and feed. Indeed, the organisms may be so abundant and active that all primary stratification is destroyed, or **'bioturbated'**. Distinctions between non-marine and fully marine environments are normally easy to make. Finer subdivisions may be possible by studying the ecology of the organisms, and whether they are found in life position, or are found to have been transported, broken and sorted by currents.

Depositional environments are constantly shifting through geological time. In order to understand thick stratigraphic sequences built up from a series of shifting environments, the sequences must be subdivided into their basic homogeneous constituent parts. This is a subjective process, and is based upon the characteristics of the rocks, the scale and purpose of any particular study, and the experience of the individual geologist. The 'homogeneous parts' are termed **facies**, from the Latin word for the aspect of the rocks. The overall aspect is a combination of all of the descriptive features of the rock; most importantly, it includes the lithology, biological features (body and trace fossils), and sedimentary structures. It is implicit that (1) each facies is different from the one stratigraphically above and below, (2) facies can and do change laterally within stratigraphic units, and (3) individual facies can be repeated in vertical or lateral succession. Facies should be defined as objectively as possible, without any implied depositional environment, although it is understood that an attempt will be made to use all of the descriptive features in building an **interpretation**.

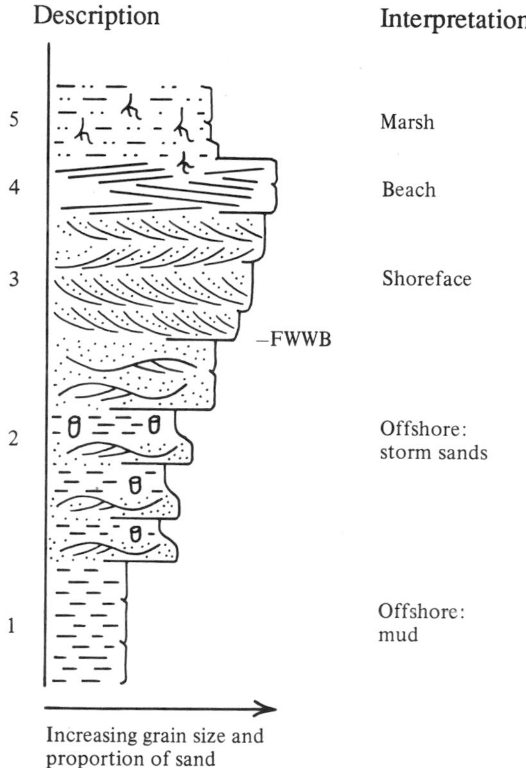

Figure 16.5 Hypothetical vertical facies sequence deposited by a prograding shoreface. In measuring this section, one would have identified five facies: (1) mudstones, (2) interbedded bioturbated mudstones (vertical tube indicates bioturbation) and sandstones with hummocky cross-stratification (undulating symbol) – note that the proportion of mud decreases upward through this facies, (3) sandstones with medium-scale cross-stratification, (4) sandstones with planar lamination inclined at a low angle, and (5) mudstones and siltstones with root traces. The facies and facies sequence are purely descriptive; interpretations are shown to the right. FWWB indicates fairweather wave base, at the bottom of the shoreface. For scale, facies 3 and 4 together might be about 10 m thick. Note how the section is drawn to illustrate the increase in grain size and proportion of sand upward. □

A single facies in the geological record can rarely be interpreted solely on the basis of the features it contains. More commonly the interpretation will depend upon two other lines of evidence: first, a *comparison with recent sediments* as exposed in trenches and excavations on land and as seen in cores from marine environments, and second, upon the context of the facies in an overall **facies sequence**.

An example of a sequence is given in Figure 16.5; each facies has a distinct lithology, set of sedimentary structures and biological features, and each indicates in a general way its possible depositional environment. But the *sequence* of facies, and the overall trend toward becoming sandier and coarser grained upward, strengthens the interpretation of each individual facies. Here, the continuous vertical sequence of facies, with gradational facies contacts, reflects a series of sedimentary environments that were originally adjacent to each other on the sea floor, with the sandier ones generally closer to shore. The sequence illustrates the fundamental importance of **'Walther's Law of the Succession of Facies'**, first stated by Johannes Walther in 1894. The law states that 'only those facies ... can be superimposed...which can be observed beside each other at the present time'. These concepts of facies, facies sequences, and the comparison of ancient and recent sediments now form the basis of environmental interpretation. Many of the concepts can be traced directly back to Walther.

Facies models

A **facies model** is essentially a summary of a given environment. It incorporates as many data as possible from many examples of recent and ancient sediments, and attempts to emphasise the features that the examples share in common. The model can be expressed as a series of vertical facies sequences (Figure 16.5) or, better, as a block diagram (Figure 16.6) that expresses the three-dimensional relationships of the facies. It is increasingly apparent that these *static* diagrams should be formulated in a more dynamic way, to illustrate how each part of the system responds to tectonic movements and relative sea-level fluctuations. The nature of the response strongly influences which parts of the system will be preserved and which destroyed during transgression and regression.

Ideally, a facies model goes beyond mere summary of an environment; it should be a working model that functions in four ways. First, it should be a *norm* to which new situations and new data can be compared and contrasted. Second, it should be a *guide* for making observations in new areas with new examples. Third, it should be a *predictor*; this is perhaps the most important aspect of the model. It suggests that given a good general model, and some preliminary observations from a new situation, the model should allow specific predictions to be made about that new situation. Sufficient information from the new situation must be available in order to apply the correct model – this is not always the case in practice. Fourth, the model forms a *basis for interpretation* – it may allow integrated hydrodynamic, environmental or tectonic interpretations to be made that could not have been done using only one or two examples. I will illustrate the construction and use of facies models in shoreline (barrier island) and shallow marine (offshore bar) settings; models are constructed and used in similar ways for other environments.

Shallow seas – introduction

Sedimentary environments can be categorised in many ways. One very simple classification would differentiate non-marine, shoreline, shallow marine (shelf), slope, and deep basin floor environments. In the rest of this chapter, I will concentrate on long, narrow sand bodies in shoreline to shallow marine settings. They exemplify relatively well understood situations (barrier islands), and also exemplify the way in which questions must be posed, answers sought, and models tentatively built for the problematic sand bodies commonly termed 'offshore bars'.

The term shallow seas will be used henceforth for both the modern continental shelves and ancient epicontinental seas. The average depth at the shelf edge today is about 130 m, and most ancient epicontinental seas appear to have been shallower than about 200 m. Shallow seas will therefore be defined as those areas between the shoreline and the 200 m isobath. Epicontinental seas include broad areas of inundation without nearby tectonic activity, such as the present Hudson Bay, Canada, as well as **foreland basins**, which are tectonically subsiding areas on the cratonic side of active mountain chains (such as the western part of the Cretaceous Western Interior Seaway of North America, and the Oligocene–Miocene north Alpine basin of France, Switzerland and Austria).

Early in this century, D. W. Johnson introduced the idea of a 'graded shelf', with gravel and sand closest to the shoreline where waves and currents are strongest, and silt and mud closer to the shelf edge where transport processes were assumed to be weaker. However, mapping and sampling the northern part of the American North Atlantic Shelf showed that even in the deepest parts there was a partial cover of very coarse material. This was emplaced by Pleistocene ice sheets and glacio-fluvial outwash processes during a low stand of sea-level, when the sea had withdrawn completely from the present shelf. During Holocene transgression, beginning about 12 000 years ago, the sea inundated the glacial deposits, leaving them as 'relict sediments'. This concept of coarse, relict sediments was introduced in the 1930s, and was current until about 1970. Beginning at that time, D. J. P. Swift and colleagues emphasised that during the transgression, the glacial material was not passively inundated and left as relict, but was constantly reworked and brought into partial or complete hydrodynamic equilibrium with processes in the transgressing sea.

The influence of relative sea-level fluctuation on shoreline and shallow marine sediments cannot be overemphasised. The term **transgression** implies a movement of the shoreline away from the centre of the basin, and hence inundation of formerly non-marine areas; **regression** implies shoreline movement toward the centre of the basin, possibly resulting in subaerial exposure of formerly marine areas.

The **shoreline** separates the sea from the land – in many areas it is characterised by beaches, particularly off barrier islands (which have lagoons behind their beach/dune systems), strand plains (which lack lagoons, and have the sand belt attached to the land) and some deltas. Seaward of the beach is the **shoreface** (Figure 16.6); it is the steeper part of a profile that gradually flattens into the **offshore** area. The shoreface is also the zone in which sand moves on a 'day-by-day' basis, driven by tidal and wind-generated flows. These flows normally become ineffective at moving sand, day-by-day, at fairweather wave base (Figure 16.6). If the base of the shoreface is defined morphologically by a change of gradient, it is impossible to recognise in the geological record. Alternatively, the base of the

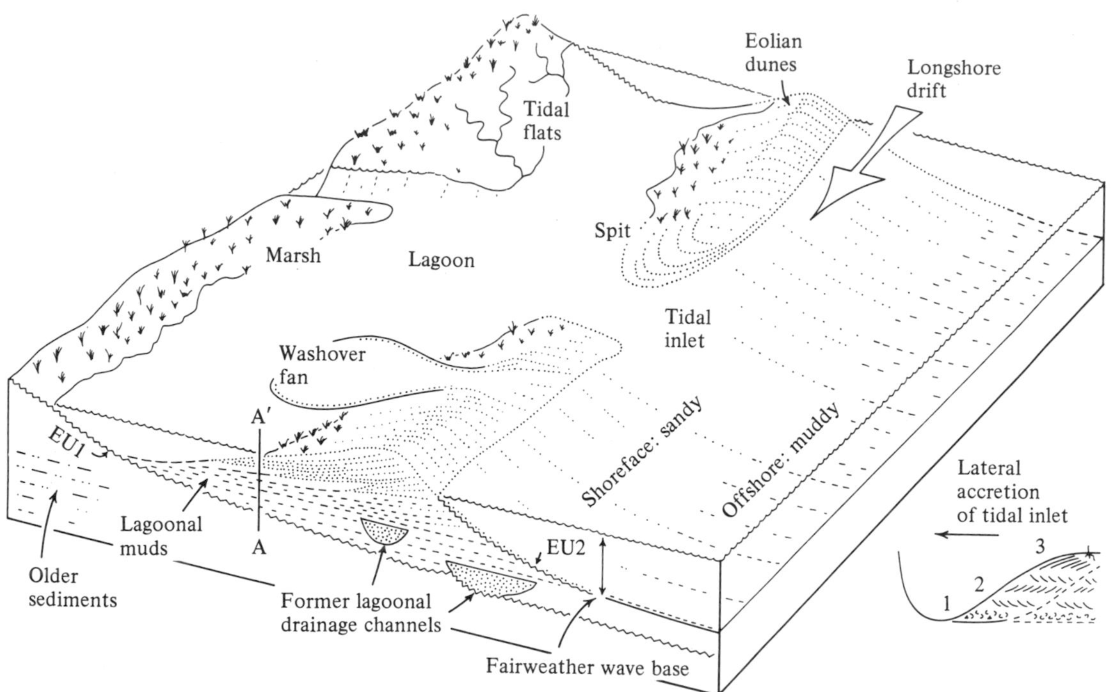

Figure 16.6 Block diagram of typical barrier–lagoon–inlet system. The barrier is being nourished by sand supplied by longshore drift; this causes the downdrift ends of the barriers to lengthen by spit elongation. This in turn causes the inlets to migrate. At the lower right, a cross-section of a laterally migrating inlet is shown – the lateral migration results in a vertical facies scheme from channel floor shell lag (1), to cross-stratified sands moved by tidal currents (2), to beach stratification formed as waves break on the spit (3). The dashed line shows the former position of the inlet. In the main block diagram, note the vegetation in marsh and tidal flats areas. The barrier geomorphology is almost independent of transgression/regression at any given time. However, the system shown is transgressive, with the barriers moving landward (to the left) as sand is washed over into the lagoon by storms. Thus there are two major erosional unconformities (EU) developing. EU1 develops as the lagoon (and particularly the channels that drain the lagoon and tidal flats) cut landward into older sediments. The vertical sequence that would be preserved, A–A', consists of lagoonal muds and former channel deposits, overlain by barrier washover sands. EU2 develops at the shoreface, as it cuts back into the barrier and the underlying former lagoonal deposits. It is unlikely that much of the barrier itself will be preserved in a transgressive setting. Finally, note the definition of the shoreface (sandy, down to fairweather wave base) and offshore (muddy) areas. □

shoreface can be taken stratigraphically at the point where dominantly muddy offshore sediments pass upward into dominantly sandy shoreface sediments – this may also approximate fairweather wave base (Figure 16.5). If sediment is constantly supplied to the shoreface, it will prograde out over the offshore area, giving rise to a facies sequence that becomes coarser and sandier upward (Figure 16.5).

In the offshore area, the most common topographic elements observed today are **linear ridges**, which are parallel or subparallel to the shoreline. They are a few kilometres wide, a few tens of kilometres long, and up to about 35 m high. They appear to have formed mostly at or very close to the shoreline, and now occur in an open marine setting due to transgression. Thus their morphology, sedimentary structures and facies sequences will reflect modification on the shelf by tidal and storm processes. Possible counterparts in the geological record are controversial.

The final point to emphasise is that on a day-by-day basis, sand movement is largely restricted to the shoreface. Longshore, tidal and fairweather wave-generated currents do not appear to transport sand across the shoreface and out into the offshore area in modern shallow seas. Only geostrophic flows generated by storms appear to move sand into the offshore, each storm achieving an 'increment' of sediment transport. This incremental distance is of the order of hundreds of metres to about 5 km per storm, with the sand moving both offshore and subparallel to shore.

Barrier island systems

Barrier islands (Figure 16.6) characterise much of the 4300 km North American coastline from New England, around the tip of Florida and into the Gulf of Mexico. In this distance, there are almost 300 individual islands. They also occur in many other parts of the world, particularly along the coasts of the Netherlands, Germany and southern Denmark. There are three main components of a barrier system – the **barriers** themselves, the **lagoons** behind the barriers, and the **inlets** that connect the lagoons with the open sea. Barrier coastlines tend to be characterised by active longshore drift of sediment, abundant sand supply, **micro-tidal** (0–2 m) or **meso-tidal** (2–4 m) ranges, and a geological history of transgression.

Transgressive barriers

Many, if not most barriers originate at times of low stand of sea-level, when older sediments are winnowed by the sea and the wind piles up sand behind the beach as beach-ridge complexes. During ensuing transgression, the sea may flood the low ground behind the beach ridges, cutting them off from the mainland and forming barriers and lagoons.

In the simplest case, the barriers migrate onshore during transgression. Sand is eroded from the beach and shoreface during storms, and is washed over the top of the barrier into the lagoon, as a **washover fan**. A facies sequence is thereby constructed which consists essentially of lagoonal facies overlain by washover facies (A–A' in Figure 16.6). Although the beach-ridge complex may locally overlie the washovers, there is little chance of long-term preservation of the beach-ridge complex in a transgressive setting. The geometry of the transgressive lagoon–washover facies sequence will be sheet-like rather than long and narrow, extending along strike for the entire length of the barrier chain, and extending out under the transgressing sea for a distance equal to that of barrier migration; this is about 100 km for the present-day Atlantic shelf of middle North America.

During transgression, some sand is lost from the barrier system by being swept out to sea during storms. It follows that, in order for the barriers to move inland continuously, they must be nourished with new sediment, either from river input, or from eroding headlands. If the barriers are not nourished, they will gradually be submerged, and the sand reworked by waves and tidal currents.

Prograding barriers

In exceptional circumstances, commonly during stillstands rather than active transgressions, a barrier that originated transgressively may be supplied with sufficient sand by longshore drift that it actively progrades. The classic example is that of Galveston Island, Texas, USA, which has prograded about 4 km in the last 3500 years. Coring through the barrier shows a facies sequence that gradationally coarsens upward from offshore into lower and upper shoreface deposits, which are capped by beach and beach-ridge sands. It is perhaps unfortunate that Galveston Island was one of the first barriers to be examined by researchers from the United States oil industry in the 1950s. Their much-publicised results gave rise to 'THE barrier island model' – a long narrow strip of sand characterised by a coarsening upward sequence with a base that grades downward into offshores muds. In fact, if a barrier does actively prograde, it is possible that the lagoon will eventually fill in, and the barrier will in effect become re-attached to the mainland. The prograding sand sheet is termed a **strand plain** rather than a barrier. One of the best examples is the Coast of Nayarit (Mexico, facing the Gulf of California), which has prograded almost 13 km as a strand plain with beach ridges.

In rare circumstances, the barrier may prograde at such a rate that regional subsidence prevents the lagoon behind the barrier from filling in. A good example is the Kakwa Member of the Cardium Formation, in the Cretaceous of Alberta, where a barrier system at least 300 km long has prograded more than 100 km seaward. The importance of this example is to emphasise that although, at any given time, barriers are characterised by long narrow sand bodies, their preserved sand body geometry in the geological record may be quite different. The differences depend on dynamic, or constantly changing factors such as relative sea-level and rate of sediment supply. Differences between present-day geomorphology and preserved sand body geometry in the geological record may be found in almost all environments, not just barriers.

Inlets

All barrier island systems are breached by inlets, which allow water supplied by rivers to drain from the lagoon into the sea. The inlets also allow tidal exchange between the lagoon and sea, and hence tend to focus relatively strong currents. There tend to be more inlets along meso-tidal coasts than micro-tidal coasts, reflecting the greater volume of water to be exchanged in each tidal cycle.

Sand supplied by longshore drift tends to form a hooked spit on the tip of the barrier (Figure 16.6). As the spit lengthens, the inlet is forced to erode on the opposite side, to maintain its width. Thus through time, the inlets migrate in the downdrift direction. In doing so, they destroy the facies sequence characteristic of the barrier, and replace it with a sequence characteristic of the inlet (Figure 16.6, inset lower right). This sequence commonly has a sharp, scoured base, overlain by a lag of coarser material winnowed and left behind by the tidal currents in the inlet. This lag consists largely of shell debris, along with the coarsest sedimentary particles in the system. As the inlet accretes laterally, a vertical facies sequence will be developed at any one point. Above the lag, this sequence will be dominated by cross-stratification, formed as tidal currents make dunes from the sand moving through the inlet. Commonly, the cross-stratification will dip in **two** opposed directions, indicating the preservation of ebb and flood oriented dunes. The top of the sequence will be characterised by flat lamination, formed by wave swash and backswash on the beach portion of the hooked spit.

Inlets can be up to about 10 m deep, and can therefore truncate much of the shoreface sequence of the barrier. It has been estimated that inlet sequences may underlie as much as 40 per cent of modern barriers.

Preservation of barrier sequences: dynamic facies models

It should now be clear that there is no such thing as a single barrier island facies sequence. One sequence is characteristic of transgressive systems, a different sequence characterises regressive systems, and yet another sequence is developed in systems dominated by inlet migration.

It follows that although a block diagram can be drawn for a modern barrier–lagoon–inlet system, the diagram is neither a model nor a predictor unless it is put into its dynamic context. The geological deposits predicted by the block diagram will differ fundamentally according to whether the system is transgressive or regressive. If transgressive, it is necessary to understand the rate of barrier nourishment. If regressive, the proportion of inlet rather than barrier sequences will depend upon the tidal range and rate of longshore movement of sediment. In general, facies models for all environments must be considered in the light of rates of sediment supply, rates of subsidence, and rates and directions of relative sea-level fluctuation.

During the Holocene transgression, many undernourished barriers became submerged. They were subsequently modified by storm and tidal processes in a shallow marine setting, and their origin as barriers is not always easy to reconstruct. These sand bodies are now termed offshore ridges or bars.

Linear offshore bars – defining a problem

The Jurassic–Cretaceous Western Interior Seaway of North America is a typical foreland basin. It was fed with clastic sediments from the actively rising cordillera to the west, and was bordered on the east by a low-relief, tectonically quiescent landmass – part of the North American craton, and the Appalachian Mountains. East of the cordillera, in the Seaway, the bulk of the sediments consist of siltstones and mudstones. Within these fine-grained marine sediments are many examples of linear ridges, or **'offshore bars'** (Figure 16.7). They average about 9 km wide, 45 km long and 14 m thick. At first examination they appear to lie many tens of kilometres seaward of their time-equivalent shorelines; if so, they cannot be interpreted as any form of drowned barrier island. Internally, the facies sequence typically becomes sandier and coarser grained upward, and many of the ridges are capped by conglomerate.

Exploration for oil and gas in these ridges has given superb three-dimensional control of their geometry, based on continuous cores and electrical and geophysical logs of the wells. This is particularly the case for their geometry in plan view, which can commonly be reconstructed in the subsurface, but which can rarely be appreciated or reconstructed in outcrop. For some of the subsurface ridges, rocks of the same stratigraphic unit can be examined in outcrop in the cordilleran fold belt. Although the ridges in the Western Interior Seaway are without doubt the best known in the world, they pose very fundamental sedimentological problems.

If the ridges *did* form way out to sea (and are not transgressed shoreface ridges or barriers), we must ask the following questions.

1. By what process or processes was the sediment transported across the shoreface and out into offshore areas of the basin?
2. Once in the basin, how was the sediment concentrated or 'focussed' into the long, narrow ridges?

3 How did the coarser- and sandier-upward facies sequences develop, and how was the gravel transported to the top of the ridge?

As well as being of academic importance, the answers to these questions have a direct economic significance; if the origin of one ridge is thoroughly understood, the location of other undiscovered ridges forming potential oil and gas reservoirs might be predicted.

Until very recently, there were no satisfactory answers to these questions. Sediment transport beyond the shoreface was believed to be dominated by storm-generated geostrophic flows. The sediments in the lower parts of many ridges consist of offshore bioturbated mudstones and siltstones, alternating with hummocky cross-stratified sandstone. The HCS, formed by storm waves, seems to support the idea of storm emplacement of the sands. However, intermittent storms randomly affecting different parts of the shoreline would be expected, through time, to give rise to broad thin sheets of sand spread evenly across the basin floor. Modern processes that might mould and focus these sheets of sand into specific long, narrow ridges are very poorly understood, and mechanisms for developing coarsening-upward, conglomerate-capped ridges in offshore settings many tens of kilometres offshore are not known. The following discussion is both a search for answers to the problems, and a demonstration of how different techniques can be used to reach for the answers.

The Cardium Formation of southern Alberta, Canada

In the subsurface, the Upper Cretaceous Cardium Formation consists of a series of long, narrow 'sand bodies' that now form oil and gas fields (Figure 16.7). The total oil recoverable from the Cardium is about two thousand million barrels (318 million cubic metres), making it an extremely important reservoir unit. The Cardium Formation is about 100 m thick, and was deposited rapidly, probably within one to two million years.

In outcrops, and in cores from the oilfields, the rocks consist of sandier-upward facies sequences that begin with burrowed marine mudstones. These are overlain by siltstones alternating with HCS sandstones, and the sequence is capped by chert pebble conglomerate with

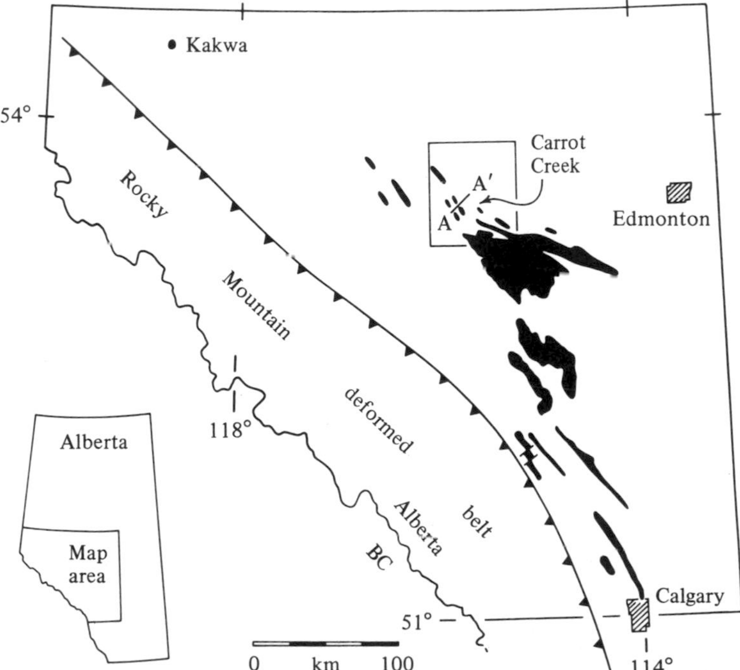

Figure 16.7 Location map of Cardium Formation oil and gas fields (shown in black) in the subsurface of Alberta. The largest field is called the Pembina field. Note their long, narrow shapes, and their trend roughly parallel to the strike of the Rocky Mountains. The Cardium is well exposed in the deformed belt, but is essentially flat lying and undeformed in the subsurface. The rectangle shows the location of cross-section A–A' (Figure 16.8) and the approximate area of the isopach map and mesh diagram (Figures 16.9 and 16.10). □

clasts commonly about 1 to 4 cm in diameter. Up to six coarsening-upward sequences can be superimposed in any one well or outcrop. The sedimentary structures consist exclusively of wave or combined flow types (HCS and wave ripple cross-stratification), without any of the unidirectional structures shown in Figure 16.1. If unidirectional flows were significant in emplacing sand into the offshore area, the sedimentary structures thus formed have been completely reworked by storm waves and combined flows. The absence of medium-scale cross-stratification is particularly significant, because it is the dominant sedimentary structure in *tidally* affected linear ridges, such as those in the southern North Sea. Its absence in the Cardium strongly suggests that tidal currents played no part in Cardium deposition.

Although the smaller scale aspects of sedimentary sequences (sedimentary structures, biological features,

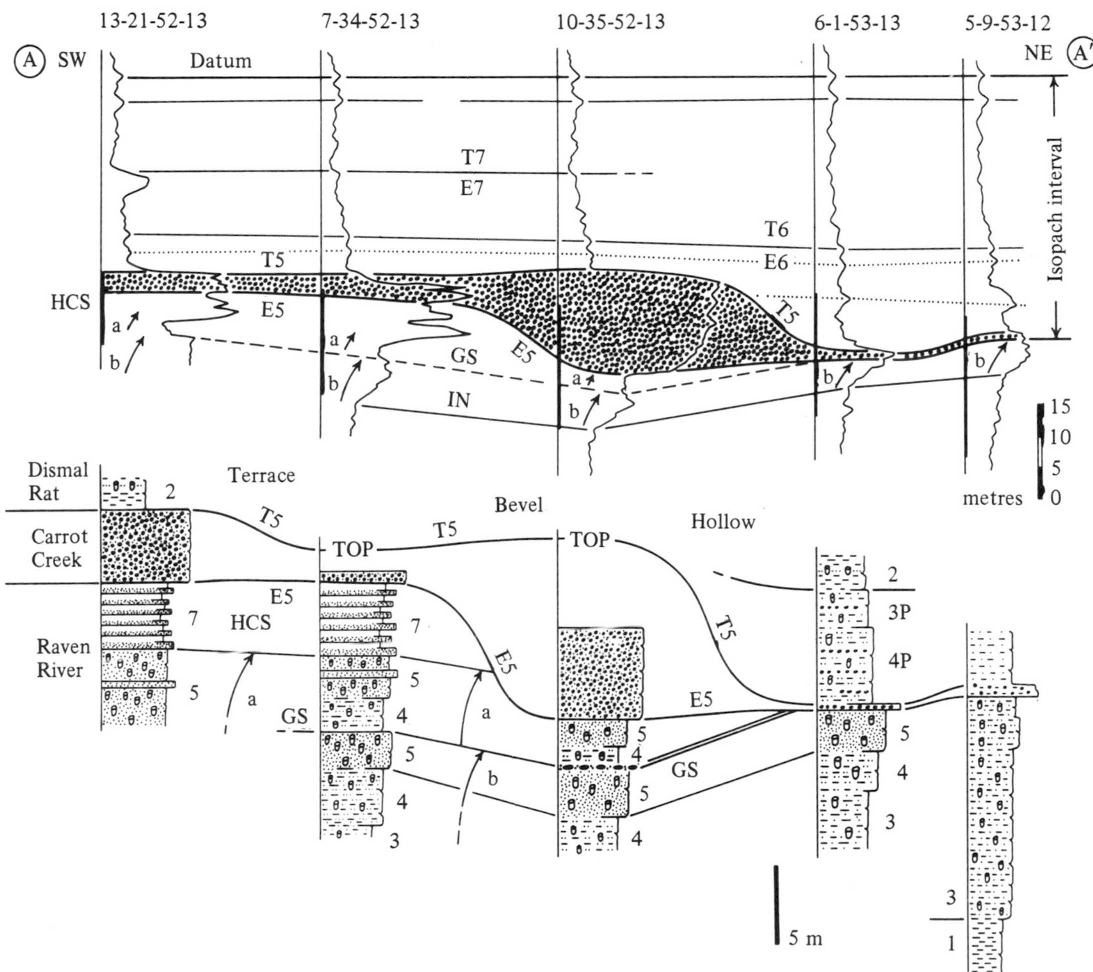

Figure 16.8 Cross-section A–A', located in Figure 16.7; it is 11.5 km long. In the upper cross-section, resistivity logs (deflections to right) are hung on a pair of upper markers. Sequences 'b' and 'a' show increasing proportions of sand upward, and are separated by a distinctive gritty siderite (GS). The change in slope or inflection point (IN) of the logs is also correlated. Note how erosion surface E5 cuts out the hummocky cross-stratification (HCS), the 'a' sequence, the gritty siderite and part of the 'b' sequence northeastward. The heavy black bars indicate cored intervals – logs of these are shown in the lower cross-section, hung on the same markers. The numbers indicate facies; 1 and 2 are dark mudstones; 3, 4 and 5 form a coarsening-upward group of bioturbated muddy facies; 7 consists of HCS sandstones and interbedded mudstones, and conglomerates are shown by the symbol of small open circles. Again note the erosion at the base of E5, and the overlying transgressive (T5) mudstones. Member names of the Cardium Formation are shown on the left. In the upper section, note the position of the isopach interval (see Figure 16.9) □

facies sequences) are very important, the larger scale sand body geometry can be equally important in contributing to an environmental interpretation. In subsurface studies, the larger scale aspects can be studied by making cross-sections that correlate electrical/geophysical logs and cores of the various wells. In Figure 16.8, a cross-section is shown (located in Figure 16.7) that correlates electrical resistivity logs. Resistivity is the resistance of pore fluids to the flow of electrical currents; large deflections to the right indicate porous rocks (sandstones or conglomerates), whereas small deflections indicate shales or mudstones with little pore space and hence little pore fluid. The logs serve two purposes; first, they indicate progressive sequences in which the rocks become more porous (and hence probably sandier) upward, such as the sequence labelled 'b' in Figure 16.8; and second, some of the small log deflections can be used as a datum for lining up the logs ('hanging' them) side by side. A prominent pair of rightward deflections can be seen at the top of Figure 16.8; the upper one is used as the datum. It is assumed to have been as close to horizontal when deposited as can reasonably be reconstructed. It is not always possible to estimate lithology from resistivity logs, and hence continuous core from the wells must be examined where available (black bars, Figure 16.8).

In the Cardium, the wells are essentially vertical and the rocks almost horizontal. The cores can be regarded as 'three inch wide outcrops', and can be divided into a succession of facies (numbered in Figure 16.8) which make up complete measured sections. If these were real outcrops, there would be no way of accurately placing them side by side in order to examine the geometry of the various facies. The advantage of *subsurface* work is that the cores can be hung on the same markers as the well logs, placing all of the measured core sections very close to their original depositional positions. The correlation of the cores (lower half of Figure 16.8) shows that the base of the conglomerate progressively cuts down into other facies toward the northeast – first the HCS sandstones are cut out (facies 7), then the upper part of the 'a' coarsening upward sequence (bioturbated siltstones and sandstones of facies 4 and 5), then the horizon labelled GS (which denotes a prominent 'gritty siderite' layer), and finally part of the 'b' coarsening upward sequence (bioturbated mudstones, siltstones and sandstones of facies 3, 4 and 5). Thus it can be demonstrated from the core correlations that the conglomerate rests on an **erosion surface** (designated E5, or the fifth of seven known Cardium erosion surfaces).

This fact must change the way we think about one of the questions posed earlier, namely the problem of how a conglomeratic cap is formed on the ridges. Because of the presence of the erosion surface, the conglomerate can no longer be regarded as the cap of a progressively coarsening-upward sequence. The underlying sequence (actually two smaller sequences, 'b' and 'a') has been erosively truncated before deposition of the conglomerate, and hence the conglomerate is not genetically part of the sequence. Walther's Law of the Succession of Facies cannot be applied if there are major erosive breaks in the facies sequence. Many geological events could have taken place after deposition of the HCS (facies 7), but the record of these events has been removed by erosion at the E5 horizon.

Cardium Formation erosion surfaces

The morphology of the erosion surface, shown in two dimensions in Figure 16.8, can be studied in three dimensions by mapping the thickness of the interval between the datum (Figure 16.8) and the E5 erosion surface. A map which shows the thickness of a designated stratigraphic interval is known as an **isopach map** (Figure 16.9). In the Carrot Creek area (Figure 16.7), about 440 cores and 960 well logs have been used to make the map. If the upper datum is regarded as the best possible estimate of an originally flat sea floor, the isopach map can be regarded as showing the palaeotopography of the E5 erosion surface (Figure 16.9). The map can also be shown in three-dimensions, as a 'mesh diagram' (Figure 16.10). Note the location of north in Figure 16.10, and the extension of the area to Township 56 (as opposed to T54 in Figure 16.9). The area labelled 'terrace' in Figure 16.10 is the stippled area (C) in Figure 16.9 that trends from the northwest to southeast corner of the isopach map.

Because of the extreme vertical exaggeration of Figure 16.10 (18 m of relief over an area of about 60 × 70 km), the topography looks very irregular, and is dominated by the features originally termed 'bumps and hollows'. However, examination of Figure 16.9 suggests the presence of three topographically high areas, termed steps. Each step has a linear topographic low immediately to the north. Step A is discontinuous along strike, and shows up on the mesh diagram as an east-trending ridge that begins at the tip of the arrow labelled 'bevel'. Step B is discontinuous, and may have been dissected. Step C is laterally the most continuous, and the steep northwestern margin of step C is termed the 'bevel' in Figure 16.10. Cores through steps A and B all show the same facies sequences as step C (and its continuation into the Pembina field; Figure 16.7),

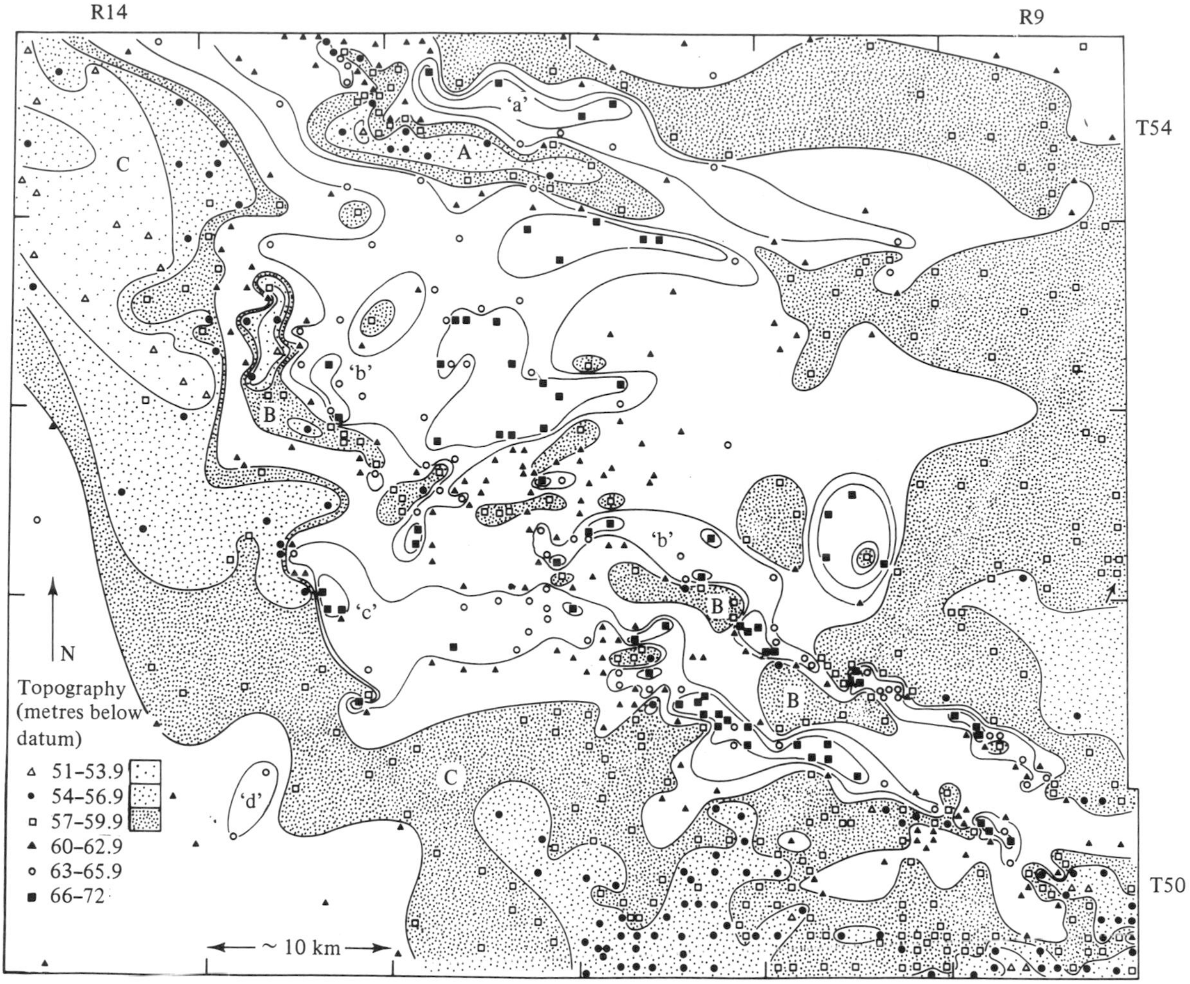

Figure 16.9 Isopach map of the datum to E5 interval. The datum (Figure 16.8) is taken to be horizontal, and the map therefore illustrates the topography of the E5 erosion surface. The various symbols (lower left) indicate depths below datum (metres). There are three main 'high' areas – step A, which is relatively short; step B, which is dissected; and step C, which is longer and wider. Note the deep areas 'a', 'b' and 'c' (black squares) immediately north of the steps. □

indicating that the steps are now part of a remnant erosional topography. They do *not* represent new bars that have grown out to sea, detached from their time-equivalent shoreline.

Seven erosion surfaces have now been identified in the Cardium Formation, although three-dimensional mapping has only been done on E5. *The morphology of this surface is the clue to its origin, and, indirectly, the clue to the problem of the linear conglomerate 'ridges'.*

Before the E5 surface was cut, storm waves controlled the deposition of fine sand, forming HCS in facies 7 below step C (the terrace of Figure 16.10). The time-equivalent shoreline lay an unknown distance to the southwest (toward the cordillera), and the sand was probably transported out across the shallow sea by storm-generated geostrophic flows. The regional strike of Cardium shorelines was roughly NW–SE, parallel to the cordillera, and the regional palaeoslope was northeastward.

In order to cut a surface with 18 m of relative relief, wave action on the bed must be greatly intensified, such that fine sand is eroded and washed away, not deposited as HCS. To increase wave effectiveness, there must be a *lowering of relative sea level.* Erosion may then take place in one of three possible ways; (1) by wave or current scour in an open marine setting, shallower than when the HCS was forming; (2) by subaerial processes; or (3) initially by subaerial processes, but with modification by erosive marine transgression. The morphology of the surface (Figures 16.9

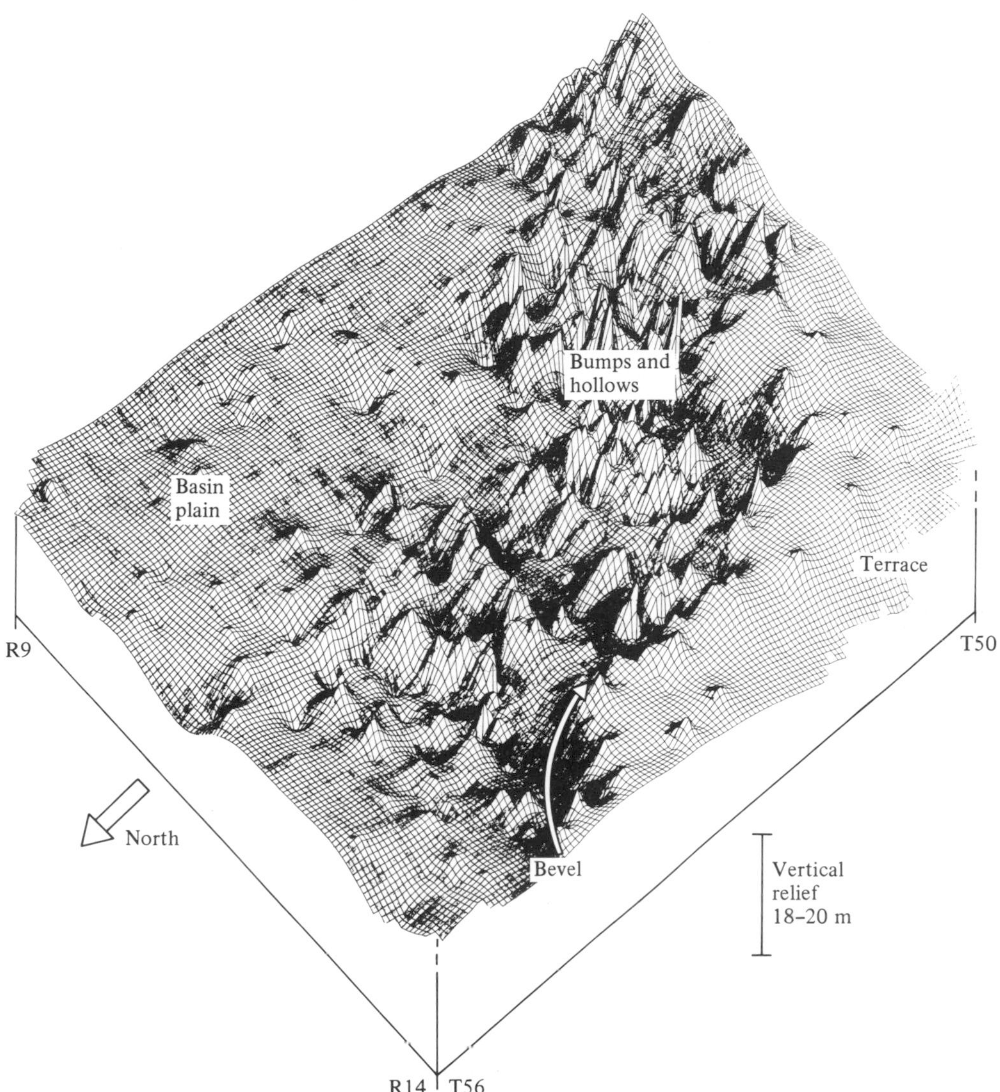

Figure 16.10 Computer-produced mesh diagram showing the topography of the surface contoured in Figure 16.9. The area is 60 × 70 km, and relief is about 18 m, so the topographic relief is extremely exaggerated. The area labelled terrace is step C in Figure 16.9. □

and 16.10) does not indicate northeastward trending relict river valleys, and there are no other suggestions of subaerial erosion. Open marine subaqueous erosion does take place during storms, bur scouring beneath the waves is unlikely to form the specific, linear shapes of the steps. Also, the scouring tends to be irregular and shallow, seldom removing even one metre of sediment; there are no known examples of elongate scours *18 m* deep on modern shelves.

The morphology of the E5 surface favours the remaining possibility, that of modification of a subaerial surface by erosion in a transgressing shoreface. The NW–SE trends of the steps are parallel to the regional strike of the cordillera, which is exactly the predicted trend of shorefaces. It is suggested that following a major relative lowering of sea-level, the shoreline and shoreface moved from an unknown position southwest of the study area to the present northern edge of step A, where the shoreface was incised in its new location. When the shoreface was at position A, all of the area to the southwest of A must have been subaerially exposed, probably with shallow gravel rivers, and trees growing on their banks. Because of the lowered sea-level, base level for the rivers would be lowered, and hence gradients increased (S in equation 1). For streams of similar hydraulic radius (R) and

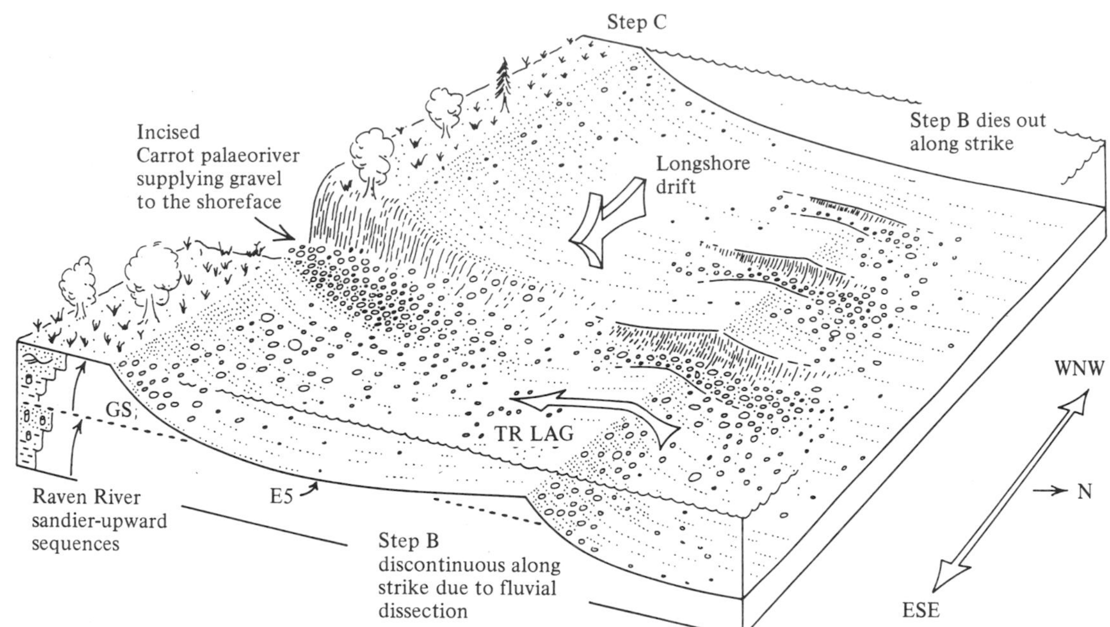

Figure 16.11 Interpretation of the isopach map (Figure 16.9) and mesh diagram (Figure 16.10). The deep areas 'a', 'b' and 'c' of Figure 16.9 are believed to be incised shorefaces, with 'b' cut before 'c'. Shoreface 'b' is shown dying-out along strike, and is dissected by a series of gravel-laden streams. During continued transgression, the beach of shoreface 'b' was eroded, and the gravel spread southwestward as a transgressive lag (TR LAG). At the next stillstand, shoreface 'c' was incised, with a new supply of gravel from palaeorivers. Note how E5 cuts into the two sandier-upward sequences, separated by a gritty siderite (GS) (see Figure 16.8). □

resistance to flow (C), mean flow velocities would be increased and hence gravel could be transported to the new, incised shoreline. It was then reworked along the shoreline by waves to form elongate gravel bodies immediately northeast of the steps. It is these gravels that were formerly regarded as capping offshore ridges.

All evidence of subaerial exposure, such as incised river channels and root traces of vegetation, has been removed. This removal probably took place during transgression, when erosion down to fairweather wave base would have removed anywhere from 5 to 15 m of section (depending on the exact depth to wave base). Gravel may have been removed from the upper shoreface or beach, and spread back across the area as a transgressive lag (wells 13-21-52-13 and 7-34-52-13 in Figure 16.8; see interpretation in Figure 16.11).

During transgression, river gradients are reduced, and much of the coarser sediment is deposited in alluvial plains and estuaries rather than being supplied to the shelf. Thus the gravels banked up against the steps were blanketed by marine muds (bioturbated siltstone and mudstone facies 4P and 3P, the P designating occasional pebble horizons, and delicately laminated mudstones of facies 2; Figure 16.8).

The final result has been to form a series of gravel shoreface deposits (Figure 16.11) as the shoreface retreated from step A via step B to step C. The bevel (Figure 16.10) associated with step C can be regarded as an exhumed shoreface. The gravels are incised into marine sandstones and mudstones, and are buried by marine mudstones. Thus they now superficially resemble long, narrow ridges in an offshore marine mud setting, with no apparent relationship to a shoreline. The details are more complicated than can be presented here, and can be found in the papers listed under 'Further reading' at the end of this Chapter.

Asking the right questions – making the right models

Before the Carrot Creek study was conducted, the gravels were regarded as linear ridges or 'offshore bars' encased in marine mudstones, and hence deposited many kilometres from their time-equivalent shoreline. It therefore seemed appropriate to ask the same questions as had been asked for other linear ridges – how was the sand and gravel transported out beyond the shoreface, how was it subsequently concentrated or focussed into long, narrow ridges, and how did the gravel-capped coarsening upward facies sequence form?

The problem was attacked using many of the techniques of clastic sedimentology – the definition and interpretation of facies and facies sequences, the construction of well log and core cross-sections, and the use of the isopach maps and mesh diagrams to give a three-dimensional view of the facies and the erosion surface. The overall reconstruction was based upon a combination of these techniques; the same methods can be used to construct interpretations of other types of clastic (and carbonate) rocks.

It is now possible to answer the questions posed at the beginning of the research (and it turns out that they were largely the wrong questions!). How was the sediment transported out across the shelf? It wasn't. The inferred relative sea-level drop caused the shoreface to advance out across the shelf, and re-incise in the area of steps A, B and C. How was the sediment focussed into long, narrow ridges? It wasn't. It was reworked along the incised shoreface by waves. How was the gravel-capped coarsening upward sequence formed out on the shelf? It wasn't. The recognition of the erosion surface demonstrated that the gravel is not genetically related to the underlying coarsening-upward sequence. Also, the shoreface interpretation avoids the problem of moving gravel onto the top of an 'offshore bar' (a process not known to occur in modern shallow seas).

The problem of modelling barrier islands was discussed above – it involves knowing the dynamic situation of the barriers in terms of transgression, regression and sediment supply. The problem of modelling 'offshore bars' is even more complex, because many if not most of the sand bodies termed offshore bars were originally deposited at or near the shoreline (for example, as barriers, or as shoreface gravels). Their character depends upon the original depositional environment, the nature and extent of erosive transgression, and the way in which the sand bodies have been modified by open marine storm, wave and tidal processes. I suggest that an 'offshore bar' is not a depositional environment waiting for its own facies model – it is part of a dynamic shelf-to-shoreline system which responds very rapidly to equally rapid relative fluctuations of sea-level; these in turn can move shoreline positions hundreds of kilometres regressively out across the shelf, or transgressively inland across the coastal plain.

Control of sea-level fluctuation

Historically, relative sea-level fluctuation has been given varying emphasis as sedimentologists have attempted to interpret stratigraphic sequences. At times, particularly in the 1960s, more emphasis was given to purely sedimentary processes in explaining phenomena such as delta switching. In the study of ancient offshore linear ridges, sediment transport mechanisms were sometimes emphasised at the expense of any consideration of sea-level fluctuation.

The Cardium has been used as an example of how depositional environments can be reconstructed, and it has been shown how the presence of erosion surfaces demands an interpretation in terms of relative sea-level changes. Currently, one of the broader goals of clastic sedimentology is to contribute to ideas on the **control of relative sea-level**, and it was emphasised above, with respect to barrier islands, that facies models must now be formulated to incorporate the effects of relative sea-level fluctuation.

Seismic stratigraphy (see Chapter 20) has demonstrated cycles or sequences of coastal onlap and sea-level fluctuation on many different time scales; megacycles, supercycles and cycles (or first, second and third-order cycles) have durations of 200 to 300, 10 to 80, and 1 to 10 million years (Ma), respectively. In the Western Interior Seaway, cycles of transgression and regression based on lithological and faunal changes have been defined. These have an average duration of about 4 to 7 Ma, and are therefore on the scale of third-order cycles as defined by seismic stratigraphers. Third-order cycles, with durations of 1 to 10 Ma, could be explained by purely **eustatic** processes, that is, sea-level changes controlled by changes in the volume of sea water or changes in the volume of the ocean basins (due to changes in the volumes of ocean ridges related to changes in rates of spreading; see Chapter 10). Because there are no known Cretaceous glaciations, eustatic sea-level changes would probably be controlled by variations in the rate of spreading.

The analysis of Cardium depositional environments, and the documentation of seven erosion surfaces within the 1 to 2 Ma span of the Cardium indicates that there are also sea-level fluctuations which are much more rapid that the the global third-order cycles. These fluctuations are probably too rapid to be explained by oscillations in the volume of oceanic spreading centres. In a foreland basin setting, the control of relative sea-level, and hence sedimentary sequences of Cardium scale and timing, is more likely to be tectonic. It could be related to loading of the crust by advancing thrust slices, although this may be too slow to explain the relative sea-level fluctuations in the Cardium.

Conclusions

The study of clastic depositional environments incorporates many of the techniques of sedimentology. Individual sedimentary structures can be interpreted in terms of their flow processes, and these structures can be grouped together, along with lithology and biological features, to define facies. Ancient facies must be compared with those forming in modern depositional environments. In that way, it may be possible to assign an environment of deposition to a single facies, but it is more likely that the overall facies sequence will give the clue to depositional environment. The clue may come from comparison with an existing well-worked-out situation or model (as with barrier islands). Alternatively, it may be necessary to deduce environments and histories of sea-level fluctuation without recourse to comparison with existing facies models.

Finally, it should be emphasised that the construction of facies models is in a state of flux. Block diagrams are good ways of illustrating geomorphological relationships between the components of a particular environment, and a carefully drawn diagram can also show the basic facies relationships below the surface. It is infinitely more difficult to understand, and then to illustrate graphically, the effects of transgression and regression, rates of subsidence, and rates of supply. Sedimentology is not only incorporating these 'outside influences' into the construction of more dynamic models; it is actively contributing information to other disciplines. The erosion surfaces in the Cardium, for example, are pointing out tectonic fluctuations in foreland basins on a much finer scale than has been suspected or modelled before.

Further reading

For more-detailed and comprehensive treatments of facies models, the reader is advised to consult the following texts, both of which require a good grounding in sedimentology:

Reading, H. G. (ed.) (1986) *Sedimentary Environments and Facies*, 2nd edition, Blackwell Scientific Publications, Oxford, 615 pp.

Walker, R. G. (ed.) (1984) *Facies Models*, 2nd edition, Geological Association of Canada, Geoscience Reprint Series 1, 317 pp. (3rd edition due to be published in June 1992.)

Readers requiring a more general introduction to sedimentary processes and models are referred to the following two texts; that by Selley is oriented towards petroleum exploration:

Leeder, M. R. (1982) *Sedimentology: Process and Product*, Unwin Hyman, 344 pp.

Selley, R. C. (1988) *Applied Sedimentology*, Academic Press, 446 pp.

Research papers discussing the Cardium Formation problem outlined in this chapter are listed below:

Bergman, K. M. & Walker, R. G. (1987) The importance of sea level fluctuations in the formation of linear conglomerate bodies: Carrot Creek Member, Cretaceous Western Interior Seaway, Alberta, *Journal of Sedimentary Petrology*, **57**, 651–65.

Bergman, K. M. & Walker, R. G. (1988) Formation of Cardium erosion surface E5, and associated deposition of conglomerate: Carrot Creek field, Cretaceous Western Interior Seaway, Alberta, in *Sequences, Stratigraphy, Sedimentology; Surface and Subsurface*, ed. D. P. James & D. A. Leckie, Canadian Society of Petroleum Geologists, Memoir 15, Calgary, Alberta, pp. 15–24.

Walker, R. G. & Eyles, C. H. (1988) Geometry and facies of stacked shallow-marine sandier upward sequences dissected by erosion surface, Cardium Formation, Willesden Green, Alberta, *American Association of Petroleum Geologists Bulletin*, **72**, 1469–94.

CHAPTER

17

LIMESTONES THROUGH TIME

Maurice E. Tucker
University of Durham

Introduction

The principle of **uniformitarianism**, that the present is the key to the past, has been advocated by most geologists since the days of James Hutton (1727–97) and Charles Lyell (1800–60) and is a common theme underlying much thinking in the Earth Sciences. However, in the past few years, it has become apparent that for limestones (or **carbonate sediments**) this line of reasoning can only be applied in a general way, and that there are many instances where the present is not the key to the past.

Carbonate sediments are deposited in a great range of environments, but as far as the geological record of limestones is concerned, the most important environments are the shallow-marine, low-latitude, tropical regions, where reefs, oolite shoals, tidal flats and lagoons are developed. In fact, in terms of volume of sediment, the deeper water shelves, slopes and ocean basins are the sites of some 80 per cent of all carbonate sedimentation at the present time. However, the pelagic oozes deposited there have a low preservation potential as a result of plate tectonic processes, notably subduction, so that deep-water limestones are under-represented in the geological record, compared with shallow-water limestones.

Carbonate rocks encompass a wide range of sediment types (**facies**), but there is much variation in the abundance of these through the geological record and some facies only occur at certain times. Carbonate facies patterns are affected by many factors, but the overriding controls are global sea-level stand and the status of organic evolution. The strong biological input to carbonates means that many limestones reflect the evolution and extinction of organisms with calcareous skeletons.

There is now good evidence to show that subtle changes took place in the chemistry of seawater through the Phanerozoic as a result of variations in the rates of sea-floor spreading and subduction. These led to fluctuations in the mineralogy of marine carbonate precipitates, and also exerted a control on the mineralogy of carbonate skeletons. The original mineralogy of a limestone is also significant in terms of its diagenesis and geochemistry. Dolomites, formed from the replacement of limestones, have an uneven distribution through time, and also reflect long-term tectonic processes.

When studying limestones, then, the principle of uniformitarianism has to be applied with care. Modern organism communities and carbonate facies *are* different from those in the Mesozoic and Palaeozoic; the mineralogy of modern sediments *is* different from some ancient limestones, and the patterns of diagenesis in limestones do vary. Indeed, it is often more useful to think that with carbonate rocks, 'the present is only the key to the Pleistocene', and that the past is more a guide to the future.

Carbonate facies through time

Thick sequences of shallow-water limestones constitute **carbonate platforms** and they are very much affected by relative sea-level changes. These may be eustatic or the result of regional tectonics. Five types of carbonate platform can be distinguished: **rimmed shelf, ramp, epeiric platform, isolated platform** and **drowned platform** (Figure 17.1). Each has a characteristic assemblage and arrangement of facies, described in more detail in the figure caption.

Assessments of the distribution of sedimentary rocks through the Phanerozoic reveal that there is a broad correlation of limestone occurrence with **global changes in sea-level** which are referred to as **eustatic** changes (Figure 17.2). Limestones are generally more common at times of higher sea-level, when continents are submerged, and so there is a very low influx of terrigenous clastic material. Shallow-water, epeiric sea carbonate platforms on a continental or craton scale are not developed at the present time because of the relatively low stand of sea-level. However, they were very widespread during the early to mid-Palaeozoic, as on the North American, Russian and Chinese cratons, when global sea-level stand was high. Carbonates were also extensively developed during the mid-Mesozoic global sea-level highstand, with many shallow-water platforms and shelves in western Europe – Middle East – North Africa and in the Gulf of Mexico region. In the late Cretaceous, somewhat deeper-water epeiric seas led to deposition of the mostly pelagic chalks over flooded cratons, such as from Ireland to the Russian platform and within the North American mid-continental seaway.

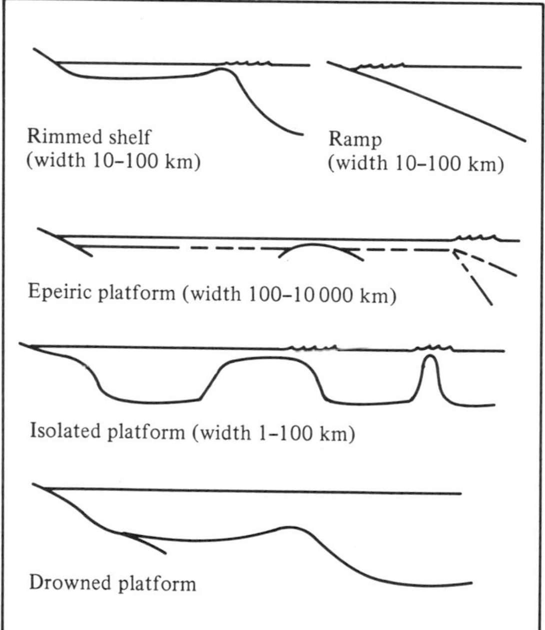

Figure 17.1 The various categories of carbonate platform. Each has a distinctive pattern of facies. In the rimmed shelf model, reefs and carbonate sand bodies occur along the shelf break, with quieter water, muddy sands and muds in the shelf lagoon behind. In the ramp model, carbonate sands of the shoreline pass down ramp to deeper water muddy sands and muds, and reefs are not important. Tidal flats and protected lagoons occur upon epeiric platforms. The facies on an isolated platform are very much affected by orientation to prevailing winds, with reefs and sand bodies along windward margins, and muddy sediments to leeward. Drowned platforms are sites of pelagic ooze deposition. □

On a more regional scale, the occurrence of smaller carbonate platforms is controlled more by the local plate tectonic regime than global sea-level. The development of rifts and passive continental margins during the Triassic – early Jurassic for example, when global sea-level was at a low stand, led to the development of many moderate-sized carbonate platforms and isolated platforms in the Tethyan region. Continued extension and passive margin subsidence led to the drowning of many of these platforms in the mid–late Jurassic, and widespread deposition of pelagic sediments.

Reef complexes and patch reefs through time

Reef complexes consist of the biologically constructed reef mass itself, together with flanking debris material and back-reef lagoonal sediments. The geological record of reef complexes is very uneven, reflecting the occurrence of organisms capable of producing reef frameworks. From studies of modern reefs, there is a well-defined facies model (Figure 17.3). A framework built by **scleractinian corals** (see Figure 17.4a), supported and cemented by crustose **coralline algae**, constitutes the reef core. Debris from the reef crest and reef front comprises the fore-reef facies of slope and talus aprons, and in the back-reef area, lime sands and muds accumulate in the protected lagoon. This general model has been applied to many ancient shelf-margin reef complexes, but in reality it is only relevant when organisms were available with rigid, massive, branching and tabular skeletons to construct solid, wave-resistant frameworks as illustrated in Figure 17.4. There were many instances in the past when these organisms did not exist or were not important, and then carbonate buildups were produced more by the baffling, binding and trapping of sediment by *in situ*

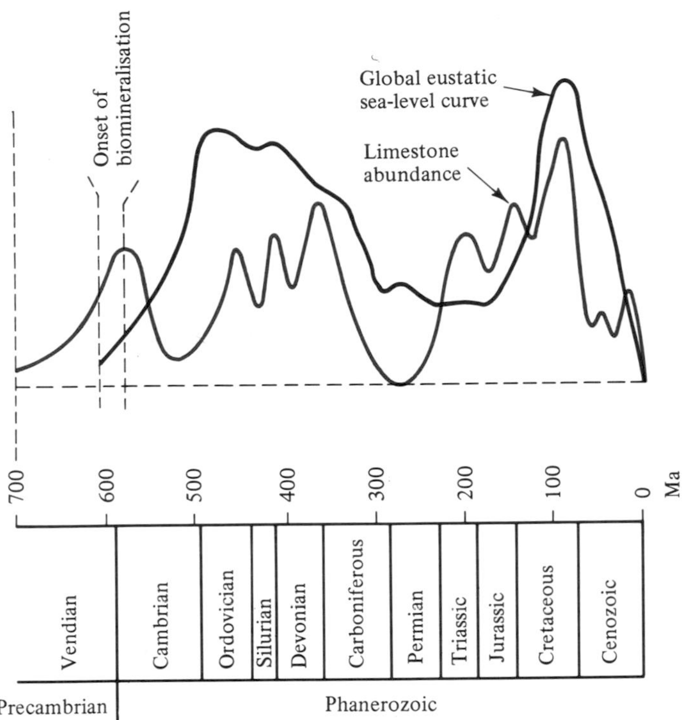

Figure 17.2 Relative abundance of carbonate rocks through the Phanerozoic compared with the first-order pattern of global sea-level change.

organisms. These buildups generally do not give rise to vast quantities of debris, to much relief or to prominent fore-reef slopes. Back-reef areas are not usually protected lagoons and marine cementation is not widespread.

Seven major periods of reef complex development with framework-constructing organisms are recognised (Figure 17.5): (1) middle–late Ordovician bryozoan–stromatoporoid–tabulate coral reefs, (2) Silurian and Devonian stromatoporoid–coral reefs, (3) late Permian sponge–calcareous algal reefs, (4) late Triassic and (5) late Jurassic scleractinian coral–stromatoporoid reefs, (6) late Cretaceous rudistid bivalve reefs and (7) Tertiary–Quaternary scleractinian coral–red algal reefs. On a broad scale, reefs developed at these times are similar to modern coralgal ones, but in detail there are still differences, such as in the amount of encrusting and binding, the extent of bioerosion especially by boring organisms, and the degree of cementation.

The geological record of **patch reefs** (relatively small reef masses surrounded by other sediment types) is much more continuous, although there is a conspicuous absence of such structures in the late Ordovician, late Devonian and early Triassic (Figure 17.5). There is a great variety of size, shape, internal structure, faunal diversity and associated facies. At one time or another, most carbonate-secreting organisms have constructed patch reefs, although many are simply accumulations of skeletal debris rather than true reefs with a framework.

One particular type of carbonate buildup with a chequered career is the **mud mound** (reef-knoll in older literature). It is a feature of Palaeozoic–early Mesozoic deeper-water ramp successions and is characterised by a dominantly fine-grained sediment, few skeletal or-

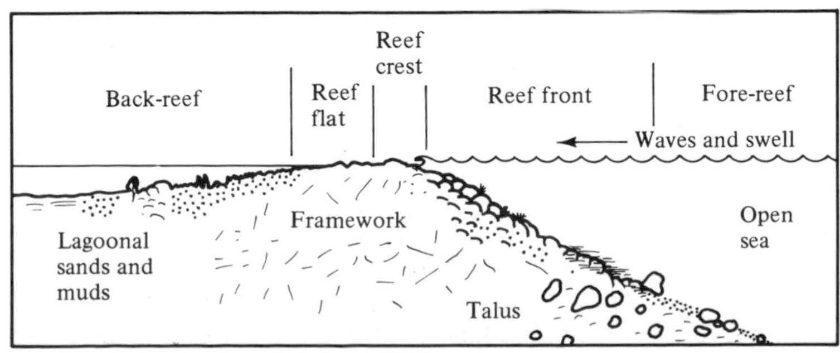

Figure 17.3 Cross-section of a modern reef showing the main sub-environments.

(a)

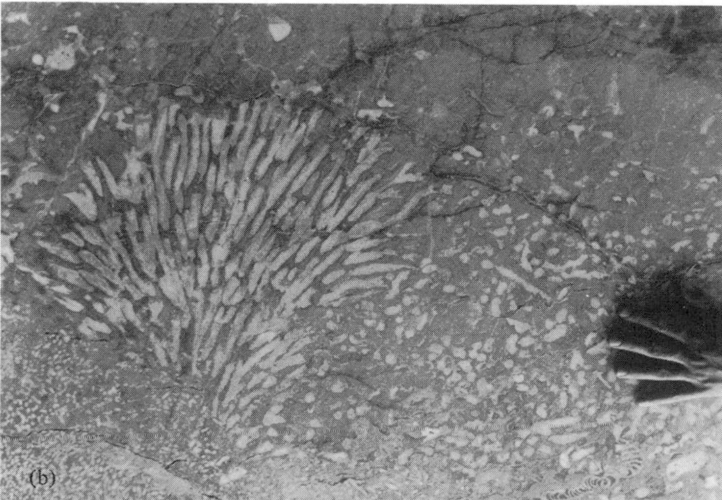

(b)

Figure 17.4 Reef corals. (a) *Acropora palmata*, the elkhorn coral, the principal frame-building coral of modern Caribbean reefs. (b) *Thecosmilia*, a large coral prominent in Triassic reefs of the Alpine–Mediterranean region. □

ganisms (although some have significant numbers of bryozoans, calcified algae or sponges), and synsedimentary cavities (i.e. formed at the time of deposition of the surrounding sediment) with fibrous cements. The cavity structures are commonly of the **stromatactis** type, which show flat, sediment-floored bases and irregular roofs. Although there has been much argument over the origin of stromatactis, a consensus now ascribes them to local sediment collapse and patchy early lithification. Mud mounds were the sites of much sediment accumulation, but mostly little was deposited in the areas between the mounds. The two major issues are the source of the lime mud and the mechanism by which it was trapped. An *in situ* origin for the mud is now generally accepted, and frequently microbial (bacterial) precipitation is invoked. One suggestion is that the mud was produced, trapped and bound largely by **coccoid cyanobacteria** so that the mounds are in effect giant **thrombolites** (unlaminated stromatolites). Certainly many mounds have a vague lamination and a clotted or peloidal texture which is typical of thrombolites.

Mud mounds are often compared to the mud banks of the inner Florida Shelf, Florida Bay and Shark Bay, Western Australia. However, these are shallow-water accumulations of skeletal debris (much of calcareous algal origin), trapped largely by sea-grass (which only evolved in the Cretaceous), with little synsedimentary cementation.

Carbonate sands

Throughout the geological record, lime sands have been deposited in the same high-energy environments of shelf-margin and inner ramp as at the present time. The former are especially well represented in the

Bahamas, occurring in windward, leeward and tide-dominated locations around the bank margin. Ramp sands are deposited in strandplain and barrier island–tidal delta complexes, as along the modern Yucatan (NE Mexico) and Trucial (Arabian Gulf) coasts, respectively.

Ancient **carbonate sands** show all the same sedimentary structures and facies sequences as the modern ones, but there are differences when the nature of the grains themselves is considered (Figure 17.6). Obviously there is much variation in the composition of bioclastic sands, since these are derived from the skeletons of the organisms living in that area at that time. Cenozoic skeletal sands are dominated by molluscan (e.g. gastropod and bivalve) debris, whereas those in the Mesozoic and especially the Palaeozoic have much more brachiopod material. Crinoidal debris is abundant in Palaeozoic carbonate sands, only locally important in Mesozoic ones and of minor significance in the Cenozoic. Precambrian skeletal sands, especially those of the later Proterozoic (Upper Riphean and Vendian) are dominated by calcified algal aggregates.

Non-skeletal sands are composed of **ooids** (spherical grains showing concentric layering due to incremental growth as they are rolled around by currents; Figure 17.7a), and less commonly **peloids** (rounded particles consisting of carbonate mud). They occur throughout the geological record, but ooids appear to have been preferentially precipitated at certain times (Figure 17.7b): during the late Cambrian, the end of theearly Carboniferous, the late Jurassic and the Holocene. These are times when, apart from relatively minor short-term fluctuations, global sea-level was either rising or falling. Ooids are conspicuously less common in mid-Ordovician through upper Devonian,

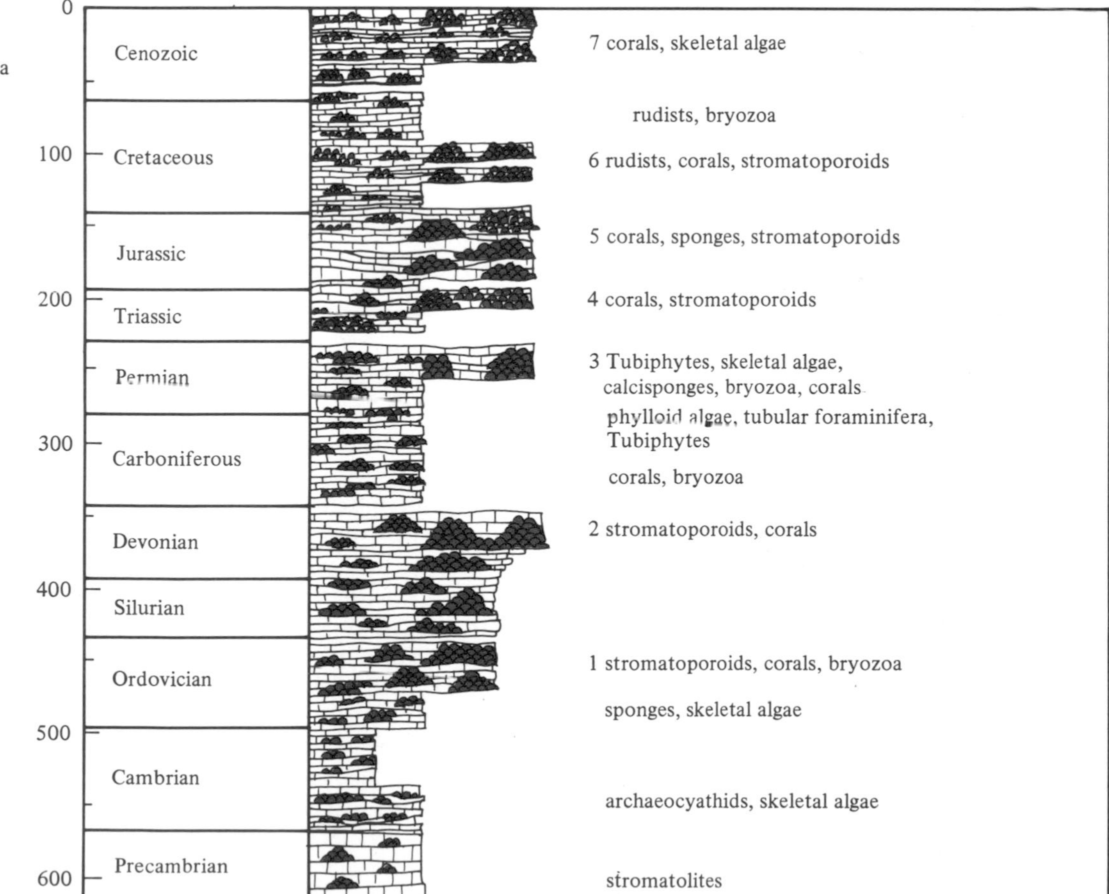

Figure 17.5 Distribution of major reef complexes and patch reefs through the Phanerozoic with the common skeletal elements involved.

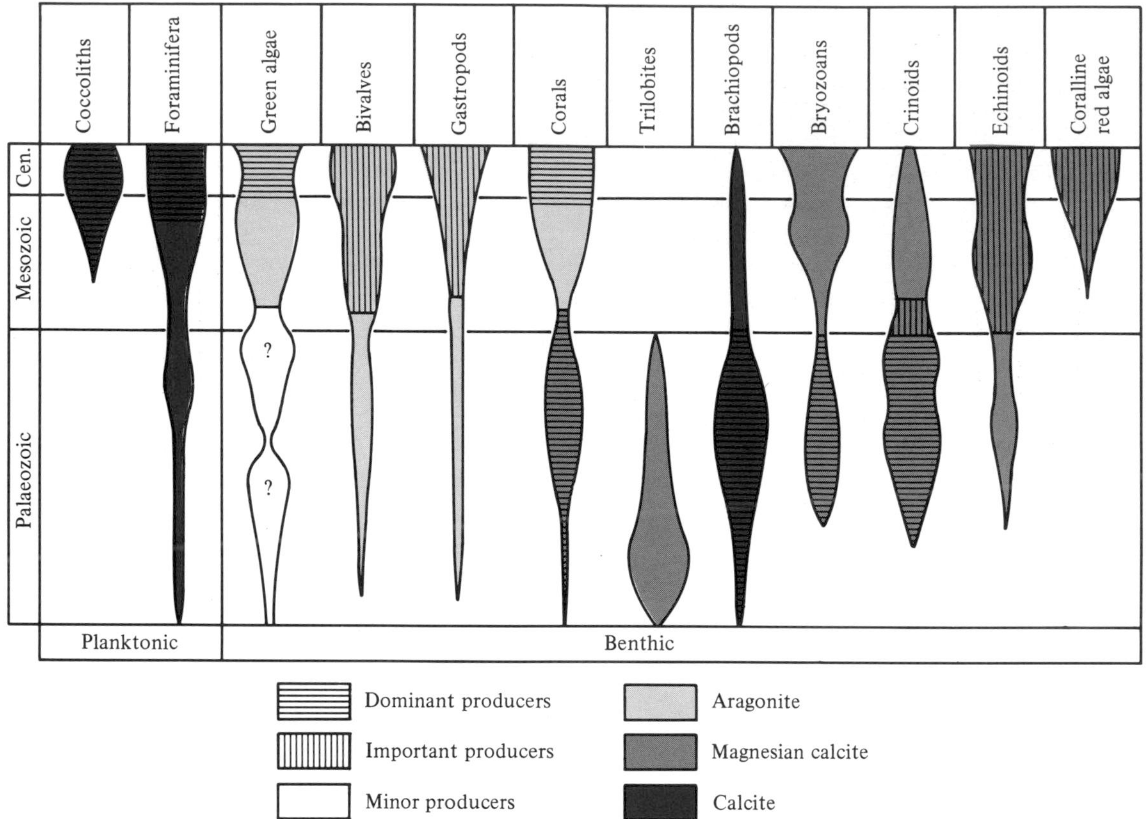

Figure 17.6 The approximate abundance and skeletal composition of the main groups of calcareous marine organisms through the Phanerozoic.

and upper Cretaceous through mid-Tertiary strata, times of global sea-level highstand. Although the reasons for these variations in ooid abundance through time are unclear, it does not appear to be for a lack of suitable high-energy environments. The most likely explanation is to do with seawater chemistry. One possibility is that carbon dioxide levels in the atmosphere may have reached sufficiently high values at times of global sea-level highstand that physico-chemical precipitation of carbonate was inhibited by a lowering of the degree of carbonate saturation.

Deep-water facies

In modern oceans, **calcareous deep-water or pelagic ooze** covers much of the sea-floor which is above the **carbonate compensation depth**; this is the depth, 3–4.5 km in low latitudes, at which the rate of carbonate deposition equals the rate of dissolution. Calcareous ooze occurs upon the flanks of ocean ridges, on oceanic plateaux, banks, seamounts and aseismic ridges, and on deeper-water continental shelves and slopes starved of clay derived from land. Back into the rock record, pelagic facies are only well-preserved when they were deposited on stable continental crust. Where they accumulated on continental margins and ocean floors, they were subsequently deformed in plate collision zones, now preserved in mountain belts, or lost down subduction zones.

Calcareous pelagic organisms do not have a continuous record through the Phanerozoic. **Foraminifera** tests and coccoliths are the dominant constituents of modern oozes, but the **coccolithophorida** (planktonic yellow-green algae) only became important from the mid-Jurassic and the planktonic foraminifera from the Cretaceous. Thus pelagic limestones become much more common from the middle Mesozoic onwards with the acme being in the late Cretaceous sea-level highstand when coccolith chalks were deposited extensively over drowned cratons. Pelagic limestones do occur in the Palaeozoic, but they are not common.

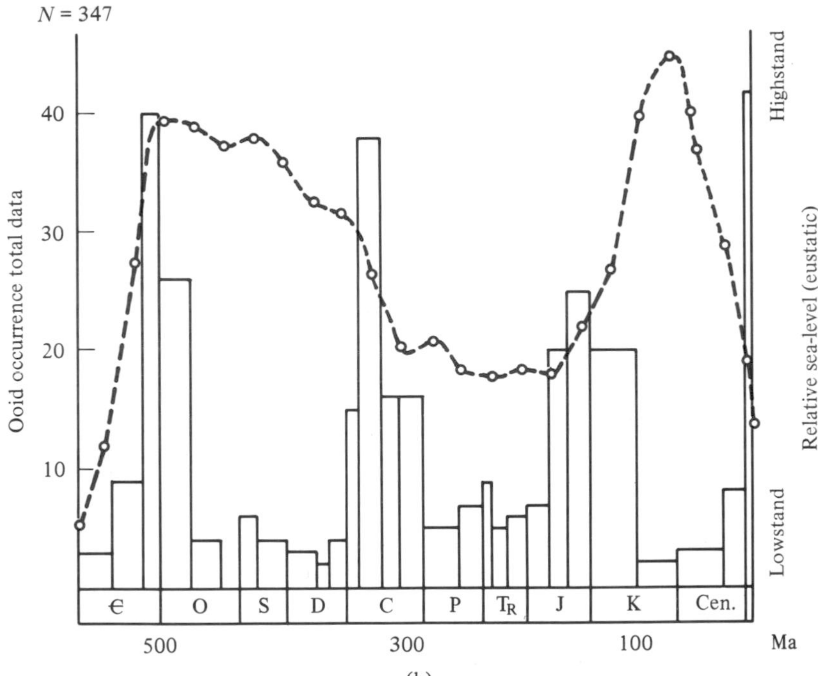

Figure 17.7 Ooid sands. (a) Modern ooids from the Bahamas. These grains, composed of aragonite today, are precipitated in shallow high-energy areas of the tropics. (b) Distribution of ooid sands through the Phanerozoic compared with the first-order global sea-level curve, showing that ooid sands were more common during the transgressive and regressive episodes compared to lowstands and highstands. □

Mineralogy of marine precipitates through the Phanerozoic

Two calcium carbonate ($CaCO_3$) minerals comprise carbonate sediments and limestones: **aragonite** and **calcite**, and two varieties of the latter are distinguished: **low-Mg calcite** with 1–4 mole% $MgCO_3$ and **high-Mg calcite** with 11–19 mole% $MgCO_3$. Aragonite and high-Mg calcite are converted to calcite (low-Mg) during diagenesis. Modern shallow-marine bioclastic sediments have a variable mineralogy depending on the composition of the skeletal grains present (see Figure 17.6). Most modern marine ooids are composed of aragonite, and marine cements, as occur in reefs, are of aragonite and high-Mg calcite. In the early days of limestone petrography, it was generally assumed that the original mineralogy of limestones was similar to modern sediments. It was widely thought that all ancient ooids were originally aragonitic, and a similar philosophy was often applied to ancient marine

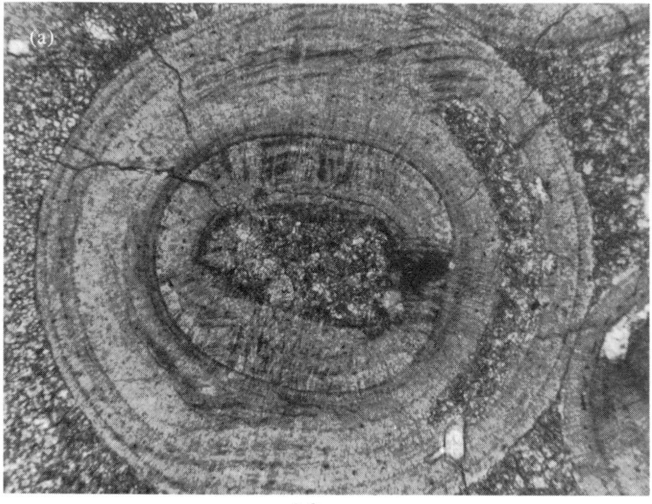

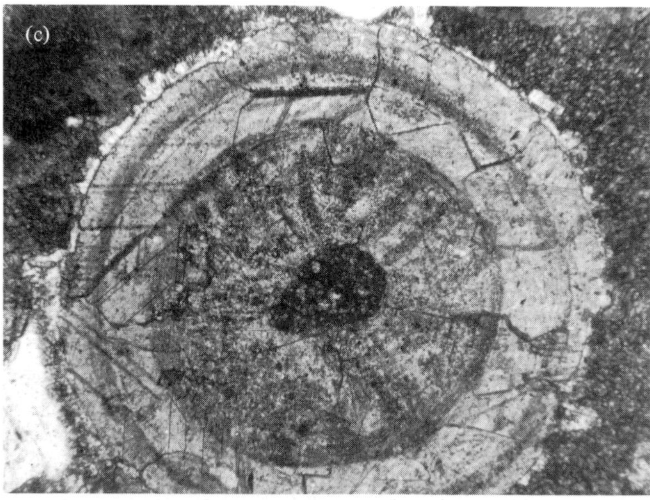

Figure 17.8 Ooid structure. (a) Ooid originally (and still) of calcite with well-preserved radial-fibrous and radial-concentric structure. (b) Ooids originally of aragonite where aragonite dissolved out and the voids later filled with calcite cement. (c) Ooid originally composed of aragonite, since replaced by calcite with some retention of the original concentric and radial fabric (calcitisation). Belt Supergroup, Late Proterozoic, Glacier National Park, Montana. □

cements. The possibility of changes in the mineralogy of marine precipitates over geological time was rarely discussed in the literature, although there were frequent references to the apparent more common occurrence of dolomites in the early Palaeozoic and Precambrian.

Ooid mineralogy through time

In 1975, Philip Sandberg (University of Illinois) argued convincingly that some ooids in the geological record were originally composed of calcite, whereas others did have an original aragonite mineralogy (since replaced by calcite), like the modern ones of the Bahamas and Trucial Coast ooid shoals. The evidence for original mineralogy of grains in ancient limestones is both petrographic and geochemical. The lines of reasoning for ooids are similar to those used to deduce the original mineralogy of bioclasts, where modern counterparts of known mineralogy can also be studied. Where ooids were originally composed of low-Mg calcite, then their original texture is normally perfectly preserved, in the same way as oyster and brachiopod shells and rugose corals for example, which were also low-Mg calcite. Ooids originally composed of high-Mg calcite mineralogy are well-preserved mostly, but in the transformation to low-Mg calcite, it is possible for some patchy dissolution to occur (this is rare) or for small crystals of dolomite (microdolomite) to form. Many ancient ooids originally (and still) of calcite have a distinctive radial-fibrous fabric, although the outer part of larger ooids may have a radial-concentric fabric (Figure 17.8a). Ooids originally composed of aragonite are preserved in ancient limestones in one of two ways, just like originally aragonitic bioclasts such as gastropods, many bivalves and scleractinian corals. Either the aragonite may be dissolved out completely to leave a mould and the void later filled with calcite spar cement (Figure 17.8b), or the aragonite may be **calcitised** (i.e. replaced by calcite) without a major void developing so that some relic structure (the concentric lamination for example) is retained (Figure 17.8c). Some ooids possess a micritic (microcrystalline calcite) fabric, and although this could be a primary feature, it could also be the result of microbial micritisation or diagenesis.

Thus, it is now generally accepted that the ancient calcitic ooids with a radial-fibrous and radial-concentric structure are primary, and that originally aragonitic ooids can be recognised by the evidence for dissolution and/or replacement by neomorphic calcite spar. Further confirmation of an original aragonite composition can be obtained in the case of calcitised ooids by searching with the scanning electron microscope for minute crystal relics of aragonite. These are commonly preserved as inclusions in the neomorphic calcite. In addition, the replacement calcite crystals may have a relatively high strontium content inherited from the aragonite. The latter typically has around 10 000 ppm Sr in abiotic marine precipitates and, on replacement by calcite, several 1000 ppm may be retained. The identification of an original high-Mg content of calcitic ooids is often not easy. The presence of **microdolomite** (very-fine-grained dolomite–CaMg $(CO_3)_2$) is generally taken to indicate an original magnesian calcite (11–19 mole% $MgCO_3$) mineralogy, but in open-system diagenesis, microdolomites may not be precipitated. Low-Mg calcite that has replaced magnesian calcite may have a slightly higher Mg content, of a few per cent $MgCO_3$, than an original low-Mg calcite ooid, and there could also be an enrichment in Fe and Mn, picked up during the stabilisation process.

It is interesting to note that although it is only quite recently that the significance of the original mineralogy of grains has been appreciated, Henry Clifton Sorby, the pioneer of microscopic studies of rocks, noticed in 1859 that the structure of Jurassic calcitic ooids was different from modern aragonitic ones, and he thought the calcitic mineralogy was primary too!

A survey of the distribution of calcitic and aragonitic ooids through the Phanerozoic by Sandberg revealed a secular variation (Figure 17.9): aragonitic ooids, which may be associated with calcitic (presumably high-Mg) ooids occur in the late Precambrian–early Cambrian, mid-Carboniferous through Triassic and Tertiary to Recent, while calcitic ooids (probably low to moderate magnesian content) were dominant in the mid-Palaeozoic and Jurassic–Cretaceous.

As is to be expected, there are exceptions to the Sandberg curve. In the much-studied Upper Jurassic Smackover Formation, for example, in the subsurface of the US Gulf Coast, both (formerly) aragonitic and (still) calcitic ooids are present. The aragonitic ooids occur in shoals which developed closer to the shoreline, whereas the calcitic ones formed in shoals farther seaward (to the south), which may also be younger. The late Jurassic is a time when only calcitic ooids would be expected from the curve. The Smackover is an important hydrocarbon reservoir and the type of porosity, as well as the diagenetic pattern are significantly determined by the original mineralogy of the ooids. **Oomoldic porosity**, through aragonitic ooid dissolution and meteoric water influx, is important in more northern reservoirs, whereas **intergranular porosity** is more common in reservoirs to the south where

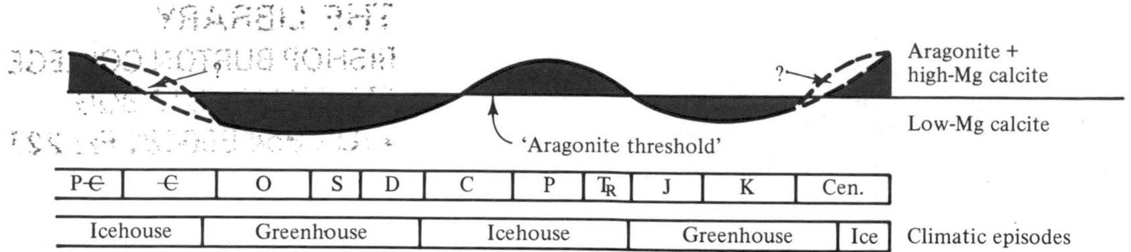

Figure 17.9 Mineralogy of abiotic marine precipitates through the late Precambrian and Phanerozoic. Calcite seas coincide with highstands and high global temperatures (greenhouse episodes) and aragonite (+ high-Mg calcite) seas with lowstands of sea-level and cold periods (icehouse episodes) (see also Figure 17.2).

little ooid dissolution has taken place because of the calcitic nature of the ooids. Thus the study of ooids is not entirely academic; it can lead to an understanding of porosity evolution in hydrocarbon provinces and a prediction of where the best reservoirs may occur.

Marine cements through time

Marine carbonate cements are precipitated directly out of seawater, like ooids, and they should show a similar pattern of mineralogical variation through the Phanerozoic (Figure 17.9). In fact, they do. Modern shallow-marine sands and reefs are being cemented by aragonite and high-Mg calcite, and one of the most conspicuous forms, especially in reefs, is **botryoidal aragonite** (Figure 17.10a). This consists of mamelons up to 50 mm radius of fibrous aragonite crystals. In the geological record, botryoidal aragonite (since replaced by calcite) is relatively common in late Palaeozoic–early Mesozoic reefs (e.g. Figure 17.10b), but absent in mid-Palaeozoic and mid–late Mesozoic buildups. In these others, primary calcite cements are ubiquitous, and mostly these are fibrous types, such as **radiaxial fibrous calcite** (Figure 17.10c). The latter is especially prominent in the mud mounds of Ordovician through Carboniferous age, filling the stromatactis cavities. Radiaxial fibrous calcite was for long regarded as a replacement of some needle-like precursor; this was mostly because of the rather unusual optical properties, but another factor was the apparent absence of this cement from modern reefs. Recently, radiaxial fibrous calcite has been reinterpreted as a primary precipitate, with the conspicuous fabrics the result of split crystal growth. Near-modern radiaxial fibrous calcite has now been found in the Pleistocene of Japan and the Miocene of the subsurface Enewetak Atoll in the Pacific. Nevertheless, the fabrics of much ancient fibrous calcite are substantially different from those of calcite cements (high-Mg) in modern reefs. In the Enewetak case, the fibrous calcite is being precipitated from seawater undersaturated with respect to aragonite, at a water depth of several hundred metres. Thus, it may be a question of more searching being required; fibrous calcite could be quite widespread in modern seas, but in deeper-water shelf-margin limestones than normally sampled.

The most common cement of all in limestones is **equant calcite spar**, usually with a drusy fabric, i.e. crystals increase in size towards the cavity centre. Again, a meteoric and/or burial origin is generally inferred, but there are now several well-substantiated examples of this spar being a marine precipitate. It occurs as the sea-floor cement which forms 'hardgrounds' of Ordovician and Jurassic age. Thus, even blocky calcite spar must be examined carefully, and its precipitational environment cannot be taken for granted. At the present time, such coarse, low-Mg calcite crystals do not precipitate in low-latitude shallow seas, although similar cements do occur in deeper, colder water, near-Recent, shelf-margin limestones.

Controls on changing mineralogy

Through the Phanerozoic, then, there were times when calcite was the abiotic precipitate from seawater, while at other times aragonite and magnesian calcite were precipitated. Although externally ooids are much the same regardless of mineralogy, internally, aragonitic ooids of high-energy environments generally have a concentric structure, whereas calcitic ooids are radial-fibrous to radial-concentric. Low-energy aragonitic ooids may also have a radial-fibrous fabric (such as those in the Great Salt Lake, Utah, and on the Abu Dhabi, Trucial Coast tidal flats). Aragonite and high-Mg calcite cements of modern seas do have exact equivalents (although replaced by low-Mg calcite) in the geological record, but many ancient marine cements originally of calcite (fibrous calcite, syntaxial crinoid overgrowths and drusy spar) are not being precipitated in shallow tropical seas at the present time. The mineralogy of these abiotic marine precipi-

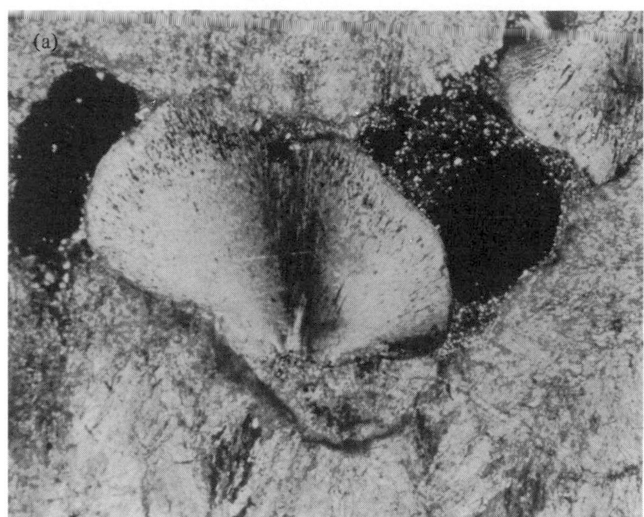

Figure 17.10 Marine cements. (a) Aragonite botryoid in coral skeleton from modern Belize reef. (b) Aragonite botryoid since replaced by calcite in brachiopod shell from Permian (Zechstein) reef, NE England. (c) Fibrous calcite, a common marine calcite cement in many Palaeozoic and Mesozoic limestones, from a Lower Carboniferous mud mound, NW England. □

tates is largely controlled by seawater chemistry, so that these deductions and interpretations coming from microscopic studies of limestones do indicate subtle fluctuations in seawater composition.

There are many factors which affect the mineralogy of carbonate precipitated from seawater, but carbon dioxide levels (**pCO_2**) and **Mg/Ca ratio** are the two most important. With pCO_2, aragonite and high-Mg calcite are precipitated when levels are low, and as pCO_2 increases, calcite is precipitated with decreasing Mg^{2+} content, and a point is reached when aragonite is no longer precipitated. Calcite with around 12 mole% $MgCO_3$ has a similar stability to aragonite, and calcite solubility increases with increasing Mg^{2+} content. It has been suggested that with very high pCO_2 in the atmosphere–hydrosphere, dolomite, or more likely a disordered Mg–Ca carbonate precursor, could be a marine precipitate. With Mg/Ca ratio as a control, a molar value in excess of five is considered to lead to aragonite and high-Mg calcite precipitation, whereas lower values will result in calcite with lower Mg content. Since modern low-latitude shallow seas have a molar Mg/Ca ratio of 5.2, aragonite and high-Mg calcite are the precipitates. Near-surface meteoric water has a molar Mg/Ca ratio of 0.5 and so low-Mg calcite is the normal precipitate there. Thus, low Mg/Ca ratio and high pCO_2 promote calcite precipitation, whereas high Mg/Ca ratio and low pCO_2 favour the precipitation of aragonite and high-Mg calcite.

The factors which determine whether aragonite or high-Mg calcite will precipitate are less clearly defined. Kinetically, aragonite is favoured over high-Mg calcite at higher seawater Mg^{2+} and SO_4^{2-} contents, at lower PO_4^{3-} concentrations, and with increasing temperature. Another important factor is the carbonate supply rate, with higher rates favouring aragonite and lower rates calcite. Thus aragonitic ooids may be expected to form in higher energy environments where CO_3^{2-} supply rate is high, whereas magnesian calcitic ooids would form in lower energy areas of low CO_3^{2-} supply rate. Such a scenario could explain the Smackover story noted earlier, with a change in energy level of ooid sand bars through time as they prograded south. An alternative explanation could be the ratio of Mg/Ca, with higher nearshore values from evaporation leading to aragonite precipitation there.

The secular trend in mineralogy of marine precipitates through the Phanerozoic (Figure 17.9) correlates in a broad way with the first-order global sea-level curve and with percentage of continental areas flooded: that is aragonite and high-Mg calcite are the marine precipitates when global sea-level is low and continents are in emergent mode, and calcite (low-Mg) is the precipitate when continents are in submergent mode during the major global sea-level highstands of the mid-Palaeozoic and Jurassic–Cretaceous. Aragonite seas correspond to the times when global temperature was relatively low (**icehouse times**) and calcite seas to the periods of higher temperature (**greenhouse times**) of Earth history (Figure 17.9). At times of calcite seas, seawater will be undersaturated with respect to aragonite so that shallow sea-floor dissolution of aragonite should take place, particularly in areas of slow sedimentation. Evidence has been found for such dissolution in Jurassic and Ordovician rocks.

It is now widely accepted that the two major global sea-level cycles of the Phanerozoic are controlled by the major plate tectonic processes of opening and closing oceans; the first cycle of the Phanerozoic reflects the formation and destruction of **Iapetus** (an ocean that separated much of north America from Europe) from the late Precambrian through to the Carboniferous, and the second cycle is the result of the formation of the Atlantic and **Tethys** (a larger ocean, of which the present-day Mediterranean is but a small relic) and the closing of Tethys. At times of high sea-level stand, rates of sea-floor spreading and subduction are high. The pCO_2 of the atmosphere–hydrosphere is raised substantially through increased rates of diagenesis and metamorphism at subduction zones and seawater Mg/Ca ratio is lowered through the preferential extraction of Mg by basalts at ocean ridges, as seawater is pumped through and chlorite and epidote are formed. Thus at times of high sea-level stand, pCO_2 can be expected to be higher and the Mg/Ca ratio lower, so that calcite should be the preferred marine precipitate. When rates of sea-floor spreading and subduction are reduced leading to a global low stand of sea-level, then pCO_2 and the Mg/Ca ratio will be lowered and raised respectively, favouring the precipitation of aragonite and magnesian calcite. Several other processes also affect pCO_2 and the Mg/Ca ratio and move them in the right direction. During low sea-level stands, increased evaporite precipitation preferentially removes Ca^{2+} so that Mg/Ca ratios are increased; weathering of silicates also releases more Mg^{2+}; increased erosion and dissolution of limestones and increased photosynthesis from more-extensive land plant cover both cause further lowering of atmospheric pCO_2. It can be clearly seen that geotectonic processes can have a marked effect on pCO_2 and the Mg/Ca ratio and the induced changes will be in concert with the major global sea-level curve. The directions of change in pCO_2 and Mg/Ca ratio through variations in rates of plate tectonic processes are consistent with those required to bring about the changes in the mineralogy of marine precipitates, with calcite at times of high sea-level stand and aragonite and high-Mg calcite at times of low sea-level stand.

Mineralogy of carbonate skeletons through the Phanerozoic

The skeletons of carbonate-secreting organisms are composed of aragonite, high-Mg calcite or low-Mg calcite; some consist of a mixture of two of these minerals. Many organisms exert a 'vital' effect which determines the skeletal mineralogy, irrespective of environmental conditions. In other cases, the mineralogy is controlled by seawater chemistry and water temperature. The latter is particularly important for magnesian calcites, such as those produced by echinoderms and calcareous red algae, which show a decreasing Mg^{2+} content with decreasing temperature. In warmer waters, some organisms have aragonitic rather than calcitic skeletons.

A consideration of the **mineralogy of carbonate skeletons** through the Phanerozoic (Figure 17.6) reveals important differences between the Palaeozoic and Mesozoic–Cenozoic. During the Palaeozoic, most benthic invertebrates produced skeletons of calcite; these include the rugose and tabulate corals, many stromatoporoids, bryozoans, echinoderms and brachiopods. Only certain molluscs had aragonitic shells. Interestingly, when **biomineralisation** (the ability of organisms to secrete hard parts) first developed in the latest Precambrian, many of the small shelly fossils were composed of aragonite. In many instances these fossils have been phosphatised, but careful examination shows the phosphate to be a replacement. The late Precambrian was a time of aragonite seas, as shown by the abundance of aragonitic ooids (now calcitised). However, just into the Cambrian, this situation changed and after a short period of bimineralic ooids, calcitic ooids are the norm. This change in mineralogy of the abiotic precipitates across the Precambrian–Cambrian boundary is matched by the change in biomineralisation from the dominantly aragonitic small shelly fossils and molluscs to the calcitic archaeocyathids, trilobites, brachiopods, echinoderms and calcified algae in the lowest Cambrian. The seawater chemistry change implied here of increasing pCO_2 and/or decreasing Mg/Ca ratio across the boundary reflects the onset of sea-floor spreading which was global in extent at this time. The development of new and extensive continental margins may have been important in the initial radiation of the metazoa (see Chapter 22), and the almost immediate development of biomineralisation may have been promoted by an excess of Ca^{2+} in seawater, a high seawater carbonate saturation, as well as the pCO_2 and Mg/Ca ratio changes. The occurrence of carbon isotopic excursions in boundary strata is consistent with phases of upwelling of deep nutrient-rich ocean water, enhanced organic productivity and organic matter burial. Higher seawater temperatures at this time may also have promoted organic evolution and skeletonisation.

In contrast with Palaeozoic skeletal limestones, Mesozoic and especially Cenozoic bioclastic sands are much more aragonitic through the dominance of aragonite-secreting molluscs, scleractinian corals and green algae. High-Mg calcite was precipitated by red algae and echinoderms, and some molluscs, such as oysters and scallops, as well as the brachiopods and many foraminifera, had low-Mg calcite shells. Thus, broadly, Palaeozoic tropical shallow-water bioclastic sands were dominantly calcitic, Mesozoic ones were a mixture of calcite and aragonite, and Cenozoic ones were more aragonitic. Modern temperature carbonate skeletal sands have a much higher calcite content from the common occurrence of red algae, echinoderms and barnacles, but high-latitude limestones are really quite rare in the geological record.

Mesozoic–Cenozoic pelagic deposits contrast with shallow-water limestones in being largely calcitic (low-Mg) through the predominance of coccoliths and planktonic foraminifera. The aragonitic pteropod (a floating gastropod) shells are only a minor component of Cenozoic pelagic oozes. As with the organisms evolving skeletons close to the Precambrian–Cambrian boundary, noted above, the low-Mg calcite mineralogy of the coccoliths and the planktonic foraminifera was very likely a function of the seawater chemistry at the time of their evolution (Jurassic), too. The marked contrast in skeletal mineralogy across the Permian–Triassic boundary from dominantly calcitic to dominantly aragonitic can also be attributed to a change in seawater chemistry. Calcitic rugose corals and brachiopods before the end-Permian extinction event are replaced by aragonitic scleractinian corals and molluscs after. There has been much discussion of the causes of the extinction event, which removed some 90 per cent of all marine species; rapid climatic changes, a sea-level fall and meteorite impact have all been suggested. However, after the event, when the new marine groups evolved, Triassic seawater had a high SO_4^{2-} content, elevated temperature and probably high Mg/Ca ratio, and pCO_2 would have been low through reduced rates of sea-floor spreading, so that aragonitic carbonate skeletons would have been greatly favoured over calcitic ones. Thus, it does seem that seawater chemistry at the time of an organism's evolution determines its skeletal mineralogy, and this is maintained, even when there is a subsequent change in seawater chemistry.

The temporal variations in skeletal mineralogy through the Phanerozoic are significant for the original composition of lime mud in fine-grained limestones, over which there has been much discussion. In modern shallow tropical seas, most of the lime mud is produced by the disintegration of carbonate skeletons, especially those of aragonitic green algae, such as *Penicillus*. The mineralogy of **micrite** (microcrystalline calcite) through the Phanerozoic could well match the trend in skeletal composition, particularly if most is of biogenic origin like modern lime muds. Although difficult to study, use of the scanning electron microscope enables micrites originally rich in aragonite to be distinguished from those largely of calcite, through the identification of minute relics of aragonite crystals.

The importance of the original mineralogy of lime mud is in its effect on the path of diagenesis. Aragonite-dominated muds have a much higher diagenetic potential than calcitic muds, through the presence of metastable grains which will be converted to calcite sooner or later during burial. Aragonite-dominated muds are much more likely to be subjected to dissolution and re-precipitation, leading to the formation of coarse microspar textures. Microporosity may also be much controlled by original mineralogy, and so this is also important in hydrocarbon reservoirs.

One particular aspect of biomineralisation which may have been controlled by seawater chemistry is the calcification of blue–green algae (cyanobacteria). They have a long geological record, evolving early in the Precambrian and being abundant today, but the calcified forms, such as *Renalcis, Epiphyton* and *Girvanella*, make a rather sudden appearance close to the Precambrian–Cambrian boundary and an equally dramatic exit in the late Cretaceous – early Tertiary. The calcification of blue–green algae appears to be mainly controlled by environmental factors, rather than vital effects, so that this temporal pattern may again reflect fluctuations in seawater chemistry. In this case, carbonate saturation state may be the controlling factor. Temperature is another possible control.

Diagenesis of limestone through time

As explained in previous sections, there is much evidence from abiotic marine precipitates and biomineralisation for subtle changes in seawater chemistry. These also account for the differences observed in the mineralogy and fabrics of marine cements through the Phanerozoic, discussed earlier, including fibrous calcite within synsedimentary cavities, drusy spar being precipitated on the sea-floor in the past, but not at the present, and aragonite cements only occurring at specific times through the Phanerozoic, notably the Carboniferous through Triassic. Since there have been changes in the mineralogy of carbonate sediments, it follows there should be variations in the processes that modify the sediments after they were deposited. These **diagenetic processes** are particularly important during the early stages of shallow burial or uplift and contact with meteoric water. The reason for this is that the carbonate minerals, aragonite, high-Mg calcite and low-Mg calcite, have different **diagenetic potentials**. Aragonite and high-Mg calcite are metastable and are converted to low-Mg calcite in time. Aragonite is commonly dissolved out, releasing $CaCO_3$ for cementation. High-Mg calcite loses its Mg^{2+}, and this may contribute towards local dolomite precipitation. Sediments dominantly composed of calcite thus have a low diagenetic potential, contrasting with sediments of mixed or largely aragonitic composition which have a high diagenetic potential. Mid-Palaeozoic bioclastic and ooid sands, and micrites would have had a lower diagenetic potential than the dominantly aragonitic Cenozoic carbonates. In the latter, grain dissolution and cementation should be much more extensive compared to the former. This should particularly be the case in near-surface meteoric environments where fresh waters preferentially dissolve out aragonite and the $CaCO_3$ is re-precipitated a little lower down in the sediment profile as calcite (low-Mg) when porewaters become supersaturated.

In near-surface diagenetic environments, the prevailing climate is very important. In the shallow-marine realm, sea-floor cementation and the formation of hardgrounds and abiotic lime mud formation are much more common in arid regions where slight to moderate hypersalinities promote carbonate precipitation. Thus along the Trucial Coast of the Arabian Gulf, an extremely arid region, inorganic precipitation of carbonate is widespread, contrasting with the more humid Bahamas and Florida regions. Evaporative dolomitisation can also be expected more commonly during more arid times, as in the Permian and Triassic. On the other hand, meteoric diagenesis moves much more quickly in humid areas, where heavier rainfall leads to more rapid carbonate dissolution and cementation, and more active mixing-zones of fresh and marine waters may lead to extensive dolomitisation.

The formation of limestone pavements, sink holes, caves, etc. is known as **karstification**. It is more important in humid than arid regions as is formation of fossil soils too. Through the geological record, one can expect to see great variations in the pattern of early diagenesis, both marine and meteoric, reflecting the palaeoclimate.

The early diagenetic history of a limestone does exert a great control on the later diagenesis, particularly with regard to the extent of compaction and pressure dissolution. Where rocks have been well lithified early, then fracture of grains and grain–grain dissolution are prevented during burial. Where little near-surface cementation has taken place, then many grains are fractured as overburden pressure increases (from tens to hundreds of metres), and sutured contacts form between grains due to dissolution at their contacts.

Most marine carbonate sediments deposited over the last few million years have been subjected to meteoric diagenesis through the dramatic falls in sea-level associated with the Great Ice Age. Thus most Pleistocene limestones have lost much original aragonite and there is extensive cementation by calcite spar. This has clouded many people's views of older limestones so that they believed that aragonite dissolution and calcitisation are near-surface phenomena and that calcite spar is only a meteoric precipitate. Now many limestones in the geological record were deposited when sea-level was rising or at a highstand, so that against background subsidence, thick sequences of limestone were gradually buried containing marine porewaters. Under these conditions, aragonite can be expected to survive until substantial depths, into the late diagenetic burial realm. There, evolving porewater chemistry may eventually lead to aragonite replacement and, in view of the quite slow fluid-flow and reaction rates at depths, especially compared to the near-surface meteoric environment, calcitisation of aragonite should be much more common than wholesale dissolution. The principle of uniformitarianism has to be applied even with care to burial diagenesis: since near-surface diagenetic reactions have varied over geological time, with near-surface meteoric diagenesis being important at some times and marine diagenesis at others, then burial diagenesis can also be expected to show variations.

One major type of limestone diagenesis is **dolomitisation**, whereby Mg ions substitute for Ca ions in calcite to produce a mixed Ca/Mg carbonate ($CaMg(CO_3)_2$). It has long been accepted that the abundance of dolomites increases back into the geological record, with dolomites being more abundant than limestones in the early Palaeozoic and Precambrian (Figure 17.11a). Several explanations have been put forward for this trend. It has been suggested that since the limestones are older, they will have had more time to come into contact with dolomitising fluids. The problem with this explanation is that most dolomitisation takes place relatively early in a limestone's history. Another possibility is the notion that dolomitising environments were more widespread in the past, or that there was something different about Precambrian seawater so that dolomite was precipitated more easily, either as a primary mineral or as a replacement. The origin of dolomites generally is still far from understood, and there are many models available. Recently, the accepted distribution of dolomites through time has been questioned and a new compilation of data (Figure 17.11b) shows that there are two maxima of dolomite abundance in the Phanerozoic, which broadly correspond to the high sea-level stands. This has been interpreted as another facet of the role of geotectonics in the hydrosphere–atmosphere system, with increased continental flooding resulting in higher atmospheric pCO_2, lower oceanic carbonate concentration and lower calcite saturation, all promoting dolomitisation by seawater.

Although there are many models available in the literature for dolomitisation, there is still much argument and little agreement on how the extensive platform dolomites are formed. In the last few years, the emphasis has turned to seawater and several new models are available for pumping it though carbonate platforms, since, after all, it is the efficient passage of the fluids through carbonate sediments which is the most important factor in dolomitisation. Seawater has an endless supply of Mg^{2+} ions. Ocean water being pumped into carbonate platform margins by tides and geostrophic currents (deep ocean currents caused by the Earth's rotation) and seawater circulation driven by rising geothermal heat under platforms are two current and popular ideas. The preferential occurrence of dolomites at high, global sea-level stands (Figure 17.11c), supports seawater as instrumental in dolomitisation.

Conclusions

The sedimentary record should be examined with an open mind and it should not be assumed that the present is always the key to the past. Studies of the Recent do help us enormously to appreciate the geological record, but it must always be remembered that things have changed. Plate tectonic processes have operated at different rates and there have been times of supercontinents as well as fragmented continents.

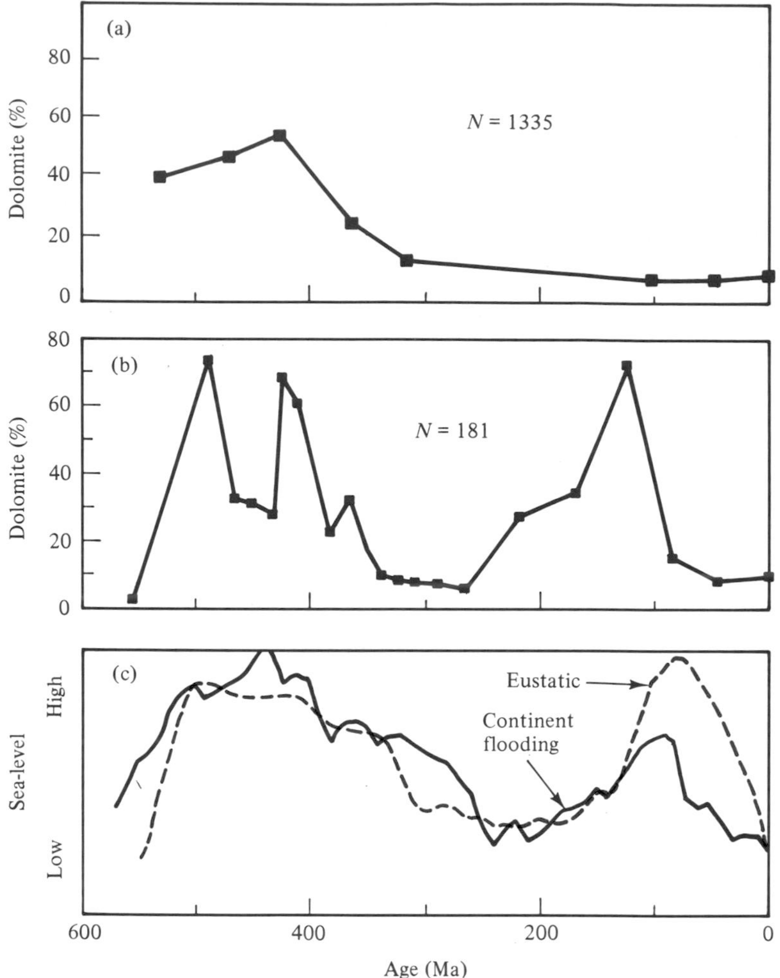

Figure 17.11 The Phanerozoic record of dolomites. (a) The long-held view of increasing dolomite with increasing age. (b) A new compilation of data showing that dolomite abundance correlates with high stands of global sea-level. (c) Changes in global sea-level and flooding of the continents during the Phanerozoic. □

Palaeoclimates have changed dramatically and commonly there is an underlying plate tectonic control here too. Evidence has been presented for subtle changes in seawater chemistry, evidence derived from simple observations down an ordinary petrological microscope. The development of carbonate skeletons and their mineralogy can also be linked to seawater composition and plate tectonic processes. Limestones hold many of the secrets of Earth history and continue to be an exciting and rewarding field of research.

Further reading

Three advanced texts dealing with carbonate rocks are listed below. They are best read after gaining some familiarity with these sediments, either through reading the appropriate parts of the books edited by Reading or Walker listed at the end of Chapter 16, or by consulting relevant chapters in: Tucker, M. E. (1991) *Sedimentary Petrology: An Introduction to the Origin of Sedimentary Rocks*, Blackwell Scientific Publications, 260 pp.

Scholle, P. A., Bebout, D. G. & Moore, C. H. (eds.) (1983) Carbonate depositional environments, *Memoir of the American Association of Petroleum Geologists*, **33**, 620–91.

Scoffin, T. P. (1987) *An Introduction to Carbonate Sediments and Rocks*, Blackie, Glasgow, 274 pp.

Tucker, M. E. & Wright, V. P. (1990) *Carbonate Sedimentology*, Blackwell Scientific Publications, Oxford, 480 pp.

Readers wishing to follow up, at a more advanced level, some of the topics introduced in this chapter, should consult the following references:

Given, R. K. & Wilkinson, B. H. (1987) Dolomite abundance and stratigraphic age: constraints on rates and mechanisms of Phanerozoic dolostone formation, *Journal of Sedimentary Petrology*, **57**, 1068–78.

Mackenzie, F. T. & Piggott, J. D. (1981) Tectonic controls on Phanerozoic non-skeletal carbonate mineralogy, *Nature*, **305**, 19–22.

Schneidermann, N. & Harris, P. M. (eds.) (1985) Carbonate cements, *Special Publication, Society of Economic Paleontologists and Mineralogists*, **36**, 379 pp.

Wilkinson, B. H., Owen, R. M. & Carroll, A. P. (1985) Submarine hydrothermal weathering, global eustasy and carbonate polymorphism in Phanerozoic marine oolites, *Journal of Sedimentary Petrology*, **55**, 171–83.

CHAPTER

18

Ralf Littke and
Dietrich H. Welte

HYDROCARBON SOURCE ROCKS

Introduction

The concept of source rocks is necessary in petroleum exploration since there is clear evidence that commercial petroleum accumulations did not form *in situ* in reservoir rocks. Most **reservoir rocks** are porous sandstones, limestones and dolomites that contain only small amounts of **primary organic matter**; i.e. finely dispersed organic matter of biological origin. Thus, oil and gas must be derived from other sources and migrated into reservoir rocks which were originally almost devoid of organic matter. Consequently, **petroleum source rocks** are defined as sedimentary rocks containing high amounts of insoluble organic material (kerogen) that can partly be transformed into liquid hydrocarbons during catagenetic heating. The term **catagenesis** describes the temperature/pressure regime in the buried sediments in which liquid petroleum is formed.

Effective source rocks are generally fine-grained sedimentary rocks with high amounts of hydrogen-rich kerogen. The classical source rock concept defines a petroleum source rock as either a siliciclastic sedimentary rock containing more than 0.5 per cent (by weight) **total organic carbon (TOC)** or a carbonate with more than 0.3 per cent TOC. This definition of a lower TOC-limit is based on empirical observations on TOC-contents in oil-bearing and non-oil-bearing basins of the Soviet Union by Soviet scientists. This lower TOC-limit for petroleum source rocks is related to the ratio of generated oil to the storage capacity of the rock: in fine-grained sediments with low amounts of organic matter, hydrocarbons will be generated during catagenetic heating just like in organic-matter-rich source rocks, but petroleum expulsion will not take place, because the storage capacity is not exceeded and conditions and processes necessary to initiate expulsion will not be fulfilled. Hence, when considering petroleum source rocks, the generation of hydrocarbons and their primary migration (expulsion out of the source rock) cannot be separated.

Source rock characteristics

Hydrocarbon source rocks are usually fine-grained shales, marlstones or limestones which appear green, grey or brown at immature stages but become dark or even black after the onset of hydrocarbon generation. Many source rocks display distinctive laminations as their most obvious sedimentary feature. Petrographically, they often contain pyrite in much higher concentrations than organic-matter-poor counterparts of similar lithology.

Recently, it has been shown that coals also act as effective sources not only for gas, but also for liquid hydrocarbons, although the efficiency and mechanism of expulsion out of impermeable coals certainly needs additional research work.

Some of the most important features of hydrocarbon source rocks with respect to their potential to provide commercial petroleum include:

- organic richness, usually measured as TOC;
- chemical composition of organic matter, especially hydrogen content;
- vertical thickness and lateral extent, i.e. the three-dimensional extension of the source rock;
- tectonic position and distance to permeable potential reservoir rocks in a basin;
- thermal history.

All but the last of these features depend on the environment in which source rocks were deposited.

Deposition of source rocks

The deposition of sediments rich in organic matter is restricted to certain boundary conditions. Such sediments are usually deposited in aqueous environments receiving a certain minimum contribution of organic matter of various origins. In subaerial environments, organic matter is easily destroyed by chemical and microbial oxidation prior to or shortly after deposition. The only notable exceptions to this rule are raised bog peats that are growing in humid climates and which are considered as important coal precursors. In raised bogs, the water-table is held at the sediment/air interface and the water is extremely acidic (pH 3–4), thus inhibiting microbial activity and remineralisation of higher plant tissues.

In aqueous environments, excellent source rocks with high concentrations of hydrogen-rich organic matter are most likely to be deposited under an oxygen-deficient water column (Figure 18.1). **Anoxic waters** are defined to contain less than 0.1 ml O_2/l water, whereas **oxygen-poor (dysaerobic) water** contains 0.1–1.0 ml/l. Such conditions do not occur at sedimentary surfaces that are affected by turbulent circulation; i.e. in fluviatile or beach environments or in very shallow shelf areas. Bottom-water oxygen deficiency is most likely to occur in stagnant seas and lakes, or in areas of high marine bioproductivity above deep shelf areas. Recent examples for these types of deposition of source rocks are, respectively, the Black Sea, Lake Tanganyika and the Peru and southwest African upwelling zones. Sediments deposited in these seas and lakes often contain more than 4 per cent TOC. However, the areas of highest organic matter concentration are usually restricted to depocentres in the individual basins (Figure 18.2) and not to the total area of deposition. The central parts of lakes and seas are often characterised by fine-grained clastic sediments. The fine-grained fraction of minerals and of particulate organic matter are of a low density and are carried away from water bodies with a high energy level to more quiet waters. There, deposition of fine-grained sediments limits the access of oxidising agents like oxygen (and sulphate) and microbial activity and therefore increases the chances for preservation of organic matter.

In modern seas, oxygen-deficient conditions in bottom waters favouring the preservation of organic particles are generally due to either restricted circulation, i.e. a restricted transfer of oxygen from surface waters to bottom waters, or to enhanced oxygen consumption below the surface waters. The first principal mechanism is found in partly silled seas, especially if waters of different densities occur within a basin without mixing (Figure 18.1a). One example is the Black Sea, in which surface waters are low in salinity and derive from freshwater runoff, whereas high-salinity marine water masses flow in from the Mediterranean. As mixing of the two different water masses is inhibited by their different physical properties, a stratified water body results in which oxygen-rich, low saline surface waters overlie saline bottom waters. The latter are anoxic due to oxygen consumption by decay of organic matter in the water and due to hydrogen sulphide (H_2S) production by microbial sulphate reduction in the sediments. Part of the hydrogen sulphide that is not fixed as iron sulphide (pyrite) escapes into bottom waters where reaction with free oxygen leads to its further consumption.

The second principal mechanism of enhanced oxygen consumption is restricted to areas of high primary productivity (Figure 18.1c). Bioproductivity is generally limited by nutrient and oxygen availability as well as by light energy necessary for the photosynthetic process. Therefore, ideal conditions for high plankton productivity occur in low-latitude areas with intense sunlight, in which nutrients, especially phosphorus, are provided either by upwelling of cold bottom water or by freshwater runoff. The resultant great mass of plankton can only partly be oxidised in the water and organic-matter-rich sediments are deposited below oxygen-poor dysaerobic bottom waters. Recent examples for this kind of sedimentation are the upwelling areas offshore the coasts of Peru, Morocco, Namibia and Oman, where oxygen-deficient water masses are in contact with deep shelf areas (100–1000 m water depth).

In recent sediments, oxygen deficiency seems to be a prerequisite for the deposition of excellent hydrocarbon source sediments in which organic matter can be preserved that otherwise is extremely labile in oxic environments. It is reasonable to assume that similar mechanisms also led to the formation of ancient source rocks, especially since the establishment of an oxygen-

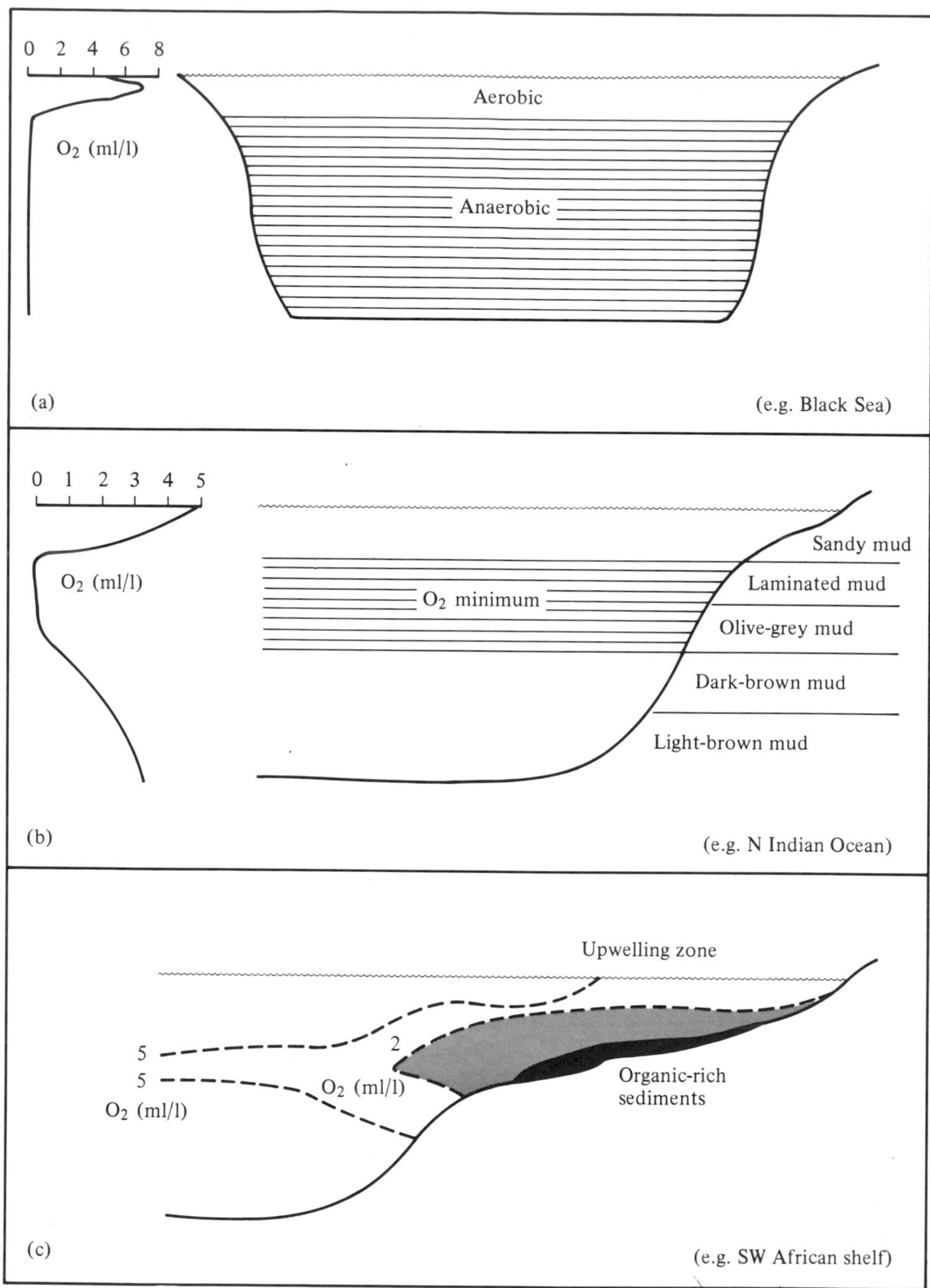

Figure 18.1 **Generalised models for the deposition of organic-rich sediments. (a) Anoxic silled basin, (b) oxygen-minimum zone impinging on the continental margin, and (c) upwelling region.** □

rich atmosphere. The well-laminated nature of most ancient and recent source rocks is due to a lack of bioturbation and current activity, both attesting to anoxic or at least dysaerobic conditions at the depositional surface. Less organic-matter-rich but still effective source rocks are deposited in more oxic environments, where the preservation of organic matter depends on sedimentation rate, i.e. at higher sedimentation rates organic particles are more rapidly buried and better preserved. As an example, turbidite layers often contain significantly more organic matter than pelagic sediments in the same oceanic realm.

Pelagic, deltaic and fluvial sediments generally do not contain much organic matter derived from aquatic organisms. However they often contain sizeable quantities of tissues of higher plants which are less easily destroyed by oxidation and bacterial degradation. This kind of organic material is less hydrogen-rich and less oil-prone than the aquatic type.

In summary, balanced optimum conditions between the energy level in a body of water, the supply of nutrients and sedimentation rate are required for deposition of first-class hydrocarbon source rocks. Less oil-prone but still effective source rocks may, however, be deposited in a variety of sedimentary environments. Near-continent silled basins or vast shelf areas are very well suited for source rock deposition, especially in view of the fact that possible reservoir rocks may be expected to occur in these same environments. Deltaic or beach sands or reefs are often deposited in the same type of basin and just separated from source rocks by a few kilometres, or a few million years in time.

Kerogen formation

Ever since photosynthesis emerged as a worldwide phenomenon on the Earth's surface about 2000 million years ago, the mass production of organic matter was initiated as a global process. However, most of this reduced (organic) carbon of biological origin is destroyed by oxidation and eventually converted back to carbon dioxide or precipitated as carbonates. Only a small portion survives in the reduced form to become fixed in sediments (Figure 18.3). The upper limit of the preservation rate of organic carbon found in certain oxygen-deficient sedimentary environments most favourable for deposition of petroleum source rocks is about 4 per cent with respect to the original biological production.

This organic material which is finely disseminated in sedimentary rocks and which escaped destruction by oxidation is termed **'kerogen'** if it is insoluble in organic solvents. Chemically, it is to a great extent derived from complicated macromolecules present in

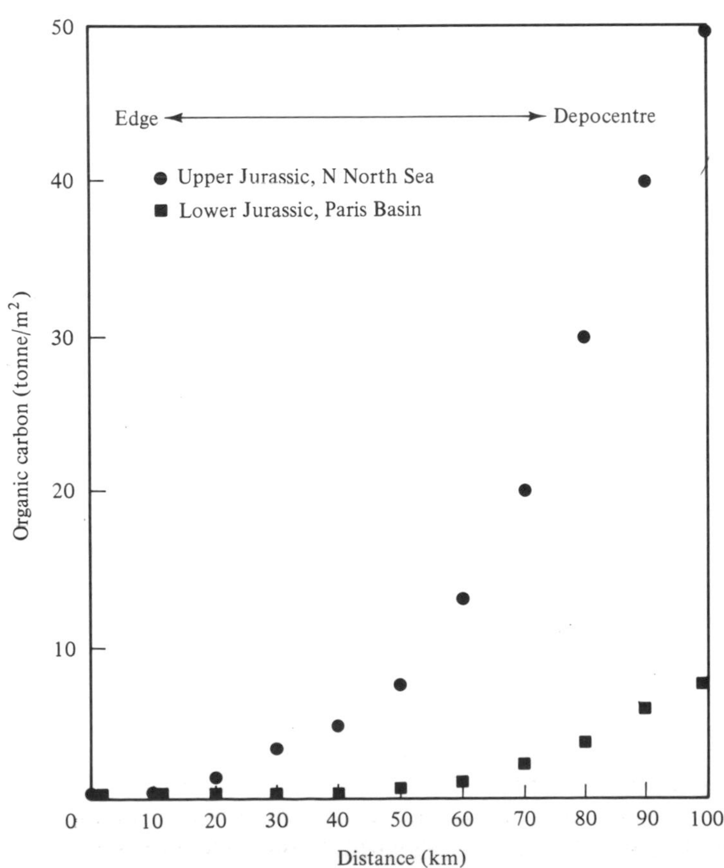

Figure 18.2 Lateral variation of accumulated organic carbon in Jurassic source rocks of the Paris Basin and the northern North Sea showing a common basinward increase. Numbers on the vertical axis are tonnes of organic carbon situated below one square metre in the respective sedimentary basin.

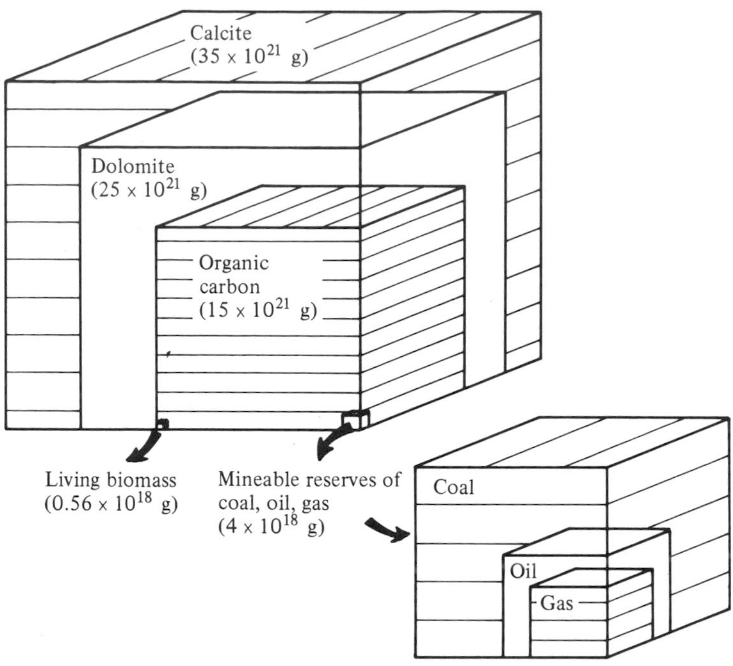

Figure 18.3 Schematic representation of masses of carbon fixed in carbonates, as organic carbon trapped in sediments, in the living biomass, and as mineable reserves. □

the lipid or lignin-rich fractions of biomass that form many resistant parts of organisms such as membranes, inner cell walls of woody material, cuticles, spores, pollen etc. These plant parts tend to be incorporated in sediments as particulate rather than dissolved organic matter. Much of the kerogen is therefore visible as microscopically small particles (macerals). Their size varies between 0.002 and 1 mm (or more) and they can often be attributed to precursor groups like wood or algae. Dependent on the vastly different chemical composition and structure of precursor materials, kerogen can be cracked to liquid or gaseous hydrocarbons or remain 'inert' during catagenesis. In organic geochemistry, the term **diagenesis** is restricted to the uppermost sediments, in which almost no thermally induced petroleum generation takes place, whereas **catagenesis** describes the temperature/pressure regime in which liquid petroleum is formed (usually at temperatures exceeding 80 °C).

A significant transformation of biologically synthesised organic matter takes place during diagenesis. This is evident from reactions leading to a loss of parts of the molecules (defunctionalisation), an addition of hydrogen (hydrogenation), or structural changes such as isomerisation and aromatisation of biological markers; i.e. of organic molecules with structures which are presumed to be characteristic of specific groups of plants. Also, tissues of higher plants which are embedded in sediments experience a transformation during early diagenesis, changing, for example, cellulose into other compounds like humic acids. This early diagenetic, *in situ* transformation is most evident from the obscured cell structure of buried wood. Of special importance is the early diagenetic incorporation of sulphur derived from hydrogen sulphide into organic molecules. Hydrogen sulphide (H_2S) is produced by sulphate-reducing, organic-matter-consuming bacteria in the uppermost sediments. Whereas the bulk of the H_2S produced is released into the overlying water or fixed as FeS_2 (pyrite), the sulphide-sulphur may also be incorporated in organic molecules. The process is quantitatively important if H_2S production is high and only little iron is available for pyrite formation. The resultant sulphur-rich kerogen is extremely susceptible to thermal stress and will become the source for sulphur-rich petroleum generated at low temperatures. Thus, a great number of biologically synthesised organic compounds undergo a significant chemical restructuring after being embedded in sediments. Among the diagenetically generated molecules only those which are either water-insoluble or protected by other resistant compounds like lignitic cell walls are able to survive as kerogen. Water-soluble compounds like amino acids and sugars are largely washed out of the sediments.

On the other hand, there is ample evidence that some biologically synthesised plant parts are not significantly changed during low-temperature diagenesis. For example, several genera of green algae are well-preserved. Their original morphology and microscopic structure only seem to be affected by compaction. These primarily hydrogen-rich particles usually contain various classes of saturated compounds, which protect them from chemical and biological degradation. These compounds are thought to remain almost unchanged during diagenesis and to become direct precursors of liquid petroleum in the course of catagenetic cracking reactions at higher temperatures. Hence, kerogen consists of a physical mixture of diagenetically *in situ* restructured biomass as well as of preserved biosynthesised compounds.

Part of the kerogen in immature sediments is labile and converted to liquid and gaseous products upon catagenetic heating, whereas the rest remains inert or is aromatised, finally grading into residual, graphitic carbon. A first and very useful classification of kerogen necessary to predict its petroleum generation potential is based on elemental composition, more precisely on the **hydrogen/carbon** and **oxygen/carbon ratios** (Figure 18.4). In recent years, additional classification schemes have been developed to describe organic facies. For example, pyrolytic methods allow an estimation of the products that may be generated from a specific kerogen, and petrologic studies give information on the variability of organic particles and their spatial distribution in source rocks. These particles are given a variety of names according to their origins. **Liptinite** (syn. **exinite**) is derived from freshwater or marine algae and land plants, and is hydrogen rich. It is subdivided into different types according to its derivation: **alginite** (from algae), **sporinite** (from spores), **cutinite** (from cuticles) etc. **Vitrinite** is derived from the ligno-cellulosic tissue of land plants; it is relatively unaltered, and is oxygen rich. **Inertinite** is carbon rich, due to the fact that it is formed by the oxidation of land plant materials in soils, during transport, or by forest fires.

The bulk of the organic matter in sediments and sedimentary rocks consists of finely disseminated kerogen particles. Kerogen is the most abundant naturally occurring organic material in the Earth's crust. As compared to kerogen, all known fossil fuels in economic deposits, being composed of gas, oil, tar sands and coals, are altogether several orders of magnitude lower in absolute amounts than kerogen. Thus compared to the global distribution of organic carbon in the Earth's crust, economic deposits of oil, gas and tar sands are rare anomalies rather than the rule. The occurrence of these anomalies does not only depend on the presence of good source rocks, but also on the right level of maturation and on the availability of migration pathways and traps. In other words, ideal geological conditions are required for the accumulation of petroleum; it is predicting and locating where such conditions have occurred that makes oil and gas exploration such a difficult exercise. However, it should be emphasised that on a worldwide basis there are still very large petroleum accumulations to be discovered. Frontier areas for future exploration are regions with poor infrastructure such as arctic and antarctic areas as well as deeper parts of already explored basins.

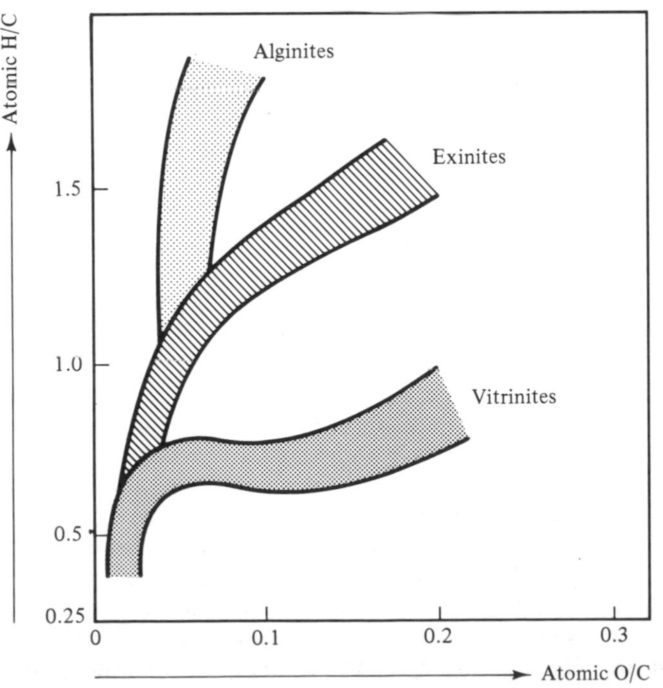

Figure 18.4 Atomic ratios of hydrogen over carbon and oxygen over carbon for three microscopically defined kerogen types. Vitrinites are mainly derived from wood, exinites from spores and pollen and alginites from algae. With increasing maturity, they evolve towards a carbon-rich residue. Plots such as this are sometimes referred to as van Krevelen diagrams, after their inventor. □

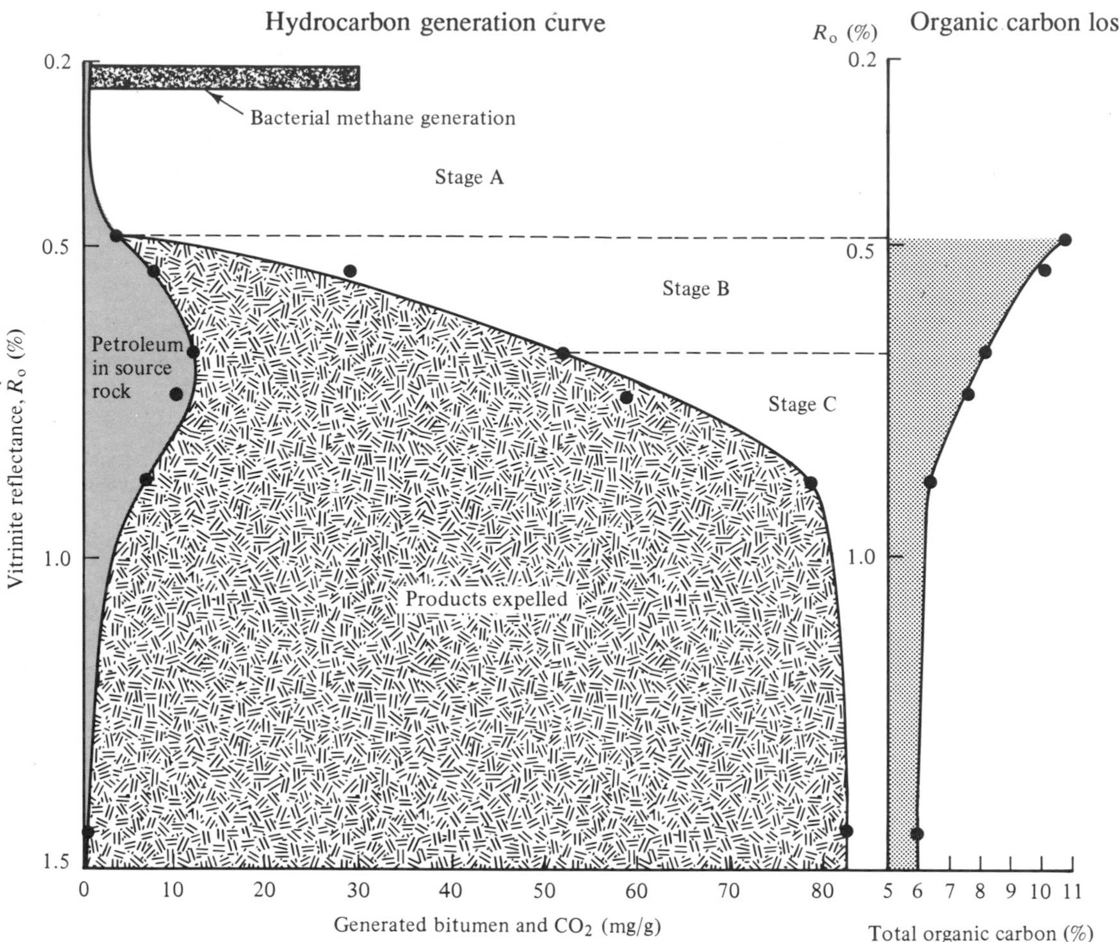

Figure 18.5 **Hydrocarbon generation curve and organic carbon loss for a series of Lower Jurassic source rocks in the Lower Saxony Basin. With increasing vitrinite reflectance (measured as the proportion of light reflected from organic matter as observed under a special microscope), total organic carbon decreases due to the conversion of kerogen into liquid and gaseous products and their expulsion. At immature Stage A, in which little oil is generated, no expulsion takes place. During mature Stage B, great amounts of liquid hydrocarbons are generated and the concentration of petroleum in the source rock is increasing. At mature–overmature Stage C, the generation of liquid hydrocarbons is exceeded by cracking from liquid to gaseous products. The concentration of bitumen in the source rock decreases due to the easier expulsion of gas.** □

Maturation and hydrocarbon generation

Oil and gas are formed by the thermal conversion of kerogen in fine-grained sedimentary rocks. With increasing burial, the temperature in these rocks is elevated, and above a given (threshold) temperature the finely disseminated kerogen begins to transform into petroleum compounds due to its inherent chemical instability at high temperatures. The generation of oil and gas in source rocks is a natural consequence of the increase of temperature over geologic time. The process of organic matter transformation with increasing temperatures is called **maturation**. With respect to petroleum generation, organic matter may be immature (= premature), mature or overmature (Figure 18.5). The reactions involved are irreversible.

In order to describe the maturation process, different **maturation parameters** like **vitrinite reflectance**, spectral fluorescence of alginite and sporinite, or ratios of specific biomarker molecules have been established. They allow a definition of the state of maturity of organic matter in a specific sedimentary rock. Although these parameters are temperature- and time-sensitive, contrary to a common belief they do not describe the wide variety of processes of petroleum generation rigorously. During petroleum generation a great number of different reactions take place. Therefore, petroleum generation is not restricted to one single temperature threshold, but extends over a range of temperatures. The reactions responsible for the change of the bulk maturity parameters mentioned above (vitrinite reflectance, fluorescence) are, however, different from those responsible for petroleum generation. In contrast, only a few reactions are involved in the change of biomarker ratios. In other words, there are a range of activation energies for petroleum generation as a bulk process which cannot accurately be described by the changes of specific maturity parameters that obey different laws of reaction kinetics. The usefulness of maturity parameters in predicting petroleum generation is further limited by the significant differences in kerogen structure and composition in different source rocks. For example, sulphur-rich kerogen will be cracked at much lower temperatures than sulphur-poor kerogen. Hence, rather precise knowledge of the chemical composition and structure of kerogen in a specific source rock is required to obtain reasonable information on temperature history and timing of petroleum generation.

The chemical and physico-chemical processes controlling the increase in vitrinite reflectance and the generation of hydrocarbons as a bulk process obey different reaction kinetics. One important consequence of this is that both temperature and time strongly influence petroleum generation. This is exemplified in Figure 18.6. At low heating rates, petroleum generation from a given kerogen will start at relatively low temperatures and extend over a long period of time until high temperatures are reached. At high heating rates, petroleum generation also starts at low temperatures but extends over a narrower time span. In the early 1960s, Jüntgen and coworkers studied the temperature and time-dependent release of gas (methane) from coals using experimental pyrolysis. During this procedure a sample was heated at a well-defined heating rate in an inert atmosphere to high temperatures and the gas generation was recorded. In more recent years, similar analytical devices were used to study petroleum generation from immature source rocks. This experimental procedure offers the principal advantage over the use of maturation parameters that petroleum generation can be predicted or reconstructed more precisely and within the framework of such important variables as type and amount of organic matter and temperature history. At the present time, the most sophisticated kinetic models describe petroleum generation from a specific source rock by a range of activation energies and frequency factors, considering the cracking of kerogen to oil and of oil to gas. In this context, however, it should not be forgotten that heating rates during laboratory pyrolysis

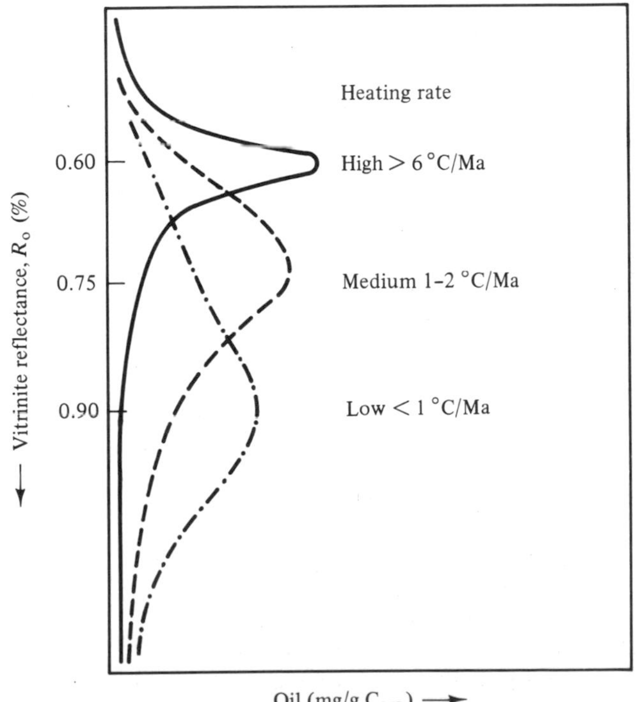

Figure 18.6 Shift of the oil generation peak on the maturity scale as measured by vitrinite reflectance due to different heating rates calculated by utilising a kinetic model for the generation of oil from kerogen.

experiments usually vary only between 0.1 and 25 °C/minute and are about 10–12 magnitudes lower under natural geological conditions (Figure 18.6). As time for petroleum generation in experiments is much shorter, temperatures are much higher, usually in the range between 300 °C and 500 °C instead of between 60 °C and 150 °C during natural petroleum generation. The reliability of any kinetic data derived from experiments clearly depends on whether natural petroleum generation obeys the same principal reactions as experimental maturation in the laboratory. At the present time, exploration concepts should include maturation assessment using *both* maturation parameters selected according to the specific geologic condition in a sedimentary basin *and* experiments on petroleum generation from possible source rocks in the study area.

The quantity of petroleum that may be generated during maturation is certainly one of the most intriguing and critical features of source rocks. It depends on the total amount of organic matter deposited as well as on its inherent chemical composition, especially on the hydrogen/carbon (H/C) ratio. Therefore, a plot of H/C versus O/C is generally used for a first geochemical characterisation of petroleum source rocks. This classification provides a method of early recognition of kerogens with a high hydrocarbon generation potential – indicated by high H/C ratios – and thereby also a recognition of source rocks capable of generating large quantities of petroleum. A real quantification of petroleum generated in a mature source rock sequence under natural geologic conditions can rarely be achieved, because a certain quantity or fraction of the products will have been expelled. Here, a special problem is posed by the more volatile products, especially for methane. A reconstruction and mass balance can, however, be performed for regions where overmature to immature source rocks with a more or less identical kerogen and similar average amounts of it are found and thoroughly analysed. Such an approach for Lower Jurassic shales in the Lower Saxony Basin revealed that about 43 per cent of the mass of the primary kerogen was converted to oil and gas and that an additional 14 per cent was converted to carbon dioxide (Figure 18.5). The rest remained as inert residue in the source rock. This degree of thermal conversion is thought to be representative of many high-quality source rocks. A significantly higher degree of conversion can only be expected for rare, extremely hydrogen-rich source rocks. Terrestrial organic matter as fixed in coals and other low-quality source rocks will produce much more carbon dioxide and less oil and gas, but the total organic matter conversion rate is also high. As most of the products generated are lost from source rocks by migration, total organic carbon contents of overmature sources are drastically lower than those of their immature equivalents.

Oil versus gas generation

With increasing depth of burial and temperature in a subsiding sedimentary basin, dispersed and concentrated organic matter, i.e. kerogen and coals, are both transformed into gases and liquids and a residual solid phase. An almost total conversion into a fluid phase, e.g. into liquid petroleum, can hardly be expected but liquids may constitute the main product for rare, primarily extremely hydrogen-rich varieties of kerogen like those derived from botryococcus-algae. The fate of the solid organic material is well established in **coals**. Peats grade subsequently into oxygen-rich lignite, sub-bituminous coal, bituminous coal, anthracite and finally graphite (which consists practically of pure carbon). In very much the same way, kerogen is eventually converted into a system of condensed carbon; i.e. condensed sheet-like aromatic systems and later into a graphitic structure, free of any oxygen and hydrogen.

The fluid products generated in a source rock at elevated temperatures consist of a variety of compounds including aromatic hydrocarbon rings and aliphatic hydrocarbon chains of vastly different molecular masses. Among these, the lightest compounds with only one to five carbon atoms – e.g. methane (CH_4), ethane (C_2H_5) etc. – are gaseous at surface pressures and temperatures. Knowledge of the products generated in a specific source rock is of utmost importance for any exploration strategy; e.g. the gas to oil ratio in a sedimentary basin is critical for its economic value.

Ratios of gas to oil generated in a source rock depend on the heating history, the degree of maturation, and on the composition and structure of organic precursor material. Generally, the gas to oil ratio increases with increasing maturity of a source rock. A notable exception to this rule is microbially produced gas, formed by methanogenic bacteria at the earliest stages of diagenesis; i.e. in immature sediments prior to the thermal conversion of kerogen.

The increasing concentration of methane in deeply buried source rocks is due to the fact that methane is the thermodynamically most stable form of reduced carbon in the Earth's crust. Only at rather shallow depths and low temperatures, i.e. at the early stages of thermal conversion of kerogen, are high-molecular weight, less stable hydrocarbons able to persist. At greater depths and temperatures, these liquid hydrocarbons are eventually converted into gas and a solid residue. This conversion can be explained using the typical hydrocarbon generation curve (Figure 18.5). Until a certain minimum maturity is reached, little oil will be generated (Stage A). During Stage B, more and more hydrocarbons are generated, and the concentration of soluble organic matter will increase significantly. At this stage, the soluble organic matter is a mixture of many different hydrocarbon molecules and additional organic molecules containing nitrogen, sulphur or oxygen atoms in minor amounts (N,S,O-compounds = heterocompounds). With increasing maturation, the ratio of hydrocarbons over N,S,O-compounds increases as well as the ratio of low-molecular-weight hydrocarbons over high-molecular-weight hydrocarbons. At Stage C, the generation process of liquid hydrocarbons is exceeded by a destructive process, i.e. liquid hydrocarbons are converted by cracking into gaseous hydrocarbons of which the end-member is methane.

The gas to oil ratio depends not only on maturation, but also on the chemical composition and structure of kerogen in a source rock. In a very general and simplified view, terrestrial-derived kerogen is more gas-prone, whereas kerogen derived from marine or lacustrine algal precursors is more oil-prone. The latter type usually contains numerous long-chain molecules which are ideal precursors for liquid oil, whereas immature kerogen derived from higher land plants mainly consists of aromatic and non-aromatic rings with attached methyl-groups which act as precursors for gas. Experimental results support this theory: coaly, terrestrial-derived material generates mainly hydrocarbons in the C_1–C_5 range (gas), whereas marine and lacustrine source rocks predominantly generate higher-molecular-weight hydrocarbons in the C_6–C_{14} range. Thus, there exists a precursor/product relationship which is extremely complex, because product (oil and gas) composition is not only controlled by the type of precursor material, but also by its degree of maturation. Also, the question whether a source rock is an open or a closed system affects petroleum composition, for oil generated but not expelled will participate in further reactions in the source rock.

Primary migration

The release of petroleum compounds from solid organic particles (kerogen) in source beds and their transport within and through the capillaries and narrow pores of a fine-grained source bed has been termed **primary migration (expulsion)** and separated from **secondary migration** through wider pores in more permeable carrier beds and reservoir rocks. The major driving forces for primary migration are pressure gradients and diffusion. Secondary migration is driven mainly by buoyancy, with a resultant general upward movement of oil and gas. During primary and secondary migration, diffusive transport takes place, especially in the case of the more-water-soluble compounds of low molecular weight such as methane. Diffusion always follows concentration gradients. In this respect, diffusion in general terms is a process negatively affecting secondary migration and hydrocarbon accumulations. Rates of diffusion are low for most hydrocarbons (except for gases) but may nevertheless be of importance for liquid petroleum migration and accumulation at elevated temperatures. Diffusion of gases is certainly an effective process in primary migration over short distances. One effect of diffusion-controlled migration is the preferential loss of gas, whereas other compounds remain in the source bed. This fractionation effect is due to great differences between diffusion coefficients of specific organic molecules controlling the rate of diffusion.

The major process in primary migration in the subsurface is pressure-driven hydrocarbon phase movement. Any petroleum phase in the narrow pores of mainly water-wet fine-grained source rocks has to overcome capillary pressures; i.e. the non-wetting oil phase when moving through the rock has to overcome the resistance of very narrow pore throats which are usually filled with water. Only a minor part of the pressure driving the petroleum phase through the narrow pores is caused by buoyancy due to density differences between water and petroleum. The less dense petroleum will move upward relative to water, thus creating a pressure against capillary force. The effectiveness of buoyancy depends on the density difference between petroleum and water as well as on the size of oil globules which are small in source rocks. A more effective source for pressure in source rock pores is created by the conversion of solid kerogen into liquid and gaseous bitumen causing a volume expansion of the fluid phase. The effectiveness of the latter mechanism depends on the depth in which hydrocarbon generation occurs. If the temperature of maximum petroleum generation is reached at a great burial depth (due to low heating rates), petroleum will be generated as liquid oil and create relatively small additional pressures. At shallower depths and higher heating rates, hydrocarbon generation will produce more gas that is not compressible and leads to great additional pressures. Another source for additional fluid-pressure in fine-grained rocks is the expansion of water at high temperatures and the dewatering of clay minerals. Generally, dewatering of compacting shales is inhibited by their low permeability and total compaction is much stronger in organic-rich source rocks than in organic-lean shales, because part of the solid framework is lost due to the conversion of kerogen to oil and gas. The resultant hydrostatic overpressures may be sufficient to fracture the source rock, thus creating new avenues for petroleum migration. Thus the effectiveness of expulsion is directly related to organic richness and hydrocarbon generation potential. In high-quality source rocks, generation of large amounts of petroleum and conversion of solid kerogen into liquid and gaseous products will cause further compaction and pore pressures sufficient to overcome capillary forces. Hence the expulsion of liquid oil can occur after a certain minimum saturation of the shale pore system (and the kerogen) has been reached (Figure 18.5). In lean source rocks, only small amounts of oil can be generated. Thus, the critical minimum saturation is not reached and oil remains trapped until it is cracked to gas which can escape by diffusion.

Fractionation effects are generally not expected to occur during pressure-driven separate phase flow of either oil or gas, but will affect the composition of expelled products if flow occurs as a solution of oil in gas. The mechanisms which are effective during petroleum expulsion can be deduced from a comparison of the molecular composition of an oil in a reservoir with the composition of extracts obtained from its source rock. Another possibility is the comparison of the molecular composition of extracts from edges of source rocks with those from central parts, where expulsion has been less effective. These studies reveal that expulsion is quantitatively more effective in source rocks adjacent to porous sediments or fractures and that depletion of aliphatic hydrocarbons is sometimes influenced by fractionation effects thus endorsing the theory that solution of oil in gas is of importance during primary migration.

Conclusions

Petroleum expulsion, like source rock deposition, kerogen formation, and petroleum (oil and gas) generation, is an extremely complex dynamic process involving a variety of mechanisms. A better understanding of all these processes affecting the formation and subsequent modification of source rocks will lead to greater success in the search for oil and gas.

Further reading

The four references below provide more detailed treatments of the topics covered in this chapter; that by Selley contains chapters that provide reading intermediate between the other references and this chapter.

Cornford, C. (1990) Source rocks and hydrocarbons of the North Sea, in *Introduction to the Petroleum Geology of the North Sea*, ed. K. Glennie, Blackwell Scientific Publications, Oxford, pp. 294–361.

North, F. K. (1985) *Petroleum Geology*, Allen Unwin, London, 607 pp.

Selley, R. C. (1985) *Elements of Petroleum Geology*, W. H. Freeman Company, New York, 449 pp.

Tissot, B. P. & Welte, D. H. (1984) *Petroleum Formation and Occurrence*, Springer-Verlag, Berlin, 699 pp.

The following paper provides an excellent summary of the depositional environments of source rocks:

Demaison, G. J. & Moore, G. T. (1980). Anoxic environments and oil source bed genesis, *American Association of Petroleum Geologists Bulletin*, **64**, 1179–1209.

CHAPTER 19

Adolf Seilacher
Universität Tubingen

EVENT STRATIGRAPHY: A DYNAMIC VIEW OF THE SEDIMENTARY RECORD

Introduction

Layered rocks are the book in which the history of our planet and its biosphere is written down. Unfortunately this album is very incomplete. Whole chapters are missing in any local section, either because no sedimentation took place in that particular period, or existing rocks were subsequently eroded away. The missing parts have to be inserted from other sections, with index fossils or radiometric dates used for correct pagination. Through the efforts of many generations of stratigraphers we now have a compound, but fairly complete geologic column, subdivided into eras, series, stages and zones and set into a framework of radiometric dates (see Chapter 7).

Resolution of the post-Jurassic stratigraphic record has also been considerably increased by the recent drilling of ocean floors (where sedimentation is slow but most continuous) and by the additional use of isotopic signatures for climatic curves (see Chapter 24) and of paleomagnetic reversals for a fine-tuned time scale. Unfortunately, the subductional recycling of oceanic plates into the melting pot of the mantle – rather than the length of drill cores – limits the backward reach in time of deep sea stratigraphy. For older rocks, discussion is still going on as to the correct correlation between different biogeographic provinces and facies realms (land, shallow and deep sea) and as to where the 'golden spike' marking a stratigraphic boundary should be most adequately put.

Event or dynamic stratigraphy has a different focus. It is not so much concerned with piecing together chapters and pages than with the less obvious, but inherent, biases in the stratigraphic account. It claims that – just as records of human history have always dwelt on revolutions and wars rather than the daily lives in between – the stratigraphic record overemphasises extraordinary perturbations. In other words, event stratigraphy is more interested in the qualitative than in the quantitative inadequacies of the geohistoric record. Since biases are caused by sedimentary processes, one should expect them to be independent of geologic age. In the present chapter I attempt to show that this is not so. We shall also see that a lot of information can be retrieved from the detailed study of the minor gaps that make up most of the stratigraphic record.

Sedimentary characteristics of individual event beds

What is an event?

In a broad sense, any change could be called an event. Still there are differences in accentuation and time scale. Daily tides – however consequential they may become for a tourist hiking over the tidal flats – are too minor and too regular to be felt as perturbations by indigenous populations, human or otherwise. They are also too short-spanned relative to general sedimentation rates to be stratigraphically registered. They simply fall below the sensitivity of the recording sedimentary system.

As another extreme, ice ages have been among the major and most consequential physical changes in Earth history (bolide impacts are another extreme!) Still they would not be classified as events in a stratigraphic sense, because they happen too gradually to show up as punctuations in the sedimentological or in the biological protocol. So what we mean by a stratigraphic event is equivalent to what we would call 'catastrophy' in the perspective of human life spans, implying suddenness and unpredictability – only in dimensions adapted to geologic time scales.

The nature of such physical perturbances is manifold. Earthquakes, volcanic eruptions, storms, floods and landslides make headlines in our newspapers. Devastating meteorite impacts are rarer and may not appear over generations. In an underwater world we should also add submarine landslides and the gravity flows and turbidity currents into which they develop.

What these events do have in common is that they are random. They also come in various sizes, without defined lower or upper limits, but ever more rarely towards the upper end of the scale. This is important because sediments also accumulate at varying rates depending on the depositional environment. Accordingly they filter out different parts of the event spectrum. For instance, a sediment pile may form in too short an interval to experience a particular kind of event. Alternatively, slow sedimentation may reduce the sensitivity of the system towards smaller events. In reality, however, the lower end of the window is usually controlled less by reactivity than by the overshadowing effect of larger events. To better understand this relationship, we must go into some details of the sedimentary processes that make the protocol. In doing so, we shall focus on subaqueous sediments because they form the major and most telling constituents of the stratigraphic record. To reduce confusion, we shall also concentrate on current-generated events and leave the wave events for a later section.

Erosional features

Except in the very centre of a meteorite impact or an earthquake, any subaquatic event would be locally registered as a more or less symmetrical bell-shaped energy (or rather turbulence) curve.

As long as turbulence levels rise, erosion will take place in forms that differ depending on energy level, nature of substrate, event duration and turbidity of the water. On a muddy substrate, it will proceed down to a level in which the sediment is compacted enough to withstand erosion – unless time is too short to reach equilibrium before the depositional phase commences. Suspended sand may push this limit because of its abrasive effect. In a unidirectional current sand abrasion leaves its signature in the form of flute casts – sedimentary structures that reflect the self-enhancing erosion by little eddies. Flute casts are always preserved in an unblurred fashion, because they become buried by the suspended sand the moment erosion stops. Shell debris and other larger particles carried in the current may also intensify erosion. The result is the production of various tool marks; but these, again, are preserved in sharp detail only because they were carved after a thin veneer of sand had already been deposited. In summary, the erosional markings observed as casts on the sole of an event bed tell us about the intensity and direction of water movement at the peak of the event. What we see, however, are largely intrasedimentary 'under-markings' that would not be exposed on modern sea bottoms or in flume experiments.

Depositional features

What happens during the waning phase can be demonstrated in a stirred glass cylinder: the suspended sediment settles in a graded fashion: larger (and denser) particles first and finer ones later. Accordingly a typical event-bed is graded with respect to grain sizes. But even if it consists of sand or silt too well sorted for this kind of grading, waning turbulence can be expressed by the **Bouma Sequence**. This term describes a succession of bed forms that reflect, first, deposition from a moving sheet of sand (even lamination), followed by a zone in which individually moving grains produced ripple cross-bedding.

In the common case that the available sediment has, during the more daily processes, been pre-sorted into a sand and a clay fraction, event sedimentation is not only graded, but also discontinuous. The interval of non-sedimentation allows the sand surface to be moulded into regular ripples, whose relief is then preserved by the muddy fallout during the same event and before burrowing animals had a chance to scar it extensively.

Another characteristic of event-deposited sands and silts results from their very rapid sedimentation: the unusually large amounts of water that are initially trapped between the loosely packed grains allow the sand to deform almost like a fluid under the tractional forces of a new surge of shooting sand (overturned cross-lamination; restricted to flood sands, see below) or of localised upward dewatering (convolute lamination).

These and many other details of sediment structures may appear to be far away from our global task of understanding the Earth. But it is by such features that we can reconstruct and distinguish past processes right in the field.

The details also convey a more general message: event deposits do not simply pile up on top of each other. By erosive phases they are highly cannibalistic and allow only the strongest and latest events to leave their signatures. This is why the stratigraphic record is full of gaps. It also looks overly dramatic, and the more so if only a few pages are provided for a given time interval – just like in history textbooks.

How to distinguish different kinds of events

After having emphasised the characteristics of event sedimentation in general, we should now pass to the distinctive features of different kinds of event deposits. With regard to sedimentary structures, the most basic divergence is between current and wave events (Figure 19.1). The deposits of floods (**inundites**) and turbidity currents (**turbidites**) occur from unidirectional currents. This is reflected in the uniform orientation of flute casts and tool marks at the base, of cross-lamination (often 'climbing') inside the bed and of asymmetric (commonly linguoid) current ripples at the top. Wave action, on the other hand, translates at the bottom into oscillating currents (as we may experience when diving in the surf zone!). This to and fro motion is expressed by bidirectional tool marks at the erosional base, hummocky cross-stratification (see Figure 16.4, p. 333) within and symmetrical oscillation ripples (straight or interfering into net-like patterns) at the top of storm beds (**tempestites**).

Other structures are also distinctive, but for less direct reasons. The brain-like load casts on sole surfaces, for instance, develop when a layer of heavy sand is suddenly deposited on uncompacted, soft mud. But why should this occur only in turbidites or flood deposits, and not in tempestites? When describing the erosional and depositional effects we have tacitly assumed that the two processes are symmetrical: strong erosion followed by strong deposition, weak erosion by weak deposition. In floods and turbidity currents, however, the energy peak shifts downcurrent during the event. Thus an unduly thick sand layer may be deposited in an area where the same event had been too weak to remove the surface layer of soft, uncompacted mud in the erosional phase.

Distinction between different kinds of **current events** is more difficult. Dynamic processes being similar, we must now resort to more circumstantial evidence. Floods (as storms) handle pre-sorted sediments, while turbidity currents, during their way down the continental slope, pick up such a variety of material that the granulometric bimodality disappears. As a result, sandy turbidites rarely show the rippled tops that are so characteristic for shallow-water event deposits, both inundites and tempestites. Biological overprints provide additional diagnostic criteria. While dinosaurs and birds could leave their footprints in a flood bed, this will hardly be the case in a turbidite formed deeper in the basin. But a distinction can also be made from the more ubiquitous, but morphologically diagnostic, traces of worms, shrimps and other burrowers between shelf and deep sea environments and their associated events.

In addition to turbidites and inundites, Figure 19.1 illustrates a number of flow-induced event deposits that form under more exceptional conditions either on land (**flash-flood deposits**) or on submarine slopes (**mass-flow deposits**). What they have in common is that particles are not completely suspended. Instead they move as a slurry, in which the interaction between variously sized components and water trapped in the pore spaces becomes a major factor.

But what about different kinds of **wave events**? Storms are a clear-cut and certainly the most common case. Since their energy source is above water, they will reach only the shallower bottoms. The uniform depth at which sedimentary shelves build out from continental margins may actually be controlled by the depth of the storm wave base. The shelf depth may be reduced in carbonate platforms if biological and diagenetic binding is fast enough to strengthen the sediments between major storms.

Shocks caused by earthquakes, volcanic eruptions and meteorite impacts also propagate as waves and can affect the sea floor at any water depth. In deep basins below storm wave base and beyond the influence of turbidity currents, they may actually become the dominant type of turbulence-inducing events. Nevertheless, due to their relative rarity, the sedimentological effects of shock events can not be readily studied in modern environments. The little that we know of shock-induced deposits (**seismites**) is derived from strange horizons in ancient quiet-basin deposits. A widespread and thoroughly mixed 'homogenite' layer in the Mediterranean, for instance, has been referred to the prehistoric explosion of the

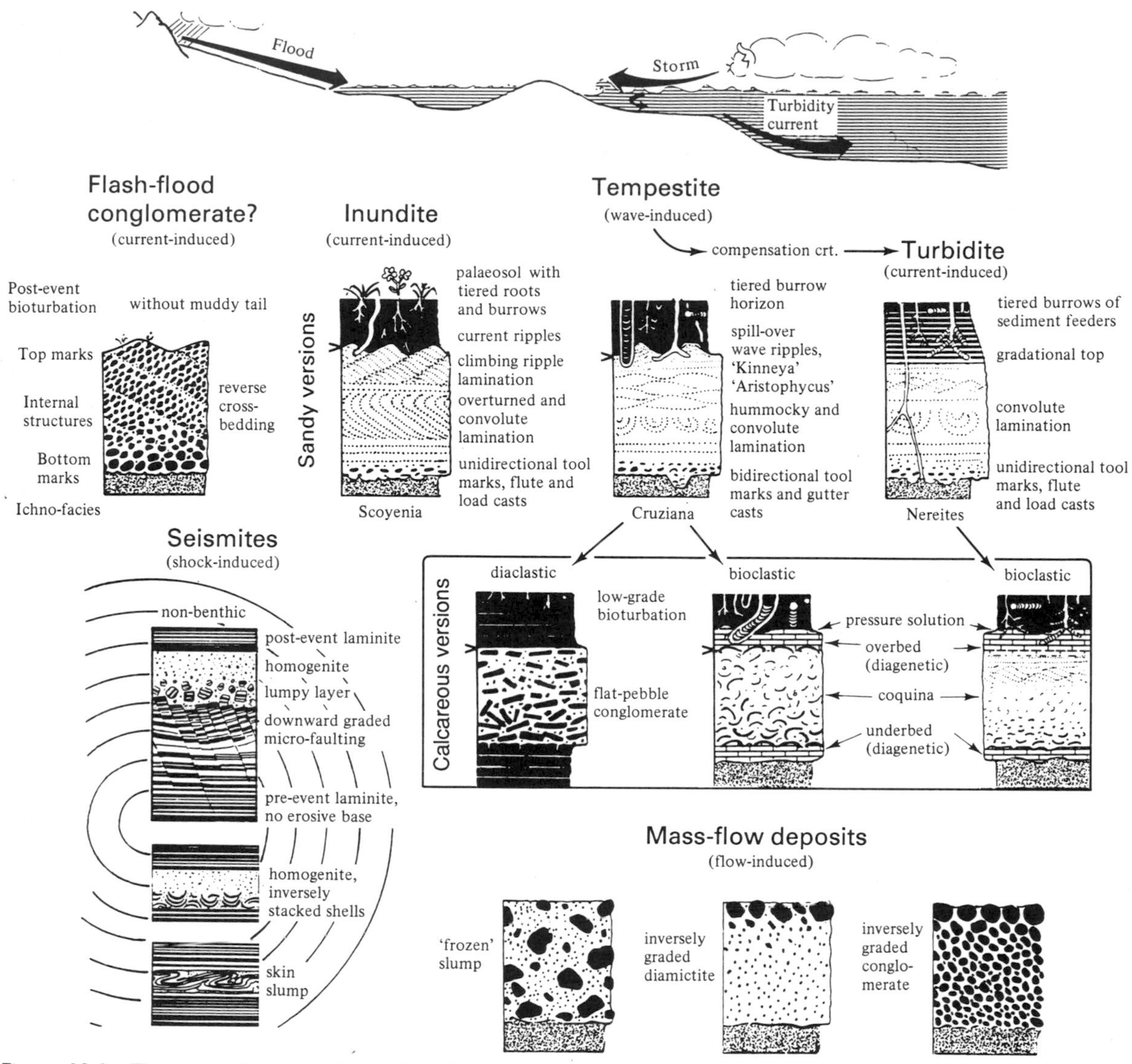

Figure 19.1 The textural characteristics of single-event deposits are related to their origin by either episodic currents (flash floods, river floods or turbidity currents), wave action (storms), gravity-induced mass flows, or seismic shocks. In carbonates, early diagenesis may considerably alter the original picture.

Santorini volcano. A similar horizon at the Cretaceous/Tertiary boundary may be caused by the supposed meteorite impact at that time. The generally well-laminated Monterey Shales of California show repeated couplets of a thin homogenite on top and a microtectonically deformed horizon underneath; they may well be seismites related to the nearby San Andreas Fault.

Shock waves are certainly intensive enough to affect a relatively thick pile of unconsolidated surface sediment. But due to their short duration they may not allow a clear succession of erosion, suspension and graded redeposition to develop. Instead the particles will become only mixed and de-stratified. But since bioturbation (the re-working of sediment by burrowing organisms) has a similar effect, this criterion is valid only in anoxic settings.

Slumping is another process that leads – at one stage – to destratification. At the onset it will crumble the layers into asymmetric slump folds. If the particles are still free to move individually, they will then mix into a structureless dough that we identify as an **olistholite** in the geologic record. In a still further stage, the fine fraction becomes suspended and travels downslope as turbidity currents, whose deposits we have already discussed. The fate of the residual coarse fraction depends on particle shapes. Angular fragments stay where they are. If, however, this fraction consists of rounded pebbles derived from glaciers, rivers or beaches, this mass will move on as a grain flow, whose mechanics lead to a particular kind of grading. Being controlled by grain-to-grain friction rather than by settling velocities in a fluid medium, grain-flow grading is inverse: like in dry wheat shaken in a container, it will be the larger pebbles that are driven to the surface. Clearly, slump events leave a diverse and distinctive record, but their effects will always be localised and lack the biological interaction that makes other event deposits so interesting.

Diagenetic overprint

So far we have described event deposits as they develop in sequences of terrigenous gravels, sands and clays. Epicontinental seas in arid climates, however, may receive so little clastic influx that coarser sediment fractions consist almost exclusively of carbonate that organisms have precipitated in their skeletons. Apart from the shape factor, biogenic sands and muds behave sedimentologically like their terrigenous counterparts. Therefore calcareous tempestites (which are the dominant type of event beds in such shallow seas) share primary features with siliciclastic ones. In fact, the distinctive internal sedimentary structures can be readily observed, but the erosional bases and rippled tops of calcareous tempestites are commonly concealed by concretionary beds above and below. The reason is that carbonate sediments are very susceptible to diagenetic alteration. The early diagenetic dissolution of aragonitic particles saturates the pore water with enough carbonate to cement adjacent mud layers, which then mummify the coarse tempestite as tightly welded concretions. In addition to the masking effect, diagenetic overprinting may also have sedimentological consequences. If it happened early enough, it could alter the transport behaviour of particles and the erodability of the whole tempestite bed upon subsequent recycling by another storm. We shall come back to this in a later section.

Taphonomic consequences of individual events

When talking about catastrophic events and their biological impacts, it is automatically the immediate victims that come to mind. The study of how the victims died and were buried is termed **taphonomy**. Such victims have actually been preserved – such as in the case of Pompeian citizens that were smothered and cast by the welded ashes of Vesuvius. Similarly, the tops of shelly tempestites may preserve the articulated skeletons of various echinoderms, and even whole parties of trilobites. They were obviously killed and buried by sudden mudfalls caused by storms. But such 'obrution deposits' (*obruere* = to smother) involve only that part of the contemporary bottom fauna which was either firmly attached or particularly susceptible to fine mud because it clogged gills or ambulacral systems. Otherwise the gruesome battlefields that physical events may leave behind have little chance to be preserved. The fish beds found at modern beaches after 'red tides' are a misleading example, because these carcasses have been secondarily concentrated by shoreward drift from a large area and have come to rest at a place where they would have minimal preservation potential.

Another kind of mass mortality has not yet been observed in modern environments. It can be deduced, however, from whole swarms of fish preserved on certain bedding planes of the Jurassic Solnhofen limestones of Bavaria and other extraordinary fossil biota (Figure 19.2). Such **'Fossil-Lagerstaetten'** owe the preservation of soft parts and articulated skeletons to the lack of oxygen (and consequently of scavengers) in the quiet bottom waters of stagnant basins. However, even under such anoxic conditions, carcasses must be mud-blanketed to be saved from eventual bacterial decomposition. So the fish swarm must have been killed and blanketed simultaneously. The alignment of the carcasses also tells us that they were laid down by a current. Storm dynamics solve this riddle. The coastal swell produced by a storm must induce a seaward compensation current. If sufficient mud has been suspended near the coast, this flow will develop into a turbidity current that hugs the slope, concentrates into submarine drainage systems and shoots beyond the storm wave base. As it approaches the anoxic zone, such a current may erode enough fetid mud to become toxic. Thus it can kill animals, carry their carcasses into the abiotic zone, deposit them in an oriented fashion and eventually provide a blanket of settling mud. In this way, storm events may become an important taphonomic factor at depths at which wave action is never felt.

Other aspects of catastrophic events are their indirect, secondary consequences. Since they act through complex ecological chains, such after-effects may extend over long periods beyond the actual event. Mass extinctions are the most famous example. Whatever kind of perturbation triggered them originally, they are too rare to be studied in terms of comparative

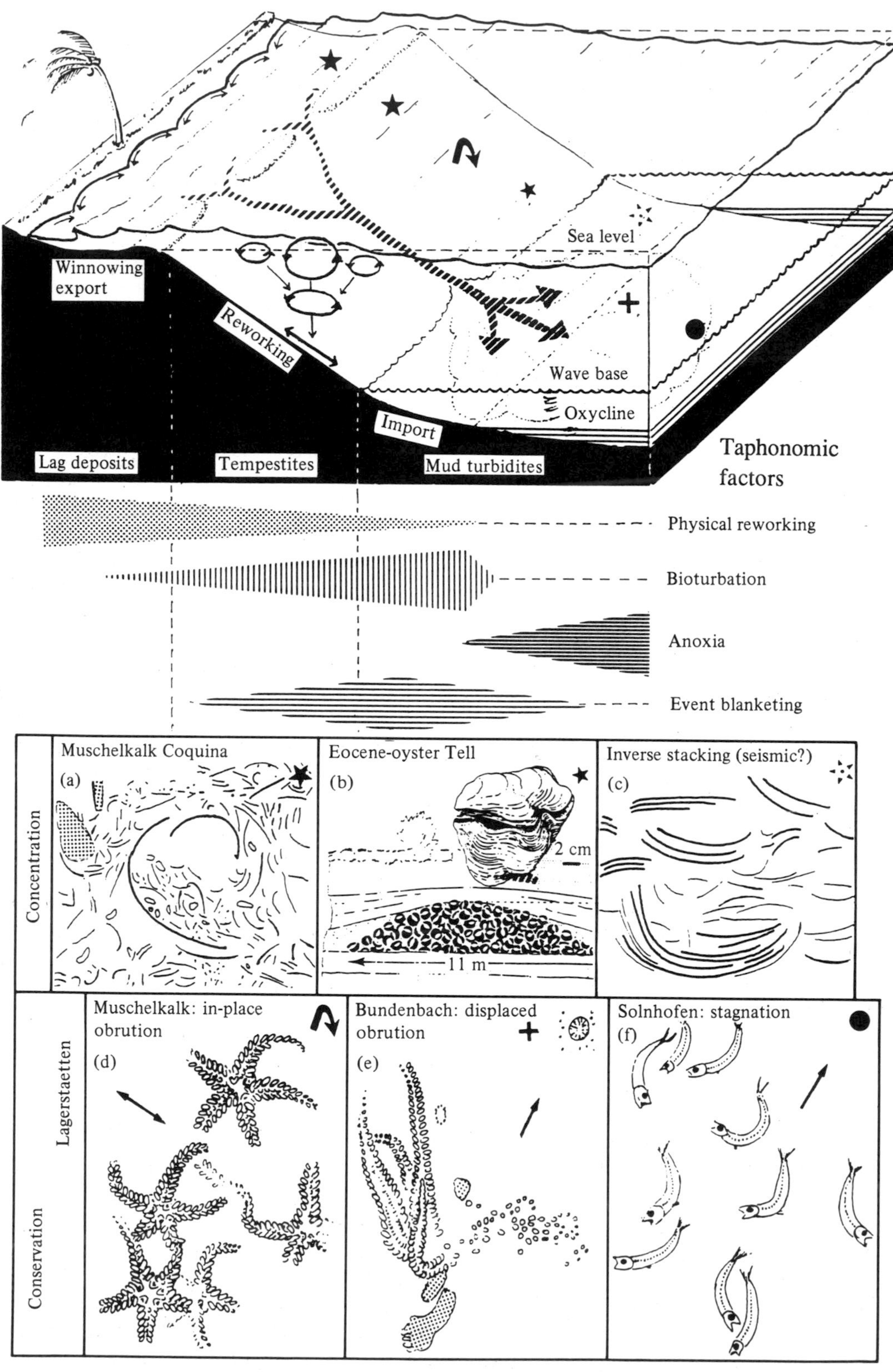

Sea level
Winnowing export
Reworking
Wave base
Oxycline
Import
Lag deposits
Tempestites
Mud turbidites
Taphonomic factors
Physical reworking
Bioturbation
Anoxia
Event blanketing
Concentration
Muschelkalk Coquina
(a)
Eocene-oyster Tell
(b)
2 cm
11 m
Inverse stacking (seismic?)
(c)
Conservation
Lagerstaetten
Muschelkalk: in-place obrution
(d)
Bundenbach: displaced obrution
(e)
Solnhofen: stagnation
(f)

sedimentology; hence they are left out from the present discussion. The same is true for other singular events, such as the drying out of the Mediterranean, the possible flushing of the Atlantic by polar fresh water or the sudden change of oceanic current patterns.

The consequences of the more modest catastrophies have been proportionately less dramatic – if only for their less global nature. Still their local impact at the ecological scale is undeniable. Minor events also have the advantage that they are common enough for us to develop and test general models. Another appeal lies in the two-sidedness of the reactions. By their repetition, ordinary turbulence events allow a variety of feedback mechanisms to evolve between sedimentary, ecological, taphonomic and diagenetic processes. This intricate system is best exemplified in storm-dominated epicontinental carbonate basins, on which we shall focus in the following section.

Boundary marker beds and event condensation

Storms would appear to be too localised to have more than regional and ecological effects. But in epicontinental carbonate sequences, such as the Jurassic of Europe, zonal boundaries (implying evolutionary change) commonly coincide with marker horizons that have the sedimentological characters of shelly storm beds. What makes the problem even more irritating is that zonal boundaries are usually based on non-benthic organisms, such as ammonites, which should be least affected by what happens at the sea bottom. Is this merely a stratigraphic convenience or a true causal relationship? Probably both, but in order to really answer this question we must go back to the possible ecological after-effects of storms – now in a shelly facies and with the additional aspect of event condensation.

Sedimentological feedback

In previous sections we have treated mollusc shells and other skeletal remains merely as another kind of sedimentary particle, though of a different mineralogical composition. This may be justified with regard to shell debris. The original shells, however, stick out from terrigenous particles both in size and shape and therefore behave quite differently under the same hydrodynamic conditions. Bivalve shells are a particularly good example. Once disarticulated, the valves will not simply roll about (as a complete shell would do), but tend to rest in the most stable convex-up position. In tempestites whole pavements of convex-up valves can thus form either on the eroded mud surface or on top of a shell bed (during the non-sedimentation interval). Even if they were covered in the course of the original storm, such pavements are likely to be swept clean of sediment during subsequent events, because they now form an armour against further erosion.

Ecologic and taphonomic feedbacks

The sedimentological feedback is often combined with a biological one. In contrast to the soft mud infauna that was there before the initial storm, the exhumed shell pavement will attract a completely different, epifaunal community consisting of encrusters or more flexibly attached bivalves, brachiopods and crinoids. This community will produce relatively large bioclasts in high concentration, because sessile organisms thrive only when background sedimentation rates are very low to begin with.

In spite of repeated winnowing, a more permanent mud cover may eventually develop and make the community of encrusters yield to a secondary one of recliners and mud stickers. The members of these two guilds (e.g. horn corals or oysters) are equally immobile, but unattached except early in their lives. Therefore they must make their skeletons large and heavy to get the necessary stabilisation. Some fossil 'boulder' oysters were in fact so out of balance with their sedimentary environment that a rare storm did

Figure 19.2 Storm effects translate to deeper parts of the water column by two effects:
(1) flattening of orbital wave motion into oscillating currents near the sea bottom, with larger wavelengths penetrating to greater depths.
(2) Dissipation of coastal swell by compensation currents that may develop into turbidity currents, hugging the slope like submarine river systems and shooting below wave base. These mechanisms define the zones in which certain types of fossil concentrations are likely to occur. The following examples are illustrated:
(a) Residual shell beds in the coastal zone;
(b) shell mounds of oversized recliners in deeper zones;
(c) stacking of shells, possibly due to seismic events;
(d) clusters of whole-body benthos smothered in life position at slope positions outside turbidity current channels;
(e) carcasses (partly disintegrating) transported and blanketed by distal storm turbidites in fan areas below wave base;
(f) fish swarms killed, transported and blanketed by fetid suspension currents in anoxic basins. □

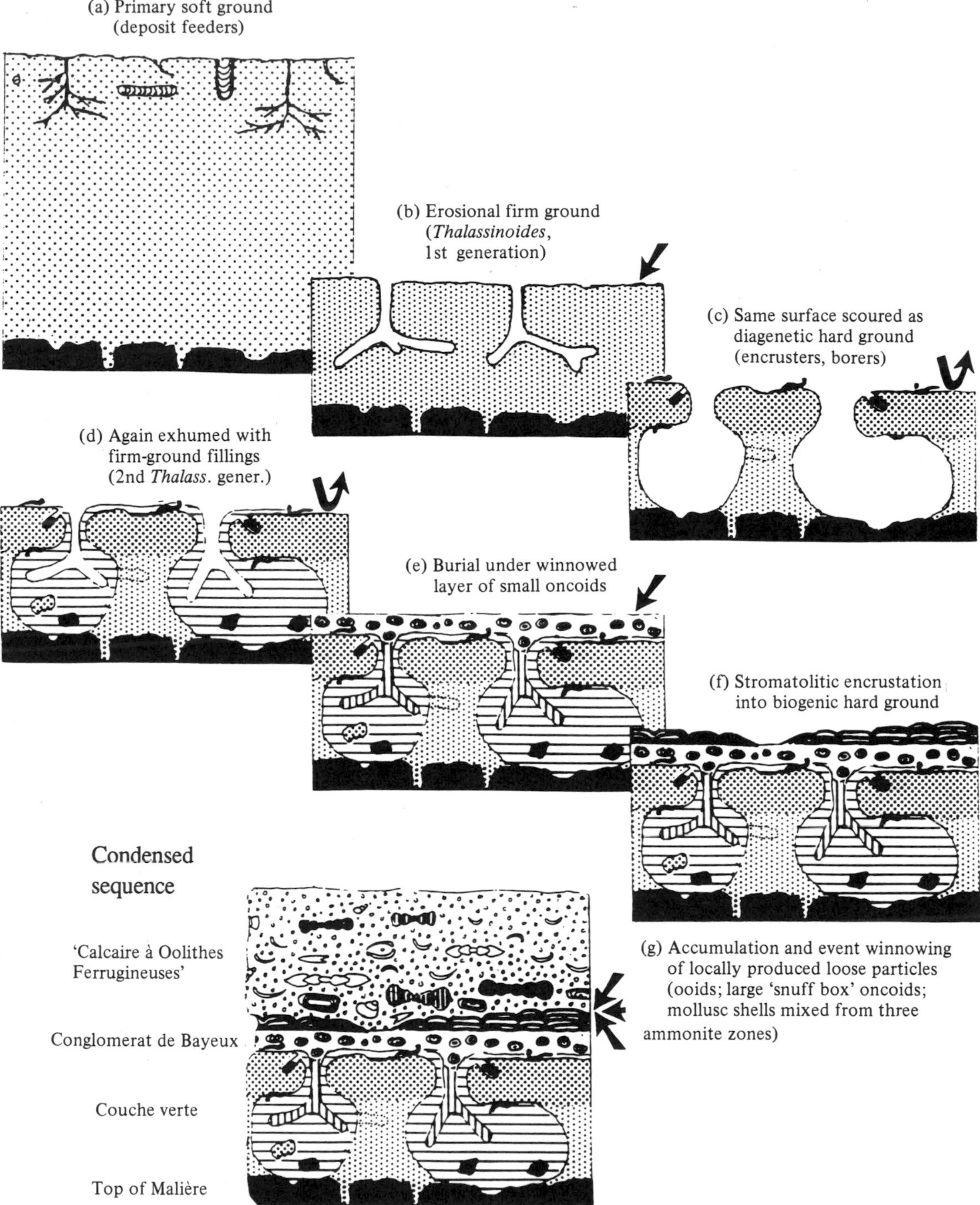

Figure 19.3 An example of a condensed marker bed from the Jurassic of Normandy. Details of this marker horizon reveal a complex history of episodic mud deposition and erosion near storm wave base. Note also the change of the diagenetic regime from a time of prevailing mud cementation (a–d) to one in which aragonitic particles were neither dissolved nor cemented and could therefore become winnowed and reshuffled into a condensed 'Concentration Lagerstaette' (g). Due to (1) reduced sedimentation rates (transgressive phase), (2) local production of biogenic and diagenetic particles or reference horizons, and (3) the telescoping effect of rare storm events (arrows), the information of the sediment pile increased dramatically, while its overall thickness (about 50 cm) remained essentially the same, during a period of more than a million years. □

not move them. Rather it winnowed the fine sediment from underneath the shells and made them sink down on generations of former storm victims before they were also buried under the resettling mud. In this way recliners, whose very existence builds on spatial isolation, could gradually build up huge mounds that have been mistaken for reefs.

Successions in multiple-event beds

What has all this to do with our initial question about storm beds and zonal boundaries? The marker horizons we talked about are probably not the products of single events, but amalgamated storm beds, in which only the last storm may have left its sedimentological signature. They correspond to long periods of time in which terrigenous background sedimentation was generally reduced. But due to the interaction of physical and biological processes, conditions during this time did not stay the same, but changed in a regular succession as shown in Figure 19.3.

1. With the long-term sedimentation rate being lowered, the erosive base of a random 'first' storm did not simply get stowed away as one of the many inconspicuous bedding planes in a pile of muddy sediments. Rather the 'memory' of this event – initially expressed only as a minor discontinuity in the compaction gradient covered by a shell veneer – acted as a reference horizon for subsequent event erosion.
2. Successive storms, being sediment-starved, exhumed this datum horizon without covering it again with loose mud. Thus it lay open as a firm ground of compacted mud into which only certain animals were able to burrow. From the Triassic onwards, firm ground communities (*Glossifungites* Association) were dominated by the gallery systems (*Thalassinoides*) of ghost shrimps, with mud-boring pholad clams of sometimes considerable sizes joining in by Cenozoic times.
3. When the initial firm ground was eventually covered by a new mud layer, the soft bottom fauna returned. But while the mud itself, along with its potential trace fossil content, had no chance to be preserved in the long run, it became a source of bioclasts, among which the massive calcitic shells of recliners and mud stickers could survive many storm re-workings.
4. At this point the former firm-ground fauna had little chance to become re-established. Still its legacy became important in the early diagenetic processes that were now increasingly fed by aragonite solution in the shell layer above. The intensive and durable perforations of the old firm-ground stage not only provided pressure-protected pockets filled with loose sediment for ready cementation, but they also increased the reactive surface to be superficially cemented. If the newly consolidated surface ever did become exhumed again for long enough periods, it was now colonised by a true hard-ground community of rock borers (*Trypanites* Community) and encrusters, some of which could even build up hummocky reeflets. But even short exposure during a major storm would be sufficient for fragments of the now cemented horizon to be ripped off and turned into large intraclasts.

Time frame and evolutionary effects

During all these ups and downs and changes of benthic communities, fallout from the open-water communities continued. In pre-Cenozoic times it contained large ammonite shells. Being aragonitic and rather delicate, they would appear to have had a low fossilisation potential compared to the benthic shell production. Still there are rare localities in which they did survive due to carbonate-saturated pore waters causing early fossilisation. Since ammonites undergo faster evolutionary transformation than benthic species, such occurrences provide us with a time frame for the cycle described above

Careful studies have shown that **condensed** ammonite **horizons** may span up to three ammonite zones, corresponding to more than a million years! Even more surprising: the ammonites are not embedded in their proper time order (as would be expected if condensation had been only a result of starved sedimentation), but mixed and rather in a size-graded succession. Also they are in much better shape than a shell could be after having been lying exposed on the sea floor for more than a very few years.

The storm explanation removes all these difficulties. A few events (in addition to reduced sedimentation) suffice to account for the mixing. They would have exposed the shells for a total of only a few hours during the whole period and covered them again with a protective mud blanket immediately afterwards.

Now to return to our initial question, does this mean that the coincidence of shelly marker beds with ammonite zone boundaries is solely an artifact of stratigraphic condensation? Not necessarily. Sedimentologists are used to judging energy levels from grain size. By this measure the coarse marker beds have been traditionally referred to regressional phases. But bioclasts do not necessarily reflect the transporting power of a current regime, but may have originated locally and be way out of balance with ambient energy levels. The same is true for 'diaclasts' (concretions, hardground fragments, cemented burrow or shell fills) formed during early diagenesis.

What all non-terrigenous large particles need, however, is time to grow. This time is available only when turbulence events are rare and sedimentation is very slow. This is not the case during the regressive, but rather during the transgressive phase, when the rising sea level holds the influx of terrigenous mud back in the coastal areas!

This sedimentological re-interpretation at long last also answers the evolutionary part of our question. If the coarse marker beds (which are less time-condensed than the extreme examples mentioned above) correspond to high rather than low sealevel stands, the appearance of new ammonite guilds at just these levels becomes a normal affair, because these were the times when better connections with the open ocean must have favoured their immigration.

Biohistoric aspects of event stratification

By now we have learned that ecological and taphonomic feedbacks are an integral part, rather than a separable epiphenomenon, of the rock-forming machinery. This insight may complicate stratigraphic procedure; but it also provides us with new geohistoric perspectives.

Skeletal production

Organic evolution is primarily recognised as a transformation of shapes and organisations. In a sedimentological context, however, such changes are rather irrelevant as long as they do not concern the general ecological and taphonomic parameters. Mud oysters, for instance, are sedimentologically fairly equivalent to Devonian mud-dwelling corals or even to Cambrian archaeocyathids. In the same sense nummulite (large foraminifera) banks may be compared to earlier crinoid limestones. But when it comes to the familiar analogy between bivalves and brachiopods we should remember that articulate brachiopods never did burrow and that their shells were always calcitic and rarely became very massive. In other words, skeletal evolution influenced event stratification mainly by providing bioclasts of different durability.

Behavioural changes

More profound changes have probably resulted from the fact that bioturbation in muddy bottoms has expanded to ever deeper levels. Before the advent of ghost shrimps, firm grounds were much less perforated. But also in soft muds the zone of bioturbational mixing appears to have been shallower in early periods than it is today. Such behavioural changes must have been more consequential than the anatomical ones, because they affected the very process of event stratification.

Shallow bioturbation gave minor event beds a better chance to survive and act as templates for selective cementation. Therefore Late Proterozoic and Early Palaeozoic carbonates tend to be thinner-bedded and more perfectly laminated than later ones. This is also the time interval in which tempestitic flat-pebble conglomerates are most abundant. Their chief constituents are variously rounded (but never deformed) limestone shards in a flat (but rarely shingled) arrangement. Alternatively the pebbles may be closely packed in edgewise orientation – a kind of arrangement that results only from continued wave action, be it on beaches or during storms in deeper water. Flat-pebble conglomerates also indicate reduced sedimentation rates, because the thin limestone beds from which they are derived needed time to become cemented, whether or not microbial activity was involved in this process. Accordingly they can be used to identify transgressive phases in continuous sections. But rare occurrences in later periods may also indicate environments in which adverse living conditions allowed the ancient pattern to persist.

On the other hand, bioturbation has not only a destructive effect. It also establishes new units of three-dimensionally structured sediment on which cementation can selectively act. Given adequate sedimentation rates in an event-dominated pile of rocks, these units may be well separated and stick out as distinctive beds, commonly in the form of nodular limestones. As bioturbation depth increased, we should expect such limestone beds to become thicker in younger periods.

Animal activity influences sedimentation not only by the production of burrows and bioclasts, but also by a less obvious effect. Sediment and suspension feeders can not help but take up, along with their food, large amounts of fine inorganic particles. But instead of being released in the same form, this fraction becomes compacted into larger, mucus-bound fecal pellets. Although it may not be readily seen in the final sediment, pelletisation must have considerably altered the hydrodynamic (and possibly also the diagenetic) behaviour of fine muds as compared to the pre-metazoan environments of the Precambrian.

Dynamic stratigraphy: the new synthesis

This review has not dealt with the grand features of Earth history, such as plate tectonics and mountain building. It has also left out the most dramatic aspect of event stratigraphy, mass extinctions. Rather we have concentrated on minor features, the kind that can be observed in any outcrop or drill core. Nevertheless, even in this everyday task new trends of understanding the Earth are being felt.

Dynamic stratigraphy, as the new approach has been appropriately labelled, implies primarily a change in perspective. Instead of only measuring and correlating sections in terms of beds and sequences, modern stratigraphers try to visualise them as three-dimensional rock bodies. The more important development is the departure from a descriptive approach that implied little more in the way of genetic principles than the one of layered succession. Sedimentary and diagenetic processes are now becoming the gauge by which the section has to be interpreted at every scale, down to the individual bed and lamina and including the many small intervals not represented by sediment. Nor is the dynamic view restricted in scope. However meticulous may be the analysis of a local sequence, it has to be related to the larger systems, with which basin analysis and global stratigraphy are concerned.

In this new framework the role of the palaeontologist is also changing. His or her original (and still continuing) task has been the relative dating of the rock record by taxonomically classified index fossils. It has been later complemented by the distinction of palaeoenvironments with the help of facies fossils. In dynamic stratigraphy, palaeobiological and palaeoecological aspects have gained an additional importance. In the present review we have stressed yet another application of palaeontological data by including the post-mortem history of the fossils. But rather than coining yet another term (such as 'taphofossils') and thereby again pushing palaeontology into a subservient function, one should recognise that at this level disciplinary pigeon-holing becomes both inadequate and dangerous. On the contrary, dynamic (or event) stratigraphy should be seen as a field, in which the efforts of sedimentologists, geochemists and palaeobiologists of broad outlook must be as integrated as the natural processes with which they are concerned.

Summary

The stratigraphic record is, like written human history, not only inherently incomplete, but also biased towards major perturbations that may be rare and catastrophic by standards of our life spans, but are common enough at geologic time scales to dominate the preserved rock record. Among the most common types of event beds are the deposits of floods, storms and turbidity currents, which can be distinguished by erosional, depositional and bioturbational structures. In calcareous rocks, such features may be masked by diagenetic overprinting, while additional information is provided by skeletal remains that preserve the memory of times between the events. By applying a combination of sedimentological, diagenetic, palaeobiological and taphonomic principles, we may not only fill the gaps in the record, but also improve our understanding of physical/biological interactions at the levels of individual events, basin history and biosphere evolution.

Further reading

Aigner, T. (1985) *Storm Depositional Systems. Lecture Notes in Earth Sciences* 3, Springer-Verlag.

Einsele, G. & Seilacher, A. (eds.) (1982) *Cyclic and Event Stratification*, Springer-Verlag.

Einsele, G., Ricken, W. & Seilacher, A. (eds.) (1991) *Cycles and Events in Stratigraphy*, Springer-Verlag.

INTRODUCTION TO CHAPTERS 20 AND 21

The titles of these two chapters – *Sequence Stratigraphy* and *Intraplate Stress and Sedimentary Basin Evolution* – seem totally unconnected, so why are they juxtaposed? There is a simple explanation. As will be explained in Chapter 20, sequence stratigraphy is considered by some (but by no means all) of its proponents to offer a means of global correlation, because they believe that the development of sequences is controlled by global changes of sea-level, which, in geological terms, are instantaneous around the Earth. However, many workers doubt there is such a control, particularly when the major ice sheets (the waxing and waning of which is the principal mechanism for producing sufficiently rapid rates of sea-level change) did not apparently exist for significant periods in the geological past (see Chapter 24). An alternative explanation for the apparent global changes of sea-level (often termed eustatic changes) deduced from sequence stratigraphic studies is that they are due to vertical movements on plate-wide scale. If this is the case, as argued by Sierd Cloetingh in Chapter 21, the global sea-level curve is an indicator of stress levels in the lithosphere, and so is related to changing plate configurations. The debate between those who believe that the dominant control on stratigraphic signatures is eustatic and those who believe it is tectonic is likely to continue, perhaps indefinitely.

CHAPTER 20

Chris Wilson
Open University

SEQUENCE STRATIGRAPHY: AN INTRODUCTION

Introduction

Stratigraphy – the study of strata and their correlation from place to place – has three distinct approaches. **Lithostratigraphy** involves the definition and correlation of units on the basis of their physical composition and features, including mineralogy, grain size and sedimentary structures. Because organisms have evolved through time, **biostratigraphy** uses the fossil content of rocks to place them in a relative time sequence and to correlate them. **Chronostratigraphy** is concerned with defining the geologic time-span of lithostratigraphic and biostratigraphic units, and utilises both absolute dating methods (see Chapter 7) and biostratigraphic data.

In the absence of *suitable* fossils, correlation of strata from place to place becomes difficult, if not impossible. However, proponents of sequence stratigraphy argue that it is a method of achieving very high resolution chronostratigraphic frameworks in which to place lithologic data obtained at outcrops and in the subsurface (well logs and reflection seismic data) which does not rely on the presence of suitable fossils. The technique is used routinely by many petroleum explorationists, and, according to its most ardent disciples, offers a means of worldwide correlation, and so has been heralded as the new global stratigraphy. Such ambitious claims are based on the premise that the overriding control on the development of sedimentary sequences is exerted by global changes of sea-level. Readers unfamiliar with the technique may be wondering what they have been missing, and what fundamental discovery underpins such seductive claims. The answer is that basically it is a new way of looking at old (and new) data. Sequence stratigraphy is a method of interpreting stratigraphic data that was formulated though the study of seismic sections, and later applied to outcrop information and well data.

This chapter can only provide a generalised introduction to sequence stratigraphy. It begins with the origin and development of the technique and then outlines the key concepts and units that are routinely used in its application. Finally, some brief comments are given on the controversy underlying the proposition that global sea-level change (eustasy) is the fundamental control on the spatial and temporal distribution of sedimentary facies.

Developing the concept

Sequence stratigraphy and, before it, **seismic stratigraphy**, were developed by researchers at the Exxon Production and Research Company in Houston, led by Peter Vail. Peter Vail's ideas concerning sequence stratigraphy were seeded during his time as a graduate student supervised by Larry Sloss. Over forty years ago, Sloss recognised major unconformity-bound sequences developed over the North American craton, and named them after Indian tribes (Figure 20.1). Sloss pointed out that each of these trangressive–

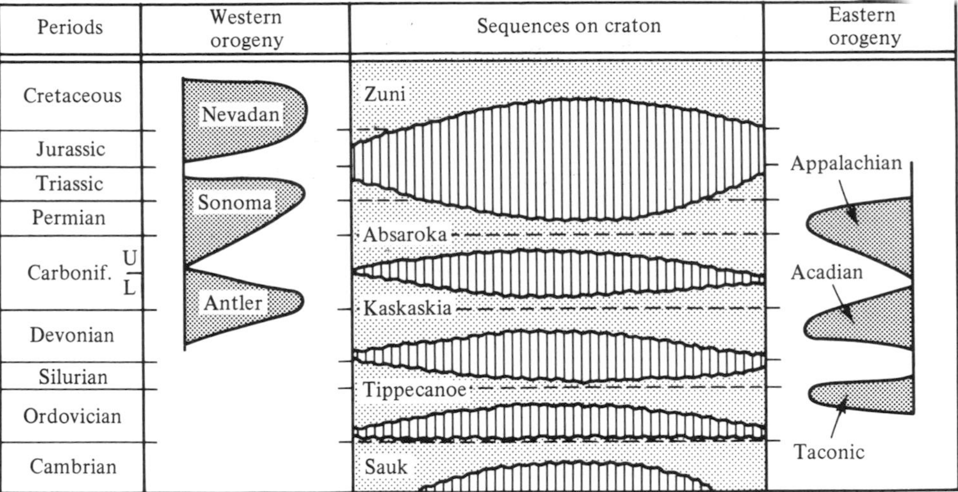

Figure 20.1 Phanerozoic cratonic sequences of North America recognised by Larry Sloss, and their relationship to orogenic episodes on the craton margins. Vertical ruling bounded by wavy lines represents unconformities, and stipple indicates sediments. From Sloss (1963).

regressive episodes alternated with phases of orogenic activity on the eastern and western margins of the craton.

Vail's sequences span much smaller periods of time than do those of Sloss, but in the Exxon global cycle charts (an extract of which is shown in Figure 20.19), the latter's sequences are termed megasequences. The fruits of the Exxon School's labours appeared in public for the first time in 1977 as Memoir 26 of the American Association of Petroleum Geologists (*Seismic Stratigraphy–Applications to Hydrocarbon Exploration*, Payton 1977). This volume rapidly became the seismic stratigrapher's bible, and – as we will see later – the Old Testament of sequence stratigraphy. It defined a sequence as 'a stratigraphic unit composed of a relatively conformable succession of genetically related strata and bounded at its top and base by unconformities or their correlative conformities'. It showed how such sequences could be identified on seismic sections, and how their development could be related to eustatic changes. The resultant 'Vail curves' of global sea-level change were based on changes of coastal onlap (see below), and had a characteristic saw-tooth shape, with abrupt initial falls followed by gradual rises. In the early 1980s a new curve was published that was much 'smoother', as by then it had been recognised that the landward encroachment of sediments across older strata (onlapping relationships, to be explained later in the chapter, see Figure 20.4) on seismic sections were related to marine *and* fluvial sediments, and that the latter would cause the saw-tooth shape of the coastal overlap curve which, if not corrected for, would result in a similar shaped curve (see Figure 20.18). The shape of the eustatic curve, and the assertion that short-term sea-level changes in the mid-Mesozoic were due to glacial effects – despite the general view that the Earth was ice free at this time – led to much criticism of the approach explained in Memoir 26. This had the unfortunate effect of detracting attention from the value of the analytical methods that were applied to seismic data which at least gave stratigraphers the chance actually to observe, rather than infer, rock relationships on a basin-wide scale.

Memoir 26 was published after the Exxon School had been using seismic stratigraphy as an exploration tool for a number of years. For example, in 1964 they were able to identify North Sea Tertiary submarine fans ahead of drilling that led to their discovery in the 1970s. A similar time lag occurred before the results of the application of the sequence concept to outcrop and well data were published in 1988 as Special Publication 42 of the Society of Economic Paleontologists and Mineralogists. Some would say that the title of this volume – *Sea-level Changes: An Integrated Approach* (Wilgus *et al.* 1988) – detracts from the analytical and interpretive techniques described within it. However, the *type* of sea-level change is not referred to in the title, and this volume is now the New Testament of sequence stratigraphy. It prompted many geologists to look at old data with new eyes, and has resulted in a plethora of sequence stratigraphy papers being given at conferences, showing how the method has stimulated new interpretations by workers who do not necessarily believe that it offers a means of global correlation.

The basic units of sequence stratigraphy

A hierarchy of units

Sequence stratigraphers recognise a hierarchy of stratal units that span may orders of magnitude in thickness from millimetres to kilometres:

Lamina
Lamina set
Bed
Bed set
Parasequence
Parasequence set
Depositional systems tract
Sequence
Supersequence
Megasequence

In this chapter, space only permits discussion of sequences and parasequences, plus another unit, the depositional systems tract, several varieties of which are developed within a sequence. The characteristics and origins of the units in the hierarchy up to bed set are summarised in Table 20.1. Figure 20.2 extends the definitions up to 'sequence' (with the exception of depositional systems tract), and shows their characteristic ranges of scales and duration of time of formation. All the units in the hierarchy 'are defined objectively by the physical relationships of the strata, including lateral continuity and geometry of the surfaces bounding the units, vertical stacking patterns,

Table 20.1. *Detailed characteristics of lamina, lamina set, bed and bed set from Campbell (1967)*

Stratal unit	Definition	Characteristics of constituent stratal units	Depositional processes	Characteristics of bounding surfaces
Bed set	A relatively conformable succession of genetically related beds bounded by surfaces (called bed set surfaces) or erosion, non-deposition, or their correlative conformities	Beds above and below bed set always differ in composition, texture, or sedimentary structure from those composing the bed set	Episodic or periodic (same as **bed** below)	(Same as **bed** below) plus • Bed sets and bed set surfaces form over a longer period of time than beds • Commonly have greater lateral extent than bedding surfaces
Bed	A relatively conformable succession of genetically related laminae or lamina sets bounded by surfaces (called bedding surfaces) of erosion, non-deposition or their correlative conformities	Not all beds contain lamina sets	Episodic or periodic. Episodic deposition includes deposition from storms, floods, debris flows, turbidity currents Periodic deposition includes deposition from seasonal or climatic changes	• Form rapidly, minutes to years • Separate all younger strata from all older strata over the extent of the surfaces • Facies changes are bounded by bedding surfaces • Useful for chronostratigraphy under certain circumstances • Time represented by bedding surfaces probably greater than time represented by beds • Areal extents vary widely from square feet to 1000s square miles
Lamina set	A relatively conformable succession of genetically related laminae bounded by surfaces (called lamina set surface) of erosion, non-deposition or their correlative conformities	Consists of a group or set of conformable laminae that compose distinctive structures in a bed	Episodic, commonly found in wave or current-rippled beds, turbidites, wave-rippled intervals in hummocky bed sets, or cross beds as reverse flow ripples or rippled toes of foresets	• Form rapidly, minutes to days • Smaller areal extent than encompassing bed
Lamina	The smallest megascopic layer	Uniform in composition/ texture Never internally layered	Episodic	• Forms very rapidly, minutes to hours • Smaller areal extent than encompassing bed

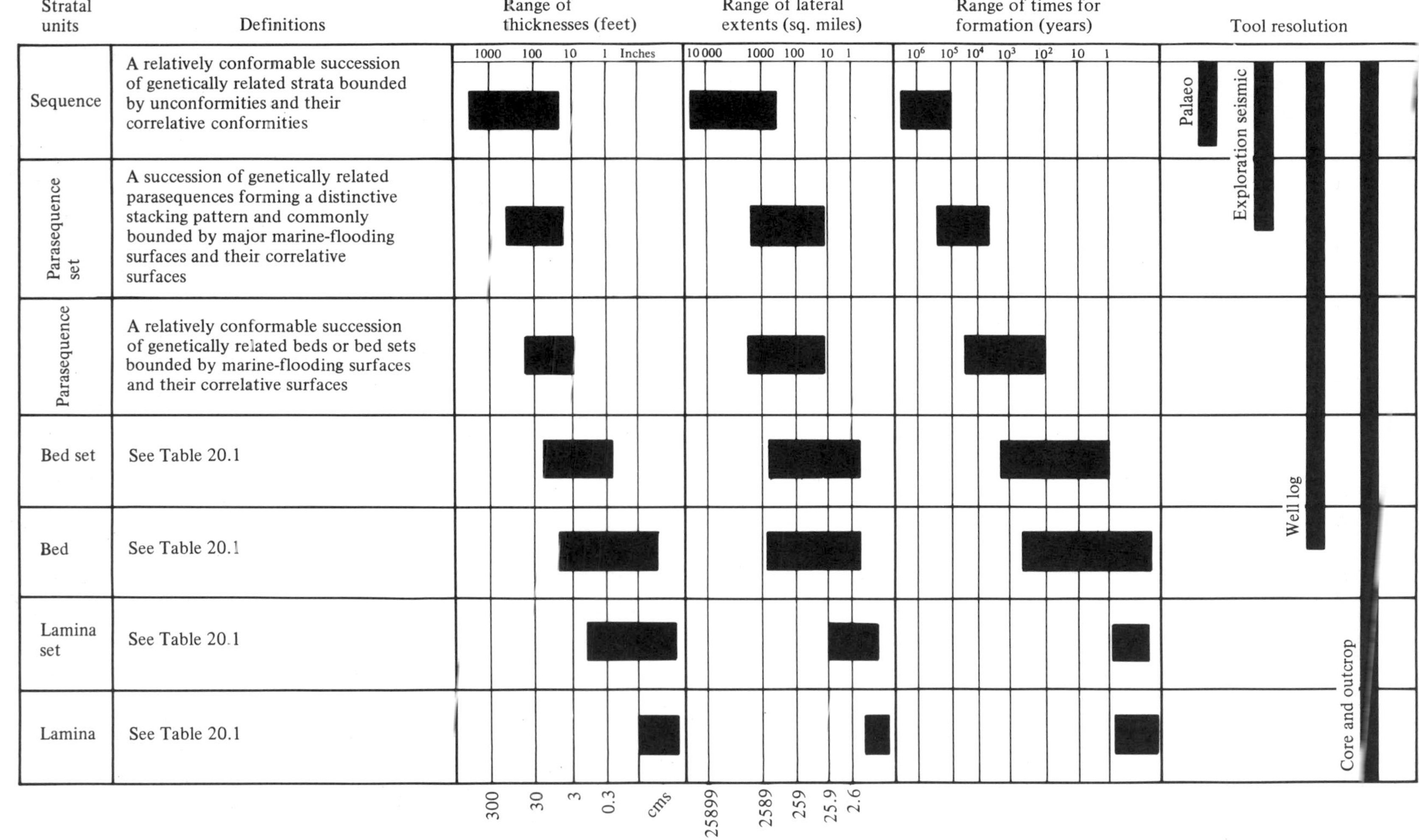

Figure 20.2 Definitions and characteristics of stratal units. The last column shows the extent to which various methods of observation can identify (or not) each unit. For an example of a well log, see Figure 20.16b. □

and lateral geometry of the strata within the units' (Van Wagoner *et al.* 1990). Furthermore, 'facies and environmental interpretation of strata on either side of bounding surfaces are critical' for the discrimination between higher order units. It must be stressed that interpretations involving regional or global origin – especially eustasy – are not used to define the units. The nature of the sequence is discussed first here, as it gives insights into how sequence stratigraphy was built on the basin-wide perspective provided by seismic reflection data.

Sequences

Discordant relations are the key criteria used to identify seismic sequences. Discordance is indicated by **reflection terminations** (see Figure 20.3). At the base of a sequence, two kinds of **baselap** may occur (see Figure 20.4); these mark the lower depositional limit of stratal units. **Onlap** occurs where originally horizontal units lap against (the term lap out may also be used in the literature) an inclined surface; it is a characteristic feature of the bases of depositional sequences. Onlap of shallow marine on non-marine sediments marks a relative rise in sea-level. **Downlap** occurs when an originally inclined stratum terminates downdip against an older surface (which may be horizontal or inclined), and it may occur at the bases of, or within, depositional sequences. **Toplap** is the updip termination of inclined units; it is characteristic of the unconformable part of the top of a sequence, and indicates non-deposition, by-passing of sediments across a shelf, or possibly some minor erosion. Toplap terminations can generally be traced downdip along reflectors into downlap. As shown in Figure 20.4, **erosional truncation** may also occur at the top of sequences; differentiating it from toplap depends on identifying the shape of the reflectors near the terminations: toplap shows reflectors that curve towards the horizontal, whereas truncation is indicated by lack of such curving and abrupt reflection terminations.

A depositional sequence, whether identified by conventional stratigraphic means or on a seismic section, can be defined as quoted earlier as 'a stratigraphic unit composed of a relatively conformable succession or genetically related strata and bounded at its top and base by unconformities, or correlative conformities'. An abstract example of three sequences is shown in Figure 20.5. Figure 20.5a shows twenty-nine stratal units displayed as a geologic cross-section, and Figure 20.5b is a chronostratigraphic section showing the same units. In other words, the first diagram shows the *spatial* distribution of the stratal units, and the second diagram emphasises their distribution in *time*.

Examine sequence boundary 1 on Figure 20.5a. At the far left, stratal unit 28 rests on unit 8; in the middle (to the left of the '1' in the box) 14 rests on 10, and as the boundary runs off the bottom of the diagram, unit 11 (in the fan) rests on unit 10. So at this last position there is no time gap, and so the unconformity has become a conformity. Therefore the stratal relationships at this point yield a very precise age for the sequence boundary: between units 10 and 11.

Sequence boundary 2 on Figure 20.5a consists of a conformity rather than an unconformity along a much greater length. In fact, unit 21 rests on unit 20 along almost the entire length of the sloping portion of the boundary. It is only along the more gently inclined part that there is a time gap (unit 22 resting on unit 20), but there is no angular discordance. So, once again, the sequence boundary can be given a precise age: between units 20 and 21. Thus the depositional sequence has chronostratigraphic significance because it was deposited during a specific interval of geologic time, although the age range of strata within it may vary within these limits from place to place.

Clearly there is a significant difference between the characteristics of unconformities 1 and 2 on Figure 20.5a. In fact two types of sequence are defined on the basis of the nature of their basal unconformities. The origins of the two unconformity types recognised are illustrated in Figure 20.6. Both types arc associated with basinward shifts in coastal onlap, which in the case of Type 1 unconformities (Figure 20.6b–d) usually moves beyond the shelf edge. This results in erosion of the shelf, including the incision of fluvial valleys and submarine canyons eroding into the shelf edge, with submarine fans being deposited at the foot of the canyons. Thus, **Type 1 unconformities** are characterised by both subaerial and submarine erosion. **Type 2 unconformities** (Figure 20.6e–g) do not show submarine erosion, because coastal onlap does not shift basinward of the shelf edge. As shown on Figure 20.6, Type 1 unconformities are associated with rapid falls of sea-level, whereas Type 2 unconformities are produced by slow falls. Recognition of these unconformity types enables rapid or relatively slower short-term sea-level changes to be recognised.

The chronostratigraphic section shown in Figure 20.5b shows how the two unconformities can be precisely dated according to their geologic age: 1 is 16 Ma, and 2 is 11 Ma. This figure also shows that most of the 'time-gaps' or hiatuses in the succession develop in both subaerial and marine settings.

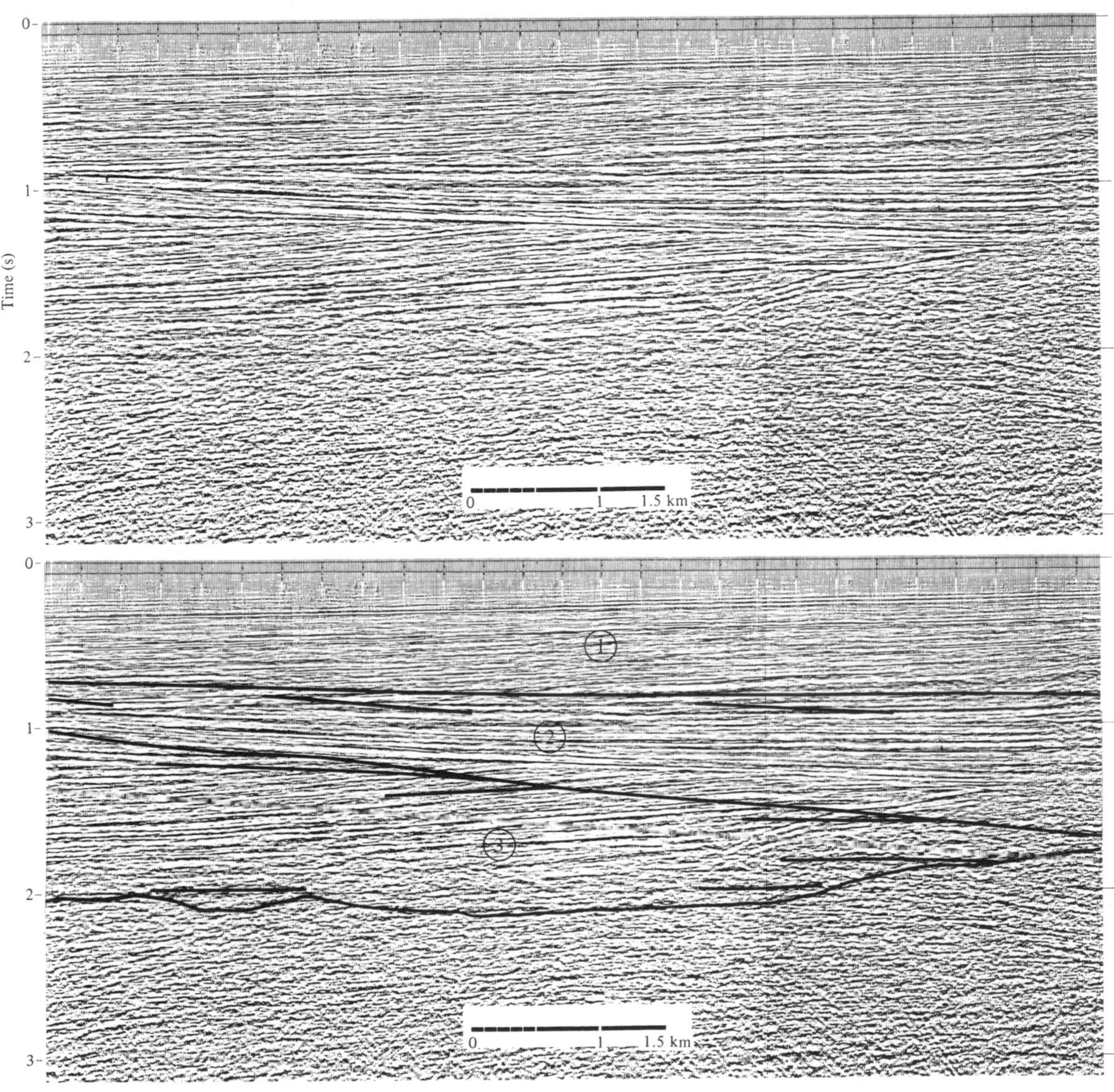

Figure 20.3 Recognising sequences on seismic sections. (a) Uninterpreted section, (b) interpreted section, showing a selection of reflection terminations which define three unconformity-bound sequence boundaries. □

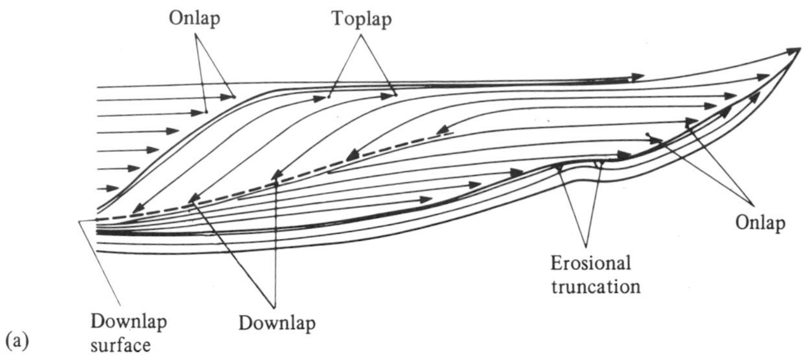

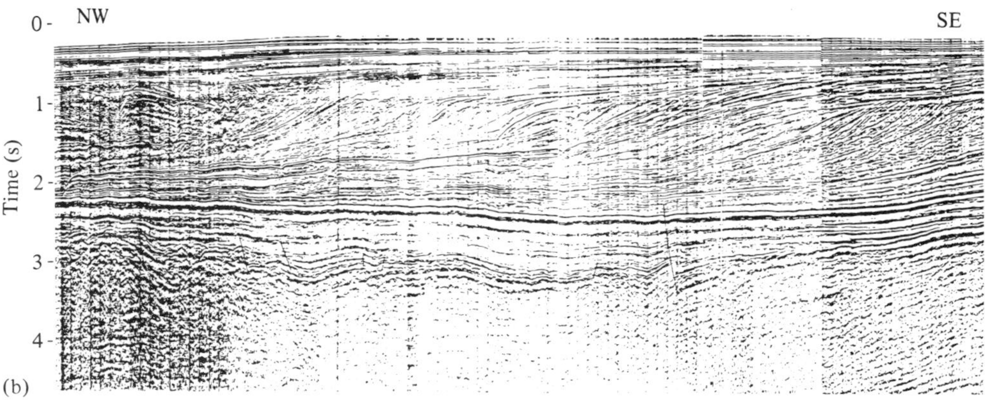

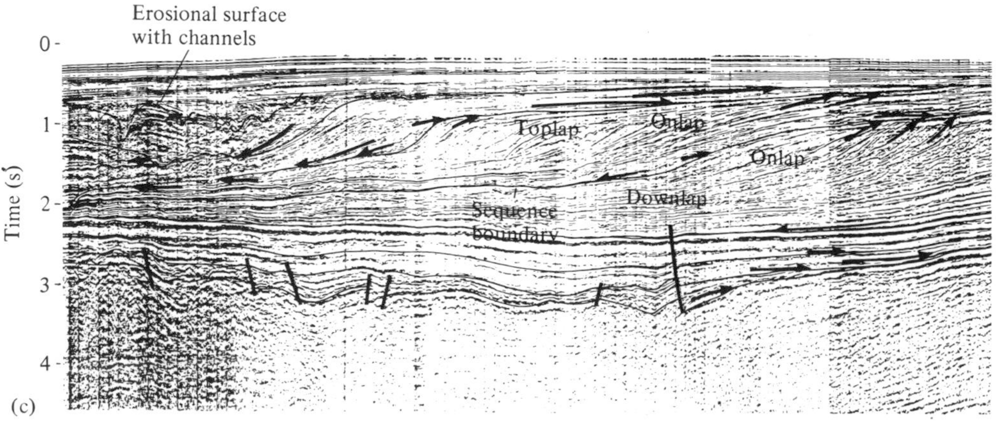

Figure 20.4 Recognising sequences on seismic sections (continued). (a) Characteristic types of reflection terminations recognised on seismic sections; (b) and (c) seismic lines, uninterpreted and interpreted, showing a series of prograding sequences and a selection of reflection terminations that help recognise them. After Vail in Bally (1987); (b) and (c). □

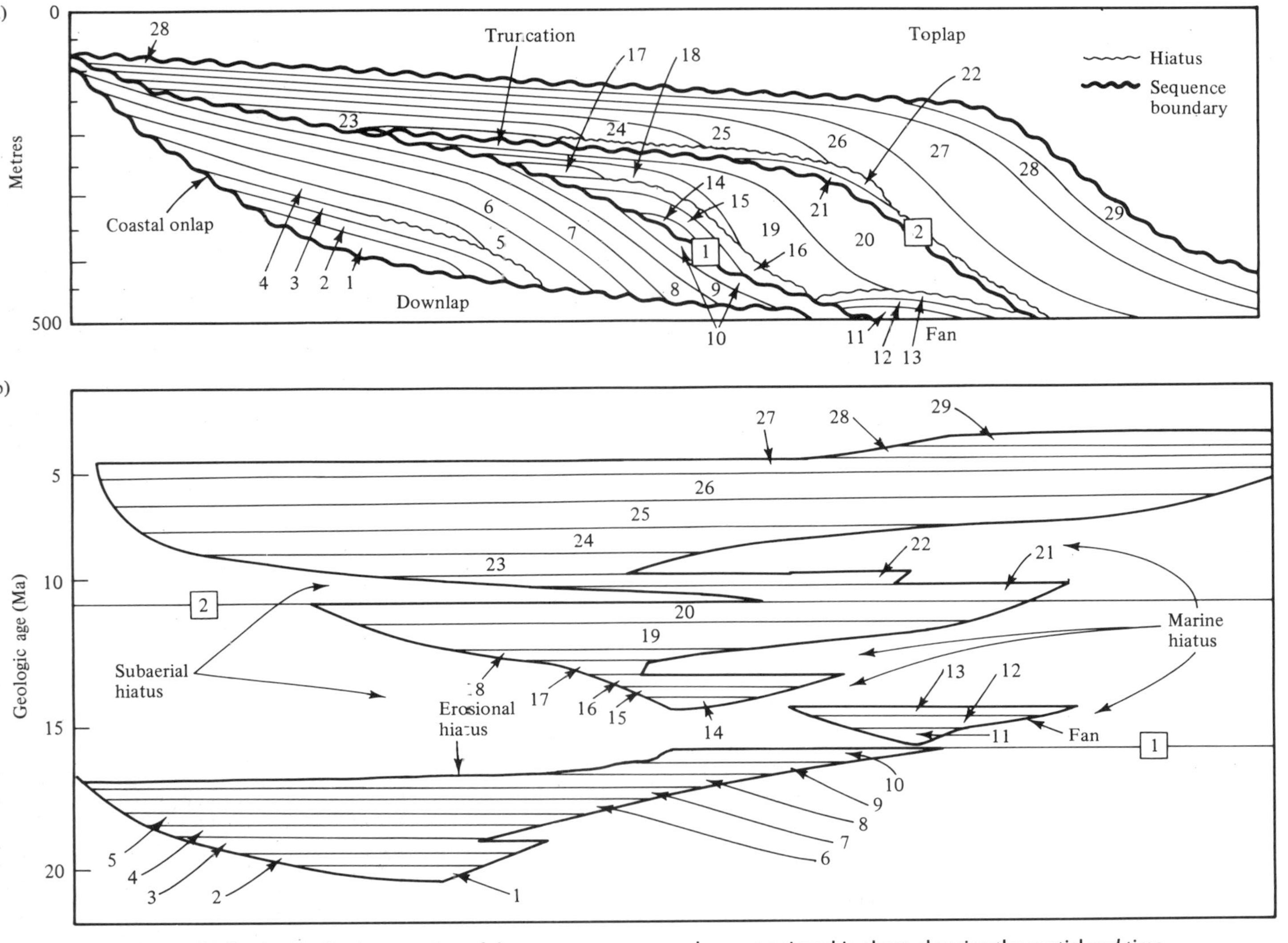

Figure 20.5 An idealised succession of three sequences, separated by boundaries 1 and 2. (a) A stratigraphic section showing the geometrical relations between the stratal units (cf. the seismic section in Figure 20.4). (b) A chronostratigraphic chart, showing the spatial *and* time relations between the stratal units. Further explanation in the text. □

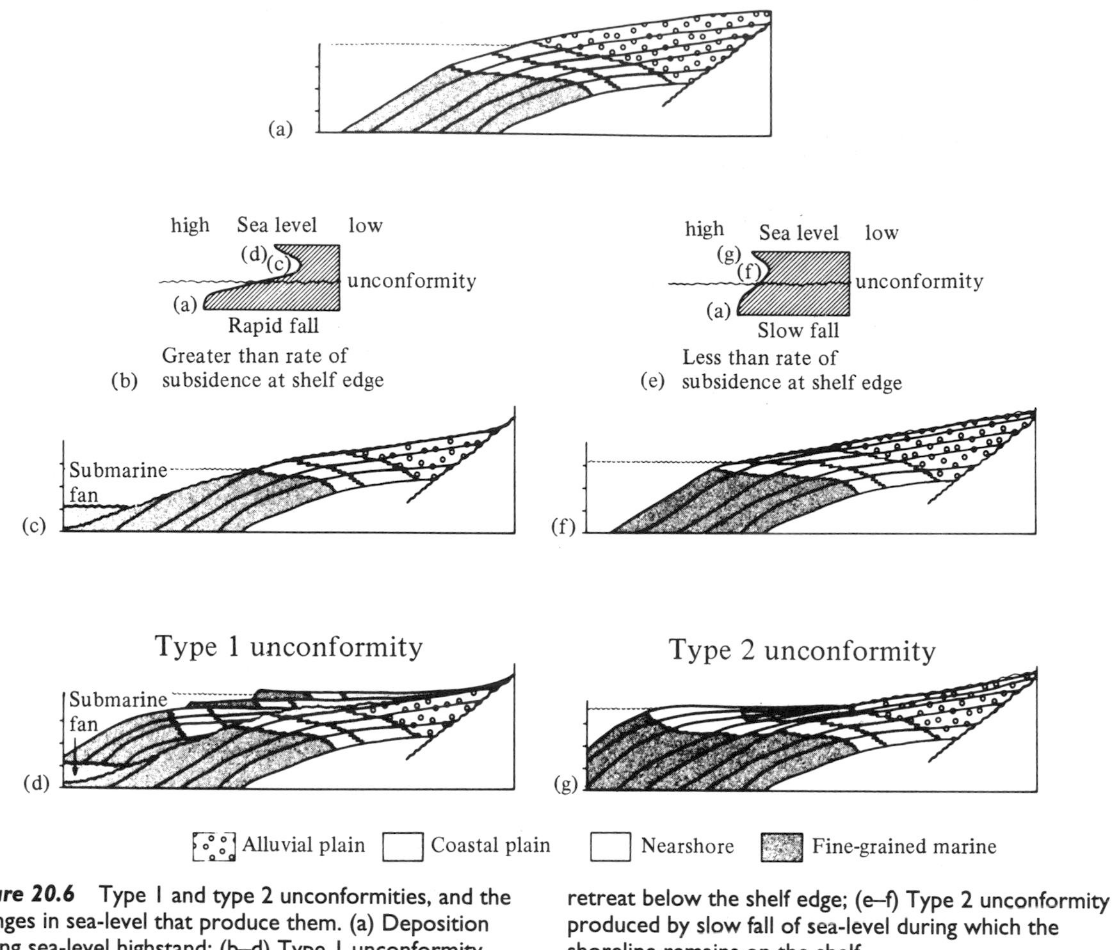

Figure 20.6 Type I and type 2 unconformities, and the changes in sea-level that produce them. (a) Deposition during sea-level highstand; (b–d) Type I unconformity produced by rapid sea-level fall causing the shoreline to retreat below the shelf edge; (e–f) Type 2 unconformity produced by slow fall of sea-level during which the shoreline remains on the shelf. □

The concept of accommodation

Shelf sedimentation progressively fills in the wedge-shaped space between the sea bottom (which is inclined seaward) and the surface of the sea (Figure 20.7). Tectonic movements, or eustatic changes of sea-level, result in changes in the space available to **accommodate** sediments transported onto the shelf. In the discussion which follows, changes of relative sea-level could be caused by either of these processes. If **relative sea-level** stays constant or rises only slowly, sediments will rapidly **prograde** seaward (Figure 20.7a). If new space is added by a rapid relative sea-level rise, progradation will be slower, and a component of sediment **aggradation** (vertical build-up) will occur (Figure 20.7b). Thus, if sediment supply remains constant, the *rate* at which relative sea-level rises – and so creates new space to be filled up – will influence the extent to which sediment aggradation and progradation occurs. Figure 20.7 shows that a slow rise (or stillstand) favours progradation, and a rapid rise favours aggradation; the stratal patterns produced by the two processes are different. *Therefore, rates of relative sea-level change influence the reflection patterns within seismic sequences.*

Downlap surfaces occur within depositional sequences (see Figure 20.4a) and are also related to changes in the rate of change of eustatic sea-level. They mark the progradation of sediments (i.e regression) as the rate of sea-level rise slows down. (On Figure 20.5a, they are shown by fine wriggly lines between stratal units 3 and 4, 16 and 18/19, and 22 and 23–26. Note that Figure 20.18a is a labelled version of Figure 20.5a.) They are often associated with condensed sections (a very thin sequence of sediments which are often iron or phosphate rich, and in which a number of biostratigraphic zones or even stages are repre-

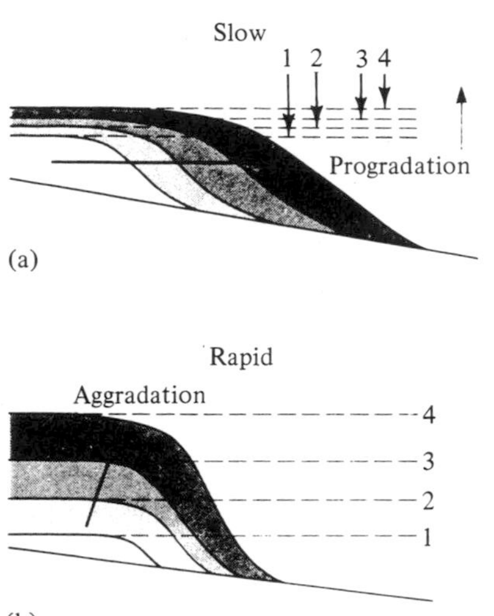

Figure 20.7 The relationship between rates of relative sea-level change (slow and fast) and sedimentary progradation and aggradation. Changes in rates of sediment input have comparable effects if relative sea-level stays constant (i.e. high sediment input results in progradation).

sented) which are the characteristic indicator of the presence of marine hiatuses. Outcrop studies show that condensed sections form when the rate of relative sea-level rise is much greater than the rate of sedimentation; this shifts the principal zone of deposition landward, resulting in very low depositional rates in the seaward direction.

The Vail School believe that eustatic sea-level changes are dominant in modulating the rate at which accommodation space is created or removed. Their reasoning is summarised in Figure 20.8. At the top of Figure 20.8 is an idealised eustatic sea-level curve. The dots on the curves are inflection points, where the rate of change is greatest. F inflection points occur on the falling limbs, R inflection points on the rising limbs. Beneath the eustatic sea-level curve on Figure 20.8 is a plot of tectonic subsidence (b); on a time-scale of 1–10 Ma, basin subsidence can be considered to be virtually linear (i.e. it is just a small part of one of the exponential thermotectonic curves shown in Figure 15.10). The eustatic and subsidence curves can be expressed in terms of rates of change (Figures 20.8c and 20.8d). Figure 20.8e shows the effect of combining the two plots of rate of change. The result is that *relative sea-level rises almost all of the time*, and so new space for sediments to fill is being created almost continuously. At F inflection points the rate at which new space is added is least – and where the eustatic fall is largest, space is removed.

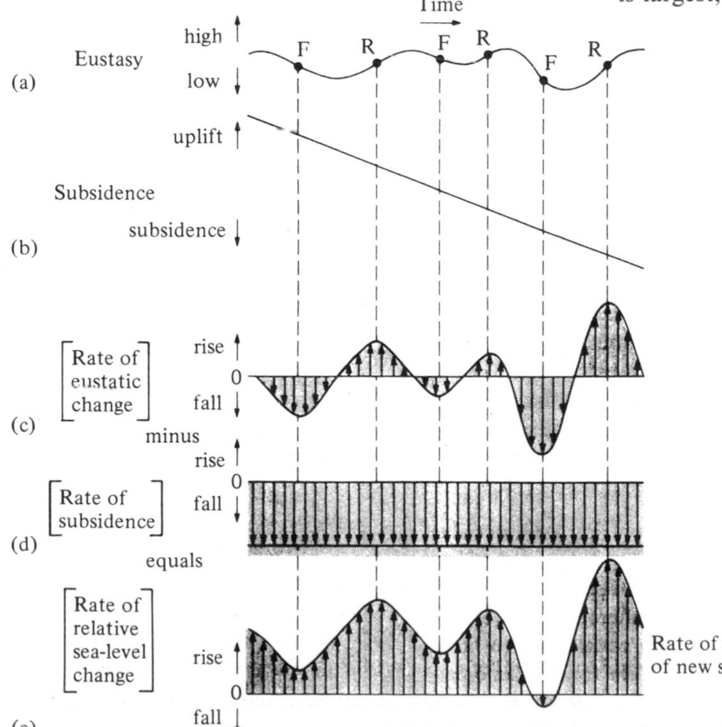

Figure 20.8 Diagram showing how eustatic sea-level changes and tectonic subsidence influence the rate of additional space to accommodate sediments.

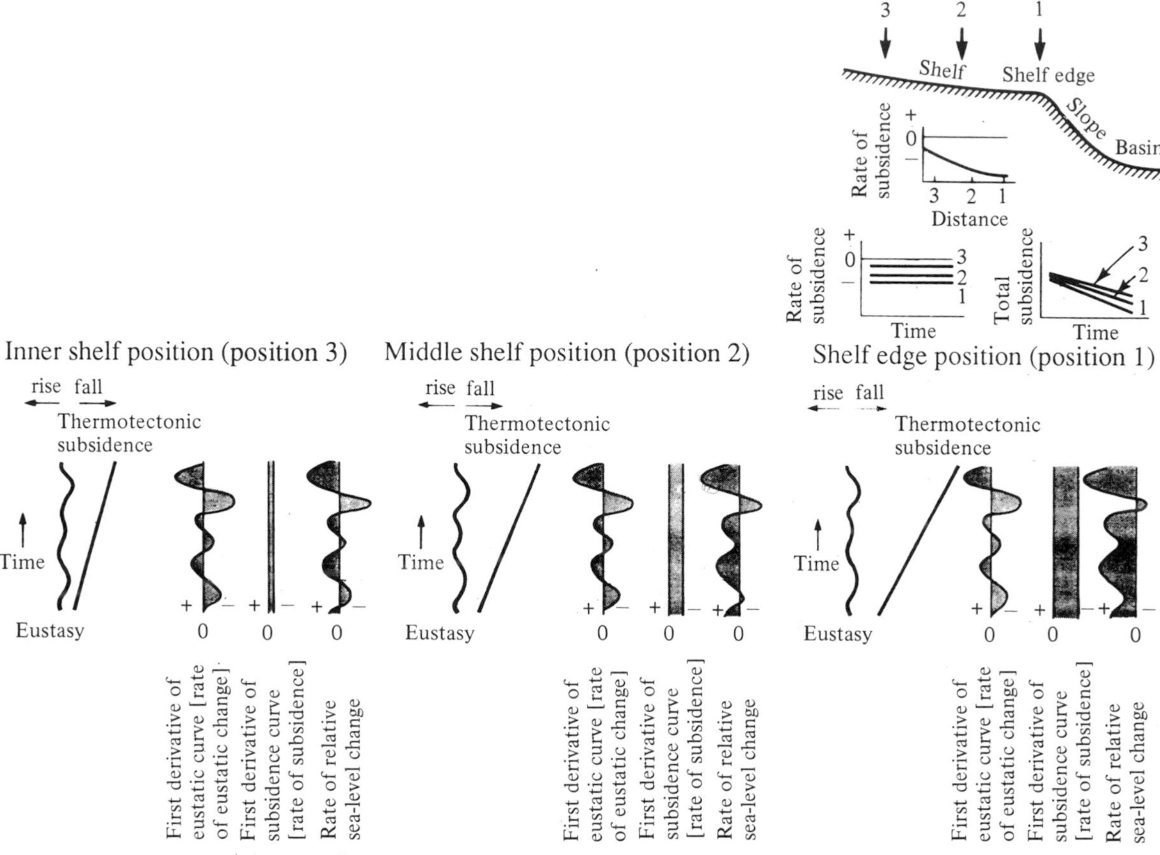

Figure 20.9 Diagrammatic summary showing how rates of formation and removal of accommodation vary across a shelf setting. Note how removal of accommodation only occurs once at the shelf edge, but is much more significant in the inner shelf location. □

Differential rates of subsidence occur around the margins of continents (and sedimentary basins) as shown on Figure 20.9. When the eustatic and subsidence curves for different locations on the shelf are combined, it can be seen that relative sea-level falls are more pronounced to landward. Therefore, in this direction non-deposition and/or erosion within sedimentary sequences becomes progressively more significant. This trend can be seen from stratigraphic sections based on outcrop studies, and on seismic sections (e.g. right side of Figures 20.4b and c).

It is unlikely that eustatic sea-level changes follow the relatively simple cyclic pattern shown in Figure 20.8a. Vail and his colleagues recognised a hierarchy of orders of change. These are illustrated in Figure 20.10. Sequences, and parasequences within them, are considered to be controlled by third to fifth-order cycles.

As shown in Figure 20.10, these three orders of sea-level change can be combined to give a composite rather more irregular curve. In turn, this can be combined with a tectonic subsidence curve to show a plot of relative change of sea-level through time, which is also a plot of changing rates in the addition of accommodation. Times when accommodation is decreasing are characterised by parts of the curve that are inclined down to the right of the diagram; these are periods when erosion will occur on shelves. We shall return to this part of Figure 20.10 later in this chapter.

Depositional systems tracts

The concept of accommodation can be used, together with assumptions about sediment input rates, to simulate the progradational filling of basins. Figure 20.11 shows the relationship between certain physical features observable in sedimentary sequences and changes in sea-level. F and R on the sea-level curves

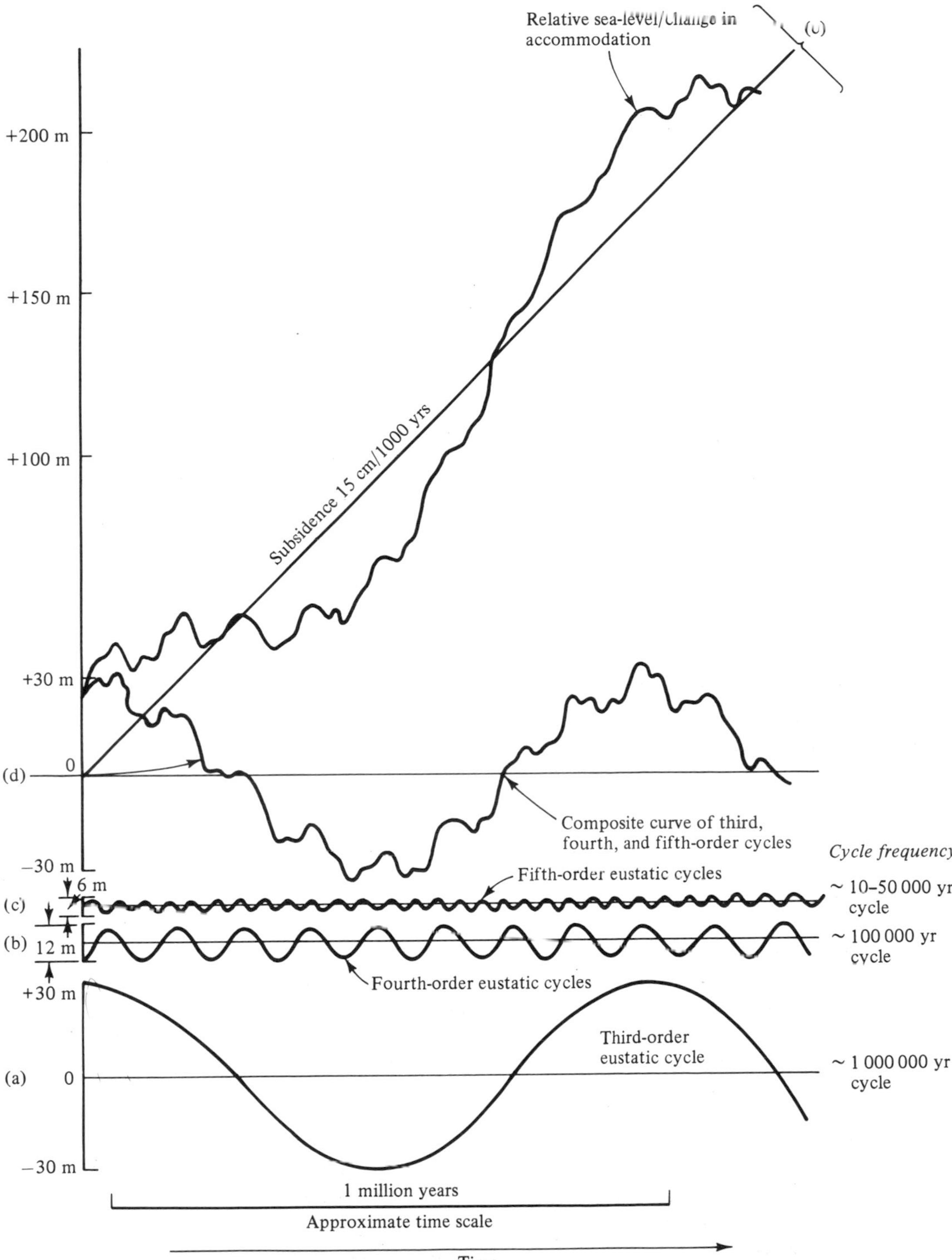

Figure 20.10 An illustration of how relative sea-level rise and changes in accommodation space are controlled by eustatic and tectonic effects. At the bottom of the diagram third, fourth and fifth-order eustatic cycles are shown (a–c). These are combined in (d) to produce a composite curve, which in (e) is added to tectonic subsidence to give a plot of relative sea-level change, or change in accommodation space. Note that relative sea-level rises (accommodation increases) almost continuously, except for brief periods when the curve slopes down to the right. During such periods, Type I sequence boundaries form. □

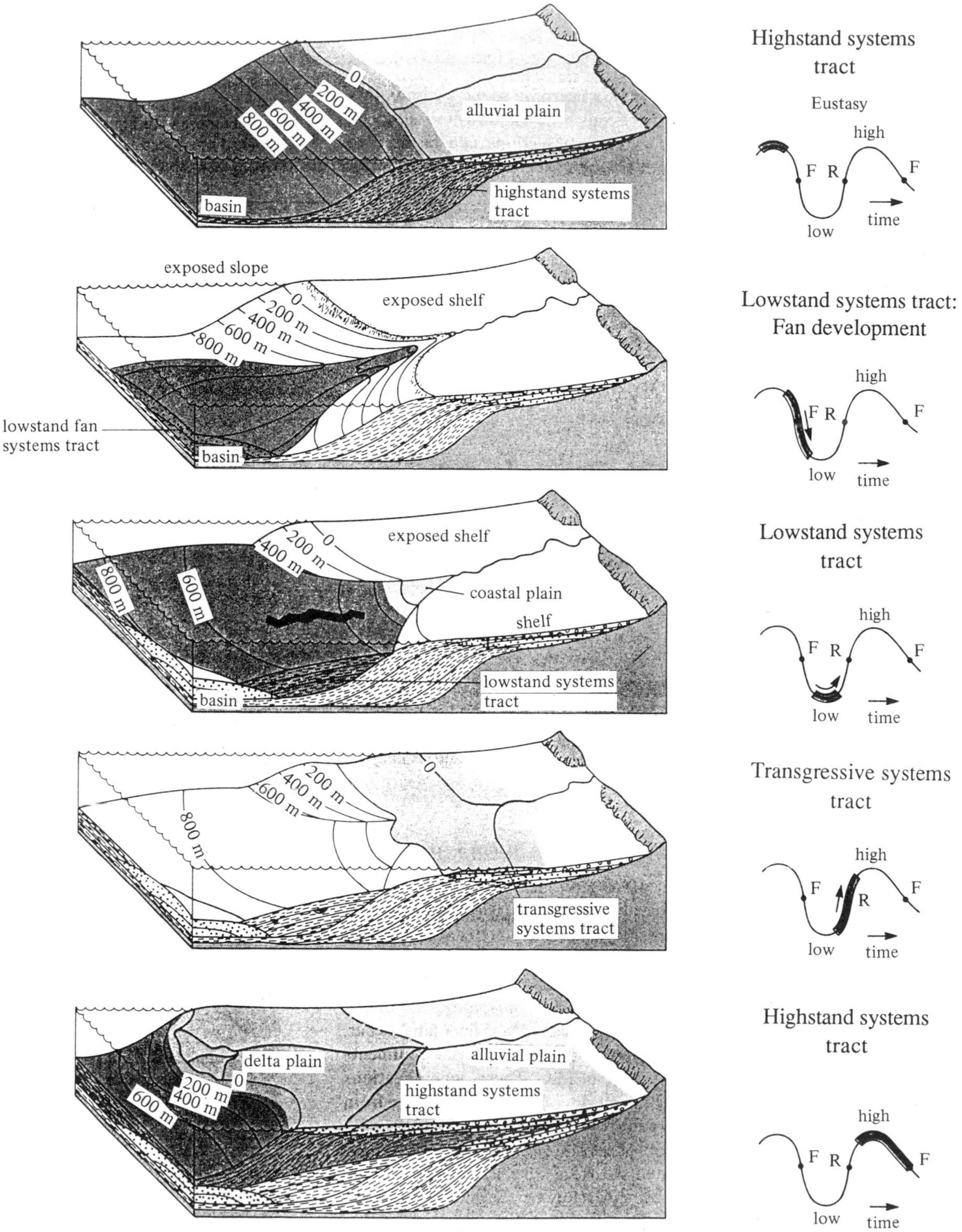

Highstand systems tract
Eustasy
high
F R
F
low
time
0
200 m
400 m
600 m
800 m
alluvial plain
basin
highstand systems tract
Lowstand systems tract: Fan development
exposed slope
exposed shelf
0
200 m
400 m
600 m
800 m
lowstand fan systems tract
basin
high
F R
F
low
time
Lowstand systems tract
exposed shelf
0
200 m
400 m
800 m
600 m
coastal plain
shelf
basin
lowstand systems tract
high
F R
F
low
time
Transgressive systems tract
0
200 m
400 m
600 m
800 m
transgressive systems tract
high
F
R
F
low
time
Highstand systems tract
delta plain
alluvial plain
0
200 m
400 m
600 m
highstand systems tract
high
F R
F
low
time

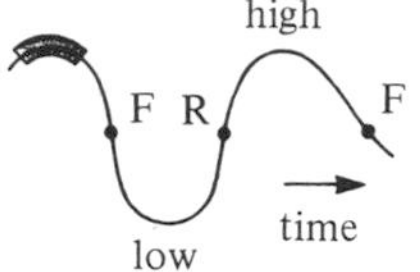

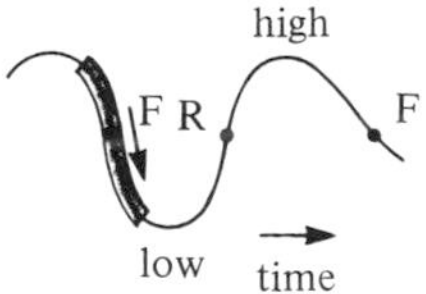

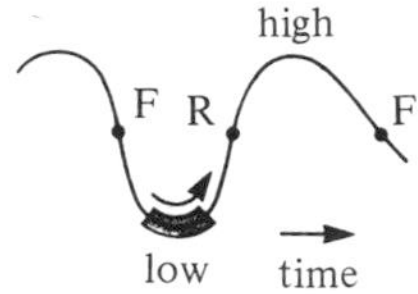

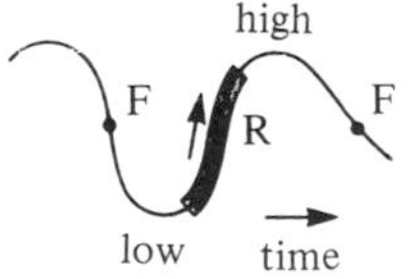

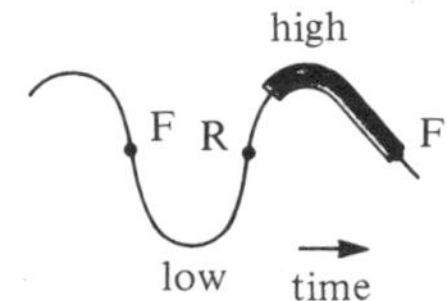

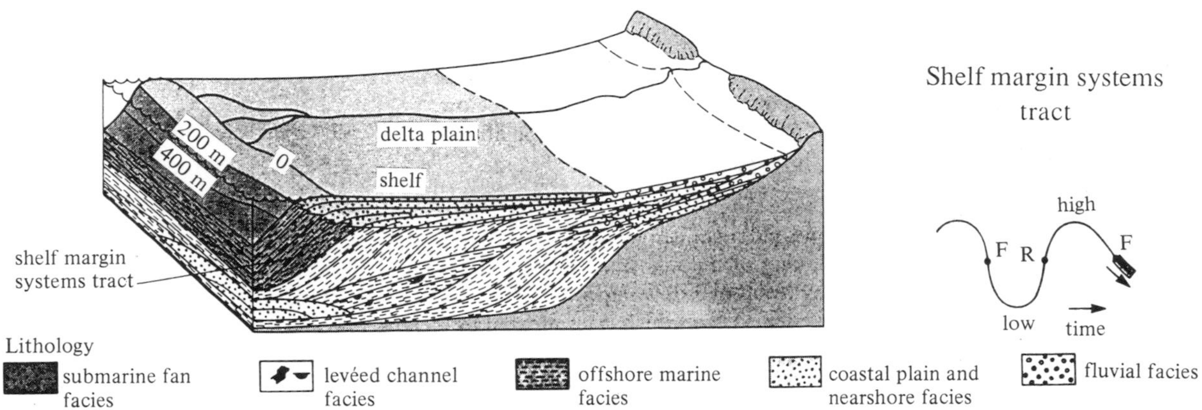

Figure 20.11 Diagrammatic summaries of the geometry of lowstand, transgressive, highstand and shelf margin deposition systems tracts. From Posamentier *et al.* in Wilgus *et al.* (1988).

on Figure 20.11 identify inflection points, where, as explained above, the rate of change is greatest. F inflection points occur on the falling limbs, and R inflection points on the rising limbs. Type 1 and 2 unconformities are linked to relative sea-level drops. Type 1 unconformities develop when accommodation space is removed (e.g. the F inflection point on the far right of Figure 20.8). Submarine fans occur above Type 1 unconformities due to sediment being eroded from the shelf and transported basinward when sea-level falls below the shelf edge. Condensed sections are associated with sea-level rises, and fluvial sediments develop in incised valleys during early sea-level rise, and during highstands and the early part of falls. These generalisations provide an additional method of interpreting the nature of sedimentary facies within sequences identified on seismic sections. This approach has been refined by researchers at Exxon, who believe that there is a predictable succession of associated deposition systems (such as fluvial, deltaic, shelf etc.) associated with different parts of a cycle of eustatic rise and fall. These associated depositional systems are termed **depositional systems tracts** – each one is associated with a specific part of the eustatic curve. Four systems tracts are recognised and illustrated (two highstand tracts are shown) and are discussed below.

Highstand systems tracts develop after the R inflection point when relative sea-level rise progressively slows, and sediment progradation begins, initiating a regression: this marks the beginning of the highstand systems tract. Progradation over earlier systems tracts may result in a downlap surface. Fluvial sediments characterise the later part of this systems tract, which is terminated by a Type 1 or 2 unconformity produced by the next eustatic fall.

Lowstand systems tracts. Two stages of development are shown on Figure 20.11. Basin floor fans develop when sea-level drops below the shelf edge, so that the shelf becomes subaerially exposed, and rivers become incised, by-passing the shelf. The sediments feed directly onto the slope, are transported downslope by turbidity currents and deposited in basin floor fans. As the rate of eustatic fall slackens, subsidence at the shelf edge again exceeds the eustatic fall, so that relative sea-level begins to rise. The incised valleys fill with sediment, and deltas may form in the upper part of the submarine canyons cut earlier. As the slope gradient in front of these deltas is high, re-sedimentation by debris flows and turbidites occurs, and a complex of levéed channels is deposited and is termed the slope fan. As relative sea-level continues to rise and the slope gradient is reduced, prograding slope fans develop. These lowstand prograding complexes persist as long as the rate of sediment supply exceeds the rate of relative sea-level rise (i.e. exceeds the rate of increase of accommodation space).

As the rate of relative sea-level rises increases, the supply of sediment cannot keep up with the rate that new accommodation space is added (i.e. eustatic rise plus subsidence), and so progradation ceases, terminating this systems tract.

Transgressive systems tracts are deposited during rapid eustatic rises, which cause a rapid rise in relative sea-level. This floods the shelf, and so the rivers are no longer incising, therefore little sediment is delivered to the shelf. Seaward of the zone of deposition, condensed sections form.

Shelf margin systems tracts. As the rate of sea-level rise slows again, another highstand systems tract will develop. The succession of systems tracts may then be repeated by the next eustatic cycle. But if the rate of

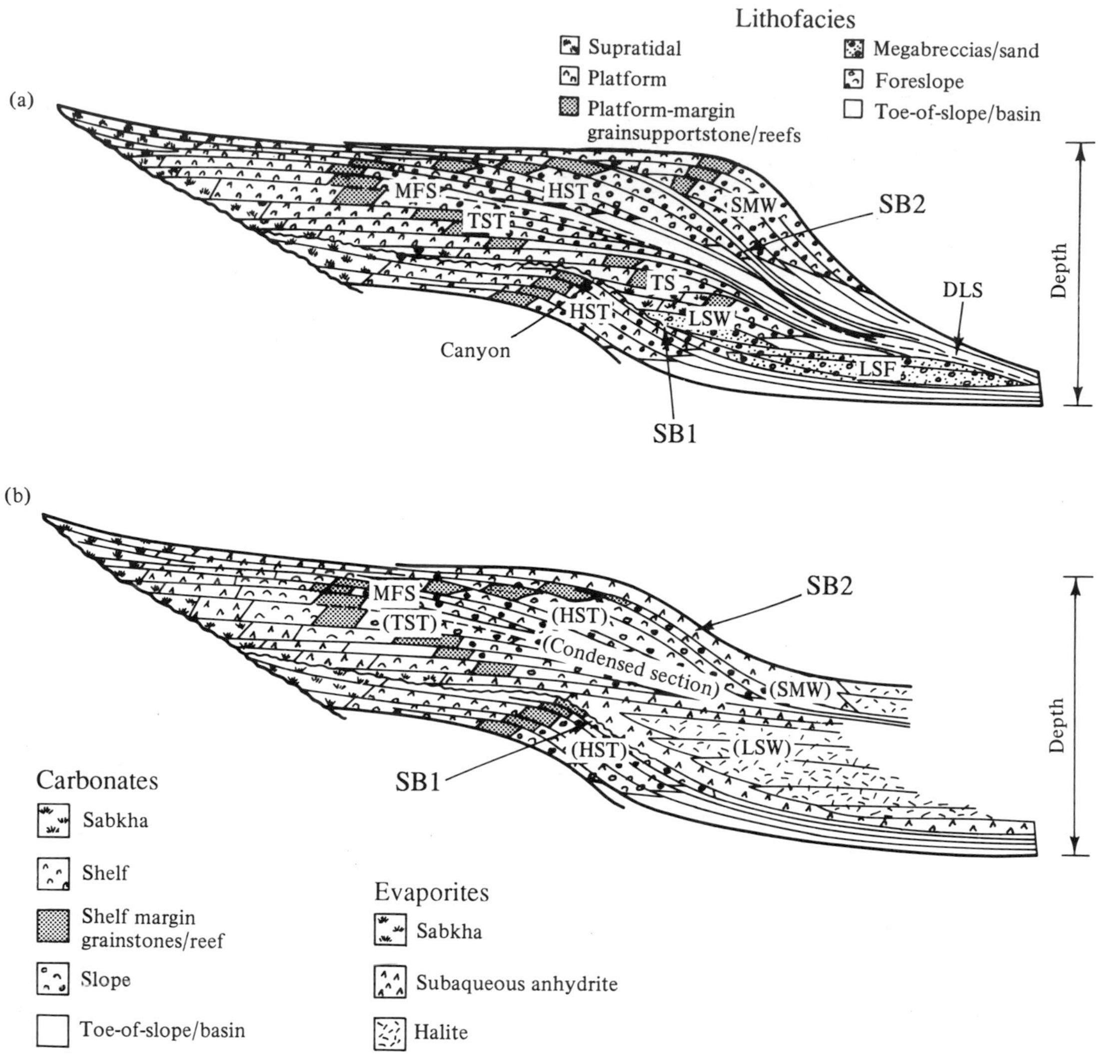

Lithofacies
Supratidal
Platform
Platform-margin grainsupportstone/reefs
Megabreccias/sand
Foreslope
Toe-of-slope/basin
(a)
MFS
HST
TST
SMW
SB2
DLS
TS
HST
LSW
LSF
Canyon
SB1
Depth
(b)
MFS
(TST)
(HST)
(Condensed section)
(SMW)
SB2
(HST)
(LSW)
SB1
Depth
Carbonates
Sabkha
Shelf
Shelf margin grainstones/reef
Slope
Toe-of-slope/basin
Evaporites
Sabkha
Subaqueous anhydrite
Halite

Surfaces

(SB) Sequence boundaries
(SB 1) = Type-1
(SB 2) = Type-2
(DLS) Downlap surfaces
(MFS) = maximum flooding surface
(TS) Transgressive surface
(First flooding surface above maximum regression)

Systems tracts

HST = Highstand systems tract
TST = Transgressive systems tract
LST = Lowstand systems tract
LSF = Lowstand fan
LSW = Lowstand wedge
SMW = Shelf margin wedge systems tract

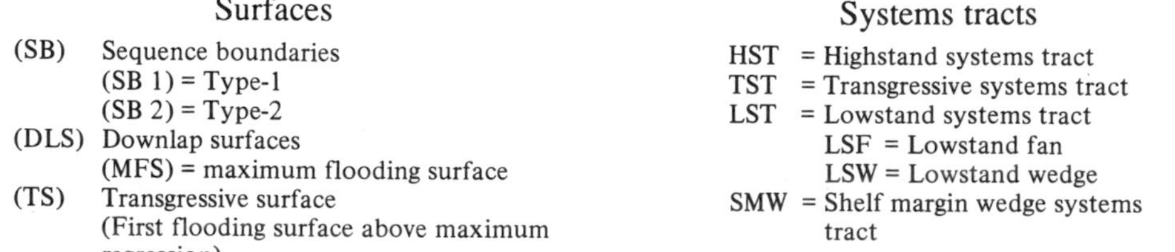

Figure 20.12 Cross-sections showing characteristic facies present in depositional systems tracts in carbonate and evaporite sequences. (a) in a pure carbonate setting; (b) in a mixed carbonate–evaporite setting. □

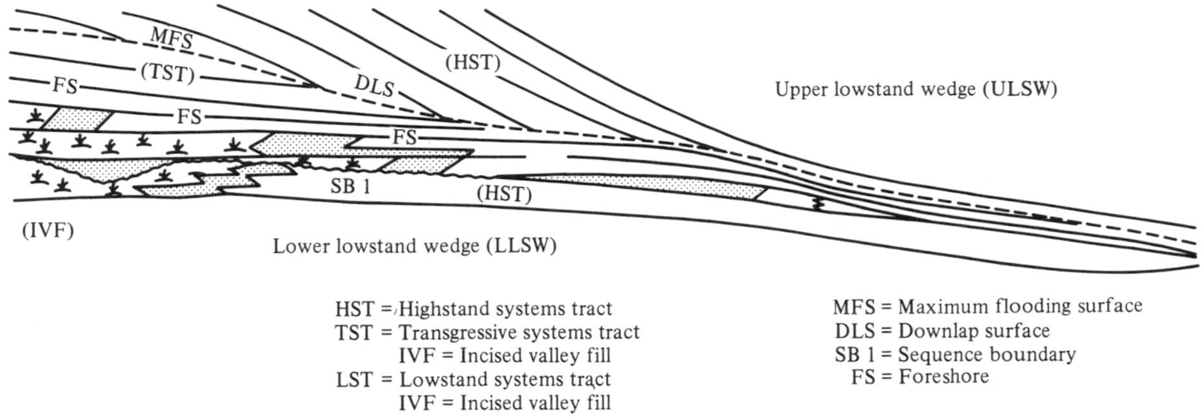

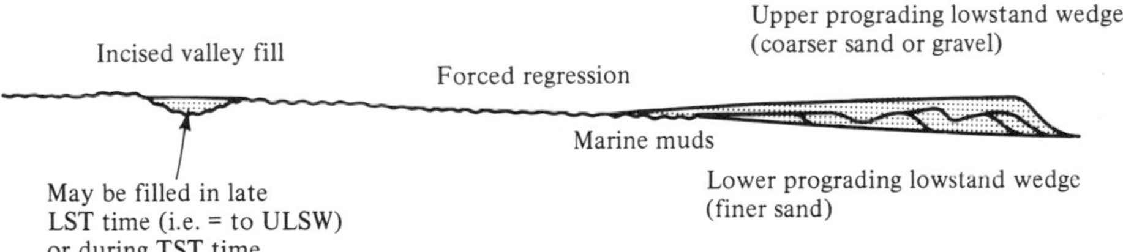

Figure 20.13 **Sequence stratigraphic features developed on a ramp. From Vail, P. R. in Bally (1987) and Vail (pers. comm.).** □

eustatic sea-level fall is lower than that associated with the lowstand systems tract (second diagram on Figure 20.11) a shelf-margin systems tract is deposited. The base of this systems tract is a Type 2 unconformity (see Figure 20.6). Aggradation rather than progradation is characteristic because the *rate* of relative sea-level rise is now higher than it was during the deposition of the previous highstand systems tract. This is because on the falling limb of sea-level change, the rate is increasing before each F inflection point, but falling after it. The shelf margin systems tract is overlain either by a transgressive systems tract, or a condensed section if the latter is not developed.

Sequence stratigraphic models have also been developed for carbonate and mixed carbonate–evaporite settings, as shown in Figure 20.12. Mixed carbonate–clastic systems also occur. In such cases, siliciclastic sediments – turbidites and shales – characterise the lowstand depositional systems tracts, due to sediment by-passing the carbonate shelf edge. The latter develops by carbonate progradation during highstands.

The Exxon group have also proposed a model for the occurrence of depositional systems tracts in basins that develop ramp profiles, and this is illustrated in Figure 20.13. However, these are only discussed briefly in publications currently available. They suggest that the lowstand systems tract is deposited in two parts. The first results from stream rejuvenation and sediment by-passing of the former coastal plain area caused by a relative sea-level fall. The second part is deposited during a slow rise, so that shoreline progradation is slowed and changes to aggradation. The resultant lower lowstand wedge sediments are finer grained than the upper lowstand wedge, and are commonly deposited in lower shoreface or offshore environments and are capped by an erosional surface. This surface does not mark the sequence boundary, which occurs at the base of the lower lowstand wedge sands.

Parasequences

Parasequences are the building blocks of sequences and depositional systems tracts. They are defined as 'relatively conformable successions of genetically related beds or bed sets bounded by marine flowing surfaces or their correlative surfaces'. Occurring in coastal plain, deltaic, beach, tidal, estuarine and shelf environments, they are commonly expressed as coarsening-up progradational units (regressive sequences) (Figure 20.14).

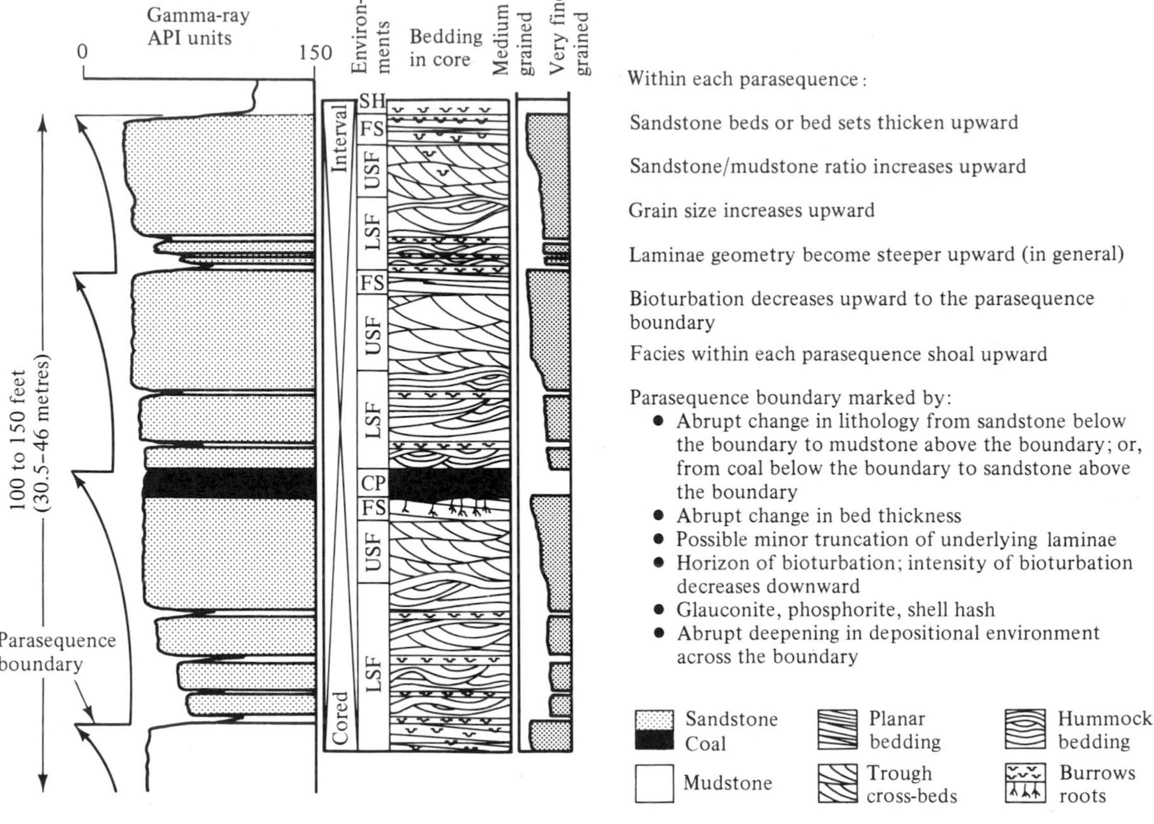

Figure 20.14 **Sedimentary log showing the characteristic features of a series of coarsening-upwards parasequences deposited in a shoreface to beach environment. The rate of deposition equalled the rate of accommodation.** □

Marine-flooding surfaces separate parasequences; there is often evidence indicative of a sudden increase in depth of deposition across the surface, and sometimes evidence of minor submarine erosion, or non-deposition. A **parasequence** set consists of a succession of parasequences deposited in related environments, and bounded by major marine-flooding surfaces and their correlative surfaces. The sets may show stacking patterns indicative of progradation, aggradation or retrogradation.

It is possible for parasequence set and parasequence boundaries to coincide with systems tract and sequence boundaries in certain circumstances. A theoretical example of how these two stratal types might occur within a sequence is shown in Figure 20.15.

The use of transgressive–regressive cycles for regional correlation is an approach familiar to many geologists. It uses as marker horizons transgressive units at the top of regressive units (e.g. coal measure marine bands, or limestones in Yoredale cycles). These are the flooding surfaces of sequence stratigraphers; they believe that though they are the best means of correlating between open marine sequences, they are not reliable in shallow marine and non-marine sequences. Sequence boundaries rather than transgressive units offer a more reliable means of regional correlation in non-marine successions for a variety of reasons, some of which are briefly given here (these are fully discussed in Van Wagoner *et al.* 1990).

1 A sequence boundary forms independently of sediment supply, whereas transgressions and regressions may be strongly controlled by it in some cases. Therefore flooding surfaces may be diachronous, as, for example, there may be local differences in rates of shoreline progradation due to variations in sediment supply.
2 Although all the points on a sequence boundary do not represent the same time span (see explanation in Figure 20.5), one instant of time is common to all points. This feature of time-

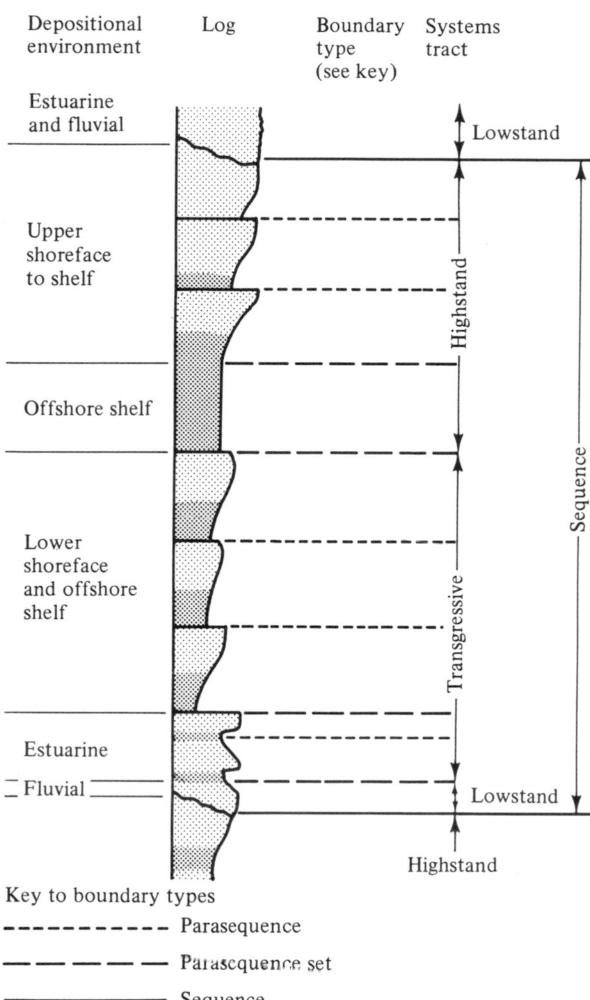

Figure 20.15 An idealised section showing how parasequences and parasequence sets might stack within a sequence. □

stratigraphic significance plus the fact that the sequence boundary is a basin-wide surface (usually separating rocks with contrasting characteristics), makes it preferable as a marker to flooding surfaces.

3 As shown in Figure 20.14, several transgressive events/flooding surfaces may occur within a sequence (at the bases of parasequences), and so might lead to confusion when attempting regional correlations of successions containing multiple transgressive events.

An exemplar sequence stratigraphic interpretation

Figure 20.16 shows how an outcrop section in the Upper Cretaceous of Utah is correlated with a succession encountered in a borehole 18 km north of the outcrop. The coarsening-up units in the lower shoreface parts of the sequences are similar to the lower part of the parasequences depicted on the log in Figure 20.14.

Sequence boundaries are marked by abrupt basinward shifts of facies type in the successions. In other words, shallower water sediments overlie deeper water ones. For example, the top of Sequence 1 at outcrop and in the borehole consists of a series of coarsening-up shoreface units (parasequences), whereas the base of Sequence 2 is marked by erosional truncation, overlain by braided river deposits at outcrop, and fluvial to estuarine sediments in the borehole. A similar change occurs across the Sequence 2–3 boundary. In both cases the lowstand deposits are more proximal (i.e. show a greater terrestrial influence) at outcrop than they do in the borehole (e.g. braided river to coastal plain versus fluvial–estuarine–coal swamp).

Detailed studies of outcrops exposed along canyons show that the Sequence 1–2 and 2–3 boundaries are very irregular, and indicate the presence or incised valleys (Figure 10.16c). The depth of the incision increases to the west – the landward direction. Correlation of the parasequences as shown on Figure 20.16c indicates a progradational pattern building eastwards – a feature typical of prograding deposits on seismic sections (cf. Figure 20.10).

Close examination of Figures 20.16a and b shows that the sequence stratigraphic units do not always correspond to the lithostratigraphic division of the successions. The boundary between Sequences 1 and 2 occurs *within* the Desert Member, and that between 2 and 3 occurs near the top of the Desert Member at outcrop, but in the borehole within the overlying Castlegate Member. Sequence stratigraphers argue that sequence boundaries record the fundamental breaks in deposition; at each sequence boundary the 'slate is wiped clean' and a new depositional record begins. Lithostratigraphic subdivisions commonly miss these fundamental boundaries, making it difficult to construct accurately a chronostratigraphic and regional framework. Once the sequence stratigraphic subdivision is made, the lithostratigraphic terminology is often so confusing that it needs to be modified substantially or abandoned (Van Wagoner *et al.* 1990, p. 40).

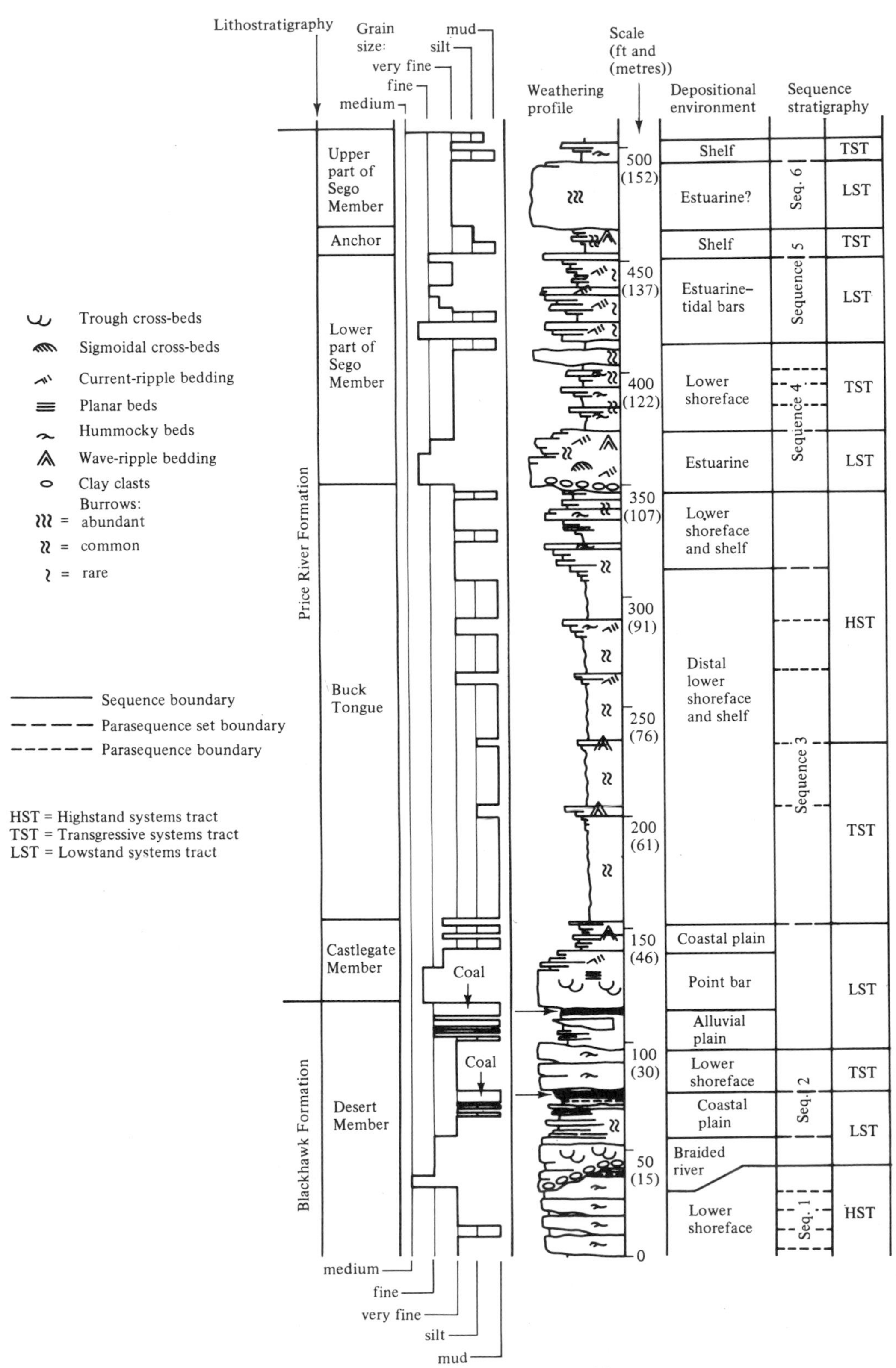
Lithostratigraphy
Grain size:
medium
fine
very fine
silt
mud
Weathering profile
Scale (ft and (metres))
Depositional environment
Sequence stratigraphy
Upper part of Sego Member
Anchor
Lower part of Sego Member
Buck Tongue
Castlegate Member
Desert Member
Price River Formation
Blackhawk Formation
Coal
Coal
500 (152)
450 (137)
400 (122)
350 (107)
300 (91)
250 (76)
200 (61)
150 (46)
100 (30)
50 (15)
0
Shelf
Estuarine?
Shelf
Estuarine–tidal bars
Lower shoreface
Estuarine
Lower shoreface and shelf
Distal lower shoreface and shelf
Coastal plain
Point bar
Alluvial plain
Lower shoreface
Coastal plain
Braided river
Lower shoreface
Seq. 6
Sequence 5
Sequence 4
Sequence 3
Seq. 2
Seq. 1
TST
LST
TST
LST
TST
LST
HST
TST
LST
TST
LST
HST
Trough cross-beds
Sigmoidal cross-beds
Current-ripple bedding
Planar beds
Hummocky beds
Wave-ripple bedding
Clay clasts
Burrows:
= abundant
= common
= rare
Sequence boundary
Parasequence set boundary
Parasequence boundary
HST = Highstand systems tract
TST = Transgressive systems tract
LST = Lowstand systems tract

(a)

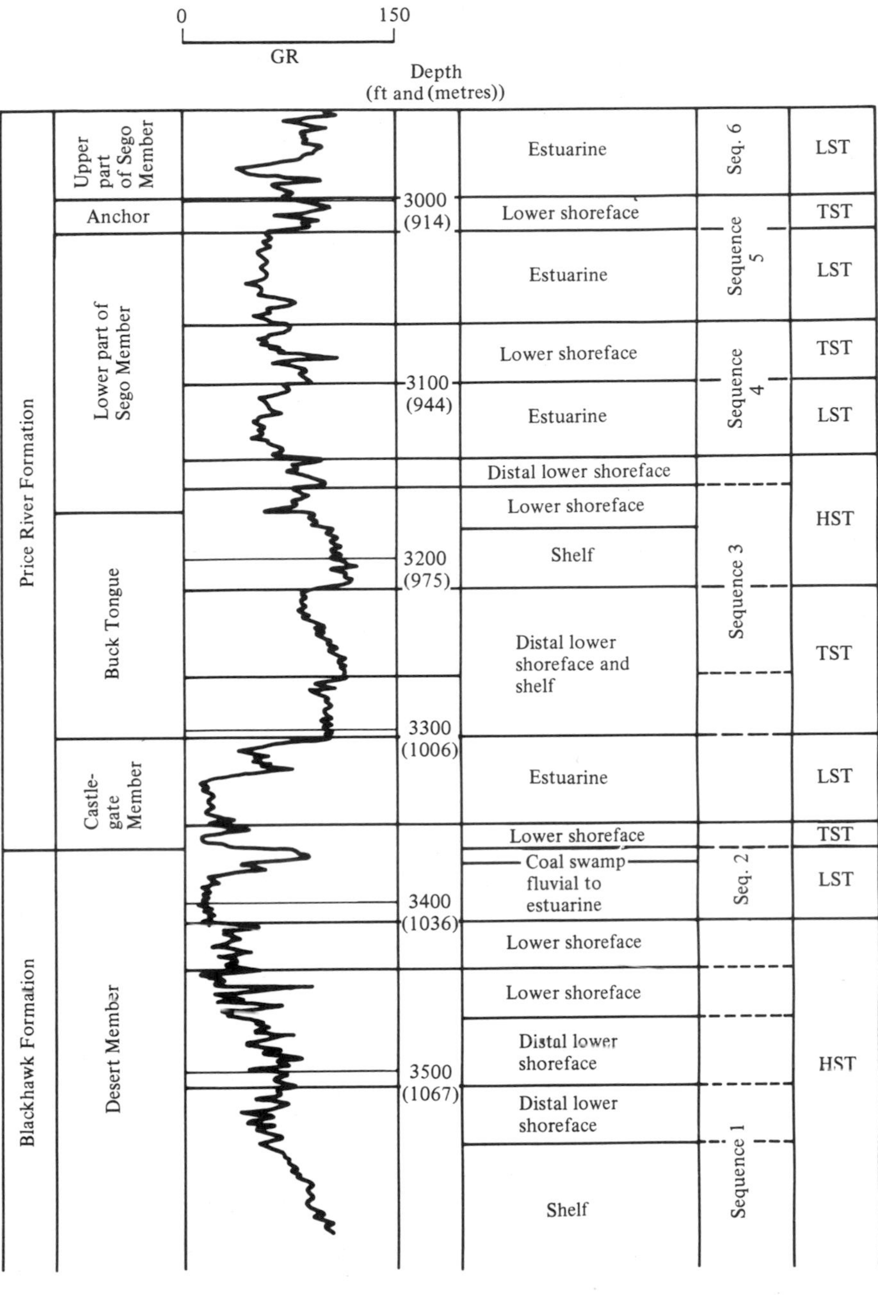

(b)

Figure 20.16 Upper Cretaceous sequences in Utah, USA. (a) Sedimentary log based on outcrop measurements, showing lithostratigraphic and sequence stratigraphic divisions. (b) Interpretation of succession encountered in a borehole located along depositional strike 18 km north of the outcrop section shown in (a). The gamma-ray (GR) log gives an indication of the shaleyness of the section, as this lithology emits relatively high amounts of gamma radiation. Thus an upward increase in this log indicates a fining-up sequence trend, and vice versa. The interpretation shown is based on this type of log, and other methods plus extensive core samples. (c) West–east (west is to the right) cross-section along outcrop belt showing correlation of Sequences 1, 2 and 3. These reveal (i) that the bases of Sequences 2 and 3 are incised valleys filled with fluvial and estuarine sediments, and (ii) that the geometry of the parasequences at the top of Sequence 1 shows eastward progradation into the basin (compare with highstand geometry shown in Figure 20.11). □

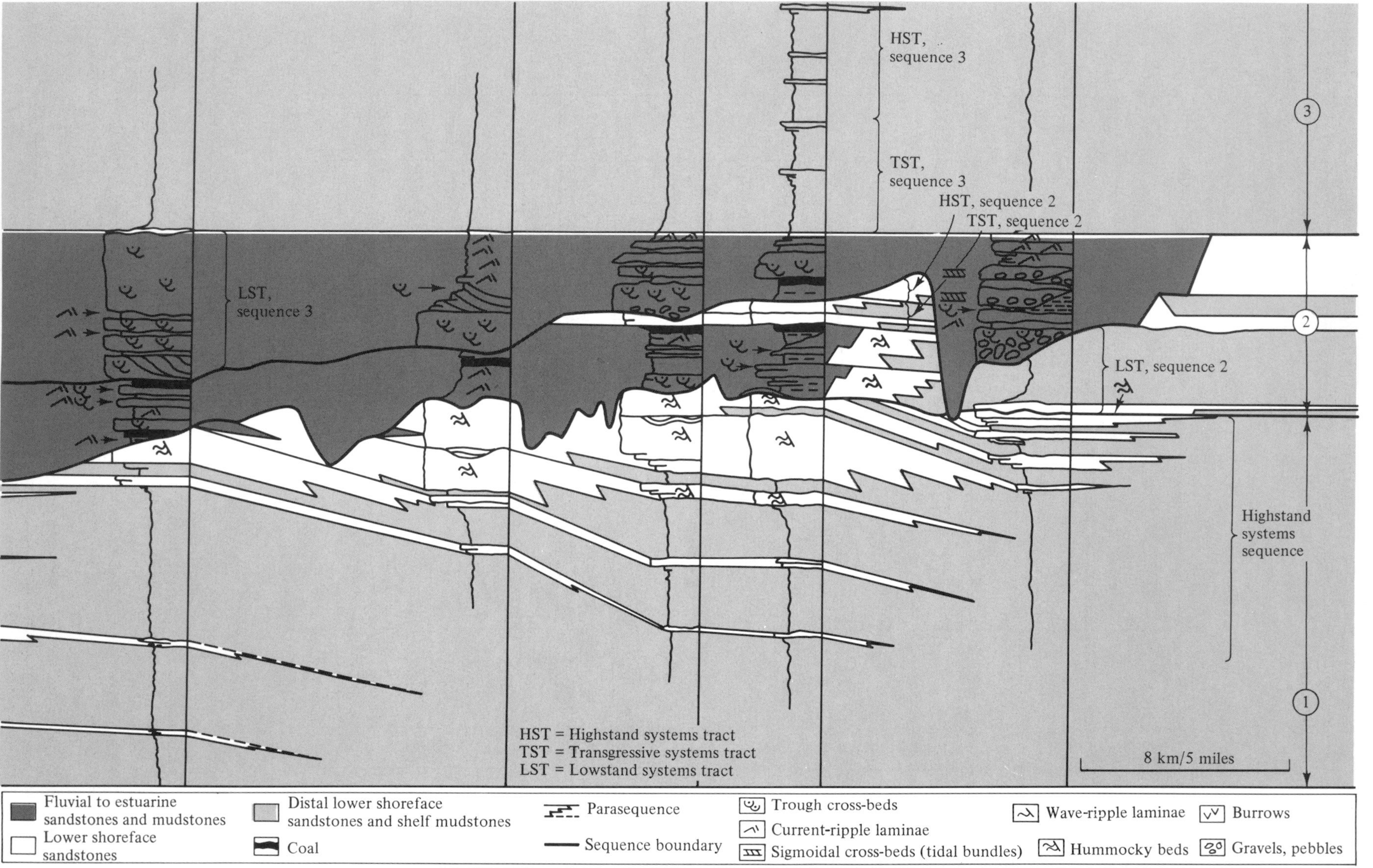
HST,
sequence 3
TST,
sequence 3
HST, sequence 2
TST, sequence 2
LST,
sequence 3
LST, sequence 2
Highstand
systems
sequence
3
2
1
HST = Highstand systems tract
TST = Transgressive systems tract
LST = Lowstand systems tract
8 km/5 miles
Fluvial to estuarine
sandstones and mudstones
Lower shoreface
sandstones
Distal lower shoreface
sandstones and shelf mudstones
Coal
Parasequence
Sequence boundary
Trough cross-beds
Current-ripple laminae
Sigmoidal cross-beds (tidal bundles)
Wave-ripple laminae
Hummocky beds
Burrows
Gravels, pebbles

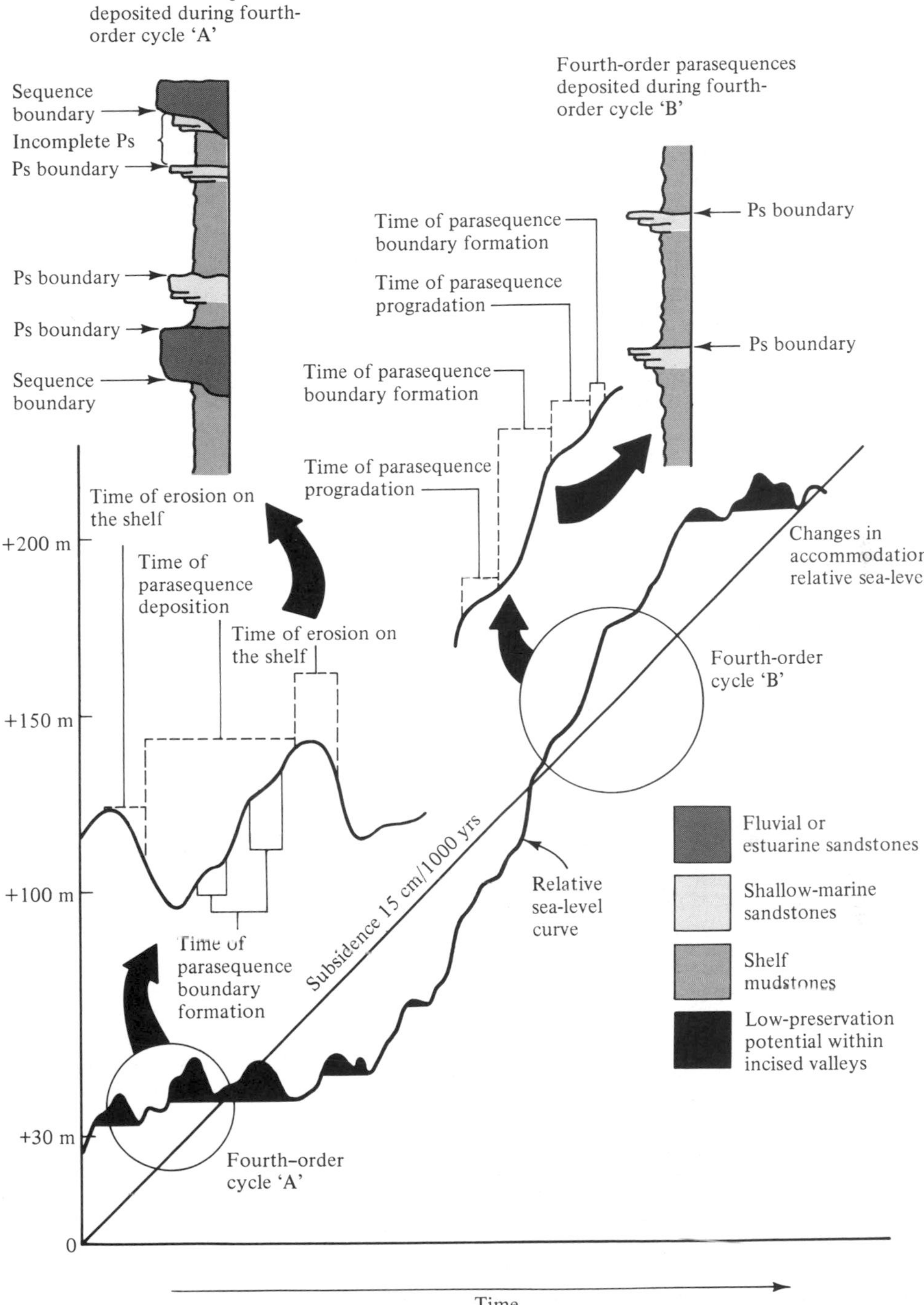

Figure 20.17 The relationship between sequences, parasequences and changes in accommodation/relative sea-level. The curve is produced by adding tectonic subsidence to eustatic changes resulting from combining third, fourth and fifth-order cycles (see Figure 20.10). Intervals on the curve where contemporaneous sediments are likely to have a low preservation potential within incised valleys are shown. Sequence boundaries form during these intervals. Parasequence boundaries mark the end of periods of low rates of rise; as the rate increases, marine-flooding surfaces result. □

The development of sequences and the parasequences observed within them can be related to the composite curve combining third, fourth and fifth-order eustatic changes of sea-level shown earlier in Figure 20.10. This figure showed how the combination of a composite eustatic curve (combining third to fifth-order changes) and tectonic subsidence caused relative sea-level to rise, and accommodation to increase almost continuously, with only brief periods of accommodation removal. The combination of eustatic and tectonic effects is shown again in Figure 20.17, with the parts of the curve that are unlikely to leave a sedimentary record due to erosion on shelves as accommodation decreases highlighted. Two parts of the accommodation curve are enlarged, and the effects of changes in rates of accommodation and loss of accommodation on the sedimentary record shown. Sequence boundaries are produced when accommodation is removed. Parasequences record cycles of change in the rate of accommodation increase, with the top of each parasequence recording the time when accommodation is rising at the lowest rate. Once accommodation space increases, a marine-flooding surface will form.

Sequence stratigraphy and eustasy

Information from studies of sedimentary succession at outcrop and in boreholes, and from seismic sections, can be used, according to the 'Vail School' to determine eustatic changes of sea-level. The timing of sea-level falls can be determined from the ages of sequence boundaries, and the maximum rate of rise from the age of transgressive surfaces indicated by condensed sections and downlap surfaces. The amount of rise and fall can be estimated from measurements of coastal onlap. The resultant short-term changes of sea-level are then superimposed on the longer term eustatic change caused by changes in the volume of oceanic ridges worldwide, as illustrated on Figure 20.18.

Continental fluvial deposits, usually developed in braided stream facies, may occur in association with Type 1 unconformities where they fill incised valleys. When fluvial deposition commences, it means that onlap visible on seismic sections shifts significantly landward. This is because fluvial sediments cannot be distinguished from deltaic or coastal sediments on seismic sections, making it impossible to define the extent to which onlap is truly *coastal* – in fact it is coastal *and* non-marine onlap. As fluvial sediments are deposited above sea-level, any onlapping relations they show are not *directly* related to relative sea-level changes, although they are directly related to base-level changes caused by relative changes of sea-level. Therefore, fluvial deposition tends to accentuate the saw-tooth nature of coastal onlap charts and, as stated previously, led to the early Vail global sea-level charts showing rapid falls followed by slow rises.

Figure 20.19 is an extract from the Exxon *Mesozoic–Cenozoic Global Cycle Chart*. This is often referred to as the 'Haq curve', after the first author of a paper in *Science* which presented it for the first time in 1987. The extensive biostratigraphic information displayed on the chart is not shown. The Exxon group claim that the proposed sequences can be traced worldwide, but as yet it is difficult to judge this claim, as the extensive database on which it is based is not in the public domain. Sequence boundaries are given absolute dates for ease of identity. This is open to the criticism that the quantitative calibration of biostratigraphic divisions is constantly under revision, and so giving absolute dates may sow the seeds of confusion. This is

Figure 20.18 Idealised lithostratigraphic and chronostratigraphic sections and charts showing how global sea-level changes are estimated by followers of the Vail/Exxon School (a and b are identical to Figures 20.5a and b, but facies types are added). Note that many actual examples from outcrops, boreholes and seismic sections are used to construct the global curve, a small part of which is shown in Figure 20.19. (a) Lithostratigraphic section, showing three sequences. Each sequence is divided into two parts by a downlap surface (which may be coincident with a condensed section at outcrop, or in boreholes). An example of how coastal onlap is measured is shown for Sequence 1, where the change in elevation of the top of coastal plain sediments through time can be seen. (b) Chronostratigraphic summary chart based on (a). The time gaps across the sequence boundaries can be seen, and the landward encroachment of sediments (i.e. onlap) in each sequence onto older strata is clearly displayed. (c) Chart showing relative coastal onlap; note how the onset of fluvial sediments accentuates the saw-tooth shape of the onlap curve. (d) Chart showing changing shoreline location, through time, and the timing of the development of downlap surfaces and condensed sections, which mark the time of maximum rate of sea-level rise. The sequence boundaries mark periods of rapid sea-level fall. The two lines on either side of the lower (Type 1) boundary show the range over which there is probably no rock record due to erosion, and across which shoreline location etc. must be extrapolated. (e) Chart showing how the short-term sea-level fluctuations deduced from (c) are superimposed on long-term estimates of global sea-level change calculated from changes in ocean ridge volumes. The amplitude of the short-term fluctuations is based on onlap measurements, with corrections being made for tectonic subsidence. □

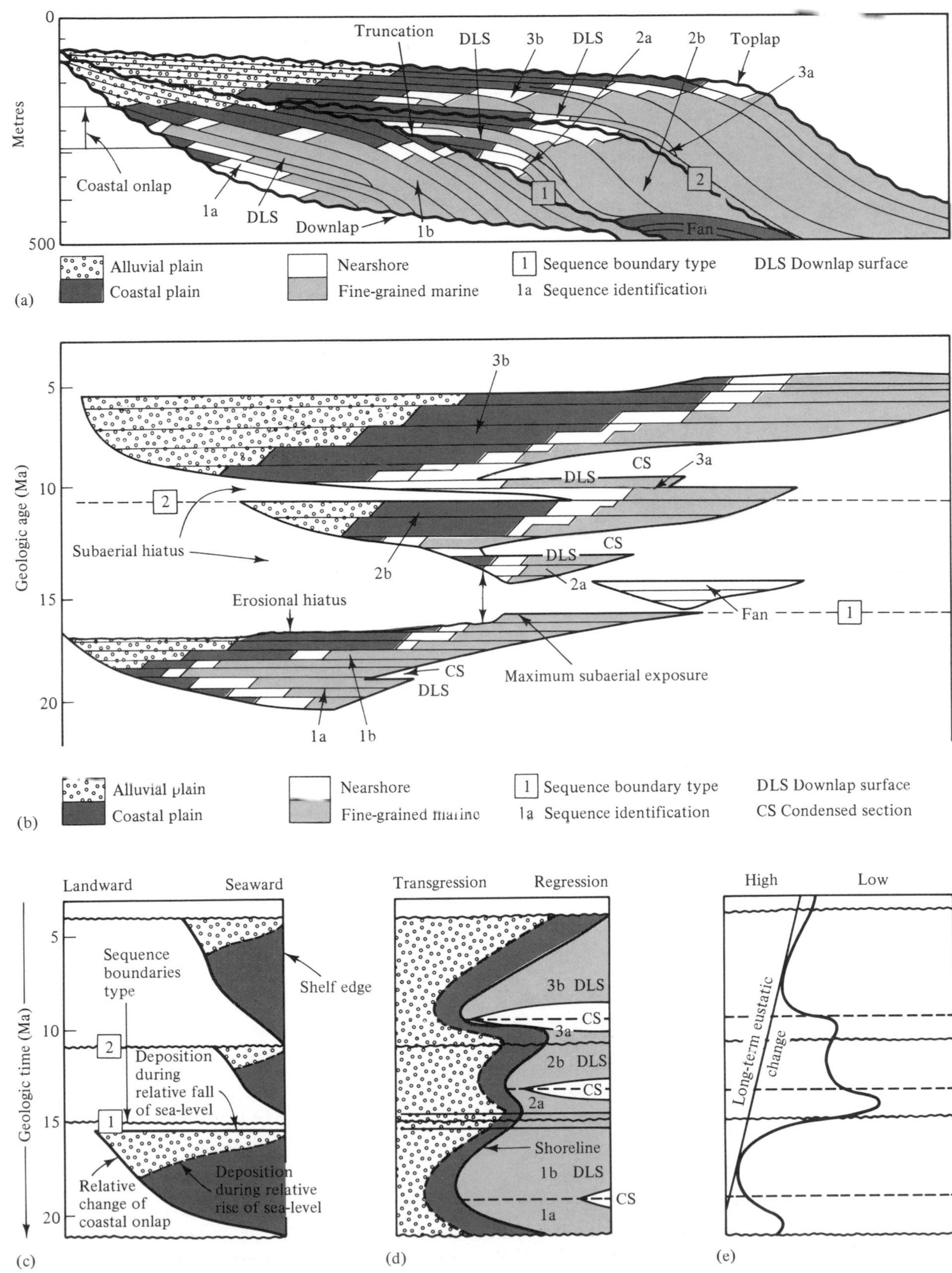
0
500
Metres
Truncation
DLS
3b
DLS
2a
2b
Toplap
3a
Coastal onlap
1a
DLS
Downlap
1b
1
2
Fan
Alluvial plain
Coastal plain
Nearshore
Fine-grained marine
1 Sequence boundary type
1a Sequence identification
DLS Downlap surface
(a)
Geologic age (Ma)
5
10
15
20
3b
CS
3a
DLS
2
Subaerial hiatus
2b
CS
DLS
2a
Fan
1
Erosional hiatus
Maximum subaerial exposure
CS
DLS
1a
1b
Alluvial plain
Coastal plain
Nearshore
Fine-grained marine
1 Sequence boundary type
1a Sequence identification
DLS Downlap surface
CS Condensed section
(b)
Landward
Seaward
Geologic time (Ma)
Sequence boundaries type
Shelf edge
Deposition during relative fall of sea-level
Relative change of coastal onlap
Deposition during relative rise of sea-level
(c)
Transgression
Regression
3b DLS
CS
3a
2b DLS
CS
2a
Shoreline
1b DLS
CS
1a
(d)
High
Low
Long-term eustatic change
(e)

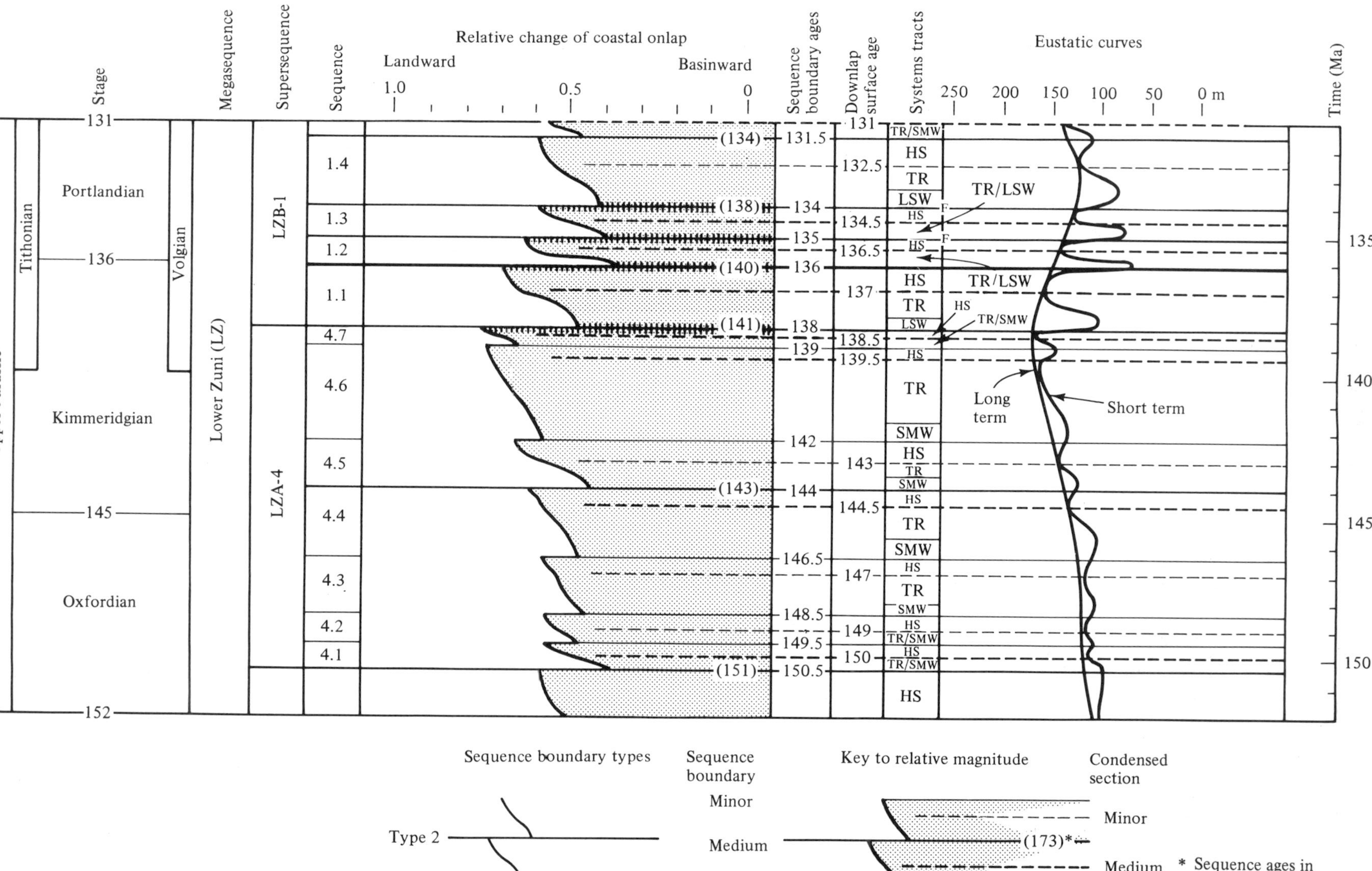

Figure 20.19 A small part of the Exxon Mesozoic–Cenozoic global cycle chart. Note that extensive biostratigraphic data presented is not shown. □

illustrated by the fact that ages given on earlier versions of the chart are given in the downlap column of Figure 20.19 in order to avoid such confusion. It must be stressed that anyone using the chart to review their own data should do so via the biostratigraphic information provided on the completed chart (this is not shown on Figure 20.19).

Confidence in the belief that the Haq cycle chart has worldwide application is based on the assumption that the dominant control on sequence development is eustatic sea-level change. At present, there is a debate in progress between those who believe that the development of depositional sequences was controlled by global sea-level changes induced by the formation and melting of ice caps, and those that consider that they were caused by widespread relative sea-level changes related to tectonics. Most geologists accept that the long-term global changes of sea-level, such as those depicted on Figure 20.18, were caused by changes in volume of the ocean 'bucket' as new ocean ridges grew in length (and hence volume) and displaced ocean water onto continents – or vice versa (see Chapter 9). But this process can only produce global sea-level changes at a rate of about 1 cm per 1000 years. The short-term eustatic changes postulated by the 'Vail School' occurred at rates of the order of 1 cm per year. Global changes at such rates can only be produced by changes in the volume of land ice (by amounts up to 150 m) or by desiccating ocean basins, such as the Mediterranean (by amounts of up to about 15 m). The problem is that coastal onlap studies indicate rapid changes of several tens of metres, and sometimes more, during the Mesozoic, at times when there appear to have been no major ice sheets.

Some workers have criticised the work of the Exxon group on the grounds that their examples are biased towards Atlantic margin basins and northwest Europe, and so the timing of changes in coastal onlap may be related to rifting and thermal subsidence of continents around this ocean. In the next chapter, Sierd Cloetingh suggests that relative sea-level changes can be induced over a wide area, at the scale of single or several tectonic plates, by changes in the stress condition of the lithosphere. He suggests that tensional stresses affecting entire plates produce an apparent rise in sea-level, followed by more rapid relative sea-level falls as the stresses are relaxed. Thus he believes that the Haq curves provide a useful source of information concerning past intraplate stress regimes.

Conclusions

Most geologists would agree that relative sea-level changes do affect the geometry and facies distribution of sediments deposited on continental margins and within sedimentary basins. Therefore, it must be emphasised that proof or disproof of the global nature of the sea-level curves produced by the Exxon group will not alter the fact that seismic stratigraphy is a powerful tool in petroleum exploration. Many companies and individuals, irrespective of whether they believe in a glacio-eustatic or a plate-wide tectonic origin for the observed sea-level curves, will continue to use the sequence stratigraphic method to identify depositional sequences using outcrop and subsurface data.

Further reading and viewing

Video

Seismic Stratigraphy (1987) running time 43 minutes. Presented by Peter Vail, and co-produced by the Open University and the American Association of Petroleum Geologists. Available from the Association in North America, and the Geological Society of London in Europe.

Texts

Bally, A. W. (ed.) (1989) *Atlas of Seismic Stratigraphy*, American Association of Petroleum Geologists Studies in Geology 27, vols. 1, 2 and 3.

These atlases contain many large-scale uninterpreted and interpreted seismic sections showing the application of seismic stratigraphic techniques in a wide range of basin types and in siliciclastic and carbonate sediments.

Berg, O. R. & Woolverton, D. G. (1986) *Seismic Stratigraphy II*, American Association of Petroleum Geologists, Memoir 39.

This book presents the advancements and refinements to the seismic stratigraphic approach that were developed after the publication of AAPG Memoir 26. It contains papers that stress the tectonic rather than eustatic controls on depositional sequences.

Campbell C. V. (1967) Lamina, Laminaset, Bed and Bedset, *Sedimentology*, **8**, 7–26.

The original paper describing the smaller scale units in the hierarchy of stratal units shown on Table 20.1.

Payton, C. E. (ed.) (1977) *Seismic Stratigraphy – Application to Hydrocarbon Exploration*, American Association of Petroleum Geologists, Memoir 26.
This is the *book on the subject, and summarises the main concepts that the Exxon School had developed over the previous decade or so. The global cycles of eustatic sea-level change proposed were subject to much criticism because their saw-tooth pattern indicated very rapid falls. Later work revised the shape of these curves when it was shown that the coastal onlap measurements did not allow for fluvial sediments onlapping earlier sequences.*

Schlee, J. (ed.) (1984) *Inter-regional Unconformities and Hydrocarbon Accumulation*, American Association of Petroleum Geologists, Memoir 36.
This contains a paper (pp. 129–44) by Vail and his co-workers explaining how sea-level curves are estimated by avoiding the problem of onlapping fluvial sediments.

Sloss, L. L. (1963) Sequences in the cratonic interior of North America, *Geological Society of America Bulletin*, **74**, 93–114.
The paper describing large-scale sequences (now the megasequences of the Haq curve).

Van Wagoner, J. C., Mitchum, R. M., Campion, K. M. & Rahmanian, V. D. (1990) *Siliciclastic Sequence Stratigraphy in Well Logs, Cores and Outcrops: Concepts for High Resolution Correlation of Time and Facies*, American Association of Petroleum Geologists, Methods in Exploration Series No. 7.
The book for those wishing to get to grips with the sequence approach and apply it to outcrops and well data. It contains excellent summaries of the components of the hierarchy of stratal units introduced in this article.

Wilgus, C. K., Hastings, B. S., Kendall, C. G. St. C., Posamentier, H. W., Ross, C. A. & Van Wagoner, J. C. (1988) *Sea-level Changes: An Integrated Approach*, Society of Economic Paleontologists and Mineralogists Special Publication 42.
The application of sequence stratigraphic techniques to studies of rock sequences is described, and the global correlation chart discussed.

Acknowledgements

The author thanks members of the 'Vail School' for numerous stimulating presentations and discussions during attendance at Esso workshop sessions for British academics and when making the video *Seismic Stratigraphy*. Without the benefit of this experience, this chapter could not have been written. Once it was, it benefited from the constructive comments of Peter Vail. Any remaining misrepresentations about the sequence stratigraphic method are the responsibility of the author.

CHAPTER

21

Sierd Cloetingh
Vrije Universiteit

INTRAPLATE STRESS AND SEDIMENTARY BASIN EVOLUTION

Introduction

The origin of sedimentary basins is a key element of the geological evolution of the continental lithosphere. During the last decade substantial progress has been made in understanding the thermomechanical aspects of sedimentary basin evolution and the isostatic response of the lithosphere to surface loads such as basins. Most of this progress has been made by the development of new modelling techniques, insights into the mechanical properties of the lithosphere (the **rheology**), and in the processing of new, high-quality data sets from previously unexplored areas of the globe.

The fundamental quantities of rheology are stress and strain respectively. **Stress** can be defined as the force per unit area acting on a surface within a solid. Once the dynamic quantity (stress) is specified, the kinematic quantity (deformation or **strain**) can be derived. Almost all basin modelling carried out so far has been in terms of lithospheric displacements, thus refraining from a full examination of dynamic controls exerted by the lithospheric stress. This is because the calculation of stress is very sensitive to the adopted mechanical properties of the lithosphere. These lithosphere rheologies have been by definition unrealistically simple, especially in view of recent advances in rock mechanics studies. This is true of models for both extensional and compressional sedimentary basins as discussed in Chapter 15. For example, most models for extensional basin formation are keyed to lithospheric strain due to an unknown and unspecified stress field rather than to the strain response of the lithosphere to a known and/or realistic stress state. However, changes in plate-tectonic regimes and associated stress fields have been shown to be quite important in controlling the subsidence record and stratigraphic architecture of extensional basins. Similarly, models of basin development in compressional environments have been conventionally related to lithospheric flexure profiles, again not involving the dynamic control of the compressional stresses intrinsic to this particular tectonic setting. A major reason that the relationship between lithospheric stresses and displacements in tectonic modelling has been largely neglected is that little has been known about the actual stress state in the lithosphere. This situation has recently changed drastically as the result of the World Stress Map Project of the International Lithosphere Project. This project has revealed the existence of regionally consistent patterns of tectonic stress in the lithosphere. Moreover, structural measurements to establish the temporal evolution of palaeo-stress have begun in a number of sedimentary basins. Simultaneously, numerical modelling has resulted in a better understanding of the causes of the observed present-day stress levels and stress directions in the various lithospheric plates. Such studies have shown a causal relation-

BOX 21.1

Stress

The state of stress of a cube of rock is usually described in terms of nine stress components of which only six are independent if the body is in equilibrium. The stress on each face of the cube (Figure 21.1a) can be separated into a stress component **normal** to the face (σ), and a **shear stress** component (τ) along the face. The shear stress itself can be resolved into two components parallel to the directions of the two other axes of the coordinate system. In general, it is possible to calculate the state of stress once the stresses on three mutually perpendicular planes are known. Three planes, called principal planes of stress, exist on which the shear stresses are zero. Stresses in the mechanically strong upper part of the lithosphere are compressive at depths larger than a few tens of metres. The principal stresses (denoted by σ_1, σ_2, and σ_3 for the largest, intermediate and minimal principal stress respectively) are in the lithosphere located in approximately horizontal and vertical planes (Figure 21.1b).

The **vertical stress** σ_v is approximately equal to the product of density of the overlying mass, gravitational acceleration and depth. The maximum horizontal stress and the minimum horizontal stress are indicated by σ_{Hmax} and σ_{Hmin} respectively. The style of deformation is determined by the relative stress magnitudes. For example, **normal faulting** occurs when $\sigma_v > \sigma_{Hmax} > \sigma_{Hmin}$, while **strike-slip faulting** occurs when $\sigma_{Hmax} > \sigma_v > \sigma_{Hmin}$, and **thrust** or **reverse faulting** occurs when $\sigma_{Hmax} > \sigma_{Hmin} > \sigma_v$. When the principal stresses are equal, shear stresses are zero and the stress state is hydrostatic. When the principal stresses are different, shear stresses τ exist. For example, as shown in the left-hand side panel of Figure 21.1b, the maximum shear stress on the plane bisecting the σ_{Hmax} and σ_v directions and coinciding with the σ_{Hmin} direction equals $(\sigma_{Hmax} - \sigma_v)/2$.

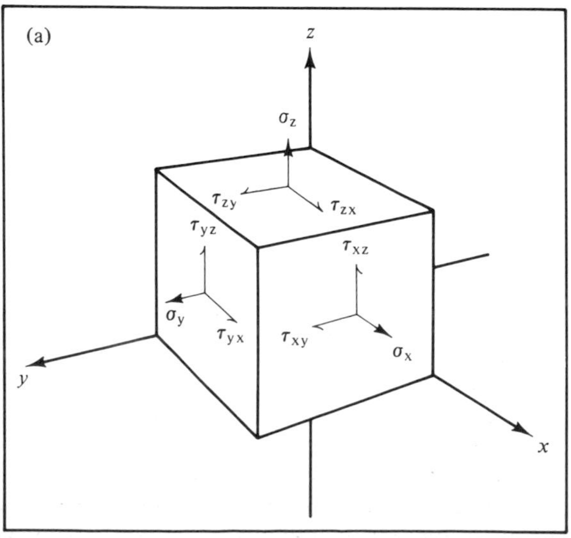

Figure 21.1 Stress conventions and orientations. (a) Stress components on three faces of an infinitesimal cube of rock volume. A stress acting on a plane can be separated into a component normal to the plane (σ_x, σ_y, or σ_z respectively) and two shear stress components acting within the plane (τ_{xy}, τ_{xz}; τ_{yx}, τ_{yz}; τ_{zx}, τ_{zy}). When in equilibrium ($\tau_{zy} = \tau_{yz}$; $\tau_{zx} = \tau_{xz}$; $\tau_{xy} = \tau_{yx}$) the stress field can be described in terms of six independent components. (b) Principal stresses in the lithosphere and planes of maximum shear stress. The principal stresses are in approximately horizontal and vertical planes. The vertical stress σ_v is approximately equal to the product of density, gravitational acceleration and depth. σ_{Hmax} and σ_{Hmin} denote maximum and minimum horizontal tectonic stress respectively. The style of deformation is determined by the relative stress magnitudes. Thrusting or reverse faulting occurs when $\sigma_{Hmax} > \sigma_{Hmin} > \sigma_v$, normal faulting occurs when $\sigma_v > \sigma_{Hmax} > \sigma_{Hmin}$, and strike-slip faulting occurs when $\sigma_{Hmax} > \sigma_v > \sigma_{Hmin}$. (c) Deformation of a horizontal circular cross-section of a bore-hole is caused by horizontal intraplate stresses σ_{Hmax} and σ_{Hmin} leading to the formation of breakouts (parts of the bore-hole wall that collapse) and hydraulic fractures. The maximum compressive stress induced around the bore-hole (and hence the breakout) is centred on the azimuth of the least horizontal tectonic stress. In a hydraulic fracturing experiment a portion of the well-bore is isolated and pressurised by injected fluids until a tensile fracture develops in the direction of the maximum horizontal stress. Down-hole measurements have confirmed that breakouts and hydraulic fractures form perpendicular to each other. □

21.1 continued

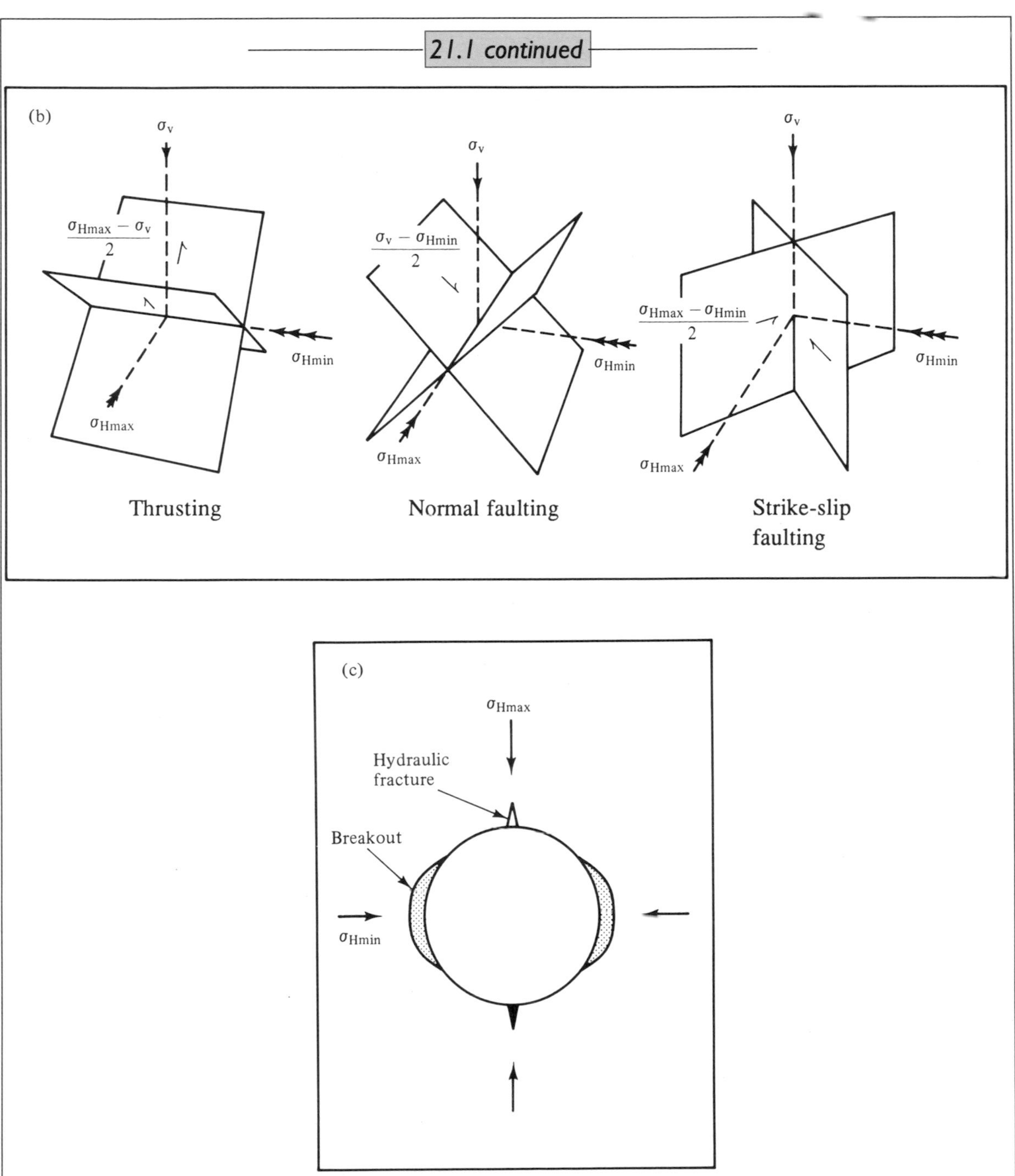

ship between the processes at plate boundaries and deformations in the plates' interiors. In models of the evolution of sedimentary basins located in the interiors of the plates, however, the role of these lithospheric stresses had been largely ignored. Recently, the first steps have been taken towards exploring the consequences of the existence of intraplate stress fields in the lithosphere for models of the formation and evolution of sedimentary basins.

Rates of tectonic subsidence following extensional rifting are dependent on stresses affecting the basin as well as on crustal cooling. This is important, as cyclic short-term changes in eustatic sea level have been regarded by many as the sole explanation for these short-term deviations from long-term subsidence. In particular, the development of quantitative stratigraphic techniques has led to the widespread use of a set of charts of cyclic changes in sea level for stratigraphic correlation. Since these key-concepts were presented by Vail and colleagues (see Chapter 20) much work has been done in testing, evaluating and developing them. In the present chapter, we discuss these basic concepts in sedimentary basin analysis in the light of recent theoretical advances in lithospheric dynamics.

We first review evidence for the existence of intraplate stress fields in the lithosphere. This is followed by a discussion of some implications of intraplate stress for quantitative modelling of sedimentary basins. Finally, the potential to separate tectonic contributions from eustatic contributions to the apparent sea-level record is explored.

Intraplate stresses in the lithosphere

The intraplate stress field in the lithosphere is the superposition of regional components and local sources of stress. The regional stresses in the lithosphere are induced by plate-tectonic forces and can be traced over large distances. On a more local scale, however, other contributors to the stress field might dominate. Examples are stresses associated with topographic anomalies and crustal thickness inhomogeneities at passive continental margins. Further, as a result of temperature variations rocks expand or contract inducing large thermal stresses in cooling lithosphere. The Earth's surface has the shape of a spheroid with polar flattening and an equatorial bulge and, consequently, plates deform as they change latitude. This deformation generates **membrane stresses** in the lithosphere. Finally, **flexural stresses** are induced due to vertical loads on the lithosphere, in particular by sedimentary sequences at passive continental margins. Stress components are described in Box 21.1. Stress magnitudes are expressed in the SI unit for stress which is the **Pascal** (Pa), equivalent to one Newton per square metre (Nm^{-2}), although it is more convenient in stress studies of the lithosphere to use the **megaPascal** (1 MPa = 10^6 Pa). At the same time, multiples of the CGS unit ($dyncm^{-2}$) are quite frequently used, in particular the **bar** (1 MPa = 10 bar) and the **kilobar**. Of the various locally induced stress sources the flexural stresses and thermal stresses stand out in magnitude (up to the order of several hundred MPa) while, as shown in Table 21.1, most of the other mechanisms produce stresses with a characteristic level of the order of a few tens of MPa.

The present stress field in the various plates has been studied in great detail by the application of a wide range of observational techniques. These include detailed analysis of earthquake focal mechanisms, *in situ* stress measurements, including hydraulic fracturing and stress-relief measurements, and analysis of stress-induced elliptical bore-hole deformations of wells or 'breakouts' (see Box 21.2).

The observed modern stress orientations show a remarkably consistent pattern, especially considering the heterogeneity in lithospheric structure in the plates. These stress-orientation data indicate a propagation of stresses away from the plate boundaries over large distances into the plate interiors. The World Stress Map Project has convincingly established the existence of these large-scale, consistently oriented stress patterns to be a general characteristic of lithospheric plates. These observations, compiled in Figure 21.2, prove that regional stress fields are usually dominated by the effect of plate-tectonic forces acting on the lithosphere.

Strong evidence exists for changes in the magnitudes and orientations of these stress fields on time scales of a few million years in association with collision and rifting processes in the lithosphere. The results of a recent compilation of stress-direction data for the European platform are displayed in Figure 21.3. In this case, the orientation and evolution of the principal **palaeo-stress** axes is inferred from **microstructure meas-**

Table 21.1 *Stress mechanisms and stress levels in the lithosphere (1 kbar = 100 MPa). Apart from regional stresses in the lithosphere induced by plate-tectonic forces, other sources might dominate on a more local scale. Examples are stresses associated with topographic anomalies and crustal thickness inhomogeneities at passive margins. As a result of temperature variations, rocks expand or contract which can induce large thermal stresses in cooling lithosphere. Deformation of the plates as they change latitude leads to membrane stresses in the lithosphere because the Earth is not a perfect sphere. Flexural stresses are generated due to vertical loads on the lithosphere, in particular by the accumulation of sedimentary sequences at passive continental margins.*

Mechanism	Stress	Order of magnitude
Ridge push	σ_{RP}	10 MPa
Drag	σ_D	1 MPa
Slab pull	σ_{SP}	10–100 MPa
Topography	σ_{TO}	10–100 MPa
Crustal thickness contrast	σ_{CT}	1–10 MPa
Thermal	σ_{TH}	100 MPa
Membrane	σ_{ME}	10 MPa
Flexure	σ_{FL}	100 MPa
Overburden	σ_{OB}	10 MPa

BOX 21.2

Techniques to measure the direction of the modern stress field

Earthquake focal mechanisms are obtained from radiation patterns of seismic waves. These define a set of two perpendicular fault planes and slip vectors, with P and T axes corresponding to the directions of maximum shortening and extensional strain for these shear faults. It is very difficult to determine the stress levels at depths of the earthquake source. However, it is generally possible to constrain relative stress magnitudes from the style of active tectonic faulting. ***In situ* stress-relief measurements** determine the strain relaxation when a rock sample is separated from the volume of surrounding rock. The change of stress after the sample is removed is then calculated from the strain relief. These measurements are only possible near the Earth's surface or on an excavated surface in mines. **Well-bore breakouts** occur spontaneously as a result of stress relaxation around the bore-hole. As predicted by simple elastic theory, the induced maximum compressive stress around a vertical bore-hole (and hence the breakout) is centred on the azimuth of the least horizontal stress (Figure 21.1c). In a **hydraulic fracturing experiment** a portion of the well-bore is isolated and pressurised by injected fluids until a tensile fracture develops in the direction of the maximum horizontal stress. Breakouts and hydraulic fractures are formed perpendicular to each other as has been shown by down-hole experiments in many wells in different tectonic settings. Breakout data are a very valuable and flexible tool in the study of the lithospheric stress field. They allow, for example, multiple determinations in a single well as well as the possibility to investigate regional consistency by the use of a large number of wells. Recently the Ocean Drilling Program has started to incorporate stress measurements in its drilling operation scheme. This is particularly important for our knowledge of the state of stress in those oceanic plates where earthquake focal mechanism studies are hampered by a low level of intraplate seismicity. The World Stress Map Project has demonstrated that the various techniques show, in general, excellent agreement.

urements in sedimentary rocks, such as pressure solution surfaces (stylolites), veins with secondary mineralisations, or small faults with a clear indication of the sense of displacement (Figure 21.4). Box 21.3 discusses in more detail techniques used to measure palaeostress.

As such, the inferred information on palaeo-stress fields is less precise than the results of studies of modern stress indicators. The study of palaeo-stress fields, however, adds geological time as a parameter crucial to understanding the temporal fluctuations of stress fields in the plates.

A complementary approach to collecting stress indicator data is the study of intraplate stress fields using numerical techniques. In the first phase of modelling intraplate stress fields resulting from plate-tectonic forces, models were tested against stress orientation data inferred from earthquake focal mechanism studies to quantify the relative and absolute importance of various possible driving and resistive forces. These studies resulted in the overall understanding that ridge push and slab pull (see Chapter 9) are the two main driving forces. Since then, a better understanding has been obtained of the age-dependence of the forces acting on the lithosphere. This development has benefited from advances in the analysis of the subduction process. Subsequent implementation of these new features and insights in stress modelling led to the successful prediction of various deformation processes within lithospheric plates.

Because of better constraints on the thermo-mechanical and tectonic evolution of the oceanic lithosphere, which is relatively well understood, these models have concentrated on oceanic plates or lithospheric plates with major oceanic parts. An example is given in Figure 21.5a which shows a comparison of the predicted and observed stress fields in the northeastern Indian Ocean. This area is at present the most seismically active oceanic intraplate region on Earth. Here stresses reach high levels due to focusing of compressional resistance associated with collision of the Indian and the Eurasian plate and subduction of relatively young oceanic lithosphere in the northern part of the Java–Sumatra Arc. Stress-orientation data

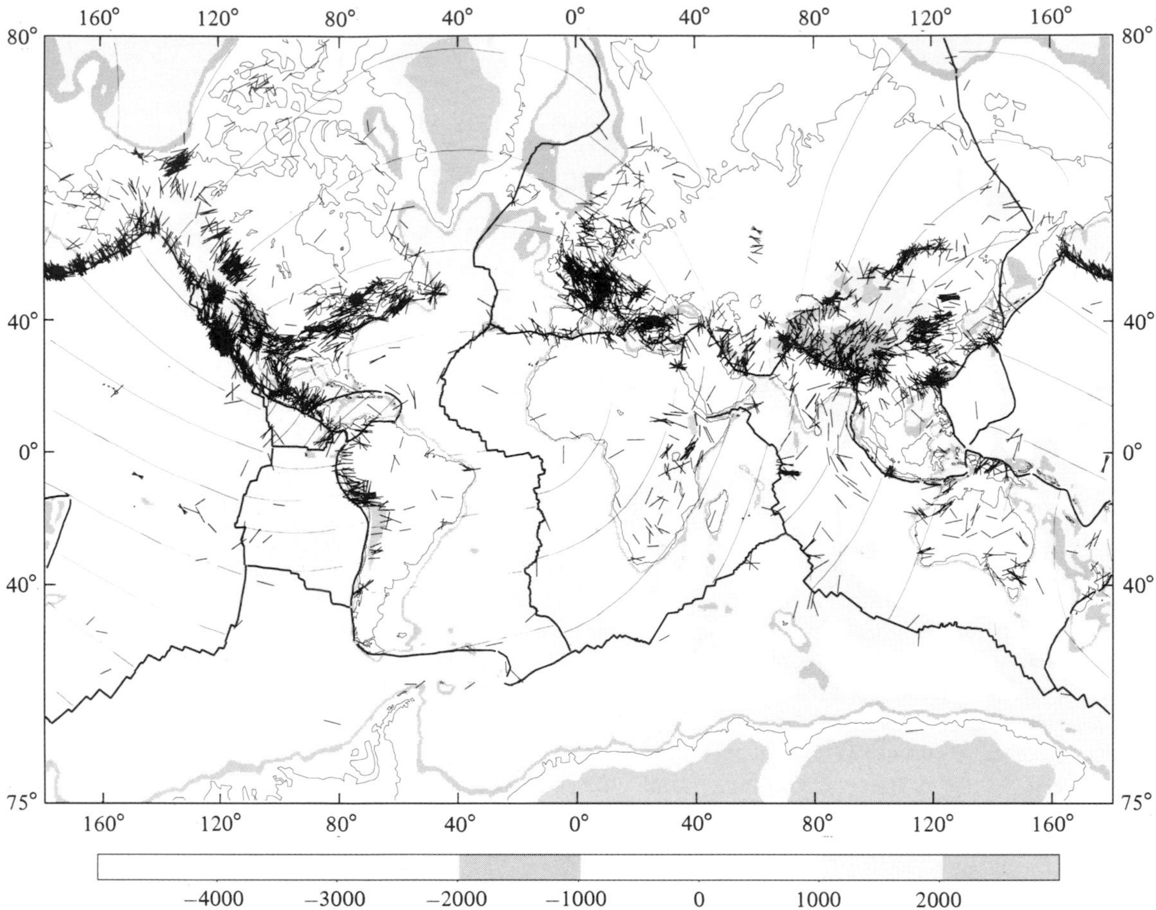

Figure 21.2 The World Stress Map. Axes give the orientation of the maximum horizontal stress field σ_{Hmax}. The line length of the data is proportional to the quality of the data, and therefore our confidence in the data. Earthquake focal mechanisms are obtained from the radiation pattern of seismic waves and lead to the definition of a set of two orthogonal fault planes and slip vectors, with P and T axes that represent the directions of maximum shortening and extension. The World Stress Map Project has demonstrated that the various techniques show, in general, excellent agreement, revealing the existence of regionally consistent patterns of tectonic stress in the lithosphere. The observed stress directions also agree quite well with the directions of absolute plate motions indicated by the dashed lines in the figure. The Indo-Australian plate, where the stress field is probably dominated by age-dependent slab pull forces, forms an exception to this pattern. Note the uneven distribution of the data as an artefact of the closer sampling of certain areas such as California and the North Sea in association with intensive studies of seismic risk and drilling for oil exploration and production respectively. Future stress measurements by the Ocean Drilling Program will forward our knowledge of the state of stress in oceanic plates such as the Pacific plate, where a low level of intraplate seismicity limits the determination of stress directions by earthquake focal mechanism studies. Shading relates to topography with top of Everest at 0 m. □

demonstrate a rotation of the observed stress from N–S oriented compression in the north to a more NW–SE directed compression in the southeastern part of the Bay of Bengal area, in agreement with the calculated stress field. The intraplate stress field as calculated also explains folding of the entire lithosphere observed in this area from seismic profiles and satellite gravity data (Figure 21.5b).

These numerical models have shown that high-magnitude stresses can be concentrated in the plates' interiors. They have also shown, in agreement with observations such as shown in Figure 21.5, that spatial variations and large-scale rotations in orientation in the stress fields may occur. These models and observations make clear that stress provinces can vary in size from an entire lithospheric plate to only portions of plates. Modelling results such as displayed in Figure 21.5 corroborate the observed stress patterns by

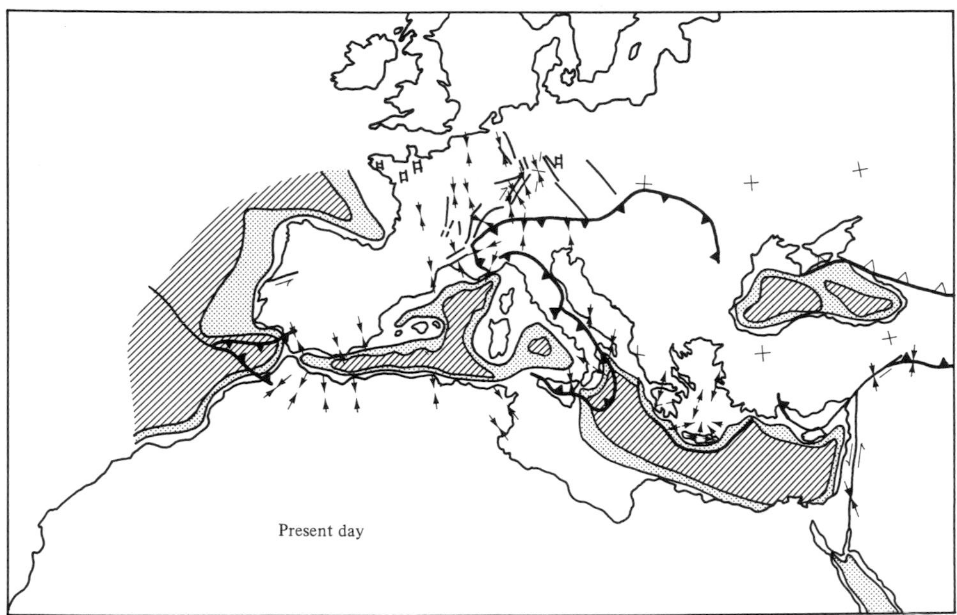

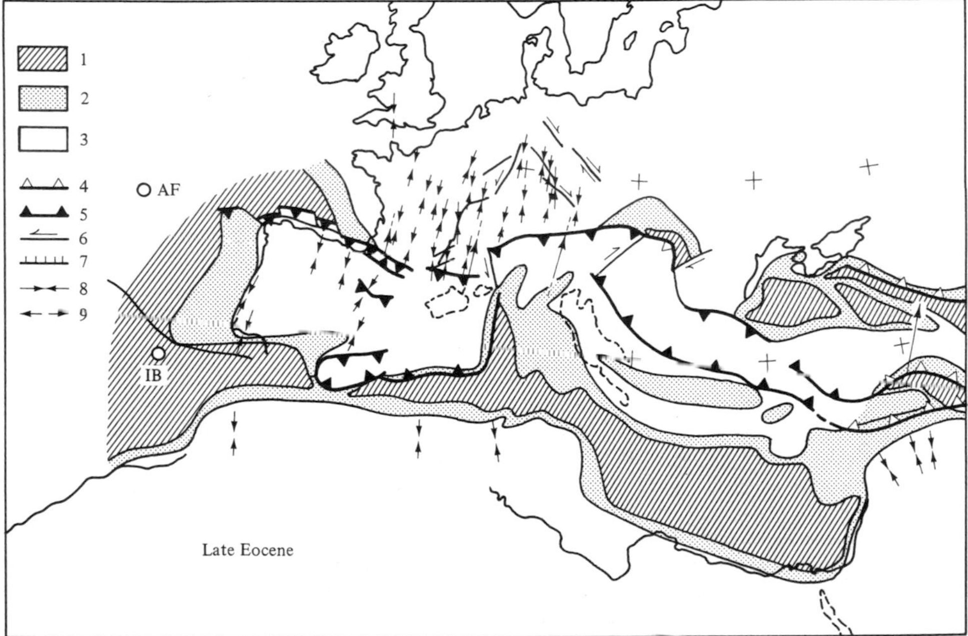

Figure 21.3 Compilation of observed maximum horizontal stress directions in the European platform. Key: 1, oceanic crust; 2, thinned continental crust; 3, continental crust; 4, subduction; 5, overthrust; 6, strike-slip fault; 7, normal fault; 8, azimuth of maximum principal stress σ_1; 9, azimuth of minimum principal stress σ_3. (a) Present-day stress field from *in situ* measurements, focal mechanism studies and geologic stress indicators. (b) Palaeo-stress field during Late Eocene times and reconstructed geodynamic evolution in a framework of Cenozoic Africa/Eurasia collision. Eurasia is fixed and AF and IB denote the Africa/Eurasia and Iberia/Eurasia rotation poles. The rotation poles and motion vectors are given for a time slice between 54 and 35 Ma. The stress data for the European platform given in a framework of Cenozoic Africa/Eurasia collision demonstrate stress propagation away from the Alpine fold belt in the platform region. □

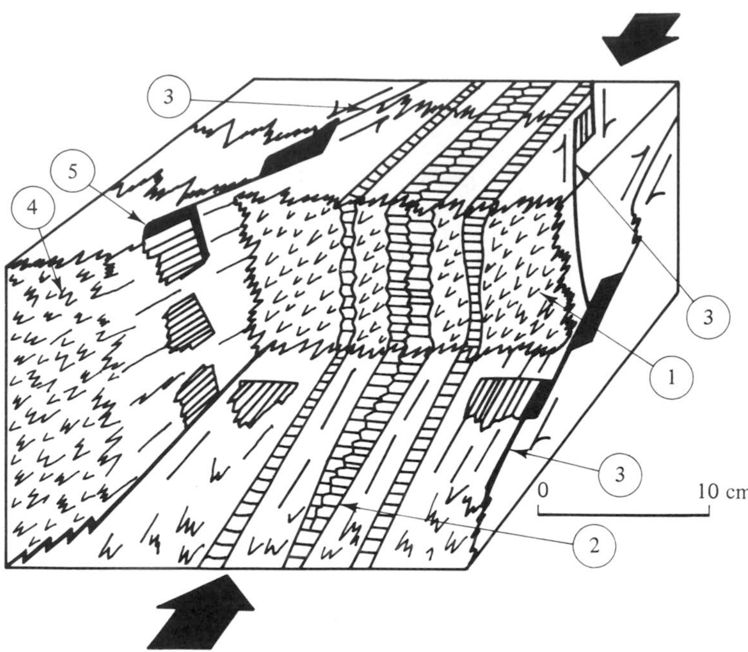

Figure 21.4 Schematic presentation of microstructures measured for the analysis of palaeo-stress patterns. 1, tectonic stylolites which are common in limestones and which are caused by pressure solution under non-hydrostatic stresses. Columns represent the direction of the local maximum principal stress axis σ_1. Stylolites form subvertically by overburden or subhorizontally due to tectonic stresses. In practice, only the orientation of tectonic stylolites with subnormal columns along the solution plane seam are measured. Although the scatter in orientation is sometimes large in each site, due to fractures and inhomogeneities, the method can provide quite consistent stress patterns on a regional scale. 2, tensional joints which are parallel veins of calcite (or quartz), with crystals growing normal to the joints. These structures are commonly associated with tectonic stylolites and form parallel to the stylolitic columns. The local minimum principal stress axis (σ_3) is inferred from the average normals to these joints. 3, measurements on a small fault plane. 4, measurements on a pressure solution surface. Reliable measurements on a small fault plane and on a pressure solution surface require the determination of the sense of displacement. 5, asymmetric steps of fibrous calcite or quartz (accretionary growth of crystal fibres during slip movement), or mechanical striation on the rocks also allow the determination of the sense of displacement. The black arrows indicate the maximum compressive stress axis. □

Figure 21.5 (a) Regional stress field in the northeastern Indian Ocean. Left: predicted stress field induced by plate-tectonic forces acting on the lithosphere. Dotted lines give the location of long seismic reflection profiles that show folding of the entire lithosphere caused by the high stress levels in the area. Right: the orientation of maximum horizontal compressive stress inferred from earthquake focal mechanism studies. Note the good agreement between theoretical modelling of the stress field and stress observation data. (b) Gravity anomalies and stresses predicted by numerical modelling. Gravity highs from satellite data corresponding to the folded oceanic lithosphere generally trend normal to the predicted stress field. Note the large-scale rotation of the fold axes(indicated by dashed lines) from an E–W trend in the area close to Sri Lanka to a NE–SW orientation in the ocean region close to the Sumatra trench. The line that marks the transition between tensional and compressional stresses in the northeastern Indian Ocean also coincides roughly with the boundary of the area where folding occurs. □

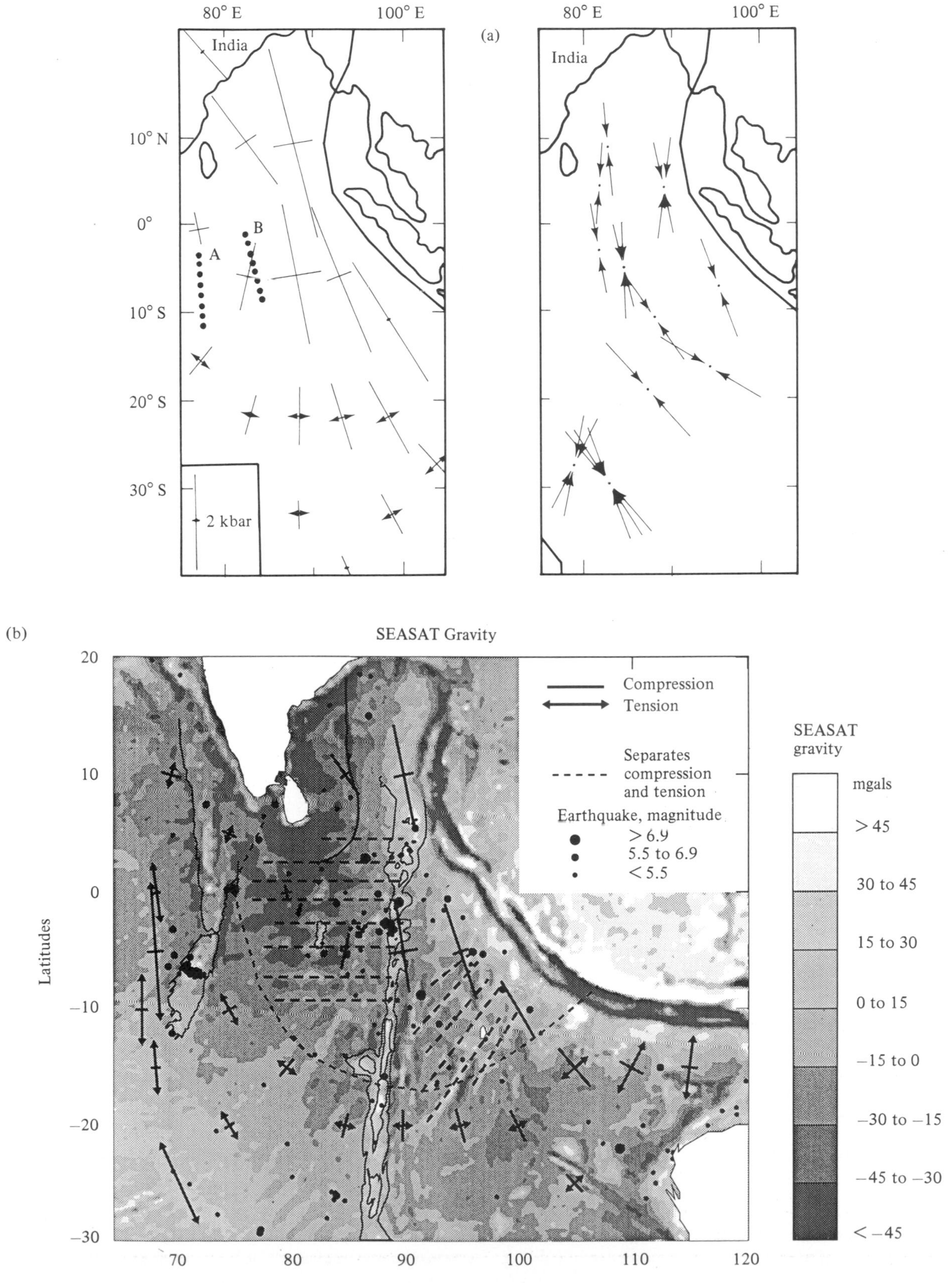

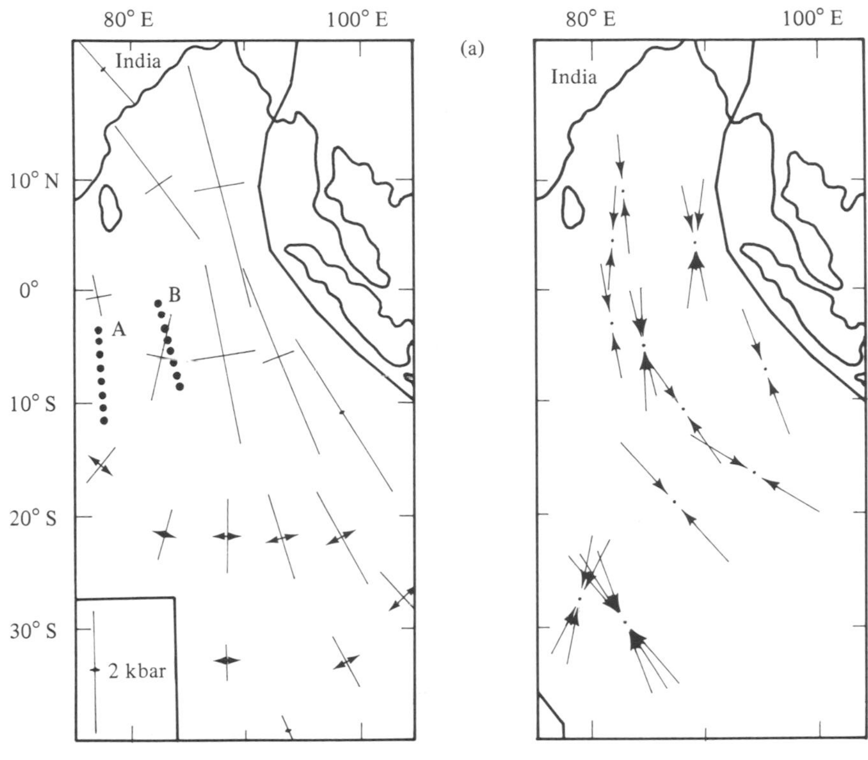

(b)

SEASAT Gravity

Compression
Tension
Separates compression and tension
Earthquake, magnitude
> 6.9
5.5 to 6.9
< 5.5

SEASAT gravity
mgals
> 45
30 to 45
15 to 30
0 to 15
−15 to 0
−30 to −15
−45 to −30
< −45

Latitudes
20
10
0
−10
−20
−30
70
80
90
100
110
120

BOX 21.3

Techniques to measure palaeo-stress fields

Stylolites are quite common in limestones and are caused by pressure solution under non-hydrostatic stresses. Columns which represent the direction of the local maximum principal stress axis σ_1 are formed subvertically due to the overburden load or subhorizontally due to tectonic stresses. In practice, only the orientation of tectonic stylolites with subnormal columns along the solution plane seam are measured. Although the scatter in orientation can be large in an individual site, for example due to fractures and inhomogeneities, usually a consistent stress pattern is observed on a regional scale.

Tensional joints are parallel veins of calcite (or quartz), with crystals growing normal to the joints. These structures are frequently associated with tectonic stylolites and are parallel to the stylolitic columns. The orientation of the local minimum principal stress axis (σ_3) is given by the average normals to the joints. Measurements on a small fault plane or on a pressure solution surface require an indication of the sense of displacement. These follow usually from tectonic stylolitic columns oblique to the surface containing them, asymmetric steps of fibrous calcite or quartz (accretionary growth of crystal fibres during slip movement), as well as from mechanical striation on fault planes.

indicating that the stress changes induced at convergent margins and collision zones can propagate over large distances through the interiors of plates, to affect passive continental margins and intracratonic basins. Temporal changes in stress levels are not limited to plate collision but also occur through rifting and fragmentation of plates. These are especially important for sedimentary basins in plates not involved in collision or subduction processes.

Stress-induced vertical motions of the lithosphere

The evolution of sedimentary basins is to a large extent controlled by the response of the underlying lithosphere to various tectonic loads. Lithospheric flexure forms an important element in determining this response. We therefore begin this section with a brief discussion of the flexural response of the lithosphere to intraplate stresses, following classical studies carried out during the first half of the 20th century. A mathematical formulation is given in Box 21.4.

The flexural response of the lithosphere is easily obtained for some simple loading cases. Assuming zero vertical load on the lithosphere, early studies of the problem made a convincing case for neglecting horizontal forces in modelling the vertical motions of the lithosphere. They showed that for reasonable levels of the compressional forces the induced vertical displacements of the lithosphere are negligible. This result, combined with lack of evidence for the existence of such horizontal forces or the occurrence of folding of the entire lithosphere such as is now known from studies of Indian Ocean tectonics led, for a long time, to the withdrawal of attention from this topic. As mentioned earlier, significant progress has been made recently in the study of horizontal stress fields in the lithosphere. Furthermore, it is now realised that vertical motions of the lithosphere at sedimentary basins are primarily the result of a variety of other tectonic processes. These include thermally induced cooling of the lithosphere amplified by the loading of sediments that accumulate in basins, isostatic response to crustal thinning, and flexural bending in response to concentrated vertical loads. The flexural response function is described in Box 21.5. Hence, it is essential to account for the presence of already existing vertical loads on the lithosphere when solving for the response of the lithosphere to horizontal stress fields, as the magnitudes of these stresses are in general far below the stresses that are required to produce folding of the entire lithosphere.

The analytical formulation of specific simple problems shows explicitly how the solution depends on various parameters. Numerical modelling techniques have the advantage of allowing more realistic geometries and variations in parameters to be handled, adding flexibility to the analysis.

Intraplate stresses and basin stratigraphy

We have seen that temporal fluctuations in stress are a natural consequence of the horizontal motions of the

BOX 21.4

Mathematical formulation of flexure

The flexural response of a uniform elastic lithosphere at a position x to an applied horizontal force N and a vertical load $q(x)$ is given by:

$$D\frac{d^4w}{dx^4} - N\frac{d^2w}{dx^2} + (\rho_m - \rho_i)gw = q(x)$$

where w is the displacement of the lithosphere, and D is the flexural rigidity ($D = E\,T^3/12(1-\varpi^2)$, with E the Young's modulus, T the plate thickness, and ϖ the Poisson's ratio. The horizontal force N is equivalent to the product of the intraplate stress σ_N and the plate thickness T. ρ_m and ρ_i are, respectively, the densities of mantle material and the infill of the lithospheric depression, usually water or sediments, and g is the gravitational acceleration.

lithosphere plates. These findings have important consequences for short-term temporal variations in the basin shape throughout geological time. Intraplate stresses, apart from being important in the formation of rifted basins, also play a critical role during their subsequent subsidence history. Intraplate stresses modulate the long-term basin deflection caused by thermal subsidence and induce rapid differential vertical motions of a sign and magnitude that depends on the position within the basin. Figure 21.6 schematically illustrates the relative movement between sea level and lithosphere at the flank of a flexural basin landward of the principal sediment load as predicted by a numerical model. The synthetic stratigraphy at the basin edge is schematically shown for three situations in Figure 21.7. In one, long-term flexure under the basin results from cooling in the absence of an intraplate stress field. Also shown is the same situation with a superimposed transition to 750 bars compression or a superimposed transition to 750 bar tension after 50 Ma. As described in Chapter 15, the thermally induced flexural widening of the basin provides an elegant explanation for long-term phases of coastal onlap. However, by their long-term nature, changes in basin shape by thermal cooling or heating, fail to produce the punctuated character of the stratigraphy of sedimentary basins. As shown by Figure 21.7, intraplate compression causes relative uplift of the basin flank, subsidence at the basin centre, and seaward migration of the shoreline. An offlap develops and an unconformity is produced. Increases in the level of tensional stress induce widening of the basin, lower the flanks and cause landward migration of the shoreline, producing a rapid onlap phase. Stress-induced vertical motions of the crust can also drastically influence sedimentation rates. Flank uplift due to an increased level of compression, for example, can significantly enhance sedimentation rates and modify the infilling pattern, promoting the development of unconformities.

During rifting phases, eventually followed by continental break-up, the tensional stress levels will be reduced. Rifting in the Atlantic, for example, rather than instantaneously, occurred at discrete rifting phases, with stepwise relaxation of tensional stresses. This process might explain sea-level fluctuations that do not correlate with accelerations in plate-spreading rates or increases in ridge lengths. Simultaneously, the correlation of short-term sea-level changes at both sides of the Atlantic may reflect rifting-related accumulation and relaxation of tensional stresses. Break-up unconformities, not successfully explained by most geodynamical models, can be explained by intraplate stresses. Similarly, the break-up of the supercontinent Pangea probably occurred with major changes in the Earth's stress field and associated changes in relative sea level.

The position of coastal onlap reflects the position where rate of subsidence equals rate of sea-level fall. During changes in the intraplate stress field the rate of subsidence is temporarily changed. Consequently, the equilibrium point of coastal onlap is shifted in position. The thermally induced rate of long-term subsidence strongly decreases with the age of the basin. Hence, the production of offlaps during late stages of passive continental margin evolution requires much lower rates of change in apparent sea level than in earlier stages of basin evolution. If these offlaps result from fluctuations in intraplate stresses, the rates of stress change required diminish with age during the post-rift evolution of the basins. This is of particular relevance for an assessment of the relative contributions of tectonics and eustasy to Cenozoic unconformities. For example, Cenozoic unconformities developed at old passive margins in association with

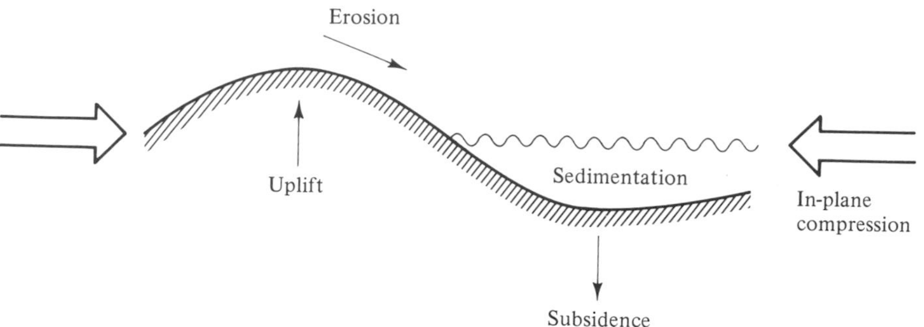

Figure 21.6 Relative movement induced by intraplate compression at the flank of a sedimentary basin landward of the principal sedimentary load. The stress-induced change in the shape of the basin affects the dynamics of sedimentation and erosion and possibly the oceanic circulation patterns. Changes in basin geometry can trigger erosion and deposition by turbidity currents perhaps in association with canyon cuttings. This could lead to the formation of unconformities traceable from the margin into the deeper part of the basin. □

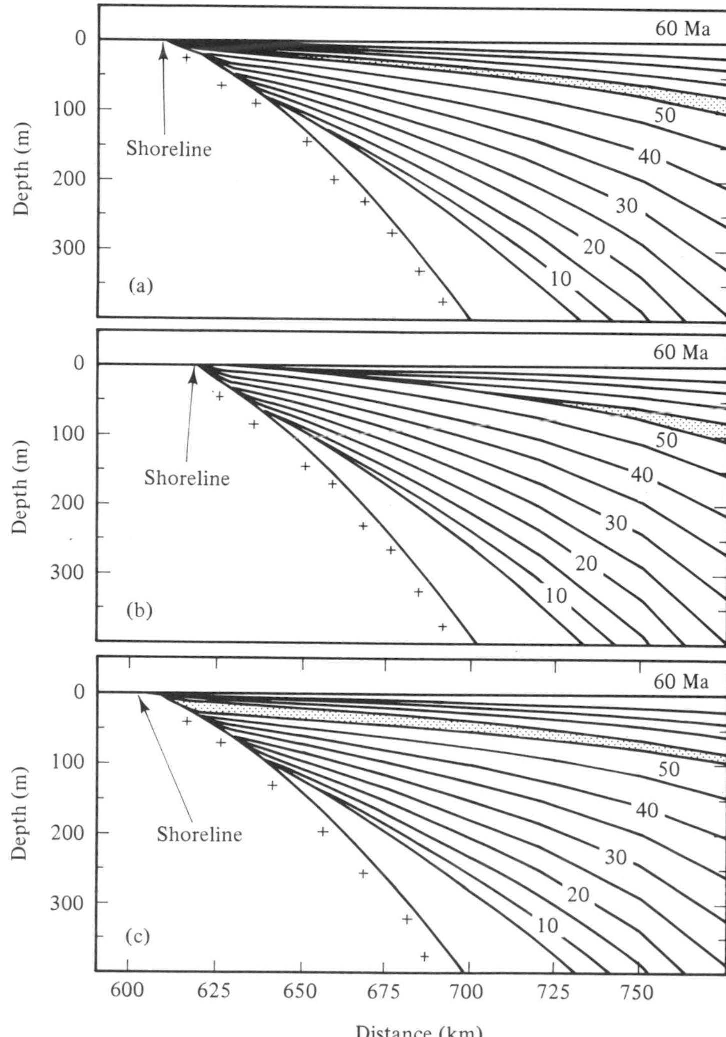

Figure 21.7 Synthetic stratigraphy for a 60-Ma-old passive margin, which is initiated by lithospheric stretching followed by thermal subsidence and flexural infilling of the resulting depression. Shading indicates the position of the

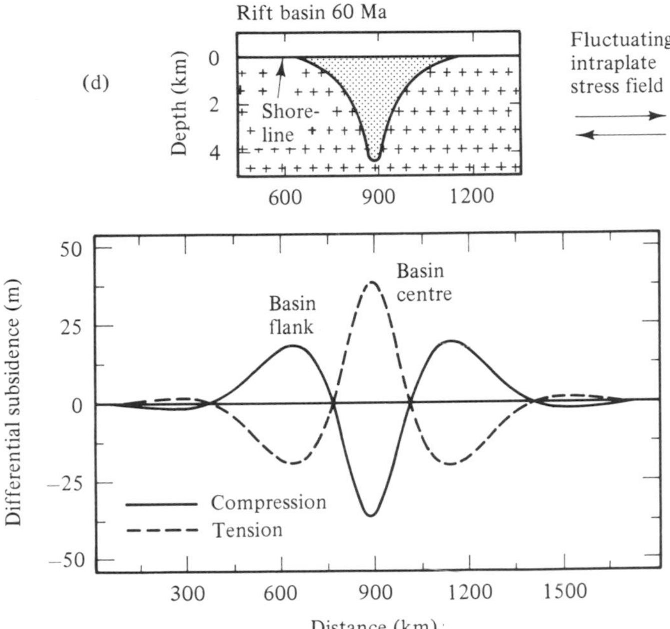

plane compression at 50 Ma induces uplift of the peripheral bulge, narrowing of the basin and a phase of rapid offlap, which is followed by a long-term phase of gradual onlap due to thermal subsidence. (c) A transition to 750 bar in-plane tension at 50 Ma induces downwarp of the peripheral bulge, widening of the basin and a phase of rapid basement onlap. (d) The differential subsidence or uplift (metres) induced by a change to 1 kbar compression (solid line) and 1 kbar tension (dashed line). □

short-term basin narrowing could be produced by relatively mild changes in intraplate stress levels. Re-narrowing and later erosion of Phanerozoic platform basins and passive margins is frequently observed, without clear evidence for active tectonism.

Figure 21.9 demonstrates that the incorporation of intraplate stresses in elastic models of basin evolution can, in principle, predict a succession of alternating rapid onlaps and offlaps observed along the flanks of basins such as the US Atlantic margin. In the figure, a two-layered stretching model for basin initiation is incorporated, as well as the effects of finite and multiple stretching phases and intraplate stresses. Inspection of Figure 21.9 shows the well-known failure of the standard elastic model of basin evolution to predict narrowing of the basin with younger sediments restricted to the basin centre. The narrowing of the basin during its late-stage evolution has often been interpreted as reflecting either the response of the basin to a phase of visco-elastic relaxation or to a long-term eustatic sea-level fall. The stratigraphic model demonstrates that although the incorporation of a long-term change in sea level enhances the Cenozoic narrowing of the basin margin, a long-term post Cretaceous decline in sea level alone cannot cause both the documented basin narrowing and the total thickness of sediments accumulated at this time. Therefore, it could be that much of the observed non-depositional or erosional character of the shelf surface is caused by stress-induced uplift of the basin flank.

Thus, modelling of the stratigraphy of the US Atlantic margin (Figure 21.9) has shown that the punctuated stratigraphy can be successfully simulated by a stress field, whose magnitude fluctuates through time, superimposed on the long-term thermal evolution of the basin. The inferred palaeo-stress is largely consistent with independent data sets on the kinematic and tectonic evolution of the northern/central Atlantic. These show a mainly tensional stress regime during Mesozoic times followed during the Tertiary by a more compressional stress field whose magnitude increases with age. A significant part of the Mesozoic unconformities of rifted basins around the Atlantic and the Arctic might be associated with large-scale, fault-controlled rifting activity, while Cenozoic sequence boundaries in the Arctic could be largely controlled by

BOX 21.5

The flexural response function

In analytical solutions of the equation describing the flexural behaviour of thin elastic plates, the loading response of the plate can be decomposed into its harmonic components by transforming the equation to the Fourier domain. The flexural response function $\Phi_e(k)$ in the presence of a horizontal load N can be written as:

$$\Phi_e(k) = \left[1 + \frac{D(2\pi k)^4 - N(2\pi k)^2}{\rho_m g}\right]^{-1}.$$

If $N = 0$, then $\Phi_e(k) = \Phi(k)$, the flexural response function of the plate in the absence of intraplate stress, as discussed in Chapter 15. In Figure 21.8a, $\Phi(k)$ is plotted as a function of the wavenumber k of the surface load, which is here simply the reciprocal of the load's wavelength (in kilometres). The curves show the relative flexural response of the elastic plate to a spectrum of surface loads of equal amplitude but different wavelengths. A value of 1 means that the flexural response is maximised (or, equivalently, isostatic compensation is 'local') and a value of 0 means that there is no flexural response (or no isostatic compensation). Figure 21.8b shows the effect of intraplate stress fields of a magnitude of a few hundred MPa on the flexural response of an elastic lithosphere. The plotted curves $\Delta\Phi_e(k) = \Phi_e(k) - \Phi(k)$ (with $\Phi_e(k)$ the response function when an intraplate stress field is applied) show the incremental changes to the flexural response functions in Figure 21.8a, in equivalent units, resulting from the application of various magnitudes of intraplate stresses to the thin elastic plate. Positive increments result from the application of compressional intraplate stresses and mean that the flexural response of the plate to a given surface load is enhanced by the presence of these stresses. The degree of the enhancement is seen to depend on the load wavelength and the flexural rigidity of the plate. The wavenumber k at which the intraplate stresses most affect the flexural response of the lithosphere is almost completely determined by the plate's flexural rigidity. The presence of intraplate stresses has a small but perceptible effect on this wavenumber, but exerts a controlling influence on the amplitude of the response for a given rigidity.

large-scale compressional activity and inversion tectonics. Similarly, simultaneous occurrence of a high frequency of faulting activity in the North Sea area and an increased intensity in the occurrence of sea-level lowerings has been reported in the Upper Jurassic of the North Sea. Intermittent phases of accumulated tensional stress, associated with rifting episodes in the northern and central Atlantic, and subsequent rapid relaxation of these stresses could explain the asymmetry in Vail's onlap–offlap charts. Both the timing and nature of Vail's second-order and third-order cycles might be to a large extent controlled by the plate-tectonic evolution and the associated changes in stress regimes of the northern and central Atlantic.

Although we have concentrated in this chapter on the relationship between tectonics and stratigraphy for rifted basins, the effect of intraplate stresses is of importance to a wider range of sedimentary environments. Another setting where the lithosphere is flexed downward under the influence of sedimentary loads is in foreland basins. Several studies have interpreted the development of unconformities in foreland basins in terms of uplift of the peripheral bulges flanking these basins. The presence of tensional or compressional intraplate stresses, the latter being more natural in this tectonic setting, can amplify or reduce the height of the peripheral bulge by an equivalent amount and thus greatly influence the stratigraphic record.

Discussion

Vail and coworkers initially interpreted their short-term cycles of sea-level change in terms of waning and waxing of ice sheets. This view was partly based on the inferred global character of the apparent sea-level cycles, partly based on the repetitious character of the cycles, and was partly due to the lack of a tectonic mechanism to explain both the rate and magnitude of the third-order cycles. Crucial in this respect has also been the basic assumption that tectonic subsidence should be slow, requiring absolute changes in sea level to explain irregularities in the subsidence record. The issue of global synchroneity has attracted major debate. Several authors have noted that Vail's cycles, although based on data from different basins around the world, are heavily weighted

21.5 continued

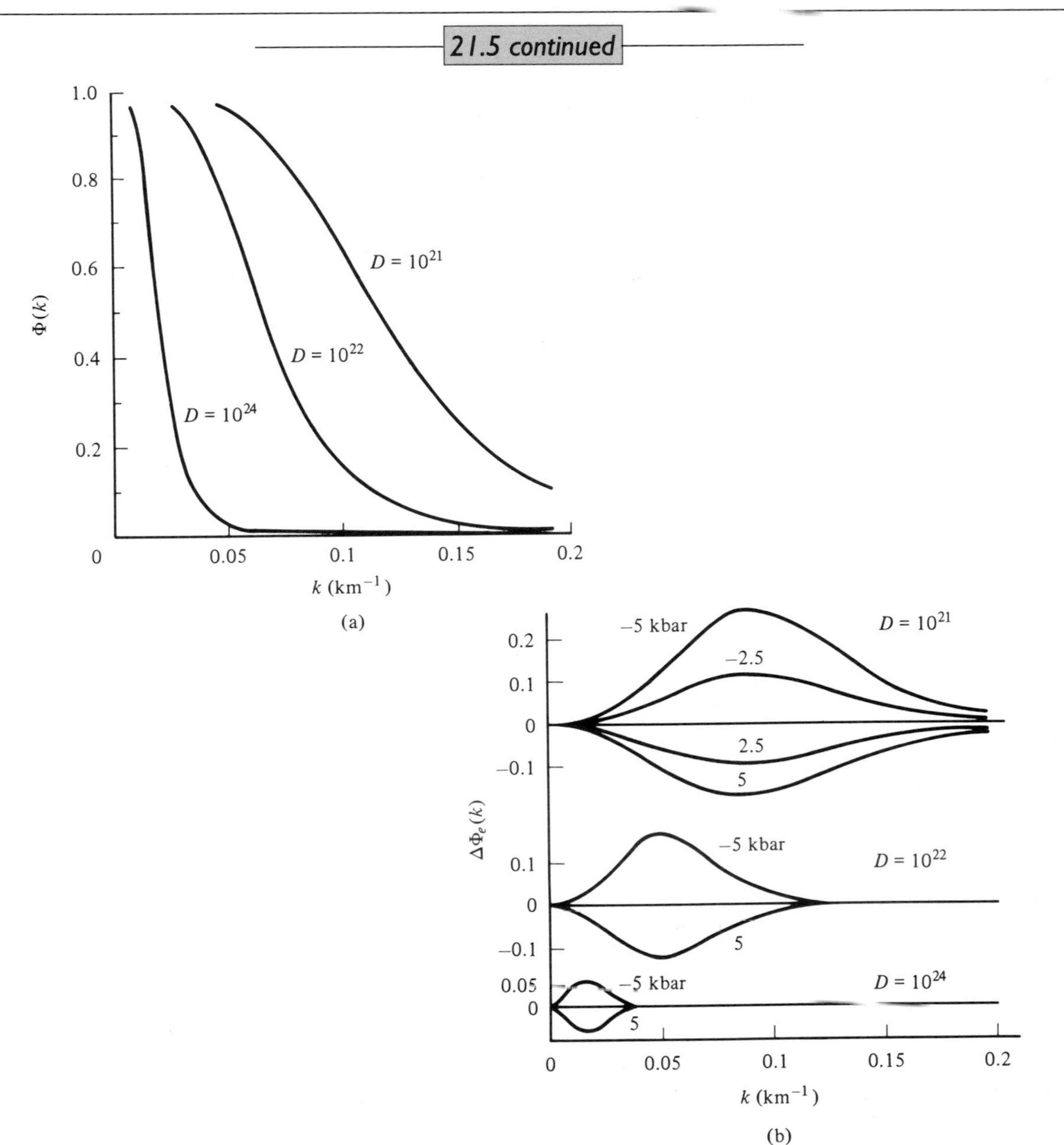

Figure 21.8 (a) The flexural response Φ(k) of an elastic thin plate, with various flexural rigidities *D* (in units of newton metres), to a surface load when there are no intraplate stresses. Φ(k) is plotted as a function of the wavenumber *k* of the surface load, which is here simply the reciprocal of the load's wavelength (in km). Each curve shows the relative flexural response of the elastic plate to a spectrum of surface loads of equal amplitude but different wavelengths. A value of 1 means that the flexural response is maximised (or, equivalently, isostatic compensation is 'local') and a value of 0 means that there is no flexural response (or no isostatic compensation). (b) The effect of intraplate stresses σ_N (in units of kbars, 1 kbar = 100 MPa; tension is positive) on the flexural response functions shown in (a). The plotted curves $\Delta\Phi_e(k) = \Phi_e(k) - \Phi(k)$ (with $\Phi_e(k)$ the response function for an applied intraplate stress field) show the incremental changes to the flexural response functions in (a), in equivalent units, resulting from the application of various magnitudes of intraplate stresses to the thin elastic plate. Positive increments result from the application of compressional intraplate stresses and mean that the flexural response of the plate to a given surface load is enhanced by the presence of these stresses. The degree of the enhancement is seen to depend on the load wavelength and the flexural rigidity of the plate. □

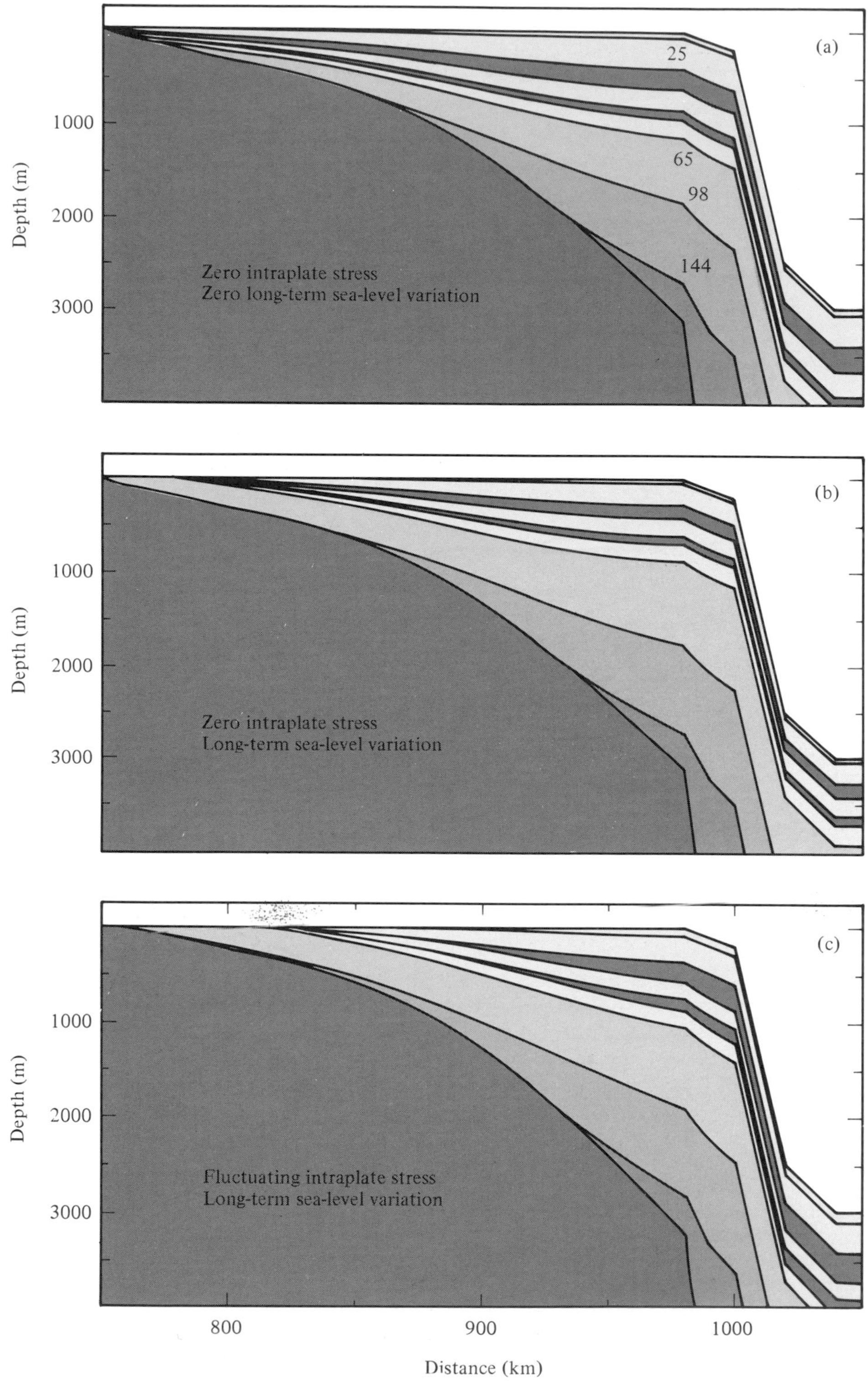

Figure 21.9 Schematic model of the stratigraphy of the US Atlantic margin at Cape Hateras. The numbers in the strata refer to stratigraphic ages in Ma. (a, b) Modelled stratigraphy for elastic rheology of the lithosphere in the absence of intraplate stresses. (c) Modelled stratigraphy showing the effect of incorporating a fluctuating intraplate stress field in the analysis. □

in favour of the North Sea and the northern/central Atlantic margins. The issue of global synchroneity is important as it obviously strongly influences present discussions on the causes of short-term changes in sea level. Tectonic mechanisms such as variations in spreading rates, hot spot activity and orogeny fail to produce changes at the rate of third-order cycles. This is because such explanations are derivatives of the thermal evolution of the lithosphere and are therefore associated with a long thermal inertia of several tens of millions of years (Table 21.2). Changes in water volume by the waning and waxing of ice sheets (**glacio-eustasy**) easily can induce both the rate and magnitude of the inferred sea-level changes but raises two basic problems. The first problem is the occurrence of third-order sea-level cycles during time intervals where there is no geological evidence in support of low-altitude glaciation. This presents a fundamental problem of explaining sea-level changes by this mechanism at times prior to the Late Cenozoic. The second problem is the inability of glacio-eustasy to cause uniform global lowerings and rises of sea level. Studies of post-glacial rebound have shown that the concept of eustasy is not valid in the case of sea-level changes resulting from substantial lateral transfer of mass, as between ice sheets and oceans. The sea level coincides with an equipotential surface and is partly controlled by the gravitational attraction of the ice sheets. For this reason, the response of sea level to the melting of an ice sheet varies strongly between sites near the ice sheets and sites farther away. Furthermore, the response of the crust to the removal of the ice load and to the loading by the meltwater causes differential motion between sea level and land. As a result, the sign and magnitude of the induced sea-level change is dependent on the distance to the location of the ice cap. This is quite important for scenarios in which glacio-eustasy is the key mechanism to explain global synchronous changes of uniform magnitude in sea level.

We have shown in the previous section that short-term changes in relative sea level can equally well be caused by rapid, stress-induced vertical motions of the lithosphere. Hence intraplate stresses, apart from being important in the formation of rifted basins, probably also play a critical role during their subsequent subsidence history. Undoubtedly, both eustasy and tectonics have contributed to the record of short-term changes in sea level. The relative contributions are, by their nature, of variable magnitude. The key question to be answered from stratigraphic analysis is related to the spatial and temporal differences in the expressions of tectonic processes and eustasy. The development of stratigraphic criteria to differentiate between tectonics and eustasy is therefore vital, and is needed both on an interbasinal and an intrabasinal scale. Recently, a number of features in the stratigraphic record of rifted basins have been recognised that are difficult to explain in terms of the eustatic framework. Among these are sediment source areas and/or sedimentary regimes that often change abruptly across sequence boundaries, intrabasinal changes in subsidence and uplift patterns that also occur at sequence boundaries, and faults that do not cross sequence boundaries.

The discrimination of tectonics and eustasy is a subtle matter, especially if biostratigraphic resolution is limited. The regional character of intraplate stresses can shed light on documented deviations from global sea-level charts. Whereas such deviations from global patterns are a natural feature of this tectonic model, the occurrence of short-term deviations does not preclude the presence of global events of tectonic origin elsewhere in the stratigraphic record. These are to be expected when major plate reorganisations in intraplate stress fields occur simultaneously in more than one plate or in time intervals prior to and shortly after the break-up of Pangea where continents and rift basins were in a largely uniform stress regime. The magnitude of the stress-induced phases of rapid uplift and subsidence varies with position within the basin, thus providing another important criterion to distinguish this contribution from eustatic effects. Similarly, differences in the mechanical properties of the lithosphere control the magnitude of the vertical motions. The presence of weak, attenuated continental lithosphere enhances the effectiveness of the stresses to cause substantial vertical motions and can explain differences in magnitude of the apparent sea-level changes such as those observed between the Tertiary North Sea and the Gippsland Basin off southeast Australia. Only the larger short-term fluctuations in sea level, with magnitudes in excess of 50 metres, require stress changes large enough to be related to major plate boundary reorganisations. This observation could explain the frequently observed correlation between the timing of plate reorganisations and rapid lowerings in the sea level. Furthermore, glacio-eustatic events sometimes occur simultaneously with a major tectonic reorganisation which could, for example, have contributed to the large magnitude of the Oligocene sea-level lowering. It appears that the repetitious character of sea-level cycles and plate-wide correlation, which are often considered to be diagnostic for a eustatic cause of the sea-level change, can equally well be explained by episodic accumulation and re-

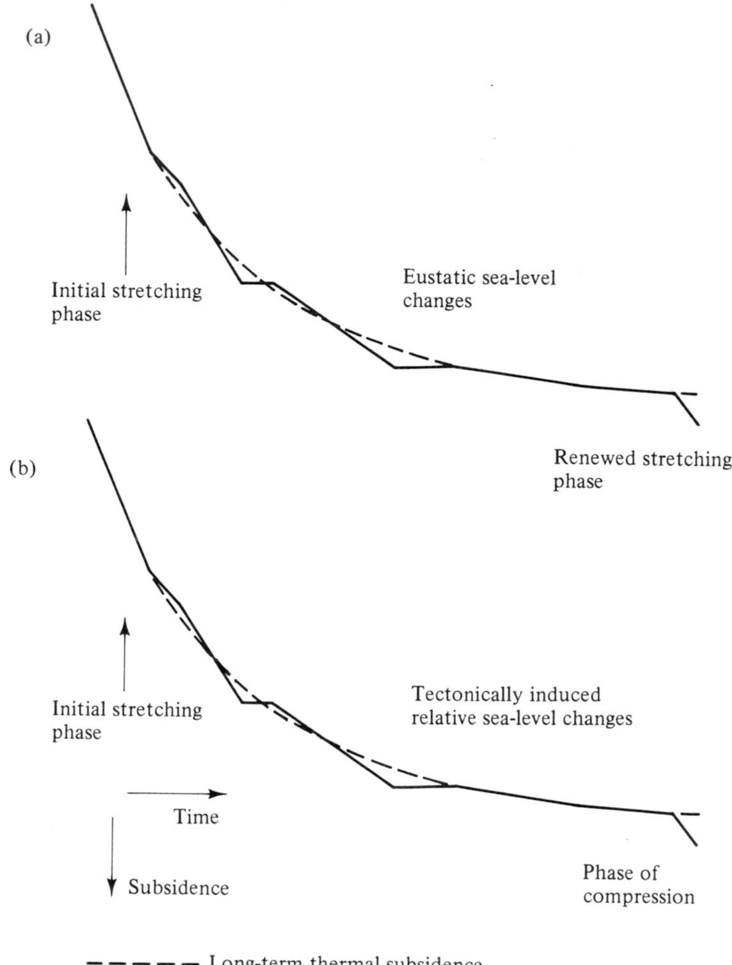

Figure 21.10 Two approaches to the analysis of tectonic subsidence. (a) Traditional interpretation: short-term deviations from long-term trends in basin subsidence are attributed to eustatic sea-level changes and renewed phases of crustal stretching. (b) Alternative interpretation: short-term deviations from long-term (thermal) subsidence are interpreted in terms of stress-induced vertical motions of the lithosphere. □

laxation of stress in the lithosphere. Similarly, as stress-induced basement subsidence and uplift phases can be of a rate and magnitude corresponding to Vail's third-order cycles (Figure 21.10), care must be taken when interpreting the record of sea-level change solely in terms of waxing and waning of ice sheets.

The stretching model of basin formation predicts a rapid phase of initial subsidence followed by long-term subsidence associated with cooling of the lithosphere. Lithospheric stretching occurs due to passive rifting of the lithosphere, after which stresses relax. An essential assumption made in this model is, therefore, that stresses are zero after basin formation. The original stretching formulation was a strictly kinematic one. More recently, work on the dynamics of stretching has demonstrated that tensional stresses of the order of several hundred MPa are required to form a sedimentary basin by this mechanism. Important in this respect have been the recent theoretical advances in lithospheric rheology based on extrapolation of rock mechanics data.

Future work has to address more fully the dynamic element of lithosphere deformation. Furthermore, present quantitative models of the origin of basins are incapable of solving problems related to subsequent structural development that may be intrinsically coupled with basin formation. For example, late-stage compression during the post-rift evolution of extensional basins

Table 21.2. *Magnitudes and rates of sea-level changes by several mechanisms that were considered in the early 1980s as potential contributors to short-term changes in relative sea level (Vail et al's third-order cycles). Apart from glacio-eustasy, the proposed (tectonic) models failed to produce both the rate (1–10 cm/1000 yrs) and the magnitude (up to the order of a 100 metres). This is primarily because of the long time constants of lithospheric thermal processes which grossly exceed the time scales (2–5 Ma) characteristic for short-term changes in relative sea level.*

	Probable maximum		Maximum maximum		
Mechanism	Magnitude (m)	Rate (cm/1000 yr)	Magnitude (m)	Rate (cm/1000 yr)	Time interval (Ma)
Glaciation	150	1000	250	1000	0.1
Ridge volume	350	0.75	500	1.2	70
Orogeny	70	0.10	150	0.20	70
Sediment	60	0.11	85	0.25	70
Hot spots	50	0.08	100	0.14	70
Flooding of ocean basins				instantaneous	

can largely explain current discrepancies between estimates of crustal thinning derived from structural analysis (obtained by measuring horizontal displacements of the normal faults active during extension) and subsidence data (where the total tectonic subsidence is used as a measure of crustal thinning). Rapid phases of basin subsidence after the initial event of basin formation and with a magnitude too large to attribute to changes in sea level are usually explained in terms of multiple stretching phases (Figure 21.10). However, care should be taken with this interpretation as an increase in the level of intraplate compression can equally well produce this type of deviation from the thermal model predictions of basin subsidence. Phases of lithospheric compression during the post-rift evolution of rifted basins can give rise to substantial deepening of the basin centre, accompanied by uplift at basin flanks, promoting the development of steershead geometries of sedimentary basins. The effect of such late-stage compressional phases is enhanced by brittle–ductile rheologies of the lithosphere, in particular where stress levels approach the lithospheric strength. Late-stage compression could, for example, explain the rapid phases of Late Neogene subsidence such as that encountered around the northern Atlantic. Therefore, current backstripping techniques, correcting only for vertical loading of the lithosphere, tend to overestimate values of crustal extension and, therefore, might result in overestimates of temperatures at depths corresponding to the hydrocarbon window.

Conclusions

Sedimentary basins form and evolve in the interiors of the plates that are subject to episodic changes in tectonic regime. Recent work on intraplate tectonics has established that there is a strong mechanical coupling between geodynamic processes at plate boundaries and deformations in the plate interiors. In fact, the record of vertical motions in sedimentary basins holds the potential for unravelling the full complexity of the interplay between the processes of basin dynamics and basin fill. Careful analysis is required to separate effects of eustatic sea-level changes from stress-induced short-term motions of the lithosphere, as both mechanisms produce rather similar short-term distortions from long-term patterns of thermal subsidence. Plate tectonics can operate on a global scale (plate reorganisations) and on a more regional scale, which is important for the discussion on global synchroneity of apparent sea-level change. Integrated studies of the structural and stratigraphic evolution of sedimentary basins, together with further development of dynamic models for basin formation and evolution, will contribute to the success of basin analysis.

Further reading

Cross, T. A. (ed.) (1990) *Quantitative Dynamic Stratigraphy*, Prentice-Hall, New York.

A report that summarises the recent developments in the new field of quantitative dynamic stratigraphy, with applications on all basin scales.

Johnson, B. & Bally, A. W. (eds.) (1986) Intraplate deformation: causes, characteristics and consequences, *Tectonophysics Special Volume*, **132**.

A collection of papers with results of studies of causes, characteristics and consequences of intraplate defor-

mation by investigators of oceanic and continental lithosphere tectonics.

Kleinspehn, K. L. & Paola, C. E. (eds.) (1988) *New Perspectives in Basin Analysis*, Springer-Verlag, New York.
A collection of papers describing current problems and frontiers in sedimentary basin analysis with special emphasis on the relation between sedimentation and tectonics.

Le Pichon, X., Bergerat, F. & Roulet, M. J. (1988) Plate kinematics and tectonics leading to the Alpine belt formation: a new analysis, *Geological Society of America, Special Publication*, **218**.
An authoritative summary of the plate-tectonic evolution and palaeo-stress field of northwestern Europe and the Alpine system.

Price, R. A. (ed.) (1989) Sedimentary basins, their formation, evolution and energy and mineral resources, *American Geophysical Union, Geophysical Monograph*, **14**.
A collection of papers on the dynamics of sedimentary basins and their importance as energy and mineral resources.

Ranalli, G. (1987) *Rheology of the Earth; Deformation and Flow Processes in Geophysics and Geodynamics*, Allen Unwin, Boston.
Excellent textbook on the mechanical properties of the lithosphere and mantle. Paperback edition available.

Tankard, A. J. & Balkwill, H. (eds.) (1989) Extensional tectonics and stratigraphy of the north-Atlantic margins, *American Association of Petroleum Geologists Memoir*, **46**.
Authoritative and extremely well-documented collection of papers on the stratigraphy and basin evolution of extensional basins in the northern Atlantic. Very useful for the wealth of high-quality seismic sections of sedimentary basins.

Wilgus, C. K., Hastings, B. S., Kendall, C. G. St. C., Posamentier, H. W., Ross, C. A. & Van Wagoner, J. C. (eds.) (1988) Sea-level changes: an integrated approach, *Society of Economic Paleontologists and Mineralogists Special Publication*, **42**.
An up-to-date integrated study of sea-level changes and the stratigraphic record with detailed description of the methodology of sequence stratigraphy. Contains a handy glossary of terms frequently used in sequence stratigraphy.

Ziegler, P. A. (1988) Evolution of the Arctic–North Atlantic and the western Tethys, *American Association of Petroleum Geologists Memoir*, **43**.
A unique and very-well-illustrated overview of the tectonic evolution of the Arctic/Northern Atlantic and western Tethys based on a compilation of industry and non-industry data.

Zoback M. L. & World Stress Map Team (1989) Global patterns of tectonic stress: a status report on The World Stress Map Project of the International Lithosphere Project, *Nature*, **341**, 291–8.
A status report summarising the results of the World Stress Map Project showing for the first time the existence of consistently oriented stress patterns on a global scale in the lithosphere.

INTRODUCTION TO CHAPTERS 22 AND 23

This volume would be incomplete if it did not include a discussion of the history of life and its evolution. In the space available, it would be impossible to cover comprehensively the full pageant of life through time, so we decided to be selective.

Popular literature and other media coverage has made the public aware of many aspects of how life developed during the Phanerozoic, or the last 550 million years (Ma) or so of Earth history during which organisms with skeletons were preserved as fossils. The fossil record of life over the period 3000 Ma before is much less familiar and is the subject of Chapter 22 by Simon Conway Morris of the University of Cambridge. Only at the very end of this long period of time did animals evolve. The early fossil record consists of simple organisms such as bacteria and algae, yet these played a crucial role in changing the composition of the Earth's atmosphere, making it possible for the more complex life forms to develop that emerged in the last tenth of geological time. Chapter 22 stops at the 'Cambrian explosion', when skeletons appeared for the first time. What caused this spectacular event in the Earth's history? As the reader will find out, there are several intriguing explanations.

The fossil record has for a long time provided much evidence for the course of evolution. But if the rock record is incomplete (as shown in Chapter 19), then the fossil record is even more so, as rarely are the soft parts of organisms preserved. However, the record we do have permits some retrospective testing of evolutionary hypotheses. Indeed, as our Open University colleague Peter Skelton explains in Chapter 23, it is impossible to conduct experiments on the long-term processes that influence evolution, simply because of the constraints of the human life span. Thus retrospective testing of evolutionary hypotheses presents an exciting challenge to palaeontologists. Chapter 23 explores this challenge.

CHAPTER

22

Simon Conway Morris
University of Cambridge

THE EARLY EVOLUTION OF LIFE

Introduction

What are the origins of our teeming biosphere, with its profligacy of organisms including such widely differing forms as deep-sea squid, ocean-cruising albatrosses, jungle orchids and hedgehogs? Much of our modern biota arose relatively recently. Its foundations were laid between about 50 and 100 million years (Ma) ago with such events as the rise of bony fish, mammals and flowering plants; however, in turn the ancestors of these organisms can be traced with reasonable facility by palaeontologists much further back in time. So it is that over a period of about 550 Ma, that is to the beginning of the Cambrian, the broad patterns of diversification and precipitous decline provide a shimmering tapestry of organic diversity. But when we move slightly further back in geological time something extraordinary happens to the fossil record: skeletons disappear and the record becomes much more cryptic and elusive; not surprisingly there are corresponding difficulties in the interpretation of the organic remains. What is even more remarkable is that this early interval, the so-called Precambrian, occupies more than 85 per cent of Earth history (Figure 22.1). One purpose of this chapter is to outline how much about the early evolution of life can be drawn from the fossil record and whether these data square with other lines of evidence from disciplines as disparate as sedimentology and molecular biology. Another aim is to explain how the abrupt appearance of skeletons at the beginning of the Cambrian is only the most obvious signal of extraordinary adaptive radiations that mark a major re-organisation of the biosphere, ushering in the **Phanerozoic**, the eon of 'visible life' and an abundant fossil record. Underlying these facts is the remarkable observation that so far as we know life is unique to this planet. It is widely agreed that life has profoundly influenced the physical evolution of the planet, certainly in terms of atmospheric composition and perhaps in more subtle ways. Some have gone so far as to suggest that life acts as a type of superorganism (Gaia) actively mediating for its own benefit conditions on the planet so as largely to isolate it from external changes, especially that of changing solar luminosity. The **Gaian hypothesis** has attracted wide attention, and even its most vigorous opponents would concede that without life the Precambrian evolution of the planet would be entirely different: one only has to look at the freezing deserts of Mars or the furnace-like surface of Venus.

The earliest Earth

Even though the Earth formed about 4600 Ma ago, the oldest sediments still preserved on the Earth's surface are only 3800 Ma old (Figure 22.1). The best known of these is the sequence at **Isua**, exposed between the waters of the Davis Strait and the Inland Ice on the west coast of Greenland. Here, despite the metamorphism, a suite of volcanic and sedimentary

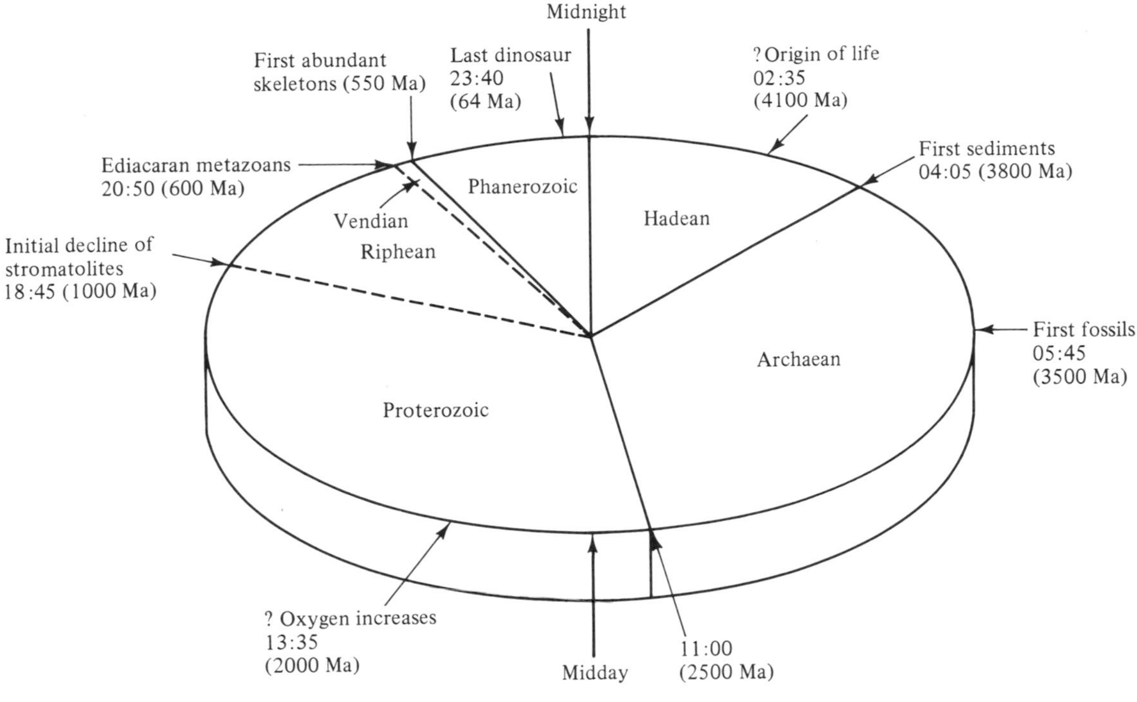

Figure 22.1 **The immensity of Earth history which stretches over 4600 Ma represented as segments of a 24-hour day. Note that the first evidence for surface processes and environments, as recorded in preserved sediments, does not occur until just after four o'clock in the morning, while an effective fossil record (represented by skeletal remains) is not available until almost nine o'clock in the evening.** □

rocks that were clearly deposited on the Earth's surface (supracrustals) are recognisable. Most significant are those sediments that record the action of running water and the proximity to exposed land whence some of the conglomerate pebbles must have been derived. In addition to these conglomerates and sandstones, there are also other sedimentary rocks such as chert and a distinctive type consisting of iron-rich bands. This is the earliest record of a banded ironstone (Figure 22.5) which later in the Proterozoic were to become much more abundant and may hold significant clues to the rise of free oxygen by the processes of photosynthesis. In addition to equivalent rocks in Labrador on the other side of the Davis Strait, that help to delimit the possible extent of the Isua 'microcontinent', other sediments that on radiometric evidence are only slightly younger than Isua are known from several areas. These include the United States (Minnesota and Michigan) and South Africa near to the border with Zimbabwe. These examples, however, are mostly metamorphosed gneisses and are correspondingly difficult to interpret. Hints of even earlier rocks are evident from zircons that, although reworked into younger material, bear a radiometric imprint of formation in excess of 4000 Ma and suggest the presence of some sort of crust.

However, for all intents and purposes the outlines of what happened in the first 800 Ma of Earth history are mostly shrouded in uncertainty. What is clear is that during this interval a major episode of degassing led to the formation of an atmosphere and ocean. Both the extent of the oceans and the composition of the atmosphere continue to be a subject of disagreement. The Isua supracrustals show at least some areas were emergent, and the intervening oceans may have reached depths similar to those of today. All are agreed that the atmosphere lacked oxygen, which only was able to accumulate to high levels by photosynthetic production of first algae (and later land plants), ultimately overwhelming the continuing supply of reducing gases from volcanoes and exposure of reduced minerals by weathering. Indeed, even today were all plants and algae to die, then within a geologically short time the planet would revert to an anoxic state, waiting to be claimed by the heirs of the earliest Archaean life which still lurk in the myriads of environments from which oxygen is excluded. If oxygen was absent from the early atmosphere, then it is agreed that nitrogen, water vapour and carbon dioxide were abundant. The last gas was probably instrumental in preventing the new Earth freezing over because, without its 'greenhouse effect', the less luminous Sun of 4000 Ma ago would have been unable to warm the planet sufficiently. Other gases present might have included methane and ammonia, but the importance of these and other constituents is less certain. The differentiation of a crust probably began well before 4000 Ma ago, while internally the descent of iron and nickel to form the Earth's core allowed a protective geomagnetic field to be generated.

The early Solar System was a violent place. For hundreds of millions of years after their formation, the planets were subject to intense bombardment by meteors and larger bodies. Just as the remorseless processes of weathering, erosion and tectonic upheavals have removed the oldest sediments, so they have obliterated all traces of this impact episode on Earth. On the Moon and other planets, however, the scars remain. The large gravitational pull of the Earth means that some of the impacts must have been colossal. Indeed, it is conjectured, on geochemical evidence, that the Moon itself is the product of a body about the size of Mars hitting the Earth and detaching a molten mass that has remained with us ever since. At first close to the Earth, its gradual retreat has been mediated by the transference of angular momentum due to tidal friction slowly reducing the rate of the Earth's rotation.

The Origin of Life

The oldest sediments, therefore, have behind them a vast period of geological time in which many of the fundamental episodes of Earth history must have taken place, including the origin of life. It is most unlikely therefore that the events involved in this momentous step will ever be detected, and even though the oldest sediments at Isua contain carbon that might be of organic origin, this tells us little of the preceding events. Concerning the time of origination, it has been suggested, however, that the likelihood of nascent life surviving is constrained by the projected diminution in the bombardment episode, the earlier intensity of which would have repeatedly sterilised the Earth's surface. If life first evolved in the deep-sea then it would have been more shielded at a somewhat earlier date than surface-dwelling forms, but it is estimated that prior to about 4100 Ma ago the position of life as a permanent tenant was decidedly precarious.

An absence of a fossil record, however, has been no barrier to speculation about how life arose. Scientists seek to explain not only the origin of the basic building blocks of organic molecules, but also their encapsulation in a cell and the ability for replication and transmission of coded genetic instructions. For many years discussion has been dominated by descriptions of experiments whereby mixtures of gases, thought to approximate to the composition of the early atmosphere, are subjected to ultraviolet irradiation, spark discharge, sudden pressure or some other energy input designed to mimic a comparable source on the early Earth. Such procedures have been remarkably successful and have led to the synthesis of a wide variety of basic molecules known to be essential for the construction of a cell. These include sugars, amino acids and nucleic acids, and some success has also been achieved in polymerisation and synthesis of larger molecules. However, there is an immense gulf between this soup of relatively simple molecules and a cell capable of division and replication. Not that attempts to produce cell-like structures have been unsuccessful. For example, by heating dry amino-acids and quenching in water, organised microspheres, with a number of cell-like properties, are readily produced. Alternatively, colloidal suspensions, of which a well-known example is gum arabic (carbohydrate) and histone (protein) in water, lead to enclosed droplets (or coacervates) which are stable and which are believed by some to suggest how primitive cells might have originated. In both cases a membrane-like structure can act as a barrier, simple chemical reactions can be run within the structure, and growth by budding can be achieved. Intriguing as these structures appear to be, there is still a huge gap between them and an operating cell with a complexity far beyond our fumbling attempts at biosynthesis.

More recently, new lines of evidence have begun to complement the existing data. There is, for example, active research into the possible involvement of clays in the origin of life. Their potential as catalysts has been long emphasised, while the alternation of different clay minerals in a clay-stack has led to the intriguing suggestion that here is a possible template for the coding of information that by subsequent transfer to living cells provided the basis for the genetic code.

The traditional line of experimentation, effectively based on Darwin's famous speculation of a 'warm, little pond' as the cradle of life, has also been extended into other environments that must have been present on the early Earth. Particular interest at the moment is focussed on hydrothermal systems, especially those associated with spreading ridges where prodigious quantities of sea-water (if submerged like the modern East Pacific Rise) or fresh-water (if exposed as in Iceland) and pumped through the cracks and fissures of the hot rock. Proponents of this scheme emphasise that not only is there an energy source, but also abundant nutrients and suitable gases, as well as clay minerals known as zeolites that could act as very efficient catalysts during primary biosynthesis. This hypothesis is not without its critics, who stress that the ambient temperatures place newly synthesised organic material in recurrent danger of destruction. It does, however, offer a coherent model, some aspects of which might be tested in modern environments.

The Precambrian fossil record

The prospects of resynthesising life are, at least at the moment, remote. The likelihood of recapitulating the sequence of events that almost certainly occurred before the oldest known sediments were deposited is even more problematic, and for the rest of the chapter we may take life as a fiat. Of what then does the Precambrian fossil record consist? Effectively there are **body, trace** and **chemical fossils**, according to whether the organism itself, a manifestation of its activity, or molecules dispersed in the sediment after its death are under consideration. However, because of the stark contrasts between the Precambrian and Phanerozoic fossil records, this tripartite division needs further explanation. Until the latest Precambrian, when macroscopic remains of animals (**metazoans**) are preserved, life may have been confined to entirely microbial forms. Not surprisingly they normally have only a minimal chance of surviving fossilisation; they are effectively soft-bodied and post-mortem decay of cell wall and contents usually is rapid. However, the long time span of the Precambrian inevitably provides sedimentary environments in which diagenesis was so rapid that microbes could be preserved. Most important in this context is very early silicification whereby entire microbial communities can be entombed by chert, but such fossils may occur in shales, dolomites and other sediments. Their interpretation, however, is not straightforward. One problem is that organisms with radically different biochemistries, tolerance of oxygen, ability to utilise substrates and so forth are often morphologically indistinguishable. Indeed, even within a particular microbial group, radical evolutionary advances in biochemical repertoire need not have any morphological expression. Aptly referred to as the 'Volkswagen Syndrome', this means an unchanged exterior may conceal revolutionary internal changes in cell machinery. We are thus faced with the dilemma that even though the great bulk of Precambrian evolution must have been the occasional, but triumphant, breakthrough in biochemical organisation, our likelihood of detecting it in among the admittedly rare microbial fossils is vanishingly small. Even distinction between the gulf of **prokaryotes** (cells without a nucleus or other organelles, and lacking sexual reproduction) and the **eukaryotes** (cells with a nucleus) is fraught with problems in terms of the Precambrian record. Eukaryotes, to be sure, spawned the great multicellular lineages leading to plants, fungi and animals (metazoans), but the origins of the unicellular ancestors are obscure. Despite the profound differences in organisation, the rapidity of decay means that of the many cellular differences only that of relative size is likely to survive. Most eukaryotic cells are substantially larger than those of prokaryotes, but the overlap is such that in itself it cannot be taken as a secure guide.

Both prokaryotes, which are generally divided into the eubacteria and archebacteria (which are distinguished on a number of profound differences, including the RNA sequences, cell wall composition and enzyme structure), and the eukaryotes are well represented by microbial forms, the former effectively exclusively so and the latter by numerous groups of unicells or simple aggregates of two to several hundred cells. Do all these microbial groups, however, stand an equal chance of preservation? It is the prokaryotic **cyanobacteria**, often known as the **blue-green algae** (Figure 22.2), that possess the highest potential for fossilisation on account of encasement by a tough gelatinous sheath that maintains a shape even if the cells within have decomposed. Cyanobacteria are largely benthic, forming mats on the floors of shallow seas, lagoons and lakes. Of the myriads of other prokaryotic groups practically nothing remains in the Precambrian fossil record. As discussed below, a few bacteria survive in the most exceptional of fossil deposits such as the Gunflint Chert, while a blip on the age curve of Precambrian carbon isotopes hints at the activities of methanogenic archebacteria. (These organisms obtain energy by reducing carbon dioxide (CO_2) and releasing methane (CH_4) during their metabolic cycle; their activities range from release of marsh gas from sediments to fermentation processes in the stomachs of cows and other ungulates.) Clues to their early diversity would seem to have been erased from any record, although we can reliably infer their presence deep into the Precambrian from the echoes of their molecular evolution, an important topic that is providing a new dimension to our grasp of Precambrian life.

Concerning the eukaryotes, it is predominantly to

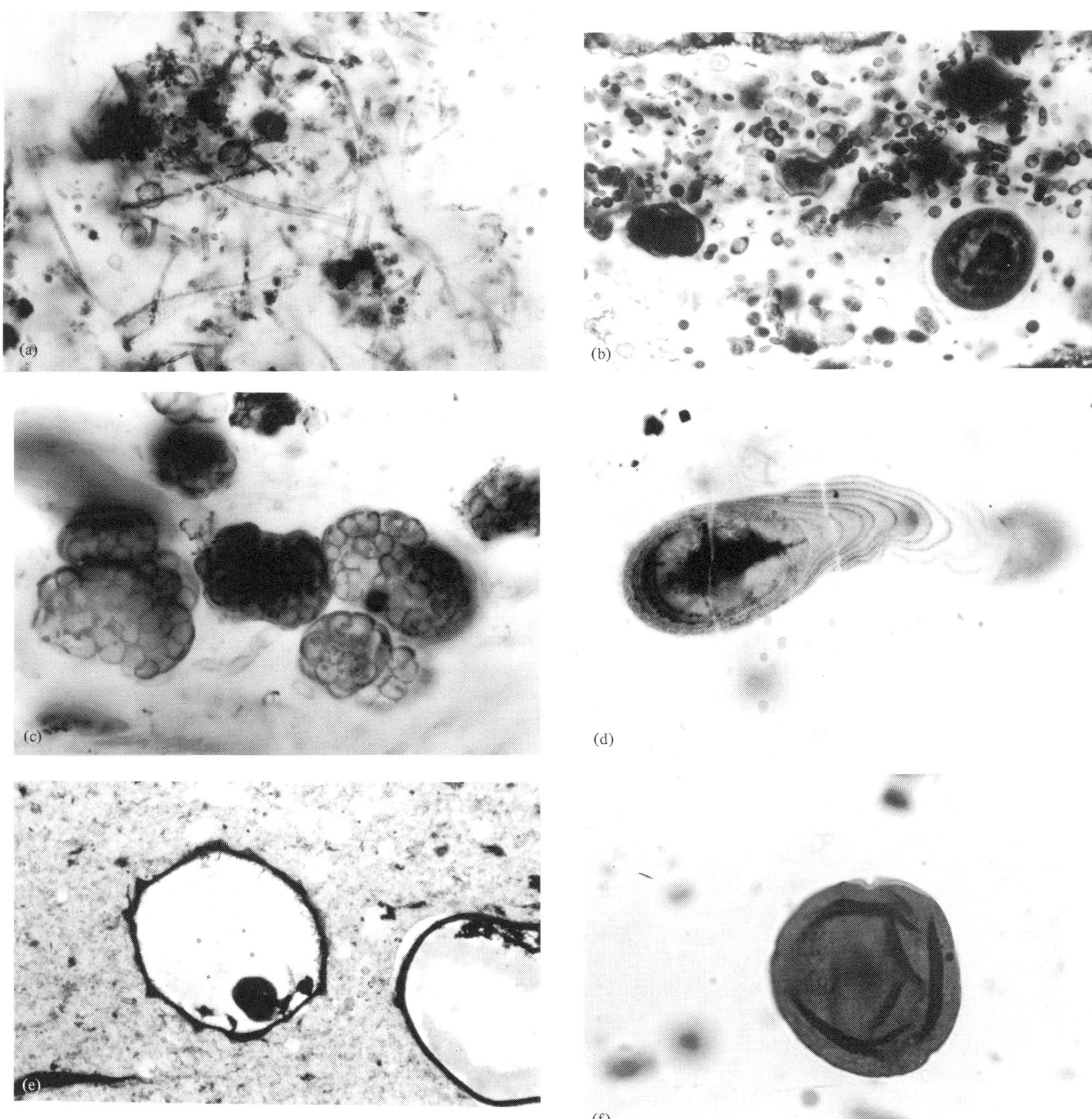

Figure 22.2 Representative cyanobacteria and acritarchs from the Precambrian. (a) Prokaryotic microbenthos from the Gunflint Iron Formation (*c.* 2000 Ma old), Ontario. Both filamentous sheaths, of which the one in the centre of the photograph has a diameter of 2 µm, and slightly larger coccoid cells are very well preserved. (b) Diverse microfossils from the Draken Formation (*c.* 750 Ma old), Spitsbergen. The large fossil (lower right) is cyanobacterial with a diameter of 75 µm. (c) A pleurocapsalean cyanobacteria (*Synodophycus euthemos*), a dweller of microbial mats in the Draken Formation, Spitsbergen. Each 'unit' within the vesicle is 4 µm diameter. (d) A morphologically distinctive cyanobacterial mat-dweller (*Polybessurus bipartitus*) from the Draken Formation, Spitsbergen. Maximum cross-sectional diameter is 35 µm. (e) Planktonic protists (*Trachyhystrichosphaera vidalii*) from the Draken Formation, Spitsbergen. Diameter of the central vesicle is 300 µm. (f) A late Riphean acritarch from the Mireodicha Formation (900 Ma old), eastern Siberia. Diameter is 60 µm. □

Figure 22.3 Fossil and modern stromatolites. (a) One of the oldest fossil stromatolites known. Replaced by chert, this 3500 Ma old stromatolite from the Pilbara region of northwest Australia shows the laminated structure. (b) Modern stromatolites exposed at low tide in Hamelin Pool, Shark Bay, on the coast of West Australia. Such a scene is little different from what most coastlines would have looked like 1000 Ma ago. □

the **acritarchs** (Figure 22.2) that we must turn in terms of the first body fossil record. Evidently planktonic, most of them are interpreted as encystment structures of algae, owing their abundance to a highly resistant coat. Typically with a diameter of about 30–100 μm, acritarchs vary widely in shape with some having smooth walls, whereas others have spinose extensions or more complex ornamentation. Although known from cherts, acritarchs are recovered in huge numbers from the disaggregation of shales. Abundant from about 1000 Ma onwards, the earliest acritarchs date back another 400 Ma. However, evidence from both chemical fossils and molecular biology suggest that this grade of organisation has yet earlier antecedents, perhaps deep in the Archaean. Apart from the chert microbiotas, mostly prokaryotic, and shale-derived acritarchs, Precambrian body fossils are mostly represented by carbonaceous films, perhaps of sea-weeds, and, in the latest Precambrian, by casts and impressions of soft-bodied metazoans belonging to the Ediacaran faunas, so named after one of their most important occurrences in the Ediacara Hills of South Australia.

What of trace fossils? In orthodox terms most palaeontologists would think of activity recorded in the sediment such as a worm burrowing or marks impressed by a strolling arthropod. Simple trails accompany the Ediacaran faunas, but the traces generated by microbes are of a somewhat different category known as **stromatolites** (Figure 22.3). These laminated structures, especially characteristic of limestones, provide the bulk of the Precambrian fossil record. They reflect the response of benthic mats to burial by sediments, a recurrent hazard in shallow waters stirred by tides and currents. The upper layers of the mats are dominated by cyanobacteria, which would be unable to photosynthesise when buried by sediment. To surmount this dilemma, the mat glides upwards through the newly deposited sediment to re-establish itself on the surface. By abandoning a layer of gelatinous sheaths and dying cells, the laminated structure of the stromatolite is developed by repeated episodes of burial and phototactic recovery. Only very rarely are the original microbial communities preserved, but study of modern stromatolitic microbiotas shows them to be composed of numerous species and also to show distinct stratification with cyanobacteria underlain by green and purple sulphur bacteria. The lower layers are more attuned to reduced oxygen levels and include decomposers that recycle some of the organic material. However, of this vertical stratification and microbial diversity little survives in the fossil record.

Stromatolites show wide morphological variability, although a broad distinction between stratiform and columnar, the latter even becoming conical, is readily distinguishable. The significance of the various stromatolite shapes and their possible relationship to the original microbial communities is largely enigmatic, but the observation that stromatolite shape and diversity change during the Precambrian suggest that links between microbial evolution and the changing Precambrian environment must have existed. Some stromatolites grew to a substantial size and, in well-preserved sequences, facies relationships may show distinction between inter-tidal and sub-tidal forms, as well as distinguishing those occupying deeper waters on the edge of the photic zone. Indeed, some stromatolites conceivably grew in dark conditions, where filamentous bacteria took the place of cyanobacteria.

Chemical or molecular fossils are the most elusive

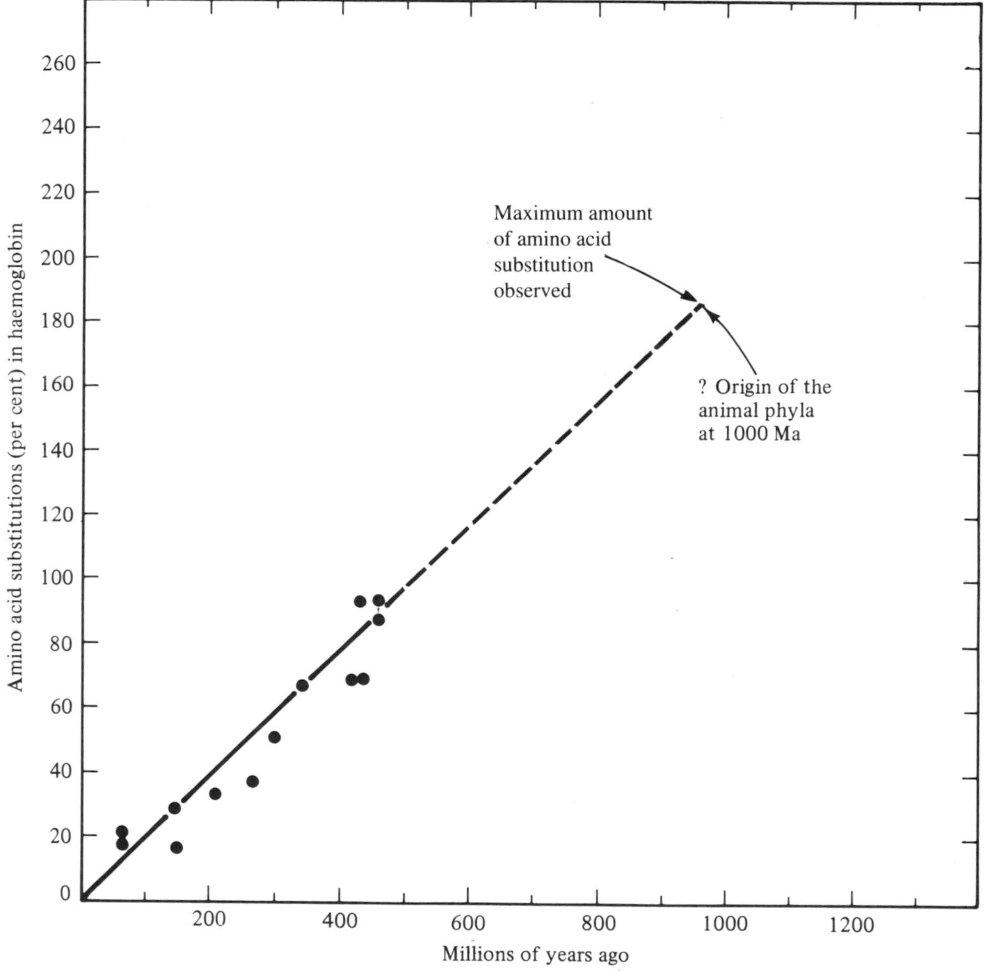

Figure 22.4 The molecular clock. The giant molecule haemoglobin, used in the blood to carry respiratory oxygen, is composed in part of long chains of amino acids. During geological time substitution of one type of amino acid for another occurs at various points along the chain. By plotting the difference in amino acid sequences in haemoglobins of different living animals against their known times of divergence as revealed in the fossil record, it is seen that the plot is a straight line. This indicates that the rates of substitution within the amino-acid chains are more or less constant, in other words the system acts as a sort of molecular clock. By extrapolating the plotted line backwards to account for the largest degree of divergence observed in amino-acid sequences of haemoglobin in living animals, one may infer that animals originated about 1000 Ma ago. This is based on the assumption, of course, that the clock-like character of substitution has proceeded at a constant pace and not been subject to periods of acceleration. □

line of evidence, partly because of their often cryptic origins, but also because contamination has been a recurrent problem. This category may be divided into insoluble residues (**kerogen**), soluble compounds such as amino acids, lipids and hydrocarbons, and **stable isotopes** (especially carbon isotopes ^{12}C and ^{13}C) whose fractionation by organisms may provide a signature that survives long after all other remains have vanished (see also Chapter 24). Apart from the problems of study mentioned above, the extraction and interpretation of chemical fossils from Precambrian rocks is also rendered difficult because few have escaped at least mild metamorphism that both degrades more complex molecules and shifts isotopic ratios. Early enthusiasm for the study of chemical fossils in the Precambrian was tempered in due course by various setbacks, but more recently advances in both the machine technology and knowledge of potential pitfalls has led to a resurgence of interest. As explained below, particular interest resides in study of isotopic ratios, although the overall persistence of values of *c*. –27‰ $\delta^{13}C$ throughout the Precambrian is widely regarded as evidence for the continuing importance of photosynthetic fractionation. This inference is based on the consistent way in which photosynthesisers 'select' the light ^{12}C in preference to ^{13}C, so leading to isotopically light carbon, i.e. negative $\delta^{13}C$. Another area that promises much in the next few years is the search for specific molecules that, by being unique to one group, provide a biomarker of existence, even if all cellular remains have entirely vanished.

In addition to the palaeontological evidence, a record survives in the modern descendants of Precambrian life in the form of the **molecular sequences** of various cellular components. Most important in this context are the nucleic acids in RNA (and DNA) and amino acids in such protein or polypeptide chains as haemoglobin and cytochrome-*c* (a respiratory protein found in mitochondria) which form the building blocks of the chains. In some cases the sequence of nucleic or amino acids is extraordinarily conservative, differing little between bacteria and vertebrates, presumably because substitution by other components would jeopardise the function. In other examples, however, not only has substitution been extensive, but there is evidence that it has taken place more or less monotonically. Thus, when two groups split the molecular systems are isolated and diverge as substitution proceeds at an approximately constant rate (Figure 22.4). Such a '**molecular clock**' provides valuable clues as to both possible times of origination and also nearness of relationship of groups that arose in the Precambrian, some of which have left no other record.

Archaean life

The potential fossil record in the Precambrian begins, of course, with the oldest sedimentary rocks. In general, the degree of metamorphism has been far too severe to offer much hope of survival. Even at Isua, where the metamorphic grade is comparatively benign, stromatolites do not occur nor have definite cellular remains been recognised, despite some claims to the contrary. Kerogen, however, does occur, and with an isotopic ratio relatively enriched in ^{12}C (i.e. negative $\delta^{13}C$) suggests that even 3800 Ma ago photosynthetic organisms may have been fractionating the carbon reservoir, although alternatively metamorphism may have led to this isotopic shift. Another 300 Ma of Precambrian time must elapse before unequivocal fossils can be recognised. These come from localities in Western Australia, in an area called North Pole on account of its remoteness, and in the Barberton Mountains of Swaziland. In these regions, limited metamorphism has allowed superbly preserved sedimentary sequences to survive, which at a few horizons yield stromatolites, although the status of a number of examples from the North Pole area continues to arouse suspicion in some quarters. Although not in direct association with these stromatolites, cherts have yielded microbial fossils, including minute filaments (about 1 μm diameter) and coccoidal spheres. The biogenicity of some of these fossils has been vigorously contested, not least because in some cases the chert cannot be shown to be penecontemporaneous with the surrounding country sediments. However, repeated scrutiny suggests at least some are genuine, although once again the Barberton Mountain examples are marginally more persuasive. What type of organism might they represent? The bias mentioned above towards fossilisation of cyanobacteria makes them a prime candidate, and at least they can be implicated in the construction of the associated stromatolites. However, the possibility that the earliest fossils represent other types of bacteria remains strong, and the North Pole examples contain particularly interesting assemblages whereby slender filaments radiate from a basal rosette. Apart from these examples, the record of Archaean microfossils is pathetically meagre. From slightly younger sediments in Swaziland, including the famous **Fig Tree Chert**, a number of possible microbial fossils have been described, mostly coccoids. Doubt attends the biogenicity of many, but selected examples are more persuasive because they show the structure of biological populations and evidence for cell fission.

Apart from another record of microbial fossils from

Figure 22.5 A typical block of banded ironstone from the Soudan Iron Formation (*c.* 2700 Ma old), Minnesota. Note in particular that banding is defined by layers of iron-poor and iron-rich chert. A number of hypotheses have been proposed to explain both this alternation and the regularity of the bedding, but most share the proposal that the iron-rich bands represent times when free oxygen was available to precipitate the soluble Fe^{2+}, dissolved in the water, into highly insoluble Fe^{3+} that then descended to the sea-bed. □

near the end of the Archaean in the Fortescue Group of Western Australia, our knowledge of the first 1300 Ma of recorded Precambrian life otherwise is almost entirely dependent on stromatolites. Even these provide only a sporadic record, which in contrast with the succeeding Proterozoic, seems conspicuously less diverse. Here is just one example of a general problem that haunts the palaeontologist. Is low diversity original or an artefact whereby once diverse assemblages are militated against by a paucity of well-preserved sequences? In the Archaean it is difficult to decide whether the scarcity of fossil evidence is due to originally inimical environments, whereby massive volcanism, an unstable and tectonically active crust and perhaps nutrient shortages quashed diversification. Alternatively, the record may appear incomplete simply because the great bulk of the sedimentary record has been destroyed and the little to survive has been variously metamorphosed and tectonically deformed. The fact that where significantly thick and unmetamorphosed sequences survive they contain quite a diversity of stromatolites is some indication that the lack of Archaean fossils is as much due to a biassed record as original impoverishment.

Proterozoic life

In contrast, the fossil record of the succeeding Proterozoic is sufficient to give confidence that at least the outlines of evolutionary history are discernible. Perhaps significantly, the transition to the Proterozoic marks a significant change in tectonic regimes. Although by no means synchronous around the globe, this shift heralds the spread of stable cratons flanked by epicontinental seas upon whose shelves clastic and carbonate sediments accumulated, superficially at least similar to much of the rock record until the present day. This is in striking contrast to the preceding Archaean where the tectonic activity is reflected in so-called greenstone belts, now generally metamorphosed and folded, with an abundance of volcanics and immature sediments, including coarse clastics. These greenstone belts typically encircle granites, but this arrangement is probably a secondary consequence of deformation and diapiric intrusion of the granites over a protracted interval. However, within the Proterozoic sequences not all depositional environments have obvious younger equivalents. Most striking are the **Banded Iron Formations** (BIFs; Figure 22.5) sometimes referred to as Banded Ironstones, which as noted above make their debut at the beginning of the sedimentary record at Isua. However, their variety of occurrences and mineralogies show they must represent a variety of depositional settings. In the Proterozoic massive BIF deposits, of considerable economic importance, occur in areas such as central Canada (Ontario), Western Australia (Hammersley Basin) and the Ukraine. These represent a complex facies that may include deposits indicative of shallow, sometimes turbulent, conditions with stromatolites and oolites. Particularly diagnostic, however, are alternating laminae of iron-rich and iron-poor chert whose regularity and basin-wide distribution is reminiscent of seasonal varves. What basinal configuration and palaeo-oceanographic conditions led to the BIFs is under active debate, but most models invoke periodic precipitation of ferrous iron (insoluble Fe^{3+}) from anoxic water rich in dissolved ferric iron (Fe^{2+}) coming into contact with oxygen. Popular suggestions have been (a) seasonal planktonic algal blooms releasing photosynthetic oxygen, or (b) periodic upwelling or turnover of deeper anoxic waters into shallower oxygenated regions. BIFs are known throughout the Archaean and for the first 500 Ma of the Proterozoic, during which period they seem to have been most voluminous and important. From about 2000 Ma, however, they are effectively absent from the rock record. Shortly afterwards **red beds**, such as desert sandstones and floodplain sediments, first become conspicuous. This transition from BIFs to red beds has long been assumed to represent the onset of an oxidising atmosphere, especially as at about the same time placer deposits (i.e. deposits where mineral grains have been concentrated by the sorting action of water, often in rivers or in shallow seas) with easily oxidisable minerals such as uraninite (UO_2, usually oxidised to U_3O_8) disappear. More or less coinciding with this interval there is growing evidence for extraordinarily large excursions in heavy carbon (positive $\delta^{13}C$) as measured in carbonates. This may represent significant changes in organic productivity, also linked to the rise of oxygen. Nevertheless, the rate of oxygen increase and its concentrations in comparison with present atmospheric levels (21 per cent) during the Precambrian excite widely divergent opinions, and even in the Archaean levels may have been more significant than sometimes conceded.

Evidence for Proterozoic life is most dramatically displayed by exquisitely preserved microbiotas in cherts (Figure 22.2), and to a lesser extent dolomites and shales. Many such horizons have been documented, but a few stand out for their quality of preservation and resultant scientific attention and publicity. Key examples are the Gunflint Chert of northern Ontario (*c.* 1900 Ma); the cherts of the Kasegalik and McLeary Formations in the Belcher Islands of northern Canada (*c.* 1900 Ma), the Bitter Springs Chert of central Australia (*c.* 800 Ma), similar aged cherts and dolomites from California, and a variety of cherts and shales from slightly younger horizons in east Greenland and Spitsbergen.

The **Gunflint Chert** forms part of a BIF exposed along the northern shores of Lake Superior, Canada. In addition to stromatolitic horizons the cherts contain superbly preserved microbiotas. Many filamentous forms can be compared reliably to cyanobacteria, although sometimes only the gelatinous sheaths survive. Such preservational artefacts have led to some taxonomic confusion, and there is still disagreement as to whether these cyanobacteria possessed certain specialised cells such as heterocysts (used for nitrogen fixation in modern species) or akinetes (resting cells capable of regeneration). Other cyanobacteria are represented by coccoid cells. As this group accounts for most Precambrian records, the occurrence of other types of bacteria in the Gunflint is all the more exciting. Some resemble manganese bacteria that inhabit Recent lakes, while peculiar forms with an umbrella-like extension are similar to certain soil bacteria. Other microbes are much more enigmatic. *Eoasphaera*, for example, has a central cell with satellite spheres adhering to its periphery. The large size of *Eoasphaera* has been used to argue it is eukaryotic, but this is not widely accepted. Recent discoveries have shown Gunflint-type microbiotas to be widespread with localities as far apart as China and Western Australia. However, this assemblage did not monopolise the Archaean seas, because similarly aged microbiotas from areas such as the Belcher Islands of northern Canada differ significantly.

Over a thousand million years later the **Bitter Springs Chert** provides another glimpse into the Precambrian world, at least as represented in shallow tropical waters where microbial mats flourished in lagoons and marine embayments. Like the Gunflint Chert, filamentous cyanobacteria abound. At first sight there seem few differences with their older congeners, but detailed study reveals the rise of new groups typified by styles of branching (or pseudo-branching) and possession of specialised cells. Coccoid cells also abound and at one time dark spots and granules within them were taken as proof of their eukaryotic status on the assumption they represented organelles. However, experiments of allowing coccoid cyanobacteria to decay showed how such features might be only a result of cytoplasmic collapse.

Fossil assemblages from the Gunflint and Bitter Springs Cherts now hold a classic place in the story of Precambrian life. However, the search for new localities has been assiduous, with particularly remarkable results from the Riphean of Spitsbergen and east Greenland. Excellently preserved microbial assemblages not only augment our knowledge of diversity, but have provided a springboard to a deeper comprehension of this early biota. Now emerging is a coherent approach to problems such as degradation sequences and life histories, the recognition of which imposes a biological order to the taxonomy, while palaeoecological investigations recognise facies control, e.g. lagoon versus offshore, and life positions such as phototactically arranged filaments in microbial mats.

In a somewhat different category to the predominantly permineralised cyanobacterial communities are the acritarchs (Figure 22.2). Typically they are extracted from shales by digestion in hydrofluoric acid, their remarkably tough organic coats ensuring survival during this robust treatment. Acritarchs are a taxonomic hodgepodge. Many probably represent encystment structures of eukaryotic algae, a few are pelagic cyanobacteria, others remains of enigmatic affinity. They first appear about 1400 Ma ago, but it is later during the Riphean that their evolutionary history is now being clarified. Two features stand out. First, evidence now exists for facies control with increasing diversity across onshore to offshore transects. Second, shortly before the Cambrian there is a precipitous decline in diversity which coincides with major ice-ages. Whatever the causal connections, this event probably deserves labelling as mass extinction, although it is unlikely to have been as abrupt as such catastrophes that heralded the end of the dinosaurs much later in Earth history. These ice-age acritarchs are depauperate and generally have simple morphologies, but in other intervals morphologically diverse acritarchs and relatively short geological ranges offer a potential role in biostratigraphy.

The final category of body fossils is carbonaceous sheets and ribbons (Figure 22.6) some of the latter sausage-shaped or forming characteristic helices. The earliest examples are about the same age as the Gunflint Chert, but abundant examples occur from about 1300 Ma onwards. The consensus states that these carbonaceous fossils are multicellular eukaryotes, such as brown or red algae. Vermiform examples like *Tawuia* have been compared tentatively to metazoans, but an absence of cephalisation or internal organs make this unlikely. Indeed it is even possible that some carbonaceous films are prokaryotic. Ultrastructural information to resolve this point could be critical.

Work on Proterozoic chemical fossils has continued to gain momentum. Numerous analyses of carbon isotopes show the continuing importance of photosynthesis as the principal agent of fractionation (*c*. –27‰ $\delta^{13}C$), but at about 2100 Ma there is a remarkable excursion into very light carbon values (*c*. –55‰ $\delta^{13}C$). This appears to represent unusual activity by methanogens, a group of archebacteria that release methane (CH_4) as an end-product of their metabolism. Very recently there have also been dramatic advances in the study of certain biomarkers, especially the detection of steranes (breakdown products from an important group of organic molecules known as sterols, of which chloresterol is probably the most familiar example). Steranes appear to be confined to eukaryotes, and so their detection in Australian

Figure 22.6 (a) Carbonaceous impressions of the fossil *Chuaria* (the circular discs) and *Tawuia* (the elongate fossil). These fossils were collected from the lower part of the Little Dal Group in northwest Canada, in sediments that are about 1000 Ma old. The relationships of *Chuaria* and *Tawuia* lie most probably with the eukaryotic algae, and they are believed to have floated near the surface of the sea. (b) A specimen of the probable alga *Tawuia*, also from the Little Dal Group. □

sediments dated at about 1600 Ma is exciting because this significantly predates the first acritarchs. Indeed, there is every reason to hope that steranes may be detected in yet older sequences. Such a result would also square with data from molecular biology which suggest the eukaryotic condition, at least in terms of its genome (the sum total of the chromosomal DNA and the genetic information it encodes), is extremely ancient, with an implied ancestry in the Archaean. Whether fossil cells from this interval could ever be reliably assigned to the eukaryotes remains speculative.

The bulk of the Proterozoic fossil record is stromatolites (Figure 22.3), which often occur in enormous abundance. In areas such as northern Canada a reef-like arrangement has been mapped with scour channels separating domal masses of stromatolites arranged along strike. Indeed few environments within the photic zone appear not to have been colonised by microbial mats and doubtless they were also present in freshwater lakes. Even the land surface, though devoid of any of our familiar plants, probably was home to hardy bacteria living in the global deserts.

In the Proterozoic stromatolites are of considerable utility in several areas. In the Riphean changes in morphology have been employed in a biostratigraphic scheme which was first erected in the Soviet Union, and which seems to have applicability elsewhere. It is unfortunate that so little is known about the possible changes in the microbial communities, although as noted below they may have been responding to the rise of grazers. Another application of stromatolites that continues to excite optimism is their use in periodicity studies, such as diurnal cycles due to tidal burial. Results, however, have been somewhat ambiguous, as have attempts to use the inclination of stromatolites as palaeolatitude indicators according to the angle of the Sun. These problems probably arise because of the innumerable local factors, e.g. currents or temporary emergence, that distort and complicate the original signal. A notable exception, however, comes from the Bitter Springs beds of central Australia where some stromatolites have a striking sinusoidal pattern along their long axes. Such a configuration appears to have arisen because of their growing near the equator, a point confirmed by palaeomagnetic work. In such a position the Sun will move from north to south and back again each year, and in doing so the successive lamellae orientate themselves towards the Sun to maximise photosynthetic potential and thus generate a sinusoidal pattern. On the assumption that the lamellae are generated on a more or less daily basis, it is possible to calculate the number of days in the Proterozoic year about 800 Ma ago. Previous studies, such as counting growth lines in Palaeozoic corals confirmed astronomical estimates that the number of days in a year was greater in the past owing to the greater speed of the Earth's rotation. This in turn is linked to tidal friction, especially on the shallow shelf seas, and the increasing distance between the Earth and Moon as the angular momentum is transferred to our daughter satellite. The Bitter Springs stromatolites provide a valuable extrapolation of these estimates, and suggest that there were then about 435 days in each year.

Whether stromatolite diversity in the Proterozoic in some way reflects the complexity and richness of the constituent microbial mats is a moot point. It is clear, however, that stromatolite variety continued to increase until about 1000 Ma ago, but that thereafter it began to decline. This trend is first noticeable in sub-tidal environments and is registered by the gradual demise of coniform stromatolites, a distinctive group in which the laminae define a series of sharp apices. By 800 Ma ago the scourge had begun to afflict even those stromatolites inhabiting inter-tidal regions in a more hostile environment subject to rapid and major fluctuations in salinity and temperature. What engendered stromatolite decline? Changes in the composition of ocean waters or basic structure of the microbial communities are possible but difficult to test. An alternative, and more popular notion, appeals to stromatolite retrogression being linked to the rise of grazers and burrowers disrupting the mats. These agents, it is argued, were the first metazoans, so heralding the dawn of animal life.

The rise of metazoans

When did animals join the cavalcade of the fossil record? In terms of body fossils they would have to be regarded as late entrants with first the Ediacaran soft-bodied assemblages (Figure 22.7) (*c.* 560 Ma ago) followed shortly afterwards by the startling irruption of skeletal remains and trace fossils that usher the 'Cambrian explosion'. On this reckoning the metazoan debut is unlikely to be older than about 600 Ma, in other words after more than 85 per cent of Earth history had elapsed (Figure 22.1). Indirect evidence, however, points to an origin at least 1000 Ma ago.

The lines of evidence that metazoans appeared much earlier than commonly thought are still tenuous, but taken together are beginning to look persuasive. The decline in stromatolite diversity may reflect metazoan activity, while the diachroneity of these changes between sub-tidal and inter-tidal environments is also consistent with the difficulties metazoans might have

Figure 22.7 Representative Ediacaran fossils from South Australia (a)–(c) and the White Sea region of the USSR (d). (a) *Dickinsonia costata*, a possible annelid worm. (b) *Tribrachidium heraldicum*, a medusoid-like animal. (c) *Cyclomedusa plana*, a medusoid, possibly related to the chondrophores. (d) *Inkrylovia lata*, a bag-like animal of uncertain systematic affinities. □

experienced in colonising the more hostile environment of the latter. Detailed studies of stromatolitic fabrics may provide evidence to clinch this point. Indeed, the development of microbial build-ups with a 'clotted' fabric, rather than the usual laminations, in the Cambrian has been interpreted as the final stage in this process of disruption. These structures, known as **thrombolites**, are often associated with skeletal debris of metazoans inferred to have grazed and disrupted them. However, new evidence suggests the thrombolitic fabric arose when coccoid cyanobacteria began to secrete calcareous walls. This is part of a much wider episode of skeletogenesis, and here it is only necessary to note that even if cyanobacterial biomineralisation arose as a response to grazing pressure, thrombolitic fabrics are not in themselves a result of metazoan disruption.

Other important clues on metazoan origins are coming from molecular biology. Just as study of ribosomal RNA (rRNA) sequences in modern microbial groups indicates that their divergence is extremely ancient and that the eukaryotic condition, at least in the sense of its genome, arose much earlier than generally supposed, so analyses of the nucleotide sequences of ribosomal RNA and amino acid sequences in haemoglobin (Figure 22.4) and collagen suggest origination times for metazoans of between 700 and 1000 Ma ago. These conclusions are based on the 'molecular clock' whereby, as noted above, random substitutions in the nucleotide or amino acid chain occur at a more or less uniform rate. That this 'clock' can keep time is shown by the good agreement between accumulated differences in the haemoglobin sequences of any two living groups and their time of divergence as seen in the fossil record and calibrated to the radiometric time scale. Although this information is derived largely from vertebrates, extrapolation to metazoans as a whole suggests that unless the 'clock' ran fast at an earlier stage initial divergences could have occurred more than 1000 Ma ago.

So, if metazoans evolved so far back in the Proterozoic, why haven't they been found? Not surprisingly there have been numerous claimants, but most transpire to be bogus pretenders. The main category are purported trace fossils, but many of these structures are patently inorganic, representing water-escape structures or shrinkage cracks. However, a stubborn core may be less easy to explain away. From 2000 Ma old rocks in Wyoming (Medicine Peak Quartzite) a variety of trace-like structures occur, but still they lack diagnostic features, e.g. backfill laminae, faecal pellets. A decision on the biogenicity of many putative Proterozoic trace fossils may come from new techniques that apply data gleaned from undoubted Phanerozoic examples. These include study of sediment fabrics whereby continual passage of an animal rotates grains into a preferred orientation and geochemical analyses of burrow margins which house cation, especially Fe and Al, anomalies that may arise originally from bacteria thriving on the mucus linings. Using these methods, burrow-like structures from 2000-Ma-old sediments from Ontario have been interpreted as genuinely organic.

However, the mystery of Proterozoic traces deepens because if they are genuine products of metazoan activity then why are they so sporadic? Empty shelf space should have been available for rapid and prolific colonisation. Indeed, it may be questioned whether our 'search-images' for Proterozoic metazoans is too strongly biassed by expectations learnt from the Phanerozoic record. The earliest metazoans were probably only a few millimetres long, incapable of producing large traces and unlikely ever to be preserved as body fossils. How then can we expect to decipher this hypothetical episode of cryptic evolution? Minute traces might survive. There are claims for faecal pellets, purportedly derived from crustacean-like animals harvesting the phytoplankton, occurring in sediments as old as the Gunflint Chert, although it may be significant that protozoans can also produce pellets. The churning by metazoans living in fine-grained sediment can impose a bioturbation fabric that at an ultrastructural level appears to be distinguishable from sediments devoid of metazoans. In principle a search for suitable Precambrian lithologies could reveal the tell-tale imprint of primitive infaunal metazoans.

The Ediacaran fauna

The search for Proterozoic metazoans remains inconclusive and it is only with the advent of the **Ediacaran assemblages** (Figure 22.7) that they emerge into geological history. By definition their origins are obscure, although their diversity has indicated to some that there must have been a prolonged period of prior evolution. The faunas have an almost worldwide distribution: curiously the earliest discoveries were made near the beginning of the century in what is now Namibia. This region has now yielded a rich haul of material, but the main impetus for the description of the Ediacaran faunas came from now-classic work in the Flinders Ranges of South Australia, including the type locality of the Ediacaran Hills. More recently Soviet geologists have been particularly active in the search for Ediacaran faunas, and have documented superb assemblages from the White Sea area, the

Ukraine and northern Siberia. Hardly a year passes without a new discovery being announced and it is now clear that Ediacaran faunas were a distinctive biotic episode and occupied a discrete slice of geological time. In a few cases tentative steps towards biostratigraphic correlations are being made, but in general these are too imprecise to indicate which faunas are older than others. In numerous cases the same sequences also contain glacial deposits, but significantly the Ediacaran faunas almost always post-date them. Even if the tillites (lithified boulder clay) and other glaciogenic sediments represent several episodes, there is evidence that the ice-ages must have been of unusual ferocity, because palaeomagnetic evidence suggests ice-sheets spread into equatorial regions with extensive glaciers at sea-level. It is tempting to link the amelioration that followed this period of global refrigeration to the rise of Ediacaran animals, but the connections may be indirect ones of changes in atmospheric composition (especially CO_2 and O_2) warming the planet and flooding of shallow shelf seas, both encouraging metazoan diversification.

With one exception described below, Ediacaran assemblages are soft-bodied, yet are widespread and abundant. Most occur in shallow-water clastics, such as those from the Flinders Ranges and White Sea, which, despite their palaeogeographic separation as inferred from palaeomagnetic data, show striking faunal similarities. A few, like those from northern Siberia, occur in carbonates. Most remarkable are evidently deep-water assemblages, best known from spectacular localities in southeast Newfoundland where it is possible to walk over bedding planes strewn with the remains of bizarre Ediacaran animals representing communities entombed by the descent of volcanic ash falls. Very similar assemblages, but much more poorly exposed, occur in Charnwood Forest, just outside Leicester in England. Their striking similarities support the widely accepted notion that formerly this part of England was attached to Newfoundland on one side of an earlier ocean. With the opening of the North Atlantic these regions were sundered, now lying thousands of kilometres apart.

This abundance of soft-bodied fossils is in marked contrast to the Phanerozoic fossil record where soft-tissues are preserved only in exceptional circumstances, usually in fine-grained sediments where anoxia or hypersalinity excluded scavengers. It is for this reason that an absence of scavengers and deep burrowers has been invoked to explain Ediacaran preservation, although even here some members of the community eluded preservation as body fossils and are only known from their traces. The reasons for this discrimination are not clear, and it is also likely that the sediments hosted a variety of smaller metazoans which, despite being absent from the fossil record, may have played key roles as ancestors to many of the Cambrian groups.

Body fossils are dominated by metazoans of an apparent cnidarian-grade, although whether there are actual **cnidarians** (i.e. relatives of the Recent sea-anemones, jellyfish, corals and so on) continues to be contentious. Particularly common are medusiform types (Figure 22.7), some of which are assigned to scyphozoans, a group of jellyfish widespread in all oceans with forms such as *Aurelia* and *Cassiopeia*. Their variable anatomy, including arrangement of gastro-vascular canals and tentacles, allows some latitude when drawing comparisons with Ediacaran **medusoids**, but it must be admitted that some investigators may have unwittingly allowed a scyphozoan 'search-image' to colour their expectations. Even so interpretation of medusoid anatomy is seldom unequivocal, especially because the attitude of burial and style of infilling or collapse of former cavities results in specimens that vary widely in appearance. Another medusoid group that may be represented are the cubozoans, better known as the box jellies and justly feared on account of their venomous sting which may be fatal. The putative Ediacaran examples display a complex internal anatomy, including possible muscles and gastric cavity, as well as traces of the trailing tentacles. Other medusoids remain in even deeper taxonomic limbo and may represent major extinct groups. Most striking are those with a prominent triradial symmetry as expressed by what appear to be internal canals or caeca. These include *Tribrachidium*, once interpreted as a primitive echinoderm, and *Albumares*. Whether these so-called trilobozoans can be accepted as cnidarians is still undecided.

Another group of cnidarians identified are the **chondrophores**, related to *Hydra* but formed of an aggregation of zooids so closely integrated that they resemble an individual animal. A modern example is the by-the-wind-sailor or *Porpita*, which may occur in vast numbers near the surface of the oceans. The feeding and reproductive zooids are suspended beneath a float with a wall resistant enough to fossilise. Most Ediacaran representatives consist of a float with concentric lines marking the former gas-filled chambers, although in rare specimens of *Eoporpita* even the zooids survive.

Stalked animals anchored to the substrate, often by a swollen hold-fast, and expanding into a leaf-like body are widely regarded as sea-pens or pennatulaceans, a group that as the **anthozoans** includes also the more familiar coral and sea-anemone. Some approach quite closely modern sea-pens and have a stalk bearing a broad foliate expanse that presumably bore the individual zooids, while at the opposite end a bulbous hold-fast acted as an anchor. Other forms such as *Charnia*, the first examples of which were discovered by a schoolboy in Charnwood Forest, also have a foliate appearance and stalked hold-fast, but differ in ways that make a pennatulacean affinity decidedly tenuous.

Not all of the cnidarian-grade benthos needs to be compared with pennatulaceans. Also occurring are various bag-like structures, superficially like sea-anemones but lacking oral tentacles and probably distantly related. Some are quite bizarre and their affinities remain problematical. Most striking are extraordinary animals from Namibia, with an apparently gaping aperture, that may have served as a mouth, and a thick wall enclosing the voluminous internal cavity.

Whatever evolutionary conundrums those animals may present, they appear to be of a relatively simple evolutionary grade and, even if their descendants survive to the present day, they are not obviously ancestral to any of the higher metazoans that appear so abruptly shortly afterwards in the Cambrian. Where then are the ancestors of such major groups as the arthropods, brachiopods, annelids or echinoderms in the Ediacaran? In part the answer hinges on the interpretation of segmented worm-like animals, but even if they are accepted as arthropods and annelids their connection with Cambrian faunas is not clear. The former group seems to be better represented, including animals with approximately triangular broad carapaces beneath which appendages may lie and more elongate creatures like *Spriggina* with multiple segments and a prominent head-shield. Among the putative annelids is *Dickinsonia*, an extraordinarily thin segmented sheet, that presumably rested on the sea-floor. Contracted specimens indicate presence of a musculature, but little is known about its internal anatomy. However, a circulatory system can be inferred because even if oxygen levels approached those of today calculations suggest that the animals would have been too thick to rely solely on diffusion of respiratory gases.

No account of the Ediacaran faunas can be complete without discussion of the only example with hard-parts, the calcareous *Cloudina* whose tubes occur in profusion in several carbonate horizons in Namibia, with the intervening clastics yielding the normal members of the Ediacaran community. *Cloudina* has also been recorded in several other regions, and there is evidence that the calcareous wall was rich in organic matter. This suggests that here we can see one of the earliest stages in the development of hard-parts which is hypothesised to have begun with the onset of isolated granules and spicules in an organic matrix that only later consolidated into robust skeletons, a point reached at the base of the Cambrian. Even if *Cloudina* prefigures this event, its affinities remain contentious and its assignment to the annelid worms is debatable. Other worm-like animals must have made the abundant, if simple, trails that occur in many Ediacaran faunas. However, their affinities are highly speculative.

Mention was made above of *Tribrachidium*, long regarded as an echinoderm but now widely interpreted as a medusoid. Recently, however, other candidates have been proffered in the form of small disc-like fossils, from South Australia, that bear a penta-radial arrangement of grooves, a type of symmetry that in younger epochs is regarded as one of the hallmarks of the echinoderms. These putative Ediacaran examples lack evidence for the diagnostic calcareous plates, but so far these animals are the best candidate for a Precambrian echinoderm.

Notwithstanding the problems of placing Ediacaran animals in a broader context of metazoan evolution, they form a distinctive assemblage. Their abrupt disappearance shortly before the Cambrian is very puzzling. The most popular suggestion is that with the rise of scavengers and deep burrowers the taphonomic balance was irrevocably shifted against soft-part preservation in normal marine environments. Some evidence suggests, however, that the Ediacaran disappearance is better regarded as a major extinction, although whether due to biological pressure of newly evolving competitors or by a changing physical environment such as shift in O_2 values cannot yet be determined. Supposition for an extinction event rests on a distinct stratigraphic interval, between the Ediacaran faunas and the first Cambrian entrants, which yields only a sparse fauna of simple trace fossils notwithstanding facies suitable for Ediacaran-type preservation. No mass extinction has extirpated all life, and indeed Ediacaran survivors can be recognised in Cambrian soft-bodied faunas. However, if a wide variety of ecological vacancies were suddenly made available could this explain the ensuing 'Cambrian explosion', a dizzying evolutionary episode which filled the ecological barrel of marine environments and out of whose vortex all the major groups of animals emerged?

The Cambrian explosion

The abrupt appearance of skeletons at the base of the Cambrian is perhaps the most remarkable event in the rock record. Appearing at first sporadically, in sections around the world, sediments are soon teeming with tubes, carapaces, shells, teeth, spicules and sclerites

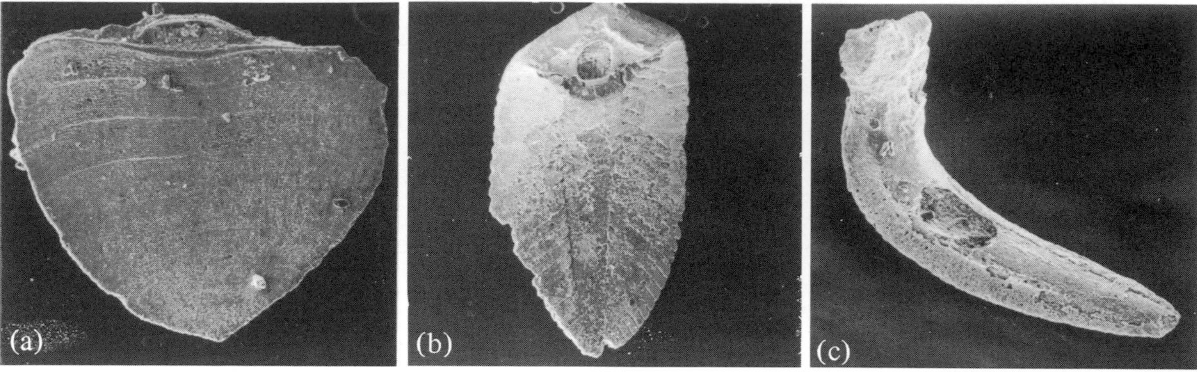

Figure 22.8 Early skeletal fossils. In the top row (a–c) are depicted electron micrographs of three isolated sclerites of a halkieriid (*Thambetolepis*) from the Lower Cambrian of Australia. The sclerites were originally composed of calcium carbonate, but were replaced after deposition in limestone sediments by a secondary mineral composed of calcium phosphate. This replacement has the advantage of making the fossils resistant to acid dissolution, so that when the limestone is digested in weak acid, the fossils are released. Although it is clear that all three sclerites come from the same type of animal, their exact distribution on the body was uncertain. Discovery of articulated halkieriids (d and e) from Lower Cambrian shales in the Buen Formation, North Greenland, now allows us to reconstruct the halkieriid animal in much greater detail. Particularly noteworthy are not only the large number of sclerites, but also a prominent shell at either end of the body. □

(Figure 22.8). But despite the exuberance of morphological designs, it must be stressed that this evolutionary irruption would still be quite evident from the corresponding adaptive radiations of soft-bodied organisms. With some important exceptions like the Burgess Shale (see below), they themselves are not preserved, but the wide variety of trace fossils show how their behavioural sophistication in excavating burrows, searching for food and so forth had advanced far beyond Ediacaran standards.

Explanations for the **Cambrian explosion** should encompass all aspects of this evolutionary breakthrough, although inevitably many emphasise controls on its most obvious manifestation, that of skeletogenesis. Broadly there are two types of hypothesis: those that stress changes in the physico-chemical environment, especially of ocean chemistry and atmospheric composition, versus those that appeal to ecological influences, in particular the rise of predators leading to the feedbacks in increasingly complex ecosystems. Disentangling cause and effect is difficult, and although it is tempting to identify a single trigger (what might be labelled the 'Assassin's Bullet Hypothesis') that unleashed the revolution, it may also be plausibly argued that a coincidence of circumstances opened the floodgates to biological diversification.

Underpinning the major physico-chemical changes that occurred close to the Precambrian–Cambrian boundary are processes involved with plate tectonics, and in particular the dispersal of the Proterozoic supercontinent. During the early break-up narrow rift oceans (Figure 22.9) formed, with their bottom waters readily stagnating because of restrictions of circulation. Such environments are particularly suitable for large-scale storage of organic carbon that sinks to the bottom sediments from the sunlit shallows. Substantial shifts in $\delta^{13}C$ near the Precambrian–Cambrian boundary might reflect such carbon accumulation in black shales, or possibly changes in oceanic productivity. Oceanic stratification of the water column may have persisted for a substantial period of geological time, but its ultimate breakdown could have been abrupt. Evidence for this comes from major changes in sulphur isotopes ($\delta^{32}S$, the proportion of ^{32}S to ^{34}S), just before the beginning of the Cambrian. Fractionation of sulphur by anaerobic bacteria in the anoxic sediments, leading to pyrite (FeS) formation, results in the overlying basinal waters becoming enriched in the isotopically heavy sulphur as the bacteria selectively precipitate lighter sulphur. Massive overturn of these isotopically heavy waters means that the isotopic signature can then be recorded in sulphates (e.g. anhydrite) accumulating in shallower waters where the rock record has survived to the present day. The

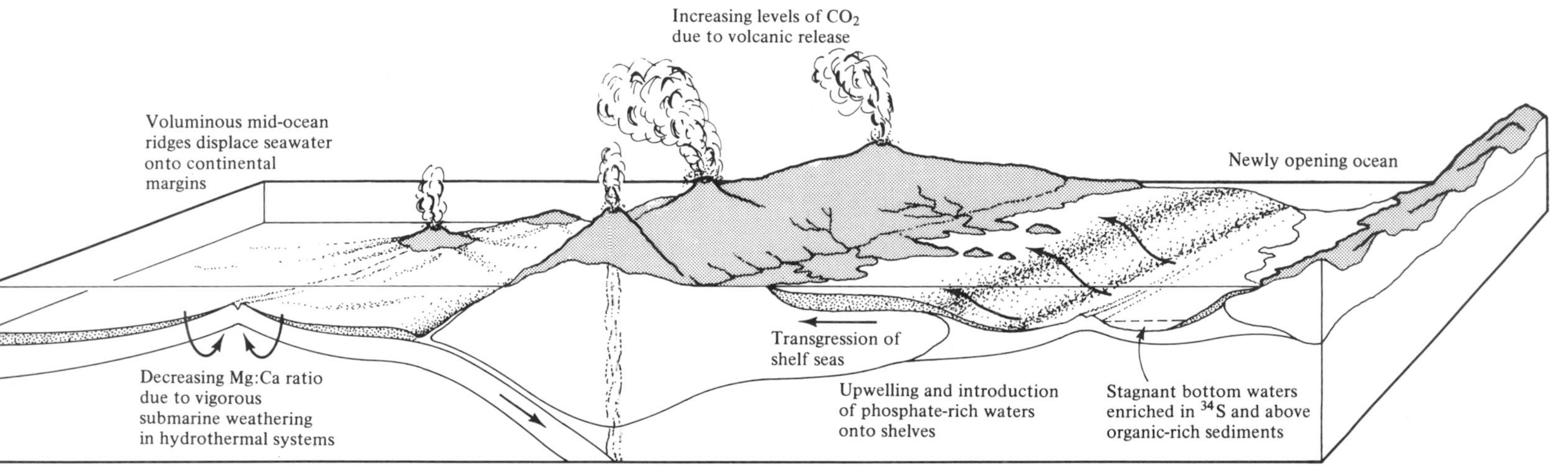

Figure 22.9 Block diagram of an idealised area of Cambrian crust to show the various physico-chemical processes that were active at the time of the 'Cambrian explosion'. Ocean chemistry was changing on account of the development of active spreading ridges, so that vigorous pumping of sea-water through the hydrothermal systems decreased the ratio of magnesium to calcium in the sea-water. Plate tectonic activity is also manifested by active subduction and major volcanism, the effect of which is to pump large quantities of CO_2 into the atmosphere. The voluminous nature of the mid-ocean ridges also leads to displacement of sea-water onto the continental margins, thereby increasing the habitable area for shallow-water creatures. Elsewhere in the oceans there is vigorous upwelling, leading to increased productivity as phosphate-rich waters are brought to the surface. In newly opened oceans restricted circulation leads to the stagnation of bottom waters, in which ^{34}S enrichment occurs while in the underlying sediments large quantities of organic matter (rich in ^{12}C) are buried. The net effects of these various physico-chemical changes on organic evolution are not well understood, but it is believed that substantial changes in the ocean–atmosphere system in the late Precambrian and Cambrian probably facilitated the evolutionary radiations that are observed in the fossil record. □

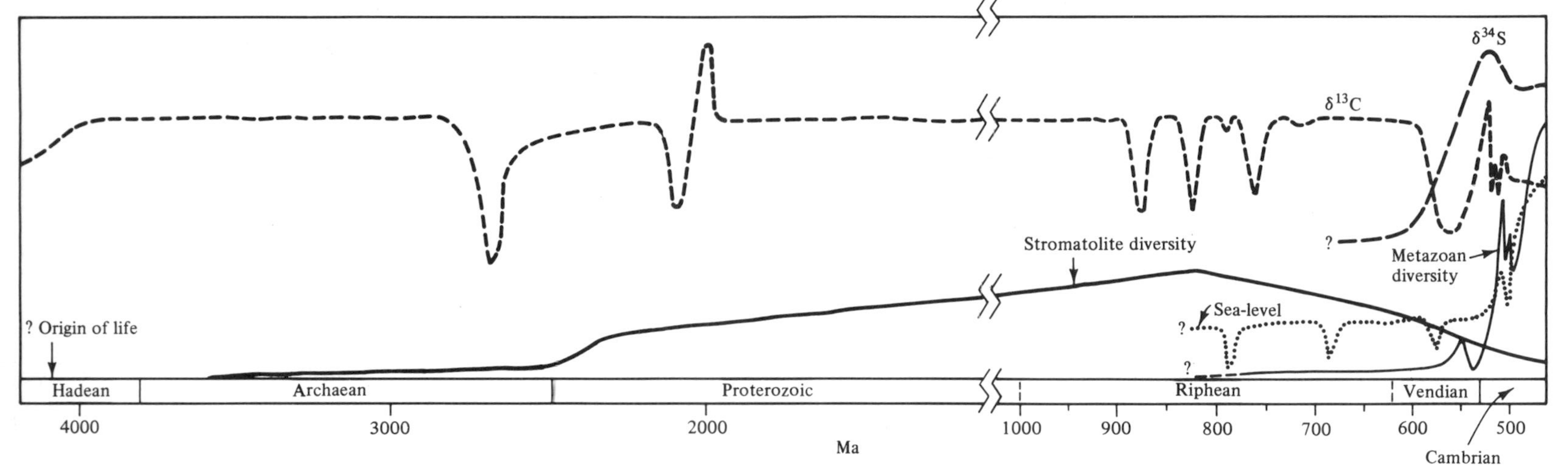

Figure 22.10 A summary of major changes in the physico-chemical environment during the first 3500 Ma of Earth history. Paucity of information for much of the Precambrian gives a meagre record, but certain trends are now becoming more clear and close to the Precambrian–Cambrian boundary there are several significant perturbations. Shown here are data on shifts in the ratios of ^{12}C to ^{13}C (notation as $\delta^{13}C$) and ^{32}S to ^{34}S (notation as $\delta^{34}S$), diversity of stromatolites and metazoans, and a sea-level curve. The changing values of $\delta^{13}C$ may indicate changes in organic productivity and/or large-scale burial of organic carbon. Less is known about $\delta^{34}S$ in the Precambrian, but the major shift near the Precambrian–Cambrian boundary has been linked to a major episode of upwelling involving deep-water brines enriched in ^{34}S. The long-term decline in stromatolites, from about 900 Ma ago, has been linked to the rise of grazing and burrowing metazoans. Metazoans may have evolved as long ago as 1000 Ma, but there is little evidence in the fossil record, until about 600 Ma ago when they diversify rapidly. They appear first as the Ediacaran faunas, then enter a mass extinction, and ultimately rediversify in the early Cambrian. The sea-level curve is rather conjectural, but depicts the major Cambrian transgression, as well as earlier periods of regression in the late Precambrian when there were several ice ages where sea-level fell because of the development of major ice-sheets. □

magnitude of this isotopic excursion suggests a type of catastrophic upwelling, and because the bottom waters were probably also nutrient rich, especially in phosphorus, their influx into shallower water may have also effected evolutionary diversifications.

Another aspect of the plate tectonics that may be relevant to the Cambrian diversifications is linked to the development of an extensive system of spreading ridges that separated the fragmenting supercontinent, and through which the hydrothermal systems pumped prodigious quantities of sea-water (Figure 22.9). Metamorphic reactions within the ridges decrease the magnesium to calcite ratio of the circulating sea-water, an effect that probably facilitated precipitation of calcium carbonate in the oceans. This process may have also been assisted by an increase in atmospheric CO_2, engendered by volcanic activity along the spreading ridges and above subduction zones pumping vast quantities of this gas into the atmosphere. Other changes in atmospheric composition that may have played an important role in mediating the 'Cambrian explosion' could have been alteration in oxygen levels. We have already seen how many geologists regard a critical threshold as being passed about 2000 Ma with the passing of BIFs and the appearance of red beds. The argument would continue that the Precambrian–Cambrian boundary marks a further increase, although it might be naive to imagine the rise in oxygen being a simple monotonic increase throughout the intervening interval. Rather, there are already clear indications of fluctuations in the Riphean (Figure 22.10), deductions based on changing values of $\delta^{13}C$ that are equated with burial of organic carbon that probably led to a rise in atmospheric oxygen owing to shifts in global redox processes. Be that as it may, rises in oxygen levels could exert powerful controls on metazoan diversification. It has long been realised that transects across the sea-floor of stagnant basins (e.g. Black Sea) from the anoxic depths to oxygenated shallows crudely parallel the events across the Precambrian–Cambrian boundary. That is, metazoans are effectively excluded when oxygen values are too low, and among the most tolerant when concentrations rise slightly are worms, while skeletal forms can only flourish in well-oxygenated water. This model is now known to be somewhat simplistic, but its general applicability has received renewed attention as the possibility of estimating oxygen levels emerges.

Biological explanations for the Cambrian explosion, on the other hand, have emphasised evidence for ecological changes, especially in predators or the deterrence of predators by potential prey. Concerning the former, remains of grasping teeth are particularly

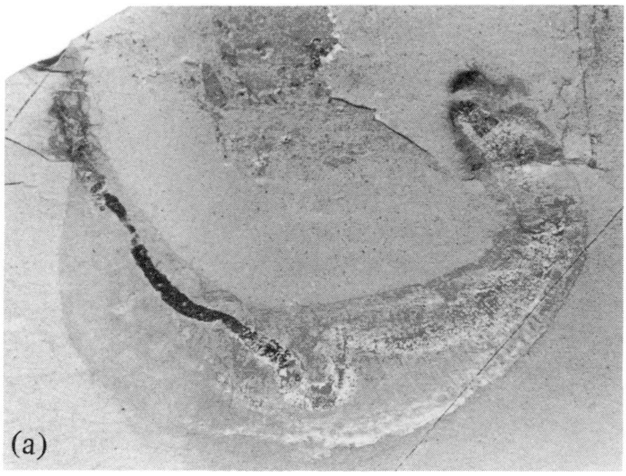

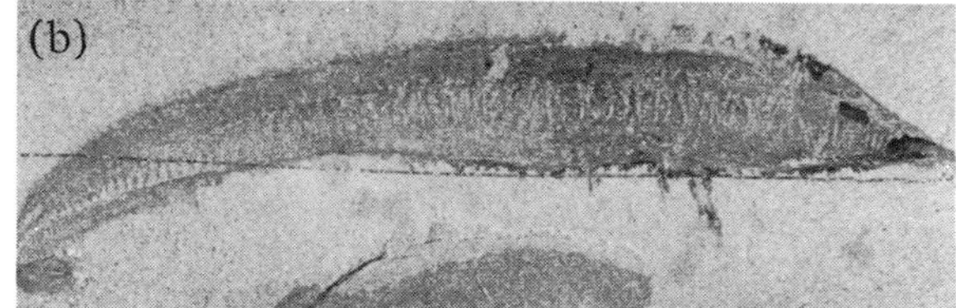

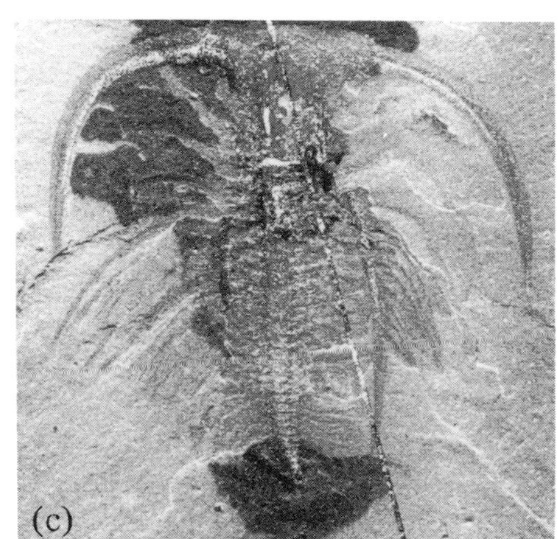

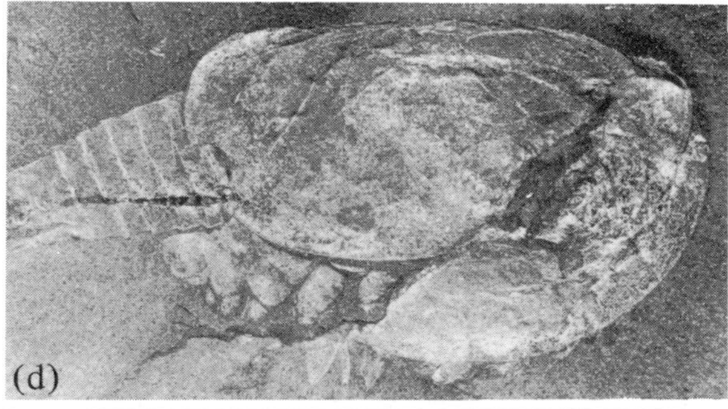

Figure 22.11 Representative Burgess Shale fossils, from the Middle Cambrian Stephen Formation of British Columbia. (a) The priapulid worm *Ottoia prolifica*. (b) The primitive chordate *Pikaia gracilens*. (c) The arthropod *Marrella splendens*. (d) The crustacean arthropod *Canadaspis perfecta*.

convincing, but in general predators were effectively soft-bodied and unlikely to be fossilised. Even here trace fossils can provide compelling evidence. Examples include burrows, apparently made by sea-anemones, stuffed with trilobite debris, arthropod excavations that intercept worm burrows, and holes drilled through skeletons. That hard-parts helped to confer a protective function is also suggested by bivalved conchs, operculate tubes (i.e. with lids that on closure protected the soft-parts) and cataphract metazoans, the last category bearing sclerites in an integrated chain-mail (cataphract) that allowed flexibility but conferred protection.

The earliest skeletal faunas are remarkable for their variety of enigmatic groups (Figure 23.8), but as the Cambrian progressed assemblages composed of trilobites, brachiopods, hyoliths (an extinct group with a conical shell, operculum and a pair of curved struts), and echinoderms become characteristic. Trace fossils hint at a diversity of soft-bodied fossils, but we are fortunate in possessing a series of extraordinary soft-bodied faunas that show how skeletal faunas by themselves can only give a fragmentary and imperfect view of Cambrian life. The cynosure of these is the **Burgess Shale fauna** from the Canadian Rockies. Here a typical Middle Cambrian skeletal assemblage is present, but is completely overshadowed by a cavalcade of soft-bodied worms (polychaetes and priapulids; the latter are a distinctive group with a spiny anterior used in burrowing), lightly skeletalised arthropods and sponges, as well as a remarkable range of bizarre metazoans (Figure 22.11) unlike any found before or since. It is understandable to label these extraordinary animals as experiments, doomed to obliteration in a relentless race to biological superiority. This may have been the case, but such was the diversity of these Cambrian faunas and so few were the types that passed the iron gates of extinction that we may have to look to chance factors weeding this evolutionary jungle into the subsequent taxonomically ordered garden of later periods. To underline this point consider one worm from the Burgess Shale, insignificant in this community but of immense evolutionary importance because as the earliest chordate it marks the beginning of the long march via many other species to Man. Who would have guessed from a Cambrian perspective its future role? However, with its appearance the early history of life may be said to have ended.

Summary

Museums of palaeontology only tell us about the history of life for the last 10 per cent or so of geological time. This is because prior to about 550 Ma ago no skeletons are known, and organisms had a minimal chance of preservation. This chapter has surveyed this immense early interval of Earth history, the Precambrian. The earliest fossils are only known from about 3500 Ma ago, more than 1000 Ma after the formation of the Earth. They consist of stromatolites and fossilised microbes. The former are abundant in the Precambrian, and their history of diversity reflects first their flourishing with increasing stability of the Earth's crust and later their demise, perhaps due to the rise of grazing metazoans. Microbial assemblages are much rarer, but key localities such as the Gunflint Chert in northern Ontario and the Bitter Springs Chert exposed in central Australia reveal much about primitive communities, especially of cyanobacteria. These lines of evidence are complemented by chemical fossils from Precambrian sediments, some of which are diagnostic of particular groups. Finally, data from molecular biology is helping to place our understanding of Precambrian evolution in new perspectives. Animals may have evolved more than 1000 Ma ago, but their fossil debut is as recent as 600 Ma in the form of the Ediacaran faunas. By 540 Ma the 'Cambrian explosion' was well under way, with a dramatic rise in the diversity of animals, both with skeletons and soft-bodied forms.

Further reading

Cairns-Smith, A. G. (1985) *Seven Clues to the Origin of Life*, Cambridge University Press.

Cloud, P. S. (1988) *Oasis in Space: Earth History From the Beginning*, Norton.

Glaessner, M. F. (1984) *The Dawn of Animal Life: A Biohistorical Study*, Cambridge University Press, 244 pp.

Gould, S. J. (1989) *Wonderful Life: The Burgess Shale and the Nature of History*, Norton.

Nisbet, E. (1987) *The Young Earth: An Introduction to Archaean Geology*, Allen & Unwin, London.

Schopf, J. W. (1983) *Earth's Earliest Biosphere: Its Origin and Evolution*, Princeton University Press.

Whittington, H. B. (1985) *The Burgess Shale*, Yale University Press.

CHAPTER

23

Peter W. Skelton
Open University

FOSSILS AND EVOLUTIONARY THEORY

Introduction

In an age when dinosaurs tumble out of cereal packets and ape-men grace our television screens (in cartoon, if not in the flesh), it is perhaps rather too easy to take for granted how much we have learnt about the life of the past from the fossil record. Two hundred years ago, these images would have been bewildering: the palaeontologists of those times were only just beginning to appreciate the extent to which the Bible had been 'economical with the truth' over the former denizens of our planet. That strange creatures unlike any alive today had once roamed the Earth and become extinct was a revolutionary idea.

Over the next half century or so, knowledge of the fossil record advanced far enough for a clear sequence of life forms to be recognised, and soon thereafter accepted by scientific consensus as evidence for evolution. Since then, unravelling the story of evolution has occupied generations of palaeontologists. Despite the notorious incompleteness of the record, there have been many notable successes. For example, the main patterns in the evolutionary history of the horses (Figure 23.1) have now been fairly extensively worked out, and paraded through many a textbook.

Lifeless lumps of stone though they are, fossils can also tell us something about the changes in life habits which have accompanied these evolutionary histories. For example, the living horses have cheek teeth with greatly extended crowns, adapted to compensate for the extensive wear that results from chewing grass, which contains highly abrasive, microscopic crystals of silica. Primitive horses, in contrast, had low-crowned cheek teeth, and palaeontologists have long argued that these animals probably browsed on soft leaves and fruit, rather than being grazers. One of the excitements in the search for fossils is that exceptional fossiliferous localities are sometimes discovered, in which the quality and detail of preservation are vastly superior to that in normal fossil assemblages. One such locality is in a Middle Eocene lake deposit in Germany, near Darmstadt. Here, numerous articulated skeletons have been found, many containing fossilised soft tissues. In the 1970s and 1980s, several fossils of primitive horses were recorded from this site, some with preserved gut contents (Figure 23.2). Confirmation of the postulated feeding habits has been provided in spectacular style: balls of leaves and even grape pips have been found in the hind-gut areas of these specimens. Some of the leaves are so well preserved that it is even possible to detect what type of tree was being browsed.

What can the fossil record tell us about the mechanism of evolution, though? For much of the 20th century, the standard answer has been 'not much'. Evolution, it has generally been felt, is a matter for the biologists to handle. But this monopoly of understanding has been challenged over the last couple of decades, and it is this exciting development which is explored in this chapter.

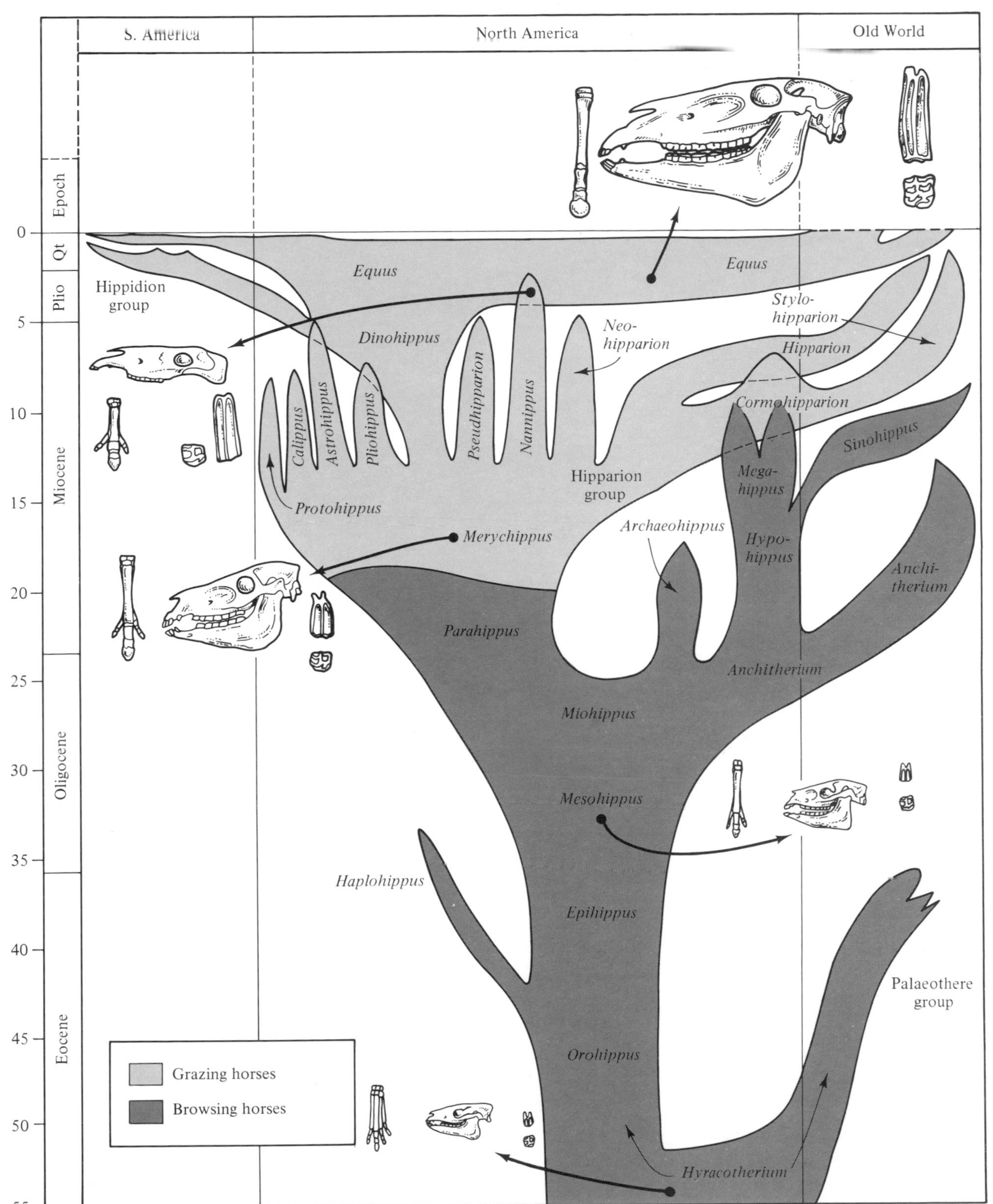

Figure 23.1 Evolutionary history of the horses. Relative skull sizes, lower forelimb bones and cheek teeth are shown for selected forms.

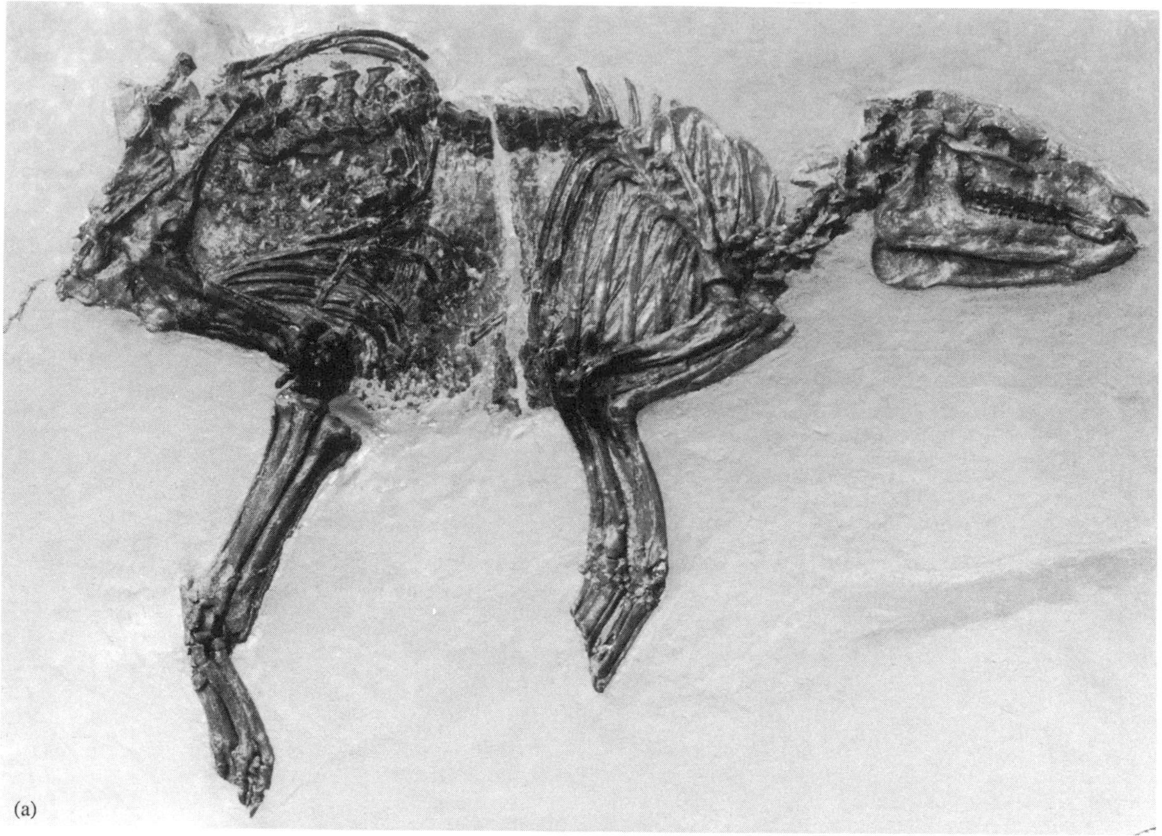

(a)

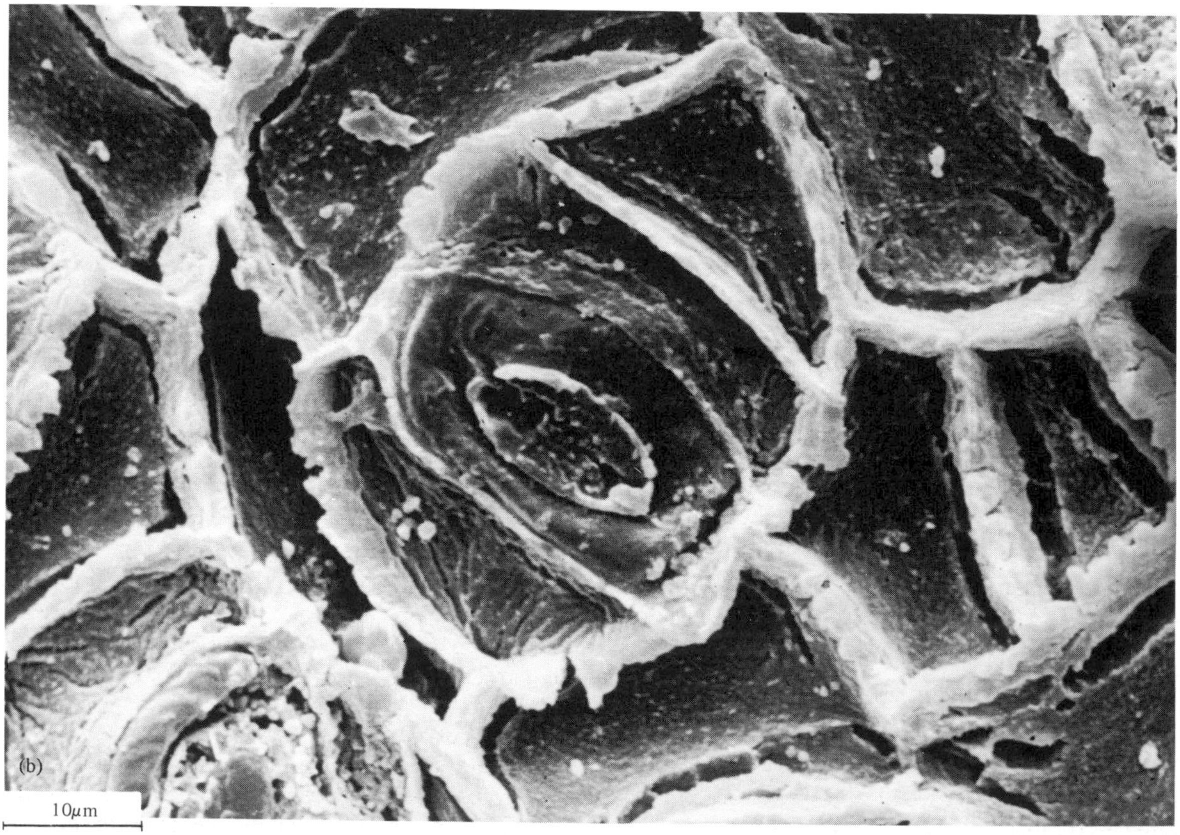

(b)

10μm

Figure 23.2 (a) Articulated skeleton plus fossilised soft tissues of the primitive horse, *Propalaeotherium parvulum*, from Middle Eocene lake deposit at Grube Messel, near Darmstadt, Germany. The dark mass in the abdominal region represents the fossilised contents of the hind-gut. (b) Scanning electron microphotograph of the surface of a leaf in the fossilised gut contents of *P. parvulum*. (c) Photomicrograph of hind-gut contents, including grape pips (arrowed). □

Pattern, process and theory in evolution

It is convenient to recognise two scales in evolution: **microevolution** concerns changes occurring within species populations and the means by which new species may arise; **macroevolution** involves the broader patterns of life mapped out by the turnover of species, and so concerns such questions as how distinct new groups of organisms arise and become extinct, and what controls the kaleidoscopic changes in their diversity and distribution through time.

The differing time scales of these phenomena have important implications for the methods by which they are studied. Many scientific hypotheses can be tested by experiments, or by direct observation of natural processes, to see if the results are as predicted by them. This is because they concern processes which can be adequately monitored in the limited time and scale of observation available to the investigator. Thus, for example, gravitational attraction can be tested by the simple experiment of throwing a ball in the air. Or, alternatively, one might chart the motions of the planets around the Sun, to see if these are as predicted by gravitational theory. Some aspects of microevolutionary theory can likewise be predictively tested. The effects on fruitfly populations of various regimes of selective mortality and breeding have been investigated in countless laboratory experiments, while scores of field studies have directly measured components of natural selection and their effects in wild populations.

Other hypotheses, in contrast, concern processes which are either too slow or too rare to be directly observed. Yet such hypotheses can still be tested, since any postulated process should have left behind a predictable pattern of effects. The expected pattern may then be confirmed or refuted by observation, and so tested with as much thoroughness as might be applied to observing the outcome of an experiment. The 'Big Bang' theory for the origin of the Universe is an obvious example: it is hardly amenable to experimental repetition, yet supporting evidence for its occurrence (such as the background radiation) is provided by astronomical observations. Such testing might be called 'retrospective', in that it seeks postulated but perhaps as yet unobserved after-effects in nature. Some longer-term microevolutionary, and all macroevolutionary ideas involve time scales for which only **retrospective testing** is possible. Models to explain mass extinctions, for example, must be tested by recourse to the testimony of the fossil record – notwithstanding our inadvertent efforts to run an appropriate experiment now.

The reliability of retrospectively tested theories of macroevolution is often underestimated by microevolutionary biologists, more used to using predictive tests in the laboratory or field. The school classroom may be somewhat to blame, with its traditional emphasis on the experimental disciplines in science teaching. Be that as it may, it has meant that there has long been an unwarranted prejudice in some quarters against any suggestion of there being evolutionary mechanisms operating above the immediately measurable microevolutionary level. Macroevolution, it is argued, is no more than microevolution writ large. Not all palaeontologists are now prepared to accept such a view. They argue instead that macroevolution is the product of a hierarchy of mechanisms operating over a wide range of time scales, and that microevolution is only one part of the story.

Evolution and palaeontology: the historical background

Two basic concepts of pattern are involved in modern evolutionary theory – descent with modification (**anagenesis**) and branching divergence of lineages (**cladogenesis**). The first evolutionary hypotheses were essentially only concerned with anagenesis. These reflected both a continuing obsession with the classical 'Great Chain of Being', which had dominated pre-evolutionary thought, and a confusion of individual development with evolutionary change in species. For example, the hypothesis of Jean Baptiste de Lamarck (1744–1829) stressed a supposed innate perfecting principle in organisms, driving species towards greater complexity, as well as the notorious notion of the inheritance of modifications acquired during growth through habitual behaviour. Lamarck envisaged myriad single evolving lineages, variously initiated by the continued spontaneous generation of simple forms. Neither branching nor extinction were accorded any significant role. Although a useful kernel from his scheme of adaptive modification in changing circumstances did survive into later evolutionary theories, the mysterious notion of a 'perfecting principle' and the appeal to the already outmoded idea of spontaneous generation won short shrift from most of Lamarck's scientific contemporaries.

Ironically, the improving knowledge of the fossil record was at first considered to yield evidence against evolution, despite testifying to the Earth's great antiquity. Lamarck unconvincingly tried to pass off fossil species as simply the evolutionary forebears of those alive today. But his brilliant younger contemporary at the French Academy of Sciences, Georges Cuvier (1769–1832), used rigorous comparative anatomy to demonstrate from fossils that many distinctively specialised vertebrates had existed for which no living counterparts could be recognised. These, he argued, must therefore have become truly extinct, and so the continuous evolution of lineages as conceived by Lamarck was wrong: successive faunas had apparently been swept away by various catastrophes and replaced by others derived from unknown sources.

With more detailed collecting, palaeontologists soon found it necessary to increase the number of 'catastrophes' needed to explain the fossil record. Then, with the rise of the **uniformitarian doctrine** – that all geological phenomena could be interpreted as the products of processes seen to operate today – the 'catastrophist' model of the history of life eventually waned. Thus, by the mid 19th century, it was recognised that the appearances and disappearances of fossil species were spread throughout sedimentary sequences. The way was now open for a cladogenetic evolutionary theory involving the continual loss of species by extinction and the compensatory rise of new ones by branching.

This integration of cladogenesis and anagenesis was achieved by the theory of **evolution by natural selection**. This was first unveiled by Charles Robert Darwin (1809–82) and Alfred Russel Wallace (1823–1913), in joint communications to the Linnean Society, in London, in 1858, and more comprehensively set out a year later in Darwin's *On the Origin of Species by Means of Natural Selection, or the Preservation of Favoured Races in the Struggle for Life*. Competition between superabundantly produced, naturally varying individuals, for limiting resources (the 'struggle for existence') was seen as the key to change. Those individuals displaying advantageous variations were predicted to be more likely to survive and reproduce than other variants. Because offspring tend to inherit parental characteristics, future generations would therefore come to be dominated by individuals possessing the advantageous variations. Thus, as populations encountered changing conditions, their composition would be progressively altered by such 'natural selection' in favour of suitably adapted individuals, so causing longer-term anagenetic evolution. In addition, widespread, variable populations could be expected to diverge according to the differing effects of selection in different environments and because of the diversity of opportunities exploited by different varieties. So, starting from the parental stock of natural varieties, first subspecies and eventually wholly distinct species would arise. Most of these, it was argued, would ultimately become extinct through competitive failure against other newly arising species with superior adaptations. Darwin was quite explicit in interpreting competition between *individuals* in the struggle for life both as the cause of adaptive evolution in populations and as the main explanation for the continuous turnover of species through time.

Darwin's consummative marshalling of the evidence (not to mention his respected position in the scientific community) soon won over a scientific consensus for evolution, though the importance of natural selection remained more contentious. Variation and inheritance were still poorly understood, and Darwin devoted much space in his later works to defending the natural selection of small, continuous variations as the principal mechanism of evolution. But it was generally supposed at that time that offspring inherited a homogenised blend of parental characteristics, and so, his critics argued, any 'favourable' variation would soon be diluted to insignificance in a few generations, by cross-breeding. These and other arguments against the efficacy of natural selection led many biologists in the latter half of the 19th century to consider the generation of variation *per se*, rather than selection, to be the principal cause of evolutionary change. But the mechanisms by which heritable variation arose remained elusive.

The Darwinian revolution in evolutionary thinking and the controversy it inspired nevertheless firmly focussed attention on individual heritable variability as the appropriate level of investigation in the quest for an explanation of how evolution comes about. With its meagre sampling of ancient populations, the fossil record was now really only seen as providing evidence that evolution had occurred: the *Origin of Species* had duly referred to the continuous turnover of species in the past, and the apparent sense of progression resulting from it. The affinities between certain ancient forms and several now distinct living forms, the increasing similarity of successively younger faunas to that of today, and the stratigraphic clustering of related species were all noted. However, the fossil record seemed to afford little illustration of the gradual transformation of populations which Darwin had expected from the theory of natural selection. He attributed this to the 'imperfection of the geological record'. For the next century or so, evolutionary palaeontology largely confined itself to the more passive exercise of unravelling the story of evolution, while attention largely moved to other disciplines for accounts of mechanism. Contemporaneous palaeontological references to mechanisms thus consisted either of lip service to the assumed operation of natural selection, or of occasional speculations concerning alternative mechanisms. The latter included innately programmed, long-term evolutionary trends towards increasing size or exaggeration of some anatomical feature (**'orthogenesis'**) and 'decadent' tendencies setting in prior to extinction (such as uncoiling in ammonites). Instantaneous changes (**'saltations'**), based on major mutations, were also hypothesised by a few.

The impasse over the fundamental mechanism of evolution was broken, around the turn of the century, by the re-discovery of some long-ignored experimental work on heredity performed several decades earlier by a little-known Austrian monk, Gregor Mendel (1822–84). Mendel had shown that certain discrete variations in the pea plants he was breeding (such as the colour and surface texture of the peas) were expressed in offspring in characteristic ratios. From these he had inferred that each individual received a randomly sorted half complement of particulate hereditary factors (or **'genes'** as they were later called) from each of its parents: blending inheritance could thus be rejected. Curiously, however, the re-discovery of this work did not immediately lead to the re-instatement of natural selection theory. Some of the early geneticists seized upon the discontinuous nature of the genetic variations studied by Mendel, and proposed that speciation was driven by mutational genetic variations, as opposed to the operation of natural selection upon small continuous variations, emphasised by Darwin's followers.

Advances in mathematical genetics during the 1920s and 1930s progressively healed this initial rift between genetics and natural selection theory. Major works by R. A. Fisher (1890–1962) and J. B. S. Haldane (1892–1964), in Britain, and Sewall Wright (1889–1987), in the USA, demonstrated that continuous variations (such as adult height in humans, for example) could also be explained by genes. Such characteristics are controlled not by single genes, but by batteries of modifier genes of additive effect. The recognition of this **polygenic inheritance** successfully integrated Mendelian genetics with the statistical study of continuous variations. Over the next couple of decades these theoretical advances in population genetics were integrated with insights on natural variability (especially geographical) in living populations, and – at a broader scale – palaeontological patterns of variability, to bring about what has been termed the **'Modern Synthesis'**. A new, **'neo-Darwinian'** consensus emerged, which re-emphasised the importance of natural selection acting on small heritable variations in populations.

The palaeontological contribution to the Modern Synthesis was notably set out by the American vertebrate palaeontologist G. G. Simpson (1902–84) in his *Tempo and Mode in Evolution* (1944) and *The Major Features of Evolution* (1953). These disposed of earlier orthogenetic and saltational hypotheses and argued for the consistency of the fossil record with the expectations of neo-Darwinian theory. In the classic case of horse evolution, for example, the fossil evidence pointed to a bushy 'evolutionary tree' of divergently branching lineages (Figure 23.1). Some lineages showed various adaptive modifications linked with changes in habit, such as the evolution of high-crowned teeth. Other lineages, however, did not show such changes, while some even showed reversals of general tendencies such as that for size increase. Such complex and inconsistent evolutionary patterns, though certainly yielding some overall 'trends' in a purely accumulative sense, contradicted the notion of orthogenesis, operating irrespective of environmental change. Nor was any support for large-scale saltational change to be found: the record yielded numerous links between genera via intermediate species differing but slightly from their relatives. From such considerations, Simpson and others inferred that macroevolutionary patterns could simply be interpreted as the scaled-up consequences of adaptation and divergence, driven by the operation of natural selection in varied and changing environments. Nevertheless, large fluctuations in rates of evolution were acknowledged, ranging from rapid bursts of considerable, though continuous change (**'quantum evolution'** – not to be confused with the essentially discontinuous saltation) to extended periods of little to no change.

In rejecting the speculative alternatives to natural selection as the fundamental mechanism of evolution, Simpson and others assured palaeontology of a place in the Modern Synthesis. But this was won at the expense of any expectation of the fossil record yielding new insights on evolutionary mechanism. Palaeontology was for the time being to remain restricted to the passive role of official chronicler of evolutionary history.

Meanwhile, the concept of species was also being revised. There was a major shift of emphasis away from the more traditional, subjective comparison of specimens with selected 'types', to the recognition of variable populations united by genetic coherence. Thus emerged the **'biological species concept'**, according to which 'species are groups of interbreeding natural populations that are reproductively isolated from other such groups' (in sexually reproducing organisms, at least). The view that, in the great majority of cases, the first step towards reproductive isolation was the geographic separation of populations (**allopatric speciation**) was especially championed by Ernst Mayr (born 1904), in a number of publications, leading up to his compendious *Animal Species and Evolution* (1963, abridged and updated in 1970). This view is now much debated.

In his discussions of allopatric speciation, Ernst Mayr particularly urged the importance of the isolation of small, localised populations peripheral to the main species range (later termed the **peripatric model of speciation**; see Figure 23.4). He suggested that the random sampling of the parental gene pool by such a small population (the **founder effect**) would yield a less diverse, perhaps imbalanced local gene pool. Mayr proposed that such isolated populations might readily undergo a **genetic revolution**, involving a rapid, selectively driven shift to a new state of genetic balance. If reproductive isolation with respect to the parental population resulted, speciation would have been effected. A large population, in contrast, would be more prone to evolutionary inertia, Mayr argued, because of the swamping of local selective effects by gene flow through interbreeding between neighbouring subpopulations. Moreover, he postulated that much of the genetic diversity would be hidden from direct exposure to selection, in any case, due to the stabilising influence on development of highly interactive gene complexes.

In a desire to bring evolutionary palaeontology fully up to date with current theory on speciation, Niles Eldredge and Stephen J. Gould published a paper in 1972, which explored the pattern that should be expected in the fossil record from Mayr's peripatric model of allopatric speciation. They noted that the morphological change that could be expected to accompany a genetic revolution, and hence speciation, would, even if thousands of generations were allowed, be accomplished by a small population in a relative moment of geological time and within a small area. Such a process, in the great majority of cases, would be unlikely to be registered in the fossil record, given the latter's statigraphically and geographically discontinuous nature (not to mention incomplete investigation by palaeontologists). Most new species could thus be expected to make sudden ('punctuational') appearances in the fossil record, due, normally, to their migration into any given area from their (unknown) local area of origin. Once a species had become well established, over an enlarged area, however, the evolutionary inertia predicted by Mayr would be reflected by a geologically long period of morphological **stasis** (or equilibrium). In other words, anagenesis should normally only be associated with episodes of localised cladogenesis (Figure 22.3). They contrasted this expected pattern for the fossil record of species, termed **'punctuated equilibrium'**, with what they viewed as the older palaeontological expectation of more or less gradual (if incompletely recorded) modification and divergence of populations to form new species (**'phyletic gradualism'**). While the latter might be represented diagrammatically by the familiar obliquely branched evolutionary trees of the older literature, the former could be expressed in the form of a rectangularly branched tree (Figure 23.3).

An important spin-off of the punctuated equilibrium model was the corollary that species could now be considered as relatively discrete units in time as well as space, with geologically precise levels of origin and termination (extinction). So, with superabundant species budding in all adaptive directions from parental populations, macroevolutionary trends could arise as a consequence of differential speciation and extinction (Figure 23.3). In such a process, the microevolutionary changes associated with each speciation would not alone explain the overall macroevolutionary patterns. The species themselves would also be units of variation for a higher order selective process, termed **'species selection'** by Steven M. Stanley. This would be analogous to the role of mutation in furnishing variation for natural selection to act upon. How these ideas have fared in subsequent years is considered in the following sections.

Microevolution and the origin of species

Biological models for the origin of species are synthesised from two main lines of evidence. First, there are the 'snapshot records' of the generally slow process of speciation yielded by the present patterns of distribution and differentiation of populations. Conjectural splicing together of examples of these can suggest how populations may progressively separate and diverge, though there is usually little evidence to hand for the timing involved. Sometimes circumstantial geographical evidence, such as the geological age of a lake or

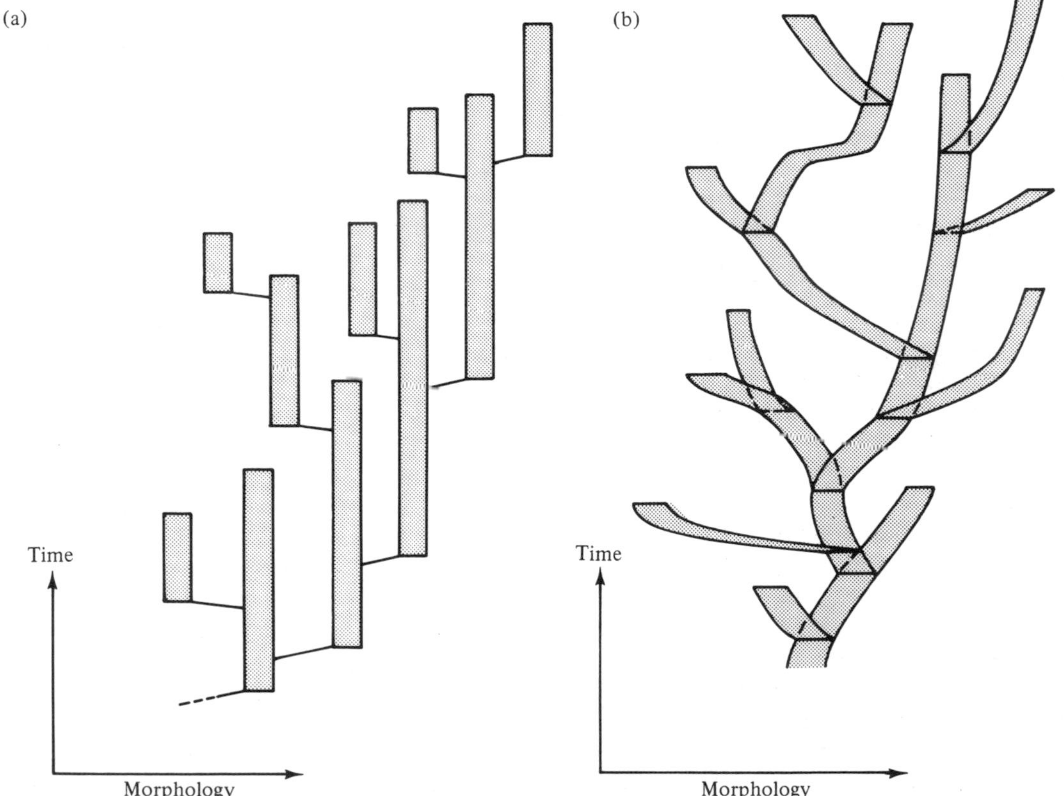

Figure 23.3 Palaeontological models for the patterns of species evolution. (a) Punctuated equilibrium. New species arise by rapid evolution in small, isolated populations; larger, well-established species populations show stasis. Note how preferential rates of speciation and extinction can yield macroevolutionary 'trends'. (b) Phyletic gradualism. Species populations evolve through time, yielding both anagenetic formation of chronospecies in single lineages, and speciation by gradual divergence. □

island, which must have been the site of some speciation, can provide constraints on the timing. The second line of approach, population genetic theory, can predict some aspects of pattern, and provide estimates for certain models of the timing (number of generations) required for given changes. What the various current models of speciation have to say about expected patterns of morphological change in space and time – the main points of interest to the palaeontologist – are summarised in Figure 23.4.

This is not the proper place to indulge in a lengthy explanation of the genetic background to these models, but before discussing the differences between them, one particular device for exploring aspects of population genetics is worth explaining, because it can yield some clues on timing. This is the diagrammatic metaphor invented by Sewall Wright, known as the **'adaptive landscape'** (Figure 23.5).

Consider a population with two alternative genes (**alleles**), A and a, for one gene locus, and another two, B and b, for a second. Possible allele combinations (**genotypes**) on the paired chromosomes inherited from the parental gametes for the first locus are AA, Aa and aa, and for the second, BB, Bb and bb. In a regime of random mating and consistent selection pressures (both very simplifying assumptions), each of these genotypes will have some average **'fitness' value**. For the population geneticist this is the product of survivorship and average fecundity for any given genotype, relative to that genotype in the population with the highest fitness. Now, it may be the case that the fitnesses of the genotypes show a simple relationship, such as AA > Aa > aa. In this case selection will invariably favour allele A at the expense of a in the gene pool of the population, until only the former remains, with all individuals of genotype AA. However, if the genotype Aa is, for whatever reason, fitter than either of the two genotypes AA and aa, selection will tend to balance the frequencies of A and a at some level determined by the relative fitnesses of the three genotypes. If a graph is now constructed, showing the average fitness of individuals in a population (i.e. the mean of the fitnesses of the genotypes weighted according to their frequencies) on the y axis, versus the frequency of, say, allele A in the population on the x axis, a convex-upwards curve like that in Figure 23.5a results. The peak of this curve represents the optimum frequency of A, for maximising production of Aa individuals, while minimising AA and aa individuals according to their lowered fitnesses. It should be clear, then, that any population with a frequency of A which is away from the peak will be drawn towards the peak by selection (Figure 23.5a), as the proportion of the fittest individuals becomes maximised.

Let us assume that a rather different situation obtains at the second locus, and that the relative fitnesses of the genotypes there are BB > Bb < bb. Plotting the mean fitness of individuals in the population against the frequency of B would now yield a concave curve (Figure 23.5b). A three-dimensional plot for both loci can now be constructed (Figure 23.5c), showing the frequency of A in the first locus, and of B in the second, as the two basal axes, and the mean fitness of individuals in the population on the vertical axis. A curved surface of mean fitness, or 'adaptive landscape' now emerges. **'Adaptive peaks'** are separated by an **'adaptive valley'** where the relative frequencies of alleles in populations tend to produce individuals of lowered fitness. Any given population can be represented as a point on this surface, showing the mean fitness of all nine possible genotypes (AABB, AABb, AaBb, etc...) weighted according to their frequencies at that position. As before, selection will tend always to pull the population uphill towards an adaptive peak, rather like a bubble of air in an upturned cup held underwater. However, the particular peak up which the population is drawn by selection clearly depends upon where it starts from (Figure 23.5c).

The adaptive peak eventually settled on by a population therefore need not be the highest in the landscape, but merely that which happens to include the allele frequencies of the population at the outset. Of course this hypothetical adaptive landscape is grossly simplistic, since in reality there would be millions of gene loci, many with complex interactive effects on fitness. For example, in the situation outlined above, the fitness values of the allele combinations at the first locus might vary according to the combination of alleles at the second locus. So a more realistic portrayal would consist of a multi-dimensional 'hypercube' with an inconceivably complex landscape. Moreover, changing circumstances would mean that the landscape would be anything but static. Nevertheless, the simple version, which can readily be visualised, is an instructive device for thinking about patterns of genetic change in populations.

If selection thus tends to 'trap' a population in an adaptive peak, how might the population ever become shifted to a new position on the grid? There are two possible ways: either the adaptive landscape itself may change (due to changing fitness values of the various allele combinations); or the population may simply transfer to a new peak because of some process of genetic change countering the effects of selection. The

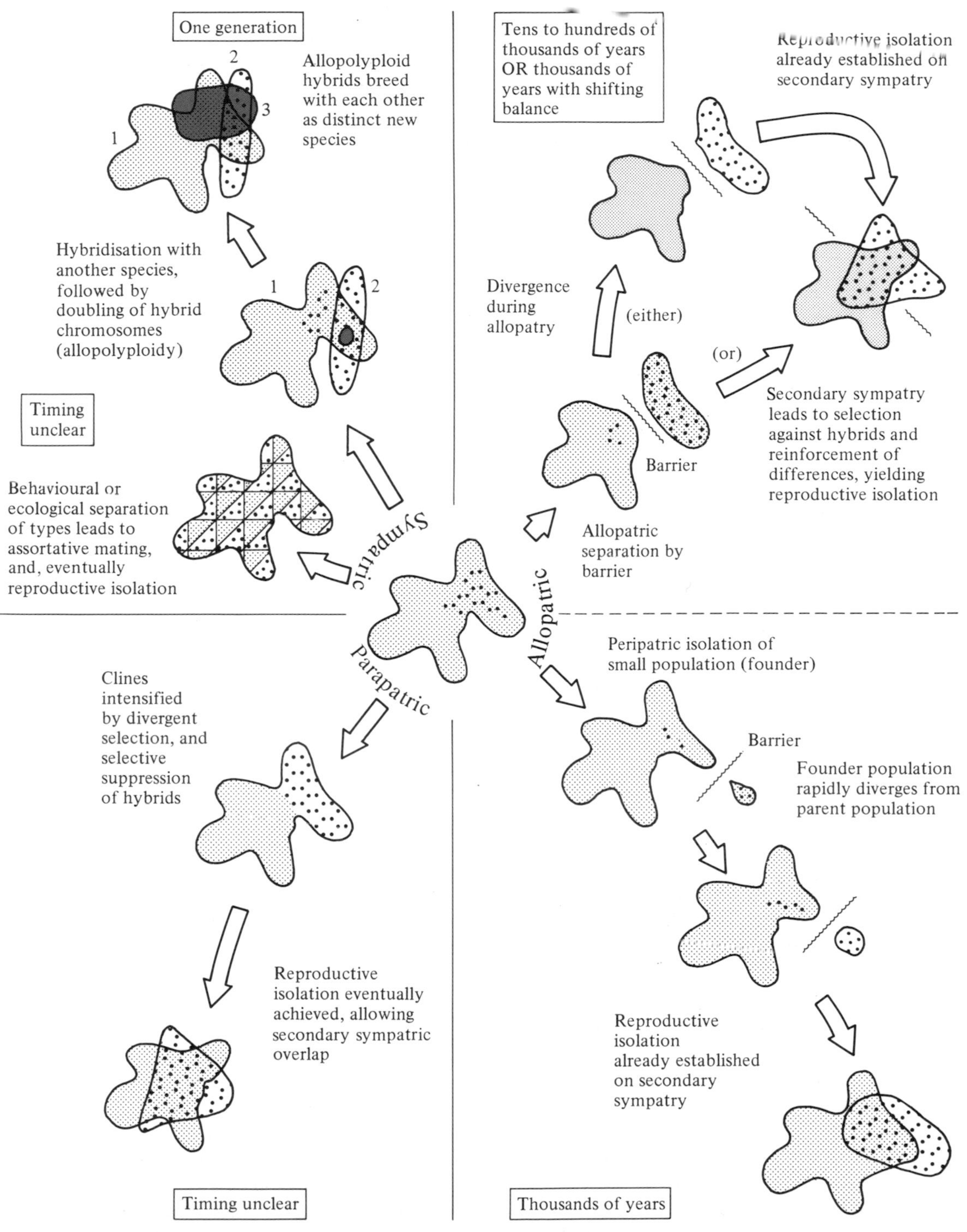

Figure 23.4 Biological models of speciation (see text for details). An ancestral population (centre) may become geographically split up (allopatry, right), and divergence may ensue (shown by change of ornament), either in large populations (above), or in a small, peripherally isolated population (peripatry, below), leading to reproductive isolation and hence speciation. Alternatively, reproductive isolation may follow selective suppression of gene flow alone a narrow hybrid zone (parapatric speciation, lower left), or the imposition of behavioural, ecological or chromosomal barriers to mating within one area (sympatric speciation). □

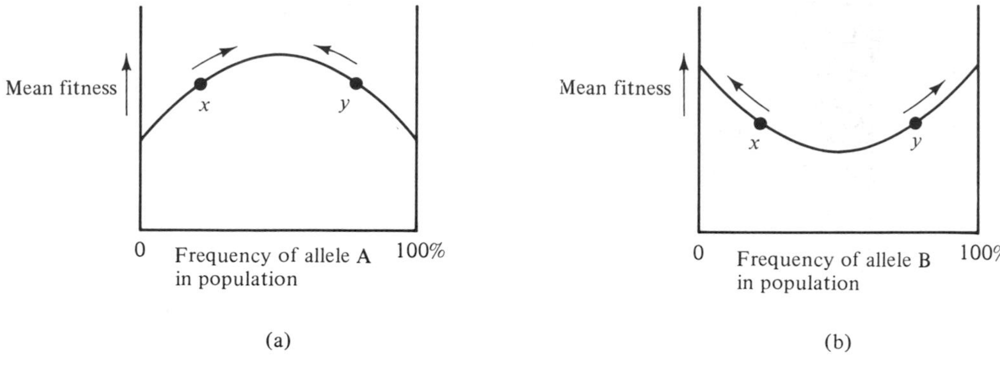

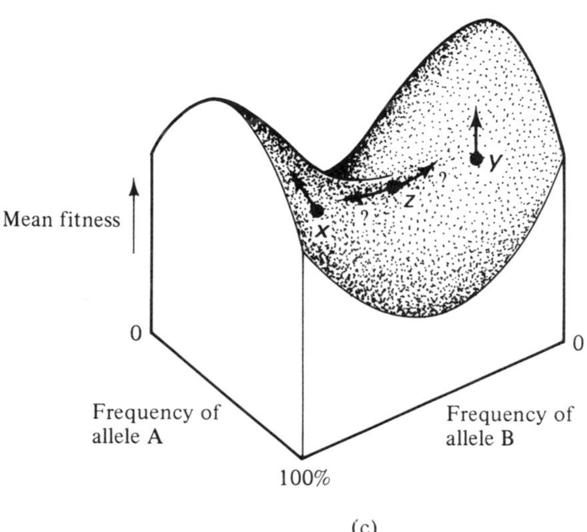

Figure 23.5 The adaptive landscape concept. (a) Because the genotype Aa is fitter than either of AA or aa, the population with the highest frequency of Aa individuals (i.e. that with 50 per cent A and 50 per cent a alleles in its gene pool) has the highest mean fitness of individuals. Populations starting with frequencies of A shown by *x* and *y* move towards the centre as selection favours Aa individuals at the expense of AA and aa individuals. (b) Because Bb individuals are less fit than BB or bb individuals, populations at *x* and *y*, in contrast, move out to the sides. (c) In the combined adaptive landscape, two adaptive peaks are separated by an adaptive valley. Populations at *x*, *y* and *z* move as shown. (NB this is a grossly simplified situation. If, for example, the alleles of the second locus affected the fitness values of those at the first locus (e.g. if Aa was only fitter than AA and aa when combined with bb, but not with BB or Bb) then the landscape would be more complex. A more realistic representation would also have vast numbers of axes, for all gene loci.) □

key to the latter is effective population size: the smaller this is, the more allele frequencies in the gene pool can be expected to fluctuate through the random sampling of parental alleles by the offspring – a phenomenon known as **genetic drift**. In extreme cases, with very small populations comprising only a hundred or so individuals, the allele frequencies may occasionally drift across an adaptive valley, there to fall under the selective regime of a neighbouring peak, within which the population would next become confined, as it previously had been in the former peak. This is termed a **peak shift**. The mathematical modelling of this is complex and not really relevant to our present concerns, but the quantitative predictions are of interest. Wright's original ideas have been extended by the American geneticist Russell Lande, who has produced a morphological version of the adaptive landscape, replacing allele frequencies on the basal axes with mean values for **phenotypic** characters (physical characters expressed in the organisms themselves). Lande's calculations show that the expected time within which a (morphological) peak shift is likely to occur increases approximately exponentially with effective population size (a number related to the numbers of breeding adults, but less than the total population size). The shift itself, when it does occur, is likely to be accomplished in just a few hundred generations. These ideas will be referred to in the following discussion.

Despite considerable debate as to its extent, there is ample evidence for allopatric speciation. The commonness of gradients of features across species populations (**clines**), and of geographical arrays of subspecies, testify to the gradual divergence of populations in different areas. That such gradual divergence may eventually lead to reproductive isolation is confirmed by examples of **'ring species'**, in which the opposite ends of a chain of subspecies geographically close on one another, but there behave as distinct species. Population genetic theory has little to offer by way of generalisations about rates of divergence in such cases, since the differentiation of the adaptive landscape between different areas is clearly bound up with the environmental differences. So, for example, American and Eurasian populations of sycamore (*Platanus*) must have been separated for millions of years, and yet they remain morphologically very similar and can form fertile hybrids. On the other hand, several distinct terrestrial species do seem to have arisen from populations which became cut off by Pleistocene climatic fluctuations. The polar bear (*Ursus maritimus*), for example, may, on the basis of fossil evidence, only date back some 40 000 years (i.e. from the beginning of the last major glaciation). Yet it has diverged markedly in habits and morphology from its probable ancestor, the brown bear (*U. arctos*). Other examples of such allopatric divergence of (presumably) relatively large populations, to yield either subspecies or distinct species, where some idea of timing could be inferred from geographical and fossil evidence, were documented by Ernst Mayr to show their generally gradual nature compared with the rates of change that could be inferred for small isolated populations. He concluded: 'Even in a species where it takes only 10 000 years to develop a well-defined island subspecies, it might well take 100 000 or perhaps 1 000 000 years for the completion of the speciation process'.

Sewall Wright, in contrast, postulated that peak shifts could provoke relatively rapid evolutionary change within large but unevenly distributed populations, in his **'shifting balance'** hypothesis. This model assumed frequent interruptions to gene flow within large but widely scattered populations, such that local subpopulations (**demes**) might at times become cut off from their neighbours. Small subpopulation sizes might then instigate genetic drift, perhaps in some demes to a higher peak on the adaptive landscape. Subsequent expansion of these demes, with migration of individuals to new sites, and resumed outbreeding, might then tip the balance of allele frequencies in neighbouring demes towards the influence of the new adaptive peak, into which they too would be drawn by selection. In this way a peak shift could spread across the whole population, or an allopatrically distinct part of it. The time scale for such a widespread shift of balance need not greatly exceed that for the original peak shifts themselves, since even low levels of selection (towards new peaks) could be expected to bring about fairly rapid change. Changes of the order of thousands of generations (hundreds to thousands of years) might not be unreasonable. An accumulation of several such events in allopatric populations could therefore conceivably lead to speciation in a series of rapid divergent shifts on a scale of thousands of years, in some cases. It is worth noting, in passing, that the shifting balance theory could also account for episodes of rapid morphological change within a single, unbranching lineage (as in Simpson's quantum evolution).

As noted earlier, Mayr's preferred model of allopatric speciation was that involving the founder effect. One of the criticisms of the founder effect was that a truly tiny population would have to be sustained through many generations in order for there to be a significant random sampling loss of any but the rarer alleles from a parental population, and that the combined chances of survival and of occupation of a new adaptive peak of such a population would anyway be slight. A modified form of the founder effect has therefore been proposed by Hampton Carson, based on his extensive studies of Hawaiian drosophilid flies. He has argued that a more effective way of bringing about random drift and fixation of alleles in an isolated population would be through a series of population explosions and crashes (**'flush–crash cycles'**). During flushes, closely interacting blocks of genes, previously assembled by selection, could be broken up, with the extensive survival of normally low fitness forms containing unusual genotypes. Then, during crashes, novel allele combinations might become randomly fixed in the residual population. Several such flash–crash cycles could progressively shift an isolated population away from its parent population. The importance of founder effects is still a matter of debate.

Again, circumstantial geographical clues can set

limits on the sort of time scale associated with inferred peripatric speciation, and these tend to indicate periods of only thousands of years. Examples tend to relate to geographically confined habitats such as lakes and islands. Lake Victoria in East Africa, for example, has undergone many changes in its approximately 750 000 year history, involving numerous fragmentations and re-connections of its waters. In common with other great lakes in the region, it has been the nursery for an explosion of species of cichlid fishes – forming veritable 'species flocks' (Figure 23.6). One small lake, Nabugabo, covering some 30 km^2, is now separated from the main lake by a narrow sand bar. This separation only dates back to about 4000 years ago, and yet five out of its seven species of the cichlid genus *Haplochromis* are endemic, presumably having evolved in the little lake since its isolation, from original subpopulations of species present in the main lake.

Beyond the fully allopatric models of speciation just discussed, there is a spectrum of models with geographical separation playing an ever decreasing role. Although there is some supporting evidence for most of these, none is yet known to match the allopatric models in importance.

In **parapatric speciation**, populations are not completely separated, but meet along a narrow 'hybrid zone' (Figure 23.4). Their differentiation is provoked by strongly divergent selection, which can overcome the diluting effects of gene flow across the hybrid zone. The model is most plausible for organisms of limited

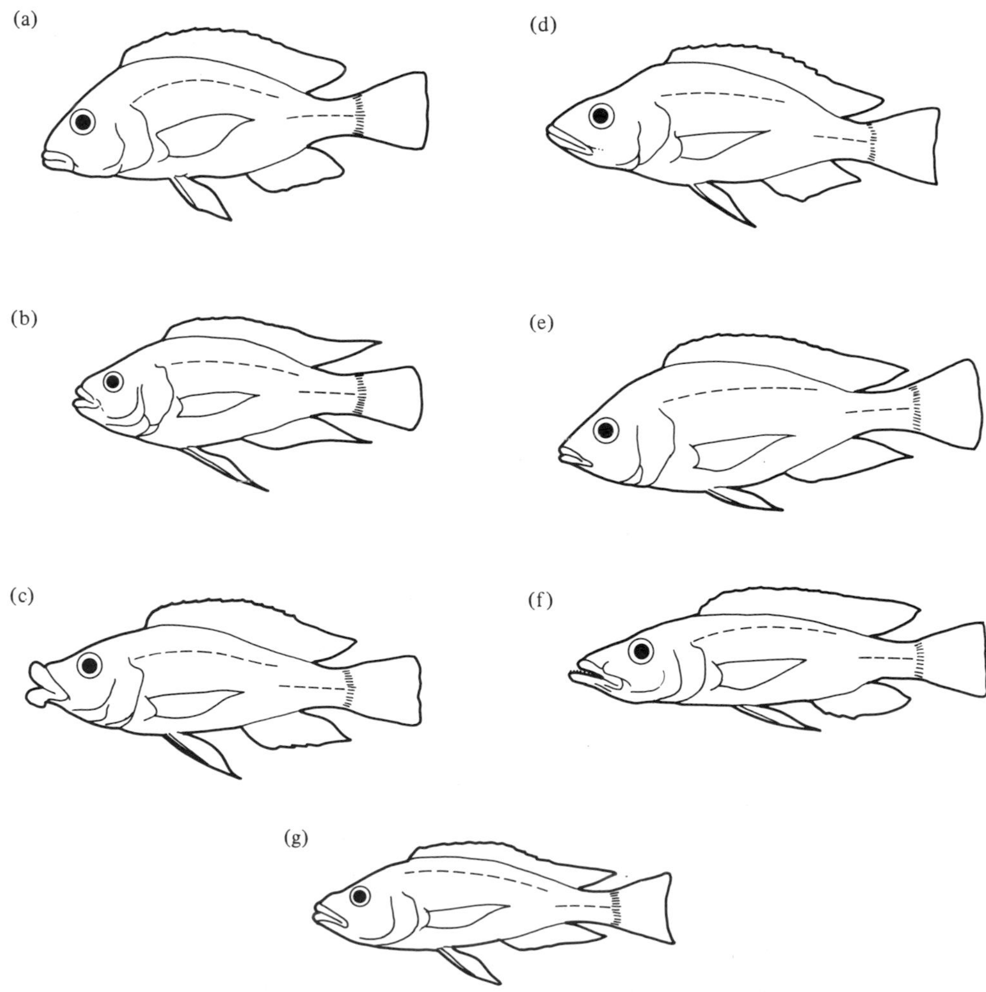

Figure 23.6 Representatives of a few of the species of the cichlid fish genus *Haplochromis* found in Lake Victoria, East Africa. Feeding habits are as follows: (a) detritus eater; (b) insectivore; (c) specialised insectivore (removes larvae and pupae from burrows; (d) feeder on juvenile fish; (e) mollusc eater; (f) fish eater; and (g) fish scale eater. □

mobility, and the main evidence for it comes from the recognition of stepped clines and/or distinct chromosomal 'races' in such organisms (e.g. flightless grasshoppers, in the latter case). Little evidence is available for the timing involved, though one rather clear example, involving the development of partial reproductive isolation of grass populations on toxic mine-waste tips in southwest Britain shows that speciation over a period of only hundreds of years may not be unreasonable. On the other hand, some hybrid zones have persisted for long periods, and so reproductive isolation is by no means an inevitable consequence of parapatric divergence. In fact, it is not at all clear how frequent speciation by this means is, nor, in many cases, can it be shown that there was not at some time an episode of discrete allopatry, sparking the divergence off.

Sympatric models of speciation, where divergence occurs in one area, imply a variety of time scales, and indeed include the only widely accepted form of 'instantaneous' speciation (operating over one generation) – that of **allopolyploid hybrids** in plants, whereby the normally infertile hybrids between different (but closely related) species may double their chromosome numbers and become fertile with each other (Figure 23.4). Otherwise, sympatric models need to explain how strong assortative mating may develop within a population, leading to the eventual separation of two (or more) distinct gene pools in one area. The most plausible mechanisms generally involve some ecological or behavioural differentiation of individuals leading to discrete patches of individuals tending to breed with each other. Postulated examples tend to involve relatively small organisms which exercise rather particular feeding habits (e.g. host-specific parasitism) and mating manoeuvres, such as (especially) many insects. Previously considered an exceptional process, there is growing evidence that reproductive isolation can indeed arise in this way.

What broad conclusions, then, may be drawn from this whistle-stop tour of biological models of speciation, as far as the palaeontologist is concerned? First the divergence of geographically separated populations is expected to have featured in a large proportion of speciations. This means that, in order to record the full pattern of morphological changes during speciation, the palaeontologist would, in most cases, need to sample successive assemblages from several different sites spread across the full geographical range of the populations in question. Most sympatric co-occurrences of sister species are likely to be post-speciational. Secondly, expected time scales are generally in the range of thousands to tens of thousands of years. Some examples (especially of gradual allopatric divergence of large populations) may range up to hundreds of thousands of years, while others may be much shorter – even 'instantaneous' in the case of allopolyploid hybrids. With the exception of the latter, such time scales are certainly 'gradual' when seen through the eyes of an evolutionary biologist. So gradual, in fact, that they are beyond direct biological observation and can therefore only be tested retrospectively.

The fossil record of species

Any sedimentary sequence is a net accumulation from a long-term budget of sediment deposition, non-sedimentation and erosion. Only where the last two factors have been absent will there be a 100 per cent record of sedimentation to represent the passage of time. This is exceedingly unusual, and most sedimentary sequences show very much lower percentages of stratigraphic completeness. The familiar short-term fluctuations in current regime (e.g. tides, and flood versus low water states in rivers etc) cause a first-order loss of completeness over the daily, seasonal and annual scale. Even what may be considered relatively quiet settings, such as deep lakes and ocean floors, are not immune from such losses: turbidity flows and, in the latter, wandering deep currents may inhibit or even cut into such condensed deposits as are allowed to settle there.

Over longer periods, sedimentary accumulation at any one point may reach a depositional limit, where stillstand or even occasional erosion may begin to take over (e.g. the point-bar crests of meandering rivers, or wave-base in seas and lakes), and this will cause yet further gaps in the long-term record. Ultimately, the only reason that any sequence will be preserved at all is the extended maintenance of a depositional receptacle, either through sea level rise, or through basement subsidence. The episodic pattern of these slower geological processes causes yet further losses of record.

The overall effect of these scaled losses to stratigraphical completeness is an inverse relationship between measured (net) rates of sedimentation and the time scale to which the measurements relate: the longer the period considered, the slower the rate of net sediment accumulation. An extensive gathering of data by P. M. Sadler establishing this relationship is shown in Figure 23.7. This is a compilation for many different depositional environments, each of which shows its own trend in the data set.

The implications of this general relationship for the palaeontological documentation of speciation events are best revealed by hypothetical illustration. Let us suppose we have a sequence containing two related

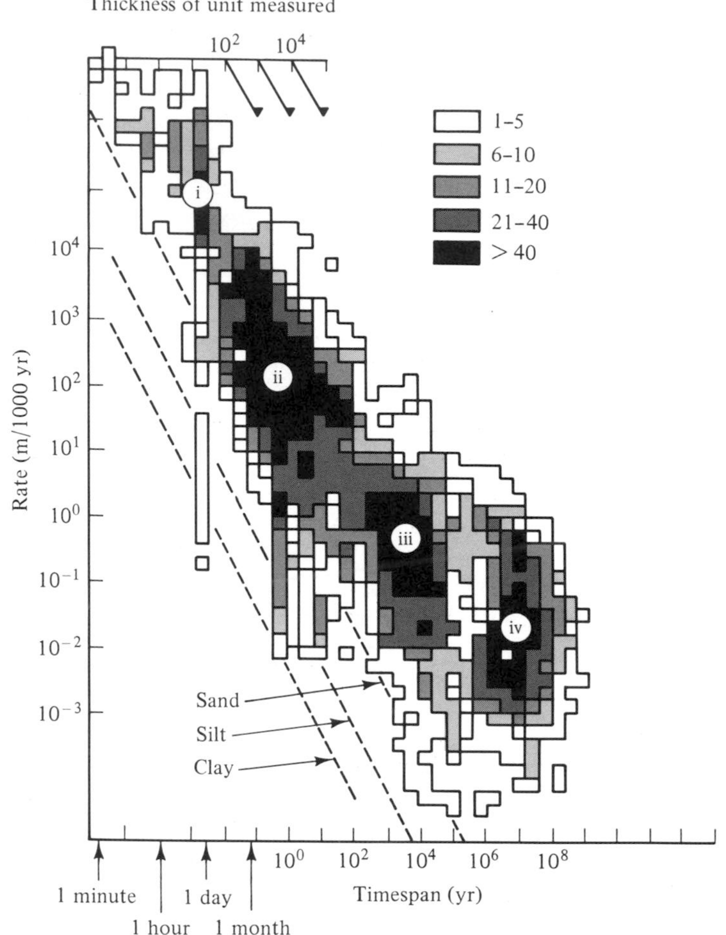

Figure 23.7 Measured net rates of sediment accumulation against the time span over which they were determined. Twenty-five thousand readings from all kinds of depositional environments surveyed are represented. The data density scale is shown at top right. The data clusters represent the four main modes of observation: (i) continuous observation; (ii) periodic monitoring at a survey station; (iii) radiocarbon dating of young sequences; (iv) older sequences dated using fossil correlation and radiometric methods. Note their approximately inverse relationship on logarithmic scales. □

species A and B, and that morphological evidence suggests that B is a probable offshoot of A. Let us further suppose that their ranges overlap, with A first appearing in the sequence studied at a level some one million years older than that at which B first appears. B might thus have evolved over any period during that million-year interval (or even earlier, if A's range is incompletely recorded). What chances are there of finding the fossil evidence for the morphological changes of this speciation?

As noted earlier, most of the speciation models involve time scales of the order of thousands to tens of thousands of years. We would therefore normally require a resolution of sampling of the fossil populations with no more than, say, a thousand years between successive samples, even to get a sketchy picture of the morphological changes involved. According to Figure 23.7, average measured rates of sedimentation at this (millennium) scale of observations are of the order of metres per thousand years. But over the whole million-year interval we are investigating, average net rates are only expected to amount to centimetres per thousand years. This means that each millennium of history, in this case, has only a 1 per cent probability of having a sedimentary record in the million-year sequence. Moreover, the probability of every one of a given sequence of millennia being represented falls off as the product of their individual probabilities: 0.01 per cent for two millennia, 0.0001 per cent for three, and so on. This is before we even consider the frequency of fossilisation when sediment is preserved. Furthermore, if a (probable) allopatric component is to be tested, we would need correlated samples from those same millennial intervals in other sequences, elsewhere within the original geographical range of the parental species. It may be readily appreciated that the chance of such a preservational coincidence becomes rapidly smaller as more widely separated areas are considered, because of the local vagaries of depositional and tectonic history.

It is clear, then, that little coherent trace of speciation can normally be expected from the fossil record. The usual pattern should comprise the 'sudden' stratigraphic appearance of any new species produced by branching, and that probably only due to its migration into the area of study. By and large, this is indeed normally the case. Rare exceptional records of speciation histories tend to involve populations originally restricted to coherent areas of deposition with unusually complete sequences, as in some lake complexes, or confined marine or continental basins in tectonically subsident areas. Just such a record has been documented by Peter G. Williamson, for freshwater molluscs of the Plio-Pleistocene deposits of the Turkana Basin in the East African Rift. Rapid changes in several lineages (over time scales of between 5000 and 50 000 years) accompanied changes in the lake level, punctuating long episodes with rather static morphology. One of the gastropods, *Bellamya unicolor*, was particularly well recorded. The local populations of this species showed three distinct episodes of rapid shift to new morphologies, presumably while isolated from the parent population, each time apparently becoming extinct a short time later with the subsequent return of individuals from the parent population, derived from some other (unknown) site. Certainly, the morphological changes recorded are consistent with allopatric divergence, though even this example has been criticised for lack of clear evidence that the changes were not simply environmentally induced deviations of growth (**ecophenotypes**).

Given the usual incompleteness of the record, punctuational change does therefore seem consistent with neo-Darwinian models of speciation, as indeed its authors originally intended, and notwithstanding the 'gradual' character (on the biological time scale) of the latter. As one commentator put it, 'one man's punctuation is another man's gradualism'. It depends upon one's time scale of observation.

Although the fossil record is thus virtually blind to most speciations, it can provide a perspective on the longer-term anagenetic history of species. It can thus go some way to testing the pattern of stasis also expected by punctuated equilibrium theory, and its corollary that most anagenesis is associated with cladogenesis.

There is no shortage of purported gradual anagenetic change in the palaeontological literature. It is much easier to pursue fossil assemblages in correct chronological order up a local sequence, than it is to monitor geographical variations and their changes through time, with the attendant problem of stratigraphic correlation. No wonder, then, that there has been so much emphasis by palaeontologists on patterns of anagenesis and so little on cladogenesis. The problem with many of the purported examples of anagenetic change, however, has been a lack of evidence that they truly represent microevolutionary change. Alternative possibilities are that the change was merely ecophenotypic (i.e. non-genetic), or that it simply reflects migration across the area of geographically variant populations such as subspecies, with, say, a shifting of climatic belts. Reviewing the situation in 1977, Gould and Eldredge concluded that none of the classic examples of supposed gradualistic change was well established since these other possibilities had not been rejected. Some more recent studies, though, did pass muster. For example, changes in the proloculus (juvenile test) diameter of a Permian larger foraminifer, *Lepidolina multiseptata*, in eastern Asia, had been shown by T. Ozawa to have transgressed the limits of its original geographical variability. The trivial nature of the change, however, was also noted. Meanwhile, of eight other measured characters, five had shown no gradual change, while the remaining three had shown very little change. This hardly amounts to the sort of radical modification of bodyplan which accompanied some major adaptive radiations, such as that of the mammals in the early Tertiary, over a similar timespan. So, Gould and Eldredge argued, this and other examples of gradualism rather reinforced the theoretical necessity for rapid speciational change to account for the latter. Subsequently, however, several palaeontologists, spurred on by this debate, have provided clear and detailed documentation of gradualistic change within lineages. And in some cases the change has exceeded the amount of difference seen between contemporaneous species so justifying (arbitrary) division of the lineages concerned into successive **chronospecies** (Figure 23.8).

On the other hand, examples of stasis have also been established. For example, nineteen Neogene lineages of bivalves were investigated by S. M. Stanley and X. Yang, who found no greater difference in shape between Pliocene specimens, four million years old, and Recent specimens, than that seen today between geographically separated populations of the species concerned. Three species, indeed, could likewise be traced back, without significant change, for 17 million years. They found that the overall stasis in these lineages resulted because 'evolution has typically followed a weak zigzag path with little net directionality'.

Fossil species thus do show both gradualism and stasis, a conclusion reached by a growing consensus. The preferred explanation for stasis in the early days, following Mayr, was that strong developmental homeostasis was imposed by strongly interactive gene complexes (p. 464). The existence of both stasis and gradualism might therefore be interpreted to reflect differing degrees of such developmental canalisation in different groups of organisms.

An alternative view of stasis, urged by several evolutionary biologists, has been gaining ground in recent years. According to this view, stasis is the result of long-term net stabilising selection, or, in other words, confinement of populations within stable adaptive peaks (p. 466). The developmental canalisation argument is criticised on the grounds that the ubiquitous existence of geographical variation in living species shows that there is frequently scope for the phenotypic expression of variable alleles. The small-scale zigzag fluctuations of features within stasis, noted by Stanley and Yang (see also Figure 23.10), can readily be interpreted as either being due to slightly fluctuating fitness values of character variations, or, in fragmented populations of many small demes, to genetic drift.

A similar tale of fine-scale zigzagging of morphology emerges in examples of gradualism when a record of high enough resolution is available. One of the most detailed studies of gradualistic change in recent years if that of Peter Sheldon, on eight trilobite lineages in the Ordovician of central Wales. Over the approximately three million years represented by the sequence, all eight lineages showed a net increase in the numbers of their pygidial ribs, though with little exact parallelism of their gradualistic trends (Figure 23.9). Sheldon interpreted the changes as indicative of selection (to give the common overall change), rather than of ecophenotypic change (for which a far greater parallelism of trends might have been expected).

A striking feature of Sheldon's study is the presence of numerous reversals in the trends. It is commonly assumed that evolution cannot be reversed (Dollo's Law), since the probability of the history of myriad influences on evolution being re-run backwards is obviously negligible. But this assumption is intended to apply to the entire phenotypic character of a species, and it certainly would be absurd to expect every jot and tittle of some previous evolutionary condition to be duplicated at a later date. No such reservation is warranted, however, when only one, or even a few phenotypic characters are considered: the circumstance promoting change in one direction might readily be reversed, so undoing the selective handiwork of former times, like some evolutionary Penelope at her loom. Thus the reversals in Sheldon's trilobites, relating only to a single character, are not in themselves alarming (although they certainly run counter to the earlier gradualistic prejudices of palaeontologists). Yet they have important implications. Had the record been poorer, or Sheldon's sampling less precise, the reversals themselves might not have been noticed. The net changes would then have seemed remarkably slight for the time elapsed. But this, of course, would have been due to the condensing of the various positive and negative trend steps of the lineages. This provides a vital clue to the evolutionary paradox of why so many gradualistic trends in fossil lineages *seem* to have been so extraordinarily slow, compared with the rates of change that may be observed in the laboratory or the field by biologists. If some of these trends were taken at face value, selection pressures of only about one selective death per 100 000 individuals per generation would be sufficient to account for them. Even random drift might be expected to swamp such rates! With reversals built into the record, however, the paradox evaporates: the apparent slowness of gradualism is just a product of poor resolution.

As with the rates of change during speciation, then, true microevolutionary rates within lineages are, evidently, generally too fast to be coherently sampled by the fossil record; they can only be glimpsed in exceptionally high resolution sequences. Both 'stasis' and 'gradualism' are but averaged patterns of such rapid zigzag change, vaguely charted by the imperfect fossil record. At the microevolutionary scale, there is therefore no qualitative difference between them. It is only over longer time scales that their overall patterns differ. So there may be no need to invoke any profound distinction between types of organism with greater or lesser extents of developmental homeostasis after all. Stasis simply implies a stable adaptive peak, while gradualism implies an altering adaptive landscape. Nor need we expect any given species to be wholly static or gradually changing. Stasis may prevail for some characters while others may be changing, at varying rates (**mosaic evolution**).

The only problem for neo-Darwinian theory still posed by examples of stasis, when seen in this light, is how an adaptive peak might remain effectively static for, perhaps, millions of years. Two main answers have been proposed. The first explanation is an extrapolation from ecological observation. Many species populations one way or another 'track' their preferred habitats in space and time. This can readily be envisaged in highly mobile organisms, such as flying insects, for example. Indeed there is good evidence from the Pleistocene fossil record that beetle species did undergo rapid geographical shifts with the movement of climatic belts associated with glacial advances and retreats. However, less obviously mobile organisms are not necessarily debarred from this. Some may have widely dispersed spores or seeds (in plants) or

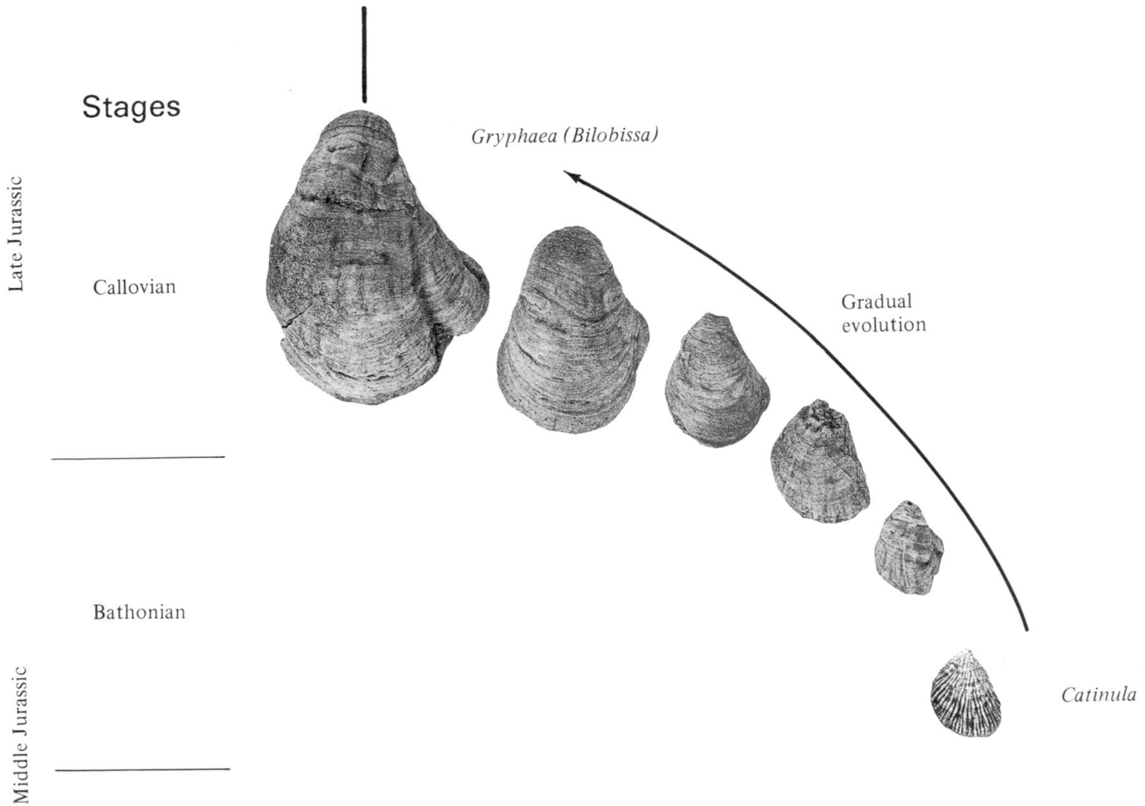

Figure 23.8 Gradual evolution of a Late Jurassic species of the oyster *Gryphaea* (*Bilobissa*) from a smaller, ribbed Middle Jurassic oyster, *Catinula*, in Europe. There is still some debate over the correct species names to be applied, but the principle of sufficient anagenetic change to warrant the recognition of distinct chronospecies is well illustrated. □

larval stages (in animals), which allow them to chance upon suitable habitats, while others may even be able, physiologically, to 'shut up shop' for extended periods of adverse conditions by, for example, encystment (e.g. in many protozoans) or hibernation. Of course these last-named life history traits are themselves adaptations to inconstant environments. But in all these cases, the organisms themselves establish the selective regime within which they carry out their life functions. Thus, so long as their ecological niche may be tracked somewhere over the course of millions of years, their morphology may remain trapped within an essentially static adaptive peak.

The second mode of stabilisation is due to the balancing of conflicting selective constraints on individual development. Thus any given feature may contribute to several different vital functions, or it may be structurally linked with other features, or it may be influenced by genes also involved in the development of other features (**pleiotropy**). So any conceivable variation which might be considered an improvement with respect to one function might prove deleterious to another. This is a notorious problem in livestock and crop breeding, where over-zealous breeding for one trait may usher in other, undesirable traits (many high-yield 'super-crops', for example, show poor resistance to waterlogging and/or disease). This essentially re-introduces the argument about developmental constraints, though in a less dogmatic form. Here, loss of fitness is the only constraint, rather than some mystical notion of rigid developmental canalisation, which can only be broken out of by major, revolutionary upheavals of developmental processes. Hence, small fluctuations in the selectively favoured phenotype may readily be envisaged with minor changes in the adaptive landscape, or with drift.

The various lines of evidence reviewed above therefore converge on the conclusion that gradualistic trends, stasis, quantum evolution and punctuational speciation are all merely different manifestations of extrapolated microevolutionary change, of the sort familiar to the evolutionary biologist. At this level of analysis there is absolutely no conflict with neo-Darwinian theory. Nor is anagenetic change especially

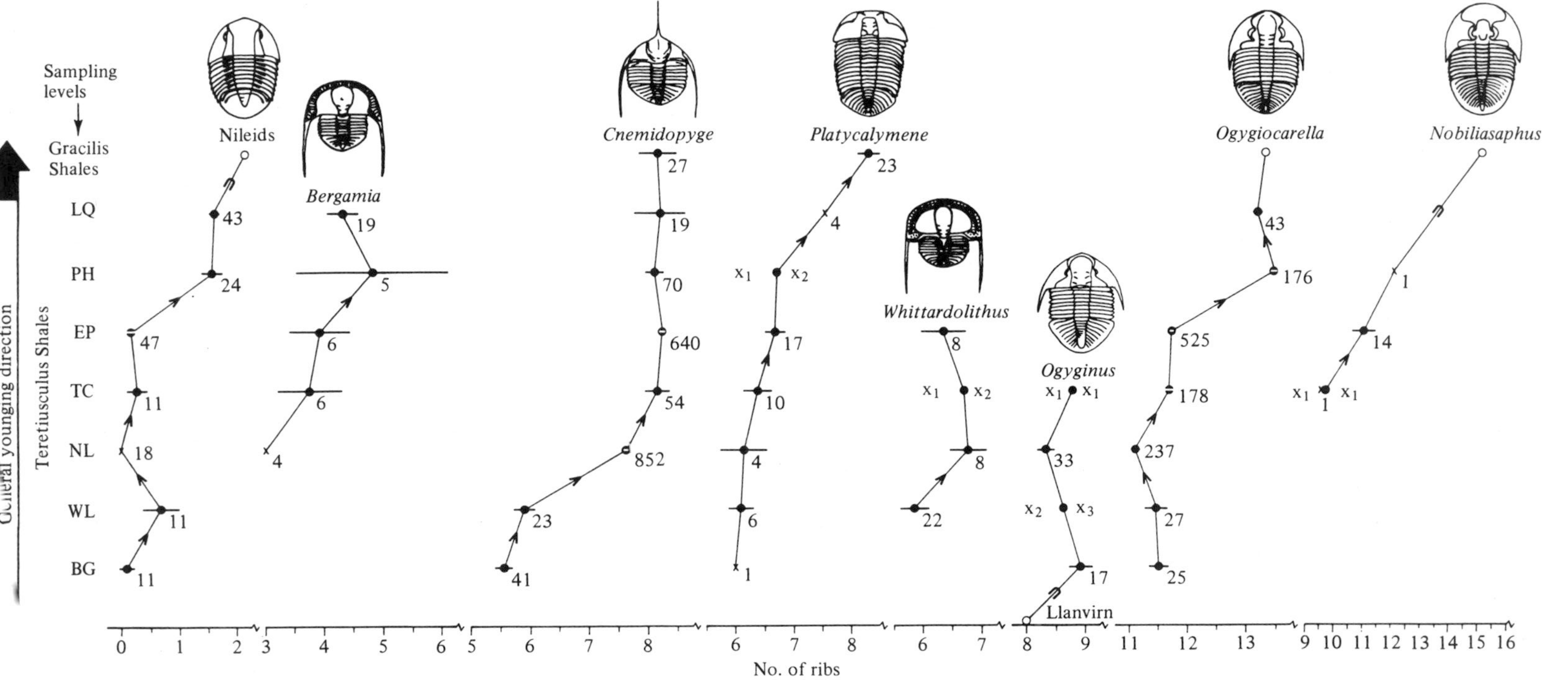

Figure 23.9 Gradualistic evolution of eight lineages of trilobites in a three million year succession in the Ordovician of central Wales. The *x* axis shows the numbers of ribs on the pygidia (tailpieces) of the trilobites. Means with 95 per cent confidence intervals are plotted, and the numbers of specimens measured are shown for successive samples (up the *y* axis). Open circles are approximate mean values and x's represent individual measurements. Arrows denote significant changes. □

associated with speciation. There are many examples of lineages even showing punctuational shifts to a new chronospecies with no evidence for another lineage splitting off. R. A. Reyment, for example, has documented just such a jump in some late Cretaceous ostracods in southwestern Morocco, whereby *Oertliella tarfayaensis*, after a long period of effective (zigzagging) stasis gave rise over a short interval to *O. chouberti* (Figure 23.10).

In summary, the proponents of punctuated equilibria seem to have scored only a partial success in their attempt to synthesise modern biological ideas on species and speciation with the facts of the fossil record. They were notably correct in recognising speciation as a usually punctuational phenomenon on the geological time scale. They were also justified in drawing attention to the common occurrence of stasis within species, even if it turned out not to be quite as pervasive as they originally supposed. Yet, they saddled themselves with the now discredited notion of evolutionary inertia in large, widespread species populations. Detailed studies of the fossil record have since shown that a full spectrum from stasis to gradualism does exist, and that, at the microevolutionary scale of observation, there is no significant difference between the busy, zigzagging histories of these. Thus, to return to the question posed earlier (p. 464), it cannot really be said that cladogenesis is necessarily the dominant source of anagenetic novelty.

Macroevolution and the sorting of species

Given the conclusion above, can the idea of species selection (p. 465) still be maintained? Or should we return to the conventional neo-Darwinian view that microevolution and time alone shape macroevolution, with species being no more than steps on the way? The answer is that the concept is still valid, for reasons which have been carefully argued throughout the 1980s by (among others) Gould, Eldredge and their co-author of various papers, Elizabeth Vrba.

Notwithstanding their changeability, species may still be regarded as 'evolutionary individuals' upon which a higher level of selection can act. To be able to serve as units for such selection, it is only necessary that species exhibit irreducible properties which are, to some extent, inherited by daughter species. Since cladogenesis multiplies these units, there is ample scope for a resulting preferential proliferation or extinction of species within different groups of organisms. Some have suggested that, in view of the existence of chronospecies, it might be preferable to talk about **'lineage selection'**, because it is the branching points which mark the birth of the evolutionary individuals concerned. However, the term 'species selection' is now well established in the literature.

It is very important to be quite clear as to what is meant by species exhibiting 'irreducible properties'. We are here concerned with the attributes of a species population as a whole, which are not themselves attributes of the individuals making up the species population. Such attributes are nevertheless inevitably derived from the characteristics of the individuals, and so are said to be 'emergent'. The idea is best explained by example. The geographical range of a species is a property of the population as a whole: no individual occupies the entire area covered by the population. Of course, the range is related to the mode of locomotion of, and the dispersal of the spores, seeds, larvae or whatever of the individuals, and so is emergent from these individual properties. Nevertheless, the range itself remains an irreducible property of the whole population.

It is easy to envisage that such a property might affect the probability of further speciation or extinction of the species. Consider, for example, some hypothetical species of seal, which is entirely confined to the North Sea. An environmental disaster in that area, such as, say, a bad oil spill during a severe winter, combined with the epidemic of distemper suffered there by seals in recent years, might signal the end of that species. The common seal, in contrast, even if likewise completely eradicated from the North Sea, would in time be able to return there by restocking from other populations elsewhere in its (much broader) geographical range. So the demise of our hypothetical species and the survival of the common seal species would then have been due to the differences in their geographical ranges, rather than to any adaptive difference between the individuals which faced the common threat of the environmental disaster.

There are two questions which now need to be addressed. First, is such a pattern of indiscriminate eradication of populations from a given area a realistic cause of extinction in nature? Second, to what extent are species properties, such as geographical range, 'heritable' by daughter species?

As with speciation, our understanding of extinction is woefully incomplete. Historically recorded examples almost inevitably involved a large component of human intervention, and many were small and vulnerable populations in any case (such as the celebrated dodo). How well-established species may eventually suffer extinction in natural circumstances has largely to be construed from theory. Obviously, what is involved is the decline in size of a population to some critically small level where random failures of reproduction and chance deaths snuff it out. So the question to be answered is, what are the main regulators of population size and how do they interact? In the broadest terms, these are reproductive capacity, on the one hand, and the host of death-dealing processes (such as starvation, predation, disease etc.) on the other. The study of the interplay of these is a huge field (most of ecology, in fact) to which justice cannot be done here. As far as extinction is concerned, however, it seems unlikely that competition between the individuals of different species would normally be a sole cause. The scope of interspecific competition has been hotly debated by ecologists in recent years and much attention has been drawn to other factors which reduce populations, such as environmental disturbance (e.g. storms, droughts etc.) and predation pressure. No two species will overlap precisely in all their ecological requirements, so that when the other regulating factors do hold off sufficiently to permit competition over shared limiting resources, it is more likely that one or other population will evolve towards uncontested niche space rather than simply become extinct. The commonplace occurrence of **niche-partitioning**, such as the coexistence of dozens of species of cone-shell gastropods (*Conus*) on some tropical shorelines, each specialising with remarkable fidelity on particular prey species, bears witness to this. Nor would any other single regulating factor probably be sufficient to cause extinction of a well-established population in most circumstances. The last few decades of history have provided numerous examples of disastrous population

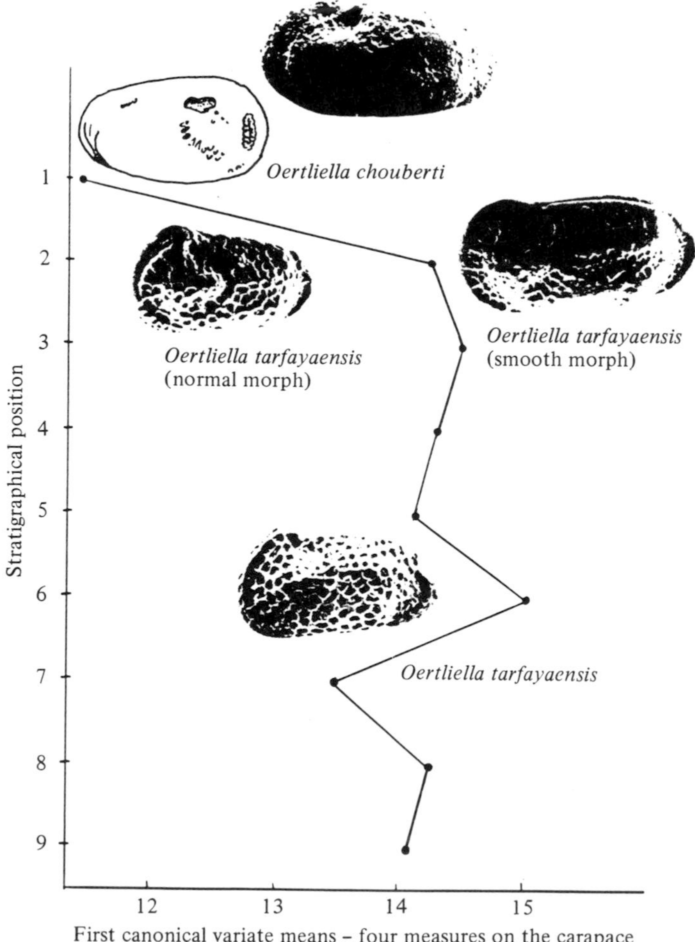

Figure 23.10 Evolution of the ostracod genus *Oertliella* in the Cretaceous of West Africa. The *x* axis shows the mean value of a statistic derived from four measurements of the carapace, for successive samples, as indicated up the *y* axis. □

crashes caused by single identifiable natural agents, such as Dutch Elm disease, distemper in seals and the ravaging of large parts of the Great Barrier Reef by the Crown of Thorns starfish, but none of these has, it seems, been the sole agent of an extinction.

It is likely, then, that many species extinctions are due to chance combinations of factors which work together to reduce populations to a critical level. Following this line of argument, it is not difficult to see how such effects may be areally bounded. The hypothetical seal example given earlier illustrates this, but one could as readily envisage a coincidence of wholly natural adverse factors, overlapping across a distinct area. The native horses of North America, for example, having radiated profusely through the Miocene, underwent a rapid decline from the beginning of the Pliocene (Figure 23.1). It is not clear why, but the coincidence of so many extinctions suggests some underlying environmental disturbance (probably related to climatic change). Even the genus *Equus* finally became extinct there some 10 000 years ago. It is only back there today as an historical (17th century) re-introduction from the Old World, to which it had earlier migrated. Thus the primary reason for the survival of *Equus*, and the contrasting extinction of other North American Pliocene genera, was the difference in their geographical ranges.

For a species property such as geographical range to be subject to any form of species selection, however, it is also necessary that it be to some degree 'heritable', in daughter species. The key to this is the emergent nature of such properties: if the individual properties which give rise to the species properties tend themselves to be conserved in the individuals of the daughter species, then there is scope for the 'heritability' of species properties. This is analogous to the role of genes in conventional natural selection. The genes confer (emergent) phenotypic properties on their individual possessors, and natural selection acts upon the latter. We are simply dealing with the same kind of hierarchical relationship at two different pairs of levels in the evolutionary hierarchy.

A highly plausible example of such selection of a heritable species property has been documented by Thor Hansen, among some Tertiary gastropods of the American Gulf Coast region (Figure 23.11). He noted a clear correlation between the geographical ranges of species and their stratigraphical durations. The geographical ranges of the various species were in turn found to be strongly linked with the mode of larval development and the environmental tolerances of the individuals comprising the species. Larval type can be recognised in these fossil gastropods from the form of the initial part, or 'protoconch' of the shell. Hansen found that planktonic larval development (where the larvae floated in the plankton, rather than developing in fixed egg masses), combined with adult tolerance of a broad range of habitat types were associated with wide geographical distribution. In so far as these individual traits tended to be passed on to daughter species, then, the emergent property of realised geographical range could also be considered 'heritable' by the daughter species.

A devil's advocate might argue that the differential survivorship of the species might have been due directly to the individual differences in mode of development and ecology. In other words, the individuals with planktonic development and broad environmental tolerance may simply have been adaptively superior to those with the opposite traits when faced with the same environmental disturbances. This is certainly a criticism which needs further testing.

Later work, by David Jablonski, has shown that this link between the distribution and duration of benthic molluscan species broke down during the mass extinction which came at the end of the Cretaceous. This makes good sense if species had indeed been the effective unit of selection: the catastrophic events associated with the mass extinction may well have been of virtually global extent: in such conditions, a mere broadening of geographical range would have provided little extra insurance against extinction.

What sort of effect on macroevolution might species selection have? While it is true that individual adaptations can only be fashioned by selection at the level of the individual, species selection could be expected to play a hand in the distribution of such features among groups of related species. Ultimately it might even suppress what might otherwise seem to be a highly adaptive feature (at the individual level) within a group, simply because this promotes species characteristics which lose out in species selection. In that sense it could be an important macroevolutionary filter, setting the scene for subsequent adaptive evolution. So when the question, 'why do the majority of species in some group possess a given feature?', is asked, it would not be sufficient to seek some adaptive explanation alone for the superiority of that feature. It would be necessary also to investigate the possibility that the feature in question was linked with a species characteristic (such as geographical range) which was itself favoured by species selection.

To return to the gastropod example, one would be tempted now to ask why non-planktonic development has not been flushed out of the group altogether, in view of its association with species of relatively lower

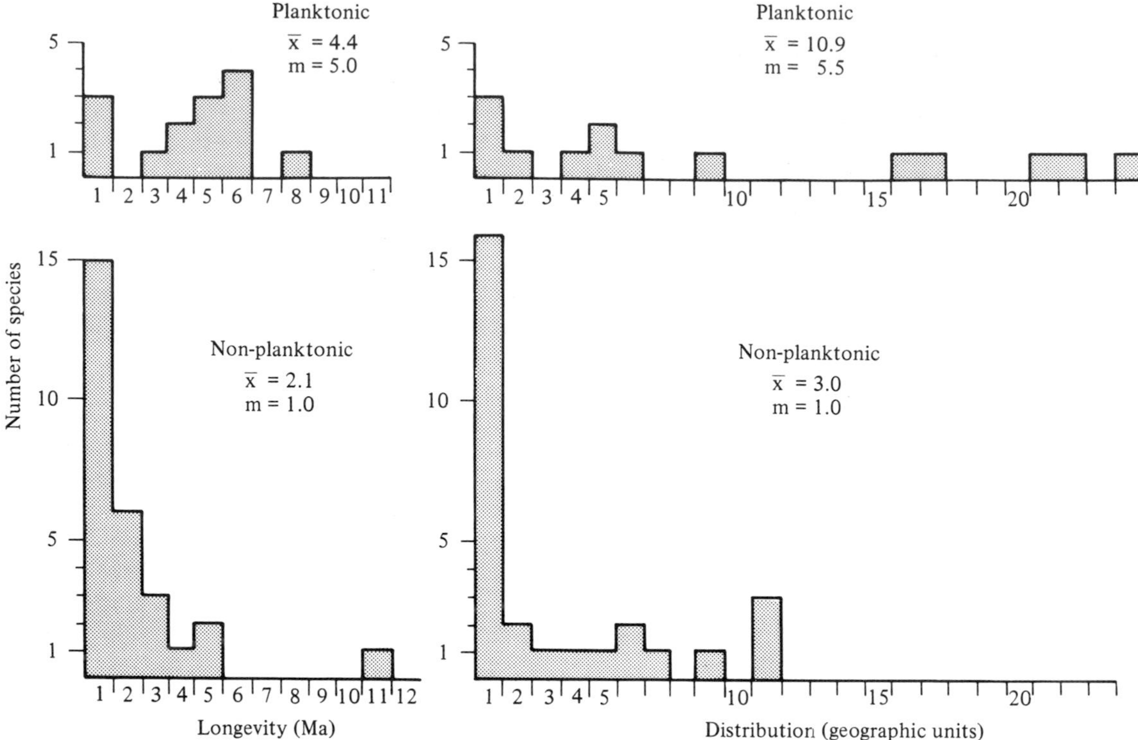

Figure 23.11 Species selection in volutid gastropods in the Palaeocene–Eocene of the Gulf Coast, USA. Histograms show: left, the stratigraphical durations of species with planktonic (above) and non-planktonic (below) larval development; and right, the relative geographical ranges (in arbitrary units) of these species. Note that the species with planktonic larvae (and also with broader ecological tolerances) tend both to occupy larger areas, and to outlast those with non-planktonic larvae. □

survivorship. In this case the answer seems to be that this mode of development also tends to be associated with increased rates of speciation, so offsetting the losses through extinction. When one is dealing with such features as population distribution and structure, it is probable that there will frequently be parallel effects on rates of speciation and extinction: aspects that promote the fragmentation, and hence tendency to speciate, of populations, are likely also to promote proneness to extinction. Nevertheless there is no intrinsic reason why such parallel effects on rates should be precisely equal and so some net species selective effect could well be expected.

Thinking along these lines is relatively new, and so there are not as yet many well-documented examples to support the idea. By the nature of its time scale – involving differences in rates of speciation and extinction over millions of years – its testing falls squarely within the palaeontologist's domain. The challenge to palaeontologists is therefore to detect further species characteristics which might have had effects on rates of speciation and extinction, and to test these proposals with data from the fossil record. For example, reproductive isolation through the appearance of behavioural or other barriers to mating is commonly (if not usually) an accidental spinoff of allopatric divergence. Individuals of some species have a strong tendency to assortative (non-random) mating, based on elaborate mate recognition devices. These species might reasonably be expected to be more prone to reproductive fragmentation of populations, and hence speciation, than those lacking such attributes. In such a case, the increased rate of speciation is itself merely an accidental spinoff of the mating behaviour, just as the increased geographical range of the gastropods discussed earlier was a spinoff from planktonic development combined with broad environmental tolerance. Yet, the increased rate of speciation could well cause net species selection in favour of species in which such mating behaviour occurs. If asked to cite examples of very 'successful' (i.e. diverse and long-lived) taxonomic groups, many students of evolution would probably include, for example, insects, spiders, birds, mammals, fish and cephalopods high on their list. As it happens, all of these have eyes, and their species commonly exhibit courtship rituals involving precise visual (as well as other forms of) mate recognition. It may therefore turn out that the evolutionary exuberance of these groups was largely caused by high rates of speciation through the accidental initiation of pre-mating barriers. The adaptive diversity of these groups would then be a secondary consequence of their rapid cladogenesis (i.e. arising through post-speciational adaptive microevolution), rather than being the main cause of it. Thus species selection would have to be recognised as an important component of macroevolution. This would mean that the 'Creator's fondness for beetles' which so impressed J. B. S. Haldane would have more to do with the tendency for their species populations to split up than with their adaptive potentials.

In conclusion, although the punctuated equilibrium theory itself only had a limited success, it brought with it the germ of an idea which promises to revolutionise the study of evolution. The time scale of the postulated process requires the sort of perspective only the fossil record can provide. There has never been a more exciting time to be an evolutionary palaeontologist! Yet Darwin has no need to turn in his grave; I'm sure he would have been delighted with the hierarchical broadening of the concept of evolution by natural selection, based, as it is, on increased emphasis on the process of cladogenesis, which he and Wallace also did so much to establish.

Summary

The fossil record has long provided evidence for the course of evolution, but more recent work has also probed it causes. Long-term processes that influence evolutionary patterns cannot be investigated experimentally, but hypotheses about them can be retrospectively tested by searching for theoretically predicted contingent outcomes in the fossil record, which differ from those of alternative hypotheses.

One important new proposal was that most morphological evolution is limited to small, geographically isolated populations, undergoing rapid evolutionary divergence (speciation) from their parent populations. Large, well-established species populations, in contrast, were postulated to remain morphologically static. Since only the latter are likely to have left a fossil record, disjointed, morphologically static species are all that the palaeontologist might be expected to find. This theory of 'punctuated equilibria', which was contrasted with earlier views of gradual change in species ('phyletic gradualism'), in turn suggested the possibility that macroevolutionary patterns might be shaped more by the factors that affect rates of speciation and extinction than by microevolutionary change within species populations – a process of macroevolutionary sorting termed 'species selection'. Biological studies of variation in species, as well as subsequent detailed analyses of the fossil record, have demonstrated a complete range of patterns between the two models: some well-established species have undergone change (without speciating), while stasis is evident in others. It is also clear, though, that much morphological evolution is too rapid to be faithfully registered in the fossil record: the latter yields only the vague outlines of highly dynamic patterns.

The collapse of punctuated equilibria as a universal model has not, however, undermined the possibility of species selection. No matter what the patterns of microevolution, there is still scope for a more narrowly defined formulation: this postulates that certain irreducible characteristics of whole species populations (such as geographical range) may both influence probabilities of speciation and extinction, and also show some degree of 'heritability' in daughter species. Macroevolutionary patterns may thus be seen as emerging from the products of microevolution (based on natural selection acting on individuals), modulated by selection operating directly on species. This relatively new idea has been little tested, but there is some evidence in its favour. It thus provides an exciting challenge for evolutionary palaeontology to contribute again to debates on the *mechanisms* of evolution, after being for so long a 'Sleeping Beauty', pricked by the bodkin of neo-Darwinian theory's exclusive stress on natural selection at the individual level.

Further reading

Briggs D. E. G. & Crowther, P. R. (eds.) (1990) *Palaeobiology: A Synthesis*, Blackwell Scientific Publications, Oxford, on behalf of the Palaeontological Association, 583 pp.
A compendium of review essays, with coverage, at an accessible level, of many aspects of evolutionary palaeontology.

Cope, J. C. W. & Skelton, P. W. (eds.) (1985) Evolutionary case histories from the fossil record, *Special Papers in Palaeontology*, **33**, 203 pp.
Papers from a conference, illustrating a range of patterns from punctuated equilibria to phyletic gradualism.

Futuyma, D. J. (1986) *Evolutionary Biology*, 2nd edition, Sinauer Associates Inc., Sunderland, MA, 600 pp.
A lucid and comprehensive modern text covering all aspects of the biological background to evolutionary theory.

Levinton, J. (1988) *Genetics, Paleontology and Macroevolution*, Cambridge University Press, 636 pp.
A comprehensive review of current thinking in macroevolutionary studies, rejecting punctuated equilibria but accepting a hierarchy of sorting processes. Somewhat dense in style.

Otte, D. & Endler, J. A. (1989) *Speciation and its Consequences*, Sinauer Associates Inc., Sunderland, MA, 679 pp.
Papers from a conference illustrating the wide range of biological models of speciation currently being tested. High-powered, but absorbing reading.

Sheldon, P. R. (1987) Parallel gradualistic evolution of Ordovician trilobites, *Nature*, **330**, 561–3.
A modern 'classic', documenting gradualistic evolution.

Stanley, S. M, & Yang, X. (1987) Approximate evolutionary stasis for bivalve morphology over millions of years: a multivariate, multilineage study, *Paleobiology*, **13**, 113–39.
The counterpart for the preceding work: a detailed documentation of stasis.

Vrba, E. S. & Gould, S. J. (1986). The hierarchical expansion of sorting and selection: sorting and selection cannot be equated, *Paleobiology*, **12**, 217–28.
A cogent essay on hierarchical processes in evolution, clearly explaining the nature of species selection.

INTRODUCTION TO CHAPTERS 24 AND 25

Today there is much concern about the intimate relationship between the human destiny and the effect of our activities on the long-term evolution of the environment. The current lack of an agreed solution to the problem of nuclear waste disposal, the detrimental effects on forests and freshwater fish stocks of acid rain, and the 'threat' of global warming, mainly due to increases of atmospheric carbon dioxide levels, are cases in point. Chapters 24 and 25 provide a geological perspective on the current concerns about the way humans may be changing the global environment.

While it was relatively easy in 1987 to gain international agreement to a 50 per cent reduction in the emission of substances that deplete the ozone layer (chlorofluorocarbons), such agreements have proved much more difficult for carbon dioxide emissions. While responsible governments may aim to cut these emissions by developing alternative energy sources, this will not be achieved globally without a dramatic change to human practices and aspirations. On the one hand this is seen as an increasingly serious problem requiring urgent action while, on the other, one might argue that it doesn't really matter to the Earth as a planet if there is a significant reduction of the land areas as ice caps melt.

Geologists know what an ice-free and ice-covered Earth is like. They know because the geological record contains evidence of successive 'greenhouse' and 'icehouse' episodes. In Chapter 24, Eric Barron reviews the evidence on which such interpretations are made, and discusses the possible causes of climatic change. Changing levels of carbon dioxide in the atmosphere may well be important, but the effects of changing distributions of land and sea, global sea-level changes caused by ocean ridge growth, and astronomical causes are also significant. The complexity of producing climatic models incorporating these variables is discussed, and the value of testing and refining such models against the recent geological record of the last advance of ice sheets is explained. In essence, this approach of retrospective testing may help refine climatic models for the present time, and so aid the evaluation of the short-term effect that anthropogenically induced changes in atmosphere carbon dioxide may have.

One might easily adopt a similar approach to earthquakes and volcanoes which are simply a product of the Earth's continuing tectonic activity, yet which represent an extreme hazard to the populations that choose to inhabit active plate margins. Our anthropomorphic answer has been to initiate studies of precursors so that natural events may be predicted and their effects mitigated, and to this end a variety of geophysical and geochemical monitoring techniques has been developed. The 1990s were designated by the United Nations and various national scientific agencies as the 'International Decade for Natural Hazard Reduction' and this has prompted renewed efforts to understand the mechanisms of volcanic eruptions, earthquakes and, of course, landslips, subsidence, tsunamis and floods. One simple but highly successful approach has been to monitor the rate of change of some parameter directly related to the hazard, such as ground movements in potential landslip zones. A forward prediction of when the rate will become infinite (or, mathematically more convenient, when 1/rate becomes zero) allows the timing of the suspected hazard to be forecasted. It can be quite another problem to move a large population from the affected zone, and there are reports that many of the three million inhabitants of Haicheng (eastern China) were evacuated at gunpoint before the massive 7.5 magnitude earthquake of February 1975. Similarly it is questionable whether many of the 22 000 sleeping residents of the Colombian town of Armero would have been willing to leave their homes if told that Nevado del Ruiz would produce a volcanic mudflow in November 1985.

No book of this kind would be complete without a summary of recent developments in the understanding of geological hazards. So we invited Michael Rampino to examine the problem in a volcanological context. As well as the short-duration or 'instantaneous' dangers they present, the more beneficial side of volcanoes is that they provide the raw materials from which the atmosphere and hydrosphere have evolved. It is salutory to reflect that volcanoes add over 10^{10} tonnes of carbon to the atmosphere each year, far more than from the burning of fossil fuels, and that even within the last million years, atmospheric levels of carbon dioxide, hence global average temperatures have, at times, been higher than they are today. It is the realisation that both particulate and gaseous emissions from volcanoes vary over long timescales that has led to some remarkable research by Michael Rampino and others at the interface between volcanology and climatology. The results show that small amounts of volcanogenic sulphur gases form persistent atmospheric aerosols which reduce solar radiation reaching the Earth's surface and so cause global cooling. This suggests that large volcanic eruptions in the past would have had a dramatic effect on the environment and hence on the progress of biological evolution.

CHAPTER 24

PALAEOCLIMATOLOGY

Eric J. Barron
Pennsylvania State University

Geology and climate

Climate is defined as a time-mean state, an average of a set of properties (temperatures, pressures, winds etc.) over a specific interval of time for a specific area. Usually, the theoretical limit of weather prediction, which is a few weeks, distinguishes climate from weather. Given this definition, the terms climatic variability and climatic change become clear. **Climatic variability** describes differences between averages of the same kind, for instance differences between two Januaries. **Climatic change** must involve a time scale which is greater than the specific time interval used to define or describe the climate of a given area.

The definition of climate is equally applicable to **palaeoclimatology**, the study of ancient climates. The only difference is the temporal reference frame. In palaeoclimatology, the averaging time is dependent on stratigraphic resolution rather than the calendar. Further, the types of observations and their interpretations are a function of the age of the record. Time is central to the scientific differences between climatology and palaeoclimatology and in large part defines the nature of problems in geology and climate.

Modern climatology is based on a large set of observations over a short interval of time. These observations are often direct measurements of the properties and characteristics of the atmosphere, oceans and ice. Because the period of extensive, direct observations is short, modern climatology lacks a sense of climatic change.

In contrast, palaeoclimatology is characterised by a tremendous sense of climatic change, but from a limited set of observations. The observations from rocks and sediments describe a rich record of global climatic change on a variety of time scales. This record may be viewed as an inventory of events and trends in Earth history which illustrates how the climate system operates and how it has evolved through time.

Three problems characterise the study of palaeoclimatology: (1) the description of past climatic states, (2) the causes of climatic change in Earth history and (3) the response of the climate system to specific factors (forcing factors) which influence climate.

First, the observations or descriptions of past climatic states define the critical problems in palaeoclimatology. However, the observations are necessarily limited and become increasingly sparse back in time. Since palaeothermometers and palaeobarometers which are the direct measures of the state of the atmosphere or ocean are unavailable, a major focus of research is to utilise physical, chemical and biological 'signatures' which can be linked to climate variables. Further, these signatures, which are recorded at the Earth's surface, must provide a three-dimensional view of the climate system. Finally, our ability to subdivide the geologic record into time intervals and to correlate events or trends across the globe has considerable implications for our perception of climate in Earth history.

Second, much of the emphasis in the study of

palaeoclimates is on the causes of climatic change because of the pervasive nature of global change recorded in geologic history. Many of the events and trends are now associated with specific causes or combinations of external and internal forcing factors. Variations in solar luminosity, changes in the Earth's orbit, changes in geography due to plate tectonics, and the link between global tectonics and geochemical cycles figure prominantly in explanations of climate change in Earth history. An equally diverse number of proposed causes remain untested or unverified.

Third, an understanding of the climatic response to a specific forcing factor is a major challenge in palaeoclimatology. This requires both a qualitative and a quantitative understanding of the complex interactions and feedback mechanisms within the climate system. Investigation of the climatic response to forcing factors has become the primary task of physically based climate models. Of considerable significance is the potential to use climate models in a predictive mode. Then, for the first time, Earth observations in the geologic record can be used as independent verification of our understanding of past climates.

The three approaches, deciphering the record of past climatic change, determining the causes of the events and trends in Earth history, and understanding the physical basis of climate change, are providing considerable information on how the climate system has evolved through time.

Tools in climate reconstruction

Ideally, the goal of climatic reconstruction is to define the conditions and characteristics of the atmosphere, oceans and cryosphere (ice sheets and glaciers) through geologic time. The challenge is to find geologic thermometers, barometers, anemometers, current meters and salinometers. In palaeoclimatology, these tools are interpretive, based on a more complex record of chemical, biological and physical signals in rocks and sediments.

Reconstructions based on chemical data

The key factor in utilising geochemical data to reconstruct palaeoclimates is that chemical and physical processes which are climate-dependent result in changes in the mineralogy, elemental composition and isotopic composition of sediments. Interpretation of these variations then depends on a knowledge of the chemical and physical processes, an understanding of how these processes influence the chemical composition of sediments, how these 'signatures' are preserved in the rock record and a knowledge of the link between the chemical characteristics and climate.

Mineralogy can be related to climate in cases where (1) mineral solubilities are controlled by temperature, (2) mineral phases are a function of salinity and can be related to the hydrologic cycle, (3) chemical alteration through weathering processes can be related to temperature and precipitation, and (4) minerals are associated with specific physical environments.

Sedimentary phosphates, evaporite minerals and **weathering products (clay minerals)** are key examples of sedimentary mineralogy as a contributor to palaeoclimatic data. The calcium-phosphate to iron-phosphate ratio is sensitive to salinity because the calcium and iron activity changes as a function of the concentration of salts. Marine phosphates are dominated by calcium, and the occurrences of iron or calcium sedimentary phosphates can be used as indicators of large differences in salinity.

Evaporite minerals are precipitated from concentrated brines. The more than eighty evaporite minerals precipitate in an order (Table 24.1) defined by brine concentration (e.g. gypsum precipitates for conditions of three to four times normal marine concentration). Stability ranges are also defined by temperatures. For example, gypsum has an upper temperature limit of 58 °C. Evaporite deposition then details conditions where evaporation exceeds precipitation and may provide some evidence for temperature limits.

Table 24.1. *Order of precipitation of common evaporite minerals*

Limestone	$CaCO_3$
Gypsum	$CaSO_4 \cdot 2H_2O$
Halite	NaCl
Magnesium sulphates	$MgSO_4$
Sylvite	KCl
Bischofite	$MgCl_2 \cdot 6H_2O$

A number of rocks are not formed in equilibrium with surface conditions. Chemical weathering then occurs ranging from oxidation to weathering products which reflect a high intensity of chemical action. For example, illite and smectite are clay minerals characteristic of little chemical weathering and are commonly associated with ice-rafted sedimentation (mechanical weathering). Kaolinite, on the other hand, is associated with weathering in humid zones. In extreme cases of chemical weathering in warm humid environments, aluminous and ferruginous soils are formed (bauxites and laterites).

The substitution of one element by another is the rule rather than the exception in the formation of minerals. Substitution is governed by ionic size, charge, crystal structure, temperature and solution composition. Minor and trace element compositions of minerals become useful climatic indices because a greater atomic substitution occurs at higher temperatures and because the concentrations of some elements are in proportion to total salt concentrations. In particular, magnesium and strontium substitution for calcium in shell material ($CaCO_3$) have been used as palaeotemperature tools. However, the greater substitution of Mg and Sr in skeletal calcites apparently is also dependent on the type of organism concerned.

Boron, gallium, rubidium, sodium and strontium are examples of useful trace elements. These elements are used as salinity indicators because the concentration is salinity dependent. However, the concentration of these elements may also depend on sediment source, clay mineralogy and grain size.

The most widely utilised geochemical methodology in climate reconstruction is **stable isotope measurement**. Isotopes are atoms with the same number of protons, but a different number of neutrons, and hence a different atomic mass. Since the kinetic and thermodynamic properties of molecules are mass-dependent, a partial segregation of **isotopes** (fractionation) occurs during chemical and physical processes so that they may be **depleted** or **enriched**. Large mass differences between isotopes, abundance and participation in important biochemical and geochemical reactions make carbon, oxygen and sulphur isotopes ideal for environmental reconstructions.

Harold Urey first noted that the **oxygen isotopic composition** of carbonates, secreted by organisms growing in equilibrium with sea water, could be used as a palaeothermometer. The oxygen isotopic composition of carbonates in molluscs (Figure 24.1) was first shown by Samuel Epstein and collaborators to be a function of temperature. The isotopic composition of planktonic foraminifera has also been demonstrated to be temperature-dependent. These organisms are the most frequently used for palaeotemperature analysis because they are widely distributed in the surface ocean.

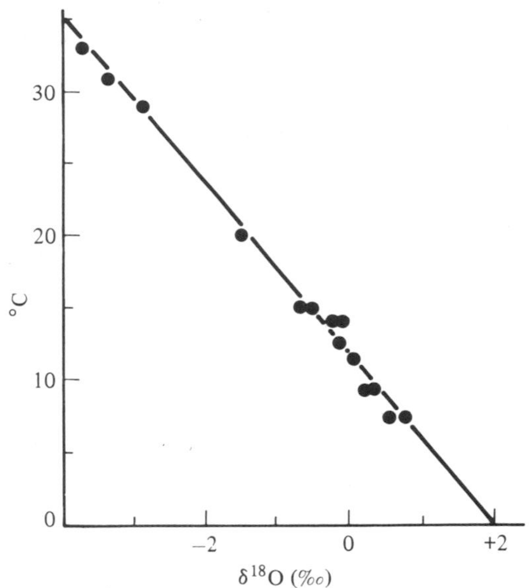

Figure 24.1 Laboratory measurements (by Samuel Epstein and collaborators) of oxygen isotopic composition of molluscs growing at different temperatures demonstrate that oxygen isotopes in carbonate shells can be used as a palaeothermometer. The notation (expressing the difference in abundance of ^{16}O and ^{18}O from a standard in parts per thousand) is the accepted mode of expressing oxygen isotope data. □

Four factors limit the use of oxygen isotopes as a palaeotemperature tool. First, the isotopic composition of sea water is dependent on evaporation, precipitation and freshwater runoff. Evaporation selectively removes the lighter oxygen-16 (^{16}O) isotope. Consequently, rain and snow are enriched in oxygen-16. One consequence of this fact is that the sea-water oxygen isotopic composition reflects the size of the ice sheets. For the Pleistocene, the isotopic composition of planktonic foraminifera is dominated by the waxing

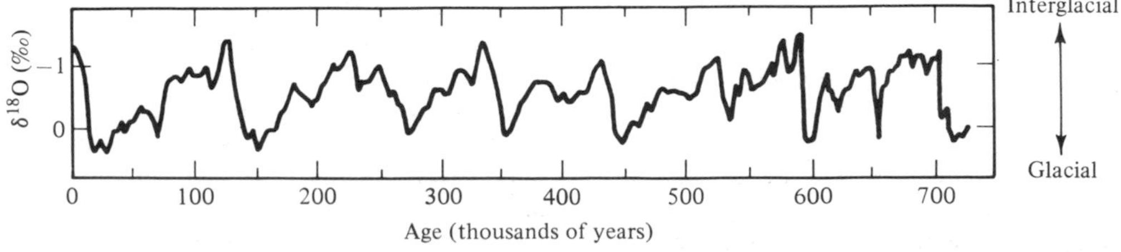

Figure 24.2 Oxygen isotopic composition of planktonic foraminifera (a floating protist with a calcium carbonate shell) measured from deep-sea cores and illustrating the waxing and waning of the ice ages. For reference, the Earth is currently experiencing an interglacial climate and the last major glaciation was 18 000 years ago. □

and waning of the ice sheets. This means that the isotopic composition can be used to reconstruct the ice ages (Figure 24.2). Prior to the Pleistocene, the history of glaciation must be known to interpret isotopic values as temperatures. The second limiting factor is organism **vital effects**; these occur within the bodies of organisms when oxygen is not secreted in isotopic equilibrium with sea water. Vital effects have been demonstrated for some organisms, such as corals and echinoderms, which then cannot be used for palaeoclimatic reconstruction with isotopes. The third limiting factor is that the use of isotopic data requires knowledge of the geographic–environmental framework. Different interpretations occur if the organism is mobile (e.g. during which season and at what location was the calcium carbonate secreted?), or lives at different depths in the ocean. Knowledge of habitat becomes essential for a meaningful interpretation. Fourth, the original isotopic composition of the carbonate material must be preserved. Both dissolution and burial diagenesis (alteration of carbonate at depth re-equilibrating the isotopic composition at a higher temperature) will influence isotopic composition. Palaeotemperature analysis requires careful examination of carbonate materials for evidence of alteration.

Carbon isotopes (^{12}C, ^{13}C, ^{14}C) are a second example of the utilisation of isotopes in palaeoclimatology. Organic carbon and carbonate are the major reservoirs of carbon. In the **organic carbon reservoir**, photosynthetic fixation of carbon preferentially incorporates ^{12}C into plant tissues. The **carbonate reservoir** involves reactions with a sequence of dissolved phases and a solid phase (calcium carbonate). Dissolved carbon in the ocean is present in three forms: bicarbonate, carbonate and carbon dioxide. Carbon isotopic fractionation (enrichment in ^{13}C) follows the species order, bicarbonate, carbonate and carbon dioxide, with bicarbonate being the more enriched in ^{13}C. Calcium carbonate is the most enriched in ^{13}C. Consequently, carbon isotopic measurements reflect the partitioning of carbon in the organic carbon (^{12}C enriched) and carbonate (^{13}C enriched) reservoirs. The partitioning of carbon isotopes between reservoirs has become useful to measure organic productivity and the ocean circulation. For example, if ocean productivity is very high then a measurement of carbon isotopes in calcium carbonate shells of organisms will reflect the greater partitioning of ^{12}C into the organic carbon reservoir because it has been incorporated into plant tissues. As a second example, the organic matter is oxidised in the ocean and the carbon is then returned to the dissolved ocean reservoir. With age of the water (the time since it has left the surface), the oxygen becomes more and more depleted. Consequently, the rate of the ocean circulation, which defines the rate at which oxygen is supplied to the deep ocean, will also influence the composition of the carbon reservoir.

The above examples are only a brief sample of the potential of chemical methods to provide substantial palaeoclimatic information.

Reconstructions based on biological data

Two basic criteria must be fulfilled in order to utilise organisms as palaeoclimatic tools. First, organisms must live in equilibrium with their environment. In this case environmental change leads to a response in the organism: either adaptation, migration or mortality. Second, the principle of **uniformitarianism**, or that the present is the key to the past, must apply. In other words, the climatic limits or tolerances of living organisms must apply throughout the time of their existence. Beyond the range of modern species, this concept is usually modified and applied to the 'closest relative' or is extended to comparisons of morphology. In general, greater confidence is derived from examining groups of organisms rather than basing conclusions on single organisms.

Fossil plants are ideal palaeoclimatic indicators based on the close correspondence between modern plants and climate (particularly temperature and precipitation–evaporation). Certainly the floras of the present-day tropics and high latitudes are distinctive both in terms of species composition and morphological character. The geologic record of plants, consisting of fossil wood, leaves, fruits, seeds, pollen and spores, is a major source of data on continental climates. The examples are numerous. For the Quaternary, pollen assemblages can be matched statistically to modern distributions to reconstruct glacial and interglacial vegetation (Figure 24.3) and hence climate. In other cases, climatic inferences can be derived from specific genera or specific families, because they appear to be highly restricted to specific modern niches. For example, members of the family Palmae are not known outside the tropics today, and hence their distribution in the geologic record is used to map the distribution of warm climates in the past. Even in cases where exact climate parameters are not known, the latitudinal shift of floral assemblages is related to climatic change (e.g. a poleward shift through time is assumed to represent global warming).

Morphological characteristics of fossil floras are also used as climatic indices based on the fact that, at

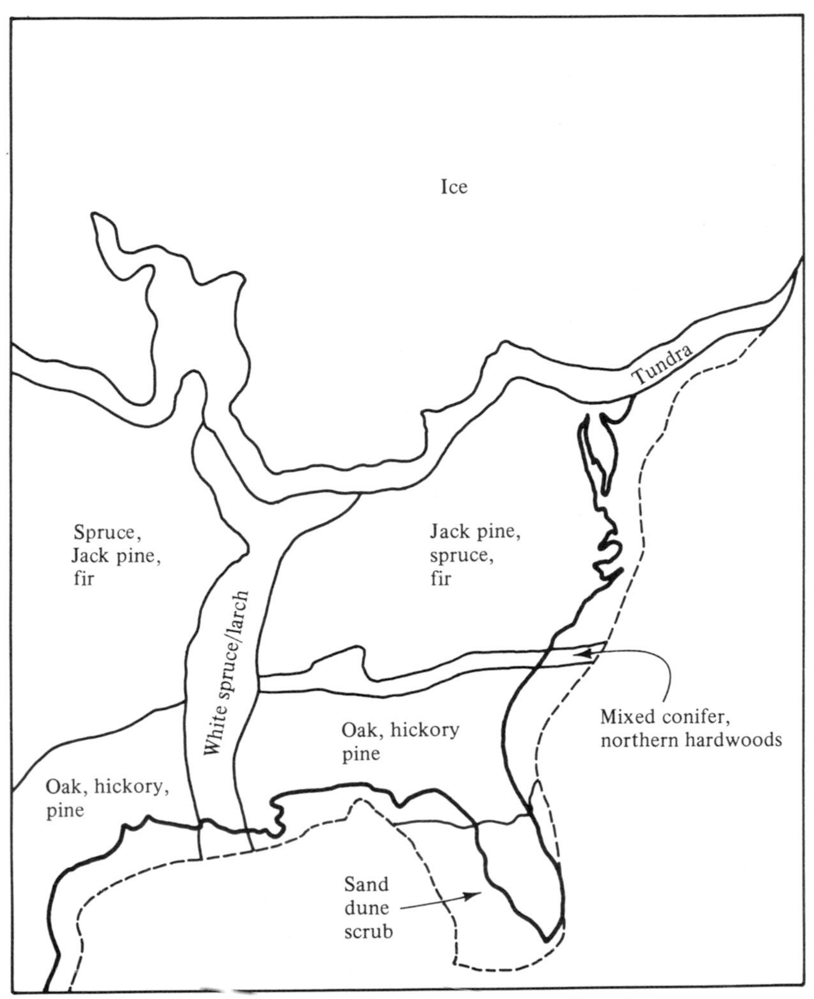

Figure 24.3 A reconstruction of vegetation patterns of the southeast part of North America for the last ice age (18 000 years ago) based on the distribution of pollen and pollen assemblages. After P. A. & M. R. Delcourt. Bold line – present day coastline, broken line – coastline 18 000 years ago.

present, geographically separated regions with similar climates often have floras which exhibit morphological similarities even if the taxa are different. For leaves, cuticle thickness, leaf size, leaf margin type, venation pattern and density, stomatal characteristics, organisation of leaves, growth habit, and leaf base type (petiole) may provide climatic data. In general, large leaves with entire margins, thick leather-like leaves, drip-tips, palmate venation, a heart-shaped petiole, vine habit and less dense venation are characteristic of warm, humid climates. Exceptions occur; for example, thick leaves may also be characteristic of cool arid climates. Small leaves with incised margins, needle leaves and pinnate venation are characteristic of cooler climates in general.

Pollen and spores, because of their morphological diversity and complexity and because of their resistance to degradation, are becoming major palaeoclimatic tools. The most direct use of pollen is by establishing the identity of the plant, although in some cases the morphology (surface characteristics designed for wind transport or reflecting an aquatic environment) may also be representative of environmental conditions.

Tree growth rings are also useful as palaeoclimatic indicators. Water availability, temperature and the length of the growing season are reflected in ring width, width of early wood, variations in cell size and variability of ring width from year to year. For example, ring width is associated with the availability of water during the growing season. Wide rings with little width variation imply that climate is not limiting to growth. Uninterrupted secondary wood implies that all the needs for growth are always present (uniform, seasonless climate).

Vertebrates are also useful to reconstruct terrestrial climates. Although new studies which relate structure or physiology to functional form are having an impact on ecological interpretations, the distribution of organisms which lack means of internal temperature regulation **(ectotherms)** provide the most frequent

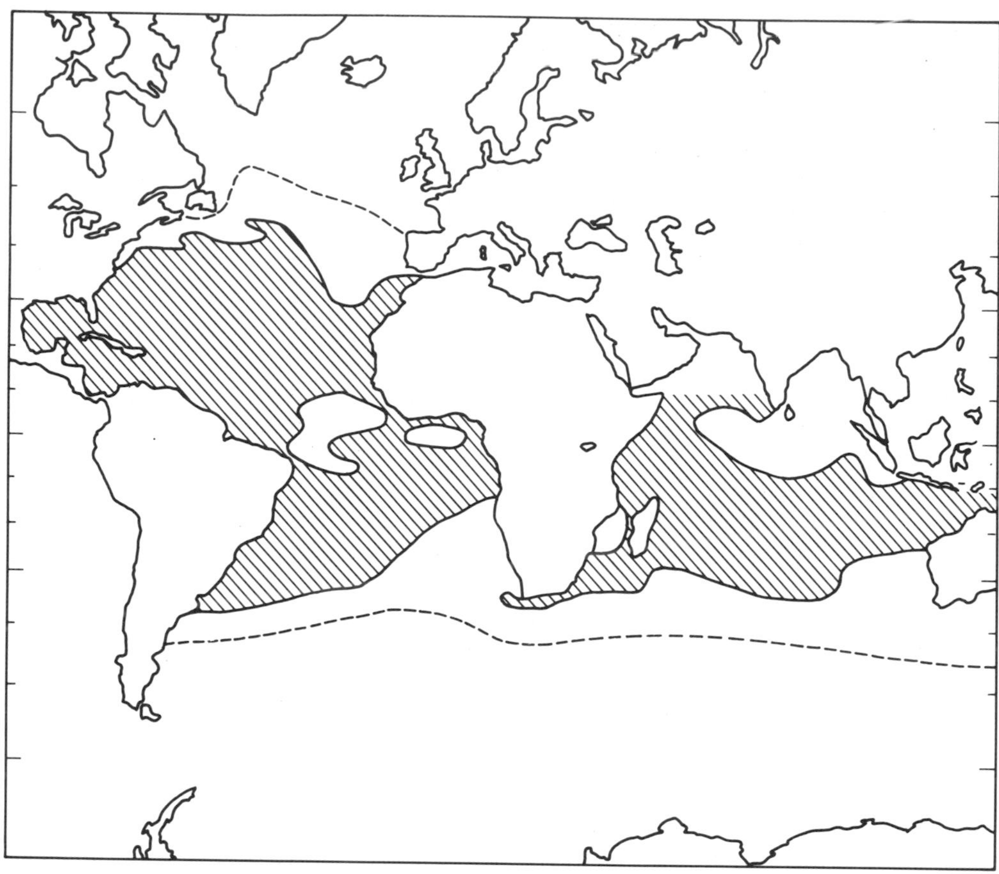

Figure 24.4 An example of the present-day distribution of one species of foraminifera (*Globigerinoides ruber*) often cited as an indicator of tropical climates. Shaded area > 20 per cent abundance; dashed line 0 per cent abundance contour.

basis of climate interpretation. For example, many large reptiles or amphibians are exclusively tropical. Crocodiles are not found poleward of a mean temperature of 15 °C today. Usually the key to accurate interpretation is size. The larger the ectotherm, the more likely that the organism is a prisoner of the climate, because it is unable to burrow or otherwise escape sub-freezing temperatures.

In the marine realm, invertebrates provide substantial palaeoenvironmental data. Perhaps the most frequently cited indication of warmth is the distribution of coral reefs and extensive carbonate buildups (see Chapter 17). In particular, hermatypic corals have narrow temperature tolerances, with an optimum range of 23–29 °C. Many modern gastropods and pelecypods have stratigraphic ranges which extend through a large part of the Cenozoic. In these cases, assemblages can be used to map climate zones. **Biogeography**, using taxa to define characteristic regions, is used to define climatic shifts or the absence of climatic zonation. Diversity of marine organisms has been related to latitude and climate, with corals, bivalves, and bryozoa illustrating substantially higher diversities in the tropics.

Many morphological characteristics of marine invertebrates also provide climatic data. Most truly large forms, such as the clam *Tridacna*, inhabit the tropics, but within a species the maximum size tends to occur in cooler climates. **Animal calcification** has been related to temperature. In many cases, cool-water pelecypods are thin-shelled and the calcification is less dense. In warmer waters, ornamentation such as spines and heavy calcification are more common. The record from the invertebrates is remarkably extensive. Growth lines of molluscs, brachiopods and corals yield data on seasonality and environmental stability. Stunting may provide evidence of environmental limits. Many of these relationships are the subject of current study.

The study of a wide variety of microscopic-sized organisms with preservable skeletons is the subject of

micropalaeontology. Such microscopic organisms are useful indicators of conditions in the marine realm. They include foraminifera, radiolaria, calcareous nannoplankton, diatoms, silicoflagellates, pteropods and ostracods. For example, **foraminifera** are unicellular or colonial unicellular protists which in most cases secrete a chambered calcium carbonate shell or test and which live as planktonic or benthonic organisms. Planktonic foraminifera assemblages have been closely tied to water mass distributions in the oceans. Particular species are often indicative of cold or warm waters (Figure 24.4). Foraminiferal morphology also provides substantial data. One species, *Neogloboquadrina pachyderma*, changes coiling direction as a function of temperature. Chamber size is correlated with optimum temperature conditions for growth. Even the growth history of the chambers of the tests provides evidence for stress or non-optimum growth conditions.

Many of the types of microfloras and microfaunas, of which foraminifera are only one example, are **good palaeoenvironmental indicators** because they are very sensitive to environmental conditions, they are abundant, they are easily identified and they are recorded in widespread environments.

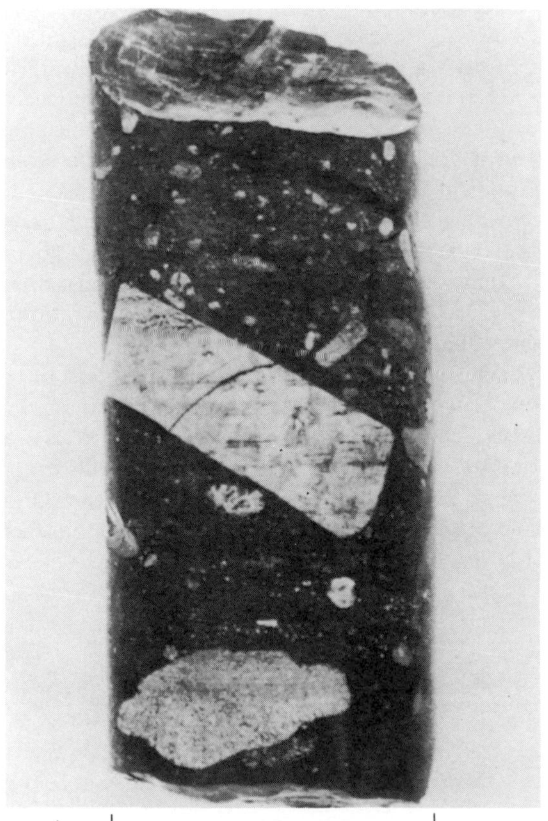

Figure 24.5 A 5 cm diameter core from a stone-rich bed, with poor sorting, characteristic of glacial deposition. □

Reconstruction based on physical data

The study of the sedimentary record provides environmental information on the presence and movement of ice and the nature of atmospheric and oceanic motions. Glaciers are a major element of erosion, sediment transport and deposition, leaving a physical record of their existence. Evidence for glacial movement ranges from minor scratches to excavated valleys. Abraded bedrock surfaces are good criteria for glaciation. The accumulation of distinctive stone-rich beds known as tills or tillites is characteristic of glacial deposition (Figure 24.5). A glacial origin is indicated by a lack of sorting, great range in grain size, preferred orientation of clasts, erratics, angular or flatiron shape of the stones, striated stones and extensive distribution. Caution must be applied in distinguishing between mass flow deposits and glacial deposits. Accumulations of till may also take distinctive form. Moraines, for example, are elongate mounds or ridges of glacial debris which form as lateral, medial or terminal deposits of a glacier.

Land ice also contributes to marine sediments where the terminus of a glacier or ice cap is at the edge of a continent. The distribution of glaciomarine sediments corresponds closely to the distribution of icebergs. **Dropstones**, outsized stones which disrupt finely laminated sediments, are characteristic of ice-rafted sedimentation. Great thickness and lateral extent of unsorted deposits in the marine realm are also associated with ice-rafted sedimentation.

Fluid motions in natural systems and the response of sedimentary particles to fluid flow yield sedimentary textures and structures which can be related to current directions and velocities throughout geologic time (see Chapter 16). Cross-bedding (the internal structure which results from the migration of ripples) and ripple marks provide evidence for current direction. Sedimentary gradients (e.g. particle size gradients), unique sources of sediment constituents and particle orientations (fossil orientations) can be interpreted as directional data. Particle size relationships are the major type of data used to interpret fluid flow velocities. Current systems can also be mapped by examining sediment dispersal patterns, shapes of sedimentary bodies, unconformities or erosional events.

Atmospheric wind direction and intensity have been reconstructed by examining the size of particles transported from desert regions and from volcanic eruptions. The arid and semi-arid regions of the Earth are major sources of atmospheric dust which is removed by settling or by rain-scavenging. Offshore, the dust size is in equilibrium with the wind intensity. The particle size, composition and mass flux of aeolian material accumulated in deep-sea sediments can be used to investigate the intensity of past atmospheric circulations, the source of atmospheric dust and the availability of sediment from the source region. On continents, **dune forms** and dune sedimentary structures are also characteristic of specific wind conditions, if an aeolian origin can be demonstrated. For example, unidirectional winds produce barchan, barchanoid ridge and transverse dunes. Barchan dunes have a crescentic shape with the steep slip face on the downwind side.

In summary, the 'signatures' recorded in the chemistry, biology and physics of sediments provide a host of opportunities to reconstruct climate in Earth history. Many of the tools provide only general information, warmer or cooler, wetter or drier, but in other cases the record is remarkably diagnostic.

The record of climate change in Earth history

On the time scale of the entire history of the Earth, the most evident aspect of the evolution of global climate is the record of extensive glaciation. The history of the Earth's climate consists of episodes of glaciation of variable duration separated by time periods for which evidence of glaciation is lacking or sparse (Figure 24.6). Along with other data, the record of glacial and non-glacial climates can be used to reconstruct a generalised global temperature history of the Earth. The glacial time periods are clearly reflected as periods of cold global climates separated by intervals of varying degrees of warmth (Figure 24.7). This record of climate change in Earth history can best be illustrated by examining an episode of climatic warmth such as during the Cretaceous, the transition between warm, non-glacial climates and glacial climates, and then examining the climate of a glacial time period in greater detail.

The earliest record for glaciation is for the middle Precambrian, near 2300 million years (Ma) ago, from North America, Africa and Australia. These middle Precambrian glacial deposits are poorly dated and could have been deposited over a period of several tens or several hundreds of million years. The second episode of extensive glaciation occurred between 950 and 615 Ma. This record of late Precambrian glaciation is found on every continent except Antarctica. The third episode of major glaciation occurred during the late Ordovician–early Silurian (the Cambrian episode shown on Figure 24.6 was not very extensive). The most beautifully exposed record is in northern Africa, but deposits are also found in South America, Europe, Asia and North America. A Devonian glaciation may have occurred, but is poorly documented. Widespread glaciation again is evident during a period of approximately 80 Ma in the late Palaeozoic (middle Carboniferous to late Permian). These glacial deposits occur over much of the area of the southern hemisphere continents, which once constituted the supercontinent of Gondwanaland. The last major episode of glaciation includes the current climate. The first evidence for ice-rafted detritus around Antarctica is from the Oligocene and extensive evidence for Antarctic glaciation is found for the Miocene. Northern hemisphere glaciation is documented from between 10 and 3 Ma.

The most extensive chemical, biological and physical records of past climate are available for the last 100 Ma. The last 100 Ma encompass the Cretaceous Period (65–140 million years ago) which is illustrated in Figure 24.7 as the warmest time period during the Phanerozoic, and includes the cooling trend from 60 Ma up to the current glacial time period.

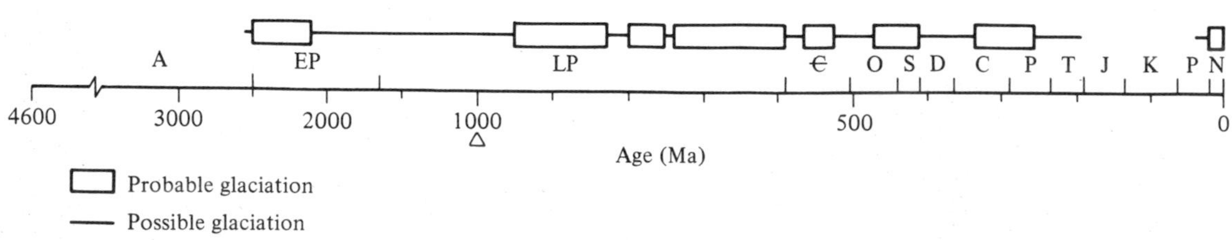

Figure 24.6 The record of the major episodes of glaciation defined as bars with times of possible record of some permanent ice indicated by a line. A – Archaean, EP – Early Precambrian, LP – Late Precambrian, C – Cambrian, O – Ordovician, S – Silurian, D – Devonian, C – Carboniferous, P – Permian, T – Triassic, J – Jurassic, K – Cretaceous, P – Palaeogene, N – Neogene. Δ – change of scale at 1000 Ma.

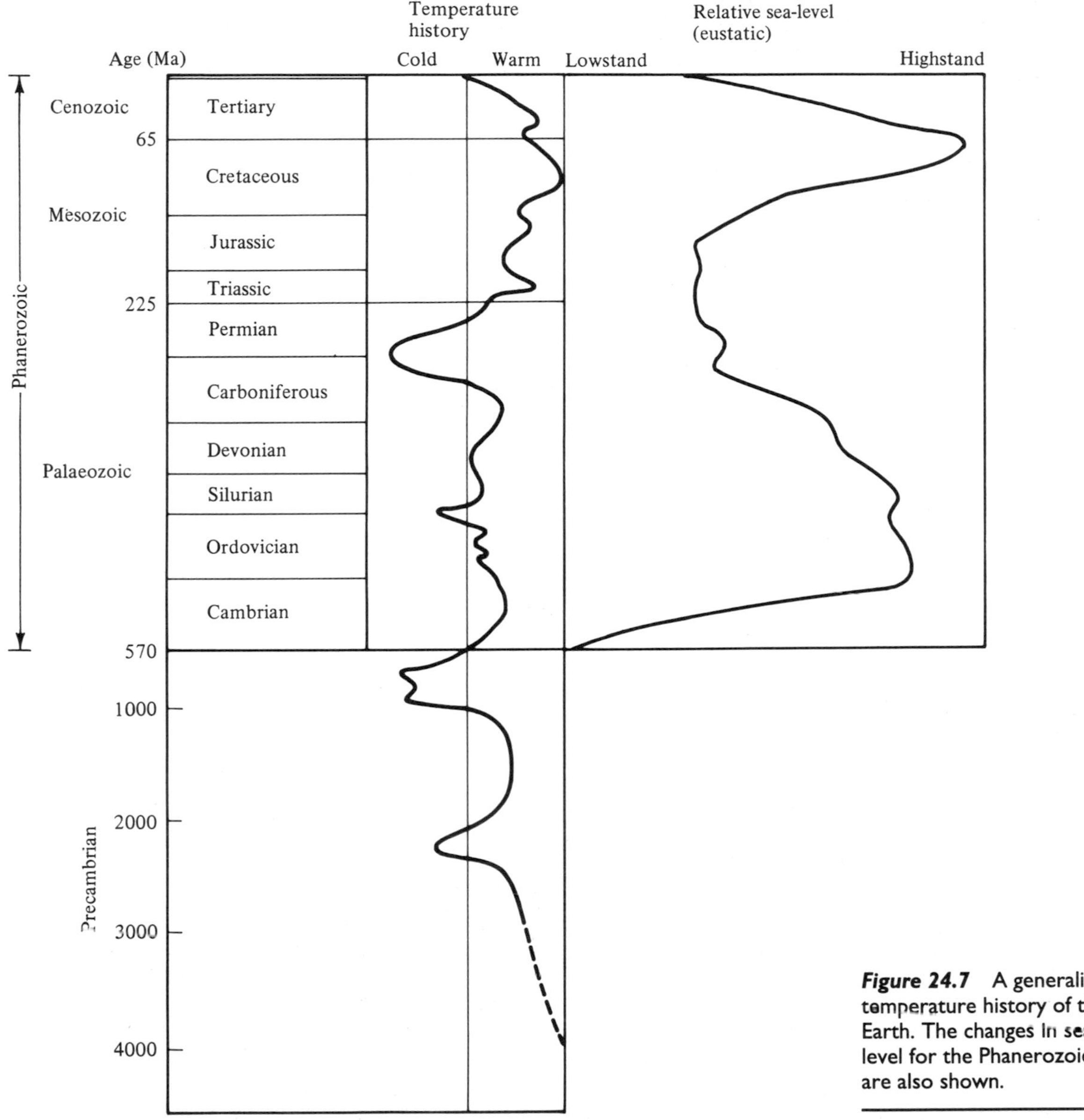

Figure 24.7 A generalised temperature history of the Earth. The changes in sea level for the Phanerozoic are also shown. □

Cretaceous warmth

The mid-Cretaceous, approximately 100 Ma, lacks any evidence for the existence of permanent ice. Warm polar temperatures are also evident from the lack of any cold-water invertebrate faunas and from the abundance of apparently warm climate floras at palaeolatitudes of 70° N and S. These data are limited only in the sense that a record from interior Antarctica, at high southern latitudes during the Cretaceous, is lacking due to the current ice cover. These data suggest a minimum polar temperature of 0 °C, but do not exclude the possibility of seasonally sub-freezing conditions. Estimates of even greater polar warmth are derived from measuring the oxygen isotopic composition of bottom waters from the Cretaceous ocean based on benthic foraminifera. Today, the ocean's deep water (~1–2 °C) is formed at high latitudes by the sinking of cold dense waters. Cretaceous deep-water temperatures are interpreted to be as warm as 17 °C. Therefore, mid-Cretaceous polar temperatures can be placed within the range of 0 to 17 °C in comparison with modern Arctic mean annual temperatures of – 15 °C.

The Cretaceous tropical temperatures may have been similar or slightly warmer than modern tropical

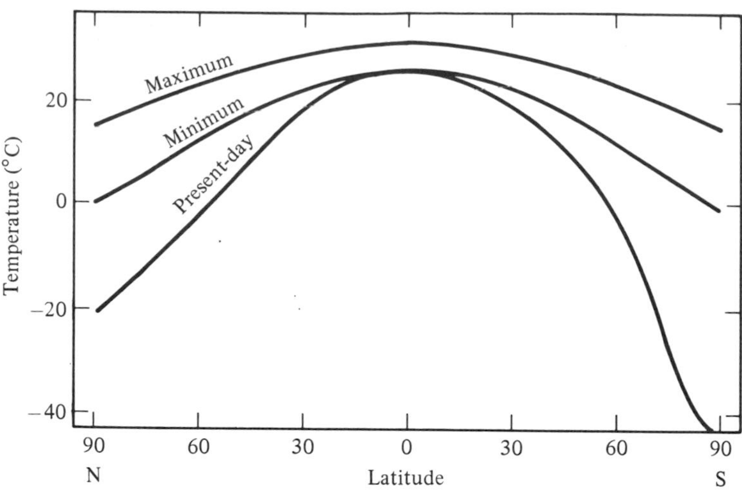

Figure 24.8 The temperature limits (maximum and minimum) with respect to latitude for the mid-Cretaceous (~ 100 Ma) based on the full spectrum of biological, chemical and physical observations. The limits are illustrated in comparison with the present day.

values. The oxygen isotopic temperatures measured from shallow-dwelling planktonic foraminifera yield temperatures of 25–27 °C. Modern foraminifera generally give results 3–5 °C cooler than the surface temperature, yielding a corrected mid-Cretaceous tropical temperature range of 27–32 °C. The warmth of the Cretaceous tropics may also be indicated by the extensive deposition on carbonate platforms and widespread occurrence of corals and other warm-water marine invertebrates.

Figure 24.8 summarises the range of temperatures with respect to latitude for the mid-Cretaceous in comparison with present-day mean annual values. On a globally averaged basis, the mid-**Cretaceous** is estimated to be 6 to 12 °C **warmer** than the present day and is probably the warmest time period in Earth history which can be well documented.

The Tertiary global cooling trend

The climatic record from 70 Ma to approximately 15 Ma is dominated by the transition between the warm Cretaceous Period and the widespread glaciation from the Miocene to the present day. The record of **Tertiary cooling** is in a sense gradual, as the decline in temperature occurred over a period of 55 Ma. However, the general trend is punctuated by a number of major events.

The most quantitative data set which describes the nature of the Tertiary cooling trend is the oxygen isotopic record from planktonic and benthonic

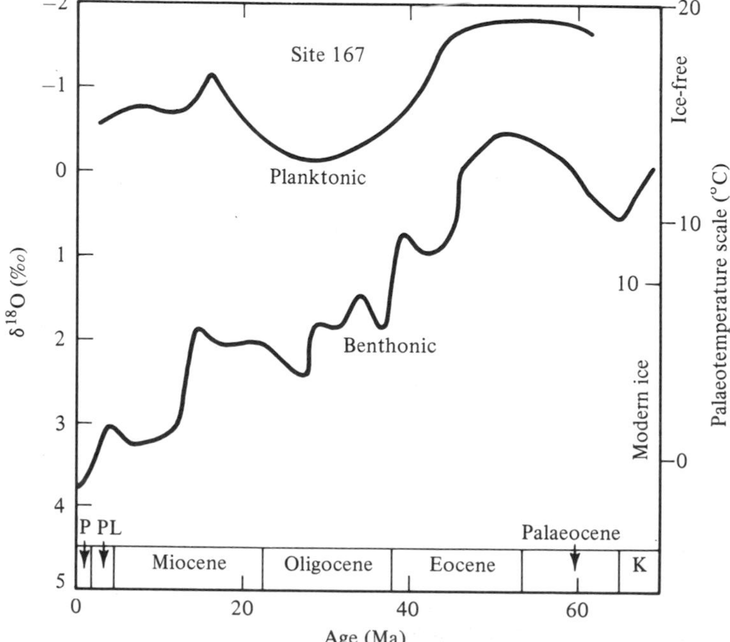

Figure 24.9 Bottom-water temperatures in the ocean, as determined from oxygen isotopes measured on benthonic (bottom-dwelling) foraminifera, and surface ocean temperatures, as determined from oxygen isotopes measured on planktonic foraminifera. The surface temperatures are for one tropical Pacific drilling site (Site 167 of the Deep Sea Drilling Project), whereas the bottom temperatures are from a global composite. In particular, the ocean deep-water temperatures illustrate a cooling trend over the last 60 Ma.

foraminifera (Figure 24.9). The record from benthonic foraminifera, usually interpreted as a measure of polar climates, is particularly informative. The isotopic palaeotemperatures illustrate a gradual decline in bottom-water temperatures from approximately 14 °C in the Paleocene–Eocene to the present-day values of 1–2 °C. The record is also characterised by major events. A variety of data is utilised to separate events related to ice-sheet growth from polar temperature changes. For example, the first significant ice-rafted deposition adjacent to Antarctica is from the Oligocene. Erosion of deep-sea sediments and changes in biogeography mark the first extensive formation of Antarctic bottom water during the major isotopic event illustrated during the Miocene.

The evidence for a global cooling from terrestrial data is equally convincing. Palaeocene beds from Ellesmere Island, within the Arctic Circle of this time, yield fossil alligators, large tortoises and relatives of the modern palms. During the Eocene–Oligocene many of the floras and faunas indicative of warm, humid conditions became restricted to the tropics and subtropics. The progressive tropical displacement of continental indicators is matched by a tropical displacement of molluscan provinces. For example, during the Eocene, tropical molluscan faunas extended to as high as 45° N palaeolatitude. This episode of a broad tropical belt was followed by periods of tropical contraction and expansion but with a marked overall contraction (cooling).

The glaciation of portions of Antarctica during the Oligocene, extensive glaciation of Antarctica during the Miocene, and northern hemisphere glaciation from the middle Miocene and the Pliocene introduce a climate characterised by glacial rhythms.

Pleistocene glacial–interglacial cycles

The continuous record of sedimentation recovered in deep-sea cores and oxygen isotopic geochemistry of foraminifera provide the opportunity to examine the record of **glacial–interglacial cycles** during the last million years in detail. The oxygen isotopic record represents both temperature variation and changes in ice volume. From a combination of biogeographic data, evidence for the isotopic composition of Greenland ice, and from comparison of benthonic and planktonic records (both should illustrate ice volume effects, but plankton should experience greater temperature variation), the oxygen isotopic record for the last million years has been interpreted as largely an indication of variations in ice volume.

The oxygen isotopic record of ice volume (Figure 24.2) illustrates a number of important characteristics of the glacial cycles. Ice ages are highly rhythmic and appear to be periodic. The amplitude of the ice ages and interglacials are roughly similar. The curves indicate a ramp-like glacial–interglacial history, with a relatively slow buildup to a full glacial episode, but rapid de-glaciation. Approximately seven major glacial episodes have occurred during the last 700 000 years. The last major ice age was approximately 18 000 years ago.

The Last Glacial Maximum

The glacial climate of 18 000 years ago can be reconstructed in detail. Most of the faunas and floras have relatives living today. More than 30 000 piston cores from all over the world ocean have penetrated sediments from the **Last Glacial Maximum** (LGM) giving a comprehensive record for isotopes and micropalaeontology. Importantly, the stratigraphic resolution far exceeds other time periods. The oxygen isotope curve of variations in ice volume and carbon-14 dating provide resolution to within a few thousand years. The data from the last ice age provide the first comprehensive and quantitative reconstruction of a past climate.

Figure 24.10 illustrates the **CLIMAP** (Climate, Long-range Investigation, Mapping and Prediction Project) reconstruction of the Atlantic Ocean surface temperature during the LGM in comparison with the present day. The CLIMAP reconstructions are the first realistic time-slice reconstruction in palaeoclimatology. The areal extent of continental ice was mapped based on dated tills and end moraines. The continental vegetation is reconstructed from the distribution of pollen and macrofloral evidence. Sea-level position is calculated from knowledge of the size of the ice caps. The seasonal extremes of sea surface temperatures are reconstructed statistically using transfer functions for the distribution of assemblages of foraminifera, radiolaria, coccolithophoroids (calcareous nannofossils) and diatoms. Each species contributes to a mathematical description of assemblage composition which is then mapped and related to physical measures of different ocean parameters (temperature, seasonal range etc.). Sea-ice distributions are also mapped, but based on the absence of subpolar surface water faunas and floras.

The results indicate a number of differences from the modern climate. The LGM was characterised by large land-based ice caps as much as 3 km in thickness and by large increases in the area of winter sea-ice. Polar water expanded to nearly 40° latitude in both hemispheres, equatorial waters cooled by approximately

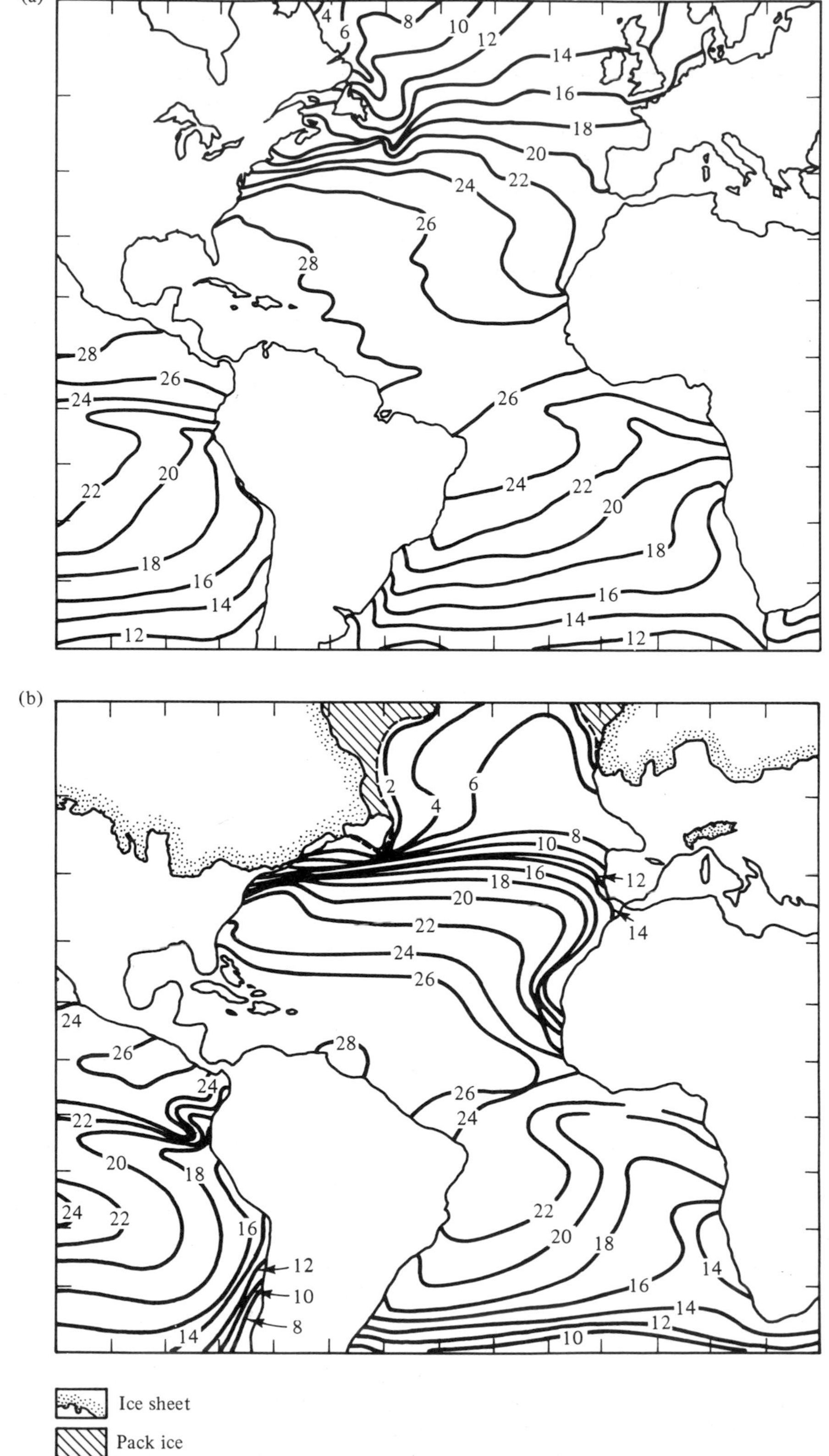

Figure 24.10 A comparison of North Atlantic sea surface temperatures (°C) in August for the present (a) and the Last Glacial Maximum 18 000 years ago (b) as measured by CLIMAP.

2 °C and there was increased penetration of cool eastern boundary currents towards the tropics. In contrast, the subtropics of the LGM were very similar to the present-day conditions. In general, thermal gradients become stronger with the expansion of polar climates. The globally averaged surface temperature was likely to have been 3–5 °C cooler than the present day.

Summary of climatic changes over the past 100 Ma

The climates of the Cretaceous, the Tertiary and the Pleistocene are interesting contrasts and provide insights into the spectrum of climate represented in Figure 24.6 of the record of major episodes of glacial and non-glacial climates throughout Earth history. Still, these time periods are only examples. Insufficient data are available to suggest that all the time periods of non-glacial or glacial climates are similar. Certainly, many data indicate that the climates of the geologic record are highly variable. Considering the large number of factors which could influence climates of the past, it is much easier to argue that palaeoclimatologists have only begun to sample the rich and diverse record of climate change in Earth history.

Causes of global climate change in Earth history

The primary problem is to identify the principal causes of climatic change which occur over the spectrum of time scales represented in Earth history. The focus of research involves changes in 'external' factors, so-called **climatic forcing factors**. In the most general terms, the climatic forcing factors fall within three categories: (1) the amount and distribution of solar radiation received at the top of the atmosphere, (2) the composition of the atmosphere and its effect on the Earth's radiation budget, and (3) the nature of the surface of the Earth. Numerous terrestrial and extra-terrestrial factors have been proposed to explain the climatic record of the Earth. Orography, (the distribution of mountain ranges), sea level, continental positions, atmospheric carbon dioxide, volcanism, orbital variations, solar luminosity and galactic forces are a few examples of factors proposed as primary causes of climatic change.

These factors may have jointly or independently influenced palaeoclimates. In most cases, for example in the cause of glacial and non-glacial climates, it is unclear whether a single cause explains the occurrence of all the glacial episodes or whether each time period of Earth glaciation resulted from a different set of factors. For most of Earth history, the climatic forcing factors are only poorly known and the specification of all of the factors which may have been important during a specific interval of time is not possible. However, the importance of a number of climatic forcing factors has been demonstrated. The climatic change examples described above are an ideal framework to illustrate some of the most significant causes of climate change in Earth history. However, the list and discussion presented below is far from comprehensive.

Causes of glacial and non-glacial climate states

Since the formulation of the hypotheses of continental drift and plate tectonics, a changing geography has become the most frequently cited explanation of the large climatic contrast between glacial climates, such as the present day, and warm, equable climates, such as the Cretaceous. In addition, hypotheses involving variations in atmospheric carbon dioxide have received considerable support from geochemistry. A combination of changes in geography and atmospheric carbon dioxide, both related to plate tectonic processes, are the leading candidates to explain the long time-scale variations in the Earth's climate.

Changes in geography are a logical explanation for climate change because the differences in geography between time periods are large. They occur over tens of millions of years, and include the extent of flooding of continents and the latitudinal distribution of continents. A number of important physical processes are influenced by land–sea distribution. The mid-Cretaceous geography is illustrated in Figure 24.11. Approximately 17 per cent of the present-day land area was covered by shallow seas during the mid-Cretaceous due to a global rise in sea level caused by an increase in the volume of ocean ridge systems (see Chapters 9 and 20). The distribution of land and sea is substantially different and the high-latitude regions are characterised by less land area in comparison with the present day.

The degree of high-latitude continentality is of importance because high-latitude land is a surface for the accumulation of high albedo (highly reflective) snow and ice. Further, high-latitude land may block the poleward penetration of ocean currents and therefore limit the poleward transport of heat. Land and ocean configuration, through the large differences in thermal inertia between the ocean and continents, also affects the amplitude of the seasonal cycle of temperatures. In particular, land–sea configurations which promote reduced summer temperatures on high-latitude continents may allow the maintenance of year-round snow fields and therefore may promote glaciation. High topography may also promote the accumulation of permanent snow.

Continental flooding is also well correlated with estimates of global warmth as shown in Figure 24.7. Ocean and land have different albedos and different thermal properties. In the absence of other factors (e.g. compensating cloud–climate feedbacks) a more oceanic Earth should absorb more incoming solar radiation. The time periods of extensive flooding, such as the mid-Cretaceous, are hypothesised to be caused by periods of more rapid sea-floor spreading and related changes in the volume of the ocean basins.

The **budget** of **atmospheric carbon dioxide**, an effective greenhouse gas, is dependent on the rate of plate tectonic movements and on changes in continental weathering rates. On the time scale of tens of millions of years, atmospheric carbon dioxide may be regarded as a balance between the rate of atmospheric input by volcanoes and the rate of removal governed by the rate of weathering of exposed silicate rocks. First, we may infer that the rate of volcanism will be a function of the rate of sea-floor spreading. Therefore, periods of high sea level should also be time periods of greater carbon dioxide injection rates. The removal of carbon dioxide involves the following general chemical weathering equation for silicate rocks:

$$CaSiO_3 + CO_2 \underset{\text{metamorphism}}{\overset{\text{weathering}}{\rightleftarrows}} SiO_2 + CaCO_3.$$

The rate of carbon dioxide removal by weathering is dependent on climate (temperature and precipitation) and area of exposed silicate rocks. Weathering rates increase for higher temperatures and precipitation. On average the area of silicate rocks should decline when sea level is high and portions of the continents are flooded. Given these basic dependencies, carbon dioxide levels which are four to ten times the present-day levels are plausible estimates for the mid-Cretaceous. Figure 24.12 portrays a pioneering calculation of carbon dioxide levels by A. Lasaga, R. Berner and R. Garrels for the last 100 Ma.

Clearly, both carbon dioxide variations over tens of millions of years and changes in land–sea distribution are a function of plate tectonic processes. Plate tectonics is considered to be a major control on the long-term climate evolution of the Earth. The importance of carbon dioxide and geography in explaining glacial and non-glacial climates is investigated in greater detail in the section on the physical basis of climate change.

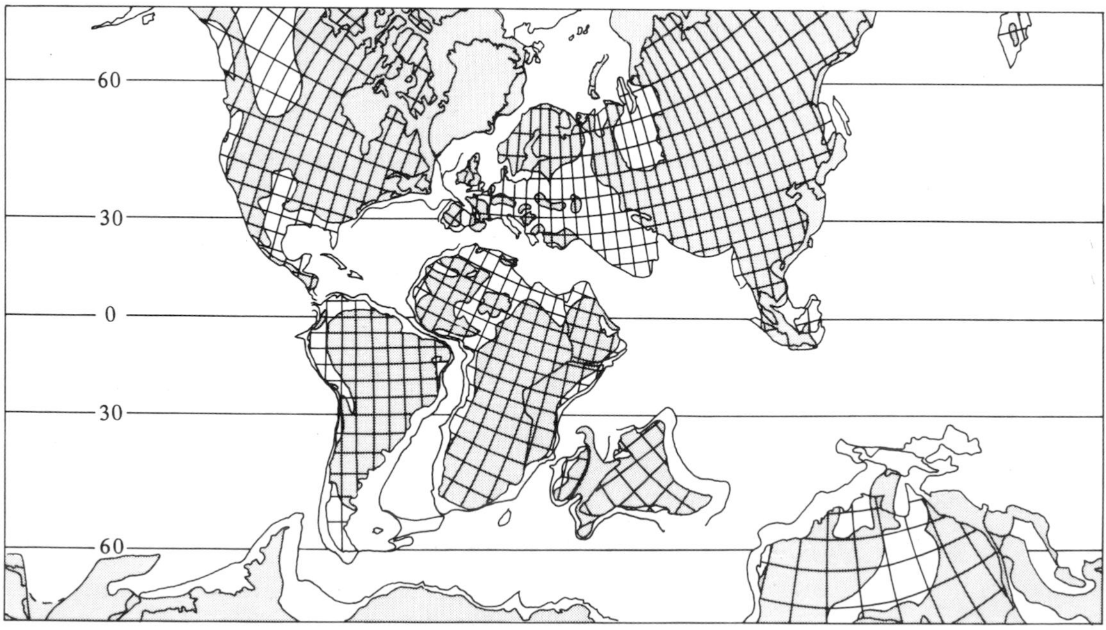

Figure 24.11 A reconstruction of the position of the continents and the area of continents above sea level for the mid-Cretaceous, approximately 100 Ma. The area above sea level is shaded. □

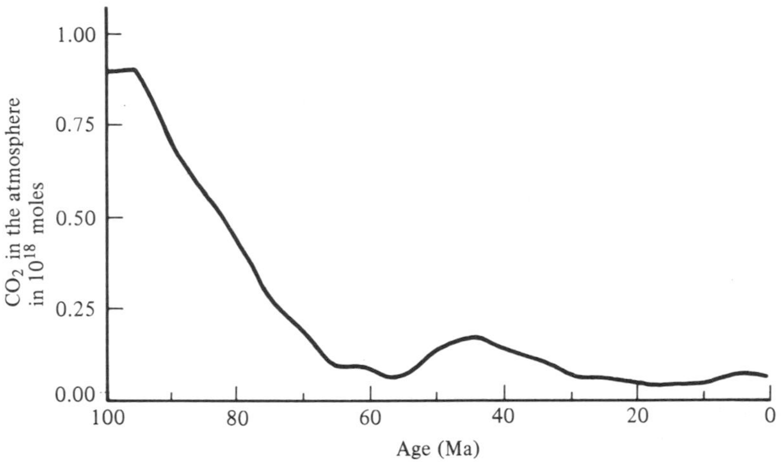

Figure 24.12 An estimate of carbon dioxide concentration in the atmosphere over the last 100 Ma based on a model of the geochemical cycles involving carbon by A. Lasaga, R. Berner and R. Garrels. □

The cause of the Tertiary global cooling trend

Following the discussion above on plate tectonics and climate change, we might deduce that the overall cooling trend during the Tertiary may reflect changes in continental palaeopositions, the decline in sea level and an overall change in atmospheric carbon dioxide levels. However, the events of the early Eocene, late Eocene, the Oligocene and the middle Miocene are a dominant part of the cooling trend. In fact, one might argue that the cooling trend can be explained entirely as a set of steps which in total yield a 50 Ma cooling. The apparent short time scale for the events illustrated in Figure 24.9 argues for different mechanisms of climate change, most notably changes in ocean circulation. In most cases these changes are initiated in response to ocean barriers or gateways or thresholds involved with the otherwise slow process of plate tectonics.

Ocean gateways or **barriers** either open or block the communication between oceans, modulating current patterns and influencing the transport of heat by the oceans (Figure 24.13). For example, the opening of the Drake Passage by mid- to late Oligocene time is

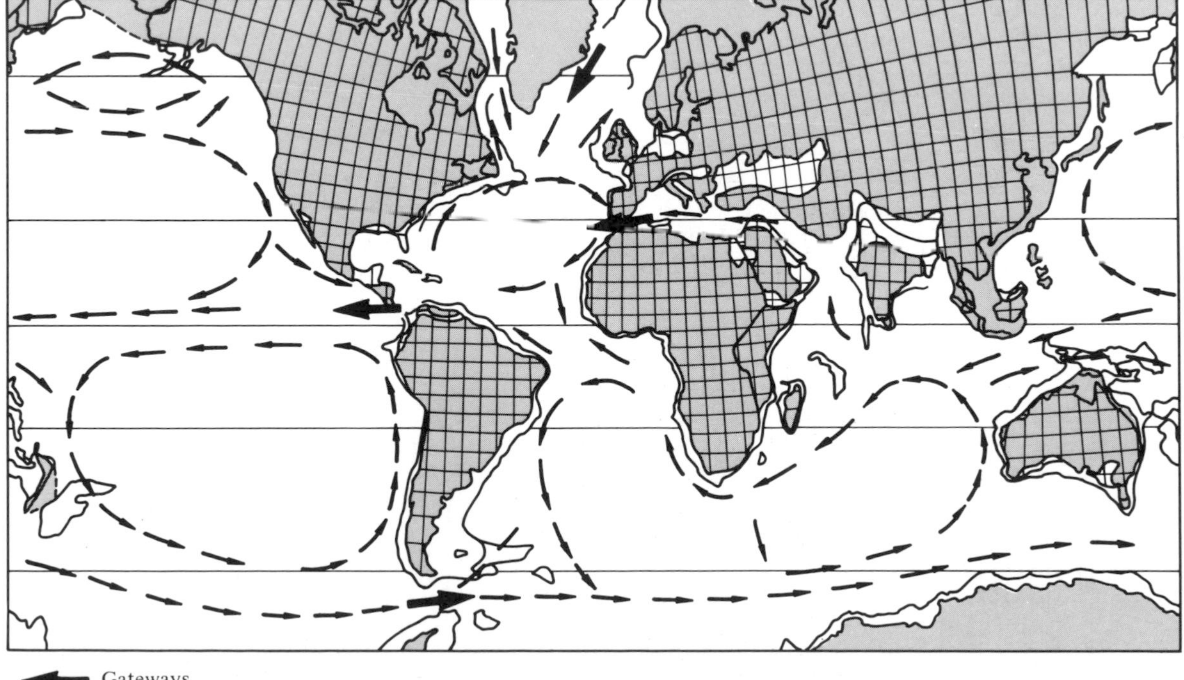

Figure 24.13 An illustration of hypothesised circulation of the ocean for the Miocene illustrating (large arrows) critical geographic gateways that may have controlled ocean circulation and climate. □

correlated with the formation of the circum-Antarctic current. This event is linked with the extension of Antarctic glaciers to the continental margin, as shown by the occurrence of ice-rafted detritus. The closure of a tropical seaway linking the Indian–Atlantic and Pacific Ocean (Tethys) and the subsidence of the Iceland–Faeroe Ridge (linking the North Atlantic Ocean to the Arctic) have been correlated with the mid-Miocene development of extensive Arctic glaciation, the development of extensive cold deep water production and the initiation of the modern ocean circulation.

The importance of ocean gateways is based primarily on time correlations between 'effects' and notable 'causes'. In general, these ideas stem from stratigraphic associations of numerous physical, biological and chemical observations in comparison with the timing of events in the tectonic evolution of ocean basins.

The cause of glacial–interglacial cycles

A wide variety of theories has been presented to explain glacial cycles, including orbital variability, volcanism, magnetic field variations, solar variability, interstellar dust, internal climatic oscillations due to the non-linearity of the climate system, Antarctic ice surges, atmospheric carbon dioxide and deep-ocean circulation. In the first half of the 20th century a Yugoslav astronomer, Milutin Milankovitch, presented a theory

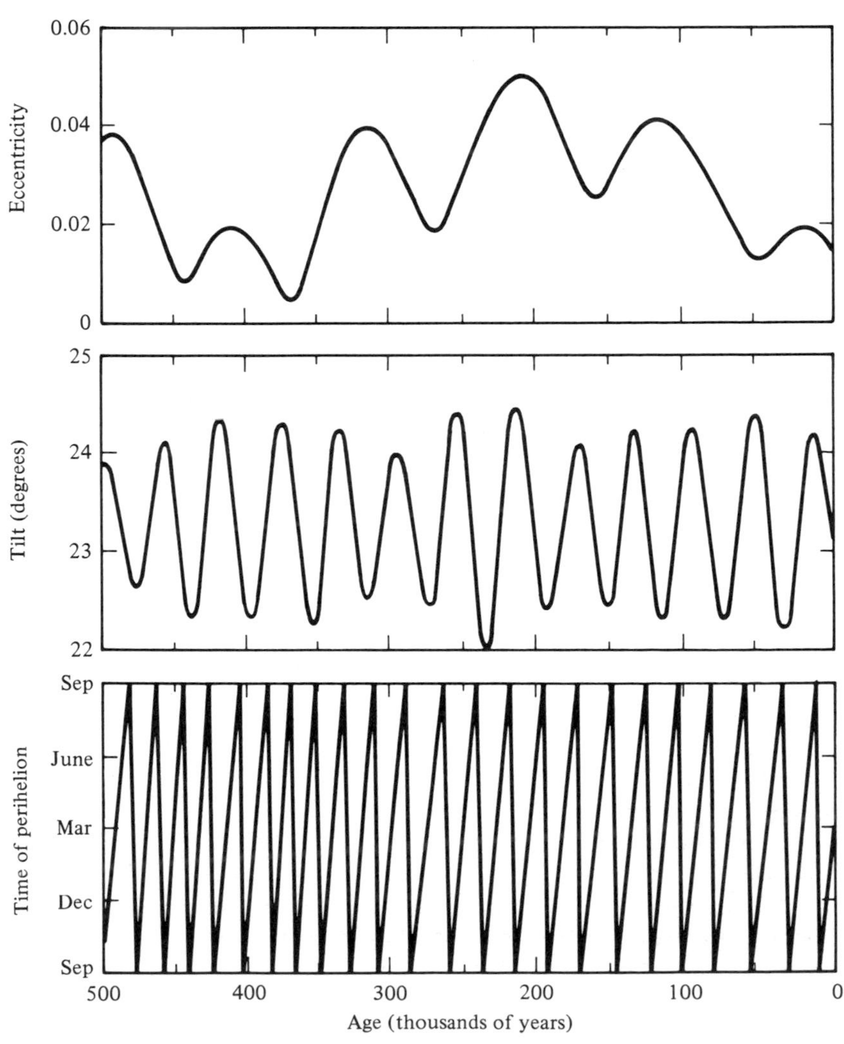

Figure 24.14 The variations in the Earth's orbital elements (orbital eccentricity, tilt of its axis and time of perihelion).

based on changes in the Earth's orbital characteristics. The Earth's orbit through time is defined by the precession of the equinoxes (the time when the Earth is nearest the Sun (perihelion) changes from January to July on a period of approximately 24 000 years), the tilt of the Earth's rotation axis (currently 23.4° but varying from 22 to 24.5° on a time scale of 41 000 years), and changes in the eccentricity of the orbit of the Earth around the Sun (period of near 100 000 years).

The variations in the Earth's orbital elements (Figure 24.14) are used to compute variations in the seasonal and latitudinal distribution of incoming solar energy. For example, a larger tilt, a more eccentric orbit and perihelion during northern hemisphere summer would produce a large amplitude of the seasonal cycle of insolation, especially at high latitudes. The resultant predicted climatic variations have become known as **Milankovitch cycles**.

Until relatively recently, Milankovitch orbital variation was dismissed as a mechanism because there is very little change in the annual global-scale solar input. The effects are dominated by changes in the seasonal distribution of solar insolation. However, the record from deep-sea exploration modified this conclusion.

The record of glacial–interglacial cycles over the last 700 000 years indicates seven well-defined glaciations which appear to be quasi-periodic (Figure 24.2). The glacial record suggests a climate response to a continual periodic forcing. This perspective revitalised the Milankovitch hypothesis. The continuous record in deep-sea cores and the excellent stratigraphic framework for the last 700 000 years lends itself to spectral analysis as a means to determine the climatic periodicities associated with glacial–interglacial rhythms. Three discrete periods characterise this record: 19 000 to 24 000 years, 43 000 years and 100 000 years. The 100 000-year period is the dominant spectral peak.

The close correspondence between the periods of orbital variations and the record in deep-sea cores has demonstrated a close causal relationship. The acceptance of Milankovitch orbital cycles as the fundamental cause of glacial cycles also provides unique insight into the climate system. In this case, glacial cycles and our ability to map glacial climates provides a 'known' response of the climate system to a 'known' and well-specified forcing factor. This provides a great opportunity to examine climatic sensitivity and climatic change. Each of the forcing factors described above provides insight into how the climate system operates and into the physical basis of climatic change in Earth history.

An understanding of the physical processes which explain past climates

The third major problem in palaeoclimatology is to demonstrate climate sensitivity to a change in a specific forcing factor, such as land–sea distribution or variations in the Earth's orbit, and to determine the climatic response to external factors. To some extent the climatic response can be demonstrated by correlation of cause and effect, but in most cases this generates only a qualitative understanding of how the climate system works. In many other cases this qualitative approach can be misleading because of the difficulty in extrapolating a larger, global view of climate from a limited set of data and because of the difficulty in describing the interactions of a complex, non-linear system.

The primary task of **climate modelling** is to replace the complex natural system by a hierarchy of simplified ones which can then be used as quantitative tools to investigate climate sensitivity and to examine the climate response to external factors. The goal is to gain physical insight into the processes which explain past climates.

Climate models also introduce a new set of uncertainties in palaeoclimatology. Every detail of the atmosphere, oceans and ice cannot be incorporated in models. The task of simplification introduces uncertainty. A full understanding of climate sensitivity implies that all of the important physical processes and feedbacks have been accurately incorporated in a model's mathematical representation of the climate system.

Given the uncertainties, the application of climate models primarily has focussed on 'sensitivity experiments' rather than an attempt to simulate or reproduce a past climate. At this point in time, only the glacial maximum at 18 000 years ago is sufficiently well documented to make a climate simulation a reasonable goal. The purpose of sensitivity experiments is two-fold. The first goal is to isolate cause and effect in the climate system by determining the model response to a change in a single factor. The second goal is to determine those processes which require further investigations, in other words to guide future investigations.

The investigation of ancient climates with mathematical models has focussed on (1) the cause of glacial versus non-glacial climates with particular emphasis on explanations of Cretaceous warmth, the role of geography in climate change and on the importance of

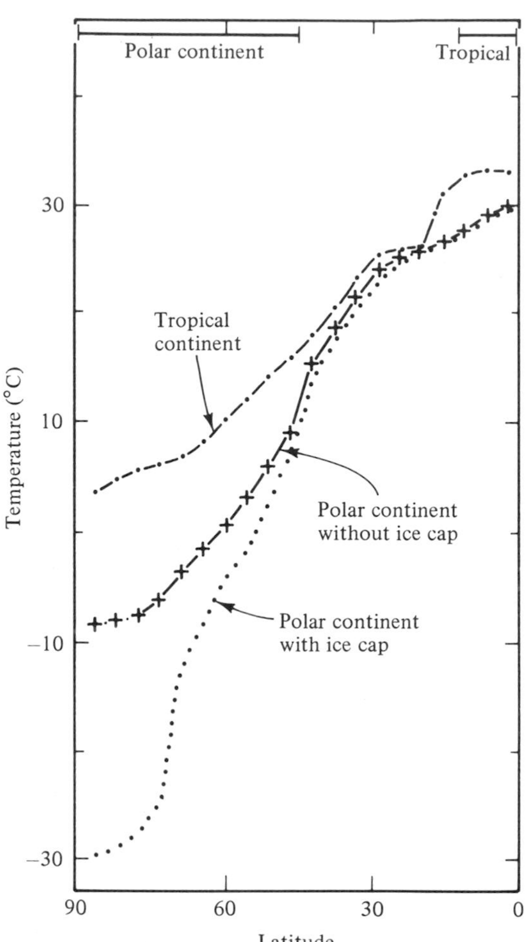

Figure 24.15 **An estimate of the range in equator-to-pole surface temperatures simulated with a climate model for extremes in the positions of the continents.** □

variations in atmospheric carbon-dioxide, (2) the climatic processes which link orbital variations with glacial–interglacial rhythms and (3) attempts to simulate the climate of the Last Glacial Maximum.

The role of palaeogeography in explanations of Cretaceous warmth and glacial versus non-glacial climates

Two types of sensitivity experiments have been performed to evaluate the role of large-scale changes in geography: (1) investigation of idealised extremes of continental configurations and (2) investigation of realistic geographic reconstructions to compare model sensitivity to the actual record of climatic change.

Extreme continental configurations have been investigated utilising a hierarchy of climate models from simple vertically and horizontally averaged energy balance models to fully resolved three-dimensional models of the climate system (**General Circulation Models, GCM**). The most interesting experiments focus on simulations with all the land in the tropics or all the land at the poles, and test the latitudinal distribution of the continents as a forcing factor for global change. In a simple energy balance model which incorporated only the differences in albedo of land and ocean and which included polar ice, the polar continent case had a globally averaged surface temperature 12 °C cooler than the tropical continent case. In the same experiments utilising a GCM, the polar continent case with a specified polar ice cap was also substantially colder (7.4 °C). The difference in polar temperatures was 34 °C. Importantly, the GCM studies further demonstrated that much of the magnitude of the polar cooling depended on the extent of polar ice. The polar cooling by geography alone was only 12 °C and the polar ice cap accounted for 40 per cent of the difference in globally averaged surface temperatures between the two cases (Figure 24.15).

The most important point of the above experiments is that palaeogeography has substantial potential to influence global warmth. Ignoring the actual development of polar ice caps in response to a changing geography, land–sea distribution may reasonably be capable of achieving as much as 4–5 °C differences in globally averaged surface temperatures. In comparison, the difference between the present and the warm, equable Cretaceous is near 6–12 °C and the difference between the present and the Last Glacial Maximum is near 3–5 °C.

A hierarchy of experiments have also been completed to examine the role of Cretaceous geography specifically. Both the area and distribution of Cretaceous continents presented in Figure 24.11 are substantially different from the present day. The most state-of-the-art climate model simulations using realistic three-dimensional geographic reconstructions yield a 4.8 °C increase in globally averaged surface temperature for Cretaceous geography using a mean annual version of a GCM with a simple energy balance ocean (Figure 24.16). A series of step-wise experiments were designed to isolate which geographic factors were responsible for the 4.8 °C global warming. Almost the entire southern hemisphere temperature change resulted from removal of the Antarctic ice cap, suggesting that large-scale continental positions had little to do with the initiation of Antarctic ice cap growth. The model showed little sensitivity to continental flooding, largely because the differences in surface albedo were

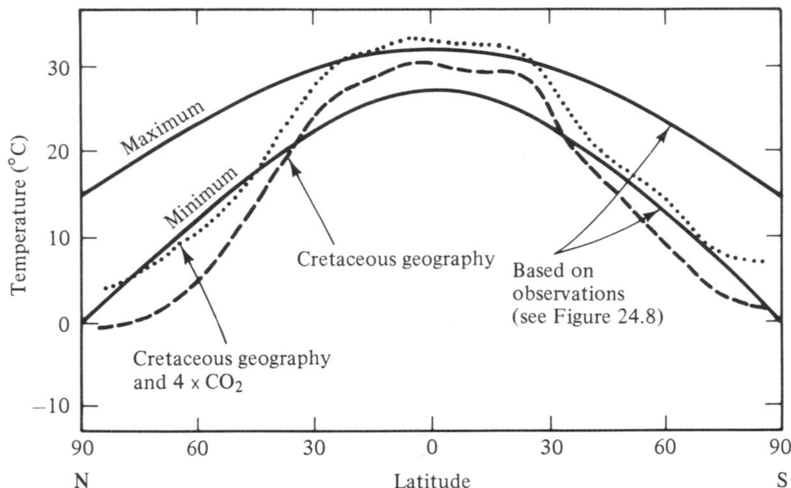

Figure 24.16 An estimate of the equator-to-pole surface temperatures simulated with a climate model for realistic mid-Cretaceous geography (Figure 24.11) with modern carbon dioxide concentrations in the atmosphere and four times modern concentrations. The results are given with respect to the limits of temperature derived from geologic observations (Figure 24.8). □

compensated for by increased cloud cover over the flooded areas of the continents. In fact, almost the entire northern hemisphere warming was found to be in response to the change in continental area above 45° N.

The results of these experiments are fascinating because they demonstrate the potentially important role of geography in climate change, yet at this stage in model development geography is insufficient to explain Cretaceous warmth and apparently insufficient to explain the initiation of the ice caps. This last point is a matter of debate. Models which incorporate the seasonal cycle of solar insolation demonstrate the role of the differences in the heat capacity between oceans and land in modulating the amplitude of the seasonal cycle. Some researchers have suggested that ocean area near polar land masses may cause lower summer temperatures which allow winter snowfall to remain through the summer. Large-scale changes in geography would then remain as a plausible mechanism for initiating ice cap growth.

The role of carbon dioxide in explanations of Cretaceous warmth

Three aspects of the previous model simulations suggest that carbon dioxide should be considered as an alternative or an accomplice in explaining Cretaceous warmth: (1) the model's inability to achieve sufficient warmth due to changes in geography to satisfy palaeoclimatic data, (2) the lack of model sensitivity to sea level despite the strong correlation between global warmth and continental flooding, and (3) the results of models of geochemical cycles which suggest that four to ten times present concentrations of carbon dioxide are plausible for the Cretaceous atmosphere.

The Cretaceous was characterised by rapid sea-floor spreading, with sea floor created nearly twice as fast as in the recent past. Increased sea-floor spreading has two effects: (1) young sea floor has a greater volume than older sea floor and so ocean volume is decreased causing continental flooding and, (2) rapid spreading results in an increase in volcanism. Hence sea level should also be correlated with atmospheric carbon dioxide levels.

To date, only one climate model experiment has included higher carbon dioxide levels for a past geography. One GCM experiment has been completed with Cretaceous geography and four times present-day carbon dioxide. Interestingly, both the globally averaged surface temperature (8 °C higher than the present day) and the distribution of temperatures with respect to latitude now fall within the limits of the Cretaceous palaeoclimatic data (Figure 24.16). Problems exist in that tropical temperatures appear to be at the upper limit of the data while polar temperatures are at the very lower limit of the data. Most importantly, in concert, changes in geography and in carbon dioxide must be considered as plausible explanations of the large differences between glacial and non-glacial climates in Earth history. Much work is required to refine this hypothesis and it may prove that carbon dioxide variations and palaeogeography do not explain the full spectrum of glacial and non-glacial climates.

Orbital elements and glacial cycles

The close correspondence of the **Earth's orbital elements**, i.e. the periods associated with orbital eccentricity, obliquity and precession with the record of ice volume recorded by oxygen isotopes provides a unique opportunity in palaeoclimatology: a known cause and

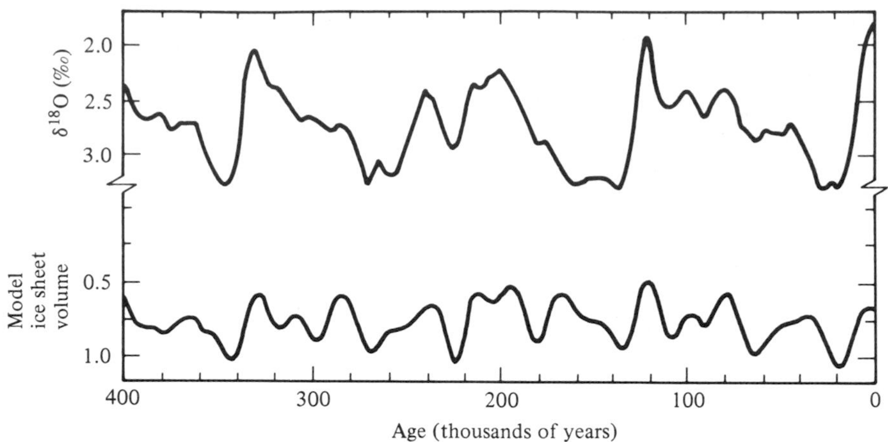

Figure 24.17 A comparison of the oxygen isotopic record of glacial cycles over the last 400 000 years with the ice volume generated by a climate model with an ice sheet. Note the close correspondence of the second-order peaks, but the lack of the 100 000 year signal in the ice-sheet model.

a well-defined climatic response. In this case, each failure or success of any model to reproduce the glacial cycles adds new insight into the physical processes required to explain past climates.

The first models which attempted to simulate glacial cycles in response to orbital variations were simple equator-to-pole models which lacked either east–west resolution or vertical resolution. The primary feedback incorporated within the models is ice–albedo feedback. Stated simply, decreased insolation results in cooler temperatures, hence more snow. Since snow has a high albedo which results in cooler temperatures, ice–albedo feedback is a positive or amplifying feedback. The result was climate sensitivity to obliquity and precession, but not to the dominant 100 000 year climate period associated with eccentricity. No ‘ramp-like’ glaciation occurred in the models. A second problem was the model's failure to simulate the real-system time lag between glaciation and orbital elements. Finally, the amplitude of the model variation was insufficient, and so excessive changes in solar insolation were required to match the extent of the ice ages.

The early model examinations of climate sensitivity to Milankovitch orbital variations directed the next area of research to ice-sheet dynamics, primarily because of the need to simulate the time lag (ice dynamics represents a component of the ‘slower’ physics of the climate system) and to evaluate the role of ice sheets in amplifying the climate response to orbital variations.

The addition of an ice-sheet model to the energy balance climate model produced an excellent representation of the second-order cycles of the changes in ice volume, but failed to generate the ramp-like 100 000 year glacial rhythm (Figure 24.17). The answer to this problem is uncertain. Ocean circulation changes may amplify ice-sheet growth or decay. Non-linear dynamics of the climate system may produce an internal climate oscillation. Time-dependent depression of the lithosphere under the loading of the ice sheet may bring the ice sheet below the snowline causing rapid deglaciation. Isostatic rebound would then place the land surface at a point poised for reglaciation. Placing a bedrock lag function for the ice sheet in climate models has resulted in a 100 000 year rhythm with reasonable estimates of the rates of isostatic adjustment. But the mechanisms of rapid deglaciation remain elusive or untested.

The results highlight the importance of including all the components of the climate system in models if realistic climate sensitivities are to be achieved. In this particular case, the geologic record of the ice ages has provided the challenge of coupling the lithosphere, cryosphere and oceans to atmospheric climate models in order to simulate climatic change. Certainly, the incorporation of ice caps has led to more reasonable ice age simulations. Evidently, we are beginning to understand the physical mechanisms required to simulate the ice ages.

Simulations of the Last Glacial Maximum

The detailed reconstructions of conditions during the Last Glacial Maximum (LGM) offer a different perspective, equally valuable in deciphering the physical processes of climate change. First, the detailed knowledge of glaciation, sea surface temperatures and continental surface characteristics can be specified in an atmospheric General Circulation Model in order to simulate atmospheric conditions during glaciation and then to 'view' the atmosphere in a different dynamic state. For example, the atmospheric conditions associated with low-latitude glacial aridity or the position of glacial storm tracks can be investigated. In addition, the atmospheric circulation response to large ice caps can be determined. One of the most interesting results in this regard is the simulation of a split jet stream flow around the North American ice sheet, quite unlike the present circulation.

Second, as climate models are developed to include more comprehensive representations of the coupled atmosphere–ocean–ice system, these models should be capable of simulating the conditions of the LGM without specification of many of the surface conditions derived from geologic data. The conditions of the LGM will become a test of the capabilities of climate models to simulate a climate very different from the present day. The first such tests are in progress. Already coupled ocean–atmosphere models are generating sea surface temperatures which are remarkably similar to the CLIMAP data set. There are some important differences in the tropics and the southern hemisphere high latitudes, but like previous efforts, these differences are leading the way to a better understanding of the physical processes needed to understand climate change throughout Earth history and into the future.

Summary

The geologic record provides a diverse and challenging inventory of climatic change during Earth history. The number of problems are tremendous, and have been barely touched upon in this chapter. Each of these events and trends is an illustration of how the climate system operates and how it has evolved through time. The challenges are clearly set forth in our goal to understand climate change. First, palaeoclimatologists must reconstruct climate from a limited data set, interpreted indirectly from biological, physical and chemical signatures. Second, the pervasive nature of global change in Earth history is a challenge. This problem draws particular attention to the large number of factors which may cause climate change. Third, an understanding of the climatic response to these forcing factors is a major challenge. This requires a qualitative and a quantitative understanding of the complex interactions and feedback mechanisms within the climate system. The examples cited here illustrate the enormous potential of the geologic record to provide information critical to understand the climate system.

Further reading

Barron, E. J. (1984) Ancient climates: investigation with climate models, *Reports on Progress in Physics*, **47**, 1563–99.

Crowley, T. J. (1988) Paleoclimate modeling, in *Physically Based Modeling and Simulation of Climate and Climatic Change – Part II*, ed. M. E. Schlesinger, Kluwer Academic Publishers, pp. 883–949.

Frakes, L. A. (1979) *Climates Throughout Geologic Time*, Elsevier, Amsterdam, 310 pp.

Imbrie, J. & Imbrie, K. P. (1979) *Ice Ages, Solving the Mystery*, Enslow Pub, Short Hills, New Jersey, 224 pp.

CHAPTER 25

Michael R. Rampino
New York University

VOLCANIC HAZARDS

Introduction

Recent volcanic disasters in the United States, Mexico, Indonesia, Colombia, the Philippines, Japan and other countries have directed scientific and public attention to the problems of forecasting and mitigating volcanic hazards. Increased awareness of natural hazards in general prompted the United States National Academy of Sciences to propose the 1990s as the International Decade for Natural Hazard Reduction. They listed cight potential projects aimed at global volcanic hazards reduction: (1) Identification and global mapping of active and potentially active volcanoes; (2) Assessment of potential hazards of these volcanoes through study of their deposits and history of past eruptions; (3) Quantitative assignment of the intensity and magnitude of all historic eruptions as a step towards establishment of a global view of volcanic energy release; (4) Baseline geophysical and geochemical monitoring of volcanoes, particularly those in densely populated areas, to provide early warning of eruptions; (5) Training and education programs for specialists in all volcanically active countries; (6) Formation of expert international volcanic crisis assistance teams to respond to developing volcanic emergencies; (7) Development of coordinated emergency warning, evacuation and response methods and techniques; and (8) Study of the environmental impacts of eruptions on the Earth's atmosphere and world climate.

New geophysical and geochemical techniques for monitoring volcanic activity, and a better theoretical understanding of how volcanoes work, are leading to improvements in our ability to predict volcanic unrest. Past experience has shown that, with adequate warning and proper planning, loss of lives during volcanic disasters can be significantly reduced. But, in order for real risk mitigation to take place, the results of scientific studies must be effectively communicated to local officials and the general public in the countries involved. There must be provisions for issuing warnings, and for evacuation, with sufficient emergency food and shelter. In addition to scientific questions, volcanic hazards reduction must also consider social, cultural, political and economic factors. The latter may be critical in underdeveloped areas.

Volcanic eruptions represent a significant worldwide natural hazard. Eruptions have claimed more than 260 000 lives in the past 400 years – about 5 per cent of all eruptions involve fatalities. The major loss of life has been concentrated mostly in a few especially destructive events such as the 1883 event at Krakatau, an island in the Sunda Straits, west of Java, which killed 35 000 persons, the 1902 eruption of Mont Pelée on the island of Martinique in the West Indies, which took the lives of 29 000 inhabitants of the city of St Pierre, or the 1985 eruption of Ruiz volcano in Colombia, in which some 22 000 lives were lost in volcanic mudflows. Property damage from 20th century eruptions alone is estimated at more than $10 thousand million (1985 value). As global population pressures continue to increase, and greater numbers of people are living near potentially active volcanoes, it can be expected that volcanic hazards reduction will become an even more pressing problem. In some areas, dense populations live in the shadow of potentially devastating volcanoes. The town of Rabaul in Papua New Guinea, with a population of 69 000, sits nestled within an active volcanic crater. The Italian city of Naples and its suburbs with several million inhabitants has grown up well within the hazardous zone around Mt Vesuvius. The potential for loss of human life, property and productive agricultural land from destructive volcanic eruptions, and the possible effects of large eruptions on global climate makes the reduction of volcanic hazards one of the most immediately relevant contributions of geological research to society.

One method of assessing volcanic risk is through geological study and historical review of the past behaviour of volcanoes, and the use of this information to outline possible future eruptive activity. This approach has worked for some well-studied volcanoes, especially those with a fairly regular cycle and style of activity. Two major problems with this approach are: (1) many of the most destructive eruptions of the past were caused by volcanoes that had been inactive for hundreds to thousands of years, and (2) even well-documented historically active volcanoes show a wide variation in the characteristics of individual eruptions, and in the time between eruptions (Figure 25.1a). Occasionally, previously unrecognised volcanic hazards such as lateral blasts or lethal gases from crater lakes occur with little or no warning.

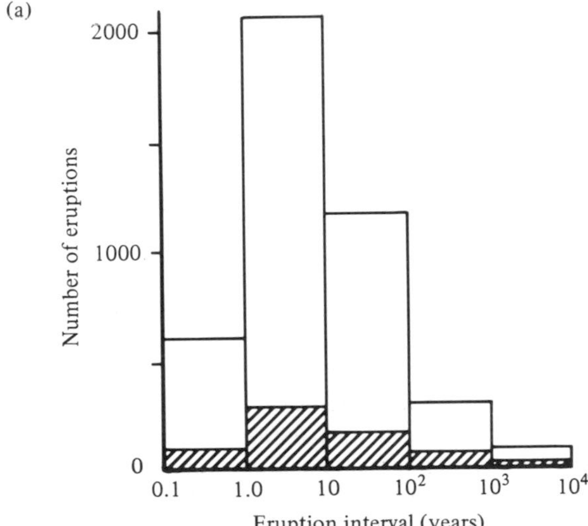

Figure 25.1 (a) Time interval between volcanic eruptions. The distribution of intervals between eruptions of the same volcano is shown for 4246 subduction zone eruptions and 629 non-subduction zone eruptions (dark pattern). The median interval for both groups is five years.

A recent catalogue of eruptions compiled at the Smithsonian Institution in Washington, DC lists more than 1300 volcanoes that have erupted during the past 10 000 years – unfortunately only about 10 per cent of these potentially destructive volcanoes have been intensively studied. Clearly, more detailed study of volcanoes that have erupted in historic times is needed, along with on-site and remote monitoring of dormant or 'inactive' volcanoes. Because of the costs involved, only a handful of volcanoes are at present well monitored. New methods of mapping volcanic hazards using data derived from satellite observations, and newly developed geophysical and geochemical techniques hold promise as ways of rapidly, systematically and relatively inexpensively covering large numbers of volcanoes. However, much basic geological fieldwork, dating, and theoretical studies of volcanic processes are also needed.

Style of eruptions and volcanic hazards

A volcano may be defined as a vent or opening in the Earth's surface through which magma (molten silicate rock) and various gases erupt. The characteristics of volcanic eruptions vary widely, from mild eruptions of lava, to paroxysmal explosions of highly fragmented gas-charged magma (the fragments of solidified magma of various sizes are called **pyroclastics**). A scale of explosivity of eruptions – the **Volcanic Explosivity Index or VEI** – was developed in the early 1980s by Christopher Newhall of the United States Geological Survey and Stephen Self of the University of Texas at Arlington. This scale is designed as a measure of the explosive magnitude of an eruption, and is a combination of the volume of the eruption and the energy released (as shown by the height of the eruption plume of ash and gases above the volcano). The VEI scale is open-ended, from 0 at the low explosivity end, for typical lava-flow eruptions, to the most explosive historic eruption, that of Tambora, on Sumbawa Island, Indonesia, in 1815, which has been assigned a VEI of 6. Figure 25.1b shows the number of eruptions of various VEI values listed in the Smithsonian catalogue.

Obviously, the hazards associated with any volcanic events are closely related to the style of the eruption. Although the largest, most explosive eruptions are

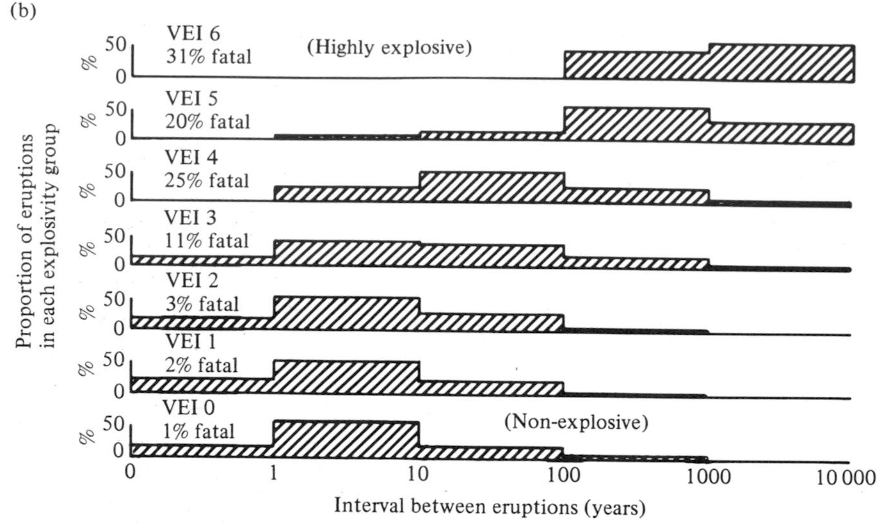

Figure 25.1 (b) Volcanic explosivity index (VEI) and the time intervals between eruptions. For each VEI group the percentage of eruptions in each time interval is shown. The number in VEI groups 0 to 6 are respectively, 354, 338, 2882, 617, 102, 19 and 8. For each group, the percentage of historical eruptions that had known fatalities is shown. □

relatively rare, they are much more likely to involve fatalities. Volcanic activity that produces only lava flows poses little danger to people, although it can be quite destructive to property. In Hawaii, lava-flow eruptions have killed only one person in the 20th century. At the same time, some 5 per cent of the island has been covered by new lava flows. Explosive volcanic eruptions, on the other hand, may be marked by powerful, rapidly moving clouds of hot ash and pumice (pyroclastics) mixed with superheated gases, devastating lateral blasts, mudflows (lahars) resulting from mixing of volcanic debris (hot or cold) and water, and destructive tsunami.

The composition of the erupted magma determines the violence of an eruption. Two primary categories of magma exist – basaltic (or mafic) magmas, which are relatively poor in silicon dioxide (40 to 48 per cent), and silicic magmas, which contain 50 to > 70 per cent silicon dioxide. Increased silicon dioxide content of the magma leads to a significant increase in viscosity (see Chapters 4 and 5 for details). Basaltic magmas are commonly somewhat hotter than silicic magma, also making for greater fluidity. Both types of magma contain dissolved gases, but silicic magmas are usually much more charged with gases. The easily flowing basaltic magma makes for relatively mild eruptions with fountains and flows of lava. By contrast, the combination of high viscosity and elevated gas content of silicic magmas leads to eruptions that are explosive in character, and the magma is usually violently fragmented in the explosion to create pyroclastic material of various sizes.

The world's active volcanoes are concentrated at the boundaries of tectonic plates and at hotspots in the oceans and within continents (see Figure 25.2 and also Chapter 9). The composition and style of volcanic eruptions is largely a function of the geological processes taking place at these active zones. Ocean ridges and hotspot volcanoes like Hawaii derive their magmas primarily from partial melting of the Earth's upper mantle. This process creates magma of basaltic composition. The magma erupted in subduction zone volcanism is derived from melting of downgoing slabs of oceanic plate and overlying sediments, and mantle. This creates a more silicic (andesitic) magma, relatively rich in gases such as carbon dioxide and water vapour derived from the subducted sediment. The important contrast for volcanic hazards is that hotspot eruptions and ocean ridge volcanism produce mostly relatively 'quiet' lava-flow activity, in contrast to subduction zone volcanoes that commonly erupt in an extremely violent manner.

Volcanic hazards

Hazards related to volcanic eruptions range from direct, and often total, destruction in the vicinity of the volcano, to regional or global perturbations of the atmosphere and climate. The direct hazards from volcanic eruptions include the effects of (1) lava flows, (2) ballistic projectiles of various sizes thrown from the volcanic vent, (3) pyroclastic fall deposits, (4) pyroclastic flows and surges (rapidly moving, ground-hugging clouds of mixed hot pyroclastics and superheated gases) including directed lateral blasts, (5) lahars (volcanic mudflows) and avalanches of volcanic debris, (6) outbursts of lethal gases from crater lakes, and (7) earthquakes related to volcanism. Indirect hazards of volcanic eruptions have also caused a great deal of damage and loss of life – these include tsunami (seismic

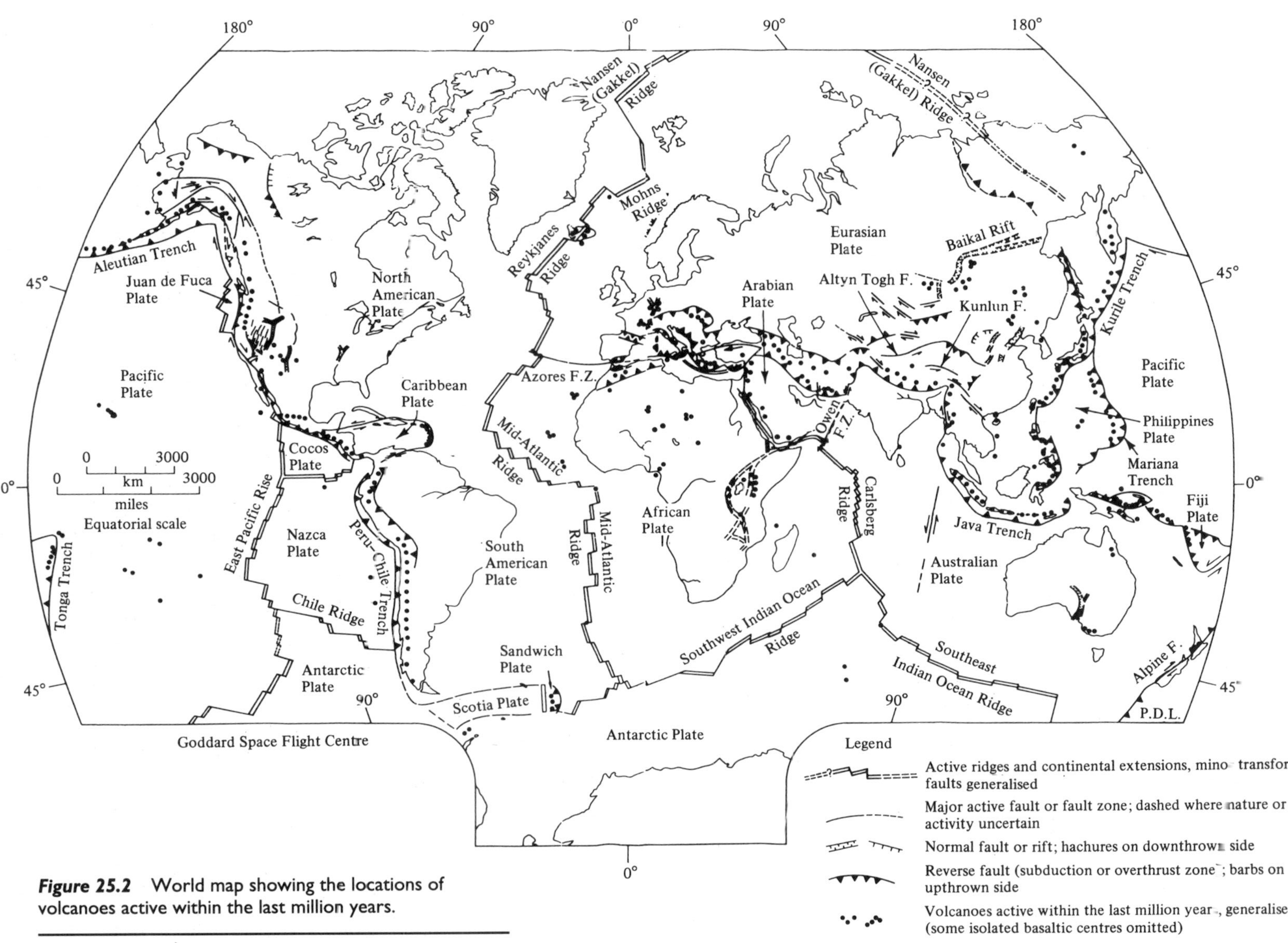

Figure 25.2 World map showing the locations of volcanoes active within the last million years.

sea waves), atmospheric shock waves, and the climatic effects of eruptions. Here we present a brief overview of the major kinds of volcanic hazards.

Lava flows

Lava flows usually originate from fissures or vents. Basaltic magma commonly loses much of its dissolved gas content as it is erupted. This process of continuous outgassing of the magma is caused by the reduction of pressure as the magma approaches the surface. In turn, outgassing can produce spectacular fountains or curtains of fire rising up to 1500 m above the volcanic vent or fissure. However, the fluidity of basaltic magma does not allow the buildup of pressures that would lead to a truly explosive eruption. Degassed basaltic lava may fall back to form lava lakes in the volcanic crater, or it may pour out forming rivers of lava flowing at velocities ranging from a few metres per hour to more than 60 km per hour on steep slopes or in narrow channels. Fresh basaltic flows have temperatures ranging from about 880 to 1130 °C, and can easily set fire to trees and wooden buildings. During the 1980s activity at Kilauea volcano, settlements on the outskirts of Hilo, Hawaii, built on lavas less than a few hundred years old, were endangered by lava flows that tended to follow the same paths as the previous flows.

Silicic lava flows usually develop only after an explosive eruption. The silicic magma, now low in gas content, is thick and slow-moving, and usually does not flow far from the volcanic vent. Such lava usually forms a dome-shaped body in the vent area. The silicic lava dome at Mount St Helens (Washington State) has been slowly and episodically extruded like thick toothpaste since a new phase of activity started with the May 1980 eruption. The dome-building has been interrupted by occasional minor explosive eruptions that have been predicted through monitoring of seismicity and the composition of gases (especially sulphurous gases) emitted from the growing lava dome.

Most recent basaltic lava flows on land are relatively small in volume, usually about 10^{-2} to 10^{-3} km^3, although much larger flows may be erupted on the ocean floor. The largest basaltic eruptions are the so-called **flood-basalt eruptions**, such as the Deccan Basalts of India (66 million years old), or the Columbia River Basalts of the Pacific Northwest (about 16 million years old), in which individual lava flows ranged up to 1200 km^3, and a total volume of about 1 to 2 million km^3 was erupted over about one million years or less. These massive eruptions may have had severe climatic consequences, which will be discussed later.

Projectiles

Projectiles of various sizes and compositions are ejected from active volcanoes (Figure 25.3). Projectiles of fresh lava greater than 64 mm in diameter are called **volcanic bombs**; similar-sized projectiles made of old volcanic rocks from the walls of the vent are called **volcanic blocks**. Volcanic ash is composed of small fragments of magma or rock less than 2 mm in diameter, but commonly much finer. The ash is primarily made up of glassy angular fragments of **pumice** – the solidified gas-bubble-rich magma. Larger fragments from 2 to 64 mm in diameter are called **lapilli** – these are usually composed of fresh pumice or pieces of old volcanic rocks from within the volcano. Ash and pumice are carried upward in the column of heated air and volcanic gases rising above the volcano, and the finest ash can be transported great distances from the site of the eruption.

Projectiles may be hurled at ejection velocities of up to 300 m per second. During the violent 1783 Asama eruption in Japan, 0.5 m diameter projectiles travelled 11 km from the vent; at Stromboli in 1930, a 30 tonne block was thrown 3 km. Volcanic bombs can easily penetrate most building materials, and hot projectiles can readily set fire to wooden buildings and trees. People in the open or in weak shelters are particularly vulnerable, and even cars do not give adequate protection. In September 1979, a 30-second explosion at Mt Etna in Sicily showered a large number of hot blocks up to 25 cm in diameter on a group of 150 tourists – nine were killed and twenty-three injured, and vehicles in the parking area provided little protection. That same month, an explosion at Aso volcano in Japan killed three people and injured eleven, as blocks up to 20 cm in diameter fell as far as 1 km from the crater. Several blocks penetrated a shelter 'protected' by a 25 cm thick concrete roof, and severely injured some of those inside.

Fallout of pyroclastics and ash

During explosive eruptions, pyroclastic materials ranging from pumice lapilli to fine ash (also called **tephra**) are carried upward by the rising plume of hot air and gases over the vent, and they eventually fall out over an extended area around the volcano. Such falls of pyroclastics are not usually particularly dangerous, except quite close to the eruption. Thick deposits of fallen ash and pumice can cause collapse of roofs, especially in rural areas where roofing materials are weak (wood or straw). During the May 1973 Heimaey eruption in Iceland, at least sixty to seventy homes

Figure 25.3 Projectiles (volcanic bombs and volcanic blocks) and volcanic ash being thrown from the volcanic vent of Anak Krakatau (Child of Krakatau), the present vent of the volcano which erupted violently in 1883. □

were buried by ash. Many strong buildings with flat roofs collapsed under loads of less than 1 m of ash.

Ash can also contaminate food and water supplies, and cause respiratory problems, creating additional hardship in the aftermath of even small explosive eruptions. Thick ash cover can kill crops and make farmland useless for years, leading to widespread famine, especially in isolated areas. The 1815 eruption of Tambora buried most cultivated land on the several islands (Figure 25.4) – the resulting starvation and disease claimed many of the 90 000 inhabitants who died as a result of the eruption.

Pyroclastic flows and surges

Pyroclastic flows and surges are rapidly moving (tens to hundreds of metres per second), ground-hugging clouds of mixed hot pyroclastics and superheated gases up to 1400 °C. Seen from a distance (the only safe way to observe them!) they appear as 'glowing clouds' or ***nuée ardente***. Flows and surges are particularly lethal, and represent the major threats to human life and structures during explosive eruptions. Relatively dense pyroclastic flows are largely confined to movement down valleys around the volcano, whereas thinner surges can move out radially from the eruption source, across valleys and over quite gentle slopes. Flows and surges are powerful enough completely to destroy most standing structures and knock down forests, and persons caught within these dense glowing clouds are killed by asphyxiation and the extreme heat.

Flows and surges originate in a partial collapse of the eruption plume of hot gases and ash that normally rises above explosive volcanoes (Figures 25.5) The collapse usually occurs because of a decrease in the gas content of the magma driving the eruption. One of the classic cases of total destruction by pyroclastic flows and surges is the 1902 eruption of Mont Pelée on the island of Martinique, where a glowing surge cloud swept through the city of St Pierre leaving some 29 000 inhabitants dead. Death was largely the result of the high temperatures and asphyxiation by inhalation of fine ash and superheated gases. Only two persons survived in the city, one was a convicted murderer locked in a jail cell with such poor ventilation that the full force of the pyroclastic surge that blasted through the town could not reach him. He was subsequently pardoned for his crimes, and became a carnival celebrity. The surges from Mont Pelée travelled across the water of St Pierre harbour, setting fire to ships and

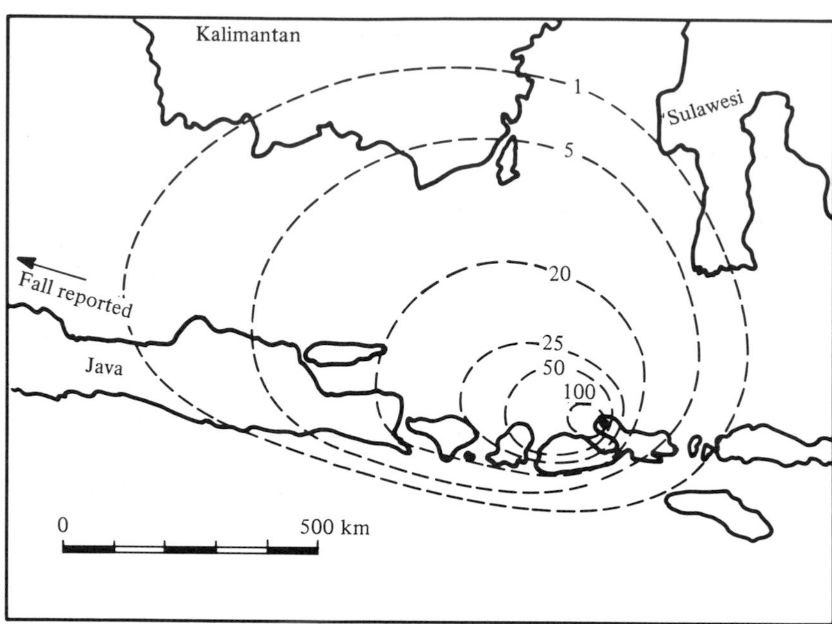

Figure 25.4 Distribution of ashfall from the great 1815 Tambora eruption. The black dot represents the position of Tambora volcano on Sumbawa Island. Ash thicknesses are given in centimetres.

leaving most of the crews and passengers dead or severely injured. In the Krakatau eruption, surges travelled more than 20 km over water and floating pumice to blast the coastal areas of Sumatra.

In the famous AD 79 eruption of Vesuvius in Italy, the residents of the Roman towns of Pompeii (population about 20 000) and Herculaneum (population about 4500) were overcome by pyroclastic surges and flows that swept down from the volcano. At Pompeii, remains of more than 2000 victims of the eruption have been discovered, and reconstructed in place through the method of injecting plaster into the natural moulds of the bodies formed in the fine surge deposits. Recent excavations at Herculaneum have uncovered the remains of hundreds of the former inhabitants of that city who took shelter in archways along the beach during the eruption. They were overcome in their hiding place by pyroclastic surges that moved rapidly through the town. These latest findings suggest that most of the inhabitants of the two buried towns did not escape by sea, as was previously thought, but actually lost their lives as they sought shelter from ashfall and destructive surges. Herculaneum was eventually buried beneath more than 20 m of pyroclastic flow deposits.

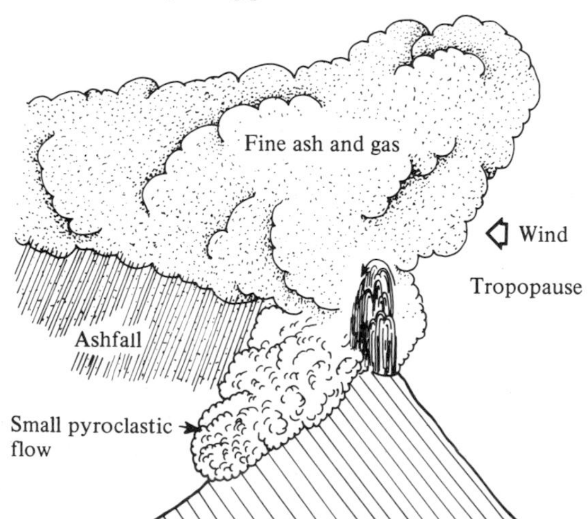

Figure 25.5 Ashfall and the production of pyroclastic flows and surges through partial collapse of an eruption column over an explosive volcanic eruption.

The 18 May, 1980 eruption of Mount St Helens in Washington State began with an avalanche of the volcano's unstable northern slope triggered by an earthquake. The collapse of the slope released a destructive lateral blast directed to the north of the volcano. The blast levelled forest and scorched trees 22 km away from its source, and was closely followed by pyroclastic flows and surges. The 1982 El Chichón eruption in Mexico, although quite small as explosive eruptions go, generated devastating surges and flows that knocked down rainforest up to 10 km from the crater, and destroyed nine villages where about 2000 people lost their lives.

Volcanic mudflows

Volcanic mudflows, or **lahars**, account for at least 10 per cent of all deaths related to volcanic eruptions. Some volcanoes, particularly those with crater lakes, or heavy snow and ice cover, are particularly susceptible to mudflow activity. For example, Kelut volcano in densely populated Java, with a large crater lake, erupted in 1919 producing lahars that killed 5110 people and damaged or destroyed 104 villages. This disaster was repeated on a smaller scale in 1966 when 210 people were killed and several villages destroyed. Ruapehu volcano in New Zealand is notorious for the large number of lahars that periodically course down the main river valleys radial to the volcano. A lahar on Christmas Eve 1953 swept away the Tangiwai railway bridge, beneath Ruapehu, a few minutes before the arrival of an express passenger train. The train dived into the river with the loss of 151 lives, mostly people on their way home for the holidays. In the 1980 Mt St Helens eruption, glacial ice was melted by the initial lateral blast, and the resulting lahars swept away trees, logging camps and bridges along the Toutle River, killing six people along the river (Figure 25.6).

Another major disaster involving volcanic mudflows was the 1985 eruption of Nevado del Ruiz volcano in Colombia. The volcano is covered by a glacier, and the small eruption of 22 November, 1985 melted a large amount of snow and ice, forming a torrent of muddy water which raced down a number of river valleys. The swift-moving water mixed with the loose volcanic muds on the upper slopes of the volcano and developed into thick mudflows that moved downstream at speeds of about 30 km per hour. The mudflows swept away homes and plantations, and buried 22 000 sleeping residents of the town of Armero to a depth of some 3 metres of ooze.

Figure 25.6 Mudflow (lahar) along the Toutle River after the 18 May, 1980 Mt St Helens eruption. Note the mud marks on the trees indicating the height of the mudflow as it passed this point. □

Outbursts of lethal gases from crater lakes

From time to time, a previously unseen volcanic hazard makes an appearance. On 15 August, 1984 a cloud of gas burst out of Lake Monoun, a volcanic crater lake in Cameroon, killing thirty-seven people by asphyxiation. This was a new kind of natural disaster, but two years later a similar event took place in another Cameroon crater lake, Lake Nyos, this time killing 1746. The gas-flows from Lake Nyos affected an area of more than 63 km^2. Volcanologist Haraldur Sigurdsson of the University of Rhode Island studied the Lake Nyos disaster and found that the cause was most likely a buildup of carbon dioxide in the oxygen-poor bottom waters of the lakes. Some disturbance of the lake water caused the rapid outgassing of bottom waters, and carbon dioxide gas escaped from the lake surface. Inhalation of air with a 10 per cent concentration of carbon dioxide causes unconsciousness and death in 10 to 15 minutes.

The cause of the disturbance of the waters of these crater lakes is not fully known. Possibly seasonal changes in wind direction caused a turnover of the density stratified waters (both disasters took place in August). A landslide or a small eruption in the depths of the lake have also been proposed, but little evidence for these exist. The bottom waters may be enriched in carbon dioxide from the Earth's mantle; indeed, the Cameroon volcanoes are particularly high in mantle carbon dioxide. Whatever the root causes, it is likely that this kind of volcanic hazard will be experienced again in areas of deep crater lakes.

Volcanic earthquakes

Some volcanic eruptions are accompanied by earthquakes. The seismic activity prior to an eruption is usually mild. So-called 'volcanic tremor' is a characteristic kind of pre-eruption seismic signal associated with the movement of magma within the volcanic plumbing system, and has been used in forecasting the timing of some eruptions. In Central America, two of the most violent eruptions in the last few hundred years (Cosiguina in 1835 and Santa Maria in 1902) occurred a few months after great earthquakes. Furthermore, significant intermediate or deep shocks precede some subduction zone eruptions. Recently, the Italian town of Pozzuoli, which sits near the active volcanic region of the Phlegrean Fields, suffered swarms of earthquakes between 1969 and 1972, and again from 1982 to 1984, that severely damaged many buildings and caused evacuation of the town. The quakes were most likely related to movement of magma beneath the region, and were accompanied by episodic uplift and tilt of the ground, but no eruption occurred.

Tsunami

Tsunami related to volcanic eruptions have been among the greatest hazards in terms of loss of human lives. For example, the Krakatau eruption on 26 and 27 August 1883 was marked by tremendous explosions and discharge of a large volume (perhaps 10 km^3) of pyroclastic flows into the sea around the island volcano. The displacement of the water as the great volume of pyroclastics entered the sea set up a series of great tsunami. The wave was 30 to 40 m high in some areas along the Sunda Straits – some 35 000 people were drowned as the waves literally swept away coastal settlements. The power of the tsunami is attested to by the fact that the great waves of 27 August carried the steamship *Berouw* more than 3 km inland from its mooring on the Sumatra coast; the crew of twenty-eight were all lost (Figure 25.7).

Volcanic tsunami are also believed by some to have been partly responsible for the fall of Minoan Crete in about 1500 BC. The explosive eruption of the Aegean volcanic island Thera (or Santorini) deposited pyroclastic flows in the sea surrounding the volcano in a manner similar to that of Krakatau. The coast of Crete, 120 km away, would have been hit by destructive tsunami, severely damaging Crete's military and trading vessels. This destruction, combined with ashfall that covered the agricultural lands of the island, is conjectured by some to have ended the great Minoan civilisation, and may have led to the flowering of the

Figure 25.7 Wave destruction along the Sunda Straits resulting from the tsunami generated by the 27 August, 1883 eruption of Krakatau volcano. The steamship *Berouw* was swept inland by the wave and marooned almost intact some $2\frac{1}{2}$ km from the sea.

□

Mycenaean civilisation, which appeared about that time in southern Greece. The Thera eruption is of interest in another light – the date is close to that assigned to the exodus of the Israelites from Egypt. Several historians have pointed out that description of the plagues of Egypt, in which darkness that could be felt spread over the land for three days, is similar to descriptions of ash clouds of historic eruptions. This idea has now been supported by the discovery of Thera ash in sediments in some of the lakes of the Nile delta region. Furthermore, the Biblical account of 'the pillar of smoke by day and fire by night' that guided the Israelites in the desert is an excellent description of a distant eruption plume of a major explosive eruption.

Forecasting volcanic eruptions

In a now famous assessment of volcanic hazards, United States Geological Survey volcanologists Dwight Crandell and Donal Mullineaux in 1975 forecast an explosive eruption of Mount St Helens before the turn of the century. Only five years later, the volcano erupted violently killing 57 people in total and causing about $1 thousand million in property damage. Crandell and Mullineaux's work was based on careful study of the record of past eruptions preserved in the deposits of ash and pumice around the volcano, which suggested that in recent centuries explosive eruptions have taken place, on average, about every 225 years. Of course, their predictions are based on probability, and they are unable to forecast the exact timing and size of eruptions.

Sometimes statistical analysis of a volcano's activity history can give clues as to the repeat time of eruptions that may be useful in forecasts. But the most direct approach is through **monitoring** of active or potentially active volcanoes by geophysical and geochemical methods, in search of some precursory phenomena diagnostic of a particular style and phase of eruption. Monitoring of volcanoes involves the recording and analysis of phenomena such as seismicity, changes in the elevation and tilt of the ground, variations in Earth magnetism and local electric fields, temperature and chemical variations in soils and hotsprings, deviations in microgravity, and the composition of magma and magmatic gases, and all of these can indicate movement of magma towards the surface, or a buildup of stresses within the volcanic magma chambers.

Long-term monitoring of volcanoes has the two primary objectives of (1) establishing patterns that will aid in forecasting, and (2) developing a broader understanding of how volcanoes work. Some eruptions have proven relatively easy to predict. For example, studies at Mauna Loa in Hawaii between 1976 and 1981 showed increases in ground elevation of 10 to 20 cm in the summit area, indicating inflation of a shallow magma reservoir. In 1983, seismicity increased and an eruption followed within a year. In another intermittently active island volcano, Rabaul in Papua New Guinea, significant increases in seismicity and inflation began in September 1983, following twelve years of slow inflation and gradually increasing seismicity. These events were interpreted as representing magma intrusion at 1 to 3 km beneath the volcano, and the situation was monitored closely. By August 1985, however, the rates of seismicity and inflation had diminished considerably, suggesting that an eruption was no longer imminent. In 1937, during the last eruption of Rabaul, 506 people lost their lives and 7500 were evacuated.

Recently, the geophysical technique of measuring the gravity field over volcanoes has provided some promise as a tool for monitoring, and perhaps predicting the timing of eruptions. Geophysicists Geoff Brown and Hazel Rymer of the Open University (England) have developed methods for determining minute changes in the Earth's gravity over volcanoes. The intensity of the Earth's gravity varies from point to point on the Earth's surface because of variations in the composition and structure of the subsurface. This connection allowed the Open University researchers to make a map of the subsurface structure of Poas Volcano in Costa Rica. They also discovered changes in the intensity of gravity over the volcanic vent prior to eruptions in 1989, probably due to magma emplacement and perhaps gas loss to the surface. Earlier in the 1980s they had noted some cyclic variations in gravity that may be related to tidal interactions. Indeed, the idea that lunar and solar tides are important in triggering volcanic eruptions and earthquakes is a topic of great interest, and if substantiated could provide an important means of predicting activity.

Forecasting volcanic eruptions requires a collaboration of scientists from several disciplines, including volcanologists, geochemists and geophysicists. Better prediction of volcanic eruptions, however, also involves the interaction of geologists with public officials, journalists and the general public. There is no easy answer as to the proper actions – this must be determined in each case. Loss of life and property can be reduced through coordinated national and international efforts in volcanic risk assessment, monitoring and warning. Once risk areas have been identified, zoning to restrict land use is probably the single most effective mitigation measure.

Even very small eruptions, such as the 22 November 1985 Nevado del Ruiz event in Colombia, can have catastrophic results when communications between volcanologists, public officials and the general populace break down. The Colombian National Institute of Geological-Mining Investigations (INGEOMINAS) had issued a report as early as 7 October, singling out Armero and the village of Chinchina as likely sites of disaster from volcanic mudflows travelling down the major river valleys from the volcano. In late October, a team of Italian volcanologists was invited to give their opinion of the possible hazards from Nevado del Ruiz. The Italians, led by Franco Barbieri, warned of an 'extremely dangerous' situation, and recommended establishment of an early warning system on the Lagunilla River valley, so that mudflows would trigger automatic warnings that would give the residents of towns downstream ample time to evacuate. Colombian and international scientists had actually prepared a hazard zoning map that accurately predicted the effects of the eruption. Although some precautions were taken by the Colombian officials, and early warning should have given the populace sufficient time to evacuate to high ground, it seems either that the word to evacuate did not get to residents in time for them to react, or that they ignored warnings because they felt the volcano was distant and that danger was not imminent. Better communication, and education regarding the nature of volcanic hazards might have prevented the massive loss of life.

Volcanoes and climate

The atmospheric perturbations caused by large eruptions can affect climate, agriculture, transportation, energy consumption, commerce and health on a global scale. Although the destruction suffered near the source of an explosive eruption is often dramatic, the effects of volcanic eruptions on the atmosphere may be the greatest potential volcanic hazard in terms of the total number of people affected. Interest in the problem of volcanic eruptions and climate goes back to the observations of Benjamin Franklin at the time of the eruption of Laki (Lakigigar) in Iceland in 1783. Franklin described what he termed a 'dry fog' in Europe during his stay there as Ambassador to France, and wrote that, the rays of sun… 'were indeed rendered so faint in passing through it, that when collected in the focus of a burning glass, they would scarce kindle brown paper.' Franklin also noted that the winter of 1783–4 in Europe and eastern North America was quite severe, and suggested that the Laki eruption and dry fog were responsible.

Several characteristic atmospheric optical phenomena produced by volcanic eruption clouds have been identified, including dimming of the Sun and Moon, blue or green-coloured Sun and Moon, enhanced sunrises and sunsets with high lavender glows, dark lunar eclipses, and Bishop's Rings (a halo around the sun produced by diffraction of sunlight by small particles). A number of studies over the last hundred years have found a correlation between some eruptions, decreases in solar radiation measured at ground observatories, and short-term climate coolings.

Sulphuric acid and the stratosphere

Sulphur dioxide (SO_2) and hydrogen sulphide (H_2S) gases injected into the upper atmosphere by volcanic eruptions form a cloud of sulphate **aerosols** (consisting of small droplets of sulphuric acid) in the stratosphere through photochemical reactions and droplet nucleation, and it is now well established that the bulk of volcanic 'dust' veils actually consisted of fine droplets of sulphuric acid (H_2SO_4). In the stratosphere, the sulphur dioxide reacts with hydroxyl (OH^-) radicals, produced by the break-up of water molecules under the influence of solar ultraviolet radiation, to form gaseous sulphuric acid. Gaseous sulphuric acid condenses on minute seed particles of dust, or on small clusters of molecules. These aerosol-forming reactions in the stratosphere may take weeks to months. The residence time of the aerosols in the stratosphere depends upon the nucleation and growth of the droplets. After the initial input of sulphur gases and the conversion to droplets, the volcanic aerosols have a typical diameter of about 0.5 μm. While most of the volcanic ash falls out of the atmosphere in a few months, these tiny droplets can take up to several years to coalesce and completely settle out of the stratosphere, and so a single volcanic eruption can affect the Earth's radiation budget and climate for several years.

Stratospheric aerosols alter the global radiation budget mainly by absorbing and backscattering incoming solar radiation, and they also absorb some outgoing infrared radiation. Absorption and backscattering should cause a cooling of the lower atmosphere and the surface, but the absorption of outgoing infrared radiation causes an increase in stratospheric temperatures. A major problem in investigating the effects of eruptions on climate is that the temperature decrease expected in hemispheric or regional temperature records after historic eruptions is about the same as the background interannual variations in temperature. Several studies have made use of the method of compositing or '**superposed epoch analysis**', a statistical technique in which the temperature records of several years before and after important eruptions are superposed, in order to stengthen the possible volcanic temperature signal (Figure 25.8). These studies have found temperature decreases in the northern hemisphere of about 0.2 to 0.3 °C after large 19th and 20th century eruptions such as Krakatau in 1883, Santa Maria in 1902 and Mt Agung in 1963.

Some scientific studies have examined individual eruptions in detail, and, through the use of computer models of the Earth's atmosphere/ocean system, have attempted to simulate the effects of these eruptions on global climate. The model results can then be compared with actual measurements of atmospheric perturbations and climate at the times of the eruptions. For example, a group at the NASA Goddard Space Flight Center Institute for Space Studies (GISS) studied the effects of the March/May 1963 eruption of Mt Agung on the Indonesian island of Bali (Table 25.1). The climate modellers calculated the expected effect of the Agung aerosol cloud on temperatures using measurements of aerosol cloud density and simulations of the Earth's climate system. They then compared these calculated temperature changes with the observed temperature perturbations. The GISS model results agreed with observations that tropospheric temperatures in the region from 30° N to 30° S showed a decrease of a few tenths of a degree over about a year. The cooling in the northern hemisphere of a few tenths of a degree Celsius is enough to reduce the growing season in northern Europe by about a week, which would be significant for agricultural production and energy consumption.

A surprising result of these studies is that relatively small volcanic eruptions, measured by the total volume of magma ejected, such as that of Mt Agung (with an estimated volume of magma of only about 0.5 km^3), lead to a relatively dense aerosol cloud totalling perhaps 10 to 20 million tonnes (megatonnes or Mt) of sulphuric acid aerosols, while the twenty times larger Krakatau eruption in 1883 (about 10 km^3 magma) produced a cloud only 2.5 to 5 times larger (50 Mt of H_2SO_4 aerosols). This shows that some eruptions are inherently richer in the sulphurous gases that become stratospheric H_2SO_4 aerosols, or are better at distributing the aerosols in the stratosphere.

The exceptional amount of sulphur released from some magmas is attested to by the small, explosive eruption of the Mexican volcano El Chichón in 1982, which sent a massive sulphur-dioxide-rich cloud up to 26 km into the stratosphere. Observations from the ground and by satellite tracked the development and subsequent spread of the aerosol cloud. The volcanic ash contribution to the atmosphere was small (*c.* 1 km^3), but the sulphuric acid aerosol contribution of 10 to 20 Mt was considerable (Table 25.1). Theoretical calculations again show that this should be sufficient to cause a decrease in surface temperatures in the northern hemisphere of a few tenths of a degree Celsius. Such a cooling was detected in 1982, but unfortunately for scientists trying to unravel the volcano/climate connection the cooling seems to have started with colder than average weather from January to March 1982 – before El Chichón erupted. The source of the excess sulphur in the magma is suspected to be deposits of sedimentary anhydrite (calcium sulphate) beneath the volcano, so that the sulphur richness of El Chichón may be a special case.

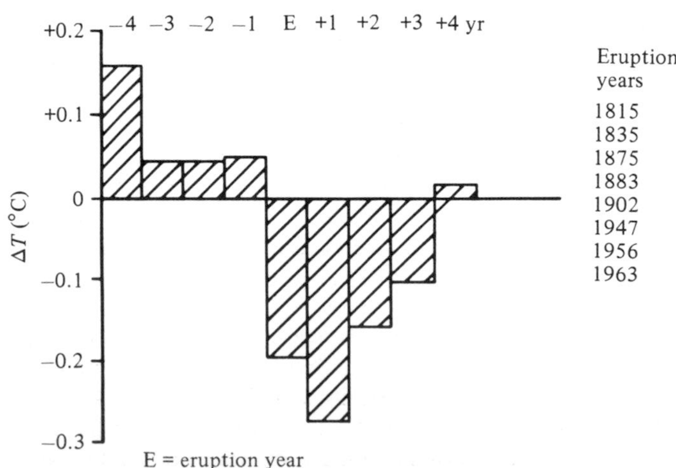

Figure 25.8 The statistical technique of superposed epoch analysis brings out the temperature signal that might be related to volcanic eruptions from the background of non-volcanic temperature variations. Temperatures for the northern hemisphere for four years prior to and four years subsequent to great eruptions are superposed to extract the net effect of eruptions on climate, in this case about 0.3 °C in the year following the eruption. □

Table 25.1. *Estimates of stratospheric aerosols and climatic effects of some volcanic eruptions*

Volcano	Latitude	Date	Stratospheric aerosols (Mt)	Northern hemisphere ΔT(°C)
Explosive eruptions				
St Helens	46° N	May 1980	0.3	< 0.1
Agung	8° S	March/May 1963	10–20	0.3
El Chichón	17° N	March/April 1982	10–20	< 0.4
Krakatau	6° S	August 1883	50	0.3
Tambora	8° S	April 1815	200	0.7
'Mystery' (Rabaul?)	4° S	March 536	300	large?
Toba	3° N	74000 years ago	1000?	large?
Effusive eruptions				
Laki	64° N	June 1783 to February 1784	100?	1.0?
Roza	47° N	14 Ma ago	6000?	large?

Proxy records of volcanic eruptions

A more or less continuous record of the clarity of the atmosphere, and hence volcanic eruption clouds, is available for the last hundred years from astronomical observations of the Sun and stars. How can one estimate the amounts of sulphur aerosols created by significant eruptions prior to that time? In the early 1980s, Claus Hammer and colleagues in Denmark developed a method of using deep cores of ice drilled from the Greenland ice cap to identify the atmospheric effects of past eruptions. The cores are marked by spikes in ice acidity that coincide with the times of historic eruptions. The acid is a result of fallout from volcanic aerosol clouds, and Hammer showed that the acid concentrations in the ice layers could be used as a so-called **proxy record** of the amounts of stratospheric aerosols. Volcanic eruptions in relatively high northern latitudes, however, can produce especially large acidity spikes because of their closeness to Greenland. Furthermore, the transport of acid aerosols to Greenland may vary with year-to-year changes in atmospheric circulation patterns. One way of correcting for these problems is by comparison with ice cores from other localities, and a number of additional ice cores has now been drilled in Antarctica, the Andes, China and the Yukon. It is now possible to correlate the acid peaks produced by fallout of volcanic aerosols from one ice core to another to get a good picture of the distribution of the aerosols in the global atmosphere.

At the same time as the work on ice cores was progressing, scientists discovered that yearly growth rings in trees from a number of places around the world commonly showed the effects of damaging frosts during the tree's growing season. Frost-damage rings in trees from the western United States were studied by Katherine Hirschboeck and Valmore C. LaMarche of the University of Arizona Tree-Ring Laboratory in Tucson, who found a close correlation between the damage rings and years of explosive volcanic eruptions. Rings in trees from other areas, such as southern Canada, show the effects of poor growing seasons in the one to two years following large eruptions. These anomalous growth patterns are indicative of outbreaks of unusual weather due to changes in regional atmospheric circulation patterns of the kind that might be set up by aerosols in the stratosphere. The frost damage rings, which could be dated by counting tree rings, were then successfully matched up with the acid spikes in ice cores, and with eyewitness accounts of the effects of volcanic eruption clouds on the visibility of celestial objects (Figure 25.9).

Some historic eruptions and their climatic effects

Through the use of proxy records, eyewitness accounts of 'dry fogs', unusual weather patterns, crop failures, and outbreaks of famine and disease it is possible to put together a picture of some of the greatest volcanic eruptions in terms of their effects on climate and society. A search of European records going back to the classic Roman and Greek periods by Richard B. Stothers of GISS turned up accounts of significant atmospheric and climatic disturbances that correlated very well with acidity peaks, tree rings and other proxy

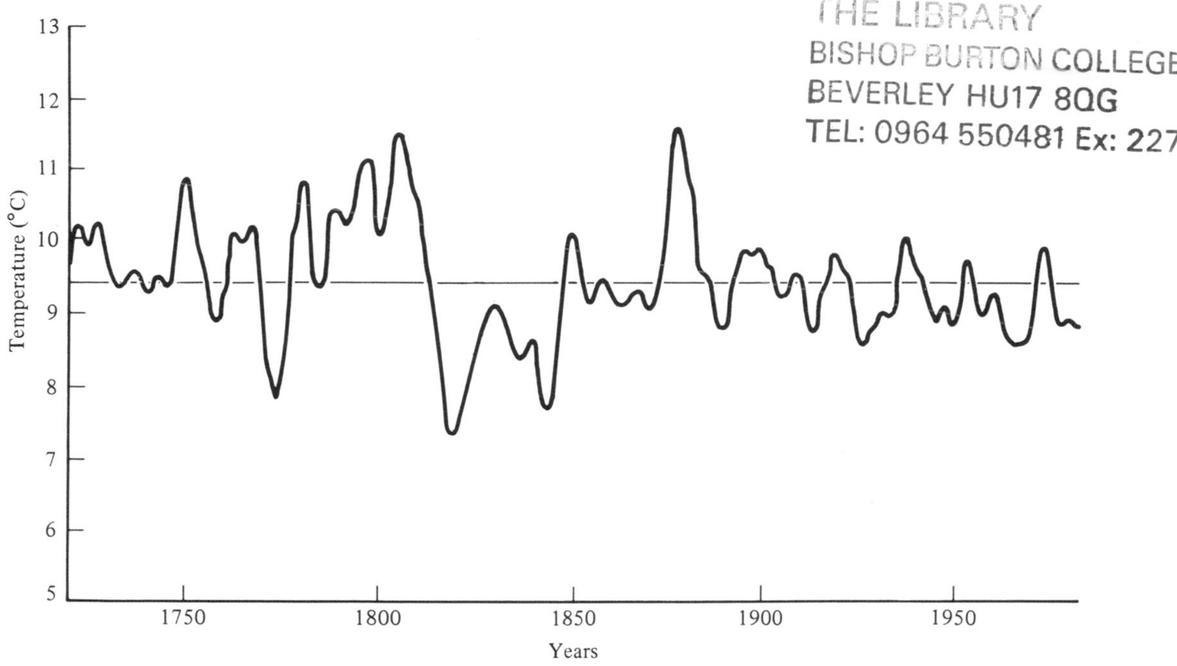

Figure 25.9 Analyses of tree-ring thicknesses from western Quebec, Canada, are used to reconstruct the climate (June to August mean temperature) for the last 250 or so years. The years following the 1815 Tambora eruption show the greatest cooling. The 3 to 6 °C regional temperature decrease in the early 1800s had severe consequences for agriculture in eastern North America. Other eruptions (e.g. Laki, 1783; Krakatau, 1883) are visible, but with much weaker effects on temperatures. □

data. Detailed information is also present in the very extensive Chinese historical annals scrutinized by Kevin Pang of the NASA, Jet Propulsion Laboratories, and his colleagues. Here we summarise some of the information put together on the aftermath of three of the greatest volcanic perturbations of the atmosphere in historic times – the Tambora (Indonesia) eruption of 1815, the Laki (S. Iceland) eruption of 1783, and the Mystery eruption (see below) of AD 536.

1816: The 'Year Without a Summer'

The year 1816 is known in climate history as the 'Year Without a Summer' and the 'Last Great Subsistence Crisis in Europe'. Although the entire decade from 1810 to 1820 was a time of noticeably decreased temperatures in the northern hemisphere, the unusual weather in 1816 followed the spectacular 10–11 April, 1815 eruption of **Tambora volcano** (Figure 25.4) – the largest known ash-producing eruption (150 km^3 of ash and pumice, equal to about 50 km^3 of magma) in the last 10 000 years. Ash fell over more than one million square kilometres, and darkness lasted up to two days at distances of 600 km from the volcano. Theoretical calculations and analysis of ash dispersal suggest that the eruption column may have reached 50 km into the stratosphere. The volcanic cloud travelled around the world, and within three months its optical perturbations of the atmosphere were observed at distant locations in Europe. For example, between 28 June and 12 July, and again later between 3 September and 7 October 1815, several observers near London, reported prolonged and brilliantly coloured sunsets and twilights. These are said to have inspired many of the subsequent landscape paintings of J. M. W. Turner.

The following year, 1816, was marked by a persistent dry fog, or dim Sun conditions reported in the northeastern United States. The summer in western Europe was cool and wet, and crop failures led to famine, disease and social unrest. Study of weather maps of the time suggest that the Tambora aerosol cloud led to a persistent marked drop in surface pressure across the mid-latitudes of the North Atlantic region, causing a southward shift in the tracks taken by middle latitude storm systems. This gave rise to a cold, wet summer. Widespread typhus in Europe added to the post-Napoleonic War misery. During this dismal European summer, a planned season of outings on Lake Geneva turned into a gloomy indoor writing session in which Mary Shelley produced *Frankenstein*, and Lord Byron his sombre poem *Darkness*.

Records of Hudson's Bay Company trading posts

show that the summers of 1816 and 1817 were the coldest of any in the modern record. Tree-ring data from central western Quebec supports these observations. The distribution and severity of sea ice in Hudson Strait in 1816 suggest prevailing northerly or northwesterly winds – atmospheric circulation patterns that would have allowed southward penetrations of Arctic air in eastern North America and western Europe, thus explaining the outbreaks of unusually cold weather during the spring and summer of 1816 in eastern Canada and the northeastern United States. From late spring through the summer, repeated frosts and snow in New England caused crop failures, resulting in poor harvests and famine.

The outbreaks of cold weather and other climatic perturbations during the summer months of 1816 are seen clearly in a number of climatic indicators from around the world, in the frost-damage rings in trees from the western United States to South Africa, in unprecedented lateness in the grape harvests in France, and possibly in the complete destruction of forests on the Atlantic island of Trindade off the coast of Brazil. On a hemispheric basis, the deviation of annual mean temperature is more difficult to assess since the station coverage in 1816 is quite patchy – an average temperature decrease for the northern hemisphere in 1816 of up to 0.7 °C is estimated, whereas the value for northern mid-latitudes is about 1.0 °C. These seemingly small average decreases are enough to shorten significantly the growing season in the northern hemisphere. This along with the episodic outbreaks of severe weather, led to the disastrous agricultural failure of 1816.

1783: The Laki fissure eruption

Franklin's observations of the 'dry fog' produced by the **Laki eruption** has focussed the attentions of climatologists and volcanologists on the events surrounding the eruption and its aftermath. The Laki eruption began in June 1783 and lasted for eight months. The eruptions were not the typical explosive eruptions of the sort that produce great amounts of pumice and ash; Laki was primarily of the fissure basalt, lava flow type, and it erupted about 12 km^3 lava from a 13 km length of fissure during June and July. From eyewitness accounts, it appears that during the first days the eruption was extremely violent, with 'enormous' Hawaiian-type lava fountains. Fine ash from the eruption fell as far away as northern Europe.

The effects of the eruption in Iceland were catastrophic. Toxic fluorine-rich volcanic gases and aerosols created a 'blue haze' that spread all across the island and led to the destruction of the summer crops. About 75 per cent of the livestock in Iceland died, and the resulting 'Blue Haze Famine' claimed 24 per cent of the Icelandic population. The dry fog reported by Franklin was also reported by others in Europe, Asia and North Africa. The haze in Europe appeared most intense during June and July, precisely the same months that Laki was most active. The eruption ceased by early February 1784, but the dry fog had already disappeared by the end of December 1783. Northern hemisphere temperature records show an abnormal temperature decline, which began in the autumn of 1783 and reached a minimum in the period from December 1783 to February 1784 (cf. Figure 25.9). Severe weather was reported in New England beginning about 20 August 1783, with the Vermont correspondent of the Newport (Rhode Island) *Mercury* reporting that 'the weather has been most extraordinary in these parts for some time past, than the oldest man among us can remember'. The below-normal temperatures persisted through the spring, autumn and winter of 1784.

AD 536: The Mystery eruption

The densest and most persistent dry fog in recorded history was observed during AD 536–537 in Europe and the Middle East and in China. The source eruption remains unknown, hence **'Mystery eruption'** but the distribution of the aerosol cloud suggests a volcano in the tropics, perhaps Rabaul in Papua New Guinea (Table 25.1). According to one contemporary writer (probably John of Ephesus) conditions in Mesopotamia (30° to 37° N) were such that 'the Sun was dark and its darkness lasted for eighteen months; each day it shone only for about four hours, and still this light was only a feeble shadow...the fruits did not ripen and the wine tasted like sour grapes.' In Rome, the Byzantine historian Procopius observed 'the Sun gave forth its light without brightness, like the Moon, during this whole year, and it seemed exceedingly like the Sun in eclipse, for the beams it shed were not clear nor such as it is accustomed to shed.' These and other passages of the time read remarkably like Franklin's and others' descriptions of the dimmed Sun and Moon seen after recent historic eruptions. Under these conditions, at maximum altitude, the Sun and Moon would have appeared about ten times fainter than normal, thus accounting for the reported darkening; scattered sunlight would have illuminated the rest of the sky.

Similar atmospheric effects were observed at the same time in China. For example, from the Chronicles of the Liang Dynasty, 'during the first of the Tai Tung years of King Wu's reign it rained dust. Again the second year yellow dust fell like snow...'. In the Chronicles of the Sui Dynasty it is stated that, 'during the Tai Tung years of King Wu of the [previous] Liang dynasty it rained ashes...'. These years correspond to AD 536 and 537. In the same years, it is reported in the Chronicles of the Sui Dynasty that: 'Frosts killed the grass during the third (April–May) and sixth (July–August) months of the reign of King Wu of Liang dynasty...', while the Kwang Wu Shing Chronicles stated: 'It snowed during the third month (April–May) of the reign of King Wu of Liang. The snow covered the ground to a depth of three feet... During the seventh month (August–September) of Tai Tung year the snow damaged seedlings in the state of Ching.' The detailed Chinese records have allowed Pang to reconstruct the distribution of early frosts, drought and famine across China – the mystery cloud and anomalously cold weather seem to have occurred throughout a large portion of the northern hemisphere.

It may be no coincidence that a severe epidemic of bubonic plague struck Africa, Europe and Asia from AD 541 to 544, wiping out from 20 to 25 per cent of the population in the stricken areas. The timing is about right for the development and spread of the plague from East Africa, and the abnormally cool weather may have facilitated the biological processes involved in passage of the plague bacteria from rats to humans through biting fleas.

Volcanic winter: the ultimate volcanic hazard?

One of the most remarkable events in geology in the 1980s was the discovery of physical evidence for the collision of a large asteroid or comet with the Earth at the time of the major **mass extinction** of life at the Cretaceous–Tertiary (K–T) boundary some 66 Ma ago. The impact of a 10-km diameter extraterrestrial object is believed to have led to a global dust cloud, and smoke clouds from widespread wildfires. These findings led several research groups to study the effects of smoke generated by fires in the aftermath of nuclear war. For example, Stephen H. Schneider and colleagues at the United States National Center for Atmospheric Research (NCAR), using sophisticated climate models, suggested that the cooling would be significant, with maximum decreases of perhaps 5 °C in low latitudes (10° to 30° N) and 10 to 15 °C at higher northern latitudes, lasting a few weeks to months.

What is the maximum climatic effect of volcanic eruptions? Historic eruptions have produced relatively small amounts of aerosols, but perhaps they can be used to estimate the atmospheric effects of the largest volcanic eruptions in the geological record. One example is the Toba eruption in Sumatra about 74000 years ago, which erupted more than 2000 km^3 of magma. Toba is estimated to have been capable of producing more than 1000 Mt of sulphuric acid aerosols. The atmospheric effects of the Toba eruption, therefore could have been more than three times that of the Mystery eruption of AD 536 (Table 25.1). The atmospheric after-effects of an explosive eruption like Toba might be comparable to some scenarios of nuclear winter.

Basaltic volcanic lava eruptions are richer in sulphur than silicic explosive eruptions such as that at Toba. Episodes of massive flood-basalt volcanism in the geological past have involved the outpouring of up to a few million cubic kilometres of basaltic magma over peak time periods of less than a million to a few million years. Flood-basalt lavas typically cover areas up to one million square kilometres with thicknesses in excess of a kilometre. Individual eruptions generated more than 500 km^3 of magma in peak eruption periods lasting days to weeks. In the past, these 'quiet' effusive basaltic lava flow eruptions were considered unlikely to produce high-altitude aerosol clouds. However, theoretical modelling of volcanic plumes now indicates that the sulphur volatiles in fissure eruptions are efficiently released, and that some proportion of these may be carried to high altitudes.

The Roza flow eruption of the Columbia River Basalt Group (about 14 Ma old) alone produced about 700 km^3 of basaltic lava. This enormous eruption would have involved a 30 m high wall of lava, over 100 km wide, at a temperature of 1100 °C, fed by huge fire fountains, and advancing at an average rate of 5 km per hour for about seven days. The quantity of atmospheric aerosols produced by such large basalt eruptions can be roughly estimated by scaling from the known amounts of aerosols generated by the largest modern fissure eruptions such as Laki. Laki erupted about 12 km^3 of magma and is estimated by various methods to have produced about 100 Mt of sulphuric acid aerosols. The sulphur release from the Roza flow eruption is estimated to have been enough to produce 6000 Mt of sulphuric acid aerosols. If only 10 per cent of these aerosols entered the stratosphere, as some recent models of flood-basalt eruptions suggest, then the stratosphere would contain about 600 Mt of aerosols, twice that of the Mystery eruption of AD 536, and about three times that of the 1815 Tambora eruption. Even the minimum atmospheric effects of a flood-basalt eruption could have dramatic global climatic consequences.

Volcanism, impacts and mass extinction

One of the current debates in geology is over the causes of mass extinctions of life. Strong evidence exists for comet or asteroid impacts at the time of the Late Cretaceous mass extinctions (66 Ma ago), which include the demise of the dinosaurs and other giant reptiles. The eruption of the massive Deccan flood basalts of India (2 million km^3 of basaltic magma) took place at the same time, and this might be more than a coincidence. During the last 250 Ma, some eleven major flood-basalts eruptions have occurred. Statistical analysis of the these eruptions shows a periodicity of about 30 Ma. This is similar to the recent findings of an approximately 30 Ma cycle in the times of mass extinctions of life, and in episodes of impact cratering on the Earth. The dates of the flood basalts agree well with the geological dates of the extinctions and impact cratering. Decompression melting studies (cf. Chapter 4) support the idea that the impact of a large asteroid or comet, capable of excavating a giant crater some 100 to 200 km across and 20 to 40 km deep, could trigger volcanism from the Earth's interior. The answer may be that both mass extinctions and flood-basalt volcanism are related to large impacts on the Earth.

Conclusions

The study of volcanic hazards and their mitigation are active areas of research, and some improvements in monitoring volcanoes and forecasting eruptions have been made in recent years. The best schemes of risk reduction involve detailed monitoring of volcanic activity, establishment of volcanic hazard zones, and development of coordinated early warning and evacuation procedures when eruptions are imminent. Volcanic eruptions are also capable of affecting the global atmosphere and climate, and the general cooling of the climate and outbreaks of severe weather that follow the production of volcanic aerosol clouds could have disastrous effects on agricultural production, especially in marginal areas, leading to hardship and famine. The recurrence of large eruptions might lead to climatic conditions that could be described as 'volcanic winter' – perhaps the ultimate volcanic hazard.

Further reading

The following are all highly accessible reviews of the subjects covered in this chapter:

Blong, R. J. (1984) *Volcanic Hazards*, Academic Press, Orlando, 424 pp.

Decker, R. W. (1986) Forecasting volcanic eruptions, *Annual Review of Earth and Planetary Science*, **14**, 267–91.

Decker, R. W. & Decker, B. B. (1991) *Mountains of Fire*, Cambridge University Press, 198 pp.

Simkin, T. *et al.* (1981) *Volcanoes of the World*, Smithsonian Institution/Hutchinson Ross, Stroudsburg, 232 pp.

Tazieff, H. & Sabroux, J-C. (eds.) (1983) *Forecasting Volcanic Hazards*, Elsevier, Amsterdam, 635 pp.

GLOSSARY

absolute time Geological time measured in years and determined by the time-dependent decay of radioactive isotopes.
accretion The process whereby small particles and gases in the solar nebula come together to form planets. Also, a gradual increase or extension of crust by natural processes, e.g. the addition of sand to a beach or the addition of basaltic rocks to oceanic crust at an ocean ridge.
achondrite A stony meteorite that lacks chondrules. Achondrites are meteorites, the appearance and composition of which are most like terrestrial basalts.
acritarch An apparently unicellular, resistant-walled algal-like organic body.
actinolite A green or greyish green mineral of the amphibole group with the formula $Ca_2(Mg,Fe)_5Si_8O_{22}(OH)_2$. Actinolite is commonly found in low-grade metabasaltic rocks.
adaptive radiation Evolutionary diversification of a group of organisms into a variety of adaptive types within a short geological time interval.
adiabatic Used in thermodynamics to describe the condition when a gas or fluid is expanded or compressed without giving or receiving heat. In an adiabatic process compression causes a rise in temperature and expansion a fall in temperature.
aerosol Solid or liquid particles suspended or dispersed in a gas.
aggradation The vertical stacking of depositional systems that results when rates of sedimentation do not keep pace with relative sea-level rise.
albite A colourless or milky white mineral. Albite ($NaAlSi_3O_8$) is the sodium end member of the plagioclase feldspar group.
alginite A component of coal within the exinite group consisting of algal matter.
alkaline Denoting igneous rock that contains greater abundances of alkali metals than the average for igneous rocks. Such rocks tend to be rich in sodium and potassium and low in calcium.
alleles The different versions of a gene that can occur at the position (locus) occupied by that gene.
allochthonous A term used to describe larger scale crustal features and rock sequences formed or produced elsewhere than in their present position.
allopatric speciation The evolutionary divergence of geographically separated populations, to form distinct species.
alteration A change in the mineralogical composition of a rock brought about by physical or chemical means. Alteration is usually distinguished from metamorphism by being more localised, and in particular is often due to the action of hydrothermal fluids in discrete zones. It can also refer to the changes in rock mineralogy and composition induced by weathering.
amino acid A compound containing both an amino ($-NH_2$) group and a carboxyl group (-COOH). A group of about twenty naturally occurring amino acids are particularly important to the biochemical structure of living things.
amphibole A group of commonly dark coloured rock-forming iron–magnesium silicates. Their struc-

ture is formed by cross-linked double chains of silica tetrahedra. Hornblende is a common form.
amphibolite A rock composed predominantly of amphiboles and plagioclase with little or no quartz.
anagenesis Evolutionary change within a single species lineage, between branching points.
analcite An isometric zeolite mineral, $NaAlSi_2O_7(OH)$, commonly found as an alteration mineral in dolerite and alkali-rich basalts.
anatexis Extreme metamorphism in which the structure of the original rock is destroyed and partial melting takes place.
andalusite A grey–purple orthorhombic aluminium silicate mineral formed at high temperature and low pressure. One of the three common polymorphs of Al_2SiO_5.
andesite A dark coloured, fine grained extrusive rock usually composed of sodic plagioclase and one of the mafic minerals, hornblende or pyroxene. Andesitic magma is generated above subduction zones.
anhydrite A white orthorhombic mineral composed of anhydrous calcium sulphate $CaSO_4$. It commonly occurs in evaporite deposits.
annelid A worm belonging to the invertebrate phylum Annelida, characterised by a segmented body, e.g., the earthworm.
anorthite The white or greyish calcium mineral of the plagioclase feldspar group. Anorthite is the calcium end member of the plagioclase feldspar group: $CaAl_2Si_2O_8$.
anorthosite A coarse-grained, light coloured plutonic rock composed predominantly of calcic plagioclase.
anthozoan A member of the group Anthozoa to which corals and sea-anemones belong.
anthracite A black hard coal with a semi-metallic lustre. Anthracite is coal of the highest metamorphic grade and is composed of between 92 and 98% carbon.
antidune A bedform similar to a dune but travelling upstream as the individual sand particles move downstream on a dense 'carpet' of sediment plus water. Antidunes form in response to higher current velocities than the velocities that would form a dune.
apatite A group of variously coloured hexagonal minerals consisting of calcium phosphate: $Ca_5(PO_4)_3(OH)$.
aquifer A body of rock that is water-saturated and is sufficiently permeable to permit ground water to move through it.
aragonite A white–grey orthorhombic $CaCO_3$ mineral. It is found in low-temperature–high-pressure metamorphic rocks as a major constituent of shallow marine carbonate sediments, and in the shells and hard parts of many invertebrates.
archaeocyathid A marine organism belonging to the phylum Archaeocyatha and characterised by a cone, goblet or vase-shaped skeleton composed of calcium carbonate.
argillaceous A term used to describe sediments with a high content of clay minerals, (e.g. clay, mudrock, shale).
artesian Ground water confined under hydrostatic pressure.
aspect ratio The ratio of width to height, commonly applied to the shape of volcanoes, sedimentary basins, etc.
asteroid A small celestial body orbiting the Sun, between Mars and Jupiter.
asthenosphere The part of the Earth's mantle, beneath the lithosphere, which flows (i.e. undergoes plastic deformation) to produce isostatic readjustments as changes in the mass of the lithosphere (above) take place.
astronomical unit A unit of planetary distance equivalent to the mean distance of the Earth from the Sun: 1.496×10^8 km
atomic number The number of protons within the nucleus of an atom. Each chemical element has a unique atomic number.
atomic weight The mean atomic weight of an element is the weighted average of the mass numbers of the isotopes of that element. (The mass number is the total number of protons and neutrons in the nucleus of an atom.)
attenuation A reduction in the amplitude of energy, used frequently for seismic waves.
augite A common black or greenish-black mineral of the clinopyroxene group $Ca(Mg,Fe,Al)(Si,Al)_2O_6$. Augite is an essential mineral in many basic igneous rocks (e.g. basalts).
aureole The zone surrounding an igneous intrusion in which the country rock shows signs of contact metamorphism due to the heating effect of the intrusion. Often called a contact aureole.
barite A white, yellow, pink or colourless mineral with the formula $BaSO_4$. It occurs in tabular crystals, and in granular and compact masses often in association with ore minerals.
basalt A general term for a dark, fine-grained extrusive igneous rock composed mainly of clinopyroxene and calcic plagioclase. Basalts consitute 90% of volcanic rocks and are, themselves, chemically varied.
basanite A group of basaltic rocks characterised by calcic plagioclase, clinopyroxene, a feldspathoid and olivine.

base-level The theoretical lowest level to which erosion on the Earth's continental surface proceeds but seldom reaches. The level at which neither erosion nor deposition takes place, i.e. an equilibrium surface.
baselap A term used in seismic stratigraphy to describe the termination of a sequence of strata along its lower boundary.
basement The undifferentiated complex of rocks that often underlies sedimentary cover in an area. Usually composed of Lower Palaeozoic or Precambrian igneous and metamorphic rocks.
basic rock An igneous rock with a silica content 44–55%. They usually contain calcic plagioclase and no quartz, and are sometimes called mafic on account of their relatively high magnesium and iron content.
batholith A large, generally discordant plutonic igneous intrusion, often elongated, and outcropping over at least 100 km^2 and extending to unknown depth. Batholiths are usually associated with orogenic belts.
bedform Any deviation from a flat bed generated by flow on the bed (e.g. ripples, dunes) and associated with the movement of sedimentary grains.
bedload Part of the sediment transported by a stream that is moved on or immediately above the stream bed by rolling and bouncing of sedimentary particles.
Benioff zone An inclined plane along which earthquake foci cluster at a subduction zone. The Benioff zone is believed to correspond to the position of the subducted slab of oceanic lithosphere.
benthic or benthonic A term used to describe bottom-dwelling marine life forms.
biostratigraphy The separation and differentiation of rocks based on the description and study of the fossils they contain.
biotite An important rock-forming mineral of the mica group. It is either dark brown or green with the chemical formula $K(Mg,Fe^{2+})_3(Al,Fe^{3+})Si_3O_{10}(OH)_2$. Biotite is a common constituent of metamorphic and igneous rocks.
bioturbation The churning of sediments by the burrowing activity of organisms.
bivalve A mollusc with a shell composed of two distinct parts, termed valves, which are joined by a flexible ligament, and which may enclose the animal. Unusually, each valve is a mirror image of the other.
blueschist A schistose metamorphic rock with a blue colour due to the presence of the blue amphibole, glaucophane. The blueschist facies is the name of the low-temperature–high-pressure metamorphic facies to which blueschists belong.
body wave A seismic wave that travels through the interior of the Earth, with a propagation mode that does not depend on any boundary surface.
botryoidal Description of a mineral having the form of a bunch of grapes, usually in the form of spherical shapes formed by radiating crystals.
boulder clay A glacial deposit consisting of boulders embedded in stiff clay; also known as till.
brachiopod A marine invertebrate animal with a two-valved shell, each valve being bilaterally symmetrical.
braided stream A stream that follows an intricate pattern of branching and re-uniting shallow channels.
breccia A coarse grained rock composed of angular broken rock fragments in a mineral cement or fine-grained matrix. Breccia is often qualified by a word indicating its mode of formation, e.g. sedimentary breccia, fault breccia or volcanic breccia.
brittle deformation The response of rocks to stress by fracturing.
brucite A layered hexagonal mineral, $Mg(OH)_2$, which commonly occurs as pearly sheets or fibres in serpentine or impure marbles.
bryozoa or bryozoan A phylum of invertebrate animals which display colonial growth with each individual housed in a tiny box-like compartment, which may be calcareous. They attach themselves to the seafloor or to other animals or rocks.
calcareous ooze A deep-sea pelagic sediment containing a high proportion of planktonic calcareous skeletal remains.
calcification Deposition of calcium salts in living tissue or the replacement of animal hard parts by $CaCO_3$ during fossilisation.
calcite A common colourless or white crystalline form of calcium carbonate, $CaCO_3$, with three cleavages at 120° to each other. Calcite is the main constituent of limestones.
caldera An approximately circular collapse structure (>1km diameter) often found above the main vent of a large volcano; usually caused by the emptying of the magma chamber at depth.
carbonaceous chondrites A group name for chondritic stony meteorites characterised by the presence of hydrated silicate minerals, such as chlorite, and considerable amounts of heavy organic compounds.
carbonate compensation depth The depth in the ocean where sea water is undersaturated with respect to $CaCO_3$ which results in the dissolution of carbonate sediments.
carbonate sediment Sediments composed of aragonite, calcite or dolomite.
carbonatite A magmatic rock composed of carbonate minerals, especially calcium carbonate, usually

associated with alkaline igneous rocks, and of restricted distribution.

cassiterite The brown or black crystalline form of tin oxide, SnO_2. Cassiterite is the main ore of tin.

chalcophile Said of an element which tends to form sulphide minerals.

chalcopyrite A brass-yellow tetragonal copper-iron sulphide mineral: $CuFeS_2$. Also known as fool's gold. Chalcopyrite is one of the main ores of copper.

chemical fossil A chemical trace of an organism that has been mostly destroyed by diagenetic processes.

chert A hard, dense, sedimentary rock composed of microcrystalline quartz, that formed either as a primary deposit or as a replacement of the pre-existing carbonate rocks.

chlorite A group of usually green platy minerals closely related to micas and clays with the general formula $(Mg,Fe^{2+},Fe^{3+})_6AlSi_3O_{10}(OH)_8$. Chlorites commonly occur in low grade metamorphic rocks or as alteration products of Fe–Mg bearing minerals.

chondrite A stony meteorite characterised by spheroidal aggregates of radially crystallised silicate minerals, known as chondrules.

chronostratigraphy The branch of stratigraphy that deals with the age of strata and their time relations.

cladogenesis The evolutionary splitting of populations, yielding a proliferation of species.

clastic Describes a rock or sediment composed principally of broken rock fragments derived from other rocks.

clay minerals A group of finely crystalline layered minerals which can take up or lose water, usually resulting from the breakdown of other minerals, such as feldspars.

cleavage The tendency of a mineral or rock to split along certain planes. In minerals cleavage results from fracturing along planes of weakness in the atomic structure. In rocks it is the tendency of a rock to split along parallel planes along which alignment of platy minerals such as micas, has occured.

cline A geographical gradation in some individual characterisitic in a population of organisms.

clinopyroxene The name for a group of monoclinic pyroxenes, of which diopside and augite are members, and with the general formula $(Ca,Mg,Fe)_2\ Si\ _2O_6$.

Coccolithophorida A group of plankton that construct a calcareous shell of round platelets known as coccoliths. They may accumulate on the ocean floor as sediment, ultimately contributing to limestones, such as chalk.

coesite A dense high pressure SiO_2 polymorph found naturally in meteorite impact craters

compaction Reduction of the pore space within sediments due to the weight of overlying rock or sediment.

compensation depth The depth at which rocks behave ductilely to compensate for changes in loading at the Earth's surface.

compressional wave See P-wave.

condensates Natural volatile liquid hydrocarbons that are produced with natural gas.

condensed horizon A thin stratigraphic sequence deposited relatively slowly and usually characterised by the presence of phosphatic and iron rich nodules, and abundant fossils.

contact metamorphism A change in the mineralogy and texture of a rock as a result of heating from an igneous body.

continental accretion The outbuilding or vertical thickening of continental crust by the addition of new material. For example, this process may happen tectonically at subduction zones where trench and ocean floor sediment is deformed or scraped onto the continental plate.

continental drift The lateral movement of continents as a result of sea-floor spreading.

continental shelf The area of the ocean floor bordering the continental land masses at a depth of 200 m or less below sea level and physically an integral part of the continent. The gradient of the shelf is approximately 0.1°.

convection Mass movement of a fluid resulting in heat transfer. The heated part expands and rises, cooler material flows in to take its place, and a convection current is established.

coralline algae Encrusting or branching red algae with a calcareous skeleton.

cordierite A blue orthorhombic silicate mineral $(MgFe)_2Al_4Si_5O_{18}$ found in thermally metamorphosed rocks and granites.

core The central zone of the Earth that lies below the major solid–liquid boundary at 2900 km depth (solid mantle above the liquid outer core). This boundary produces a major seismic discontinuity. It consists of an inner core, which is thought to be solid and a liquid outer core.

country rock The host rock surrounding an igneous intrusion, vein or orebody.

covalent bonding A type of chemical bonding in which the bonds are represented by electron pairs shared between atoms.

crater A volcanic vent or depression, usually approximately circular and <1km diameter.

crater lake A lake formed in a volcanic vent or crater by the accumulation of rain water and ground water.

craton An old, stable region of the Earth's crust characterised by low heat flow and the absence of volcanic or seismic activity.
creep Gradual plastic deformation of a rock as a result of continuous stress.
critical point A point representing the conditions of pressure and temperature at which two phases become indistinguishable in all properties. In a one component system this would be when liquid and vapour become indistinguishable.
cross stratification An arrangement of strata lying at an angle to the main stratification. Cross stratification is produced by the migration of bedforms like ripples or dunes.
crust The outer skin of the Earth, overlying the mantle, which varies in thickness from 6 km to 90 km.
cryosphere The part of the Earth's surface consisting of ice and frozen ground.
crystal settling The sinking of crystals in a magma due to their greater density.
cumulate An igneous rock formed by the accumulation of crystals settling from a magma.
cyanobacteria Prokaryotic organisms with chlorophyll which occur in a variety of environments, marine, freshwater, soil, etc.
D'Arcy's law A formula that describes the flow of fluids assuming that flow is laminar.
dacite A fine grained extrusive rock with a slightly more silicic composition than an andesite.
declination The horizontal angle at any location on the Earth's surface between true geographic north and magnetic north.
décollement A detachment plane in rocks that separates zones of contrasting deformation styles; found, for example, at a transition from brittle to ductile deformation.
dehydration reaction A metamorphic reaction that results in the liberation of water from a mineral.
deme A local population of interbreeding organisms.
detrital mineral A mineral (usually a heavy mineral) that has been eroded from pre-existing rocks and deposited unaltered into new sediment.
devolatisation The loss of volatile components (H_2O, CO_2, CH_4 and other gases and hydrocarbons) from a rock during metamorphism, or from a magma during depressuring.
diachronous Used to describe a sedimentary rock unit that, when traced over a large area, is found to contain fossils of different ages. The age of deposition of the bed was different in different places.
diagenesis The processes whereby a sediment is converted to a rock. These processes include compaction, cementation and recrystallisation.
diapir A large magma body which intrudes the upper crust because it is less dense than its surrounding rocks.
diatom A microscopic single celled plant. Diatoms secrete a wide variety of silica walled structures.
diffusion The process by which a solute (or energy) moves from a region of high concentration (or high energy) to a region of low concentration. The solute may be a component in the fluid phase or ions in a crystal.
diffusivity A measure of the rate at which diffusion occurs through a material.
diopside A green clinopyroxene with the formula $CaMgSi_2O_6$, which is found in some igneous rocks and in metamorphosed limestones.
diorite A plutonic rock intermediate in composition between a gabbro and a granite usually containing amphibole, plagioclase, pyroxene and some quartz. The intrusive equivalent of andesite.
DNA (deoxyribonucleic acid) An organic polymer composed of two strands wrapped round each other in a double helix. DNA carries the genetic information of living organisms.
dolomite A common white or yellow rhombohedral carbonate mineral: $CaMg(CO_3)_2$. It is also the term for a sedimentary rock composed dominantly of the mineral dolomite. The process by which a limestone is converted to dolomite is known as dolomitisation.
downlap A discordant relation in which inclined strata terminate downwards against an initially horizontal or inclined surface.
ductile deformation Deformation of a solid by mechanisms such as creep. A ductilely deformed rock appears to have flowed.
dune A migrating bed form formed by wind or water currents. Dunes are similar to ripples but are greater than 5 cm high.
dunite A basic rock composed of more than 90% olivine and often accompanied by the mineral chromite.
dyke A sheet-like igneous body intruded into other rocks. Dykes in sedimentary rocks cross-cut bedding, unlike sills which are parallel to bedding.
earthquake A sudden movement or trembling of the Earth caused by the abrupt release of slowly accumulated stress on a fault.
echinoderm A group of bottom-dwelling marine invertebrates mostly characterised by five-fold radial organization and a skeleton composed of crystalline calcite plates.
eclogite A rare metamorphosed basic rock composed of Fe–Mg garnet (almandine-pyrope) and sodic pyroxene (omphacite), formed by regional metamor-

phism under conditions of very high pressure and temperature. Also the descriptive term for rocks belonging to the metamorphic eclogite facies. Rocks which belong to this facies have been metamorphosed under the same high pressure conditions that would produce an eclogite.
ecophenotype Non-genetic variation in organisms due to difference in the environment in which growth has occurred.
ejecta blanket A deposit surrounding an impact or explosion crater, composed of material ejected during the formation of the crater.
elastic Having the ability to return to its original shape after being deformed by an applied force.
elastic modulus The ratio of the magnitude of the applied stress to the magnitude of the resulting strain produced when a material is deformed.
electron microprobe An analytical instrument used to determine the composition of minerals. The machine works by focusing an electron beam on the mineral of interest and measures the X-rays produced. These X-rays are then used to determine the elemental composition of the mineral.
element A substance that consists of atoms each with the same atomic number.
endothermic Said of a chemical reaction during which heat is transferred to the reacting phases from their surroundings.
enstatite A common orthopyroxene, $Mg_2Si_2O_6$, which is an important constituent of intermediate and basic igneous rocks.
epeiric sea See epicontinental sea.
epicontinental sea A shallow sea on the continental shelf or within a continent.
epidote A pistachio-green mineral, $Ca_2(Al,Fe)_3Si_3O_{12}(OH)$, which is a common constituent of low grade metamorphic rocks.
epigenetic Used to describe a mineral deposit which is younger than its enclosing rocks.
epithermal Used to describe a mineral deposit occurring within 1 km of the Earth's surface which formed in the temperature range 50–200 °C.
erosion The mechanical process of removing material by wearing away the Earth's surface through the action of wind, water and ice.
eukaryote Organisms whose cells contain a nucleus within which lie the chromosomes.
Euler's theorem The theory that describes a displacement of points over a spherical surface in angular coordinates relative to a single fixed point, termed the Euler pole.
eustasy The worldwide global changes of sea-level caused by changes in water volume due to the formation and melting of ice sheets, or changes in the volume of ocean basins induced by changing volumes of ocean ridges.
eutectic Used to describe the minimum melting temperature in a system composed of two or more solid phases. Also used to describe the bulk composition (eutectic composition) formed by the mixture at that temperature.
evaporite A sedimentary rock composed of minerals precipitated from an evaporating saline solution.
exhalative Used to describe minerals precipitated from hydrothermal solutions venting (exhaling) onto the ocean floor.
exinite A hydrogen-rich constituent of coal derived from spores, cuticular matter and resins.
exothermic Used to describe a chemical reaction that liberates heat.
exotic terrane A group of related rocks occurring in a lithologic association formed in a location foreign to that in which the rocks are now found.
fault A fracture of the Earth's crust along which relative movement of the rocks on either side takes place.
fault plane The more or less planar surface of a fault.
fayalite A brown-black iron silicate, Fe_2SiO_4, of the olivine group found rarely in acid igneous and some metamorphic rocks.
feldspar A group of white to pink abundant rock-forming silicate minerals with a composition varying between $CaAl_2Si_2O_8$ and $NaAlSi_3O_8$ (plagioclase feldspars) or between $KAlSi_3O_8$ and $NaAlSi_3O_8$ (alkali feldspars)
felsic Used to describe an igneous rock with abundant light coloured minerals like quartz, feldspar and muscovite (e.g. granite).
ferrosilite An iron component in the orthopyroxene mineral group, $Fe_2Si_2O_6$. Ferrosilite is only a component and does not occur naturally as a separate mineral phase.
flood basalt Basalt which occurs in large accumulations of horizontal or sub-horizontal lava flows.
fluid inclusion A tiny inclusion in a mineral containing a liquid and/or a gas trapped during the crystallisation or recrystallisation of the mineral.
fluorite The cubic crystalline form of calcium fluoride CaF_2, commonly purple in colour.
flute cast The cast of a scour mark formed by a sediment-laden turbulent current moving over a muddy bottom. The cast is preserved on the underside of the sandstone deposited from the turbulent current.
Foraminifera Unicellular organisms characterised by chambered shells composed of calcite (or more

rarely silica or aragonite).
foreland basin A sedimentary basin formed in front (in the foreland) of an orogenic belt as the crust is loaded by advancing thrust sheets.
formation water The water present in a sedimentary water-bearing formation.
forsterite The pale-green magnesium end-member of the olivine group of minerals: Mg_2SiO_4. Forsterite is common in basic igneous rocks.
fractional crystallisation The segregation of minerals in a body of cooling magma due to their crystallisation at different temperatures. Early-formed crystals may sink through the magma and accumulate at the base of the intrusion, resulting in a residual liquid with a different composition from the original melt.
fractionation The separation of chemical elements in nature by preferential concentration into particular mineral phases.
free oscillation An oscillation of a body that, once initiated, occurs without an external influence.
fusion A nuclear reaction involving the combination of two lighter atoms to form a heavier atom. This type of reaction produces a large amount of energy when the elements produced are lighter than iron, and is the basis of the hydrogen bomb.
gabbro A coarse-grained igneous rock, the plutonic equivalent of basalt, principally consisting of calcic plagioclase and clinopyroxene (augite).
galena A cubic, grey metallic crystalline mineral with the formula PbS. Galena is the main ore of lead.
garnet A silicate mineral group with cubic symmetry and the general formula $A_2B_3(SiO_4)_3$ where A can be Ca, Fe^{2+}, Mg and Mn and B can be Al, Fe^{3+}, Cr and V^{3+}. Garnets are common constituents of metamorphic rocks.
gastropod A mollusc characterised by a distinct head with eyes and tentacles, and a single, often spiralled, shell closed at its apex, e.g., a snail.
gene A sequence of DNA (either continuous or divided) which produces a given protein, and hence serves as an inherited instruction for the development of characters.
genotype The genetic constitution of an organism.
geobarometry The use of trace elements in a group of minerals with compositions that reflect sensitively the pressure at which they were in equilibrium.
geochemical cycle The sequence of stages in the exchange of chemical elements between major geochemical reservoirs.
geochronology The study of time in relation to the history of the Earth and the development of age dating techniques for that purpose.
geomagnetic field The Earth's magnetic field.
geotherm The variation of temperature with pressure (or depth) in the Earth.
geothermal fluid A fluid heated within the Earth, e.g., fluid emanating from hot springs.
geothermometry The use of trace elements in a group of minerals whose compositions are sensitive to the temperature at which they were in equilibrium.
glacio-eustasy Worldwide changes in sea level related to the formation and melting of ice sheets.
glacio-fluvial outwash Sediment deposited by water emanating from a glacier.
glaucophane A blue monoclinic sodic amphibole; $Na_2(Mg,Fe^{2+})_3Al_2Si_8O_{22}(OH)_2$. characteristic of the high-pressure–low-temperature metamorphic blueschist facies.
gneiss A coarse-grained, high-temperature metamorphic rock composed of alternating light and dark bands. The pale minerals are usually quartz and feldspars and the dark minerals are usually micas and amphiboles.
goethite A brownish-red iron hydroxide mineral: FeO(OH).
gossan The highly oxidised, usually red-brown coloured iron-bearing weathering products that overlie a sulphide deposit.
granite A coarse-grained (plutonic) intrusive igneous rock usually composed of quartz, feldspar and micas. It is formed from a silica-rich magma and is the intrusive equivalent of the volcanic rock rhyolite. Granites can be produced by fractionation of a basic magma or melting of the crust, and modern examples have formed above destructive plate margins.
granodiorite A coarse-grained (plutonic) intrusive igneous rock intermediate in composition between a granite and a diorite. The rock is typically composed of plagioclase, potassium feldspar, biotite and/or hornblende and quartz.
granulite A rock belonging to the high temperature metamorphic granulite facies which formed at temperatures in excess of 650 °C. Granulite facies rocks are characterised by the presence of pyroxenes in basic rocks, and sillimanite, garnet and alkali feldspar in pelitic rocks.
graphite A soft, grey hexagonal mineral which is one of the two naturally-occurring crystalline forms of carbon, the other being diamond.
gravity The phenomenon of attraction between any two masses in the Universe. Usually used to describe the force acting on objects at the Earth's surface which attracts them towards the centre of the Earth.
gravity anomaly The amount by which the measured acceleration due to gravity on the Earth's surface

departs from the calculated theoretical value which is based on the Earth being radially symmetrical and slightly flattened at the poles. Gravity anomalies are measured in metric units of acceleration called milligals ($1x10^{-5}$ ms^{-2}).

great circle A circle on the Earth's surface, the plane of which passes through the centre of the Earth.

greenschist A green schistose (foliated) metamorphic rock whose green colour is due to the minerals epidote, chlorite or actinolite (Fe-rich calcic amphibole) which are typical minerals of the metamorphic greenschist facies. The greenschist facies covers rocks regionally metamorphosed at low grades in the temperature range 300–500 °C.

greenstone A term applied to a green metamorphosed basic igneous rock (usually massive rather than schistose, see above) typified by the green minerals chlorite, epidote and amphibole, and often found in discrete regional zones, termed greenstone belts.

ground water Underground water found located in pores and fissures within rocks. It is derived mainly from some combination of rainwater percolating downwards, from water trapped in sediments during deposition and from magmatic sources.

groundmass An igneous term used to describe the fine-grained matrix in which larger crystals (phenocrysts) are contained.

gypsum White or colourless monoclinic crystalline form of hydrous calcium sulphate: $CaSO_4.2H_2O$. Gypsum is usually found associated with evaporite deposits.

haematite A metallic or red iron oxide mineral: Fe_2O_3. Haematite is the main ore of iron.

half life The time taken for half of the parent atoms of a radioactive isotope to decay.

halogen One of the elements in group VII of the Periodic Table: fluorine, chlorine, bromine, iodine and astatine.

harzburgite A mantle-derived peridotite rock composed principally of olivine with small amounts of orthopyroxene.

heat flow This is the amount of heat leaving the Earth which varies from place to place, measured in $mW\ m^{-2}$ (also known as heat flux).

hornblende The commonest mineral of the amphibole group. It is black, dark green or brown with variable composition and the general formula $Ca_2Na(Mg,Fe^{2+})_5(Al,Fe^{3+},Ti,Si)_8O_{22}(OH)_2$.

hornfels A metamorphic rock composed of equidimensional grains produced as the result of contact metamorphism.

hot spot A volcanic centre 100–200 km in diameter which is persistent for tens of millions of years and is thought to be the surface expression of a long-lived rising plume of hot mantle.

humite A white-yellow orthorhombic mineral: $Mg_7Si_3O_{12}(F,OH)_2$. Humite is found in altered volcanic rocks. It is also the name for a group of fluorine-bearing minerals related to humite.

hummocky cross stratification Provided by storm weather wave action, during which convex-up bed forms (hummocks) are produced, separated by concave-up swales (qv).

hydrocarbon A mixture of organic compounds composed of hydrogen and carbon: e.g. crude oil and natural gas.

hydrostatic pressure The pressure due to the weight of overlying water in the zone of water saturation.

hydrothermal Used to describe the action of hot water or the results of hot water circulation in the Earth's crust, e.g., hydrothermal alteration or hydrothermal mineral deposits.

hydrothermal chimney A vent on the ocean floor that produces hot solutions rich in base metal sulphides. Hydrothermal chimneys are usually associated with ocean ridges.

hydrous Used to describe a mineral with water in its crystal structure usually in the form of OH groups.

igneous texture An interlocking rock or mineral texture associated with the crystallisation of minerals from a melt.

ignimbrite A volcanic rock formed by the deposition and consolidation by welding of hot ash flows or nuées ardentes, from an explosive volcanic eruption.

impact structure A circular crater-like structure produced by the impact of an extraterrestrial body such as a meteorite on a planetary surface.

incompatible element An element which tends to concentrate in the liquid rather than in the solid phases of a crystallising melt. During melting incompatible elements are usually concentrated in the first melt to form.

index mineral A mineral characterising the particular pressure and temperature of metamorphism.

inertinite A carbon rich component of coal which is rich in carbon and which is relatively unaffected by burial (i.e. it is inert to diagenetic processes).

intergranular porosity The porosity between grains or particles of rock.

intracratonic basins Sedimentary basins developed on a cratonic area of the Earth's crust.

ion An atom or molecule that has lost or gained electrons leaving it positively or negatively charged. Ions are characterised by their radius and charge.

isobaric Said of a reaction or group of reactions (usually metamorphic) that have occurred at constant pressure.

isochemical Said of a reaction (usually metamorphic) that happens without changing the bulk chemical composition of a rock.
isopach map A contour map showing the thickness of a bed or formation. Isopachs are lines connecting points of equal thickness of a stratigraphic unit.
isopleth A line connecting points of specified constant numerical value on a map or graph.
isostasy The condition of equilibrium whereby areas of rigid lithosphere tend to 'float' in the plastically-deforming weak asthenosphere.
isotherm A line connecting points of equal temperature.
isotopes Atoms of the same element with different numbers of neutrons in their nuclei, i.e. the same atomic numbers but different mass numbers.
jadeite A sodic clinopyroxene, $Na(Al,Fe^{3+})Si_2O_6$, found in high-pressure metamorphic rocks.
kaolin A group of clay minerals with a two layer structure and the approximate formula $Al_2Si_2O_5(OH)_4$.
kerogen Fossilised insoluble organic matter found in sedimentary rocks.
komatiite An ultramafic (i.e. ultrabasic) lava, char-containing olivine, garnet and phlogopite. Kimberlites are intruded in roughly conical bodies, widening upwards, known as kimberlite pipes. Kimberlites are one of the main sources of diamonds.
komatiite an ultramafic (i.e. ultrabasic) lava, characterised by a high MgO content, silica less than 44% by mass, and olivine displaying a spinifex texture (elongate branching olivines).
kyanite A blue aluminium silicate mineral one of the three polymorphs of Al_2SiO_5. Kyanite is typical of high pressure metamorphic rocks.
laccolith A concordant igneous intrusion with a flat floor and a dyke-like feeder system.
lagoon A shallow stretch of seawater separated or partially separated from the sea by a long narrow barrier, such as beaches and dunes, or reefs.
lapilli Pyroclastic clasts in the size range 2–64 mm.
leucite A white or grey mineral of the feldspathoid group: $KAlSi_2O_6$. Leucite is a mineral found in silica-undersaturated alkaline igneous rocks.
levée A ridge or embankment of sand built up on either bank of a river on its flood plain. Levées are formed by coarse material being deposited close to the river in times of flood.
lherzolite A mantle rock composed of olivine, clinopyroxene and orthopyroxene with or without plagioclase (shallow upper mantle), spinel or garnet (deeper upper mantle).
lignite A soft, brown-black coal intermediate in coalification between peat and bituminous coal.
lipids A general term for oils, fats and waxes found in living organisms.
liquid immiscibility The ability of two different liquid phases to separate from a magma and co-exist as two discrete phases rather than one single phase.
liquidus The locus of points in diagrams of temperature-pressure-composition that mark the boundary between the liquid field and the solid plus liquid field.
lithification The process of conversion of an unconsolidated sediment to a solid rock.
lithophile Said of an element that is concentrated in silicate phases rather than metal or sulphide phases.
lithosphere The outer part of the Earth down to about 100-200 km which lies above the asthenosphere and is essentially rigid. It includes the Earth's crust and the upper part of the mantle.
lithostatic pressure Pressure in a rock due to the weight of the overlying rock.
load cast A sole structure formed on the base of sand bed due to underlying mudrocks flowing upwards (and/or the sand moving downwards) and deforming the base of the sand.
mafic minerals Dark-coloured minerals like pyroxenes and amphiboles. Mafic, literally magnesium–iron bearing, is also used to describe dark coloured igneous rocks. Igneous rocks with 44-55% silica are also said to be mafic (or basic).
magma Naturally occurring molten or partially molten rock formed within the Earth from which igneous rocks crystallise. Magma commonly contains dissolved gases which are released on crystallisation and/or depressuring.
magma chamber A reservoir of magma in the crust from which volcanic material is derived.
magnesite A white to brown crystalline form of magnesium carbonate: $MgCO_3$.
magnetite A black magnetic iron oxide mineral: Fe_3O_4. Magnetite commonly occurs as octahedral crystals and is a common mineral in igneous rocks. It forms an important iron ore.
majorite A garnet mineral thought to be present at high pressures, equivalent to over 400km depth in the Earth's mantle: $Mg_3(Fe,Al,Si)_2Si_3O_{12}$.
mantle The part of the interior of the Earth that lies between the Mohorovicic seismic discontinuity at 6 to 90 km depth and the core–mantle boundary at 2900 km depth. The mantle has an overall peridotite (ultramafic)composition, with the types of minerals present varying with depth.
mass spectrometer An instrument that separates atoms according to their charge-to-mass ratio and thus enables the masses and relative abundances of isotopes

to be determined. Mass spectrometers are fundamental instruments in isotope geochemistry.
mélange A mapable body of rock composed of fragments and blocks of all sizes, both exotic and native (locally derived), embedded in a sheared matrix. Mélanges are normally tectonic in origin.
metabasalt A metamorphosed basalt usually composed of amphibole and plagioclase.
metamorphic facies Metamorphic rocks of any origin formed within certain limits of pressure-temperature conditions (e.g. greenschist facies), reflecting the environment or area in which a rock was formed.
metamorphic grade A term indicating the temperature and pressure of metamorphism. High grade corresponds to high pressures and temperatures and low grade to low pressures and temperatures.
metamorphism The process by which the mineral composition of a rock is changed by the action of heat and/or pressure. Metamorphic rocks can be derived from pre-existing sedimentary, igneous or other metamorphic rocks.
metasomatism The process by which the chemical composition of a rock is changed by addition and/or removal of an element or elements carried by a fluid penetrating the rock.
metazoan A multicellular animal belonging to the group Metazoa, in which cells are organized into distinct tissues. The largest group in the animal kingdom, including all animals except single-celled animals and sponges.
meteoric water Surface water (rainwater, riverwater, etc.) that penetrates rocks from above.
meteorite Extraterrestrial material that falls to Earth. Most meteorites are thought to be composed of relatively primitive matter, similar to that which formed the Earth, and to be derived from the asteroid belt between the solar orbits of Mars and Jupiter.
mica A group of hydrated aluminium silicate minerals including biotite and muscovite, characterised by a sheet structure with one perfect cleavage. Micas are important rock-forming minerals in many igneous and metamorphic rocks.
migmatite A high-grade metamorphic composite rock formed of new igneous and old metamorphic parts which can be visually distinguished.
migration The process by which events on a seismic reflection profile are mapped in their approximately true position. The term is also used in the context of hydrocarbon migration from source rocks to reservoir rocks.
Milankovitch cycle Cycles of variation in the amount of solar energy falling on the Earth caused by variations in the Earth's orbit around the sun, and the tilt of its axis of rotation.
mineral assemblage The minerals that compose a rock; particularly used in relation to igneous or metamorphic rocks.
Mohorovicic discontinuity (Moho) A seismic discontinuity thought to represent the boundary between the crust and the mantle. Typically 35 km below the surface of the continents and 5–11 km below the ocean floor.
mollusc An invertebrate animal belonging to the phylum Mollusca and characterised by a soft unsegmented body, a 'muscular foot' and commonly a calcareous shell. Gastropods and bivalves are both types of mollusc.
moment of inertia Applies to a body rotating (spinning) about an axis. The moment of inertia is the sum, for all component parts within the body, of the product $m \times r^2$ where m is the mass of each part and r the distance of each part from the specified axis of rotation.
muscovite A transparent hydrous aluminium silicate of the mica group: $KAl_2(AlSi_3)O_{10}(OH)_2$. Muscovite is a common constituent of metamorphic rocks in the greenschist facies, and of granites.
mylonite A fine-grained banded rock which has formed as a result of grain size reduction due to shearing.
natural selection The preferential survival and/or reproduction of individuals displaying certain characteristics. Where the characteristics are heritable, this may lead to evolutionary change, or stabilization of a population.
nebula An interstellar diffuse cloud of dust particles and gases.
nepheline A pale green or grey mineral of the feldspathoid group: $NaAlSiO_4$ which is frequently present in igneous rocks with a high sodium content and a low percentage of silica, i.e. the undersaturated rocks.
nephelinite Igneous extrusive or shallow intrusive rock composed largely of the minerals nepheline and clinopyroxene. It typically occurs in stable continental areas as part of intracontinental magmatism.
neutrons Uncharged particles found in the nucleus of an atom.
noble gas One of the gaseous elements helium, neon, argon, krypton, xenon and radon in group VIII of the Periodic Table.
normal fault A fault caused by extension, in which the rocks on the upper side of the fracture have moved down relative to the rocks on the under side of the fracture.
nucleotide One of the molecules that condense

together to form nucleic acid like DNA and RNA.
nuée ardente A fast-flowing cloud of hot turbulent glowing gas erupted from a volcano containing solid or liquid (molten) pyroclastic material.
obduction The process by which oceanic lithosphere may be thrust onto less dense continental crust at destructive plate margins.
ocean ridge A raised ridge sometimes taking a median line within an ocean basin along which new ocean lithosphere is formed. The ridge is topographically high because high heat flow causes thermal expansion of the lithosphere.
olivine A group of silicate minerals including forsterite and fayalite with the general formula $(Mg,Fe,Mn,Ca)_2SiO_4$ and usually dark green. Olivines are common minerals in basalts, peridotites and some other igneous rocks. Olivine is the dominant mineralogical constituent of the upper mantle.
oolite A rock composed of spherical carbonate particles termed ooliths or ooids which are formed of concentric layers of calcite or aragonite deposited around a nucleus (shell fragment, quartz grain etc.).
ophiolite A group of basic to ultrabasic igneous rocks and deep sea sediments (cherts) thought to represent oceanic lithosphere (principally crustal) material, which is tectonically emplaced onto an island arc or continental margin.
orogeny The processes that form structures within mountainous areas, including faulting and folding, metamorphism and igneous activity. Elevation of mountains is due to uplift (or epeirogeny) of thrusted continental crust during isostatic adjustment.
orthogenesis A now discredited hypothesis involving innate evolutionary tendencies amoung organisms (e.g. for the elaboration of horns and other stuctures) divorced from the effects of natural selection.
orthopyroxene A group of orthorhombic pyroxene minerals including enstatite and hypersthene, found in igneous or metamorphic rocks, and with the general formula $(MgFe)_2Si_2O_6$.
ostracod A microscopic aquatic crustacean, with a bivalved calcareous carapace.
oxidation A chemical reaction involving the addition of oxygen atoms, and/or the removal of electrons from a solid, such as a mineral.
oxygen fugacity A measure of the concentration of O_2 in a fluid. Similar to the partial pressure of oxygen in the fluid phase but applied to non-ideal fluids.
palaeomagnetism The study of the remanent magnetisation crystallised at their formation in minerals to determine the history of the Earth's magnetic field and the relative positions of the continents.
pegmatite A very coarse-grained igneous rock found as dykes, veins or lenses in plutonic rocks, usually granites.
pelagic Used to describe marine organisms which live in the open ocean rather than near the shore or on the ocean bottom.
pelagic ooze A deep ocean sediment formed from the hard parts of pelagic organisms and very fine suspended sediment.
pelite (metapelite) A metamorphosed rock formed from mud-grade sedimentary rocks.
peloid A carbonate particle in a sediment composed of carbonate mud (micrite).
peridotite A dense, coarse-grained crystalline rock composed of mainly olivine with some pyroxene. Peridotite is the principal rock forming the Earth's upper mantle.
permafrost Permanently frozen soil or subsoil found in the arctic regions.
permineralisation A fossilisation process whereby the remains of organisms are preserved due to the growth of mineral material in their tissues.
perovskite A calcium–titanium silicate mineral: $CaTiO_3$. Perovskite is often rich in rare earth elements, and the structure of perovskite is thought to be adopted by some mantle silicates at high pressures.
petiole The stalk of a leaf.
petrogenesis The origin and mode of formation of a rock.
petrogenetic grid A pressure-temperature phase diagram used in metamorphic petrology to determine the sequence of reactions by which a rock formed. On a petrogenetic grid the equilibrium reaction boundary curves between minerals or groups of minerals are shown.
pH A measurement on a scale which provides a measure of the acidity (inversely related to the concentration of hydrogen ions) in a solution. Low pH corresponds to high acidity.
phase diagram A graphical diagram on which the boundaries between different solid phases and between solids, liquids and gases, representing chemical reactions are shown.
phenocrysts Large crystals enclosed in the fine-grained groundmass of an igneous rock. They are generally believed to be of the first minerals to form as magma cools and solidifies.
phenotype A particular characteristic or all the characteristics an organism possesses.
phonolite An extrusive igneous rock composed of alkali feldspar, mafic minerals and nepheline or another feldspathoid mineral. Most phonolites are porphyritic.
phyletic gradualism The model of evolutionary

change whereby species lineages are expected to show long-term anagenetic change in their characteristics, to the extent of becoming classifiable as new species.
phyllite A cleaved metamorphic rock often with a finely wrinkled silky sheen, intermediate between a slate and a schist.
picrite An extrusive igneous rock composed mainly of olivine with pyroxene and possibly some amphibole or biotite.
pillow lava Lava extruded under water which, due to rapid chilling of the propagating front, has developed a pillow-like shape in cross-section. Pillow lavas occupy much of the top layer of the oceanic crust beneath the sediment cover.
placer A mineral deposit formed by the mechanical concentration of the heavy minerals usually by the action of water.
plagioclase feldspar The calcium and sodium-bearing members of the feldspar mineral group: $CaAl_2Si_2O_8$ - $NaALSi_3O_8$
plankton Microscopic organisms that float near the ocean surfaces.
plate tectonics The theory that the outer layer of the Earth consists of rigid plates of lithosphere in motion relative to each other and to the interior of the Earth.
pleiotropy The condition when one gene affects more than one character.
pluton A mass of coarse-grained intrusive igneous rock.
polymer A large molecule formed by the linking together of smaller molecules.
polymorph One of a group of minerals with the same chemical formula but different internal atomic structures.
porosity The percentage of a rock that is occupied by interstices, pore spaces between grains. The pore spaces may be filled by water, oil, natural gas, or near the surface, by air.
porphyritic Used to describe an igneous rock which contains large crystals (phenocrysts) in a fine-grained groundmass.
porphyroblast A large crystal in a finer grained matrix within a metamorphic rock. Similar to phenocryst but refers to metamorphic rather than igneous rocks.
porphyry An igneous rock of any composition that contains phenocrysts. Often qualified by the name of the phenocryst mineral phase, e.g. quartz porphyry.
pressure vessel A container in which experiments on small rock samples are conducted at high temperatures and pressures. Used extensively in experimental petrology and geochemistry.
progradation The lateral stacking pattern of sedimentary units that results when sediment supply exceeds the rate of sea-level rise, causing sediment to build seawards.
prograde metamorphism The series of metamorphic changes that occur in response to an increase in temperature and/or pressure.
prokaryote A single-celled organism with no distinct nucleus, such as bacteria and blue–green algae.
pumice A light coloured, highly vesicular, sometimes glassy volcanic rock often with the same composition as rhyolite. Pumice is usually formed by the explosive volcanic eruption of silica-rich magma.
punctuated equilibrium The model of evolution whereby morphological change is expected to occur in brief episodes associated with speciation, and the resulting species are expected to remain more or less static thereafter.
pure shear Strain in a body which does not involve rotation. The body is elongated in one direction and shortened at right angles to that direction.
P-wave A compressional or dilational seismic wave that involves particle motion in the direction of wave propagation. P-waves are so called as they are the first waves (primary) waves to arrive after an earthquake.
pyrite A metallic, yellow mineral composed of iron sulphide: FeS_2. Pyrite is the most abundant sulphide mineral.
pyroclastic rock A clastic (fragmental) rock formed by the explosive ejection of volcanic material. Such deposits include volcanic ash, pumice and ignimbrites.
pyrophyllite A white–grey, mica-like, aluminium silicate mineral found in metamorphic rocks: $AlSi_2O_5(OH)$.
pyroxene A group of silicate minerals in which the silica tetrahedra are in single-chain structures, giving the minerals two cleavages at approximately 90° to each other. Pyroxenes have variable compositions but are usually rich in iron and magnesium, with formulae like $(Mg,Fe)Si_2O_6$ (orthopyroxenes) or $Ca(Mg,Fe)Si_2O_6$ clinopyroxenes. Pyroxenes are most commonly found in basic igneous rocks.
pyrrhotite An iron sulphide mineral in which some of the iron ions are lacking: $Fe_{1-x}S$. Pyrrhotite is darker and softer than pyrite and often contains a proportion of nickel sulphide (pentlandite) for which it is mined.
quantum evolution Rapid evolutionary change in a species lineage.
quartz A common rock forming silica mineral: SiO_2. Quartz is usually white, or clear and glassy (pale grey); internally it has a three dimensional framework structure and is resistant to weathering. It is a common constituent of many igneous, metamorphic and sedi-

mentary rocks (e.g. quartzites and sandstones).

radiaxial Used to describe crystals growing in a cavity in a fan-like pattern approximately perpendicular to the cavity wall.

radioactive decay The spontaneous decay of an unstable atomic nucleus into a lighter nucleus with the emission of a-, b- or g-radiation.

radiogenic isotope An isotope that was produced by radioactive decay of a radioactive isotope.

radiolaria Marine pelagic micro-organisms with a siliceous skeleton.

radiometric date The age of a rock in years determined by the relative proportions of radioactive isotopes and their decay products present within a rock.

rare earth elements A set of chemical elements that resemble each other very closely with atomic numbers 57–71. High concentrations of these elements in the Earth's crust are not common, but as trace elements they are valuable in determining petrogenetic processes.

red bed A siliciclastic sedimentary rock with a red colour due to the presence of ferric oxide formed in an oxidising environment.

reduction The opposite chemical process to oxidation which involves the removal of oxygen or the addition of electrons during chemical reactions.

reflection (seismic) The bouncing of a wave off a surface in such a way that it obeys the law of reflection, i.e. the angle of incidence is equal to the angle of reflection measured relative to the normal (at right angles) to the surface.

refraction (seismic) The process by which a wave's direction of propagation changes as it travels across a boundary between two materials. This bending is caused by the speed of the wave being different in the two materials.

refractory A material which can resist high temperatures. Refractories are used to line furnaces and, in the Earth, refractory minerals are the least easily melted.

relative time Geological time determined by placing events in a chronological order.

reservoir rock A rock with sufficient porosity and permeability to permit the accumulation of oil, gas or water.

residual liquid The molten part of the magma that remains in a chamber after some crystallisation and fractionation has taken place.

retrograde metamorphism Metamorphism that results in a high-grade mineral or assemblage changing to a lower grade mineral or assemblage. Retrograde metamorphism usually results from the infiltration of water at lower temperatures than the maximum metamorphic temperature attained.

reverse faulting A steeply inclined fault caused by compression in which rocks on the upper side of the fault move up relative to rocks on the lower side of the fault.

rheology The study of the deformation and flow of matter.

rhyolite A light coloured, extrusive silica-rich igneous rock. Rhyolite is the extrusive equivalent of a granite. Rhyolites are very viscous and tend to be erupted explosively forming ash flows, pumices etc. but may form banded lava flows.

ribosome The site of protein synthesis in a cell.

Richter scale A logarithmic scale of earthquake magnitude devised by Charles Richter. The scale ranges from negative values for very small earthquakes up to a maximum magnitude of about 9 as a function of the absolute strength of the Earth's crust.

ripple A bedform which is a small ridge of sand, resembling a ripple of water, formed on the bedding surface of a sediment. Ripples form perpendicular to the current flow direction.

RNA (Ribonucleic acid) A molecule involved in the synthesis of protein.

rock mechanics The science of the physical behaviour of rocks.

rocksalt A massive coarsely crystalline form of the cubic mineral, halite (which is the crystalline form of sodium chloride: NaCl).

salinometer An instrument which uses electrical conductivity to measure the salinity of seawater.

saltation The movement of sedimentary particles by bouncing along a surface.

scanning electron microscope A microscope in which a scanning beam of electrons is used to generate a three-dimensional image of a sample.

scarp A straight line of steep cliffs or slopes. If the steep slope has been formed by a fault it is known as a fault scarp.

schist A strongly foliated crystalline metamorphic rock. The foliation results from the alignment of micas. The term is often qualified by the name of the major mineral present in the rock e.g. biotite-schist, hornblende-schist.

seafloor spreading The process by which the lithospheric plates either side of an ocean ridge grow by the addition of new material as the plates either side of the ridge move apart.

seismic discontinuity A distinct boundary separating two rock units within which seismic waves travel at different speeds.

seismic stratigraphy The interpretation of seismic reflection data in terms of infered stratal relationships and rock types in order to construct a geological history of the region surveyed.

seismic wave An elastic wave generated in the Earth by earthquakes or produced artificially by explosions.

sericite A fine-grained white mica which forms as an alteration product of aluminium silicate minerals.

serpentinite A rock composed of serpentine, a group of hydrous magnesium silicates $(Mg,Fe)_3Si_2O_5(OH)_4$ that form as an alteration product of ferromagnesium (mafic) minerals such as olivine. Usually greenish and greasy to the touch.

shadow zone A zone at some distance from the epicentre of an earthquake which does not receive direct seismic waves due to refraction at a boundary where the seismic velocity of the materials present decreases downwards across the boundary (e.g. in the Earth, the core mantle boundary at 2900 km depth causes a shadow zone from 103 to 142° away from the epicentre).

shear wave See S-wave.

shear zone An originally planar zone of rock that has been deformed or fractured by shearing motions. Shear zones often act as zones of fluid flow and therefore have associated hydrothermal alteration or mineralisation.

shoreface The sloping zone between the seaward limit of the shore and the flatter offshore zone. The lower limit is usually taken to be at storm wave-base, which is the depth to which wave activity impinges on the sea bottom.

siderophile Said of an element which prefers to be in its metallic form rather than bound in silicates or sulphides within meteorites.

signal to noise ratio The ratio of the amplitude of the desired energy to that of the unwanted energy.

silica The oxide of silicon: SiO_2. It occurs naturally as the mineral quartz.

silica saturation This refers to the excess or deficiency of SrO_2 in an igneous magma which remains when all the main minerals have crystallised. An *oversaturated* basalt contains silica glass in the matrix, a *saturated* basalt contains both orthopyroxene $(Mg,Fe)_2Si_2O_6$ and olivine $(Mg, Fe)_2SiO_4$ in proportions that reflect the silica content of the magma, while an *undersaturated* basalt will contain no orthopyroxene, but olivine and a feldspathoid mineral such as nepheline.

siliciclastic sediment A sediment composed of quartz and other silicate minerals as opposed to a sediment composed of carbonate minerals.

sill An igneous body intruded parallel to the bedding plane of the host sediments.

sillimanite A white to silvery orthorhombic aluminium silicate. Sillimanite is typical of high grade metamorphic pelites and occurs as prismatic and needle-like crystals. One of the three polymorphs of Al_2SiO_5.

simple shear A type of constant volume deformation similar to shearing a deck of cards in one direction.

skarn A term used to describe silicate minerals that have replaced a limestone or dolomite as a result of the infiltration of, typically, Si, Al, Mg, and Fe-bearing solutions usually derived from an intrusive body.

slate A fine-grained cleaved metamorphic rock formed by the recrystallisation of a mud rock. The cleavage forms as a result of the alignment of platy minerals.

slumping The sliding downslope of a mass of sediment or rock by rotary motion along a concave upwards surface. The process can occur on land to produce landslips, or beneath the sea to produce submarine slumps.

solidus On a temperature–pressure–composition diagram the solidus is the locus of points separating a zone in which all phases are solid from one in which there is a total or partial melt.

spinel A group of closely-related oxide minerals with cubic structure and similar formulae and structure, literally $MgAl_2O_4$.

sporinite One of the organic particles, composed of spores, which make up coal.

stable isotope An isotope which does not undergo radioactive decay.

standing wave A regular oscillating wave that exists between two fixed (nodal) points.

staurolite A brown silicate mineral typical of pelitic rocks that have undergone medium-grade metamorphism. Staurolite is usually iron rich: $(Fe,Mg)_2Al_9Si_4O_{23}(OH)$.

stishovite A very high-pressure tetragonal form of silica: SiO_2. Stishovite is found with coesite in shocked rocks associated with meteorite impact craters and is believed to occur at great depth in the Earth's mantle.

Stokes' Law A formula which expresses the rate of settling of a particle in a fluid as a function of its radius, the particle density, the fluid density, the fluid viscosity and the acceleration due to gravity.

stoma A pore in a leaf through which oxygen, carbon dioxide and water vapour within the leaf are exchanged.

strain When a material is subject to stress, the strain is the deformation produced; for example, longitudinal compressional strain is the change in length of the sample divided by the original length of the sample.

strand plain A progradеd shore built seawards by waves and currents, which is continuous for some distance along the coast.

stratigraphy The study of the sequence and correlation of rock strata in order to interpret the geological history on local and global scales.

stratovolcano A steep-sided volcano that is composed of alternating layers of lava and pyroclastic deposits.

stress The force acting on a unit area of a surface within a solid.

strike-slip fault A high angle or vertical fault on which the movement is parallel to the strike of the fault, sometimes called transverse or transcurrent motion.

stromatolite An organic sedimentary structure formed by sediment trapped by filamentous organisms such as cyanobacteria. This results in mounds of finely layered sediment.

subduction The process by which oceanic lithosphere is conveyed down into the mantle at destructive plate margins. The area in which this takes place is called a subduction zone, the surface expression of which is a deep trench on the ocean floor.

supergene enrichment A process in ore bodies whereby surface-derived fluids leach metals at shallow levels in the ore body and reprecipitate them at deeper levels in the ore body, hence enriching the ore at deeper levels.

superheated Heated above the temperature required to cause melting or another phase transition.

suspended load Sediment in a current that is carried suspended in the water without coming into contact with the stream bed. It consists mainly of clay, silt and sand.

swale A bed form consisting of a concave-up depression formed by fair weather wave action. If sediment is deposited during the wave activity, swaley cross stratification is produced.

S-wave A transverse seismic wave produced by shearing motion. S-waves are so called because they are the second set of waves to arrive from an earthquake.

sympatric speciation The evolutionary divergence of populations living in the same area, to form distinct species.

syngenetic Said of an ore formed at the same time as its enclosing rocks.

talc An extremely soft sheet silicate: $Mg_3Si_4O_{10}(OH)_2$. Talc is found in altered basic igneous rocks and in metamorphosed dolomites.

taphonomy The branch of palaeontology concerned with the manner of burial and the origin of plant and animal remains.

tempestite A sedimentary storm deposit showing violent disturbance of pre-existing sediments followed by rapid re-deposition.

tephra A general term for all pyroclastic deposits produced by explosive volcanism.

thermal conductivity A measure of the ability of a material to conduct heat, numerically equal to the heat flow divided by the thermal gradient.

thermodynamics The mathematical treatment of the relation of heat to other forms of energy. In petrology the thermodynamics of minerals is widely used to interpret the equilibrium between different mineral phases and the temperatures and pressures of mineral/mineral assemblage formation.

tholeiitic basalt A silica saturated or silica oversaturated (i.e. having free, uncombined silica) basalt. Typical of continental plateaux and ocean ridge basalts.

threshold flow velocity The minimum velocity under which wind or water will start to move particles of sediment.

thrust fault A low angle reverse (compressional) fault.

thrust sheets Sheet-like units of crustal rocks bounded by a thrust fault.

tillite A sedimentary rock formed by the lithification of poorly-sorted glacial debris known as till.

titanomagnetite A titanium-bearing variety of the mineral magnetite.

tonalite A coarse-grained igneous rock of intermediate silica content, and with up to 10% quartz. Tonalite can be considered as a quartz-rich variety of diorite.

trace fossil A sedimentary structure resulting from the activities of living organisms, such as feeding trails, burrows, foot prints, etc.

trachyte A fine-grained extrusive igneous rock composed of alkali feldspar and minor mafic minerals often showing flow alignment of minerals.

transcurrent boundary A boundary between two rock units, comprising a strike-slip fault (a transcurrent fault).

transform faults A fault laterally displacing ocean ridges. Superficially they appear to be strike-slip faults, but lateral movement in opposite directions to either side of the fault only occurs in the zone between the offset segments of ocean ridge.

transgression An advancement or spread of sea over land (marine transgression), due to a rapid rise in relative sea level.

tremolite A magnesium amphibole associated with metamorphosed carbonates: $Ca_2Mg_5Si_8O_{22}(OH)_2$.

trench A narrow elongate depression of the ocean

floor with a crescent shape in plan view and usually parallel to volcanically active continental or island arcs.

trilobite An extinct group of arthropods with three-lobed, segmented bodies. Trilobites are confined to the Palaeozoic Era.

triple junction A point where three lithospheric plates meet.

troilite A variety of pyrrhotite, with no ferrous iron deficiency, found in meteorites and on the moon: FeS.

tuff A general term for all consolidated pyroclastic rocks, but often used for the finer grained variation.

turbidite A sediment or rock deposited from a turbidity current, characterised by a fining up grain size distribution and changes in cross stratification type indicating an upward waning of current velocity.

turbidity current A bottom flowing density current of suspended sediment moving swiftly down a subaqueous slope and spreading out at the base of the slope. Such currents occur in lakes and on the continental shelf where they carry sediment on to the ocean floor.

ultrabasic (ultramafic) rock An igneous rock with a silica content less than 44%. Such rocks are dominated by the mineral olivine and are the dominant constituent of the Earth's mantle. Undersaturation (of silica) - see silica saturation.

uniformitarianism The principle that geological events can be explained by processes observable today. This assumes that these processes have not changed during geological time.

variation diagram A diagram that shows the variations in the chemical compositions of a suite of igneous rocks, usually based on two chemical components with concentrations that form the two axes of a graph.

viscosity A measure of the ease with which a fluid will flow.

vitrinite A constituent of coal composed of humic material.

volatile Readily vaporisable. Volatiles dissolved in magma become gaseous as the pressure is released or as crystallising proceeds.

volcanic ash Fine pyroclastic material.

washover fan A fan-like deposit of sand washed across barrier islands into the lagoons during storms.

water-table The surface of an unconfined body of ground water at which the pressure is equal to the atmospheric pressure.

wavefront A surface representing the position of a travelling seismic disturbance at any given time.

weathering The chemical or physical breakdown of rocks at the surface by atmospheric agents and physical processes.

X-ray diffraction A technique for determining crystal structure which uses the diffraction of X-rays by a crystal lattice to determine the arrangement of atoms in a crystal.

X-ray fluorescence A method of using X-rays to determine the chemical composition of a sample by irradiating it with short-wavelength X-rays and measuring the longer wavelength fluourescent X-rays emitted by the sample.

xenolith A foreign inclusion in an igneous rock.

zircon A tetragonal mineral: $ZrSiO_4$ which is a common accessory phase in many igneous and some sedimentary rocks.

FIGURE ACKNOWLEDGEMENTS

The figures in this book have been gathered from the work of many Earth scientists worldwide. The publishers gratefully acknowledge the following sources from which the figures have been redrawn, modified and inspired.

Chapter 1

Figures 1.2, 1.4, 1.6, 1.7, 1.8, 1.9, 1.10, 1.12, 1.13, 1.14, 1.15 NASA.

Chapter 2

Figure 2.1 E. Anders and N. Grevesse; figure 2.2 J. A. Wood; figure 2.3 R. N. Clayton; figures 2.4, 2.6 NASA/Johnson Space Center; figure 2.7 N. E. Newsom; figure 2.8 A. G. W. Cameron and W. Benz.

Chapter 3

Figure 3.1 E. Anders; figure 3.3. Bruce Bolt © 1988 W. H. Freeman and Company; figure 3.5 S. Grand and D. Helmberger; figure 3.6 Dziewonski and Anderson; figure 3.11 Toshiro Tanimoto.

Chapter 7

Figure 7.1 D. Williams; figure 7.8 J. C. Claoué-Long © 1990 Elsevier Science Publishers.

Chapter 8

Figure 8.2 A. Hofmann © 1988 Elsevier Science Publishers; figure 8.11 J. E. Lupton and H. Craig © 1981 American Association for the Advancement of Science; Figure 8.12 E. R. Oxburgh and R. K. O'Nions © 1987 American Association for the Advancement of Science.

Chapter 9

Figures 9.1, 9.2, 9.3, 9.4, 9.5, 9.6, 9.7, 9.9, 9.10, 9.11b T. H. van Andel; figure 9.11a D. L. Jones © 1982 Scientific American Inc.; figure 9.13b W. C. Pitman; figure 9.14 T. H. van Andel © University of Chicago Press; figure 9.15 F. Press and R. Siever © 1978 W. H. Freeman.

Chapter 10

Figures 10.1, 10.3, 10.4, 10.8 A. G. Smith © 1971 Vision Press Ltd; figure 10.5 D. F. Argus © 1989 American Geophysical Union; figure 10.6 C. DeMets © 1990 Royal Astronomical Society; figure 10.7 J. Jacobs © 1987 Academic Press; figure 10.9 D. H. Tarling © 1971 Chapman and Hall; figure 10.12 M. Stefanik and D. M. Jurdy © 1984 American Geophysical Uniion; figure 10.13 D. M. Jurdy and R. G. Gordon © 1984 American Geophysical Union.

Chapter 11

Figure 11.1 Dudley Foster, Woods Hole Oceanographic Institute; figure 11.2 R. W. Hutchinson and D. L. Searle; figure 11.4 E. Horikoshi; figure 11.5 H. Ohomoto; figure 11.6a R. H. Sillitoe and H. F. Bonham; figure 11.7 K. C. Dunham and H-J. Schneider; figure 11.8 G. Garven and R. A. Freeze; figure 11.9 Derry, Clark and Gillatt; figure 11.10 M. J. Russell; figure 11.11 R. C. Larter, A. J. Boyce and M. J. Russell; figure 11.12 H. Mills.

Chapter 12

Figure 12.1b F. J. Turner © 1968 Hemisphere Publishing Corporation.

Chapter 13

Figure 13.3 F. A. Cook © 1980 Scientific American Inc; figure 13.5 J. Sharry © 1986 American Geophysical Union; figure 13.6. C. Coruh © 1987, Royal Astronomical Society; figure 13.7 K. D. Nelson *et al.* © 1985 Geological Society of America; figure 13.11 C. E. Keen and B. DeVoogd © 1988 American Geophysical Union/European Geophysical Society; figure 13.12 S. McGeary © 1989 Geological Society of London; figure 13.13 C. Flack and M. Warner © 1990 Elsevier Science Publishers; figure 13.14 C. Peddy © 1989 Geological Society of London.

Chapter 14

Figure 14.9 P. Molnar and P. Tapponnier © American Association for the Advancement of Science; figure 14.11 D. A. Rothery; figure 14.14 P. England and G. Housman © Geological Society of London; figure 14.16 Ben Avraham © 1981 American Association for the Advancement of Science.

Chapter 15

Figure 15.1 St John © 1980 American Association of Petroleum Geologists; figure 15.2 B. E. Parsons and J. G. Sclater; figure 15.3 L. Montadert and A. Mauffret; figure 15.5 N. White and D. P. McKenzie; figure 15.6 B. Wernicke; figure 15.7 C. E. Keen.

Chapter 16

Figure 16.6 G. E Reinson; figure 16.7 Plint, Walker and Bergman © 1987 Canadian Society of Petroleum Geologists; figure 16.8 Bergman and Walker © 1987 Society for Sedimentary Geology; figure 16.9 Bergman and Walker © 1988 Canadian Society of Petroleum Geologists; figures 16.10, 16.11 Canadian Society of Petroleum Geologists.

Chapter 18

Figure 18.1 R. C. Thunnel, D. F. Williams and P. R. Belyea © 1984 Elsevier Science Publishers; figure 18.2 A. Y. Huc © 1984 Pergamon Press; figure 18.4 B. P. Tissot and D. H. Welte © 1984 Springer-Verlag; figure 18.6 M. N. Yalcin and D. H. Welte © 1989 Turkish Association of Petroleum Geologists.

Chapter 19

Figure 19.1 A. Seilacher © 1990 Springer-Verlag; figure 19.2 Brett and A. Seilacher © 1990 Springer-Verlag; figure 19.3 F. Fürsich.

Chapter 20

This chapter was based on an article in *Geoscientist*, Vol. 1, No. 1, and thanks are given to the Geological Society of London for permission to reproduce text and figures. Figure 20.1 L. L. Sloss © 1963 Geological Society of America; figures 20.2, 20.10, 20.14, 20.16, 20.17 J. C. Van Wagoner, R. M. Mitchum, K. M. Campion and V. D. Rahmanian © 1990 American Association of Petroleum Geologists; figures 20.3, 20.4 A. W. Bally © 1989 American Association of Petroleum Geologists; figures 20.5, 20.18 P. R. Vail, J. Hardenbol and R. G. Todd © 1984 American Association of Petroleum Geologists; figures 20.6, 20.7 R. C. L. Wilson © 1987 Open University Press; figures 20.8, 20.9 H. W. Posamentier and P. R. Vail © 1988 Society of Economic Paleontologists and Mineralogists; figure 20.13 P. R. Vail © 1987 American Association of Petroleum Geologists; figure 20.19 Haq © 1988 Society of Economic Paleontologists and Mineralogists.

Chapter 21

Figure 21.2 M. L. Zoback and M. Magee © 1991 Royal Society of London; figure 21.3 X. Le Pichon © Geological Society of America; figure 21.4 J. Letouzey © 1986 Elsevier Scientific Publishers; figure 21.5 Stein, Cloetingh and Wortel © American Geophysical Union; figure 21.8 Stephenson and Lambeck © 1985 American Geophysical Union; table 21.2 Pitman and Golovchenko © 1983 SEPM.

Chapter 22

Figure 22.2 A. H. Knoll; figure 22.3 K. J. McNamara; figure 22.4 B. Runnegar © Norwegian University Press; figure 22.6 H. J. Hofmann.

Chapter 23

Figure 23.1 B. J. MacFadden © 1985 The Paleontological Society; figure 23.2 Photography by D. Amy, BBC, OUPC, © Forschungsinstitut und Naturmuseum Senkenberg; figure 23.6 P. H. Greenwood © National History Museum; figure 23.7 P. M. Sadler © 1981 University of Chicago Press; figure 23.8 A. L. A. Johnson © 1987 Macmillan Magazines; figure 23.9 P. R. Sheldon © 1987 Macmillan Magazines; figure 23.10 R. A. Reyment © 1982 CNRS; figure 23.11 T. A. Hansen © 1980 The Paleontological Society.

Chapter 24

Figure 24.1 S. Epstein; figure 24.12 A. Lasaga, R. Berner and R. Garrels; figure 24.13 B. Haq.

Chapter 25

Figure 25.1 P. Simkin; figure 25.2 P. D. Lowman, NASA; figure 25.3 Photograph taken on 9 September 1979 by M. R. Rampino; figures 25.4, 25.6 S. Self; figures 25.5, 25.8 M. R. Rampino and S. Self; figure 25.7 E. Cotteau; figure 25.9 G. C. Jacoby.

INDEX

Index entries for whole chapters are indicated by bold type chapter numbers. Entries referring to whole sections of chapters are indicated by italic type.